AEROSOL CHEMICAL PROCESSES IN THE ENVIRONMENT

AEROSOL CHEMICAL PROCESSES IN THE ENVIRONMENT

Edited by
Kvetoslav R. Spurny

Special Editorial Consultant
Dieter Hochrainer

LEWIS PUBLISHERS
Boca Raton London New York Washington, D.C.

Library of Congress Cataloging-in-Publication Data

Aerosol chemical processes in the environment/ K.R. Spurny, editor.
p. cm.
Includes bibliographical references and index.
ISBN 0-87371-829-1 (alk. paper)
1. Aerosols--Environmental aspects. I. Spurny, Kvetoslav.

QC882.42.A32 2000
541.3′.4515—dc21 99-089288
CIP

Direct all inquiries to CRC Press LLC, 2000 N.W. Corporate Blvd., Boca Raton, Florida 33431.

Lewis Publishers is an imprint of CRC Press LLC

International Standard Book Number 0-87371-829-1
Library of Congress Card Number 99-089288
Printed in the United States of America 1 2 3 4 5 6 7 8 9 0
Printed on acid-free paper

Dedication and Acknowledgment

This volume is dedicated to its editor, Kvetoslav R. Spurny, whose untimely death on November 3, 1999 shocked and saddened all associated with its publication.

CRC Press/Lewis Publishers is privileged to have had a long-standing relationship with Dr. Spurny, whose deep interest in science, drive to initiate and complete tasks, and kind personality are as integral to his publishing file as the tangible documents therein. The Publisher sincerely regrets that aerosol science has lost one of its founding fathers, a great scientist, and a renowned contributor to the literature in this field.

Furthermore, the Publisher gratefully acknowledges Dr. Dieter Hochrainer of Boehringer Ingelheim, Dr. Spurny's former colleague at the Fraunhofer Institute, for his assistance in the final production of this book.

Acknowledgments

The editor is extremely grateful to the authors for their excellent contributions. They have not only contributed to broader information on the progress, state of the art, and new concepts in the field of aerosol chemical processes, but they also have recognized the need for a monograph in aerosol chemistry.

The editor would like to thank the staff of Lewis Publishers, a division of CRC Press LLC, for their substantial help with the realization of this book. The authors are globally distributed and, for the majority of them, English is not their mother tongue. Without the great, patient, and highly professional support of the publishers, the realization of this book would not have been possible. All — authors and publishers — have done an excellent job for aerosol science and aerosol scientists.

Preface

Aerosol science today is an interdisciplinary branch of science that incorporates several environmental, biological, and technological research fields. In a conference organized in Prague (Czechoslovakia) in 1962, we were able to bring together lecturers from different scientific fields — physicists, chemists, meteorologists, biologists, physicians, hygienists, agrochemists, astrophysicists, etc. — and from several different countries. The conference showed the important role of aerosols and aerosol research in many basic and applied scientific and technological fields (Spurny, K., Ed., *Aerosols, Physical Chemistry and Applications*, Academia Publ. House, Prague, 1965).

Furthermore, the interdisciplinary cooperation was found to be very useful and necessary. I also remember the important role of chemistry in basic and applied aerosol research in a contribution, published in 1971 (Spurny, K.R., A note on the development of the chemistry of aerosols, *J. Aerosol Sci.*, 2, 389, 1971). Now, being retired and 75 years old, I still feel that there is a need for more synthetic work in aerosol chemistry, consisting of summarizing and evaluating the very many aerosol chemical publications dispersed in the various journals among several disciplines.

Nevertheless, such a task is not easily realized. It is perhaps beyond the feasibility of one or more editors to organize and compile a monograph like the *Aerosol Chemistry Handbook*, which would include and describe all or the most important aerosol chemical processes involved in already known scientific and technological areas.

We consider this book a partial contribution to such a task. We have picked up several examples that show the impact of aerosol chemistry in several fields, mainly in basic and atmospheric research.

American, European, and Japanese colleagues have substantially contributed to the realization of this book. I would like to thank them very much and hope their contributions will be helpful and useful to the readers.

Kvetoslav R. Spurny

Editor

Prof. Dr. Kvetoslav R. Spurny was Head of the Department of Aerosol Chemistry at the Fraunhofer Institute for Environmental Chemistry and Ecotoxicology in Germany from 1972 to 1988. After his retirement, he continued to work as an aerosol chemist. Prior to this, he was an environmental chemist at the Institute for Occupational Hygiene in Prague (1952 to 1956) and Head of the Department of Aerosol Sciences at the Czechoslovak Academy of Sciences in Prague (1957 to 1972). He was a Visiting Scientist at the National Center for Atmospheric Research, Boulder, Colorado, (1966 to 1967) and Visiting Scientist at the Nuclear Research Center, Fontenay aux Roses, France, in 1969.

Dr. Spurny obtained his diploma in Physics and Chemistry from Charles University, Prague, in 1948, a Ph.D. in chemistry at the same university in 1952, and a C.Sc. as a Candidate of Chemical Sciences at the Czechoslovak Academy of Sciences in Prague in 1964.

Professor Spurny was a member of the American Chemical Society, American Association for the Advancement of Science, American Association of Aerosol Research, British Occupational Hygiene Society, the New York Academy of Sciences, and was president of the Association for Aerosol Research from 1983 to 1984. He wrote three books on aerosols and over 150 original publications in aerosol physics and chemistry. In 1989, he was the recipient of the American David Sinclair Award in Aerosol Sciences.

Contributors

Kai Bester
Institute of Organic Chemistry
Hamburg, Germany

János Bobvos
Municipal Institute of the State Public
Health Officer Service
Budapest, Hungary

Mikhail V. Buikov
Institute of Radioecology
Ukrainian Academy of Agricultural Sciences
Kiev, Ukraine

Miroslav Chomát
Institute of Radio Engineering and Electronics
Academy of Sciences of the Czech Republic
Prague, Czech Republic

I. Colbeck
Department of Biological Sciences
University of Essex
Colchester, U.K.

Marco Del Monte
Dipartamento di Scienze della Terra e
Geologico-ambientali
Bologna, Italy

K. Hang Fung
Department of Applied Science
Brookhaven National Laboratory
Upton, New York

Mario Gallorini
CNR
Centro di Radiochimica e Analisi per
Attivazione
Universitá di Pavia
Italy

Alexandra Gogou
Department of Chemistry
Division of Environmental and Analytical
Chemistry
University of Crete
Heraklion, Crete, Greece

A. Hachimi
Laboratoire de Spectrométrie de Masse et
Chimie Laser
IPEM
Metz, France

Heinrich Hühnerfuss
Institute of Organic Chemistry
University of Hamburg
Hamburg, Germany

Mark Z. Jacobson
Department of Civil and Environmental
Engineering
Stanford University
Stanford, California

Satoshi Kadowaki
Aichi Environmental Research Center
Nagoya, Japan

Ivan Kašík
Institute of Radio Engineering and Electronics
Academy of Sciences of the Czech Republic
Prague, Czech Republic

G. Krier
Laboratoire de Spectrométrie de Masse et
Chimie Laser
IPEM
Metz, France

Markku Kulmala
Department of Physics
University of Helsinki
Helsinki, Finland

Ari Laaksonen
Department of Applied Physics
University of Kuopio
Kuopio, Finland

Claude Landron
Centre de Recherches sur les Matériaux à Haute
Temperature
Orléans, France

Willy Maenhaut
University of Gent
Gent, Belgium

Vlastimil Matějec
Institute of Radio Engineering and Electronics
Academy of Sciences of the Czech Republic
Prague, Czech Republic

P. Mériaudeau
CNRS
Institut de Recherches sur la Catalyse
Villeurbanne, France

Javier Miranda
Department of Experimental Physics
Instituto de Física
Universidad Nacional Autónoma de México
México, D.F., Mexico

J.F. Muller
Laboratoire de Spectrométrie de Masse et Chimie Laser
IPEM
Metz, France

Anders G. Nord
National Heritage Board
Satens Historska Museer
Stockholm, Sweden

S. Nyeki
Department of Biological Sciences
University of Essex
Colchester, U.K.

Vincent Perrichon
LACE–CNRS
Université Claude Bernard
Villeurbanne, France

E. Poitevin
Laboratoire de Spectrométrie de Masse et Chimie Laser
IPEM
Metz, France

Jan Rosinski
Clouds and Precipitation Group
Istituto delle Scienze dell´Atmosfera e dell´Oceano
Bologna, Italy

P. Rossi
Agenzia Regionale Prevenzione e Ambiente
Modena, Italy

Glenn O. Rubel
Department of the Army
R&T Directorate
SBCCOM-ECBC
Aberdeen Proving Ground, Maryland

Cristina Sabbioni
CNR
Instituto delle Scienze dell'Atmosfera e dell'Oceano
Bologna, Italy

Imre Salma
KFKI Atomic Energy Research Institute
Budapest, Hungary

Kvetoslav R. Spurny
Aerosol Chemist
Schmallenberg, Germany

Euripides G. Stephanou
Department of Chemistry
Environmental Chemical Processes Laboratory
University of Crete
Heraklion, Crete, Greece

Ignatius N. Tang
Department of Applied Science
Brookhaven National Laboratory
Upton, New York

Timo Vesala
Department of Physics
University of Helsinki
Helsinki, Finland

Giuseppe Zappia
CNR
Scienze dei Materiali e della Terra
Universitá di Ancona
Ancona, Italy

Éva Zemplén-Papp
Hungarian Academy of Sciences
Budapest, Hungary

Aerosol Chemistry

WHAT IS AEROSOL CHEMISTRY DEALING WITH?

An aerosol is a collection of fine and very fine particles dispersed in the gas phase. While aerosol physics tries to describe the mechanical and dynamical behavior of this system and the movement of particles in several force fields, and considers the single particle to be chemically inert, aerosol chemistry involves the physicochemical and chemical properties of the particles, in the chemical processes of particle generation, gas-to-particle and particle-to-particle reactions, interface interactions, and — on a large scale — the chemical effects of particles in several environmental fields and situations. Single aerosol particles are rarely inert; they are chemically varied and reactive.

Editor's Introduction

The physics and chemistry of aerosols have become generally adopted and are commonly in use today. They fall under aerosol science. In general, aerosols, as dispersed systems, have the same historical beginning and development as colloid chemistry. Both colloid chemistry and aerocolloids (aerosols) originated approximately in the same time period, in the second half of the 19th century. Thomas Graham, who was the first to distinguish between crystalloids and colloids in the 1860s, is considered the founder of "classical" colloid chemistry. The first observations and identifications of finely dispersed particles in gases, especially in air, were also made during this time period and recorded by the most important scientists such as J. Tyndall (1870), M. Coulier (1875), J. Aitken (1880), and L.J. Bodaszewsky (1881) et al.

The generic term, **aerosol**, evidently was coined by the well-known physicochemist F.G. Donnan near the end of World War I (see Green, H. and Lane, W. *Particle Clouds*, Van Nostrand, Princeton, NJ, 1969). A. Schmauss, a German meteorologist, was the first to introduce this term into the literature (Schmauss, A., Kolloidchemie und meterologie, *Meteorol. Zschr.*, 37, 1–8, 1920). He had compared the properties and behavior of colloidal systems in liquids with dispersed systems in gases. As an analogy to the term "hydrosol," he used the designation "aerosol" for the aerocolloids. In the publication mentioned, he concluded that: "Between the aerocolloidal solutions and the systems of solid and liquid particles in the atmosphere, large analogies do exist and therefore the latter can be named aerosols, while the first ones are known as hydrosols."

Physical investigations of aerosols were the domain of the classical period of aerosol science. Greater practical interest in the properties of and processes in aerosol chemistry began during the 1950s. "Photochemical smog" was a new term at that time, and started to be used for the designation of highly dispersed aerosols in the atmosphere of cities, which were heavily polluted by volatile organics and gaseous and particulate emissions produced primarily by motor vehicles. This aerosol is formed by several photochemical mechanisms and chain reactions in the air after intensive UV irradiation. The first important observations and measurements date back to the 1950s (e.g., see Haagen-Smit, A.J. et al.).

The importance of physicochemical processes — such as gas-to-particle conversion (or vice versa), heterogeneous chemical reactions, etc. — was recognized. Important laboratory and atmospheric investigations, including the development of mathematical models, have been undertaken since the 1970s. Several laboratories and institutes have taken part in such research. In my opinion, the most basic results were published at that time by S.K. Friedlander and his "school of aerosol physics and chemistry" established initially at Cal Tech (California Institute of Technology) in Pasadena and continued later at UCLA (University of California at Los Angeles). (See also Friedlander, S.K., *Smoke, Dust and Haze*, John Wiley & Sons, New York, 1971). Friedlander, his pupils, and coworkers were very successful in describing homogeneous and heterogeneous reactions under laboratory and atmospheric conditions. John Seinfeld, Friedlander's successor at Cal Tech, and his staff are continuing with aerosol formation studies by developing, describing, and verifying mathematical models. (See also Seinfeld, J.H., *Atmospheric Chemistry and Physics of Air Pollution*, John Wiley & Son, New York, 1985.)

Another important event dealing with the development of aerosol chemistry was the Conference of Multiphase Processes in Albany, New York, in 1981 (Schryer, D.R., Ed., *Heterogeneous Atmospheric Chemistry,* Geophysical Monograph 26, *Am. Geophys.,* Union, Washington, D.C., 1982). The published presentations of this conference showed that heterogeneous or multiphase chemical processes play an important role in general atmospheric chemistry.

The role of chemistry in the polluted atmosphere, as well as in several effects which atmospheric anthropogenic aerosols produce in the total environment (e.g., deterioration of human health, ecosystems, materials, etc.), is perhaps of greatest importance. Nevertheless, aerosol chemistry is significantly involved in several technological fields as well.

A very important one is the technology of the production of new materials by means of aerosol chemical reactions and processes. Virtually any material in the form of fine particles with controlled compositions, microstructures, morphologies, and particle size (ranging from nanometer to micron size) can be produced by means of this technology (Kodas, T., Ed., Aerosols in material processing, *J. Aerosol Sci.,* 24, 271, 1993.).

A further field in which aerosol chemistry is involved is the technology of combustion processes. Heterogeneous and particle surface chemical reactions are a very important part of the formation and modification processes by aerosol production in burning gases, liquids, and powders (Kauppinen, E.I., Ed., Combustion aerosols, *J. Aerosol Sci.*, 29, 387, 1998).

It is clear that there exists no sharp dividing line between aerosol physics and aerosol chemistry. Generally speaking, both are involved in the dynamics of any aerodispersed system and in its effects. As mentioned, the literature dealing with aerosol chemistry is still rather widely dispersed, and monographs summarizing and evaluating chemical studies in aerosols seem to be very desirable.

Kvetoslav R. Spurny

Table of Contents

PART III
AEROSOL SYNTHETIC CHEMISTRY

PART IV
AEROSOLS AND BUILDINGS

PART V
AEROSOLS IN THE ATMOSPHERE

Part I

General Aspects

1 Aerosol Chemistry and Its Environmental Effects

Kvetoslav R. Spurny

CONTENTS

INTRODUCTION

Two kinds of aerosols can be distinguished: "good" aerosols and "bad" aerosols. While the good aerosols are useful in their applications and effects, the bad aerosols produce negative, harmful effects on the environment and the human population.

Aerosol synthesis of nanoscale particles and powders belongs among the good aerosols. These aerosol processes are currently used for the large-scale production of several modern materials,[1-4] such as alumina, silica, carbon black, uranium dioxide, titanium dioxide powders,[1] ceramic superconductors,[2] magnetic semiconductors,[3] etc.

Another field of application of good aerosols is their usage in the field of medicine (i.e., for the production and application of diagnostic and therapeutic aerosols).[5]

Although the distinction between good and bad aerosols is sharp, there do exist aerosols, for which the effects can be positive as well as negative. *Agroaerosols* are such an example. Their application in the protection of plants, forest, crops, etc. has a very positive impact on agricultural development. On the other hand, agroaerosols can also produce negative effects on the health of farms, on atmospheric and aqueous environments, on soils, etc.

Generally speaking, some of the good aerosols can also be harmful. The nano-sized particles and powders produced in aerosol synthesis can be toxic when inhaled by humans. Unfortunately, the majority of anthropogenic aerosols in the ambient air have negative impacts; for example, on human health, on the living as well as nonliving environment, and on several atmospheric processes (climate changes, ozone depletion, cloud formation, and other atmospheric processes). This short

0-87371-829-1/00/$0.00+$.50

chapter attempts to summarize the effects of the bad aerosols to mainly the effects produced by *secondary atmospheric anthropogenic aerosols* (SAAA).

SECONDARY ATMOSPHERIC ANTHROPOGENIC AEROSOLS

Atmospheric anthropogenic aerosols are aerodispersed systems of solid and liquid particles with different sizes, forms, and chemical compositions. Only a small portion of these particles — the primary ones — which are dispersed into the atmospheric environment by wind from soils, weathered rocks, deserts, sea spray, volcanoes, etc., are single inorganic or organic compounds. The majority of atmospheric anthropogenic aerosol particles, especially the secondary ones, represent chemical mixtures. Also, the chemical composition of single particles can be anisotropic; this means that the chemical composition is different, for example, on the surface and inside of particles. Furthermore, the very fine particles can form agglomerates and, because of relatively high specific surfaces, may adsorb or absorb volatile substances or gases. It will be shown later that the single-particle chemistry is also a function of the particle size.

The size as well as chemical composition of the atmospheric anthropogenic aerosols are further determined by emission sources. Therefore, a correlation exists between emission sources and ambient air particle characteristics.[6-9]

FORMATION, TRANSFORMATION, AND CHARACTERIZATION

The processes that impact species once they are released into the atmosphere involve a full spectrum of chemistry and physics. The emitted gas and vapor molecules can react chemically in the gas phase, can be absorbed into a particle or droplet where they might chemically react, can be transported into the stratosphere, or can be removed by interactions with the Earth's surface. Emitted particles can coagulate with other particles, grow by absorption of vapor molecules, or be removed at the Earth's surface or incorporated into a water droplet.[10-12]

The chemistry of air pollutants in the atmosphere — the formation, growth, and dynamics of aerosols — the meteorology of air pollution, and the transport, diffusion, and removal of species in the atmosphere are processes by which the total resulting atmospheric aerosol is characterized (Figure 1.1).[13] A significant fraction of SAAA, such as sulfates, nitrates, ammonium, organics, etc., is produced in the atmosphere by combination of several physico-chemical processes: gas-phase chemistry, aqueous-phase chemistry, condensation, adsorption, etc.[14]

The formation of secondary organic aerosols is one of the most complicated processes in heterogeneous atmospheric chemistry, involving oxidation of the precursor volatile organic component, accumulation of the species of the secondary organic aerosol, and reversible transport to the aerosol phase.[14-16]

Useful and verified physico-chemical models are able to describe the formation of the SAAA and predict their behavior and characterization at urban and regional levels.[11] The same is also valid for modeling the physical processes that are involved in model aerosol dynamics. They are able to determine aerosol size distribution functions in correlation with time and space.[17] The processes affecting aerosol behavior are summarized in the Table 1.1 and Figure 1.2. The entire process can then be expressed by the General Dynamics Equation first formulated by Friedlander in 1977 and modified by Whitby and McMurry in 1997 (Figure 1.3).[17,18]

PARTICLE SIZE-DEPENDENT CHEMISTRY AND TOXICOLOGY

The mass distribution of particle sizes found in the atmosphere includes several modes. The "nucleation mode" includes those particles with diameters less than 0.1 μm; these are termed "ultrafine particles" or "highly dispersed aerosols." This fraction includes "nanometer particles," defined as particles with diameters less than 50 nm.[19]

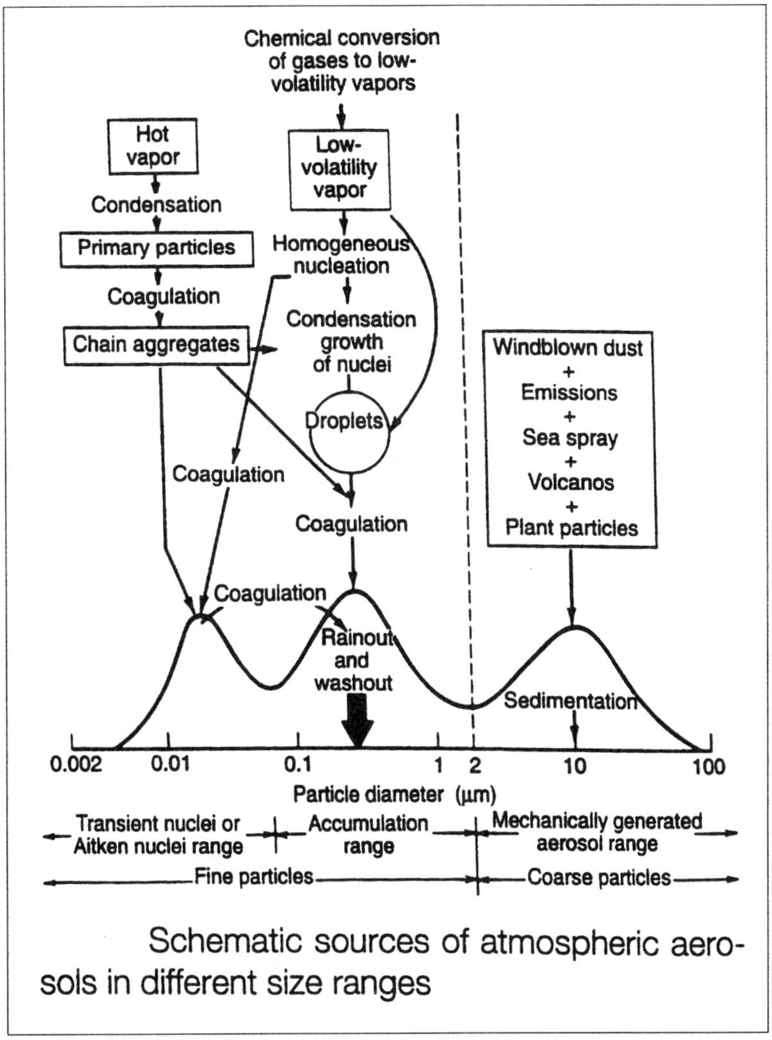

FIGURE 1.1 Schematic picture showing the physical and physico-chemical processes in the development of atmospheric aerosols. (From Reference 13. With permission.)

TABLE 1.1
Processes Affecting Aerosols

Internal Processes	External Processes
Coagulation	Diffusion
Fragmentation	Convection
Particle Growth	Particle migration due to external forces
Condensation/evaporation of vapors	Deposition to surfaces in a closed system
Adsorption/desorption of gases	Convection-like fluxes in an open system
Absorption of gases	
(reactions on or within particles)	
Internal sources/nucleation	
Fraction of bulk-phase material	
Reentrainment of particles from surfaces	

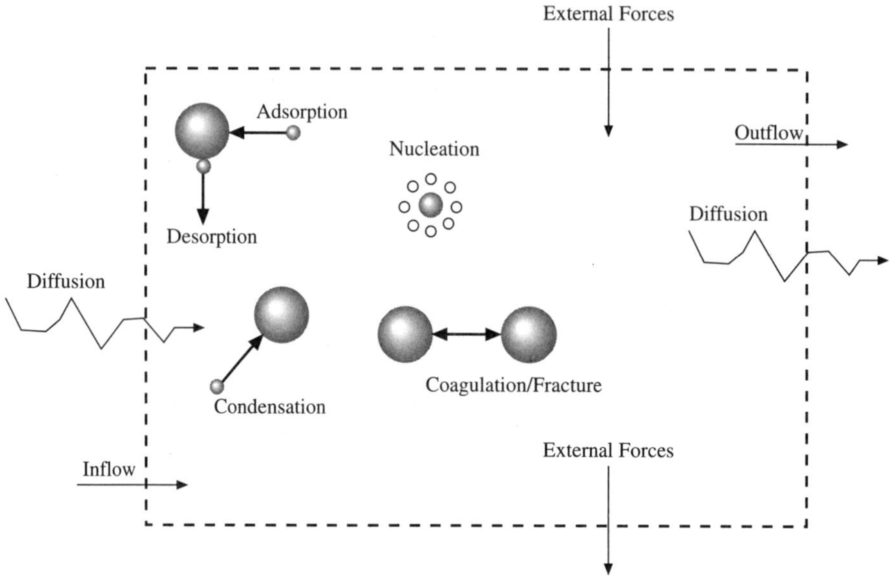

FIGURE 1.2 Aerosol dynamic processes for a control volume. (From Reference 17. With permission.)

Electron microscopical and atomic force microscopical investigations of the atmospheric particle fraction less than 0.1 μm (PM-0.1) can be used to confirm the existence of "nano-particulates" in the polluted atmosphere (Figure 1.4).[20]

As particles less than 0.1 μm in diameter coagulate relatively fast (lifetime less than 1 h), from the standpoint of sampling and measurement as well as from a toxicological standpoint, the "ultrafine fraction" was recently defined as a fraction with particle diameters less than 1 μm. This fraction then includes the "nucleation" mode and a large portion of the "accumulation" mode. It is also designated as PM-0.1 (particulate matter). The fraction now designated as "fine particles" is the fraction with particle diameters less than 2.5 μm; this fraction is also called the PM-2.5 fraction, and includes "nucleation" and "accumulation" modes. All particles with diameters below 10 μm are designated as the "respirable" or "thoracic" fraction and also as the PM-10 fraction. It includes the "nucleation," "accumulation," and a portion of the "coarse" modes. TSP (total suspended particles) is then designated as the fraction that includes the "nucleation," "accumulation," and "coarse" modes.

TSP and PM-10 were first introduced and standardized in 1971.[21] The fractions PM-2.5 or PM-1 are now considered standardized.[22] Data exists regarding the relationships among the measured fractions PM-2.5, PM-10, and TSP.[23]

TSP (particles with aerodynamic diameters less than 50 μm) represents a very broad size range, including PM-10, PM-2.5, and PM-1.[24] TSP, therefore, has a very poor toxicological definition. PM-10 has been defined from a physiological standpoint only and represents the "thoracic" particle fraction. TSP and PM-10 do not account for, correctly and sufficiently, the health risk potential of particulate air pollutants. Their definitions neglect the role of particle chemistry and toxicity.

The PM-2.5 fraction, recently proposed and already used in routine measurements, seems to be a much better standard that appears to account for the toxic and carcinogenic potential of particulate air pollutants. PM-1 and PM-0.1, with particle sizes below 1 μm and 0.1 μm, respectively, represent the "most toxic modes" of atmospheric anthropogenic aerosols. Both fractions are characterized by high particle numbers and specific surfaces, while their mass concentrations are very low.

From a toxicological standpoint, the chemical composition of the fine and ultrafine airborne particles is of basic importance. Nitrates, sulfates, ammonia, and inorganic and organic carbon are

I.

$$\frac{\partial}{\partial t}n(v_p) = \underbrace{-\nabla\cdot \mathbf{v}n(v_p)}_{\text{convection}} \underbrace{-\nabla\cdot \mathbf{c}(v_p)n(v_p)}_{\text{external forces}} + \underbrace{\nabla\cdot D(v_p)\nabla n(v_p)}_{\text{diffusion}}$$

$$+\underbrace{\frac{1}{2}\int_0^{v_p}\beta(\bar{v}_p, v_p-\bar{v}_p)n(\bar{v}_p)n(v_p-\bar{v}_p)\mathrm{d}\bar{v}_p - \int_0^{\infty}\beta(\bar{v}_p,\bar{v}_p)n(v_p)n(\bar{v}_p)\mathrm{d}\bar{v}_p}_{\text{coagulation}}$$

$$+\underbrace{\left[\frac{\partial}{\partial t}n(v_p)\right]_g}_{\text{particle growth}} + \underbrace{\dot{n}_s(v_p)}_{\text{internal sources}},$$

II.

$$\frac{\partial}{\partial t}(M_k) = \underbrace{\nabla\cdot \mathbf{v}M_k}_{\text{convection}} \underbrace{-\nabla\cdot\int_0^{\infty}d_p^k\mathbf{c}(d_p)n(d_p)\mathrm{d}d_p}_{\text{external forces}} + \underbrace{\nabla\cdot\int_0^{\infty}d_p^kD(d_p)\nabla n(d_p)\mathrm{d}d_p}_{\text{diffusion}}$$

$$+\underbrace{\frac{1}{2}\int_0^{\infty}\int_0^{\infty}\left(d_{p_1}^3+d_{p_2}^3\right)^{k/3}\beta(d_{p_1},d_{p_2})n(d_{p_1})n(d_{p_2})\mathrm{d}d_{p_1}\mathrm{d}d_{p_2} - \frac{1}{2}\int_0^{\infty}\int_0^{\infty}\left(d_{p_1}^k+d_{p_2}^k\right)\beta(d_{p_1},d_{p_2})n(d_{p_1})n(d_{p_2})\mathrm{d}d_{p_1}\mathrm{d}d_{p_2}}_{\text{coagulation}}$$

$$+\underbrace{\int_0^{\infty}\frac{\mathrm{d}d_p^k}{\mathrm{d}v_p}\frac{\partial}{\partial t}(v_p)n(d_p)\mathrm{d}d_p}_{\text{particle growth}} + \underbrace{\int_0^{\infty}d_p^k\dot{n}_s(d_p)\mathrm{d}d_p}_{\text{internal sources}}.$$

FIGURE 1.3 General dynamic equations for aerosols formulated by Friedlander (I) and by Whitby and McMurry (II). (From Reference 17. With permission.)

the most abundant species in the PM-2.5 fraction. This fraction predominantly incorporates metals such as V, Cr, Mn, Zn, Se, Pb, Ni, Cd, Pt, Pd, Rd, etc. In a recent study,[25] ultrafine atmospheric particles (PM-0.1) sampled in the Los Angeles area were characterized by physical and chemical parameters. The measured particle number concentrations were found to be in the range of 10^4 particles per milliliter of air. The mass concentrations were in the range of 1 μg m^{-3}. Organic compounds were the largest contributors to ultrafine mass concentrations. A small amount of sulfate was present in these particles. Iron was the most prominent transition metal found in ultrafine particles (Figure 1.5). The toxic and carcinogenic organic particulate compounds are highly dispersed aerosols. The polycyclic aromatic hydrocarbons (PAHs) belong among the most common ubiquitous organic air pollutants. They are present in the atmosphere as volatile and semivolatile, but mainly as fine and very fine particulates. It can be demonstrated that the PAHs and their derivatives are bounded practically on fine and very fine particulates.[26,27]

The majority of such particles with PAHs and other organic toxicants are produced by combustion processes. Diesel vehicles are one of the major sources of fine, atmospheric toxic particulate matter in urban environments. Submicron diesel-particle mass size distributions display three lognormal modes that are centered at 0.09, 0.2, and 0.7–1.0 mm of particle aerodynamic diameter (Figure 1.6). The size distributions of elemental carbon (EC) and organic carbon (OC) are quite different: EC peaks at 0.1 mm, and OC peaks between 0.1 and 0.3 mm.[28]

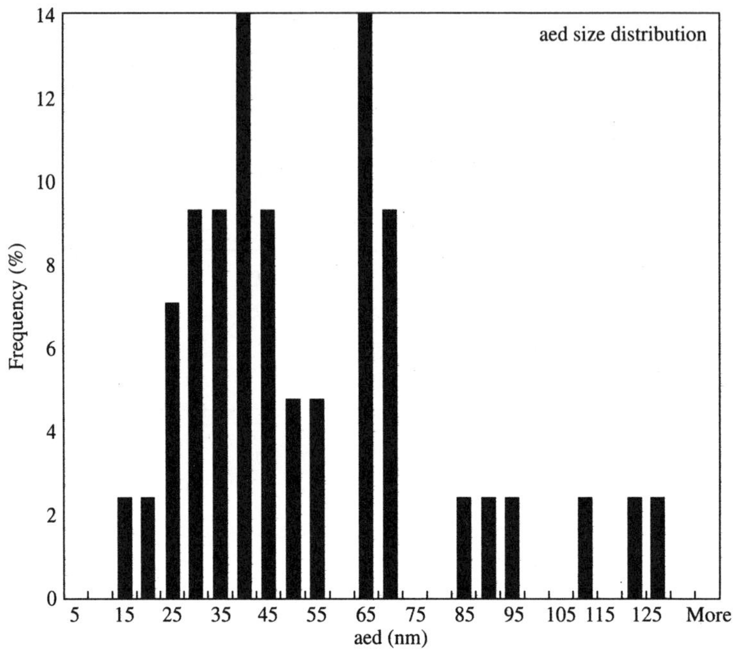

FIGURE 1.4 An example of the aerodynamic (aed) distribution of atmospheric aerosols (Vienna, Austria) in the nanometer size region. (From Reference 20. With permission.)

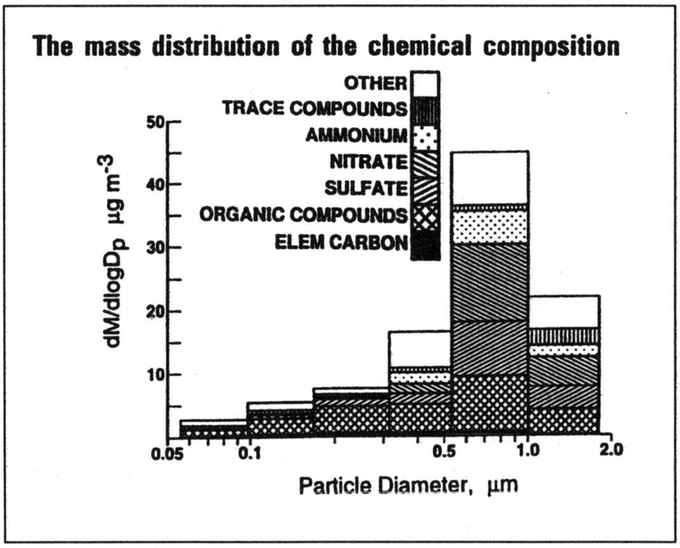

FIGURE 1.5 The mass distribution of the composition of wintertime fine and ultrafine particles measured at Pasadena, CA. (From Reference 25. With permission.)

In vitro cyto- and genotoxicity tests have been used to evaluate the inhalation hazard from airborne particulates.[29] Furthermore, the application of *in vitro* procedures makes it possible to correlate the genotoxicity, mainly the mutagenicity, to physical (e.g., particle size) and chemical (e.g., PAHs, etc.) properties of air particulate samples.

Investigations of basic importance were realized in the urban air of the city of Bologna (Italy).[30] The mutagenicities of total (PM-10) and of particle size fractions-particle diameters, in mm:

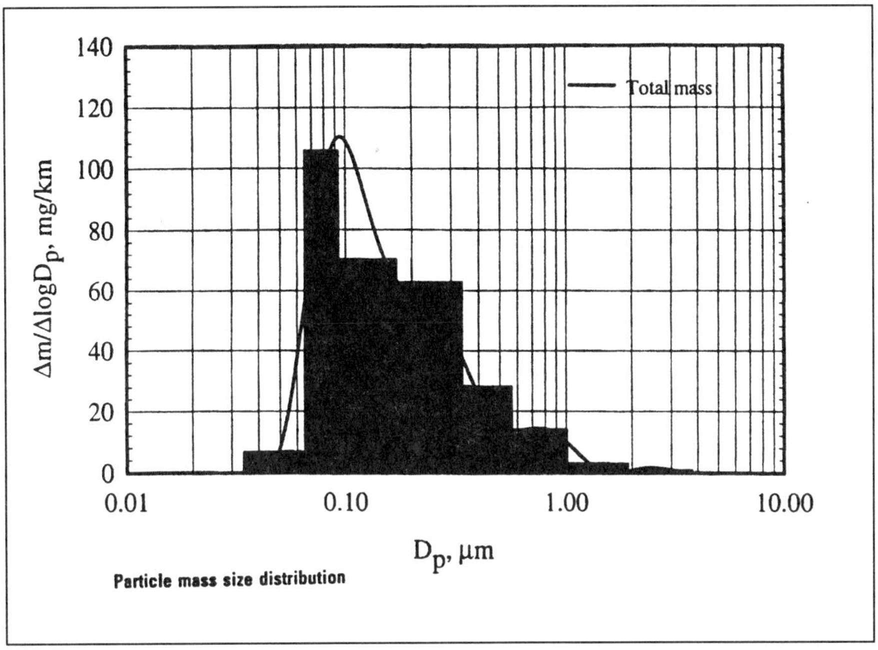

FIGURE 1.6 The size distribution (particle diameter D_p) of particles emitted in diesel engine exhaust. (From Reference 27. With permission.)

0.4: 0.4–0.7; 0.7–1.1; 1.1–3.3

of urban air particulates were identified using tests on *Salmonella typhimurium.* There was no correlation between total or coarse-particle concentrations in air and the mutagenic activity. The correlations increased as particle size decreased; moreover, the finer the particles, the greater the mutagenicity (Figure 1.7). Therefore, the PM-10 concentrations do not appear to be representative of air quality, at least with regard to mutagenicity. PM-2.5 seems to combine a better air quality concept with effective health risk.

Increasing Fine Particulate Emissions

Since about 1980, important changes in the physical and chemical characterizations of air pollutants have been observed and quantitatively determined.[31-34] The previous heavy pollutions by coarse dusts and dark smokes disappeared from the skies of cities in highly developed countries. The emission situation has changed considerably in these countries, mainly due to improved burning and dust abatement technologies. Today, particulate emissions from transportation fuel combustion, etc. contain fine aerosols, with particle sizes less than 2 μm. The coarse particulates in industrial emissions are very efficiently removed by modern air-cleaning equipment, but fine and very fine particulates penetrate into the atmospheric environment.

While the concentrations of coarse particles (approximate size range of 2 to 10 μm) have considerably decreased, an increasing tendency of the concentration of fine (particle sizes less than 2.5 μm) and ultrafine (particle sizes less than 1 μm) particles has been confirmed in several more recent observations and measurements.[33,34]

Therefore, the decrease in emitted particulates from different sources does not imply a simultaneous reduction in the fine particle fraction. For example, the newer diesel engine (1991) produces less mass of soot particles, but has increased emissions of ultrafine soot particles in comparison with the previous (1988) diesel engine. This could occur if the newer engine causes incomplete

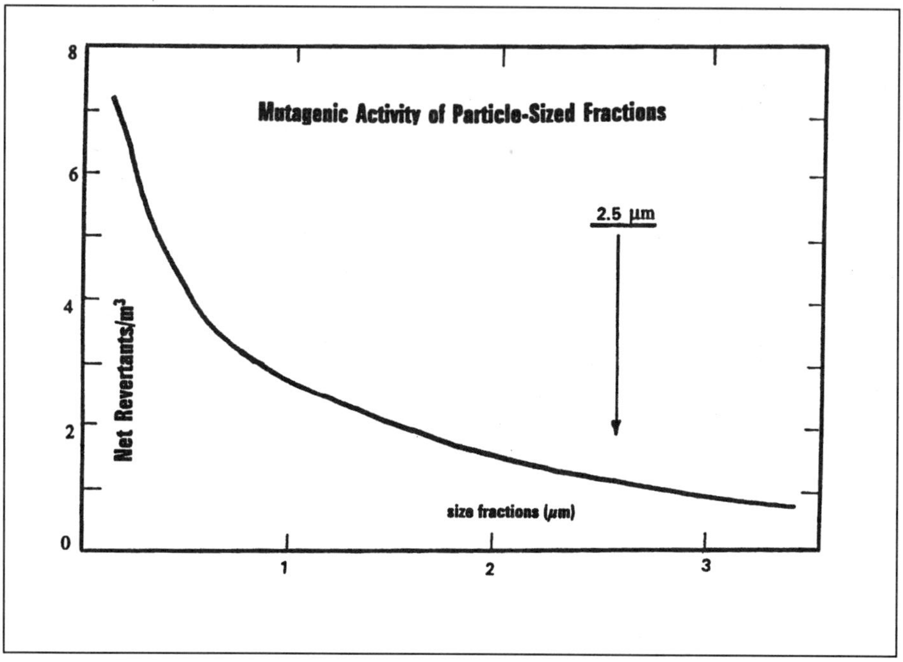

FIGURE 1.7 Mutagenic activity of air particulates as a function of particle diameter, indicated in net revertants per cubic meter air. (From Reference 30. With permission.)

burn-out of soot particles, causing particle agglomerates to disintegrate.[25] An increased production of fine and very fine particulate emissions seems to be the characteristic development in atmospheric pollution at the present time.

HEALTH EFFECTS

The results of existing toxicological studies (*in vitro* and *in vivo*) suggest that both physical (particle size, shape, surface, and biopersistence) and chemical (solved and leached toxic chemicals, and surface catalytic reactions) properties of fine (PM-2.5) particulate fractions are involved in several health effects.

Air Particulate Epidemiology

Recent epidemiological studies that have examined particulate air pollutant concentrations in relation to health statistics conclude that elevated fine-particle matter concentrations (PM-2.5) are associated with increased mortality and morbidity in the general population, and especially in vulnerable children and elderly persons.[34-37]

Figure 1.8 summarizes the results of such investigations. The ambient air concentrations of fine particulates (PM-2.5) correlate well with excess mortalities of six U.S. cities.[37] Correlation with the concentrations of the PM-10 particulate fraction was not satisfactory.

Toxicology of Fine Particulates

The mechanisms responsible for the observed increases in mortality and morbidity are virtually unknown. Nevertheless, there seems to be clear evidence that specific components of the air particulates alone, or as components of mixtures with other pollutants, produce adverse biological

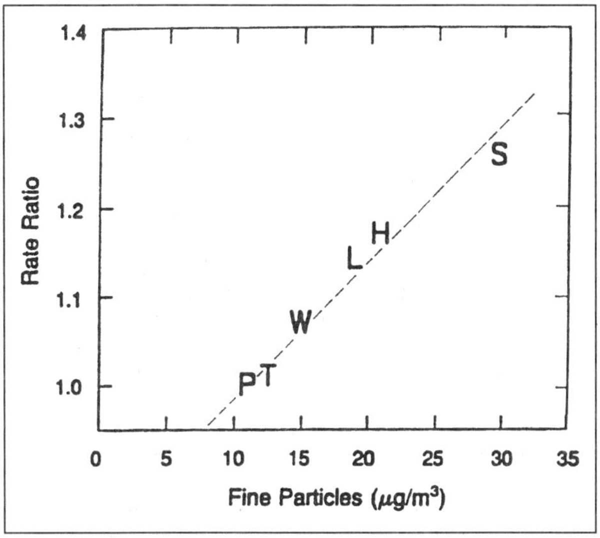

FIGURE 1.8 Estimated adjusted mortality rate ratios and pollution levels (PM-2.5) in six U.S. cities: Portage, WI (P); Topeka, KS (T); Watertown, MA (W); St. Louis, OH (L); Harrison, TN (H); and Steutenville, OH (S). (From Reference 37. With permission.)

responses consistent with human morbidity findings.[34,36] Results of some recently published epidemiological studies, as well as animal inhalation experiments, support the hypothesis that physical and chemical properties of single particles are involved in the toxic, genotoxic, and carcinogenic health effects of inhaled atmospheric particulates. Particle size, shape, and electric charge, particulate surface, and solubility are the most important physical parameters that have been correlated with the observed toxic effects. These parameters also influence substantially the particle lung deposition and lung clearance rates. Fine and very fine particles penetrate into the deep lung compartments, and their deposited fraction depends strongly on particle diameter and breathing rates. Furthermore, the electric charge of fine particles enhances airway deposition. Similarly, particle lung clearance depends on particle size. The hypothesis that fine and very fine particles, when inhaled, can be very toxic to the lung is therefore supported by their high deposition efficiency in the lower respiratory tract, by slow clearance rates, by their large numbers per unit mass, and by the increased surface areas available for interactions with cells. Fine and very fine particles have been shown to be highly toxic to rats. They are poorly taken up by lung macrophages and are capable of penetrating the pulmonary epithelium into the interstitium.[38-44]

In several investigations, the dependency on chemical composition of inhaled air particulates was also proved.[44] Particle–cell interactions potentially could be amplified by the presence of radicals on their surfaces. Transition metals (e.g., Fe, Mn, V, Ni, Ti, etc.), which are the most important inorganic constituents of air particulates, release in interactions with cells metal ions; for example, ferric ions, which catalyze the production of hydroxyl free radical via the Fenton reaction.[45] The free radical activity of particle surfaces may have several "damaging" effects on cells with which they make contact: lipid peroxydation, protein oxidation, DNA strand breaks, antioxidant depletion.

There also exists a direct production of short-lived free radicals (OH, HO_2, H_2O_2) in particulates in the air by several chemical reactions. Their concentrations are approximately one order of magnitude higher in polluted urban regions than in rural and remote areas.[46]

Particle toxicity and carcinogenicity are also enhanced by numerous organic constituents (e.g., PAHs, oxy- and nitro-PAHs, etc.) that are adsorbed or absorbed on and in fine and ultrafine insoluble particulates.[47-49] The majority of such particles with organic toxicants are produced by combustion

California considers tougher diesel exhaust rule

The California Air Resources Board (CARB) is expected to vote in January on a proposed rule that would identify diesel exhaust as a toxic air contaminant (TAC), paving the way for the state to establish regulations that are more stringent than existing federal and state diesel exhaust limits.

After four years of study, CARB released a draft report in May showing that lifelong exposure to the state's current ambient concentrations of diesel exhaust significantly increases the risk of cancer in humans.

Diesel exhaust is a complex mixture of gases, vapor, and fine particles composed of arsenic, benzene, and nickel, which are known to cause cancer in humans. At least 40 other components of the exhaust, including suspected human carcinogens benzo[*a*]pyrene, 1,3-butadiene, and formaldehyde, have been listed as TACs by CARB and as hazardous air pollutants by EPA, according to the draft report.

California has adopted a series of mobile source controls that are expected to reduce diesel exhaust particulate matter (PM) emissions from on-road mobile sources by 75% between 1990 and 2010. In July, EPA came out with a National Ambient Air Quality Standard for particulate matter with an aerodynamic diameter of 2.5 micrometers or less. California will consider how this standard will affect its diesel exhaust particle controls, according to the proposal, which noted that a larger percentage of the fine $PM_{2.5}$ inventory is attributable to diesel fuel combustion sources.

VOL. 31, NO. 8, 1997 / ENVIRONMENTAL SCIENCE & TECHNOLOGY / NEWS ■ 355 A

FIGURE 1.9 Proposal of tougher diesel exhaust rules in California. (With permission.)

processes. Diesel vehicles are a major source of fine, atmospheric toxic particulate matter in urban environments. Therefore, special diesel engine exhaust rules will be necessary (Figure 1.9).

CLOUDS, GLOBAL, CLIMATIC EFFECTS, AND OZONE DEPLETION

Cloud formation as well as cloud properties are substantially influenced by the physical and chemical characteristics of the aerosols present in the air.[50-56] Since cloud condensation nuclei (CCN) and ice nuclei (IN) impart initial cloud characteristics, and consequently influence the evolution of clouds, essentially all cloud properties are dependent on these aerosols. In turn, clouds modify the preexisting aerosol, by removal and by aqueous reactions.

The ocean is the main source of hygroscopic CCN, through the bursting of bubbles in breaking waves. Nevertheless, the continental air is also rich in natural and man-made CCN. The importance of man-made CCN is increasing. Both physical and chemical properties of the anthropogenic CCN are involved in their role in cloud formation. Several particulate air pollutants are suitable nuclei for water condensation; for example, the diesel exhaust aerosol has proven to be a very effective CCN.

Generally speaking, fine dispersed aerosols produced by anthropogenic processes are often good or very good CCN. The increasing concentration of anthropogenic CCN will be able to influence climatic changes, especially through the cloud albedo. Furthermore, the continental anthropogenic aerosol may also influence maritime clouds. Nitrate and sulfate concentrations in oceans are increasing, probably via transport of continental particulate air pollutants. The anthropogenic CCN significantly affect clouds and modify the greenhouse effect, forcing increased concentrations of CO_2.[51,52] Furthermore, the particle size, chemical composition, and reactivity of emitted anthropogenic aerosols substantially influence the optical properties of the global atmosphere. Increasing concentrations of a number of atmospheric chemicals, including aerosols, lead to concerns about the possibility of resulting climate changes.

Another role played by aerosol chemistry involves atmospheric chemical reactions. For example, organosulfur species (like DMS, dimethylsulfide) are oxidized in the marine atmosphere and form CCN.

In order to estimate the direct effects on climate, physical, chemical, and global radiative transfer models have been developed.[53,54] To validate such models, ambient aerosol properties (such as complex refractive index, particle size, and shape) as well as the chemical composition of particles need to be quantified. These aerosol properties need to be integrated with processes that influence aerosol formation, transformation, and transport.[53] Aerosol chemistry play a very important role in such processes. The role of aerosols in tropospheric and stratospheric ozone photochemistry has also been partially quantified.[57-59] It could be shown that heterogeneous reactions on aerosols can play a significant role in ozone depletion mechanisms. In the stratosphere, low temperature reactions on nitric acid trihydrate, on water ice particles, or on particles consisting of ternary supercooled solutions of sulfuric acid, nitric acid, and water, lead to rapid conversion of HCl and $ClONO_2$ species into active chlorine species (Cl, ClO, Cl_2O_2), which act catalytically to rapidly destroy ozone.[57]

It is quite possible that several aerosols produced by anthropogenic sources (e.g., carbonaceous aerosols emitted by diesel engines and jet aircraft) are also involved in the heterogeneous chemistry of ozone depletion mechanisms.[59]

AEROSOLS IN THE FOREST ATMOSPHERE

Several air pollutants, including inorganic and organic aerosols, are involved in the observed deterioration and decline of forests in Europe and North America.[60-65] The aerosols in the atmospheric environment of the forest are mainly imported into the forest areas from different industrial and automobile traffic emission sources. However, they are also produced or modified by the vegetation inside the forest atmosphere.

The forest smog periods are well-known. These are formed in the forest atmosphere by chemical and photochemical, as well as by gas-to-particle conversion, processes.[63] These aerosols — organic as well as inorganic, solid as well as liquid — are then deposited on tree leaves and needles by means of different separation mechanisms. The forest is very effective at filtering particles out of the atmosphere.[64] The physical and chemical interactions with the leaf and needle surfaces and cells are responsible for their different phytotoxic effects. Air pollutants, both gaseous and particulates, produce harmful effects on plants. The stomates are major portal entries for these pollutants. Deposited aerosol particles block the stomates and/or, after dissolution or evaporation, diffuse into plants through them. Figure 1.10 illustrates the stomates (ST) and deposited particles on spruce needles sampled in a declined forest. Changes in the physiological and biochemical parameters of the plant cell are a possible explanation for current forest diseases.

"Forest Aerosol"

Qualitatively, inorganic aerosols in the forest atmosphere differ very little from urban aerosols. The local emission sources of inorganic particulates or of their precursors in a forested area are limited. The majority of inorganic particulates are imported into the forest atmosphere from urban and industrial anthropogenic sources and are therefore already altered. Particulates of PM-2.5 constitute the most important fraction of this aerosol type. Sulfate and ammonium, especially ammonium acid sulfates, account for more than 50% of the fine-fraction mass. The acidity of a forest aerosol is higher — often much higher — than the acidity of the aerosol in urban areas. The fine aerosol fraction contains the majority of nitrates, organic and inorganic carbon, and trace metals. On needles and leaves, deposited trace metals are involved in microbial metabolism. The capacity of foliage

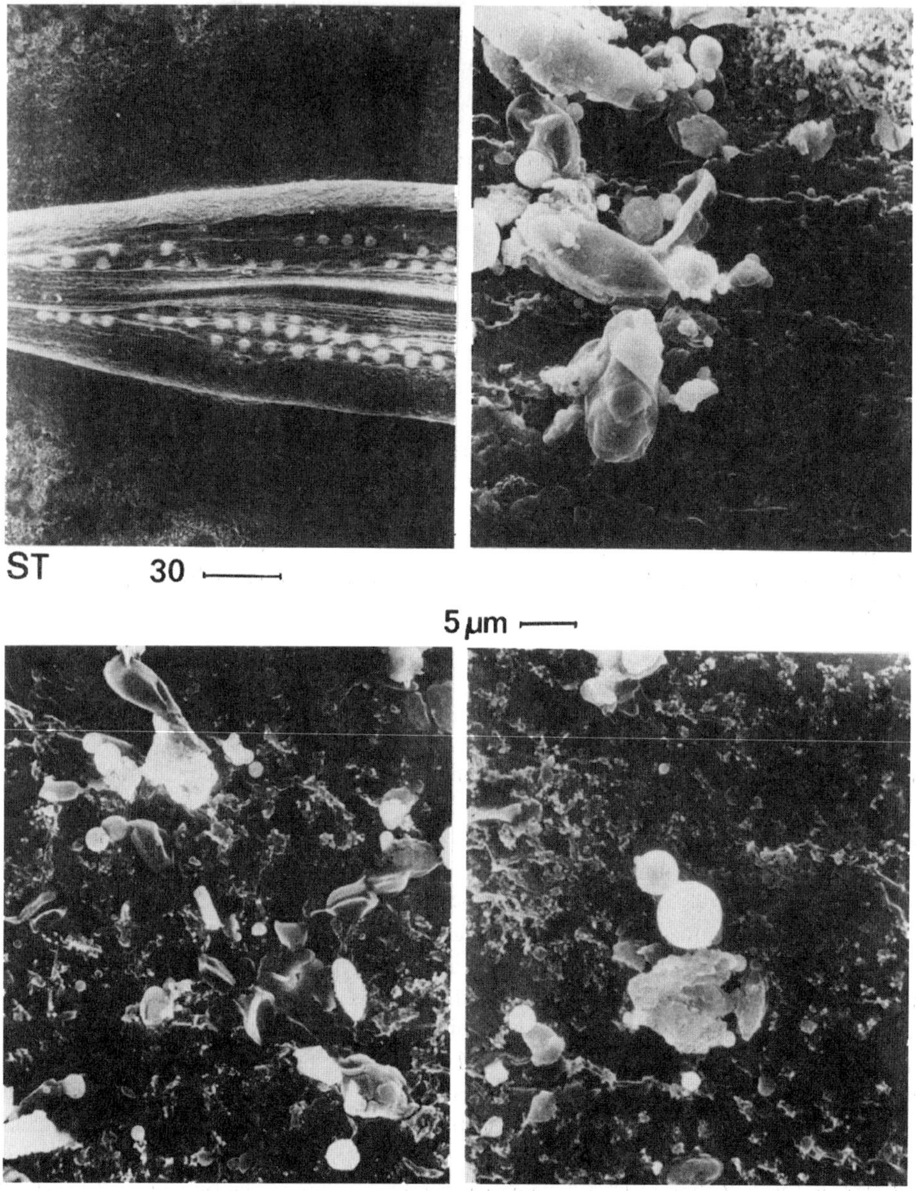

FIGURE 1.10 Scanning electron micrographs of spruce needles and deposited mineral particles.

to accumulate trace-metal particles is considerable. The metal cations interact with the microorganisms of the phyllosphere and are capable of causing or influencing those that function as pathogens.[60] The important differences between urban and forest pollution occur in the field of organic vapors, gases, and aerosols. In some forested areas, organic compounds are the major component of atmospheric aerosols with carbon representing as much as 80% of the total aerosol mass.

The organics adsorbed onto the particles can be divided into a group containing substances of biogenic origin; a group of fatty acids, fatty acid esters, and *n*-alkanes; and a group containing anthropogenic substances such as phthalic acid esters and polycyclic aromatic hydrocarbons (PAHs). The *n*-alkanes seem to be of biogenic origin from plant waxes.

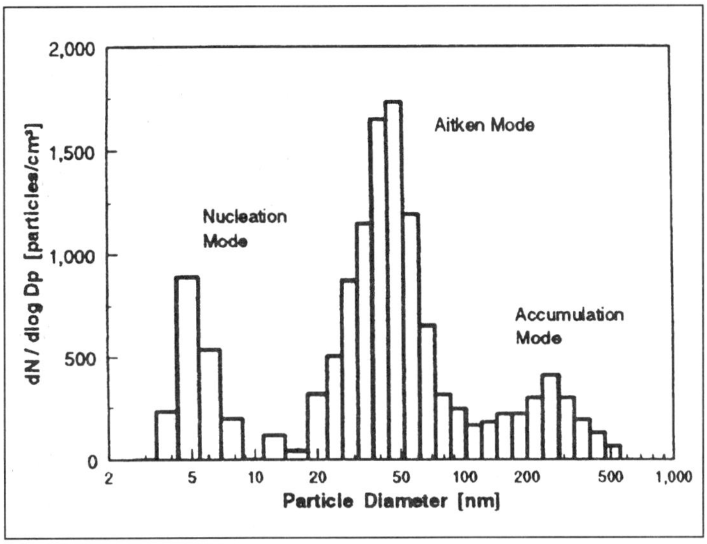

FIGURE 1.11 Three-modal structure of submicron number size distribution of the aerosol in forest. (From Reference 62. With permission.)

Monoterpenes with a concentration level of the order of 100 $\mu g\ m^{-3}$ seem to be important precursors for the formation of highly dispersed secondary aerosols in the atmosphere of forested areas.[63] This secondary organic aerosol of biogenic origin lies in the range of nano-sized particles and is often trimodal. Figure 1.11 shows such a size distribution, which was measured in a boreal forest in Finland. The measurements were carried out inside the forest, and there were no local sources of pollution near the site.[62]

Total worldwide emissions of terpenes plus isoprene are approximately 9×10^8 metric tons per year. In the presence of NO_2, SO_2, PAN (peroxyacetyl nitrate), and UV-irradiation, these gaseous substances are oxidized and produce aldehydes, ketones, etc., which form finely dispersed organic aerosols. Some products of the oxidation by ozone are phytotoxic. The reaction of β-pinene with ozone in the presence of SO_2 leads to the formation of sulfur-containing acidic aerosols.

Assuming that terpene emissions close to tree surfaces (needles and bark) are relatively high, a fast reaction with ozone and subsequent oxidation of SO_2 to sulfuric acid can be expected. In this way, forest trees actively promote acidic depositions. Thus, there is no question that aerosol chemistry contributes substantially into the mechanisms of "forest morbidity."

BUILDING DETERIORATION

Deterioration, corrosion, and weathering of buildings and monuments are produced by complex interactions of physical, chemical, and biological processes, with an important contribution due to air pollutants.[66-74]

Urban stone is damaged primarily by the process of sulfatation. Airborne SO_2 interacts with wet stone, for the most part calcite ($CaCO_3$), to form gypsum ($CaSO_4 \cdot 2H_2O$). The resulting mixture of calcite and gypsum crystals forms a surface crust that slowly consumes the stone (Figure 1.12).[71] Furthermore, different types of microorganisms and bacteria were identified in and under the corrosion crust of deteriorated facades.[75] The bacteria generate, within the stone, slime films that are suitable as a living environment for many microorganisms. These films promote adsorption and absorption of different atmospheric gaseous and aerosol pollutants. Among these, NH_3, $(NH_4)_2SO_4$, as well as ammonium salts, play an important role. The nitro-bacteria living and reproducing in

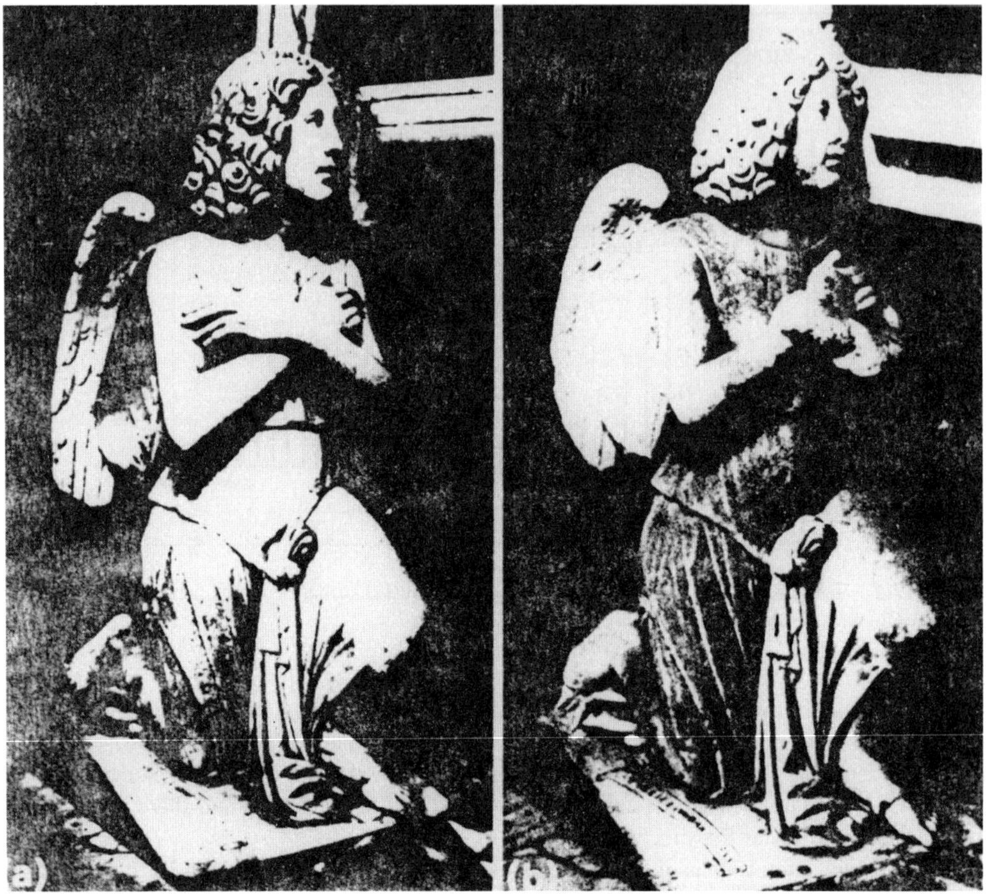

FIGURE 1.12 Photographs of the same marble sculpture: before (left) and after (right) deterioration by air pollutants. (From Reference 71. With permission.)

the lime, consume CO_2, NH_3, and other ammonium compounds and transform them (by oxidation) into NO_2 (HNO_2, HNO_3, etc.). These biogenic acids strongly promote stone deterioration and also catalyze the conversion of SO_2 into sulfates. In this way, the nitro-bacteria and deposited aerosols substantially influence the production of gypsum.

There are observations and measurements showing significant amounts of atmospheric aerosol pollutants being deposited on building facades and monument surfaces.[76-78] Highly dispersed aerosols of sulfates, nitrates, chlorides, fluorides, metals, black carbon, etc. can react in the presence of water with the stone material (e.g., with $CaCO_3$ and $MgCO_3$). The slightly soluble calcium and magnesium carbonates are transformed into more soluble calcium and magnesium sulfates, nitrates, chlorides, fluorides, etc. The salts are then dissolved and transported from the corroded surfaces by rainwater.[68,74]

Different studies have shown that soot and fly-ash particles deposited on the solid stone surface accelerate SO_2 oxidation in the presence of moisture.[79-81]

The residual acidity of fly-ash (Figure 1.13) and soot particulates enhances and catalyzes chemical reactions, and the presence of transition metals, probably as oxides, might accelerate the oxidation of SO_2 to MSO_4^{2+N}. Such a hypothesis was confirmed with laboratory experiments.[82,83] The presence of atmospheric aerosols embedded in the damage layers found on monuments and historical buildings has also been documented and confirmed. Figure 1.14 illustrates the results of such measurements done in the damage layers of stone monuments in Rome.[84]

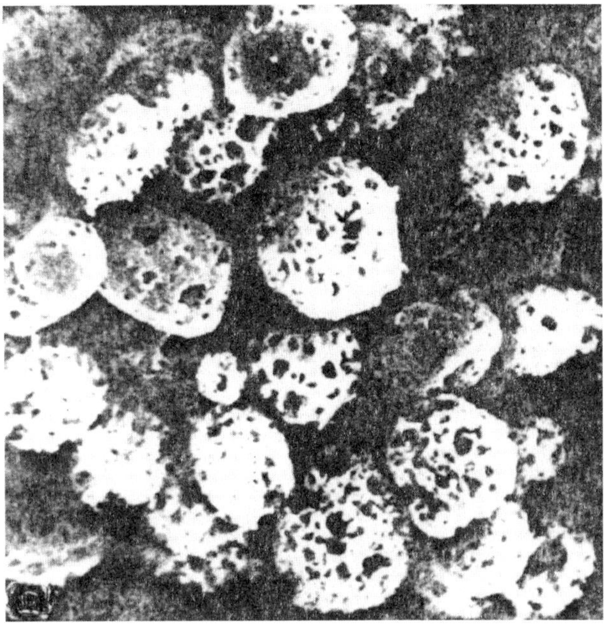

FIGURE 1.13 Scanning electron micrograph of ash particles deposited on a sandstone wall.

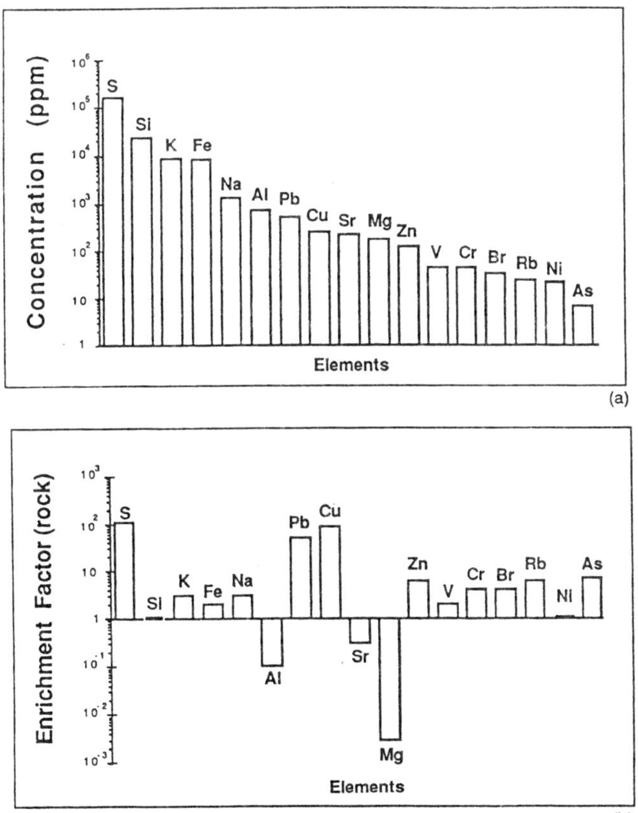

FIGURE 1.14 Mean elemental concentration (a) and rock enrichment factors (b) of elements measured in the damage layers of stone monuments in Rome. (From Reference 84. With permission.)

SAAA, mainly carbonaceous and metallic aerosols, are involved in the complex physicochemical and biological mechanisms that result in the deterioration and damages of stones and other inorganic building materials.

Many pioneer investigations and important contributions relevant to this problem have been done by Italian researchers (see also other chapters in this book).

REFERENCES

1. Pratsinis, S.E. and Kodas, T.T., Manufacturing of materials by aerosol processes, K. Willeke and P.A. Baron, Eds., *Aerosol Measurement,* Van Nostrand Reinhold, New York, 1993, 721.
2. Kodas, T.T., Emgler, E.M., and Lee, V.Y., Generation of thick $Ba_2YCu_3O_7$ films by aerosol deposition, *Appl. Phys. Lett.,* 54, 1923, 1989.
3. Pankov, V., Modified aerosol synthesis for nanoscale hexaferrite particles preparation, *Mater. Sci. Eng.,* A224, 101, 1997.
4. Haas, V., Birringer, R., Gleiter, H., and Pratsinis, S.E., Synthesis of nanostructured powders in an aerosol flow condenser, *J. Aerosol Sci.*, 28, 1443, 1997.
5. Clarke, S.W. and Pavia, D., *Aerosols and the Lung,* Butterworths, London, 1984.
6. Venkataraman, C. and Friedlander, S.K., Source resolution of fine particulate PAH using a receptor model modified for reactivity, *J. Air Waste Management Assoc.,* 44, 1103, 1994.
7. Kao, A.S. and Friedlander, S.K., Chemical signatures of the Los Angeles aerosol, *Aerosol Sci. Technol.,* 21, 283, 1995.
8. Kao, A.S. and Friedlander, S.K., Frequency distribution of PM-10 chemical components and their sources, *Environ. Sci. Technol.,* 29, 19, 1995.
9. Watson, J.G., Chow, J.C., Lowenthal, H.D. et al., Differences in the carbon composition of source profiles for diesel- and gasoline-powered vehicles, *Atmos. Environ.,* 28, 2493, 1994.
10. Seinfeld, J.H., *Atmospheric Chemistry and Physics of Air Pollution,* John Wiley & Sons, New York, 1986.
11. Seinfeld, J.H., Dynamics of urban and regional atmospheric aerosols, *J. Aerosol Sci.,* 28, S417, 1997.
12. Meng, Z.Y., Dabdub, D., and Seinfeld, J.H., Size-resolved and chemically resolved model of atmospheric aerosol dynamics, *J. Geophys. Res. Atm.,* 103, 3419, 1998.
13. Whitby, K.T. and Svendrup, G.M., California aerosols: their physical and chemical characteristics, *Adv. Environ. Sci. Tech.,* 10, 477, 1980.
14. Pandis, S.N., Formation and properties of secondary atmospheric aerosols: from the laboratory to the supercomputer, *J. Aerosol Sci.,* 28, S367, 1997.
15. Odum, J.R., Hoffmann, T., Bowman, F. et al., Gas/particle partitioning and secondary organic aerosol yields, *Environ. Sci. Technol.,* 30, 2580, 1996.
16. Barthelmie, R.J. and Pryor, S.C., Secondary organic aerosols: formation potential and ambient data, *The Sci. Total Environ.,* 205, 167, 1997.
17. Whitby, E.R. and McMurry, P.H., Modal aerosol dynamics modeling, *Aerosol Sci. Technol.,* 27, 673, 1997.
18. Friedlander, S.K., *Smoke, Dust and Haze,* John Wiley & Sons, New York, 1977.
19. Pui, D.Y.H. and Chen, D.R., Nanometer particles, *J. Aerosol Sci.,* 28, 539, 1997.
20. Kölensperger, G., Friedbacher, G., Grasserbauer, M., and Dorffner, L., Investigation of aerosol particles by atomic force microscopy, *Fresenius J. Anal. Chem.,* 358, 268, 1997.
21. *Federal Register,* EPA national primary and secondary ambient air quality standards, FR 36-8186, 1971.
22. Raber, L.R., EPA's air standards, *Chem. Eng. News,* April 14, 10, 1997.
23. Brook, J.R., Dann, T.F., and Burnett, R.T., The relationship among TSP, PM-10, PM-2.5 and inorganic constituents of atmospheric particulate matter at multiple Canadian locations, *J. Air Waste Manage. Assoc.,* 47, 2, 1997.
24. Chow, J.C., Measurement methods to determine compliance with ambient air quality standards for suspended particles, *J. Air Waste Manage. Assoc.,* 45, 320, 1995.

25. Hughes, L.S., Cass, G.R., Gone, J.E.C., Ames, M., and Olmez, I., Physical and chemical characterization of atmospheric ultrafine particles in the Los Angeles area, *Environ. Sci. Technol.,* 32, 1153, 1998.
26. Venkataraman, C. and Friedlander, S.K., Size distribution of PAH and elemental carbon. 1, 2, *Environ. Sci. Technol.,* 28, 555, 1994.
27. Miguel, A.H., Kirchstetter, T.W., Harley, R.A., and Hering, S.V., On-road emissions of particulate PAH and black carbon from gasoline and diesel vehicles, *Environ. Sci. Technol.,* 32, 450, 1998.
28. Kerminen, V.M., Mäkela, T.E., Ojanen, C.H. et al., Characterization of particulate phase in the exhaust from a diesel car, *Environ. Sci. Technol.,* 31, 1883, 1997.
29. Seemayer, N.H., Hadnagy, W., Behrend, H., and Tomigas, R., Inhalation hazard from airborne particulates evaluated by *in vitro* cyto- and genotoxicity testing, *Exp. Pathol.,* 37, 228, 1989.
30. Pagano, P., De Zaiacomo, T., Scarcella, E., Bruni, S., and Calamosca, M., Mutagenic activity of total and particle-sized fractions of urban particulate matter, *Environ. Sci. Technol.,* 30, 3512, 1996.
31. Winkler, P. and Kaminski, U., Increasing submicron particle mass concentrations in Hamburg, *Atm. Environ.,* 22, 2871, 1988.
32. Hitzenberger, R., Fohler-Norek, C., Dusek, U., Galambos, Z., and Sidla, S., Comparison of recent (1994) black carbon data with those obtained in 1985 and 1986 in the urban area of Vienna, Austria, *The Sci. Total. Environ.,* 189/190, 275, 1996.
33. Trier, A., Submicron particles in urban atmosphere, *Atm. Environ.,* 31, 909, 1997.
34. Wilson, R. and Sprengler, J.P., *Particulates in Our Air,* Harvard University Press, Cambridge, MA, 1996.
35. Pope, C.A., Particulate air pollution and health. A review of the Utah Valley experience, *J. Exposure Analysis and Environ. Epidemiol.,* 6, 23, 1996.
36. Schwartz, J., Dockery, D.W., and Neas, L.M., Is daily mortality associated specially with fine particles?, *J. Air and Waste Manage. Assoc.,* 46, 927, 1996.
37. Dockery, D.W., Pope, C.A., Xioing, Xu, et al., An association between air pollution and mortality in six U.S. cities, *N. Engl. J. Med.,* 329, 1753, 1993.
38. Seaton, A., McNee, W., Donaldson, K., and Godden, D., Particulate air pollution and acute health effects, *Lancet,* 345, 176, 1995.
39. Ferin, J., Oberdörster, G., Penney, D.P., Soderholm, S.C., et al., Increased pulmonary toxicity of ultrafine particles, *J. Aerosol Sci.,* 21, 381, 1990.
40. Cheng, K.H. and Swift, D.L., Calculation of total deposition of ultrafine aerosols in human extrathoracic and intrathoracic regions, *Aerosol Sci. Technol.,* 22, 194, 1995.
41. Oberdörster, G., Gelein, R.M., Ferin, J., and Weiss, B., Association of particulate air pollution and acute mortality: involvement of ultrafine particles, *Inhalation Toxicol.,* 7, 111, 1995.
42. Cohen, B., Xiong, J., and Li, W., The influence of charge on the deposition behavior of aerosol particles with emphasis on singly charged nanometer sized particles, J.M.C. Marijnissen and L. Gardon, Eds., *Aerosol Inhalation,* Kluwer Academic, Dordrecht, 1996, 127.
43. Godleski, J.J., Hatch, V., Hauser, R. et al., Ultrafine particles lung macrophages of healthy people, *Am. J. Respir. Crit. Care Med.,* 151, A264, 1995.
44. Godleski, J.J., Sioutas, C., Katler, M. et al., Death from inhalation of concentrated ambient air particles in animal models of pulmonary disease, *Am. J. Crit. Care Med.,* 153, A15, 1996.
45. Donaldson, K., Beswick, P.H., and Gilmour, P.S., Free radical activity associated with the surface of particles, *Toxicol. Lett.,* 88, 293, 1996.
46. Kao, A.S. and Friedlander, S.K., Temporal variations of particulate air pollution: a marker for free radical dosage and adverse health effects, *Inhalation Toxicol.,* 7, 149, 1995.
47. Finlayson-Pitts, B.J. and Pitts, J.N., Tropospheric air pollution: ozone, airborne toxics, polycyclic aromatic hydrocarbons and particles, *Science,* 276, 1045, 1997.
48. Finlayson-Pitts, B.J. and Pitts, J.N., Eds., *Atmospheric Chemistry,* John Wiley & Sons, New York, 1986.
49. Lioy, P.J. and Daisey, J.M., Eds., *Toxic Air Pollutants,* Lewis Publishers, Boca Raton, FL, 1987.
50. Gerber, H.E. and Deepak, A., Eds., *Aerosols and Their Climatic Effects,* A. Deepak Publishers, Hampton, VA, 1984.
51. Lodge, J.P., Ed., Global climatic effects and aerosols, *Atm. Environ.,* 25A, 2433, 1989.
52. Kulmala, M. and Wagner, P.E., Eds., *Nucleation and Atmospheric Aerosols,* Pergamon Press, Oxford, U.K., 1996.

53. Rood, M.J., Ambient aerosols and their influence on climate, *Newsletters of the Chinese Assoc. Aerosol Res.,* 15, 4, 1997.
54. Zhang, Y., Chen, L.L., Carmichael, R., and Dentener, F., The role of mineral aerosols in tropospheric chemistry, Gryning and Schmiermeier, Eds., *Air Pollution Modeling and Its Application,* Plenum Press, New York, 1996, 239.
55. Siefert, R.L., Johansen, A.M., and Hoffmann, M.R., Measurements of trace metals oxidation states in fog and stratus clouds, *Air Waste Manage. Assoc. J.,* 48, 128, 1997.
56. Ohta, S., Hori, M., Yamagata, S., and Murao, N., Chemical characterization of atmospheric fine particles in Sapporo with determination of water content, *Atm. Environ.,* 32, 1021, 1998.
57. Crutzen, P.J., The role of particulate matter and ozone photochemistry, M. Kulmala and P.R. Wagner, Eds., *Nucleation and Atmospheric Aerosols,* Pergamon Press, Oxford, U.K., 1996.
58. Dentener, F.J., Carmichael, G.R., Zhang, Y., Lelieveld, J., and Crutzen, P.J. Role of mineral aerosol as a reactive surface in the global troposphere. J. Geophys. Res. 101, 22, 869, 1996.
59. Schurath, U., Atmospheric aerosols: sources and sinks of reactive trace gases, *J. Aerosol Sci.,* 28, S31, 1997.
60. Smith, W.H., *Air Pollution and Forests,* Spring Publishers, Heidelberg, Germany, 1981.
61. Spurny, K.R., Physico-chemical characterization of aerosols in a forest environment, *J. Aerosol Sci.,* 20, 1103, 1989.
62. Mäkelä, J.M., Aalto, P., Jokinen, V. et al., Observations of ultrafine aerosol particle formation and growth in boreal forest, *Geophys. Res. Lett.,* 24, 1219, 1997.
63. Forstner, H.J.L., Flagan, R.C., and Seinfeld, J.H., Secondary organic aerosol from the photooxidation of aromatic hydrocarbons, *Environ. Sci. Technol.,* 31, 1345, 1997.
64. McLachlan, M.S. and Horstmann, M., Forest as filter of airborne organic pollutants, *Environ. Sci. Technol.,* 32, 413, 1998.
65. Welke, B., Ettlinger, K., and Riederer, M., Sorption of volatile organic chemicals in plant surfaces, *Environ. Sci. Technol.,* 32, 1099, 1998.
66. Del Monte, M., Sabbioni, C., and Vittori, O., Airborne carbon particles and marble deterioration, *Atm. Environ.,* 15, 645, 1981.
67. Amoroso, G.G. and Fassina, V., *Stone decay and conservation,* Elsevier, Amsterdam, 1983.
68. Malissa, H., Hoke, E., and Grasserbauer, M., Erarbeitung mirkoanalytischer Charakterisierungs-Kriterien für Schäden der Bausubstanz in Wohnbaubereich als Folge von Mikrophasenumwandlungen, *Bericht. Bundesministerium für Bauten und Technik,* Wien, 1984.
69. Camuffo, D., Del Monte, M., and Angaro, A., The pH of the atmospheric precipitation in Venice, *Sci. Total Environ.,* 38, 234, 1984.
70. Del Monte, M., Sabbioni, C., and Vittori, O., Urban stone sulfatation and oilfired carbonaceous particles, *Sci. Total Environ.,* 36, 369, 1984.
71. Del Monte, M. and Vittori, O., Air pollution and stone decay: the case of Venice, *Endeavour,* 9, 117, 1985.
72. Del Monte, M., Sabbioni, C., Ventura, A., and Zappia, G., Crystal growth from carbonaceous particles, *Sci. Total Environ.,* 36, 347, 1986.
73. Spurny, K.R., Marfels, H., Schörmann, J. et al., Atmospheric aerosols and building corrosion, *J. Aerosol Sci.,* 18, 605, 1987.
74. Spurny, K.R., Marfels, H., and Schörmann, J., Atmospheric anthropogenic aerosols and building deterioration, *Bautenschutz und Bausanierung,* 11, 150, 1988.
75. Lübbert, E., Ed., Building conservation and protection of monuments, *Zschr. Bauinstandhaltung und Denkmalschutz,* 1986, 1.
76. Horvath, H., Pesava, P., Toprak, S., and Aksu, R., Technique for measuring the deposition velocity of particulate matter to building surfaces, *Sci. Total Environ.,* 189/190, 255, 1996.
77. Lammel, G. and Metzing, G., Pollutant fluxes onto the facades of historical monuments, *Atm. Environ.,* 31, 2249, 1997.
78. Pio, C.A., Ramos, M.M., and Duarte, A.C., Atmospheric aerosol and soiling of external surfaces in an urban environment, *Atm. Environ.,* 32, 1979, 1998.
79. Hutchinson, A.J., Johnson, G.E., Thompson, G.E. et al., The role of fly-ash particulate material and oxide catalysts in stone degradation, *Atm. Environ.,* 26, 2795, 1992.

80. Torfs, K. and Van Grieken, R., Chemical relations between atmospheric aerosols, deposition and stone decay layers on historic buildings at the Mediterranean coast, *Atm. Environ.,* 31, 2179, 1997.
81. Maropoulou, A., Bisbikou, K., Torfs, K., et al., Origin and growth of weathering crusts on ancient marbles in industrial atmosphere, *Atm. Environ.,* 32, 967, 1998.
82. Sabbioni, C., Zappia, G., and Gobbi, G. Carbonaceous particles and stone damage in a laboratory exposure system, *J. Geophys. Res. Atm.,* 101, 196, 1996.
83. Ausset, P., Crovisier, J.L., Del Monte, M. et al., Experimental study of limestone and sandstone sulfatation in polluted realistic conditions, *Atm. Environ.,* 30, 3197, 1996.
84. Sabbioni, C., Contribution of atmospheric deposition to the formation of damage layers, *Sci. Total Environ.,* 167, 49, 1995.

2 Physical Chemistry of Aerosol Formation

Markku Kulmala, Timo Vesala, and Ari Laaksonen

CONTENTS

INTRODUCTION

The formation and growth of aerosol particles in the presence of condensable vapors represent processes of major importance in aerosol dynamics. The emergence of new particles from the vapor changes both the aerosol size and composition distributions. The size distribution of the aerosol at a given time is affected by particle growth rates, which in turn are governed partially by particle compositions. Aerosol deposition, which transfers chemical species from the atmosphere, is influenced by particle size. It is easy to see, therefore, that the physical and chemical aspects of aerosol

0-87371-829-1/00/$0.00+$.50

dynamics are very closely coupled. In short, chemical reactions determine particle compositions and modify their dynamics significantly, while the number, size, and composition of aerosol particles determine conditions for heterogeneous and liquid-phase chemical reactions.

The formation of aerosol particles by gas-to-particle conversion (GPC) can take place through several different mechanisms, including (1) reaction of gases to form low vapor pressure products (e.g., the oxidation of sulfur dioxide to sulfuric acid), (2) one- or multicomponent (in the atmosphere generally with water vapor) nucleation of those low pressure vapors, (3) vapor condensation onto surfaces of preexisting particles, (4) reaction of gases at the surfaces of existing particles, and (5) chemical reactions within the particles. Steps 1, 4, and 5 affect the compositions of both vapor and liquid phases. Step 2 initiates the actual phase transition (step 3) and increases aerosol particle number concentration, while step 3 increases aerosol mass.

The purpose of this chapter is to focus on Steps 2 and 3 of the rather generalized picture of GPC given above. In actuality, the formation of new particles from the gas phase is only possible through homogeneous nucleation, or through nucleation initiated by molecular ion clusters too small to be classified as aerosol particles. Heterogeneous nucleation on insoluble particles initiates changes in particle size and composition distributions, but does not increase particle number concentration. Soluble aerosol particles may grow as a result of equilibrium uptake of vapors (mostly water), but only when the vapor becomes supersaturated can significant mass transfer in the form of condensation take place between the phases.

The driving force of the transition between vapor and liquid phases is the difference in vapor pressures in gas phase and at liquid surfaces. For a species not dissociating in liquid phase, the vapor pressure ($p_{l,i}$) at the surface of aerosol particle is given by:

$$p_{l,i} = X_i \Gamma_i(T,X) Ke_i(T,X) p_{s,i}(T) \tag{2.1}$$

Here, X_i is the mole fraction of component i, Γ_i is the activity coefficient, Ke_i is the Kelvin effect (increase of saturation vapor pressure because of droplet curvature), and $p_{s,i}$ is the saturation vapor pressure (relative to planar surface). For more detailed discussion of liquid phase activities, refer to the chapter subsection "Vapor Pressure and Liquid Phase Activities."

If the partial pressure of species i in the gas phase ($p_{g,i}$) is higher than $p_{l,i}$, a net mass flux may develop from gas phase to liquid phase. A prerequisite is the existence of (enough of) liquid surfaces; this is the case if the aerosol contains a sufficient amount of soluble particles that are able to absorb water and other vapors at subsaturated conditions (i.e., grow along their Köhler curves). Condensation will then start as soon as the vapor becomes effectively supersaturated. However, if liquid surfaces are not present, supersaturation may grow until heterogeneous nucleation wets dry particle surfaces and triggers condensation. If the preexisting soluble and insoluble particle surface area is not sufficient to deplete condensable vapors rapidly enough, supersaturation may reach a point where homogeneous nucleation creates embryos of the new phase.

The phase transition between vapor and liquid phases is often made easier by the presence of more than one condensing species. The reason for this can be understood from Equation 2.1: mixing in the liquid phase tends to lower the equilibrium vapor pressure $p_{l,i}$ of species i compared with $p_{s,i}$, and therefore effective saturation takes place at lower vapor densities in multicomponent vapor than in vapor containing a single species.

Homogeneous nucleation may create new particles in air with low aerosol concentration, but a high effective supersaturation is needed. Therefore, homogeneous nucleation is always a multicomponent process in the atmosphere, involving a vapor such as sulfuric acid, which has a very low saturation vapor pressure and can form droplets with water even at low relative humidities. Homogeneous nucleation of pure water requires relative humidities of several hundred percent, and is thus out of the question in the atmosphere. Heterogeneous nucleation can take place at significantly lower effective supersaturations than homogeneous nucleation. Atmospheric heterogeneous nucleation of water is, in principle, possible; the required relative humidities (R.H.) would be just a few percent over one hundred (depending on the surface characteristics of the particles) and lower still if some other vapor were to participate. However, usually the atmospheric R.H. does not reach values high enough for heterogeneous nucleation of water to take place because rapid condensation on soluble aerosol particles depletes the vapor already at relative humidities below 101%. This is the process predominantly responsible for the generation of clouds and fogs in the atmosphere. Here, the starting point of condensation is not a genuine nucleation process, and can be called activation of soluble aerosol particles. Note also that in the case of activation, other vapors besides water may have an effect: by depressing the equilibrium vapor pressure of the particles ($p_{l,i}$), they may lower the threshold R.H. at which activation takes place.

This chapter focuses on the various aspects of aerosol formation by gas-to-particle conversion. Subsequent chapter sections are devoted to a review of theoretical investigations on one- and two-component homogeneous nucleation; heterogeneous nucleation; activation of soluble particles; and condensational growth of aerosol particles, respectively.

HOMOGENEOUS NUCLEATION

To date, several different theories have been proposed to explain homogeneous nucleation from vapor (for review, see Reference 1). These theories can be roughly divided into microscopic and macroscopic ones. From a theoretical point of view, the microscopic approach is more fundamental, as the phenomenon is described starting from the interactions between individual molecules. However, microscopic nucleation calculations have thus far been limited to molecules with relatively simple interaction potentials, such as the Lennard-Jones potential, and are therefore of little practical value to aerosol scientists who usually deal with molecules too complex to be described by these potentials. The macroscopic theories, on the other hand, rely on measurable thermodynamic quantities such as liquid densities, vapor pressures, and surface tensions. This enables them to be used in connection with real molecular species; and although certain assumptions underlying these theories can be called into question, their predictive success is in many cases reasonable. We shall therefore focus on the macroscopic nucleation theories below.

ONE-COMPONENT NUCLEATION

Classical Theory

The first quantitative treatment that enabled the calculation of nucleation rate at given saturation ratio and temperature was developed by Volmer and Weber,[2] Farkas,[3] Volmer,[4,5] Becker and Döring,[6] and Zeldovich,[7] and is called the classical nucleation theory. The classical theory relies on the *capillary approximation:* it is assumed that the density and surface energy of nucleating clusters can be represented by those of bulk liquid. According to the classical theory, the reversible work of forming a spherical cluster from n vapor molecules is equal to the Gibbs free energy change and can be written as:

$$W = \Delta G_n = n\Delta\mu + A\sigma \tag{2.2}$$

where the chemical potential change between the liquid and vapor phases is given by $\Delta\mu = -kT \ln S$, S is the saturation ratio of the vapor, k is the Boltzmann constant, T is temperature, σ is the surface tension of bulk liquid, A denotes the surface area of the cluster with a volume of $V = nv$, and v is the liquid-phase molecular volume. The equilibrium number concentration of n-clusters is given by:

$$N_n = q_0 \exp(-\Delta G_n / kT) \tag{2.3}$$

In the classical theory, the constant of proportionality q_0 is assumed to be equal to N, the total number density of molecular species in the supersaturated vapor (often approximated by the monomer density).

In supersaturated vapor, the first term of Equation 2.2 is negative and proportional to the number of molecules in the cluster, whereas the second term is positive and proportional to $n^{2/3}$. Consequently, the Gibbs free energy will exhibit a maximum as a function of cluster size. The cluster corresponding to the maximum is called critical, as it is in unstable equilibrium with the vapor; clusters smaller than the critical one will tend to decay, whereas clusters larger than the critical one will tend to grow further. Thus, the term "nucleation rate" refers to the number of critical clusters appearing in a unit volume of supersaturated vapor in unit time. Below, the properties of the critical cluster are denoted by an asterisk.

The radius r^* of the critical cluster can be located by setting the derivative of ΔG with respect to n zero, resulting in the so-called Kelvin equation:

$$r^* = \frac{2\sigma v}{kT \ln S} \tag{2.4}$$

The critical work of formation is then given by

$$W^* = \frac{16\pi\sigma^3 v^2}{3(kT \ln S)^2} \tag{2.5}$$

To derive an equation for the nucleation rate, one has to consider the kinetics of cluster formation; that is, rates at which clusters of various sizes grow because of addition of monomers from the vapor (condensation), and rates at which they shrink because of evaporation. The details of the kinetics are bypassed here (for more information, see e.g., Reference 8), noting just that the steady-state nucleation rate is given by:

$$J = RNZ \exp(-W^*/kT) \tag{2.6}$$

Here, the condensation rate $R = (kT/2\pi m)^{1/2} NA^*$ is the number of molecules impinging on a unit surface per unit time, multiplied by the surface area of the critical cluster, m is the mass of a vapor molecule, and the so-called Zeldovich factor

$$Z = \sqrt{\frac{\sigma}{kT} \frac{2v}{A^*}} \tag{2.7}$$

accounts for the difference between the steady-state and equilibrium concentrations, and for the possibility of re-evaporation of supercritical clusters. It is assumed here that the sticking probability of molecules hitting the critical cluster is unity. In the steady-state, the cluster size distribution remains constant as a function of time, which can result either from constant monomer concentration (which is a generally used approximation), or from monomer production during nucleation. Note that uncertainties in the pre-exponential factors of Equation 2.6 have a much smaller effect on the value of the nucleation rate than uncertainties in W^*.

Self-Consistency

Later investigators (Courtney,[9] Blander and Katz,[10] and Reiss et al.,[11]) have pointed out that with $q_0 = N$, Equation 2.3 does not obey the law of mass action (see also Reference 12), and have argued that a correct treatment of nucleation kinetics results in multiplication of I in Equation 2.6 by a factor of $1/S$ (i.e., q_0 in Equation 2.3 should equal the number concentration of molecules in saturated vapor). Relatedly, Girshick and Chiu[13] and Girshick[14] considered the limiting consistency problem caused by the fact that the classical distribution of n-clusters does not return the identity $N_1 = N_1$. They proposed a self-consistency corrected (SCC) model in which the work of nucleus formation is calculated from

$$W^*_{SCC} = W^*_{CNT} - kT \ln S - A_1\sigma \tag{2.8}$$

where A_1 is the surface area of a (spherical) monomer in liquid phase and CNT denotes classical theory. The nucleation rate is then

$$J_{SCC} = \frac{\exp\left(A_1\sigma/kT\right)}{S} J_{CNT} \tag{2.9}$$

Note that, although the SCC approach offers a correction for both the mass action consistency and limiting consistency problems, the choice of Equation 2.8 must be regarded as somewhat arbitrary. Wilemski[12] argued that the mass action consistency problem is more serious than the limiting consistency problem because mass action consistency is fundamentally necessary, while limiting consistency is not a fundamental property that must be satisfied by a distribution.

The predictive powers of the classical theory and the SCC model appear quite similar, although the classical theory predicts lower nucleation rates than the SCC model. Both theories predict the critical supersaturations S_{cr} (supersaturation at which the nucleation rate reaches a certain level) of some substances rather well and others not so well; the classical theory seems to succeed especially with butanol[15] and the SCC model with toluene.[13] The prediction of correct nucleation rates is usually more difficult than that of critical supersaturations because J is generally a very steep function of S, and thus both of the above theories predict in some cases nucleation rates differing from the experimental ones by several orders of magnitude. A common problem with both theories is the incorrect temperature dependence of the predicted S_{cr} found with many substances. In any case, the fact that almost-correct critical supersaturations are predicted by theories relying on the capillarity approximation is quite remarkable in itself.

Nucleation Theorem

An important new development in nucleation studies is the rigorous proof of the so-called Nucleation Theorem, given by Oxtoby and Kashchiev.[16] The Nucleation Theorem relates the variation of work of formation of the critical cluster with its molecular content:

$$\left[\frac{\partial W^*}{\partial \mu_{ig}}\right]_{\mu_{jg},T} = -n_i \tag{2.10}$$

Here n_i denotes the number of molecules belonging to species i in a multicomponent vapor, and μ_{ig} is the gas-phase chemical potential of species i in a multicomponent vapor. This result was first proposed for one-component systems by Kashchiev,[17] who assumed that the surface energy of the nucleus is only weakly dependent on supersaturation. Viisanen et al.[18] derived the Nucleation Theorem using statistical mechanical arguments that assume that the critical cluster and the surrounding vapor can be treated as if decoupled. This approach was generalized to binary systems by Strey and Viisanen,[19] and Viisanen et al.[20] extended Kashchiev's original derivation to the two-component case. However, only with the work of Oxtoby and Kashchiev[16] was it was realized that the Nucleation Theorem is a completely general thermodynamic statement free of any specific model-related assumptions, and holds down to the smallest nucleus sizes. The Nucleation Theorem is particularly useful because it allows the measurement of numbers of molecules in critical clusters. This is possible because the rate of nucleation depends on the work of nucleus formation and on a pre-exponential kinetic factor, which in turn is only weakly dependent on supersaturation. It can be shown that

$$\left[\frac{\partial(kT\ln J)}{\partial\mu_{ig}}\right]_{\mu_{jg},T} = n_i + m \tag{2.11}$$

where m is between 0 and 1. Measurements of molecular content of critical clusters have been performed by Viisanen and Strey,[15] Viisanen et al.,[18,20,21] Strey and Viisanen,[19] Strey et al.,[22,23] and Hruby et al.[24] These studies have shown that with one-component nuclei, the Kelvin equation predicts the critical nucleus size surprisingly well, down to about 40 to 50 molecules.

Scaling Correction to Classical Theory

Applying the Nucleation theorem to a general form of reversible work of critical nucleus formation $W^* = W_{CNT} - f(n^*, \Delta\mu)$, where the function f gives the departure from classical theory, McGraw and Laaksonen[25] derived the following differential equation:

$$\Delta\mu\frac{dn^*}{d\Delta\mu} + 3n^* = 2\frac{d}{d\Delta\mu}\left(fg^*\Delta\mu\right) \tag{2.12}$$

In the classical theory, $f = 0$, and the equation can be solved to give $n^*_{CNT} = C(T)\Delta\mu^{-3}$. The temperature-dependent function $C(T)$ is identified with the help of the Kelvin relation to be $C(T) = (32\pi\sigma^3 v^2)/3$.

In general, f is non-zero; and without additional information, Equation 2.12 cannot be solved. However, in the special case that each side of Equation 2.12 vanishes separately, a class of homogeneous solutions is obtained for n^* and the product fn^*:

$$n^* = C'(T)\Delta\mu^{-3} \tag{2.13}$$

$$fn^* = D(T)\Delta\mu^{-1} \tag{2.14}$$

As $\Delta\mu \to 1$, one must have $n^* \to n^*_{CNT}$ and, hence, $C'(T) = C(T)$. Thus, in the generalized theory, the number of molecules in the critical nucleus is the same as in the classical theory, in agreement with experiments. The work of nucleus formation, on the other hand, becomes

$$W^* = W^*_{CNT} - D(T) \tag{2.15}$$

that is, the difference from the classical theory being given by a function that depends on temperature only and not only supersaturation. This is also in accord with experiments, as the classical theory predicts with many substances the supersaturation dependence of nucleation rate reasonably well. (The predictions of the supersaturation dependence and the molecular content of the nucleus are of course linked by the Nucleation Theorem.)

McGraw and Laaksonen[25] showed that the scaling relation (Equation 2.15) is supported by results[26] from the density functional theory of nucleation[26]; but they did not present a general theory for calculating the temperature-dependent function $D(T)$. However, Talanquer[27] pointed out that $D(T)$ can be obtained by requiring that the work of nucleus formation, W^*, goes to zero at the spinodal line (see Reference 28). The function $D(T)$ then becomes

$$D(T) = C(T)\Delta\mu_s^{-2}/2 \tag{2.16}$$

Talanquer obtained the chemical potential at the spinodal, $\Delta\mu_s$, from the Peng-Robinson equation of state, and showed that this scaling correction (Equation 2.16) substantially improves the classical nucleation rate predictions for several nonpolar and weakly polar substances.

Binary Nucleation

Classical Theory

Classical binary nucleation theory (extension of the Kelvin equation to two-component systems) was first used by Flood,[29] Volmer,[5] and Neumann and Döring.[30] Unaware of the earlier work, Reiss[31] considered binary nucleation, and noted that the growing binary clusters can be thought of as moving on a saddle-shaped free energy surface, the saddle point corresponding to the critical cluster. Building on Reiss' work, Doyle[32] derived the so-called generalized Kelvin equations for binary critical clusters. These equations contained derivatives of surface tension with respect to particle composition. In 1981, Renninger et al.[33] noted that the equations of Doyle were thermodynamically inconsistent, and that the correct binary Kelvin equations do not contain any compositional derivatives of surface tension. This is, incidentally, in accord with the early German investigators. Wilemski[34] showed how the derivatives are removed by the correct use of the Gibbs adsorption equation, and Mirabel and Reiss,[35] and Nishioka and Kusaka[36] later argued that there are even more fundamental thermodynamical reasons for the derivative terms not to appear (see also Reference 28). The discussion below, however, follows Wilemski's derivation.

The change of Gibbs free energy of formation of a spherical binary liquid cluster from the vapor phase is expressed as (e.g., see Reference 31)

$$\Delta G = n_1\Delta\mu_1 + n_2\Delta\mu_2 + 4\pi r^2\sigma \tag{2.17}$$

where n_i denotes the number of molecules of the ith species in the cluster, $\Delta\mu_i$ is the change of the chemical potential of species i between the vapor phase and the liquid phase taken at the pressure outside of the cluster, r is the radius of the cluster, and σ is the surface tension. The properties of the cluster are assumed to be the same as for macroscopic systems with plane surfaces; and possible effects on density and surface tension caused by the curvature of the cluster are neglected (the capillarity approximation). Following Wilemski[34,37] and Zeng and Oxtoby,[26] one can write the total number of molecules of species i in the cluster as:

$$n_i = n_i^s + n_i^b \tag{2.18}$$

Here, n^s_i and n^b_i are the numbers of surface and interior ("bulk") molecules of the ith species in the cluster, respectively. The above-mentioned thermodynamic quantities are determined using the bulk mole fraction X_b.

The saddle point on the free energy surface can be found by setting

$$\left(\frac{\partial \Delta G}{\partial n_i}\right)_{n_j} = 0 \tag{2.19}$$

Applying these conditions to Equation 2.17 and making use of the Gibbs-Duhem equation and the Gibbs adsorption isotherm (see, for example, References 34 and 37),

$$n_1^b d\mu_1^l + n_2^b d\mu_2^l = 0 \tag{2.20}$$

$$n_1^s d\mu_1^l + n_2^s d\mu_2^l + A d\sigma = 0 \tag{2.21}$$

one obtains the binary Kelvin equations

$$\Delta\mu_i + \frac{2\sigma v_i}{r^*} = 0 \tag{2.22}$$

The partial molecular volumes v_i are related to the cluster radius as follows:

$$\frac{4}{3}\pi r^3 = n_1 v_1 + n_2 v_2 \tag{2.23}$$

Note that the v_i values depends on composition, and they are determined at X_b. From the Kelvin equations, one obtains

$$\Delta\mu_1 v_2 = \Delta\mu_2 v_1 \tag{2.24}$$

which can be solved numerically to find the bulk mole fraction X^*_b of the critical cluster at given gas-phase chemical potentials. For the radius and the free energy of formation of the critical cluster, one has

$$r^* = -\frac{2\sigma v_i}{\Delta\mu_i} \tag{2.25}$$

$$\Delta G^* = \frac{4}{3}\pi r^{*2}\,\sigma \tag{2.26}$$

Finally, as noticed by Laaksonen et al.,[38] the total number of molecules in the critical cluster can be calculated using Equations 2.23 and 2.27

$$n_1 d\mu_1^l + n_2 d\mu_2^l + A d\sigma = 0 \tag{2.27}$$

(which follows from the addition of Equations 2.20 and 2.21).

The predictions of binary classical nucleation theory have been found to be qualitatively correct in the case of nearly ideal mixtures.[19] However, for systems in which surface enrichment of one of the components takes place (marked by considerable nonlinear variation of surface tension over the mole fraction range), the predictions of the theory become unphysical. For example, with constant alcohol vapor concentration in a water/alcohol system, addition of water vapor will suddenly result in *lowering* the predicted nucleation rate, associated with a prediction of *negative* total number of water molecules in the critical cluster.

Explicit Cluster Model

An alternative (classical) way to find the critical cluster would be, instead of using the Kelvin equations, to construct the free energy surface ($\Delta G - n_1 - n_2$) with Equation 2.17, and locate the saddle point, for example, with the help of a computer. This can be readily done with systems exhibiting no surface enrichment, that is, if $X_b = X = n_2/(n_1 + n_2)$. If this is not the case, a method is needed to calculate X_b for a general (n_1, n_2)-cluster. Flageollet-Daniel et al.[39] proposed to treat water/alcohol clusters in terms of a microscopic model, allowing for enrichment of the alcohol at the surface of the cluster, and at the same time depleting the interior of alcohol. Laaksonen and Kulmala[40] have proposed an alternative explicit cluster model and demonstrated[41] that for a number of water/alcohol systems, the agreement with the cluster model and experiments is rather good.

The cluster model describes a two-component liquid cluster as composed of a unimolecular surface layer and an interior bulk core with

$$n_i = n_i^s + n_i^b \tag{2.28}$$

The volume of a cluster is calculated assuming a spherical shape. The numbers of molecules in the surface layer are determined from

$$A = 4\pi r^2 = n_1^s A_1 + n_2^s A_2 \tag{2.29}$$

where A_i is the partial molecular area of species i. The surface composition is assumed to be connected to the surface tension of a bulk binary solution via a phenomenological relationship:

$$\sigma(X_b) = \frac{(1 - X_s)v_1\sigma_1 + X_s v_2 \sigma_2}{(1 - X_s)v_1 + X_s v_2} \tag{2.30}$$

Here, $X_s = n^s_2/(n^s_1 + n^s_2)$, and σ_i denotes the surface tension of pure i. This description is, in effect, an approximation to the Gibbs adsorption isotherm. The cluster size is allowed to affect the distribution between surface and interior molecules as the partial molecular areas are taken as curvature dependent (for details, see Reference 41). The cluster model predicts, for a given set of total numbers of molecules at fixed gas temperature, the numbers of interior and surface molecules in the cluster. The surface tension, liquid phase activities, and density are calculated using the interior composition. These quantities and the total number of molecules are then used to determine the binary nucleus by creating a saddle surface in three-dimensional ($\Delta G - n_1 - n_2$) space and searching the saddle point.

The principal difference between the cluster model and the classical theory is that in the former, the cluster size is allowed to affect the relative fractions of the molecules at the surface and in the interior, and thereby also the mole fraction of the critical cluster. Surprisingly enough, it seems that this is sufficient for correcting the unphysical predictions of the classical theory, at least qualitatively (although one should bear in mind that the approximate nature of Equation 2.30 and the equations describing the molecular areas might contribute). Laaksonen[41] found that the theory produced well-behaved activity plots (plots of vapor phase activities at which the nucleation rate is constant at given temperature), and that the predicted nucleation rates were within 6 orders of magnitude of the measured rates in several water/alcohol systems. Furthermore, Viisanen et al.[20] showed that the explicit cluster model predicts almost quantitatively correct numbers of molecules in critical water/ethanol clusters over the whole composition range.

Hydration

The association of molecules in the vapor phase can significantly affect their nucleation behavior. It is known that methanol, for example, has a considerable enthalpy of self-association, and one should therefore treat both the theoretical predictions and experimental results of the methanol nucleation with caution. Another system with a tendency to associate is the binary sulfuric acid/water mixture, which is important in ambient aerosol formation. The very high enthalpy of mixing of these species causes them to form hydrates in the gas phase. The hydrates consist of one or more sulfuric acid and several water molecules, and have a stabilizing effect on the vapor. In other words, it is energetically more difficult to form a critical nucleus out of hydrates than out of monomers (although from a kinetic viewpoint, hydration does make nucleation a little bit easier).

Jaecker-Voirol et al.[42] deduced a correction for the classical free energy of cluster formation, taking into account the effect of hydration. The hydrates were assumed to contain one sulfuric acid and one or more water molecules. Expressing the chemical potential difference of species i with the help of liquid and gas phase activities, one obtains

$$\Delta\mu_i = -kT \ln \frac{A_{ig}}{A_{il}} \tag{2.31}$$

where the activities are given by $A_{il} = p_{i,sol}/p_{i,s}$ and $A_{ig} = p_i/p_{i,s}$; and p_i, $p_{i,s}$, and $p_{i,sol}$ denote the partial pressure, saturation vapor pressure, and vapor pressure over the solution, respectively.

The correction for the acid activities has the following form:

$$\left(-kT \ln \frac{A_{ag}}{A_{al}}\right)_{cor} = -kT \ln \frac{A_{ag}}{A_{al}} - kT \ln C_h \tag{2.32}$$

The correction factor C_h due to hydration is given by:

$$C_h = \left[\frac{1 + K_1 p_{w,sol} + \ldots + K_1 K_2 \times \ldots \times K_h p_{w,sol}^h}{1 + K_1 p_w + \ldots + K_1 K_2 \times \ldots \times K_h p_w^h}\right]^{n_a} \tag{2.33}$$

where the subscripts w and a refer to water and acid, respectively, K_i is the equilibrium constant for hydrate formation, and h is the number of water molecules per hydrate. Jaecker-Voirol et al.[42] noted that an approximate expression is obtained for the equilibrium constants by taking the derivative of ΔG of a hydrate with respect to the number of water molecules. Kulmala et al.[43] extended the classical hydration model into systems where the gas phase number concentrations of acid and water molecules may be of the same order of magnitude. They also showed that the fraction of free molecules to the total number of molecules in the vapor can be solved numerically, rendering the equilibrium constants unnecessary. However, the resulting sulfuric acid hydrate distributions were shown to be similar to those calculated by Jaecker-Voirol et al.[42]

Nucleation Rate

The nucleation rate in a binary system is:[44]

$$I = R_{AV} FZ \exp(-\Delta G^*/kT) \tag{2.34}$$

Here, F is the total number of molecular species in the vapor, and R_{AV} is the average condensation rate. For nonassociating vapors, one obtains

$$R_{AV} = \frac{R_1 R_2}{R_1 \sin^2 \Phi + R_2 \cos^2 \Phi} \tag{2.35}$$

where Φ is the angle between the n_2-axis and the direction of cluster growth at the saddle point of the free energy surface. The Zeldovich nonequilibrium factor Z can be obtained from second derivatives of ΔG at the saddle point (see Reference 44 for details). An approximate expression for Z was given by Kulmala and Viisanen,[45] who derived the classical binary equations starting from the "average monomer" concept. In this case the Zeldovich factor is reduced to that of the one-component case (Equation 2.7 with $v = (1 - X)v_1 + Xv_2$) and $S = (A_{1g}/A_{1l})^{1-X} (A_{2g}/A_{2l})^X$. The corresponding approximate growth angle is the steepest descent angle with $\tan \Phi = X/(1 - X)$.

Although it is true that the nucleation rate is governed primarily by the exponent of ΔG, one should be careful when using the kinetic expressions. Kulmala and Laaksonen[46] showed that various kinetic expressions presented in the literature produced H_2O/H_2SO_4 nucleation rates differing from each other by several orders of magnitude. In a recent study, Vehkamäki et al.[47] used numerical methods to solve the binary kinetic equations exactly. They found that in the water/sulfuric acid system the exact rates were within one order of magnitude of those produced by the analytical expressions of Stauffer.[44]

HETEROGENEOUS NUCLEATION

The quantification of heterogeneous nucleation is even more difficult than that of homogeneous nucleation. This is due to the complexity of interactions between the nucleating molecules and the underlying surface. The heterogeneous nucleation rate is strongly dependent on the characteristics of the surface, and it is extremely difficult to produce well-defined surfaces for experimental investigations. The lack of experimental data, on the other hand, makes it difficult to verify any theoretical ideas. It seems probable that in the future more information of the details of heterogeneous nucleation phenomena will be acquired through molecular dynamics or Monte Carlo simulations, rather than through laboratory experiments (see Reference 48).

A further complication, compared to laboratory conditions, emerges when one tries to carry out calculations of heterogeneous nucleation at ambient conditions: in the lab, the surface materials at least are known, but this is usually not the case in the atmosphere. The uncertainties associated with atmospheric heterogeneous nucleation calculations can therefore by very large. However, some guidance can be acquired about the conditions at which heterogeneous nucleation can take place using the classical nucleation theory, which is reviewed below. Also considered are some factors that may be important, but are left out of the classical description.

BINARY HETEROGENEOUS NUCLEATION ON CURVED SURFACES

Free Energy of Embryo Formation

In classical nucleation theory, the Gibbs free energy of formation of a liquid embryo from a binary mixture of vapors onto a curved surface is given by the expression[49]:

$$\Delta G = -n_a kT \ln \frac{A_{ag}}{A_{al}} - n_b kT \ln \frac{A_{bg}}{A_{bl}} + S_{12}\sigma_{12} + (\sigma_{23} - \sigma_{13})S_{23} \tag{2.36}$$

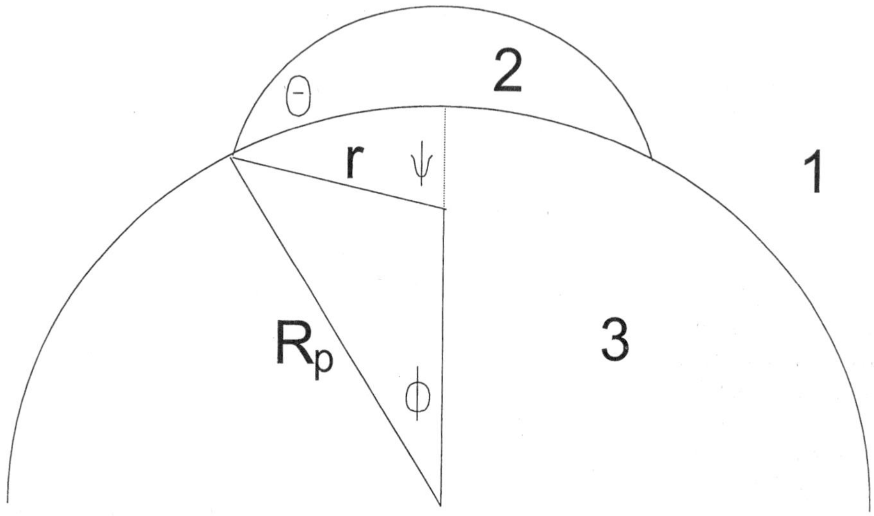

FIGURE 2.1 A cluster (2) on aerosol particle (3) in gas phase (1).

σ_{ij} and S_{ij} are the surface free energy and surface area of the interface, respectively, between phases i and j. The gas phase is indexed by 1, the liquid phase embryo by 2, and the substrate by 3. The contact angle θ is given by $\cos\theta = m = (\sigma_{13} - \sigma_{23})/\sigma_{12}$, and the values of S_{12} and S_{23} by

$$S_{12} = 2\pi r^2 (1 - \cos\psi) \tag{2.37}$$

$$S_{23} = 2\pi R_p^2 (1 - \cos\phi) \tag{2.38}$$

Here,

$$\cos\psi = -\frac{(r - R_p m)}{d} \tag{2.39}$$

$$\cos\phi = \frac{(R_p - rm)}{d} \tag{2.40}$$

$$d = \left(R_p^2 + r^2 - 2rR_p m\right)^{1/2} \tag{2.41}$$

where r is the radius of the embryo and R_p the radius of the solid surface (see Figure 2.1).

One should notice that the heterogeneous nucleation theory gives the same value for the critical radius as the homogeneous theory. However, the differentiation demands some calculus. After it is completed, one finds that the critical value for the Gibbs free energy is given by (see, for example, Reference 49):

$$\Delta G^* = \frac{2}{3}\pi r^{*2}\, \sigma_{12} f(m, x) \tag{2.42}$$

where

$$f(m,x) = 1 + \left(\frac{1 - mx}{g}\right)^3 + x^3\left[2 - 3\left(\frac{x - m}{g}\right) + \left(\frac{x - m}{g}\right)^3\right] + 3mx^2\left(\frac{x - m}{g} - 1\right) \tag{2.43}$$

and

$$g = \left(1 + x^2 - 2mx\right)^{1/2} \tag{2.44}$$

and

$$x = \frac{R_p}{r^*} \tag{2.45}$$

Nucleation Rate

The nucleation rate can be given as (see References 44 and 46):

$$I = R_{av} FZ \exp\left(-\frac{\Delta G^*}{kT}\right) \tag{2.46}$$

Here, F denotes the total number of nucleating molecules, clusters, particles, etc., depending on the system in question. (For example, in homogeneous nucleation, F would be the total number of molecules in the vapor, and in ion-induced nucleation the number of ions). In the case of heterogeneous nucleation, the identification of F is not straightforward. Several different expressions for the heterogeneous nucleation rate can be found in the literature. These differ from each other in the way of counting F; one can use the binary sulfuric acid/water system as an example.

When very small solid particles act as condensation nuclei, the nucleation rate is defined as[50]:

$$I_1 = R_{av} N_{par} Z \exp\left(-\frac{\Delta G^*}{kT}\right) \tag{2.47}$$

Here, N_{par} is the number concentration of the solid particles.

Another formula includes the adsorption mechanism through the quantity N^{ads} (the number of water and acid molecules adsorbed per unit area on the solid nuclei). In the case of atmospheric H_2SO_4, the number of acid molecules is several orders of magnitude smaller than the number of water molecules, and it is sufficient to count only the adsorbed water molecules $N_w^{ads} = \beta\tau$. Here, β is the impinging rate of molecules on the surface of the solid particle, and τ is the time that a molecule spends on the surface of the solid particle, given by $\tau = \tau_o \exp(E/RT)$, where τ_o is a characteristic time and E is the heat of adsorption. Hamill et al.[51] used the value 2.4×10^{-16} s for τ_o, and the value 10,800 cal mol^{-1} for E. Lazaridis et al.,[52] on the other hand, made use of the fact that τ_o corresponds to $1/\nu_o$,[53] where ν_o is the characteristic frequency of vibration. The vibration between two molecules can be calculated using the nearest-neighbor harmonic oscillator approximation. The angular frequency (ω) of the oscillator is:

$$\omega = 2\pi\nu = \sqrt{\frac{d^2V}{dr^2} \cdot \frac{1}{m_\mu}} \tag{2.48}$$

where m_μ is the reduced mass of the two molecules. For V, Lazaridis et al.[52] used the modified Lennard-Jones potential of polar molecules, resulting in $\tau_o = 2.55 \times 10^{-13}$ s, which corresponds to water–water interaction. For E, they used the latent heat of condensation (see Reference 53).

The minimum nucleation rate is now given by[51]

$$I_2 = \pi r^{*2} N_w^{ads} R_{av} Z N_{par} \exp\left(-\frac{\Delta G^*}{kT}\right) \tag{2.49}$$

and the maximum nucleation rate by[51]

$$I_3 = 4\pi R^2 N_w^{ads} R_{av} Z N_{par} \exp\left(-\frac{\Delta G^*}{kT}\right) \tag{2.50}$$

The minimum nucleation rate has been derived assuming the nucleation takes place when the condensation nucleus is covered with critical clusters. The expression for maximum rate assumes that a new particle is produced instantaneously when one critical cluster is formed on the condensation nucleus.[51]

The effect of sulfuric acid hydration can be included in the expressions for the nucleation rate. For details, see Reference 52.

Nucleation Probability

The heterogeneous nucleation rate is a somewhat arbitrary concept. In some cases, it would be better to consider the number of aerosol particles that have nucleated. This number depends on the nucleation time (or duration of the experiment), nucleation rate (per unit time and unit area), and the surface area of the pre-existing particle. Denote the initial number concentration of aerosol particles by N_0. The number of non-nucleated particles can be obtained by solving the following differential equation.

$$\frac{dN}{dt} = -NI4\pi R_p^2 \tag{2.51}$$

After integration, one has, for non-nucleated particles,

$$N = N_0 \exp\left[-4\pi R_p^2 I\tau\right] \tag{2.52}$$

The probability of an embryo appearing on an aerosol particle with radius R_p is then:

$$P = 1 - \exp\left[-4\pi R_p^2 I\tau\right] \tag{2.53}$$

and τ is the duration of the experiment, or nucleation time.

The Effect of Active Sites, Surface Diffusion, and Line Tension on Heterogeneous Nucleation

Lazaridis et al.[54] have developed a model for heterogeneous nucleation on aerosol particles with so-called *active sites,* following the work of Gorbunov and Kakutkina.[55] Active sites refer to areas at which nucleation is easier compared to the surroundings. Active sites may be caused by variations in surface composition, curvature, roughness, etc. In the model of Lazaridis et al.,[54] the interaction between an embryo and an active site is described by a contact angle smaller than that between the embryo and the surrounding substrate. The embryo formation occurs via three successive stages. In the beginning, the surface area of the interface between the liquid embryo and the substrate (S_{23}) is smaller than the surface area of the active site (S). When the embryo has grown enough, so that $S_{23} = S$, the contact angle, and thereby the form of the embryo, start to change. Finally, after the transformation stage, the embryo continues to grow with a new contact angle. As expected, active sites enhance the nucleation rate. Furthermore, Lazaridis et al.[54] found that the curve describing nucleation probability as a function of particle size has a steeper slope in the active site model compared with a model employing uniform surfaces.

Another important feature missing from the classical theory is surface diffusion. Pound et al.[56] has pointed out that surface diffusion delivers molecules more efficiently to the embryo than impingement from the vapor. The two-dimensional diffusion coefficient can be determined as (see Reference 57):

$$D = \frac{\delta^2}{4\tau_D} \tag{2.54}$$

Here, δ is the mean jump distance, and τ_D the average jump time from site to site:

$$\tau_D = \tau_0 \exp(-U / UT) \tag{2.55}$$

with τ_0 denoting the vibration period of an adatom and U the activation energy for diffusion. The rate of molecules arriving to the embryo is given by the following expression (see, for example, Reference 58):

$$R_{aa} = 2\pi R_p \sin\phi \left(\frac{N_a^{ads} \bar{v}}{\phi} \right) \tag{2.56}$$

$\bar{v}$ is the average velocity of the adsorbed molecules ($\bar{v} = \delta/\tau_D$). The nucleation rate per unit area and unit time is given by:

$$I = R_{aa} Z N_{ads} \exp(-\Delta G^*/kT) \tag{2.57}$$

Lazaridis[58] found several orders of magnitude enhancements in theoretical nucleation rates of water at 273K when surface diffusion was allowed for.

The discontinuity of two or more volume phases is connected to surface tension. Line tension, on the other hand, arises from the discontinuity between two or more surface phases. The Gibbs free energy, which takes into account the line tension (σ_t), is

$$\Delta G^* = \frac{2}{3}\pi r^{*2} \sigma_{12} f(m,x) - \frac{\sigma_t}{R_p \tan\phi} S_{23} + 2\pi R_p \sigma_t \sin\phi \tag{2.58}$$

The inclusion of a positive line tension increases the critical value for the Gibbs free energy and decreases the corresponding nucleation rate and nucleation probability. However, experimental results show higher nucleation rates than the classical theory.[59] The concept of negative line tension, which has been studied by Scheludko et al.,[60] will decrease the Gibbs free energy and increase the nucleation rate. Lazaridis[58] has studied the effect of line tension on heterogeneous nucleation in more detail.

ACTIVATION

Recent studies have shown (see, for example, Reference 61) that atmospheric aerosol particles are very often internally mixed (i.e., they are at least partly soluble). This section examines the activation process, which will follow the hygroscopic growth of soluble aerosol particles. In the atmosphere, the most important solvent is water; however, other vapors may take part in the process.

Traditionally, the activation of aerosol particles into cloud droplets has been described by the Köhler theory (see, for example, Reference 62); as the saturation ratio S of water increases, hygroscopic salt particles take up water so that they stay in equilibrium with the environment. When S exceeds 100% (usually by a fraction of 1%, the exact number depending on the particle size and on its composition), the particles start to grow spontaneously. A Köhler curve plots the particle radius versus the saturation ratio; a maximum in S is seen at the activation radius of the particle in question.

In the ordinary Köhler theory, a solution droplet consisting of water and some nonvolatile salt is considered. The Kelvin equation for the droplet reads

$$\ln\left(\frac{p_1}{p_{1,sol}}\right) = \frac{2\sigma v_1}{kTr} \tag{2.59}$$

Here, p_1 is the partial vapor pressure of water. The vapor pressure of water above the solution is determined by the liquid phase activity.

When other condensable vapor (typically some acid) is present in the system and condenses on the droplet simultaneously with water, Equation 2.59 is still valid. However, all thermodynamic variables now refer to the ternary solution; and to complete the calculation, one has to find out the acid mole fraction. One can therefore write a second Kelvin equation that describes the equilibrium of the acid:

$$\ln\left(\frac{p_2}{p_{2,sol}}\right) = \frac{2\sigma v_2}{kTr} \tag{2.60}$$

Equations 2.59 and 2.60 can now be solved simultaneously with a simple iterative method to yield the equilibrium radius of the ternary solution droplet at any (p_1, p_2). One thus obtains a Köhler surface rather than a Köhler curve, the axes being the particle radius and the saturation ratios of the two vapors. The question that arises is: what is the route that a droplet follows in crossing this surface? Kulmala et al.[63] presented equilibrium curves for droplets at various constant nitric acid vapor concentrations. Such conditions could be realized, for example, in flow diffusion chambers, where the walls act as a source for vapor. However, in atmospheric situations, condensation depletes the water-soluble vapor; that is, in equilibrium growth calculations, it is necessary to account for the conservation of number of moles:

$$n_{2g} + n_2 = n_t \tag{2.61}$$

where $n_{2g} = N_2/C$, N_2 is the gas phase concentration of species 2, and C is the concentration of droplets, n_2 is number of moles of species 2 in a droplet, and the total number of moles n_t is a constant that usually equals n_{2g} at low R.H.

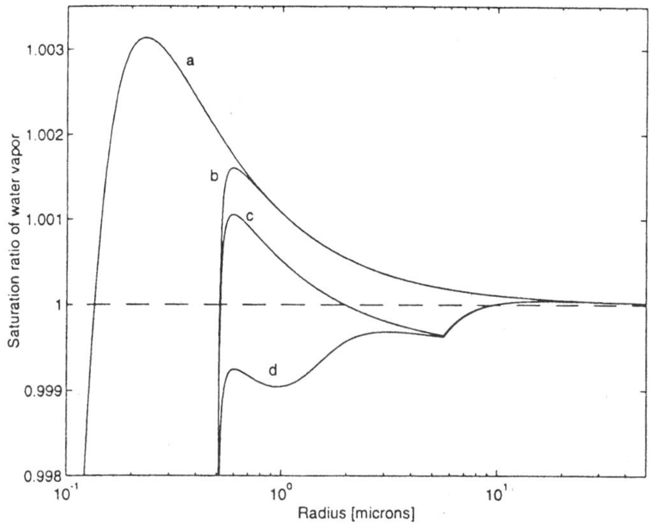

FIGURE 2.2 Curve (a) is the conventional Köhler curve for an initial 30-nm dry particle consisting of ammonium sulfate at 298K. Curve (b) shows the effect of an insoluble 400-nm core; the amount of ammonium sulfate is the same as with curve (a). The effect of insoluble material was studied in detail by Hänel.[86] Curve (c) shows what happens when the 400-nm core consists of slightly soluble rather than insoluble material. The solubility used (0.00209 g cm^{-3}) corresponds to that of $CaSO_4$.[87] The sharp minimum of the curve shows the point at which all of the core is dissolved. $CaSO_4$ was chosen as the example because it occurs commonly as gypsum dust or as the product of reaction of CaO in fly ash with H_2SO_4 in the air. $CaSO_4$ has been found in fogwater collected in Po Valley, Italy.[88] Other possibilities exist, including many slightly soluble organics. In curve (d), the effect of a highly soluble gas, nitric acid, has been added. The initial gas phase concentration of HNO_3 is 1 ppbv, and the Henry's law constant used[89] (mole fraction scale) is 853.1 atm^{-1}. Because nitric acid is allowed to deplete from the gas as it is absorbed by the droplets, its effect depends on the aerosol number concentration, which in this case was assumed to be 100 cm^{-3}; the aerosol size distribution was taken to be monodisperse (note that a qualitatively similar curve would result if the aerosol population was 1000 cm^{-3} and the initial HNO_3 concentration 3 ppbv.) The smooth minimum in the curve is caused by the depletion of the acid from the gas.

To determine the equilibrium growth curve in the presence of a highly water-soluble gas such as nitric acid by solving Equations 2.59 to 2.61 numerically, one needs thermodynamic data such as equilibrium vapor pressures as functions of mole fractions and temperature for the three-component system. (Actually, the system has more than three components, as far as the thermodynamic data is concerned, because the salt and the acid are decomposed into various ionic species in the liquid phase.) When the droplets are dilute enough (R.H. above 99%) and the dry particles are larger than about 30 nm, Equations 2.59 to 2.61 can be cast in the form of a single equation.[64] Furthermore, the effect caused by a slightly soluble aerosol species[65] that contributes solute into the liquid phase throughout droplet growth can be accounted for (see Reference 64 for details). Figure 2.2 shows what happens to an ordinary Köhler curve calculated for 30-nm ammonium sulfate particles when the effects of a slightly soluble 500-nm calcium sulfate core and absorption of 1 ppb of nitric acid are taken into account (droplet concentration 100 cm^{-3}).[66] The figure indicates that nonactivated droplets can grow to a 10-μm size range, forming a visible pollution fog. (Note that the fog has to be formed relatively slowly for nonactivated droplets to grow to large sizes; otherwise, the ambient R.H. might exceed 100% and activation may take place.)

With most clouds, the activation of aerosol particles is a very dynamic process as the ambient parameters (temperature, vapor concentrations) change constantly while the droplets are growing, and water vapor depletion is no longer an equilibrium process. It is therefore advisable to use condensation theories in studies of cloud droplet activation. Kulmala et al.[63,67] have studied the activation and growth of cloud droplets using an air parcel model that describes the formation of a convective cloud. Simultaneous condensation of water and nitric acid vapors were allowed for. The study showed that enhanced nitric acid concentrations increased the number of pre-existing aerosol particles that were able to activate. Consequently, the cloud droplet concentration was increased and average droplet size decreased.

CONDENSATION

Condensation follows the first step of the phase transition, be it nucleation or activation. Condensation causes an increasing amount of the new phase to accumulate on liquid or solid particles suspended in the gas. Unlike with nucleation, the theory of condensation is well-established, and agreement with experimental results is rather good (see, for example, Reference 68). The main difficulty in the determination of mass and heat fluxes is the lack of accurate, experimentally verified liquid phase activities. Finding data is not an easy task, especially in the case of multicomponent mixtures, and quite approximate semi-empirical expressions are often used. This section considers first the determination of liquid activities and vapor pressures at the droplet surface; and then provides two useful expressions for mass fluxes in the transition and continuum regime. By means of these expressions, the rates of the droplet composition and mass, the quantities that fully describe the droplet evolution, can be formulated. In evaluation of the growth of droplet population, these rate equations are coupled with expressions for variables of surrounding atmosphere (temperature, pressure, partial pressures, etc.). For details, see References 68 to 70.

VAPOR PRESSURES AND LIQUID PHASE ACTIVITIES

To determine the partial vapor pressures just above the droplet surface, one must know the corresponding equilibrium vapor pressures (these two are the same only if the droplet is large enough for the mass transport to be wholly diffusion controlled). To calculate the equilibrium vapor pressures of species in a liquid mixture, one needs their saturation vapor pressures and activities. The vapor pressure of species i for a flat surface is given by (see Reference 71):

$$p_{ia,f} = A_i p_{is,f} \tag{2.62}$$

where $p_{is,f}$ is the saturation vapor pressure of pure species i for a flat surface, and A_i is the activity of species i in the liquid mixture. The saturation vapor pressure varies exponentially with the temperature. The activity of species i at some temperature, pressure, and composition is defined as the ratio of the fugacity of species i at these conditions to the fugacity of species i in some standard state (see, for example, Reference 72).

For nondissociating species (and approximately for weak acids and alkalis, including water), a convenient way of expressing the activity is to introduce the activity coefficient Γ_i of species i by the definition

$$\Gamma_i \equiv \frac{A_i}{X_{il,a}} \tag{2.63}$$

where $X_{il,a}$ is the mole fraction of species i in the liquid. The activity coefficient depends on temperature, composition, and pressure. However, far from critical conditions, and unless the pressure is large, the effect of pressure on the activity coefficient is usually small.[73] The activity is commonly defined so that the activity of the exceeding component approaches its mole fraction as its $X \rightarrow 1$. Then, according to this definition and Equation 2.63, $\Gamma_i \rightarrow 1$ as $X_{il,a} \rightarrow 1$. Consequently, the expression for the equilibrium vapor pressure of species i above a flat surface is given by:

$$p_{ia,f} = \Gamma_i X_{il,a} p_{is,f} \tag{2.64}$$

This is now the general formula for nondissociating species and it reduces to two useful expressions for trace species. At the limit $X_{il,a} \to 0$, the activity coefficient approaches a finite, temperature-dependent value, and the product $\Gamma_i p_{is,f}$ is called Henry's law coefficient (or absorption equilibrium constant). In this definition, Henry's law coefficient has a dimension of pressure, but there are also other, physically equivalent definitions for which one can refer to Reference 72. At the limit $X_{il,a} \to 1$, by definition $\Gamma_i \to 1$, and consequently, $p_{ia,f} = X_{il,a} p_{is,f}$. This is called Raoult's law.

For completely dissociating species (salts, strong acids, and alkalis), a convenient way of expressing the equilibrium vapor pressure over a flat surface is:

$$p_{ia,f}^{dis} = X_{il,a}^{+} X_{il,a}^{-} f_{\pm}^{*2} / K_{Hx} \tag{2.65}$$

where $X^{+}_{il,a}$ and $X^{-}_{il,a}$ indicate mole fractions of positive and negative ions, respectively, forming the species concerned. They are calculated on the basis of the total number of components present in the solution. $f_{\pm}^{*2}$ is the mean rational activity coefficient (i.e., infinite dilution is taken to be the standard state), and K_{Hx} is the Henry's law constant (now in units of Pa^{-1}). For further information on dissociating species, refer to works of Clegg and Brimblecombe[74] and Clegg et al.[75]

Finally, one must account for the effect of surface curvature. The curvature modifies slightly the attractive forces between surface molecules, with the net result that the vapor pressure is higher than that for a flat surface. Now, the expression for the vapor pressure of species i at the droplet surface is given by (see References 76 and 77):

$$p_{ia} = p_{ia,f} \exp\left(\frac{2\sigma v}{kTr}\right) \tag{2.66}$$

or, for dissociating species, by:

$$p_{ia} = p_{ia,f}^{dis} \exp\left(\frac{2\sigma v}{kTr}\right) \tag{2.67}$$

where σ is the surface tension, v is the molecular volume of the liquid mixture, and r is the droplet radius. The effect of the curvature is called the Kelvin effect, and it becomes significant for particles smaller than about 1 μm.

For the final goal of determining the mass fluxes, one must know the partial vapor pressures just above the droplet surface. In the continuum, regime, the droplet surface can be assumed to be saturated and the equilibrium achieved; that is, the vapor pressure of species i just above the droplet surface is the same as at the droplet surface. Thus, in the continuum regime, the mole fraction X_{ia} of species i just above the droplet surface is found by the ideal gas assumption as:

$$X_{ia} = \frac{p_{ia}}{p} \tag{2.68}$$

where p is the total gas pressure.

In the transition regime, the saturation equilibrium is disturbed, and the vapor pressures just above the droplet surface are lower than the equilibrium vapor pressure. In calculating the mass fluxes, the disturbance of the saturation equilibrium can be formally taken into account by a correction factor (see Equation 2.72).

Mass Flux Expressions

Several mass flux expressions and growth equations can be found in the literature. The well-known Mason equation is based on inaccurate approximations (see Reference 78). Presented here are two useful, physically sound expressions (see References 68, 79, and 80).

Presented first are the mass flux expressions that explicitly include the droplet temperature. In the second approach, the effect of the droplet temperature is taken into account implicitly. In practice, the uncoupled approach can be applied in all atmospheric aerosol problems and the semi-analytical approach is valid when the sum of the saturation ratio is near 1.

Uncoupled Solution

Consider first the so-called uncoupled approach because it is frequently applied and typically gives accurate enough results when compared with experiments (see References 81 and 82). It is somewhat simplified compared to the exact expression for binary condensation,[79] as the Stefan flow contribution for the flux I_i is assumed to be caused by diffusion of species i alone, rather than by diffusion of both species. However, contrary to the exact solution, the uncoupled solution allows for the exact temperature dependencies of molar density and diffusion coefficients.

The most rigorous uncoupled expressions have been derived by Kulmala and Vesala[80] and are given by:

$$I_1 = C_1 \frac{8\pi r M_1 D_{1l} p}{RT_\infty} \ln\left(\frac{1 - X_{1\infty}}{1 - X_{1a}}\right) \tag{2.69}$$

or

$$I_2 = C_2 \frac{8\pi r M_2 D_{2l} p}{RT_\infty} \ln\left(\frac{1 - X_{2\infty}}{1 - X_{2a}}\right) \tag{2.70}$$

where M_i is the molecular weight (g mole^{-1}) of vapor i, R is the universal gas constant, T_∞ is the (ambient) temperature far from the droplet, and X_{ia} are estimated using Equation 2.68. The binary diffusion coefficients (D_{il}) between the vapor i and the carrier gas are calculated at the temperature far from the droplet. The correction factors C_i are needed because of the temperature profile around the droplet. The following form of C_i takes into account the temperature dependence of the diffusion coefficients more rigorously than the commonly used geometric mean:

$$C_i = \frac{T_\infty - T_a}{T_\infty^{\mu_i - 1}} \frac{2 - \mu_i}{T_\infty^{2-\mu_i} - T_a^{2-\mu_i}} \tag{2.71}$$

Here, T_a is the droplet temperature and μ_i is the exponent for the temperature dependencies of the binary diffusion coefficients ($D_i \propto T^{\mu_i}$). The exponent varies in most cases from 1.5 to 2.0,[73] and at the limit $\mu \rightarrow 2.0$, the value for the correction factor is obtained in a straightforward manner.

Note that at the limits $X_i \rightarrow 0$ the logarithmic terms are reduced to ($X_{ia} - X_{i\infty}$). This form does not take into account the effect of the convective-like Stefan flow, which is toward the droplet surface in the case of condensation. Physically, the carrier gas (air) and the vapor diffuse in opposite directions; and because the carrier gas is not released from the droplet, the total molar density would tend to decrease near the droplet surface were it not for the Stefan flow. In atmospheric applications, the Stefan flow is typically insignificant because the vapor mole fractions are very small.

In the transition regime, the transport of mass and heat is partly under mass diffusion and heat conduction control, and partly under mass kinetic and heat kinetic control. To take this into account, the above mass flux expressions are commonly multiplied by a semi-empirical correction factor. One can adopt the following correction factor from Fuchs and Sutugin[83]:

$$\beta_{M,i} = \frac{1 + Kn_i}{1 + \left(\frac{4}{3\alpha_{M,i}} + 0.337\right)Kn_i + \frac{4}{3\alpha_{M,i}}Kn_i^2} \tag{2.72}$$

where a dimensionless group, the Knudsen number, is:

$$Kn_i = \frac{\lambda_i}{r} \tag{2.73}$$

λ_i is the mean effective free bath of the vapor molecules i in the gas and, thus, the Knudsen number is the ratio of two length scales — a length scale λ_i characterizing the gas with respect to the transport of mass, and a length scale r characterizing the droplet. From simple kinetic theory of gases, the zero-order approximation for λ_i can be expressed by means of a measurable macroscopic property D_{il} and the average absolute velocity of the vapor molecules (e.g., see Reference 84). The proper value of the mass accommodation coefficient $\alpha_{M,i}$ for various substances has recently been under discussion. The condensational growth of binary and unary droplets has been measured with high accuracy in an expansion cloud chamber to determine accommodation coefficients.[81,82,85] The size range where these measurements are valid is from 0.5 to 10 μm. The experimental results show that the accommodation coefficient is 1 or near 1 for water, n-alcohols, and n-nonane molecules impinging on a respective liquid interface, and also for some mixtures like water/n-propanol and water/nitric acid. If there is no experimentally verified information available, it is suggested that unity be used for the accommodation coefficient — this is consistent with recent theoretical work.[86]

In order to apply the expressions for the mass fluxes, the droplet temperature should be known. In the case of a flat internal temperature profile and insignificant droplet thermal capacity, the algebraic, but implicit expression for the droplet temperature can be derived using the droplet energy balance. In practice, the most useful expression is (see, for example, Reference 85):

$$T_a = T_\infty - \frac{L_1 I_1 + L_2 I_2}{2\pi a \beta_T (k_a + k_\infty)} \tag{2.74}$$

where L_i is the latent heat of vaporization and k_a and k_∞ are the gaseous thermal conductivities at the droplet and ambient temperatures, respectively. β_T is the transitional correction factor for the heat transfer, and it can be estimated by means of Equation 2.72, where the mass accommodation coefficient is replaced by the thermal accommodation coefficient (commonly set to unity), and the mean free path used for the Knudsen number is expressed by a gaseous heat conductivity and the average absolute velocity of vapor molecules (see Reference 84). If the thermal capacity is included, the droplet temperature can be estimated using a first-order differential equation. For various ways to analyze the droplet temperature, refer to Reference 80.

Semi-Analytical Solution

For several applications, an analytical form of mass flux expressions without the dependence on the droplet temperature is desirable. Expressions in this form can be derived by replacing the explicit droplet temperature dependence by the dependence on the products of latent heats and mass fluxes. Following this approach, linearizing with respect to mass fluxes and neglecting the effects of the Stefan flows and the radiative heat transport, mass fluxes can be found as solutions for the equations[63]:

$$I_1 = -4\pi r \frac{M_1 \beta_{M1} D_1 p_{1s,f}(T_\infty)}{RT_\infty}\left[A_{1g} - A_1\left(1 - \frac{L_1 M_1 (L_1 I_1 + L_2 I_2)}{4\pi r \beta_T k_\infty RT^2}\right)\right]$$
$$I_2 = -4\pi r \frac{M_2 \beta_{M2} D_2 p_{2s,f}(T_\infty)}{RT_\infty}\left[A_{2g} - A_2\left(1 - \frac{L_2 M_2 (L_1 I_1 + L_2 I_2)}{4\pi r \beta_T k_\infty RT^2}\right)\right] \tag{2.75}$$

where A_{ig} is the gas phase activity.

The solution for the preceding set of equations is[63]

$$I_1 = \frac{(1 - B_{22}) b_1 (A_{1g} - A_1) + B_{12} b_2 (A_{2g} - A_2)}{1 - B_{22} - B_{11} + B_{11} B_{22} - B_{12} B_{21}}$$
$$I_2 = \frac{(1 - B_{11}) b_2 (A_{2g} - A_2) + B_{21} b_1 (A_{1g} - A_1)}{1 - B_{22} - B_{11} + B_{11} B_{22} - B_{12} B_{21}} \tag{2.76}$$

where

$$b_i = \frac{-4\pi r M_i \beta_{M,i} D_i p_{is,f}}{RT_\infty}$$
$$B_{ij} = \frac{A_i L_i L_j M_i}{4\pi r \beta_T k_\infty RT_\infty^2} b_i \tag{2.77}$$

Note that although the mass fluxes now depend on saturation deficits of both species, these expressions do not take into account the diffusive coupling arising from the Stefan flows, but here the coupling arises from the elimination of the explicit dependence on the droplet temperature. The fact that the droplet temperature is not equal to the gas temperature is described approximately in the above equations, by means of terms containing latent heats of vaporization. This gives rise to the coupling of mass fluxes. In the uncoupled approach, the temperature coupling is taken into account by the explicit temperature dependence.

ACKNOWLEDGMENTS

The authors would like to acknowledge financial support from the Academy of Finland (project 44278) and from the Petroleum Research fund of the American Chemical Society.

REFERENCES

1. Laaksonen, A., Talanquer, V., and Oxtoby, D.W., *Annu. Rev. Phys. Chem.,* 46, 489, 1995.
2. Volmer, M. and Weber, A., *Z. Phys. Chem.,* 119, 277, 1926.
3. Farkas, L., *Z. Phys. Chem.,* 125, 236, 1927.
4. Volmer, M., *Z. Elektrochem.,* 35, 555, 1929.
5. Volmer, M., *Kinetik der Phasenbildung,* Verlag Von Theodor Steinkopff, Dresden und Leipzig, 1939.
6. Becker, R. and Döring, W., *Ann. Phys. (Leipzig),* 26, 719, 1935.
7. Zeldovich, J., *Soviet Phys. JETP,* 12, 525, 1942.
8. Zettlemoyer, A.C., *Nucleation,* Marcel Dekker, New York, 1969.
9. Courtney, W.G., *J. Chem. Phys.,* 35, 2249, 1961.
10. Blander, M. and Katz, J., *J. Stat. Phys.,* 4, 55, 1972.
11. Reiss, H., Kegel, W.K., and Katz, J.L., *Phys. Rev. Lett.,* 78, 4506, 1997.
12. Wilemski, G., *J. Chem. Phys.,* 103, 1119, 1995.
13. Girshick, S. and Chiu, C.-P., *J. Chem. Phys.,* 93, 1273, 1990.
14. Girshick S., *J. Chem. Phys.,* 94, 826, 1991.
15. Viisanen, Y. and Strey, R., *J. Chem. Phys.,* 101, 7835, 1994.
16. Oxtoby, D.W. and Kaschiev, D., *J. Chem. Phys.,* 100, 7665, 1994.
17. Kashchiev, D., *J. Chem. Phys.,* 76, 5098, 1982.
18. Viisanen, Y., Strey, R., and Reiss, H., *J. Chem. Phys.,* 99, 4680, 1993.
19. Strey, R. and Viisanen, Y., *J. Chem. Phys.,* 99, 4693, 1993.
20. Viisanen, Y., Strey, R., Laaksonen, A., and Kulmala, M., *J. Chem. Phys.,* 100, 6062, 1994.
21. Viisanen, Y., Wagner, P.E., and Strey, R., *J. Chem. Phys.,* 108, 4257, 1998.
22. Strey, R., Wagner, P.E., and Viisanen, Y., *J. Phys. Chem.,* 98, 7748, 1994.
23. Strey, R., Viisanen, Y., and Wagner, P.E., *J. Chem. Phys.,* 103, 4333, 1995.
24. Hruby, J., Viisanen, Y., and Strey, R., *J. Chem. Phys.,* 104, 5181, 1996.
25. McGraw, R. and Laaksonen, A., *Phys. Rev. Lett.,* 76, 2754, 1996.
26. Zeng, X.C. and Oxtoby, D.W., *J. Chem. Phys.,* 94, 4472, 1990.
27. Talanquer, V., *J. Chem. Phys.,* 106, 9957, 1997.
28. Debenedetti, P.G., *Metastable Liquids*, Princeton University Press, 1996.
29. Flood, H., *Z. Phys. Chemie A,* 170, 280, 1934.
30. Neumann, K. and Köring, W., *Z. Phys. Chem. A.,* 186, 203, 1940.
31. Reiss, H., *J. Chem. Phys.,* 18, 840, 1950.
32. Doyle, G.J., *J. Chem. Phys.,* 35, 795, 1961.
33. Renninger, R.G., Hiller, F.C., and Bone, J., *J. Chem. Phys.,* 75, 1584, 1981.
34. Wilemski, G., *J. Chem. Phys.,* 80, 1370, 1984.
35. Mirabel, P. and Reiss, H., *Langmuir,* 3, 228, 1987.
36. Nishioka, K. and Kusaka, I., *J. Chem. Phys.,* 96, 5370, 1992.
37. Wilemski, G., *J. Phys. Chem.,* 91, 2492, 1987.
38. Laaksonen, A., Kulmala, M., and Wagner, P.E., *J. Chem. Phys.,* 99, 6832, 1993.
39. Flagelleot-Daniel, C., Cao Dinh, and Mirabel, Ph., *J. Chem. Phys.,* 72, 544, 1980.
40. Laaksonen, A. and Kulmala, M., *J. Chem. Phys.,* 95, 6745, 1991.
41. Laaksonen, A., *J. Chem. Phys.,* 97, 1983, 1992.
42. Jaecker-Voirol, A., Mirabel, P., and Reiss, H., *J. Chem. Phys.,* 87, 4849, 1987.
43. Kulmala, M., Lazaridis, M., Laaksonen, A., and Vesala, T., *J. Chem. Phys.,* 94, 7411, 1991.
44. Stauffer, D., *J. Aerosol Sci.,* 7, 319, 1976.
45. Kulmala, M. and Viisanen, Y., *J. Aerosol Sci.,* 22, S51, 1991.
46. Kulmala, M. and Laaksonen, A., *J. Chem. Phys.,* 93, 696, 1990.
47. Vehkamäki, H., Paatero, P., Kulmala, M., and Laaksonen, A., *J. Chem. Phys.,* 101, 9997, 1994.
48. Zapadinsky, E.I., *Report Series in Aerosol Science,* 27, 1994.
49. Fletcher, N., *J. Chem. Phys.,* 29, 572, 1958.
50. Hamill, P., Kiang, C.S., and Cadle, R., *J. Atmos. Sci.,* 34, 151, 1977.
51. Hamill, P., Turco, R.P., Kiang, C.S., Toon, O.B., and Whitten, R.C., *J. Aerosol Sci.,* 13, 561, 1982.
52. Lazaridis, M., Kulmala, M., and Laaksonen, A., *J. Aerosol Sci.,* 22, 823, 1991.
53. Adamson, A., *Physical Chemistry of Surfaces*, John Wiley & Sons, New York, 1982.

54. Lazaridis, M., Kulmala, M., and Gorbunov, B.Z., *J. Aerosol Sci.,* 23, 457, 1992.
55. Gorbunov, B. and Kakutkina, N., *J. Aerosol Sci.,* 13, 21, 1982.
56. Pound, G.M., Simnad, M.T., and Yang, L., *J. Chem. Phys.,* 22, 1215, 1954.
57. Frenkel, J., *Kinetic Theory of Liquids,* Clarendon Press, Oxford, 1946.
58. Lazaridis, M., *J. Colloid Int. Sci.,* 155, 386, 1993.
59. Mahata, P. and Alofs, D., *J. Atmos. Sci.,* 32, 116, 1975.
60. Scheludko, A., Chakarov, V., and Toshev, B., *J. Colloid Interface Sci.,* 82, 83, 1981.
61. Svenningsson, I.B., Hansson, H.-C., Wiedensohler, A., Ogren, J.A., Noone, K.J., and Hallberg, A., *Tellus,* 44B, 556, 1992.
62. Pruppacher, H.R. and Klett, J.D., *Microphysics of Clouds and Precipitation*, D. Reidel, 1978.
63. Kulmala, M., Laaksonen, A., Korhonen, P., Vesala, T., Ahonen, T., and Barrett, J.C., *J. Geophys. Res.,* 98, 22949, 1993a.
64. Laaksonen, A., Korhonen, P., Kulmala, M., and Charlson, R.J., *J. Atmos. Sci.,* 55, 853, 1998.
65. Shulman, M.L., Jacobson, M.C., Charlson, R.J., Synovec, R.E., and Young, T.E., *Geophys. Res. Lett.,* 23, 277, 1996.
66. Kulmala, M., Laaksonen, A., Charlson, R.J., and Korhonen, P., *Nature,* 338, 336, 1997.
67. Kulmala, M., Korhonen, P., Vesala, T., Hansson, H.-C., Noone, K., and Svenningsson, B., *Tellus,* 48B, 347, 1996.
68. Vesala, T., Kulmala, M., Rudolf, R., Vrtala, A., and Wagner, P.E., *J. Aerosol Sci.,* 28, 565, 1997.
69. Nikmo, J., Kukkonen, J., Vesala, T., and Kulmala, M., *J. Hazardous Mat.,* 38, 293, 1994.
70. Kulmala, M., Vesala T., and Wagner, P.E., *Proc. R. Soc. London, A* 441, 589, 1993b.
71. Atkins, P.W., *Physical Chemistry,* 3rd ed., Oxford University Press, 1986.
72. Prausnitz, J.M., Lichtenthaler, R.N., and de Azevedo, E.G., *Molecular Thermodynamics of Fluid-Phase Equilibria,* Prentice-Hall, Englewood Cliffs, NJ, 1986.
73. Reid, R.C., Prausnitz, J.M., and Poling, B.E., *The Properties of Gases and Liquids,* 4th ed., McGraw-Hill, New York, 1987.
74. Clegg, S.L. and Brimblecombe, P., *J. Phys. Chem.,* 94, 5369, 1990.
75. Clegg, S.L., Pitzer, K.S., and Brimblecombe, P., *J. Phys. Chem.,* 96, 9470, 1992.
76. Seinfeld, J.H., *Atmospheric Chemistry and Physics of Air Pollution,* John Wiley & Sons, 1986.
77. Seinfeld, J.H. and Pandis, S.N., *Atmospheric Chemistry and Physics. From Air Pollution to Climate Change*, John Wiley & Sons, 1998.
78. Kulmala, M., Majerowicz, A., and Wagner, P.E., *J. Aerosol Sci.,* 20, 1023, 1989.
79. Vesala, T. and Kulmala, M., *Physica A,* 192, 107, 1993.
80. Kulmala, M. and Vesala, T., *J. Aerosol Sci.,* 22, 337, 1991.
81. Rudolf, R., Majerowicz, A., Kulmala, M., Vesala, T., Viisanen, Y., and Wagner, P.E., *J. Aerosol Sci.,* 22, S51, 1991.
82. Rudolf, R., *Ph.D. thesis*, University of Vienna, Department of Experimental Physics, 1994.
83. Fuchs, N.A. and Sutugin, A.G., *Highly Dispersed Aerosols,* Ann Arbor Science, Ann Arbor, MI, 1970.
84. Hirschfelder, J.O., Curtiss, C.F., and Bird, R.B., *Molecular Theory of Gases and Liquids,* John Wiley & Sons, New York, 1954.
85. Wagner, P.E., *Aerosol Microphysics,* W.H. Marlow, Ed., Vol. II, Springer Verlag, Berlin, 1982, 129.
86. Clement, C.F., Kulmala, M., and Vesala, T., *J. Aerosol Sci.,* 27, 869, 1996.
86. Hänel, G., *Adv. Geophys.,* 19, 73, 1976.
87. *CRC Handbook of Chemistry and Physics,* 74th ed., (Eds. Lide, D.R. and Frederikse, H.P.R.), CRC Press LLC, Boca Raton, FL, 1993.
88. Del Monte, M. and Rossi, P., *Atmos. Environ.,* 31, 1637, 1997.
89. Brimblecombe, P. and Clegg, S.L., *Atmos. Environ.,* 24A, 1945, 1990.

3 The Estimation of Time-Dependent (Relaxation) Processes Related with Condensation and Evaporation of Liquid Drop

Mikhail V. Buikov

CONTENTS

INTRODUCTION

The theory of condensational growth and evaporation of drops consisting of pure liquid or solution is a complicated thermodynamical problem that concerns many aspects of kinetic gas theory. A complete review of the theory of growth and evaporation of drops can be found elsewhere.[1-3] Here, only one problem will be investigated: the time-dependent processes as a result of which steady-state temperature, vapor, and salt concentration are reached. Usually, the adopted drop temperature is constant inside the drop volume and is determined by heat balance between phase transition and thermal conduction; the salt concentration equal to the average volume value, vapor concentration, and temperature outside drop obey steady-state distributions in the drop vicinity. This approach is applicable if the rate of change of droplet size is small enough and the temperature of the environment varies slowly with time. The drop temperature can be defined as a psychometric one because it the same as that of aspiration psychrometer in the case of evaporation. The investigation of the transition to steady-state may be of some interest from a general point of view, and can find application if the environment temperature is changing fast enough and in the case of intensive drop growth and evaporation when the deviation of salt concentration near drop surface from mean

0-87371-829-1/00/$0.00+$.50

value results in substantial influence on the rate of change of drop size. Temperature relaxation has been investigated[4,5]; the influence of inhomogeneous salt concentration on drop growth has also been studied.[6-8] Intensive evaporation of solution drop was considered by Buikov and Sigal.[9] Below are presented the main results of these research efforts. The exact formulation of the problem of growth (evaporation) of a drop of solution is presented below, the application of the heat potentials to the solution is described, and an analysis of the time-dependent process resulting in establishing the steady state is considered. In conclusion, the intensive evaporation of solution drop when the formation of a salt crust is possible at the drop surface is also considered.

GENERAL EQUATIONS

The equations that describe diffusion of some non-volatile salt in volatile dissolvant (e.g., water), heat conduction inside and outside the drop, and vapor diffusion in a gas-vapor environment in a spherical coordinate system are:

$$\frac{\partial c}{\partial t} = D_1\left(\frac{\partial^2 c}{\partial r^2} + \frac{2}{r}\frac{\partial c}{\partial r}\right) \tag{3.1}$$

$$\frac{\partial T_1}{\partial t} = K_1\left(\frac{\partial^2 T_1}{\partial r^2} + \frac{2}{r}\frac{\partial T_1}{\partial r}\right) \tag{3.2}$$

$$\frac{\partial T_2}{\partial t} = K_2\left(\frac{\partial^2 T_2}{\partial r^2} + \frac{2}{r}\frac{\partial T_2}{\partial r}\right) \tag{3.3}$$

$$\frac{\partial q}{\partial t} = D_2\left(\frac{\partial^2 q}{\partial r^2} + \frac{2}{r}\frac{\partial q}{\partial r}\right) \tag{3.4}$$

The rate of change of drop radius is given by the conventional formula

$$\frac{dR(t)}{dt} = \frac{D_2}{\rho}\left.\frac{\partial q}{\partial r}\right|_{r=R(t)} \tag{3.5}$$

but the effects under consideration will be displayed through the gradient of the vapor concentration. If a drop at $t = 0$ is placed into a gas-vapor environment, then the initial conditions are:

$$c(r,0) = c_0,\quad T_1(r,0) = T_{10},\quad T_2(r,0) = T_\infty,\quad q(r,0) = q_\infty,\quad R(0) = R_0. \tag{3.6}$$

The boundary conditions are:

$$q(r,t) \to q_\infty,\quad T_2(r,t) \to T_\infty,\quad r \to \infty. \tag{3.7}$$

$c(r,t)$ and $T_2(r,t)$ must be finite at $r = 0$.

Temperature, heat flux, and vapor concentration should be continuous at the drop surface ($r = R(t)$); that is,

$$T_1(r,t) = T_2(r,t), \quad \text{at} \quad r = R(t), \tag{3.8}$$

$$LD_2 \frac{\partial q}{\partial r} + k_1 \frac{\partial T_1}{\partial r} + k_2 \frac{\partial T_2}{\partial r} = 0, \quad \text{at } r = R(t); \tag{3.9}$$

$$q(r,t) = q_s(T)\{1 - Ac(R(t))\}, \quad \text{at } r = R(t) \tag{3.10}$$

Equation 3.10 takes into account the effect of dissolved salt in the simplest form. The total amount of salt inside the drop should not change with time:

$$4\pi \int dr\, r^2 c(r,t) = \frac{4\pi}{3} R_o^3 c_0. \tag{3.11}$$

Taking the derivative of both sides of Equation 3.11 and using Equation 3.1 we have

$$\frac{dR}{dt} c(R,t) + D_1 \left.\frac{\partial c}{\partial r}\right|_{r=R} = 0 \tag{3.12}$$

In a growing (evaporating) drop, the salt concentration decreases (increases) with r and does not change if $R = const.$

THE SOLUTIONS OF THE EQUATIONS

There are some specific (characteristic) time intervals in the problem. The first one (t_l) is the time interval during which the steady-state field of the vapor is established in the vicinity of the drop. The time interval t_2 describes the relaxation of salt concentration. The establishment of the psychometric temperature can be reached after an elapsed time t_3. The last time interval is connected with a substantial increase in drop size (t_4); for example, for a solution drop growing in the saturated environment, it can be taken as the time when the salt concentration will be smaller than the initial value. It is well-known[1,3] that $t_l << t_4$ and $t_l << t_2$ ($D_2 << D_l$) and it will be clear later that $t_l << t_3$). So, as conventionally adopted, one can use the simplified formula for q:

$$q(r,t) = q_\infty + \left(q_s - q_\infty\right) R(t) r^{-1}. \tag{3.13}$$

The solutions of Equation 3.1 to 3.4 can be presented as thermal potentials[10] (dimensionless variables):

$$x\Sigma(x,\tau) = \int_0^\tau d\partial \sigma(\partial) K(\tau, \partial, x)$$

$$K(\tau,\partial,x) = \frac{1}{2\sqrt{\pi(\tau - \partial)}} \left[\exp\left(-\frac{(x - Z(\partial))^2}{4(\tau - \partial)} \right) \right] - \exp\left(-\frac{(x + Z(\partial))^2}{4(\tau - \partial)} \right) \tag{3.14}$$

$$xY_1(x,\tau) = \int_0^\tau d\partial\sigma_1(\partial)K_1(\tau,\partial,x)$$

$$K(\tau,\partial,x) = \frac{1}{2\sqrt{\pi\beta_1(\tau-\partial)}}\left[\exp\left(-\frac{(x-Z(\partial))^2}{4\beta_1(\tau-\partial)}\right)\right] - \exp\left(-\frac{(x+Z(\partial))^2}{4\beta_1(\tau-\partial)}\right) \quad (3.15)$$

$$xY_2(x,\tau) = \int_0^\tau d\partial\sigma_2(\partial)K_2(\tau,\partial,x) \quad (3.16)$$

$$K_2(\tau,\partial,x) = \frac{1}{2\sqrt{\pi\beta_2(\tau-\partial)}}\left[\exp\left(-\frac{(x-Z(\partial))^2}{4\beta_2(\tau-\partial)}\right)\right] \quad (3.17)$$

$$\Pi(x,\tau) = \frac{(\Pi_\infty - \Pi_s)Z(\tau)}{x} \quad (3.18)$$

$$\frac{dZ}{d\tau} = \delta\frac{d\Pi}{dx}\bigg| x = Z(\tau) \quad (3.19)$$

These solutions satisfy the equations and the initial and boundary conditions at great distances:

$$\Sigma = Y_1 = Y_2 = 0, \quad \tau = 0 \quad (3.20)$$

$$Y_2 \to 0, \quad x \to \infty. \quad (3.21)$$

The heat balance equation is transformed into:

$$\Gamma_0\frac{\partial\Pi}{\partial x} + \Gamma_1\frac{\partial Y_1}{\partial x} + \frac{\partial Y_2}{\partial x} = 0, \quad x = Z(\tau). \quad (3.22)$$

The temperature continuity and salt conservation equations are, respectively,

$$Z(\tau)Y(\tau) = \int_0^\tau d\partial\,\sigma_1(\partial)K_1(Z,\tau,\partial) = \int_0^\tau d\partial\,\sigma_2(\partial)K_2(Z,\tau,\partial), \quad (3.23)$$

$$\int_0^Z dx\,x^2\,\Sigma(x,\tau) = -\frac{Z^3-1}{3}. \quad (3.24)$$

The subsidiary functions $\sigma(\partial),\sigma_1(\partial),\sigma_2(\partial)$ can be determined using other boundary conditions.

RELAXATION OF SALT CONCENTRATION

Concentration and thermal relaxations are not connected because the heat of dissolution is not taken into account. This is in agreement with Equation 3.24, which will be turned into the equation for $\sigma(\theta)$: the dependence on temperature is only through drop size, but not directly. If the dependence of Σ (x, τ) on x in Equation 3.24 is neglected, then we obtain the conventional formula for salt concentration as:

$$\Sigma(x,\tau) \cong Z^{-3}(\tau) - 1 \tag{3.25}$$

or, $c(r,t) \cong R^{-3}(t)$.

To get the next approximation taking into account the difference between bulk and surface concentrations, the kernel K (x, τ, θ) is expanded into a series on x, keeping the terms $\sim x^3$:

$$\Sigma(x,\tau) \cong \Sigma_1(\tau) + x^2\,\Sigma_2(\tau) \tag{3.26}$$

Equation 3.25 is applicable if the following inequality is true:

$$\Sigma_1 >> x^2\,\Sigma_2 \quad \text{or } \Sigma_1 >> Z(\tau)\,\Sigma_2\,. \tag{3.27}$$

Because

$$\Sigma_1(\tau) \cong Z^{-3}(\tau) - 1, \tag{3.28}$$

and

$$\Sigma_2(\tau) \cong \frac{1}{6}\frac{d\Sigma}{d\tau} \tag{3.29}$$

then, instead of Equation 3.27, we have that it is possible to neglect salt concentration relaxation if

$$1 - Z^{-3}(\tau) >> \frac{1}{Z^2(\tau)}\frac{dZ(\tau)}{d\tau} \tag{3.30}$$

This inequality cannot be satisfied for small τ; but because $\tau_4 >> \tau_2$, then

$$\frac{dZ(\tau)}{d\tau} \cong 1 + \text{const. } \tau, \tag{3.31}$$

$Z(\tau)$ can be expanded for small τ and we obtain, from Equation 3.30,

$$\tau >> \frac{1}{6} \text{ or } t_2 >> \frac{D_1}{6R_0^2} \tag{3.32}$$

It is a criterion of the applicability of the bulk concentration approximation, when only a small gradient of salt concentration exists inside a drop, that

$$\frac{d\Sigma}{dx} \cong -\frac{x}{Z^2(\tau)}\frac{dZ}{d\tau}. \tag{3.33}$$

The gradient is negative for growth and positive for evaporation. For a 10-μm drop, t_2 is about 10^{-2} s; for a 100-μm drop, t_2 is equal to some seconds; and for a 1-mm drop, t_2 is some minutes.

In the opposite case, when $\tau << 1/6$, it is possible to derive an approximate formula for the salt concentration, introducing Equation 3.14 into Equation 3.25, after integration, expanding for small τ, and adopting $Z = 1$, one can derive:

$$\int_0^\tau d\partial\,\sigma(\partial) = -\frac{\sqrt{\pi}}{3}\left[Z^3(\tau) - 1\right]. \tag{3.34}$$

This means that

$$\sigma(\partial) = -\varsigma; \quad \varsigma = \sqrt{\pi}\left.\frac{dZ}{d\tau}\right|\tau = 0. \tag{3.35}$$

So, for the salt concentration, one has

$$x\,\Sigma(x,\tau) \cong -\varsigma\int_0^\tau d\partial\frac{\exp\left(-\frac{(1-x)^2}{4\partial}\right)}{2\sqrt{\pi\partial}} \tag{3.36}$$

The deviation of the salt concentration from the initial value takes place only in a very thin layer near the drop surface:

$$x \cong 1 - 2\sqrt{\tau}, \tag{3.37}$$

and deeper inside the drop volume $c = c_0$.

THERMAL RELAXATION

Leveling of the temperature inside a drop is a more complicated process than salt concentration relaxation because it involves simultaneous heat exchange inside and outside the drop and is described by the heat balance equation (Equation 3.24), which can be transformed to the following form:

$$\Gamma_0\Pi(Z,\tau) + (1+\Gamma_1)Y(\tau) + \Gamma_1\sigma_1(\tau) + \sigma_2(\tau) + \Psi(\tau) = 0, \tag{3.38}$$

where

$$\Psi(\tau) = \int_0^\tau \frac{d\partial\,\sigma_1(\partial)}{4\sqrt{\pi}\beta_1^{3/2}(\tau-\partial)}\exp\left[-\frac{1}{\beta_1(-\partial)}\right] \tag{3.39}$$

Two more equations for subsidiary functions σ_1 and σ_2 should be added to Equation 3.38 to find the steady-state temperature $Y(\tau)$:

$$Y(\tau) = \int_0^\tau \frac{d\partial\sigma_2(\partial)}{2\sqrt{\pi\beta_2(\tau-\partial)}} = \int_0^\tau \frac{d\partial\sigma_1(\partial)}{2\sqrt{\pi\beta_1(\tau-\partial)}} \exp\left[1 - \frac{1}{\beta_1(\tau-\partial)}\right] \tag{3.40}$$

There are two complications in solving Equations 3.38 through 3.40: (1) non linearity due to the presence of $\Pi_s(Y, \tau)$, and (2) time dependence through Σ, which depends on $\Pi_s(Y(\tau), \tau)$. The first difficulty can be easily overcome because temperature Y is, as a rule, small and Π_s can be expanded using the first approximation. Because thermal relaxation is much faster than concentration relaxation, and because drop size change is very small during thermal relaxation, one obtains:

$$\Pi_s(Z,\tau) \cong \Pi_\infty - \Pi_1 Y(\tau) \tag{3.41}$$

So, instead of Equation 3.38, one has

$$\left(1 + \Gamma_1 - \Gamma\Pi_1\right)Y + \Gamma_0\Pi_\infty + \Gamma_1\sigma_1(\tau)\beta_1^{-1} + \sigma_2(\tau)\beta_2^{-1} + \Psi(\tau) = 0 \tag{3.42}$$

Laplace transformation can be used to solve Equations 3.40 to 3.42; and for the Laplace transforms, one obtains

$$\sigma_1(s) = \frac{2\sqrt{\beta_1 s}Y(s)}{1 - \exp\left(-2\sqrt{s\beta_1^{-1}}\right)}, \tag{3.43}$$

$$\sigma_2(s) = 2\sqrt{\beta_2 s}Y(s), \tag{3.44}$$

$$Y(s) = \frac{\Pi_\infty\Gamma_0}{\Phi(s)s}. \tag{3.45}$$

The complex roots of the equation

$$\Phi(s) = 0 \tag{3.46}$$

can be found for two cases: (1) $\Gamma_0 << \Gamma_1$ and (2) $\Gamma_0 >> \Gamma_1$. The parameter $\lambda = \Gamma_0/\Gamma_1$ in usual variables is

$$\lambda = \frac{\Pi_\infty L D_2 q_s(T_0)}{R_0}\frac{R_0}{T_0 k_1} \tag{3.47}$$

It is the ratio of two fluxes: condensation heat flux and thermal conductivity flux inside the drop. So, the inequality $\lambda << 1$ means that the real heat flux of phase transition heat is much smaller than the potential amount of heat that can be transferred by thermal conductivity in the drop. If $\lambda << 1$, there are branch point and two poles with small real parts in Equation 3.46, so the asymptotic formula for the surface temperature is

$$Y(\tau) = Y_p\left(1 - \exp\left(-\frac{\tau}{\tau_3}\right) - \frac{\Phi_0}{2}\sqrt{\frac{\pi}{\beta_2\tau}}\right) \tag{3.48}$$

Y_p is steady-state (psychometric) temperature of the drop surface:

$$Y_p = \frac{\Pi_\infty \Gamma_0}{1+\Gamma}, \text{ and} \tag{3.49}$$

$$\tau_3 = \frac{\Gamma_1}{3\beta_2(1+\Gamma)}. \tag{3.50}$$

It can be shown that a more simple formula (Equation 3.51) is also applicable because it is possible that $\tau_3 >> \tau >> 1$. Temperature relaxation in this case is slow and regular enough. The primary reason for this is the slow growth or evaporation of the drop, which is determined by the low value of supersaturation Π_∞. In the opposite case, when $\lambda >> 1$, there are no small poles in Equation 3.46 and thus,

$$Y(\tau) = Y_p\left(1 - \exp\left(-\frac{\tau}{\tau_3}\right)\right) \tag{3.51}$$

$$Y(\tau) = Y_p\left(1 - \frac{\Phi_0\sqrt{\pi}}{\sqrt{2\beta_2\tau}}\right) \tag{3.52}$$

because the exponential term is absent in this case (the steady-state temperature is reached more quickly). In both cases, thermal relaxation inside the drop is slower than that outside because $\beta_2 >> \beta_1$. To derive the conventional formula for the outside temperature from Equation 3.40, one obtains:

$$\sigma_2(\tau) = 2\sqrt{\frac{\beta_2}{\pi}}\int_0^\tau \frac{d\partial}{\sqrt{\tau-\partial}}\frac{dY(\partial)}{d\partial}. \tag{3.53}$$

By substituting Equation 3.53 into Equation 3.16 after some transformation, it is easy to derive the following expression.

$$xY_2(\tau) = \frac{2}{\sqrt{\pi}}\int_{\frac{|1-x|}{2\sqrt{\beta_2\tau}}}^{\infty} dz\, e^{-z^2}\, Y\left(\tau - \frac{(1-x)^2}{4z^2\beta_2}\right) \tag{3.54}$$

If $\beta_2\tau >> (1 - x)^2$, then a formula similar to that for vapor concentration (Equation 3.18)) follows.

$$Y_2(x,\tau) = \frac{Y(\tau)}{x} \tag{3.55}$$

The steady-state temperature Y_p is obtained from Laplace transform (Equation 3.46) when $s \to 0$; but the same formula can be obtained from Equation 3.42 when $t \to \infty$; the last three terms vanish and the first two terms represent the conventional heat balance equation of the drop. From Equation 3.43, it follows that $\sigma_2(\tau)$ can be connected with dY/dt by an integral equation similar to Equation 3.53, but with a different kernel. This kernel at large τ will be proportional to $\tau^{1/2}$, so asymptotically, this will be similar to $\Psi(\tau)$ and both these terms will be exponentially small ($\lambda << 1$) or as τ^{-1} when $\tau \to \infty$. More troublesome is the derivation of the formula for the temperature gradient inside the drop. An approach similar to that used in deriving Equation 3.27 for the salt concentration can be applied. Expansion of the kernel $K_2(\tau, \theta, x)$ in Equation 3.16 leads to:

$$Y_1(x,\tau) = \Pi_{11} + x^2\Pi_{12}, \tag{3.56}$$

$$\Pi_{11}(\tau) = \frac{2}{\sqrt{\pi}} \int\limits_{\frac{1}{4\beta_{1r}\tau}}^{\infty} ds\, e^{-s^2} Y\left(\tau - (4\beta_1\tau)^{-1}\right), \tag{3.57}$$

$$\Pi_{12}(\tau) = \frac{2\beta_1}{\sqrt{\pi}} \int\limits_{\frac{1}{4\beta_{1r}\tau}}^{\infty} ds\, e^{-s^2} s^2 \left(\frac{2s^2}{3} - 1\right) Y\left(\tau - (4\beta_1\tau)^{-1}\right). \tag{3.58}$$

It is natural that the second term that determines the gradient is proportional to temperature conductivity. For large τ, from Equation 3.58, one can obtain

$$Y_1(x,\tau) = Y(\tau)\left(1 + \frac{1}{96\sqrt{\pi\beta_1^2\tau^3}}\right). \tag{3.59}$$

The temperature gradient inside the drop is positive (negative) for the growing (evaporating) drop, smaller near the drop center, and very rapidly decreases with time.

THE RATE OF CHANGE OF DROP SIZE

Thermal relaxation can influence the rate of drop growth only through the saturatation vapor density in Equation 3.5 and Equation 3.13, which depend on surface temperature. If vapor supersaturation is small, the deviation of surface temperature from the steady-state value will result in some retardation of the rate of growth during relaxation ($T_{10} < T_\infty$) and will be greater than at steady-state. The mirror-reflected situation will take place in the case of evaporation. If the environmental temperature varies with time for the period (t_∞) much greater than the relaxation time ($\lambda << 1$), then the drop temperature will follow it. For the period smaller than the relaxation time, the drop will grow under average environment temperature. In the case $\lambda >> 1$, when there is no characteristic time, the latter case corresponds to $t_\infty << R_0^{-1} D_2$ and the former to $t_\infty >> R_0^{-1} D_2$.

Unlike thermal relaxation, deviation in the salt concentration can directly influence drop growth. For a saturated vapor environment and neglecting thermal relaxation, the following formula is derived for small time (usual variables).

$$\frac{dR}{dt} = R_0\left[1 + \frac{\omega t}{R_0^5} - \frac{2\omega^2 R_0 t^{3/2}}{3\sqrt{\pi D_1}}\right] \tag{3.60}$$

The second term in the brackets describes the growth at the initial salt concentration and does not depend on salt diffusivity. The third term decelerates the growth due to the decrease in salt concentration near the drop surface. The smaller the salt diffusivity, the greater the salt concentration gradient. This formula is valid during the first moments of concentration relaxation ($t << t_2$); during the last moments ($t >> t_2$), we have the equation to determine $Z(\tau)$:

$$Z^5(\tau) - 1 + \frac{5g}{3}\left[Z^3(\tau) - 1\right] = 5g\tau. \tag{3.61}$$

For weak solutions, $g << 1$, and one obtains for the zero approximation:

$$Z_0(\tau) \cong (1 + gt)^{1/5}. \tag{3.62}$$

The next iteration gives

$$Z(\tau) \cong Z_0(\tau)\left(1 - g\frac{Z_0^2(\tau) - 1}{Z_0(\tau)}\right) \tag{3.63}$$

INTENSIVE EVAPORATION OF THE SOLUTION DROP

An example of the application of the theory of concentration relaxation is intensive evaporation of the solution drop. Salt concentration enhancement near the drop surface can be large and may result in the formation of a solid crust. The treatment of this problem was considered by Buikov and Sigal. Intensive evaporation takes place under high undersaturation in the environment and blowing of the drop with dry air, so it is possible to assume that the drop radius is a linear function of time; that is,

$$R(t) = R_0 - bt. \tag{3.64}$$

Using Equations 3.13 and 3.21, it is possible to obtain the integral equation for the subsidiary function $\sigma(\theta)$ as:

$$\int_0^\tau d\partial\,\sigma(\partial) K_4(\tau, \partial)\, Z(\partial) = -\frac{2}{3}\left(Z^3(\tau) - 1\right); \;\; Z(\tau) = 1 - a\tau. \tag{3.65}$$

The salt concentration can be calculated from Equation 3.25. The following formula was derived for salt concentration at the drop surface:

$$\Sigma(\tau) = \frac{a\sqrt{\tau}}{\sqrt{\pi Z(\tau)}}\left(1 + \frac{1}{2\sqrt{\pi D(a)}}\right). \tag{3.66}$$

Crystallization of salt will start when the solution near the surface is saturated ($c(R(t_s), t_s) = c_s$) and then the solid crust can grow at the drop surface. For the time for the formation of crust of thickness δ the following formula can be applied:

TABLE 3.1
The Time to Start Crystallization

Velocity of Blowing ($cm\ s^{-1}$)	Experimental Value (s)	Calculated Value (s)
40	0.390	0.350
90	0.155	0.145
160	0.100	0.046

TABLE 3.2
The Time to Form Crust

Substance	Experimental (s)	Calculated (s)
Na_2SO_4	210	175
NH_4NO_3	235	215

$$t_c = t_s + \sqrt{\frac{2\delta\rho_s R_c^2}{b^2 R_s^2 \left(\frac{dc}{dr}\right)\bigg|_{r = R_s}}} \tag{3.67}$$

Values of t_s and t_c calculated for the experimental conditions are given in Tables 3.1 and 3.2.[11,12]

SUMMARY AND CONCLUSIONS

The classical formula for the condensation or evaporation rate of a liquid drop derived by Maxwell and modernized by Fuchs[1] is based on some hypotheses of complete physical lucidity. This formula is widely used in many branches of aerosol science. Giving up the hypotheses leads to more complicated mathematical problems and complicates the solution. In the research work reviewed in this chapter, a new approach is introduced: a heat balance equation on the drop surface that connects thermal processes inside and outside the drop with phase transition heat. The application of thermal potentials to solving the system of equations of heat, vapor, and salt diffusion resulted in the integral equations for subsidiary functions. This system of equations may be more appropriate for use in numerical methods than the primary system of differential equations. The approximate analytical analysis of the processes of drop growth carried out using these integral equations makes it possible to penetrate more deeply into heat and salt transfer inside the drop, as well as to follow the transition from the initial state of the drop to steady-state growth and steady-state fields of temperature, vapor, and salt. The asymptotical formulas for salt and temperature gradients are also derived. Analytical studies resulted only in some corrections to the conventional formulas, but nevertheless it has been demonstrated that the procedure developed can be useful for intensive processes of drop evaporation. It is hoped that the theory developed can find more wide application, namely for intensive growth and evaporation — especially if the drop or environmental temperature varies with time rapidly enough.

REFERENCES

1. Fuchs, N.A., *Evaporation and Droplet Growth in Gaseous Media*, Pergamon Press, New York, 1958.
2. Mason, B.J., *The Physics of Clouds*, Clarendon Press, London, 1957.
3. Pruppacher, H.R. and Klett, I.D., *Microphysics of Clouds and Precipitation*, D. Reidel, Dordrecht, 1978.
4. Buikov, M.V. and Dukhin, S.S., Diffusional and heat relaxation of evaporating drop, *Eng. Phys. J.*, 5(3), 1962 (in Russian).
5. Buikov, M.V., Diffusional and heat relaxation of evaporating drop. Part 2, *Eng. Phys. J.*, 5(4), 1962 (in Russian).
6. Buikov, M.V., Time dependent growth of solution drop. I. Relaxation of concentration, *Colloidny J.*, 24(6), 1962 (in Russian).
7. Buikov, M.V., Time-dependent growth of solution drop. II. Heat relaxation, *Colloidny J.*, 25(1), 1963 (in Russian).
8. Buikov, M.V., Some Problems of Growth and Evaporation of Drops in Gaseous Media, Thesis of candidate dissertation. Kiev (in Russian), 1963.
9. Buikov, M.V. and Sigal, V.I., Intensive evaporation of solution drop, P*roblems of Evaporation, Combustion and Gas Dynamics of Disperse Systems*, Editor, V.A. Fedoseev, Naukova Dumka Publ. House, Kiev (in Russian), 1967.
10. Smirnov, V.I., *Course of Higher Mathematics*, Vol. 2, Gostechizdat, Moscow (in Russian), 1948.
11. Ranz, W.E. and Marshall, W.R., *Chem. Eng Progr.*, 48(3), 1952.
12. Charlesworth, D.E. and Marshall, W.B., A.I.Ch. E.J., 6(1), p. 1959, 1960.

NOMENCLATURE

A	$M_v/M_s\ \rho$
$c(r,t)$	Concentration of salt
D_1	Diffusivity of salt
D_2	Diffusivity of vapor
D(a)	$= 1 - \text{v.p.}\int_0^\infty ds\, e^{-s^2}\left(a + 4s^2\right)\left(a - 4s^2\right)^{-1}\pi^{-1/2}$ v.p. means the main value.
g	$= \dfrac{Ac_a\delta}{1+\Gamma}$
$\kappa_1(K_1)$	Heat (temperature) conductivity of liquid
$\kappa_2(K_2)$	Heat (temperature) conductivity of gas-vapor mixture
L	Phase transition heat
m	Mass of salt in drop
$M_s(M_d)$	Molecular weight of salt (dissolvant)
$q(r,t)$	Vapor concentration
$q_s(T)$	Density of saturated vapor
r	Radial coordinate
$R(t)$	Radius of drop
t	Time
t_1	$= R_0^2 D_2^{-1}$
t_2	$= R_0^2 D_1^{-1}$
$T_1(r, t)$	Temperature of liquid inside drop

$T_2(r, t)$ Temperature of gas-vapor mixture

$T_p = T_{10} + \dfrac{LD_2\left(q_\infty - q_s(T_\infty)\right)}{\kappa_2(1+\Gamma)}$ Steady-state drop surface temperature

x $= rR_0^{-1}$

$Y_1(x, \tau)$ $= \left(T_1(r,t) - T_{10}\right)T_{10}^{-1}$

$Y_2(x, \tau)$ $= \left(T_2(r,t) - T_\infty\right)T_\infty^{-1}$

$Y(\tau)$ $= (T(R(t)) - T_{10})T_{10}^{-1}$ Temperature of drop surface

Y_p $= (T_p - T_{10})T_{10}^{-1}$ Steady-state surface temperature

$Z(\tau)$ $= R(t)/R_0$

β_1 $= \kappa_1 D_1^{-1}$

β_2 $= \kappa_2 D_1^{-1}$

γ $= D_2\, D_1^{-1}$

δ $= \dfrac{D_2 q_s(T_0)}{D_1\rho}$

Γ $= \Gamma_0\, \Pi_\infty$

Γ_1 $= \kappa_1\, \kappa_2^{-1}$

Γ_0 $= LD_2 q_s(T_\infty)T_\infty^{-1}\kappa_2^{-1}$

λ $= \dfrac{\Pi_\infty LD_2 q_s(T_\infty)}{R_0}\left(\dfrac{T_0\kappa_1}{R_0}\right)^{-1}$

$\Pi(z(\tau), \tau)$ $= \Pi_s(Y) - AC_0\left[1 + \Pi_s(Y)\right](1 + \Sigma(x, \tau))$

$\Pi_s(Y)$ $= \left(q_s(T) - q_s(T_\infty)\right)q_s^{-1}(T_\infty)$

Π_∞ $= \dfrac{q_\infty(T_\infty)(1 - Ac_0) - q_\infty}{q_\infty}$

Π_1 $= \dfrac{1 - Ac_0}{q_\infty}\dfrac{dq_s(T_\infty)}{dT}$

ρ Liquid density

ρ_s Density of salt

$\sigma(\theta), \sigma_1(\theta), \sigma_2(\theta)$ Kernels of thermal potentials

$\Sigma(x, \tau)$ $= \left(C(r,t) - C_0\right)C_0^{-1}$

$\Sigma_1(\tau)$ $= \displaystyle\int_0^\tau \frac{d\theta\, \sigma_2(\theta) Z(\theta)}{\sqrt{2\pi}(\tau - \theta)^{3/2}} e^{-\frac{Z^2(\theta)}{4(\tau-\theta)}}$

$\Sigma_2(\tau)$	$= \int_0^{\tau} \frac{d\theta\, \sigma_2(\theta) Z(\theta)}{6\sqrt{\pi}(\tau-\theta)^{3/2}} \left[\frac{Z^2(\theta)}{6(\tau-\theta)} - 1 \right] e^{-\frac{Z^2(\theta)}{4(\tau-\theta)}}$
τ	$= t\, D_{1.} R_0^{-2}$
$\Phi(s)$	$= 1 + \Gamma - \Gamma_1 + s^{1/2}\beta_1^{-1/2} + \Gamma_1 s^{1/2} \beta_2^{-1/2} cth\left(s^{1/2}\beta_2^{-1/2}\right)$
Φ_0	$= \frac{2}{\pi\sqrt{\beta_1}(1+\Gamma)}$
ω	$= \frac{q_s(T_\infty) A m D_2}{\rho(1+\Gamma) R_0^5}$

4 Phase Transformation and Growth of Hygroscopic Aerosols

Ignatius N. Tang

CONTENTS

INTRODUCTION

Ambient aerosols play an important role in many atmospheric processes affecting air quality, visibility degradation, and climatic changes as well. Both natural and anthropogenic sources contribute to the formation of ambient aerosols, which are composed mostly of sulfates, nitrates, and chlorides in either pure or mixed forms. These inorganic salt aerosols are hygroscopic by nature and exhibit the properties of deliquescence and efflorescence in humid air. For pure inorganic salt particles with diameter larger than 0.1 micron, the phase transformation from a solid particle to a saline droplet occurs only when the relative humidity in the surrounding atmosphere reaches a certain critical level corresponding to the water activity of the saturated solution. The droplet size or mass in equilibrium with relative humidity can be calculated in a straightforward manner from thermodynamic considerations. For aqueous droplets 0.1 micron or smaller, the surface curvature effect on vapor pressure becomes important and the Kelvin equation must be used.[1]

In reality, however, the chemical composition of atmospheric aerosols is highly complex and often varies with time and location. Junge[2] has shown that the growth of atmospheric aerosol particles in continental air masses deviates substantially from what is predicted for th growth of pure salts. He explained this difference by assuming a mixture of soluble and insoluble materials within the particle, thus introducing the concept of mixed nuclei for atmospheric aerosols. Subsequent investigation by Winkler[3] led to an empirical expression for the growth of continental atmospheric aerosol particles. Tang[4] considered the deliquescence and growth of mixed-salt particles, relating aerosol phase transformation and growth to the solubility diagrams for multi-component electrolyte solutions.

In this chapter, an exposition of the underlying thermodynamic principles on aerosol phase transformation and growth is given. Recent advances in experimental methods utilizing single-particle levitation are discussed. In addition, pertinent and available thermodynamic data, which are needed for predicting the deliquescence properties of single- and multi-component aerosols, are compiled. Information on the composition and temperature dependence of these properties is

0-87371-829-1/00/$0.00+$.50

required in mathematical models for describing the dynamic and transport behavior of ambient aerosols. Such data, however, are very scarce in the literature, especially when dealing with aerosols composed of mixed salts as an internal mixture.

SINGLE-PARTICLE LEVITATION EXPERIMENTS

Numerous methods have been employed by investigators to study aerosol phase transition and growth in humid air. Thus, Dessens[5] and Twomey[6] conducted deliquescence experiments with both artificial salt and ambient particles collected on stretched spider webs. They examined the particles with a microscope and noted phase transition in humid air. Orr et al.[7] investigated the gain and loss of water with humidity change by measuring the change in electrical mobility for particles smaller than 0.1 μm. Winkler and Junge[8] used a quartz microbalance and studied the growth of both artificial inorganic salt aerosols and atmospheric aerosol samples collected on the balance by impaction. Covert et al.[9] also reported aerosol growth measurements using nephelometry. Finally, Tang[10] constructed a flow reactor with controlled temperature and humidity and measured the particle size changes of a monodisperse aerosol with an optical counter. Although these methods suffer from either possible substrate effects or some difficulties in accurate particle size and relative humidity measurements, they have provided information for a clear understanding of the hydration behavior of hygroscopic aerosols.

In recent years, however, new experimental techniques have been developed for trapping a single micron-sized particle in a stable optical or electrical potential well. These new techniques have made it possible to study many physical and chemical properties that are either unique to small particles or otherwise inaccessible to measurement with bulk samples. An earlier review by Davis[11] documented the progress up to 1982. Since then, many interesting investigations have appeared in the literature. In particular, thermodynamics[12-14] and optical properties[15,16] of electrolyte solutions at concentrations far beyond saturation that could not have been achieved in the bulk, can now be measured with a levitated microdroplet. This is accomplished by continuously and simultaneously monitoring the changes in weight and in Mie scattering patterns of a single suspended solution droplet undergoing controlled growth or evaporation in a humidified atmosphere, thereby providing extensive data over the entire concentration region. Other interesting works on the physics and chemistry of microparticles have been discussed in the recent review by Davis.[17] In this section, the experimental methods used by Richardson and Kurtz[18] and Tang et al.[13] are described in some detail.

Single particle levitation is achieved in an electrodynamic balance (or quadrupole cell), whose design and operating principles have been described elsewhere.[19-22] Briefly, an electrostatically charged particle is trapped at the null point, of the cell by an ac field imposed on a ring electrode surrounding the particle. The particle is balanced against gravity by a dc potential, U, established between two endcap electrodes positioned symmetrically above and below the particle. All electrode surfaces are hyperboloidal in shape and separated by Teflon insulators. When balanced at the null point, the particle mass, w is given by

$$w = \frac{qU}{gz_o}, \tag{4.1}$$

where q is the number of electrostatic charges carried by the particle, g the gravitational constant, and z_o the characteristic dimension of the cell. It follows that the relative mass changes, w/w_0, resulting from water vapor condensation or evaporation can be measured as precisely as measurement of the dc voltage changes, U/U_0, that are necessary for restoring the particle to the null point. Here, the subscript, o, refers to measurements for the initial dry salt particle.

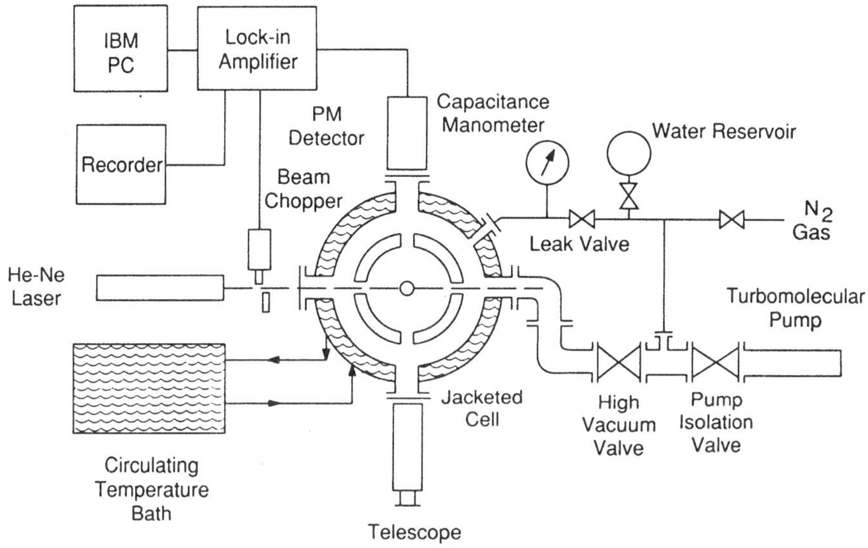

FIGURE 4.1 Schematic diagram of the single-particle levitation apparatus.

A schematic diagram of the apparatus is shown in Figure 4.1. The single-particle levitation cell is placed inside a vacuum chamber equipped with a water jacket that can maintain the cell temperature within ±0.1°C. A linear, vertically polarized He-Ne laser beam, entering the cell through a side window, illuminates the particle, 6 to 8 μm in diameter when dry. The particle position is continuously monitored by a CCD video camera and displayed on a TV screen for precise null point balance. The 90° scattered light is also continuously monitored with a photomultiplier tube. The laser beam, which is mechanically chopped at a fixed frequency, is focused on the particle so that a lock-in amplifier can be used to achieve high signal-to-noise ratios in the Mie scattering measurement.

Initially, a filtered solution of known composition is loaded in a particle gun; a charged particle is injected into the cell and captured in dry N_2 at the center of the cell by properly manipulating the ac and dc voltages applied to the electrodes. The system is closed and evacuated to a pressure below 10^{-7} torr. The vacuum is then valved off and the dc voltage required to position the particle at the null point is now noted as U_0. The system is then slowly back-filled with water vapor during particle deliquescence and growth. Conversely, the system is gradually evacuated during droplet evaporation and efflorescence. The water vapor pressure, p_1, and the balancing dc voltage, U, are simultaneously recorded in pairs during the entire experiment. Thus, the ratio, U_0/U, represents the solute mass fraction and the ratio, $p_1/p^o{}_1$, gives the corresponding water activity, a_1, at that point. Here, $p^o{}_1$ is the vapor pressure of water at the system temperature. The measurement can be repeated several times with the same particle by simply raising the water vapor pressure again and repeating the cycle. The reproducibility is better than ±2%.

HYDRATION BEHAVIOR AND METASTABILITY

A deliquescent salt particle, such as KCl, NaCl, or a mixture of both, exhibits characteristic hydration behavior in humid air. Typical growth and evaporation cycles at 25°C are shown in Figure 4.2. Here, the particle mass change resulting from water vapor condensation or evaporation is plotted as a function of relative humidity (RH). Thus, as RH increases, a crystalline KCl particle (as illustrated by solid curves) remains unchanged (curve A) until RH reaches its deliquescence point (RHD) at 84.3% RH. Then, it deliquesces spontaneously (curve B) to form a saturated solution droplet by water vapor condensation, gaining about 3.8 times its original weight. The droplet

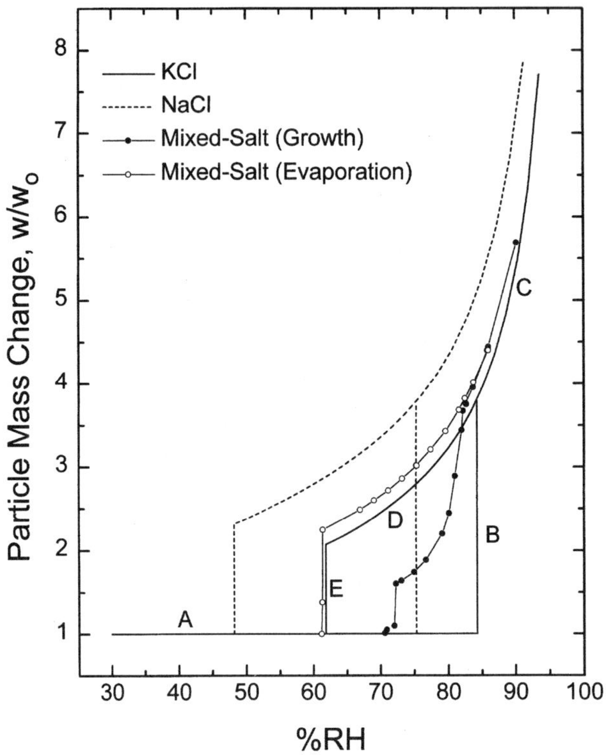

FIGURE 4.2 Growth and evaporation of KCl/NaCl particles in humid environment at 25°C.

continues to grow as RH further increases (curve C). Upon decreasing RH, the solution droplet loses weight by water evaporation. It remains a solution droplet even beyond its saturation point and becomes highly supersaturated as a metastable droplet (curve D) at RH much lower than RHD. Finally, efflorescence occurs at about 62% RH (curve E), when the droplet suddenly sheds all its water content and becomes a solid particle. Similar behavior is illustrated in Figure 4.2 as dashed curves for an NaCl particle, which deliquesces at 75.4% at 75.4% RH and crystallizes at about 48% RH. Note that, for a single-salt particle, the particle is either a solid or a droplet, but not in a state of partial dissolution.

In a bulk solution, crystallization always takes place not far beyond the saturation point. This happens because the presence of dust particles and the container walls invariably induce heterogeneous nucleation at a much earlier stage than what would be expected for homogeneous nucleation to occur. On the other hand, in a solution droplet where the presence of an impurity nucleus is rare, homogeneous nucleation normally proceeds at high supersaturations. Thus, the hysteresis shown in Figure 4.2 by either the KCl or NaCl particle represents a typical behavior exhibited by all hygroscopic aerosol particles. The observations reported by Rood et al.[23] also revealed that in both urban and rural atmospheres, metastable droplets indeed existed more than 50% of the time when the RH was between about 45 and 75%. Since solution droplets tend to become highly supersaturated before efflorescence, the resulting solid may be in a metastable state that is not predicted from the bulk-phase thermodynamic equilibrium. In fact, some solid metastable states formed in hygroscopic particles may not even exist in the bulk phase.[24] It follows that the hydration properties of hygroscopic aerosol particles cannot always be predicted from their bulk solution properties.

A case of interest is Na_2SO_4 aerosol particles. In bulk solutions at temperatures below 35°C, sodium sulfate crystallizes with ten water molecules to form the stable solid-phase decahydrate,

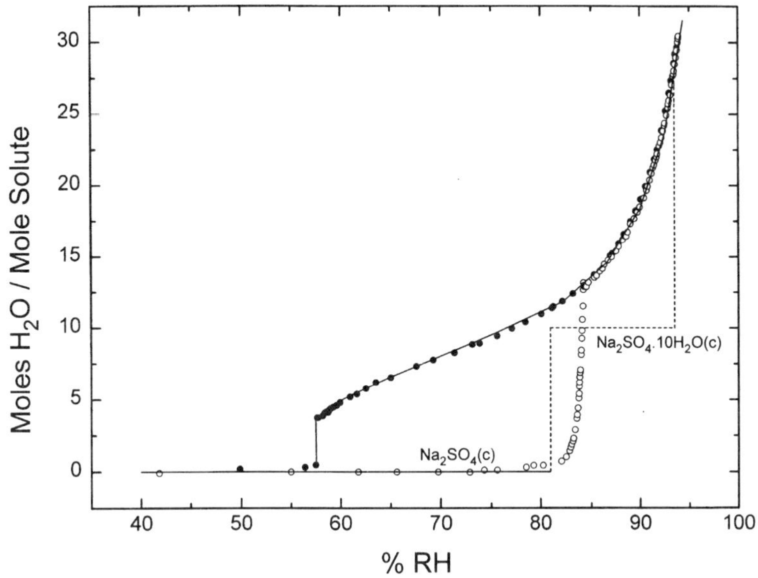

FIGURE 4.3 Growth and evaporation of a Na_2SO_4 particle in humid environment at 25°C.

$Na_2SO_4 \cdot 10H_2O$.[25] In suspended microparticles, however, it is the anhydrous solid, Na_2SO_4, that is formed most frequently from the crystallization of supersaturated solution droplets. This fact is established both by particle mass measurements[14] and by Raman spectroscopy.[24] Figure 4.3 shows the growth (open circles) and evaporation (filled circles) of an Na_2SO_4 particle in a humid environment at 25°C. The hydration behavior is qualitatively very similar to that of the KCl or NaCl particle shown in Figure 4.2. Thus, as the RH increases, an anhydrous Na_2SO_4 particle deliquesces at 84% RH to form a saturated solution droplet containing about 13 moles H_2O per mole solute (moles H_2O/mole solute). Upon evaporation, the solution droplet becomes highly supersaturated until, finally, crystallization occurs at about 58% RH, yielding an anhydrous particle.

At high supersaturations, the decahydrate is no longer the most stable state. The relative stability between anhydrous Na_2SO_4 and the decahydrate can be estimated from a consideration of the standard Gibb's free energy change, ΔG^o, of the system:

so that,

$$Na_2SO_4(c) + 10H_2O(g) = Na_2SO_4 \cdot 10H_2O(c),$$

$$\Delta G^o = \Delta G_f^o[Na_2SO_4 \cdot 10H_2O] - \Delta G_f^o[Na_2SO_4] - 10\Delta G_f^o[H_2O] = -RT\ln(1/p_1^{10}). \tag{4.2}$$

Here, c and g in the parentheses refer to the crystalline state and gas phase, respectively. Taking the tabulated[26] ΔG_f^o values –871.75, –303.59, and –54.635 kcal mol^{-1} for $Na_2SO_4 \cdot 10H_2O(c)$, $Na_2SO_4(c)$, and $H_2O(g)$, respectively, we obtain a value of –21.81 kcal mol^{-1} for ΔG^o, which leads to 19.2 torr as the equilibrium partial pressure of water vapor, or 81% RH at 25°C. It follows that, instead of the decahydrate, the anhydrous Na_2SO_4 becomes the most stable state below 81% RH. Thus, as depicted by the dashed lines shown in Figure 4.3, a solid anhydrous Na_2SO_4 particle would have transformed into a crystalline decahydrate particle at 81% RH, which would then deliquesce at 93.6% RH, to become a saturated solution droplet containing about 38 moles H_2O/mole solute, according to solution thermodynamics.[27] However, the observed hydration behavior of the particle, as shown in Figure 4.3, is quite different from what is predicted from bulk-phase thermodynamics.

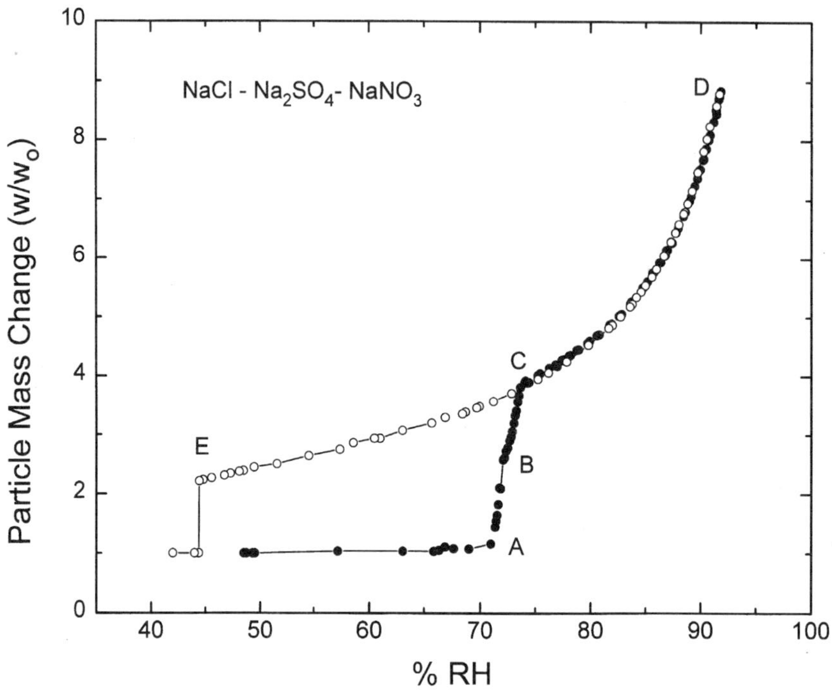

FIGURE 4.4 Growth and evaporation of a mixed-salt particle composed of NaCl, Na_2SO_4, and $NaNO_3$ in humid environment at 17.5°C.

The hydration behavior of a mixed-salt particle is more complicated in that partially dissolved states may be present. This is illustrated again in Figure 4.2 by the growth (filled circles) and evaporation (open circles) of a mixed-salt particle composed of 80% KCl and 20% NaCl by weight. The particle was observed to deliquesce at 72.5% RH, followed by a region where excess KCl gradually dissolved in the solution as the RH increased. The particle became a homogeneous solution droplet at 82% RH. Upon evaporation, the solution droplet was observed to crystallize at about 61% RH. Figure 4.4 shows the growth and evaporation of another mixed-salt particle composed of equal amounts of NaCl, Na_2SO_4, and $NaNO_3$. At 17.5°C, the particle was observed to deliquesce at 72% RH.[16,28] There was also a region following deliquescence where excess solids were gradually dissolving in the solution. At 74% RH, this mixed-salt particle became a homogeneous solution droplet, which would then grow or evaporate as RH was increasing or decreasing, respectively, as shown in Figure 4.4. Upon evaporation, the particle was observed to persist as a metastable solution droplet and finally crystallized at about 45% RH. Thus, the general hydration characteristics are similar for multi-component aerosol particles.

Tang[4] has considered the phase transformation and droplet growth of mixed-salt aerosols. The particle deliquescence is determined by the water activity of the eutonic point, E, in the solubility diagram, as shown in Figure 4.5 for the KCl–NaCl–H_2O system. Here wt% NaCl is plotted vs. wt% KCl for ternary solutions containing the two salts as solutes and H_2O as the solvent. The solid curves, AE and BE, shown here for 25°C, are solubility curves constructed from data taken from Seidell and Linke.[25] Each point on the solubility curves determines the composition of a saturated solution in equilibrium with a specific water activity. Thus, point A represents the solubility of NaCl at a concentration of 26.42 wt% and a_1 of 0.753, and point B is the solubility of KCl at 26.37 wt% and a_1 of 0.843. The solution is saturated with NaCl along the curve AE and with KCl along BE. The eutonic point, E, is the composition (KCl/NaCl ≐ 11.14/20.42%) where both salts have reached their solubility limits in the solution at the given temperature. This is usually the compo-

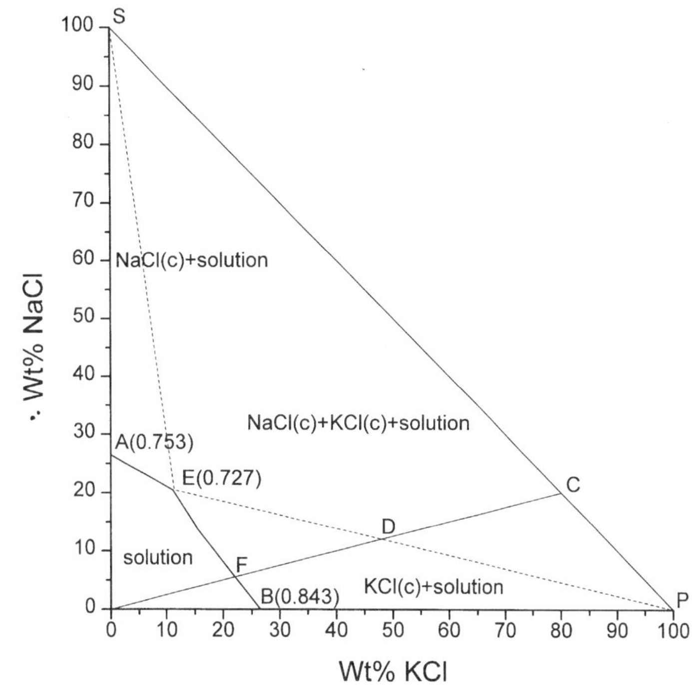

FIGURE 4.5 Solubility diagram for the system KCl-NaCL-H_2O at 25°C.

sition at which the water activity is the lowest among all compositions.[4,29] It is, therefore, the composition of the solution droplet formed when a solid particle of any composition (e.g., KCl/NaCl = 80/20%, as represented by point C) first deliquesces. Wexler and Seinfeld[30] have shown theoretically that the RHD of one electrolyte is lowered by the addition of a second electrolyte, essentially explaining why the RHD of a mixed-salt particle is lower than that of either single-salt particles.

EQUILIBRIUM DROPLET SIZE AND WATER ACTIVITY

The equilibrium between an aqueous salt solution droplet and water vapor in humid air at constant temperature and relative humidity has been considered by many investigators since the earlier work of Koehler.[31] A thorough account of the thermodynamics of droplet-vapor equilibrium can be found in books by Dufour and Defay[32] and by Pruppacher and Klett.[33] For a solution droplet containing nonvolatile solutes, the equation

$$\ln \frac{p_1}{p_1^o} = \ln \gamma_1 y_1 + \frac{2\upsilon_1 \sigma}{RTr} \tag{4.3}$$

is quite general and applies to both single- and multi-component systems, provided that the solution properties are determined for the system under consideration.[4,34] Equation (4.3) relates the equilibrium radius r of a droplet of composition y_1 (mole fraction) to RH, namely, $\%RH = 100\, p_1/p_1^{\circ}$, and to the solution properties such as the activity coefficient γ_1, partial molar volume υ_1, and surface tension σ. Here, the subscript 1 refers to water as the solvent. p_1 is the partial pressure and p_1° the saturation vapor pressure of water at temperature T (°K). R is the gas constant. For a droplet 0.1 μm in diameter, the contribution of the second term on the right-hand side of Equation (4.3) is about 2%. Consequently, for larger droplets, the droplet composition agrees closely with that of a bulk solution in equilibrium with its water vapor at given T, and the water activity of the solution droplet is simply

$$a_1 = \gamma_1 y_1 = \frac{p_1}{p_1^o} = \frac{\%\text{RH}}{100}. \tag{4.4}$$

The change in particle size at a given relative humidity can be readily deduced from a material balance on salt content before and after droplet growth to its equilibrium size. The following equation is obtained:

$$\frac{d}{d_o} = \left(\frac{100\rho_o}{x\rho}\right)^{1/3} \tag{4.5}$$

Here, d and ρ are, respectively, the diameter and density of a droplet containing x% by weight of total salts. Again, the subscript, o, refers to the dry salt particle. It follows that, in order to calculate droplet growth as a function of RH, it is essential to have water activity and density data as a function of droplet composition.

The simplest measurements that can be made with the single-particle levitation technique are water activities of electrolyte solutions over a large concentrated range, especially at high supersaturations that could not have been done with bulk solutions. For highly hygroscopic inorganic salts such as NH_4HSO_4, $NaHSO_4$, and $NaNO_3$, the solution droplets may persist in the liquid form to such a degree that one solvent molecule is shared by five or six solute molecules.[16] Such data are not only required in modeling the hydration behavior of atmospheric aerosols, but also crucial to testing and furthering the development of solution theories for high concentrations and multicomponent systems. Indeed, some efforts have begun to modify and extend Pitzer's semiempirical thermodynamic model for relatively dilute electrolyte solutions to high concentrations.[35-37]

$(NH_4)_2SO_4$ is one of the most important constituents of the ambient aerosol. A large effort has been made to obtain thermodynamic and optical data for modeling computations. Thus, Richardson and Spann[12] have made water activity measurements at room temperature with $(NH_4)_2SO_4$ solution droplets levitated in a chamber that can be evacuated and back-filled with water vapor. Cohen et al.[14] have employed an electrodynamic balance placed in a continuously flowing gas stream at ambient pressures and made water activity measurements for a number of electrolytes, including $(NH_4)_2SO_4$. The two sets of data show some discrepancies, which amount to 0.04 to 0.05 in water activities, or 5 to 6 wt% at high concentrations. Chan et al.[38] have repeated the measurements in a spherical void electrodynamic levitator (SVEL) and obtained results consistent with those of Cohen et al. The SVEL is a variation of the electrodynamic balance with the inner surfaces of the electrodes designed to form a spherical void.[39] Tang and Munkelwitz[16] have also made extensive measurements in their apparatus, which is closer in design to that of Richardson and Spann but butter thermostatted. Their results, together with those of previous studies, are shown in Figure 4.6. It appears that, although the agreement among all data sets is acceptable for aerosol growth computations, there is a need for more intercomparison studies to reduce the variability before the method can become standardized for precise thermodynamic measurements. The discrepancies could be due to experimental uncertainties in balancing the particle at the null point, adverse effects of thermal convection in the cell, and/or unavoidable measurement errors in humidity and temperature.

Because of space limitations, as well as the specific purpose of this review, water activity and density are given only for a few selected inorganic salt systems, most of which are of atmospheric interest. Both water activity and density are expressed in the form of a polynomial in x, the solute wt%, namely,

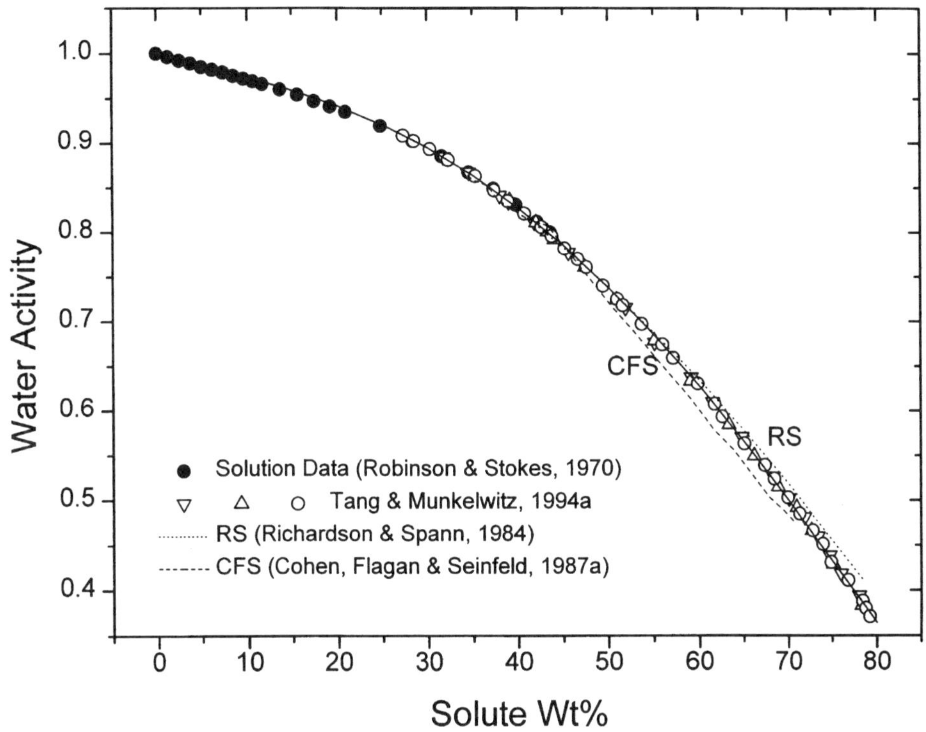

FIGURE 4.6 Water activities of aqueous $(NH_4)_2SO_4$ solutions as 25°C.

TABLE 4.1
Summary of Polynomial Coefficients for Water Activities and Densities

	$(NH_4)_2SO_4$	NH_4HSO_4	$(NH_4)_3H(SO_4)_2$	Na_2SO_4		$NaHSO_4$	$NaNO_3$	NaCl
x (%)	**0–78**	**0–97**	**0–78**	**0–40**	**40–67**[a]	**0–95**	**0–98**	**0–48**
C_1	–2.715 (–3)	–3.05 (–3)	–2.42 (–3)	–3.55 (–3)	–1.99 (–2)	–4.98 (–3)	–5.52(–3)	–6.633(–3)
C_2	3.113 (–5)	–2.94 (–5)	–4.615 (–5)	9.63 (–5)	–1.92 (–5)	3.77 (–6)	1.286 (–4)	8.624 (–5)
C_3	–2.336 (–6)	–4.43 (–7)	–2.83 (–7)	–2.97 (–6)	1.47 (–6)	–6.32 (–7)	–3.496 (–6)	1.158 (–5)
C_4	1.412 (–8)						1.843 (–8)	1.518 (–5)
A_1	5.92 (–3)	5.87 (–3)	5.66 (–3)	8.871 (–3)		7.56 (–3)	6.512 (–3)	7.41 (–3)
A_2	–5.036 (–6)	–1.89 (–6)	2.96 (–6)	3.195 (–5)		2.36 (–5)	3.025 (–5)	–3.741 (–5)
A_3	1.024 (–8)	1.763 (–7)	6.68 (–8)	2.28 (–7)		2.33 (–7)	1.437 (–7)	2.252 (–6)
A_4								–2.06 (–8)

[a]For this concentration range, $a_w = 1.557 + \sum C_i x_i$.

$$a_1 = 1 + \sum C_i x^i \tag{4.6}$$

and

$$\rho = 0.9971 + \sum A_i x^i, \tag{4.7}$$

where the polynomial coefficients, C_i and A_i, are given in Table 4.1.

TABLE 4.2
Predicted and Observed %RHD for Some Pure-Salt Particles

Salt	Solution Phase	Particle Phase	Pred. %RHD	Obs. %RHD
NaCl	Anhydrous	Anhydrous	75.3	75.3 ± 0.1
KCl	Anhydrous	Anhydrous	84.3	84.2 ± 0.3
$(NH_4)_2SO_4$	Anhydrous	Anhydrous	80.0	79.9 ± 0.5
NH_4HSO_4	Anhydrous	Anhydrous	39.7	40.3 ± 0.5
Na_2SO_4	Decahydrate	Anhydrous	93.6	84.5 ± 0.5
$NaNO_3$	Anhydrous	Anhydrous	73.8	74.1 ± 0.5
NH_4NO_3	Anhydrous	Anhydrous	61.8	61.2 ± 0.5
$Sr(NO_3)_2$	Tetrahydrate	amorphous	85.0	69.1 ± 0.5

Data for mixed-salt solutions are very limited. Tang et al.[40,41] measured the water activity of bulk solutions of $(NH_4)_2SO_4/NH_4HSO_4$ (molar ratio 1/1) and $(NH_4)_2SO_4/NH_4NO_3$ (3/1; 1/2). Spann and Richardson[42] measured the water activity of $(NH_4)_2SO_4/NH_4HSO_4$ ($1.5 \leq [NH_4^+]/[SO_4^{2-}] \leq 2$) solution droplets, using the electrodynamic balance. Cohen et al.[43] used the electrodynamic balance to measure the water activity of mixed-electrolyte solution droplets containing NaCl/KCl, NaCl/KBr, or NaCl/$(NH_4)_2SO_4$. Chan et al.[38] used the SVEL to measure the water activity of solution droplets containing various compositions of $(NH_4)_2SO_4/NH_4NO_3$. Recently, Kim et al.[64] again used the SVEL to measure the water activity of solution droplets for the $(NH_4)_2SO_4.H_2SO_4$ system. All investigators seem to agree that the simple empirical relationship, known as the ZSR relation (Zdanovskii,[44] Stokes and Robinson[45]), is capable of predicting with satisfaction the water activity of mixed-salt solutions up to high concentrations, although other, more elaborate methods may perform better at low concentrations.

For a semi-ideal ternary aqueous solution containing two electrolytes (designated 2 and 3) at a total molality $m = m_2 + m_3$, the ZSR relation

$$\frac{1}{m} = \frac{y_2}{m_{02}} + \frac{y_3}{m_{03}} \tag{4.8}$$

holds when the solution is in isopiestic equilibrium with the binary solutions of the individual electrolyte at respective molalities m_{02} and m_{03}. Here, $y_2 = m_2/m$ and $y_3 = m_3/m$. Semi-ideality refers to the case where the two solutes may interact with the solvent but not with each other. It is also conceivable that a solution behaves semi-ideally when the solute–solute interactions are present but canceling each other. Systems showing departure from semi-ideality are common.[46] For such systems, a third term, by_2y_3, can be added to the right-hand side of Equation 4.8, where b is an empirically determined parameter for each system.

PARTICLE DELIQUESCENCE

As discussed earlier, for single-salt particles larger than 0.1 μm, the deliquescence point corresponds to the saturation point of the bulk solution. Thus, %RHD for a single-salt aerosol particle is, in principle, equal to $100a_1^*$, where a_1^* is the water activity of the saturated electrolyte solution. In Table 4.2, the observed %RHD of some inorganic salt particles are compared with predictions from bulk solution data, which are available in the literature (e.g., see References 47 and 48). Note that, within experimental uncertainties, the comparison is reasonably good only for those inorganic salts whose stable crystalline phase in equilibrium with the saturated solution is identical to the observed particle phase.

TABLE 4.3
Predicted and Observed %RHD for Some Mixed-Salt Particles

System	Eutonic Composition	Solution Phases	Obs. %RHD	Pred. %RHD K-M Method	Pred. %RHD ZSR Method
KCl(A) NaCl(B)	2.183 5.106	A + B	72.7 ± 0.3	71.7	72.1
$NaNO_3$(A) NaCl(B)	6.905 4.161	A + B	68.0 ± 0.4	65.7	67.1
Na_2SO_4(A) $(NH_4)_2SO_4$(B)	1.057 5.494	A · B · $4H_2O$ + B	71.3 ± 0.4	76.4	76.4
Na_2SO_4(A) NaCl(B)	0.708 5.530	A + B	74.2 ± 0.3	75.5	74.7
Na_2SO_4(A) $NaNO_3$(B)	0.413 10.28	A · B · $2H_2O$ + B	72.2 ± 0.2	74.6	74.1

For a ternary system consisting of two salts as solutes and water as solvent, it is possible to compute the water activity at the eutonic point using the ZSR method. Other estimation methods, such as those by Meissner and Kusik,[49] Bromly,[50] and Pitzer[51] are also available in the literature. Stelson and Seinfeld[52] used the M-K method to calculate the water activities for the NH_4NO_3–$(NH_4)_2SO_4$–H_2O system and found a good agreement between the theoretical predictions and the experimental measurements of Tang et al.[41] Koloutsou-Vakakis and Rood[53] also presented a salient description of a thermodynamic model for predicting RHD for the $(NH_4)_2SO_4$–Na_2SO_4–H_2O system. They compared their %RHD predictions with field measurements by temperature- and humidity-controlled nephelnometry, assuming the aerosol sample to be internally mixed.

Table 4.3 shows the comparison of the predicted %RHD by the M-K and ZSR methods with experimental measurements for a number of mixed-salt particles. It is shown that for simple mixed-salt systems, where no crystalline hydrates or double salts are present in the solid phases, the predictions are in good agreement with the measurements. However, for more complicated systems such as the Na_2SO_4–$(NH_4)_2SO_4$ and the Na_2SO_4–$NaNO_3$ solutions, where the eutonic composition is in equilibrium with a double salt, the predicted %RHD is somewhat off. Also note that, since in an aerosol particle the solid phase may not be what is expected from the bulk solution, the observed %RHD may also be different from what is predicted on the basis of the bulk-solution eutonic composition.

Klaue and Dannecker[54,55] investigated the deliquescence properties of the double salts $2NH_4NO_3 \cdot (NH_4)_2SO_4$ (2:1) and $3NH_4NO_3 \cdot (NH_4)_2SO_4$ (3:1), using a humidity-controlled X-ray diffractometer to observe changes in the crystalline phase. They concluded that %RHD for 2:1 was 68% RH, instead of 56.4% RH as reported by Tang,[34] who made the measurement in a continuous-flow aerosol apparatus. Subsequently, Tang et al.[41] reported water activity measurements for mixed-salt solutions of NH_4NO_3–$(NH_4)_2SO_4$ and showed that the water activity at the eutonic composition was 0.66, clearly indicating that the earlier measurement was too low. The measurement error could have resulted from water adsorption on aerosol particles due to the presence of NH_4NO_3, which obscured the deliquescence point, just as what might have happened in the case of pure NH_4NO_3 aerosol particles, using the continuous-flow method.

The temperature and composition dependence of the deliquescence humidity has been investigated by Tang and Munkelwitz.[16,56] Consider, for example, a solid KCl particle surrounded by humid air at a temperature *T*. At its deliquescence humidity corresponding to a water vapor partial pressure of p_1 atm, the particle transforms into a droplet by condensing, on a molar basis, one mole of water vapor, $H_2O(g)$, onto n moles of crystalline KCl(c) to form a saturated aqueous solution of molality m_s. Assume again the diameter of the droplet to be larger than 0.1 μm so that the Kelvin effect due to surface tension can be ignored. The vapor-liquid equilibrium can be expressed by the following reactions:

$$H_2O(g) = H_2O(l) \tag{4.9}$$

$$H_2O(l) + n\ KCl(c) = KCl(aq,\ m_s) \tag{4.10}$$

Here, the symbols in the parentheses have the following meanings: g denotes vapor, l liquid, c crystalline, aq aqueous solution. The heat that is released in Reaction (4.9) is the heat of condensation of water vapor, which is equal to its heat of vaporization, $-\Delta H_V$. The heat that is absorbed in Reaction (4.10) is the integral heat of solution, ΔH_S, which can be calculated from the heats of formation tabulated in standard thermodynamic tables.[26] The overall heat involved in the process is the sum of the two heats:

$$\Delta H = n\Delta H_S - \Delta H_V. \tag{4.11}$$

Thus, applying the Clausius-Clapeyron equation to the phase transformation, one obtains

$$\frac{d\ln p_1}{dT} = -\frac{\Delta H}{RT^2} = \frac{\Delta H_V}{RT^2} - \frac{n\Delta H_S}{RT^2}. \tag{4.12}$$

Since by definition,

$$\frac{d\ln p_1^0}{dT} = \frac{\Delta H_V}{RT^2}, \tag{4.13}$$

it follows that, by combining Equations 4.4, 4.12, and 4.13, one obtains

$$\frac{d\ln a_1}{dT} = -\frac{n\Delta H_S}{RT^2}. \tag{4.14}$$

Here, n is the solubility in moles of solute per mole of water, which can be found either in International Critical Tables[47] or in the compilation by Seidell and Linke.[25] For the convenience of integrating Equation 4.14, n is expressed as a polynomial in T

TABLE 4.4
Thermodynamic and Solubility Data of Electrolyte Solutions

Systems	%RHD	ΔH_S (cal mol^{-1})	A	B	C
$(NH_4)_2SO_4$	79.9 ± 0.5	1510	0.1149	–4.489 (–4)	1.385 (–6)
Na_2SO_4	84.2 ± 0.4	–2330	0.3754	–1.763 (–3)	2.424 (–6)
$NaNO_3$	74.3 ± 0.4	3162	0.1868	–1.677 (–3)	5.714 (–6)
NH_4NO_3	61.8	3885	4.298	–3.623 (–2)	7.853 (–5)
KCl	84.2 ± 0.3	3665	–0.2368	1.453 (–3)	–1.238 (–6)
NaCl	75.3 ± 0.1	448	0.1805	–5.310 (–4)	9.965 (–7)

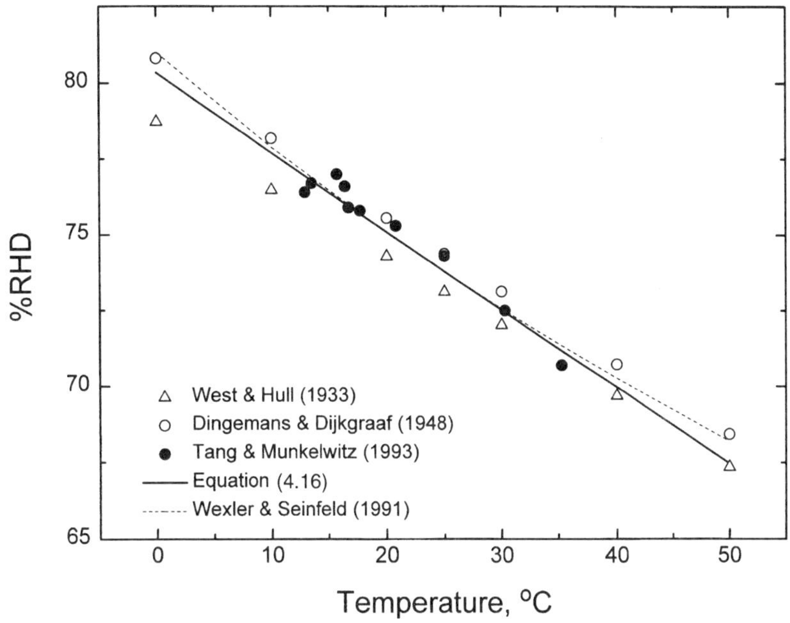

FIGURE 4.7 Deliquescence humidities as a function of temperature for $NaNO_3$ particles.

$$n = A + BT + CT^2. \tag{4.15}$$

Upon substituting n from Equation 4.15 into Equation 4.14, rearranging and integrating the resulting equation from a reference temperature, T^*, one obtains

$$\ln \frac{\%RHD(T)}{\%RHD(T^*)} = \frac{\Delta H_S}{R}\left[A\left(\frac{1}{T} - \frac{1}{T^*}\right) - B\ln\frac{T}{T^*} - C(T - T^*)\right]. \tag{4.16}$$

Since for the most electrolyte solutions the thermodynamic properties at 25°C are well documented, 298.2K is a convenient choice for T^*.

The derivation of Equation 4.16 for a single-salt particle is straightforward. Edger and Swan,[57] in considering the vapor pressure of saturated aqueous solutions, used the Van't Hoff equation relating the solubility to the integral heat of solution and obtained an equation essentially showing that $\ln a_w$ is a linear function of n over a limited temperature increment. Recently, Wexler and Seinfeld[30] derived a similar but simplified equation by assuming both constant latent heat and constant saturation molality over a small temperature change. Thus, the derivation of Equation 4.16 here is more rigorous, assuming only that the integral heat of solution is constant.

Equation 4.16 shows that the effect of temperature on %RHD is predominantly governed by the sign and magnitude of the integral heat solution. In Table 4.4, the parameters required for computing %RHD by Equation 4.16 are given for a few inorganic salts of atmospheric interest. Figure 4.7 shows a comparison of %RHD between the bulk solution data (open symbols) and the single-particle measurements (filled circles) for the $NaNO_3$–H_2O system. A comparison is also shown between characteristics by Equation 4.16 (solid curve) and by a simpler formula (dashed curve) given by Wexler and Seinfeld.[30] It is apparent that, while in general the agreement between measurements and theory is good, the single-particle data show less scatter than the bulk-solution data and agree better with theoretical predictions. The two theoretical models also agree with each other in the limited temperature range 10 to 30°C, but start to show some departure at other temperatures as a result of different assumptions used in the solubility data.

For mixed-salt systems, particle deliquescence is determined by the water activity at the eutonic point. Consider, therefore, the deliquescence of a mixed-salt particle at the eutonic composition represented by n_2 moles of NaCl, n_3 moles of KCl and 1 mole of H_2O:

$$H_2O(l) + n_2NaCl(c) + n_3KCl(c) = \text{Solution}. \tag{4.17}$$

Because of a lack of experimental data for multi-component systems, the heat that is absorbed in Reaction 4.17 can only be estimated from the respective integral heats of solution for the binary solutions, NaCl–H_2O and KCl–H_2O, namely,

$$\Delta H_S = n_2\Delta H_{S2} + n_3\Delta H_{S3} - \Delta H_1. \tag{4.18}$$

Here, the subscript, 1, refers to the solvent and the other subscript numbers refer to the solutes. The last term in Equation 4.18 accounts for the fact that ΔH_1, the differential heat of solution due to the solvent, has been included in each of the two integral heats of solution and, therefore, should be subtracted once from the total heat of solution. This is usually a small correction term and can be neglected in most cases.

The solubilities n_2 and n_3 can be obtained from the eutonic composition and expressed as a function of temperature, as in Equation 4.15. Sometimes, polynomials higher than the second order may be needed. Substituting Equation 4.18 into Equation 4.14, rearranging, and integrating lead to the final equation.[56]

$$\begin{aligned}\ln\frac{\%RHD(T)}{\%RHD(T^*)} &= \frac{\Delta H_{S2}}{R}\left[A_2\left(\frac{1}{T}-\frac{1}{T^*}\right) - B_2\ln\frac{T}{T^*} - C_2(T-T^*)\right] \\ &+ \frac{\Delta H_{S3}}{R}\left[A_3\left(\frac{1}{T}-\frac{1}{T^*}\right) - B_3\ln\frac{T}{T^*} - C_3(T-T^*)\right] - \frac{\Delta H_1}{R}\left(\frac{1}{T}-\frac{1}{T^*}\right)\end{aligned} \tag{4.19}$$

Equation 4.19 was derived strictly for the case of simple two-component mixtures forming a single eutonic composition in saturated solutions. Further work is needed for more complex aerosol systems.

Figures 4.8 and 4.9 show, respectively, the results obtained for aerosol particles containing various compositions of KCl–NaCl and $NaNO_3$–NaCl. The two lines shown for the single-salt particles are computed from theory, using tabulated parameters given in Table 4.4. The corresponding line for mixed-salt particles is computed from Equation 4.19 and pertinent data in Table 4.5. It is clear that the agreement between theory and experiment is good. The slight but noticeable departure at either end of the theoretical line may be due to our assumption of additive heats of solution made in Equation 4.18. Since there is no experimental heat of solution data available for the multi-component systems of atmospheric interest, Equation 4.19 derived on the basis of additive properties can still be used to provide a reasonable estimate in any ambient aerosol modeling studies, at least in a limited temperature region. It is also worthwhile to point out that, for salt mixtures having simply solubility properties, the deliquescence humidity is governed only by the water activity at the eutonic composition and is thus independent of the initial dry-salt composition. The temperature dependence of the mixed-salt particle usually more or less follows the direction of the component salt whose eutonic solubility is the higher of the two.

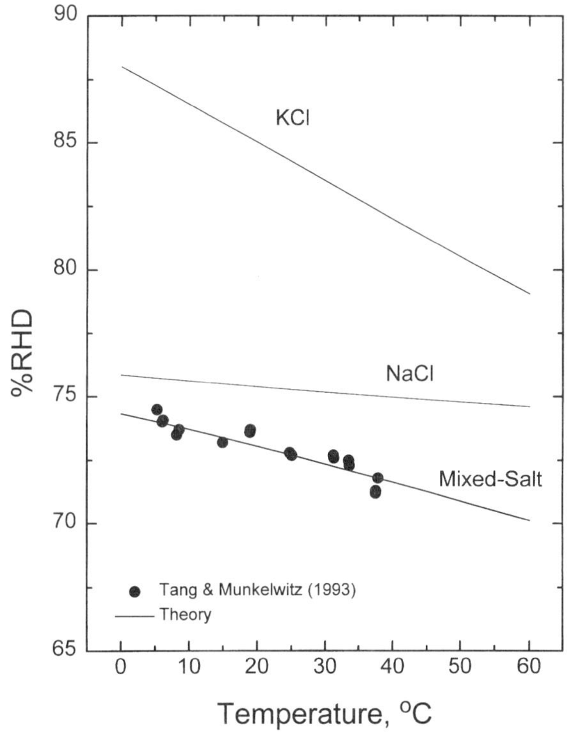

FIGURE 4.8 Deliquescence humidities as a function of temperature for mixed KCl-NaCl particles.

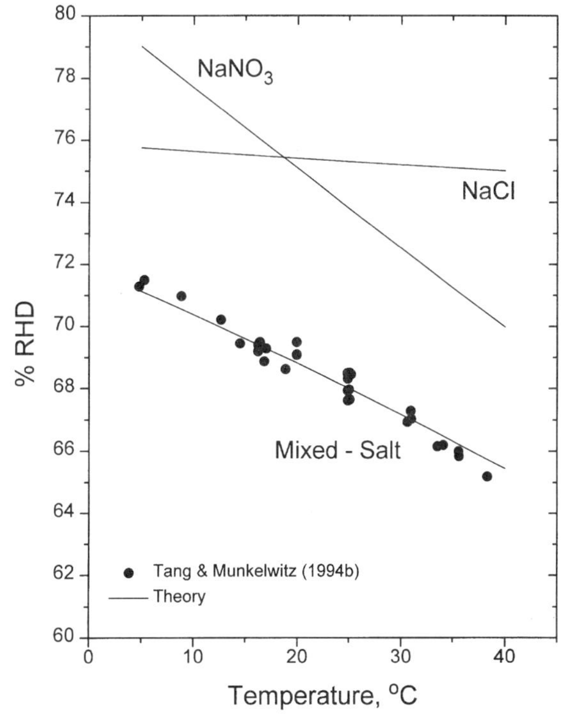

FIGURE 4.9 Deliquescence humidities as a function of temperature for mixed Na_2SO_4-$NaNO_3$ particles.

TABLE 4.5
Thermodynamic and Solubility Data of Aqueous Mixed-Salt Solutions

System	%RHD at T^*	Δh_{si} (cal mol^{-1})	A_i	B_i	C_i	D_i
NaCl	72.7 ± 0.3	448	2.618 (–1)	–9.412 (–4)	1.254 (–6)	
KCl		3665	–6.701 (–2)	1.394 (–4)	7.225 (–7)	
Na_2SO_4	72.2 ± 0.2	–2330	–4.591	4.413 (–2)	–1.407 (–4)	1.489 (–7)
$NaNO_3$		3162	6.134	–5.847 (–2)	1.852 (–4)	1.879 (–7)
$(NH_4)_2SO_4$	71.3 ± 0.4	1510	1.977 (–2)	2.617 (–4)		
Na_2SO_4		–2330	–2.187	2.343 (–2)	–8.411 (–5)	1.017 (–7)
NaCl	68.0 ± 0.4	448	5.957 (–1)	–3.745 (–3)	9.134 (–6)	–8.173 (–9)
$NaNO_3$		3162	4.532 (–1)	–4.106 (–3)	9.909 (–6)	5.552 (–10)
NaCl	74.2 ± 0.3	448	–5.313 (–1)	5.477 (–3)	–1.631 (–5)	1.689 (–8)
Na_2SO_4		–2330	–4.584 (–1)	5.000 (–3)	–1.723 (–5)	1.933 (–8)

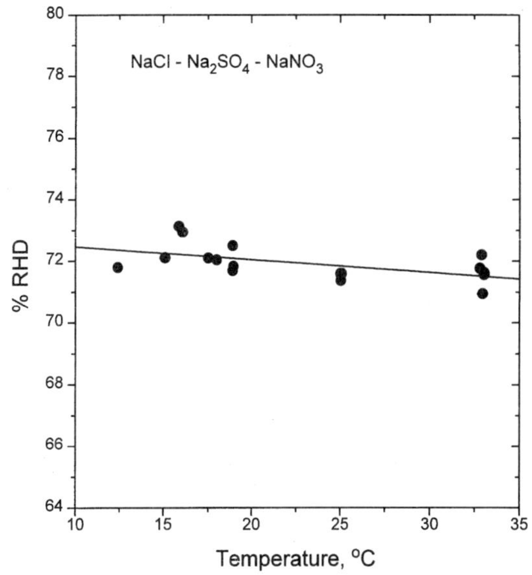

FIGURE 4.10 Deliquescence humidities as a function of temperature for mixed Na_2SO_4-$NaNO_3$-NaCl particles.

As discussed earlier, no simple mathematical analysis is yet possible at the present time for mixed-salt particles containing more than two deliquescent salts. The deliquescence properties of the three-salt system, NaCl–Na_2SO_4–$NaNO_3$, whose growth curve is shown in Figure 4.4, was studied in the limited temperature range 12 to 33°C. The results shown in Figure 4.10 indicate that, within experimental error, the deliquescence humidity can be considered constant at 71.8 ± 0.5%. A least-squares line drawn through the data points shows only very slightly, if any, temperature dependence. Because ambient aerosols are likely multi-component systems composed of more than two inorganic salts, further work to elucidate the hygroscopic properties of these complex aerosols is needed in order to predict their transport and light-scattering behavior in a humid environment.

SOLUTE NUCLEATION AND DROPLET EFFLORESCENCE

The persistence of a solution drop during evaporation to high degrees of supersaturation with respect to the solute is typical of suspended hygroscopic aerosol particles, which are free of the presence of foreign substrates. While the droplet is in equilibrium with the surrounding water vapor, it is metastable with respect to the solid-phase solute. Therefore, solute nucleation is expected: the higher the degree of supersaturation, the larger the nucleation rate.[58] According to the classical nucleation theory, the net rate of embryo formation, J, per unit volume per unit time is given by

$$J = K\exp\left(-\Delta G_c/kT\right), \tag{4.20}$$

where ΔG_c is the maximum free-energy barrier to transition to the more stable phase and k the Boltzmann constant. K, an undetermined kinetic factor, is either estimated from the binary collision frequency to the reaction rate theory[59] or expressed by some complex formula derived from various theories as discussed by Tamara et al.[60] Theoretical estimates of its value range from 10^{24} to 10^{36} $\text{cm}^{-3}\ \text{s}^{-1}$. An intermediate value that has been commonly used is $10^{30}\ \text{cm}^{-3}\ \text{s}^{-1}$. For a given rate of critical nucleus formation, J, the expected induction time, t_i, before a nucleation event happens in a droplet of volume, V_d, is given by[61]

$$t_i = \frac{1}{V_d J}. \tag{4.21}$$

Substituting Equation 4.21 into Equation 4.20 and rearranging, one obtains

$$\Delta G_c = kT\ln\left(V_d t_i K\right). \tag{4.22}$$

Assume that the nucleation embryos are crystallites formed by density fluctuations in the supersaturated solution droplet. The free-energy barrier to nucleation of a given-size crystalline embryo is

$$\Delta G = A\sigma + V\Delta G_v, \tag{4.23}$$

where A and V are, respectively, the total interfacial area and volume of the embryo, σ is the average interfacial free energy based on A, and ΔG_V is the excess free energy per unit volume of the embryo over that of the solution. For simplicity, the embryo is usually assumed to be spherical in shape so that A and V can be expressed in term of its radius, r. Other shapes consistent with the unit cells specific to given crystalline habits have also been considered, using an appropriately defined characteristic length.[14,62,63]

If the solute in the saturated solution is chosen as the references state and the definition of the solute mean activities is invoked, then, ΔG_V is given by

$$\Delta G_v = -\frac{\nu\rho_0 RT}{M}\ln\frac{a_\pm}{a_\pm^*}, \tag{4.24}$$

where $a_\pm$ and $a_\pm$* are, respectively, the solute mean activities in the supersaturated and saturated solutions. M is the solute molecular weight, ρ_o is the density of the crystalline phase, and ν is the number of ions produced by the dissociation of a salt molecule.

TABLE 4.6
Properties of Nucleation Embryos in Aqueous Salt Solutions

Salt	m (critical)	S (critical)	σ (ergs cm^{-2})	r (critical)	N (# molecules)
NaCl	13.8[a]	5.15[a]	104.0	6.81	30.0
	13.0	5.23	103.0	6.84	30.0
KCl	12.3[a]	3.64[a]	70.4	8.26	38.0
	12.6[b]	3.4[b]	67.9	8.41	40.0
	12.5	2.91	62.0	8.81	46.0
$(NH_4)_2SO_4$	17.5[a]	2.52[a]	46.6	10.2	35.0
	30.0	3.05	52.8	9.55	29.0
Na_2SO_4	13.2[a]	3.71[a]	74.1	8.06	25.0
	14.0	2.7	61.6	8.83	33.0
$NaNO_3$	78.0	2.97	62.9	8.74	45.0
	380.0	3.45	68.5	8.38	39.0

The critical size of the embryo corresponding to the maximum free-energy barrier is obtained, in the case of a spherical embryo, by letting $(\partial \Delta G/\partial r) = 0$. Hence,[63]

$$r_c = -\frac{2\sigma}{\Delta G_v} \tag{4.25}$$

and, consequently,

$$\sigma^3 = \frac{3kT\ln\left(V_d t_i K\right)}{16\pi}\left[\frac{\nu\rho_0 RT}{M}\ln S_{\pm}\right]^2, \tag{4.26}$$

where $S_{\pm}$ is the critical supersaturation at the onset of crystallization and is given by the ratio, $a_{\pm}/a_{\pm}^*$. Using the Gibbs-Duhem equation, ln $S_{\pm}$ can be calculated from the water activity measurement according to the following equation[45]:

$$\ln S_{\pm} = \int_{a_1^*}^{a_1} \frac{55.51}{\nu m} d\ln a_1. \tag{4.27}$$

Here, lna_1 is usually expressed as a polynomial in solute molality for the convenience of carrying out the integration.

In droplet crystallization experiments, $S_{\pm}$ can be measured with much higher precision than what would be possible in bulk solution studies. Thus, the uncertainties in σ determination by the single-particle levitation experiment lie largely in estimating the product $(V_d t_i K)$. Taking a typical droplet of 15 μm in diameter, an induction time about 1 s, and 10^{30} for K, the estimate of $\ln(V_d t_i K)$ is about 49, a representative value for ionic solution droplets. A change in the product by two orders of magnitude results in about 3% change in the value of σ, whereas a 15% change in S would lead to about 7% change in σ.

In Table 4.6, the estimated interfacial energy, σ, critical embryo size, r_c, and number of molecules, N, in the spherical embryo are given for some common inorganic salts. The calculation is based on the solute concentration in molality, m, and supersaturation, S, measured at the onset of solute nucleation in droplets. It is worthwhile to note that, although for each system there are discrepancies in the observed critical supersaturations, the estimated embryo properties show reasonable agreement. In addition, the nucleation embryo properties for $NaNO_3$, a highly hygroscopic salt, do not vary much, despite the fact that the critical solute concentration may span a wide range from 78 to 380 m. The invariance appears to give credence to the embryo properties determined from studies of homogeneous nucleation in suspended aqueous solution droplets.

ACKNOWLEDGMENTS

The author is indebted to his colleague, Harry R. Munkelwitz, who designed and constructed the single-particle levitation apparatus and performed the experiments reported through the years. This research was performed under the auspices of the U.S. Department of Energy under Contract No. DE-AC02-98CH10886.

REFERENCES

1. La Mer, V.K. and Gruen, R., *Trans. Faraday Soc.,* 48, 410, 1952.
2. Junge, C.E., *Ann. Met.,* 5, 1, 1952.
3. Winkler, P., *J. Aerosol Sci.,* 4, 373, 1973.
4. Tang, I.N., *J. Aerosol Sci.,* 7, 361, 1976.
5. Dessens, H., *Q. J. R. Meteor. Soc.,* 75, 23, 1949.
6. Twomey, S., *J. Meteor.,* 11, 334, 1954.
7. Orr, C, Jr., Hurd, F.K., and Corbett, W.J., *J. Colloid Sci.,* 13, 472, 1958.
8. Winkler, P. and Junge, C.E., *J. Rech. Atm.,* 6, 617, 1972.
9. Covert, D.S., Charlson, R.J., and Ahlquist, N.C., *J. Appl. Metero.,* 11, 968, 1972.
10. Tang, I.N. and Munkelwitz, H.R., *J. Aerosol Sci.,* 8, 321, 1977.
11. Davis, E.J., *Aerosol Sci. Technol.,* 2, 121, 1983.
12. Richardson, C.B. and Spann, J.F., *J. Aerosol Sci.,* 15, 563, 1984.
13. Tang, I.N., Munkelwitz, H.R., and Wang, N., *J. Colloid Interface Sci.,* 14, 409, 1986.
14. Cohn, M.D., Flagan, R.C., and Seinfeld, J.H., *J. Phys. Chem.,* 91, 4563, 1987.
15. Tang, I.N. and Munkelwitz, H.R., *Aerosol Sci. Technol.,* 15, 201, 1991.
16. Tang, I.N. and Munkelwitz, H.R.,. *J. Geophys. Res.,* 99, 18801, 1994.
17. Davis, E.J., *Advances in Chemical Engineering,* Vol., 18, Academic Press, San Diego, CA, 1992.
18. Richardson, C.B. and Kurtz, C.A., *J. Am. Chem. Soc.,* 106, 6615, 1984.
19. Straubel, H., *Z. Elektrochem.,* 60, 1033, 1956.
20. Wueker, R.F., Shelton, H., and Langmuir, R.V., *J. Appl. Phys.,* 30, 342, 1959.
21. Frickel, R.B., Shaffer, R.E., and Stamatoff, J.B., *Chambers for the Electrodynamic Containment of Charged Particles,* U.S. Department of Commerce, National Technical Information Service, Report #AD/A056 236, 1978.
22. Davis, E.J., *Langmuir,* 1, 379, 1985.
23. Rood, M.J., Shaw, M.A., Larson, T.V., and Covert, D.S., *Nature,* 337, 537, 1989.
24. Tang, I.N., Fung, K.H., Imre, D.G., and Munkelwitz, H.R., *Aerosol Sci. Technol.,* 23, 443, 1995.
25. Seidell, A. and Linke, W.F., *Solubilities of Inorganic and Metal Organic Compounds,* 4th ed., *Am. Chem. Soc.,* Washington, D.C., 1965.
26. Wagman, D.D., Evans, W.H., Halow, I., Parker, V.B., Bailey, S.M., and Schumm, R.H., *Selected Values of Chemical Thermodynamic Properties,* National Bureau of Standards Technical Note 270, U.S. Department of Commerce, Washington, D.C., 1966.
27. Goldberg, R.N., *J. Phys. Chem. Ref. Data,* 10, 671, 1981.
28. Tang, I.N. and Munkelwitz, H.R., *J. Appl. Meteor.,* 33, 791, 1994.

29. Kirgintsev, A.N. and Trushnikova, L.N., *Russ. J. Inorg. Chem.*, 13, 600, 1968.
30. Wexler, A.S. and Seinfeld, J.H., *Atmos. Environ.*, 25A, 2731, 1991.
31. Koehler, H., *Trans. Faraday Soc.*, 32, 1152, 1936.
32. Dufour, L. and Defay, R., *Thermodynamic of Clouds,* Academic Press, New York, 1963.
33. Pruppacher, H.R. and Klett, J.D., *Microphysics of Cloud and Precipitation,* D. Reidel, Dordrecht, Holland, 1978.
34. Tang, I.N., *Generation of Aerosols*, Willeke, K., Ed., Chap. 7, Ann Arbor Sci., Ann Arbor, MI, 1980.
35. Clegg, S.L. and Pitzer, K.S., *J. Phys. Chem.*, 96, 3513, 1992.
36. Clegg, S.L., Pitzer, K.S., and Brimblecombe, P., *J. Phys. Chem.*, 96, 9470, 1992.
37. Clegg, S.L. and Brimblecombe, P., *J. Aerosol Sci.*, 26, 19, 1995.
38. Chan, C.K., Flagan, R.C., and Seinfeld, J.H., *Atmos. Environ.*, 26A, 1661, 1992.
39. Arnold, S. and Folan, L.M., *Rev. Sci. Instr.,* 58, 1732, 1987.
40. Tang, I.N., Munkelwitz, H.R., and Davis, J.G., *J. Aerosol Sci.*, 9, 505, 1978.
41. Tang, I.N., Wong, W.T., and Munkelwitz, H.R., *Atmos. Environ.*, 15, 2463, 1981.
42. Spann, J.F. and Richardson, C.B., *Atmos. Environ.*, 19, 819, 1985.
43. Cohn, M.D., Flagan, R.C., and Seinfeld, J.H., *J. Phys. Chem.*, 91, 4575, 1987b.
44. Zdanovskii, A.B., *Trudy Solyanoi Laboratorii Akad. Nauk SSSR,* No. 6, 1936.
45. Stokes, R.A. and Robinson, R.H., *J. Phys. Chem.*, 70, 2126, 1966.
46. Sanster, J. and Lenzi, F., *Can. J. Chem. Eng.*, 52, 392, 1974.
47. West, C.J. and Hull, C., *International Critical Tables,* McGraw-Hill, New York, 1933.
48. Robinson, R.A. and Stokes, R.H., *Electrolyte Solutions*, Butterworth, London, 1970.
49. Meissner, H.P. and Kusik, C.L., *AIChE J.*, 18, 294, 1972.
50. Bromley, L.A., *AIChE J.*, 19, 313, 1973.
51. Pitzer, K.S., *J. Phys. Chem.*, 77, 268, 1973.
52. Stelson, A.W. and Seinfeld, J.H., *Atmos. Environ.*, 16, 2507, 1982.
53. Koloutsou-Vakakis, S. and Rood, M.J., *Tellus,* 46B, 1, 1994.
54. Klaue, B. and Dannecker, W., *J. Aerosol Sci.*, 25, S189, 1993.
55. Klaue, B. and Dannecker, W., *J. Aerosol Sci.*, 25, S287, 1994.
56. Tang, I.N. and Munkelwitz, H.R., *Atmos. Environ.*, 27A, 467–473, 1993.
57. Edger, G. and Swan, W.O., *J. Am. Chem. Soc.*, 44, 570, 1992.
58. Walton, A.G., *The Formation and Properties of Precipitates*, Interscience, New York, 1967.
59. Turnbull, D. and Fisher, J.C., *J. Chem. Phys.*, 17, 71, 1949.
60. Tamara, D., Snyder, T.D., and Richardson, C.B., *Langmuir,* 9, 347, 1993.
61. Cohn, M.D., Flagan, R.C., and Seinfeld, J.H., *J. Phys. Chem.*, 91, 4583, 1987c.
62. Enustun, B.V. and Turkevich, J., *J. Am. Chem. Soc.*, 82, 4502, 1960.
63. Tang, I.N. and Munkelwitz, H.R., *J. Colloid Interface Sci.*, 98, 430, 1984.
64. Kim, Y.P., Pun, B.K.-L., Chan, C.K., Flagan, R.C., and Seinfeld, J.H., *Aerosol Sci. Technol.*, 20, 275, 1994.
65. Dingemans, P. and Dijkgraaf, L.L., *Rec. Trav. Chim.*, 67, 231, 1948.

5 On the Role of Aerosol Particles in the Phase Transition in the Atmosphere

Jan Rosinski

CONTENTS

INTRODUCTION

The dry atmosphere of the earth consists mostly of nitrogen and oxygen. In addition to the two permanent gases, there is one variable one: water vapor. Water vapor concentration varies from close to 0 to nearly 3%. Water is the only constituent of air that, in the range of temperatures present on Earth, can exist as vapor, liquid, or solid. The lowest concentration can be found over polar regions where temperatures are mostly far below 0°C. The highest concentrations exist in the equatorial region. The heat required to vaporize or condense water or water vapor, respectively, is equal to 595.9 gram-calories (15°C) per gram at T = 0°C. The heat of fusion at T = 0°C is 79.7 cal g^{-1} and, consequently, the heat of sublimation of ice is 675.6 cal g^{-1}. During the vapor → liquid phase transition, the latent heat is released; it is also released during the liquid → solid phase transition. The latter constitutes 13.4% of the former.[1] Most of the water vapor enters the lower atmosphere through evaporation of liquid water from the surface of the Earth and, to a very small extent, through sublimation of ice. At 0°C, water vapor pressure over a flat surface of water is equal that over ice; that is, 4.579 mmHg. At a temperature of –15°C, the saturated water vapor

0-87371-829-1/00/$0.00+$.50

pressure over supercooled liquid water is 1.436 mmHg and 1.241 mmHg over ice. The vapor pressure over water is always larger than over ice for all temperatures below 0°C. Because of this difference, liquid water evaporates in the presence of ice, resulting in the ice growing. Under such conditions, the surface temperature of the evaporating water drop decreases and the temperature of the growing ice surface increases due to condensation. The largest difference in vapor pressures, $P_{water} - P_{ice} = 0.20$ mmHg is for the temperature range between –11°C and –12°C (–11.8°C). This is the basis of the Wegener-Bergeron-Findeisen mechanism of formation of precipitation in the Temperate Zone.

If one starts cooling an air parcel (e.g., in an updraft), eventually at some altitude or some lower temperature, one will discover the presence of first cloud droplets. The first cloud droplets form just below water vapor saturation on aerosol particles that are hygroscopic. With subsequent lowering of the temperature, the water vapor will become supersaturated and more cloud droplets will form. They will form on cloud condensation nuclei (CCN) that constitute a fraction of the population of aerosol particles.[2,3] When the temperature of a rising and cooling parcel of air reaches temperatures below 0°C, ice can form within a cloud. Ice is formed on aerosol (or hydrosol) particles that can act as ice-forming nuclei (IFN).[4]

The CCN initiate a phase transition of water vapor to liquid water; this is called the V → L phase transition. This process has been thoroughly treated in many textbooks, and will be discussed in this chapter only when it constitutes an integral part of the formation of ice. Ice can be formed during the vapor–solid (ice) transition (V → S phase transition) or during the liquid–solid (L → S) phase transition. In all cases, aerosol (or hydrosol) particles are necessary to make the phase transition possible in the atmosphere at temperatures above ~ –40°C; below this temperature, homogeneous nucleation of ice may take place.

It should be pointed out that the L → S phase transition taking place at temperatures below ~–40°C in the atmosphere consists of freezing liquid water suspensions (cloud droplets) of hydrophilic hydrosol particles in a water solution of different chemical compounds present in the CCN. In view of this, the L → S phase transition occurring at very low temperatures should be called spontaneous freezing of droplets; the term "freezing by homogeneous nucleation" should be reserved for freezing of pure water in laboratory experiments. Pure water droplets do not exist in the atmosphere.

There are two major sources of aerosol particles. The first one is the Earth's surface and the second one is oceans. Particles differ in chemical composition, in solubility in water, and in the structure of their surfaces and their density. Aerosol particles are lifted from the surfaces of the Earth and the oceans by turbulence associated with winds. An example of the mass lifted from the two surfaces is given in Figure 5.1.[5–7] Large water drops settle rapidly to the ocean surface, controlling the mass concentration of aerosol produced by oceans. The size distribution of marine aerosol particles is governed by two mechanisms. The larger, 0.5- to 10-μm diameter dry sea salt particles are produced by bursting bubbles, and the very small ones ($d < 0.01$ μm) through gas-to-particle conversion. Over land, soil particles up to 250 μm in diameter are suspended in the atmosphere by strong updrafts associated with storms, and their lifetime is also controlled by gravitational forces. Aerosol particles from, for example, Mainland China (115°E) have been observed to travel eastward over the Pacific Ocean as far as 170°E longitude (Figure 5.2). During their residence time over the Pacific, they coagulate with aerosol particles generated by the ocean and produce terrestrial/marine mixed aerosol particles.[8] Aerosol particles formed by the Pacific Ocean travel eastward in the Northern Hemisphere and produce marine/terrestrial mixed aerosol particles while they cross the North American continent. As a result, practically all aerosol particles are mixed aerosol particles.[9–11] Contribution from the oceans consists mostly of sulfates; these particles are soluble in water and can act as CCN.[12,13] Aerosol particles of soil origin consist mostly of water-insoluble clay minerals, and of water-soluble sodium chloride and sulfates; the latter two compounds act as CCN.

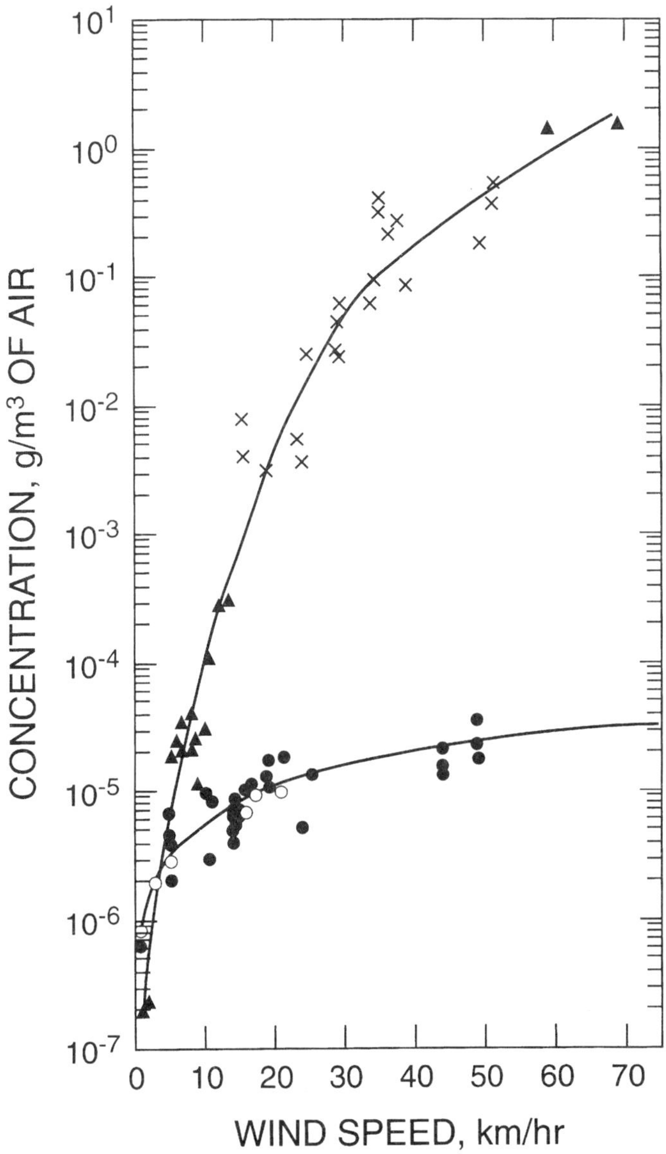

FIGURE 5.1 Concentration of soil particles (× – Chepid, 1957; △ – Rosinski et al., 1973) and sea salt particles (●, Reference 115; ○, Reference 9) as a function of wind speed.

Air parcel trajectories must be known in studies of aerosol particles participating in phase transitions in the atmosphere; they will establish the origin and contribute to the knowledge of the life history of aerosol particles under investigation.[14]

MODES OF ICE NUCLEATION

There are three basic modes of ice nucleation: freezing, contact, and sorption. The same aerosol particle present in a cloud may nucleate ice by any of the three mechanisms. Usually, the difference will be in the temperature at which phase transition into solid (ice) takes place.

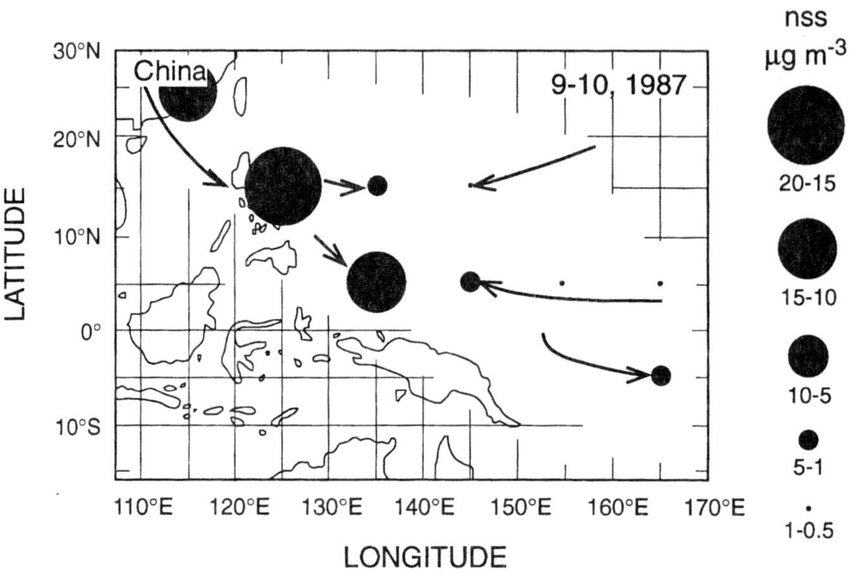

FIGURE 5.2 Transport of non-sea-salt sulfate particles over the Pacific Ocean.

LIQUID → SOLID (ICE) PHASE TRANSITION: FREEZING NUCLEI

Different-sized aerosol particles will act as cloud condensation nuclei (CCN) at different water vapor supersaturations (S_w). A dry particle, for example, ammonium sulfate of 6×10^{-2} μm diameter will act as CCN at critical supersaturation of 0.2%; a supersaturation of 1% is required to activate a particle of 1.6×10^{-2} μm diameter.[15] Liquid water droplets formed in the atmosphere are therefore water solutions of portions of an aerosol particle that acted as CCN. The water-insoluble part of that particle may be wetted, transferred into the interior of a droplet, and later act as an IFN. Transfer of aerosol particles into the liquid phase (aerosol particles become hydrosol particles) takes place during the following processes active in the atmosphere:

1. Transfer of aerosol particles through condensation of water vapor on aerosol particles active as CCN. This process can be subdivided into three separate groups:
   a. Condensation of water vapor at subsaturations with respect to saturation over liquid water, S_w, (hygroscopic particles); $S \leq S_w$.
   b. Condensation of water vapor at conditions of slight supersaturation, $S > S_w$; this takes place at and just above cloud bases.
   c. Condensation of water vapor at high supersaturations, $S >> S_w$, that are present in the vicinity of freezing drops or wet hailstones (freezing water).

   In the above three cases of V → L phase transition, the liquid phase consists of a solution of the water-soluble part of the CCN and of the water-insoluble particles present as hydrosol (hydrophilic) particles; if particles are not wetted, hydrophobic particles will remain floating on the surface of a drop. A water solution droplet can freeze at higher temperatures than the freezing temperature of pure water, or can be frozen at different temperatures with the help of a hydrosol (hydrophilic) or a hydrophobic particle.
2. Transfer of aerosol particles into cloud droplets and raindrops. This transfer takes place in the atmosphere by means of several different mechanisms, including:
   a. Brownian diffusion of submicron aerosol particles.
   b. Phoretic forces associated with condensation and evaporation of droplets: submicron- and micron-sized aerosol particles are affected by this scavenging mechanism.

c. Aerodynamic capture: larger particles will be captured by this process.
d. Turbulent diffusion: this mechanism is responsible for bringing different-sized particles together.
e. Electrostatic forces: these forces act on particles of all sizes.

The above-listed scavenging mechanisms will act in the atmosphere simultaneously. Not all of the mechanisms act at the same time, but in different combinations most probably with electrostatic forces always present.

3. Formation of hydrosol particles in the liquid phase of clouds. In addition to the above processes that transfer existing aerosol particles into the liquid phase, there are some additional mechanisms taking place in a cloud that introduce newly formed hydrosol particles directly into the condensed water. They can be grouped into two major categories. The first category consists of:
a. Formation of solid hydrosol particles during cooling of a water solution of dissolved salts (CCN); this takes place in an updraft
b. Formation of solid hydrosol particles in evaporating droplets
c. Formation of solid particles through chemical reactions between different chemical compounds supplied by the CCN and other scavenged water-soluble salts

The second category consists of submicron- and micron-sized hydrosol particles that are shed from the surfaces of larger particles when they are transferred into the condensed water drop. These particles are shed upon contact with water droplets larger than 40 μm in diameter. Concentration and size of the shed particles vary with the size of the parent particle and type of soil (Figure 5.3). This process is not the breaking of aggregates; it is a separate process.[5-7,16]

4. Hydrosol particles as IFN. Most aerosol particles consist of aggregates of water-soluble and water-insoluble particles. Aerosol particles that can act as CCN are generally water soluble; they consist of some water-insoluble particles found together in the water-soluble matrix. A droplet formed on a CCN particle consists of a water solution of water-soluble salts and a suspension of water-insoluble particles. Particles that will not be wetted (hydrophobic particles) will float on the surface of a droplet; they may nucleate ice by delayed-on-surface nucleation. Concentrations of salts and suspended (hydrophilic) particles will decrease during the growth of a droplet growing by condensation in an updraft. As the parcel rises, it will eventually pass through the 0°C temperature level. Above this altitude, cloud droplets become supercooled suspensions of hydrosol particles in solutions of CCN in water. The liquid → solid (ice) phase transition can now take place. It was found from experiments performed over the years that for all modes of ice nucleation, each particle size — even if monodispersed and chemically and physically homogeneous — is always associated with a freezing temperature spectrum.[17] Hydrosol particles and dissolved chemical compounds participate in the initiation of the L → S (ice) phase transition. To see if there is any relation between aerosol particles (aerosol particles transferred into liquid water), Rosinski introduced the concept of a water-affected fraction of aerosol particles.[14] The water-affected fraction (by number) in a given size range i is

$$f_i = 1 - N_i / L_i \tag{5.1}$$

where L is the concentration of aerosol particles and N is the concentration of water-insoluble hydrosol particles.

Transfer of aerosol particles from and into the i size range when they become hydrosol particles is shown in Figure 5.4. Group A consists of aerosol particles that are insoluble in water. The water-affected fraction for aerosol particles in Group A is equal to zero; they are transferred into water without changing size. Another extreme is when the aerosol population consists of water-soluble

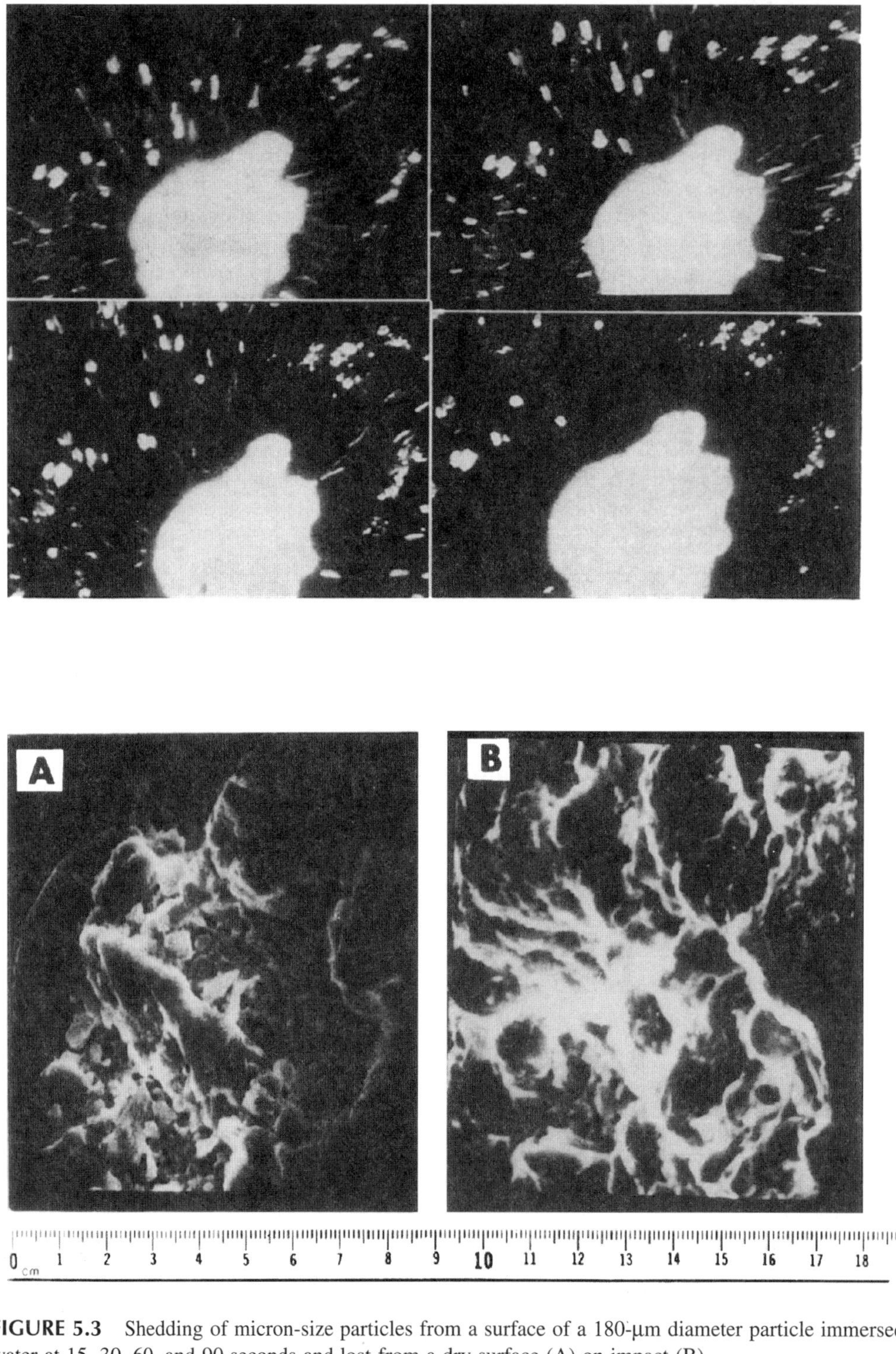

FIGURE 5.3 Shedding of micron-size particles from a surface of a 180-μm diameter particle immersed in water at 15, 30, 60, and 90 seconds and lost from a dry surface (A) on impact (B).

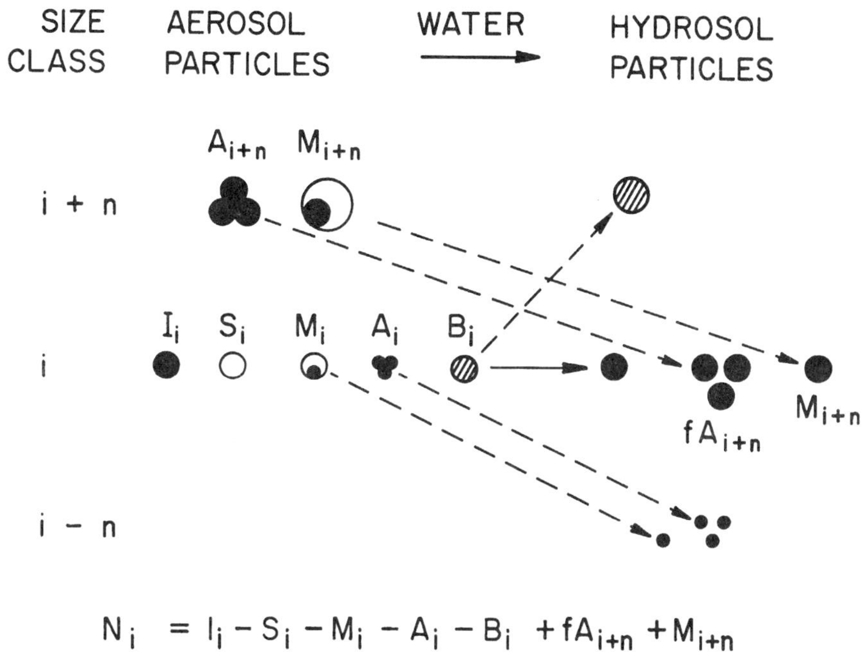

FIGURE 5.4 Transfer of aerosol particles into water.

particles only. The water-affected fraction is equal to 100%, indicating the complete absence of hydrosol particles for Group B of the aerosol particles. Group C consists of mixed aerosol particles, that is, of particles that are aggregates of water-soluble and water-insoluble particles. When the soluble part dissolves in water, the insoluble particle becomes a hydrosol particle. It can remain in the i size range, it can be transferred into the $(i - 1)$, or even into the $(i - n)$ size range and be completely lost if that lower size range is outside the size range under investigation. Category D consists of aggregates of smaller particles that may produce even larger numbers of hydrosol particles. For large concentrations of aerosol particles in the D category, the water-affected fraction becomes a negative number. Some of the results from experiments performed during 1969 and 1970 are presented in Figure 5.5. The negative values of f_i were found in experiments in which liquid impinger was used; they were for the lower size ranges of i equal to 1.5–3 and 3–5 μm diameter size ranges (experiments I, 0–0). However, there were aerosols that did not produce negative values of f_i (I, x–x), indicating the presence of aerosol particles that did not consist of aggregates that could be broken either during the contact of particles with water or the mechanical force present in the impinger that is exerted on particles. That force does not exist in nature when aerosol particles are transferred into the liquid phase of a cloud. The f_i values determined on filters clearly show the presence of two different classes of aerosol particles. The f_i values for aerosol particles of marine origin were found to be around 99% (II, ●). For pure continental air masses, the f_i values were around 1 to 5% (II, –). Aerosol particles present in mixed air masses have f_i values between the extreme values. For continental–marine air (II, ▲) f_i values were about 30% and for marine–continental air (II, –x) they were 72 to 90%. Generally, f_i values were higher (55 to 90%) in the presence of southerly winds; for westerly winds, they were from 1 to 83% in Colorado.

Part of the aerosol population acts as CCN; if the water-insoluble parts of mixed particles can act as IFN, then there should exist a direct relation between IFN and CCN. The ratio of CCN to IFN concentrations is about 10^6 in an unpolluted atmosphere. An example of that relation is shown

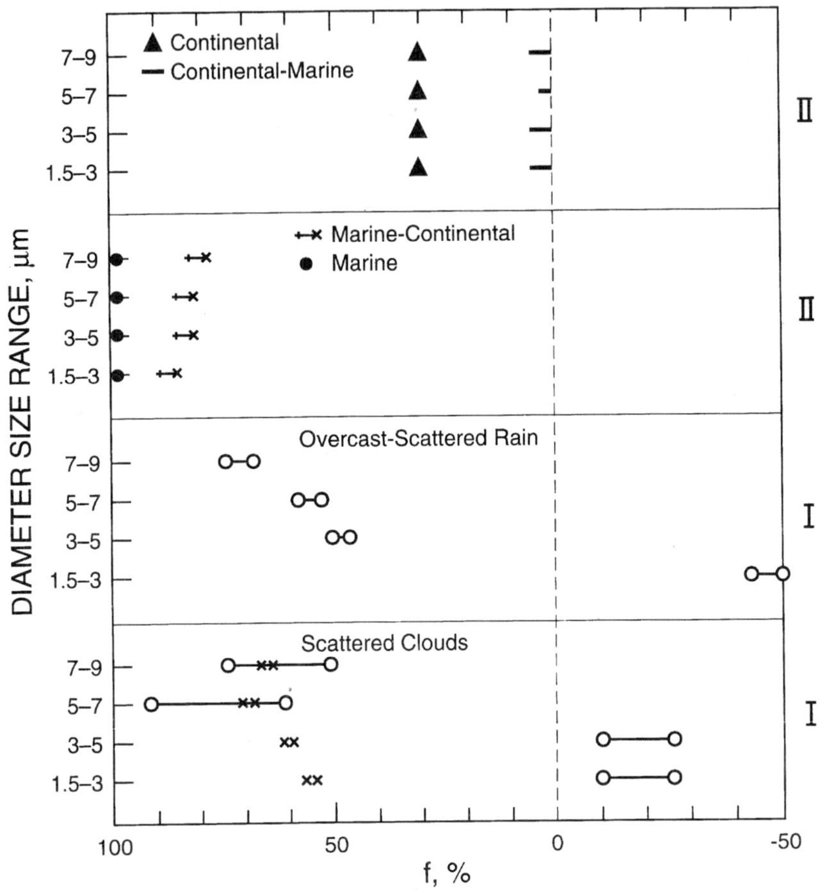

FIGURE 5.5 Water affected fraction for aerosol particles present in different air masses (I, f_i from liquid impinger; II, f_i from filters).

in Figure 5.6.[18] Khorguani, et al.[19] found a correlation between concentrations of CCN and IFN in 40% of measurements made over the North Caucasus Mountains. Results of these measurements strongly suggest a relation between CCN and IFN; they also suggest that, at first, condensation takes place on aerosol particles active as CCN and, after cloud droplets have formed, ice particles (frozen droplets) are produced through ice nucleation. The liquid phase is the solution phase, which is generally more difficult to nucleate than pure supercooled liquid water. The molal depression of the freezing point was found to be proportioned to the molality of a solution; this is known as Blagden's law. It was published in 1778, but R. Watson discovered the depression of the freezing point in 1771; his findings somehow went unnoticed. Junge[9] pointed out that the salt concentrations in cloud droplets just formed on CCN are too high for the L → S phase transition to take place at cloud temperatures. Experiments by Sano et al.[20] completely changed the understanding of freezing of droplets formed on CCN. They showed, in experiments using 8 µm average diameter water solution droplets, the existence of temperature maxima at which L → S transitions take place; this temperature was a function of the concentration of dissolved chemical compounds in water. In nature, the temperature at which droplets freeze will depend not only on the concentration of dissolved CCN, but also on the type of insoluble particle or particles that were part of the aerosol particle acting as CCN and remained within a droplet as hydrosol particles. The consequence of this finding is that not only can a droplet growing by condensation reach critical dilution and freeze, but also an evaporating droplet can come to the same critical concentration and also freeze. This

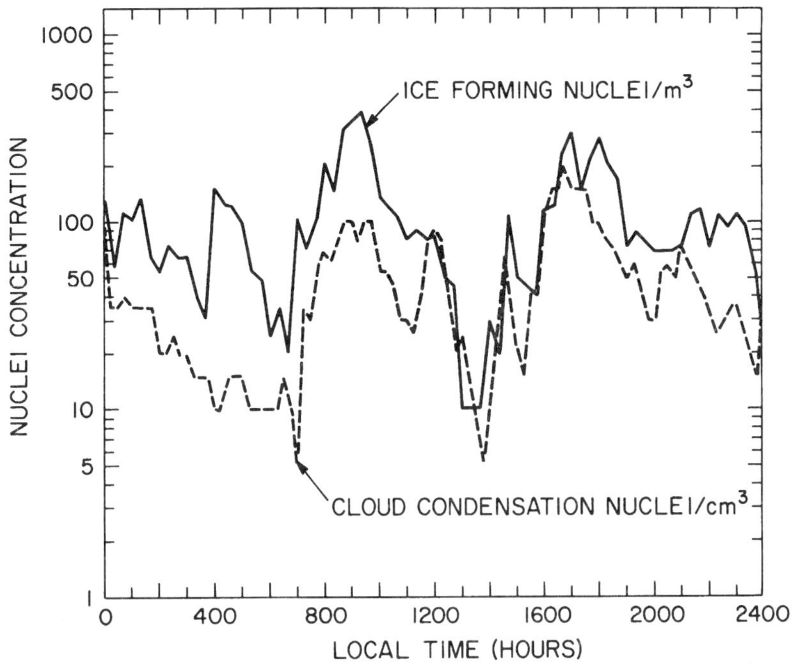

FIGURE 5.6 Concentration of IFN (–20°C) and of CCN (S_w = 1.5%) at an altitude of 3000 m in Colorado.

is shown in Figure 5.7; the hypothetical curve represents actual data but cannot be used to determine the temperature of the L → S phase transition taking place in different clouds.[21,22] The role of hydrosol particles of different origin on this freezing phenomenon is shown in Figure 5.8.[8] The maximum freezing temperatures at a given concentration of ammonium sulfate in water solution were –4°, –9°, and –12°C for marine and continental aerosol particles acting as IFN in a pure salt solution. They were all recorded at an ammonium sulfate water solution of 10^{-4} *M*. At that salt concentration, the difference between the highest temperature of drop freezing of a water solution of pure ammonium sulfate and a solution of IFN present (aerosol particles) of marine and continental origins is 8° and 3°C, respectively. For 10^{-1} *M* solutions, the difference was 10° and 6°C; these differences were the largest observed. It is clear that there is a difference between aerosol particles of marine and continental origin active as IFN through freezing.

Cloud condensation nuclei consist mostly of sulfates and chlorides.[10,11,23–25] Sulfate-bearing aerosol particles are predominant in the marine atmosphere. The ratio of sulfate-bearing aerosol particles to the number concentration of aerosol particles in the 0.1 to 0.3 μm diameter size range was found to be between 0.99 and 1.0. Sulfates, most probably ammonium sulfate, are therefore present in practically all cloud droplets in the marine atmosphere. The sulfate ion constitutes an integral part of IFN of marine origin. Results of experiments performed with aerosol particles of continental and marine origins are shown in Figures 5.9 and 5.10.[26,27] It was found that the concentration of IFN present in marine air masses increases with increasing S_w at constant temperature. On the other hand, the concentration of IFN of continental origin remained constant over a wide range of S_w at constant temperature. This suggests that the marine atmosphere contains aerosol particles with a wide size range. Larger aerosol particles will act as CCN at lower water vapor supersaturation; smaller particles will nucleate liquid water (vapor → liquid phase transition) at higher S_w. An aerosol particle of 0.1 μm diameter (9.26×10^{-16} g) acting as CCN will initiate a water droplet that will grow in an updraft. The concentration of ammonium sulfate in water solution will reach the critical concentration of 10^{-4} *M* when the growing droplet reaches ~4 μm in diameter. The critical concentration is the concentration of the solute at which the L → S phase transition

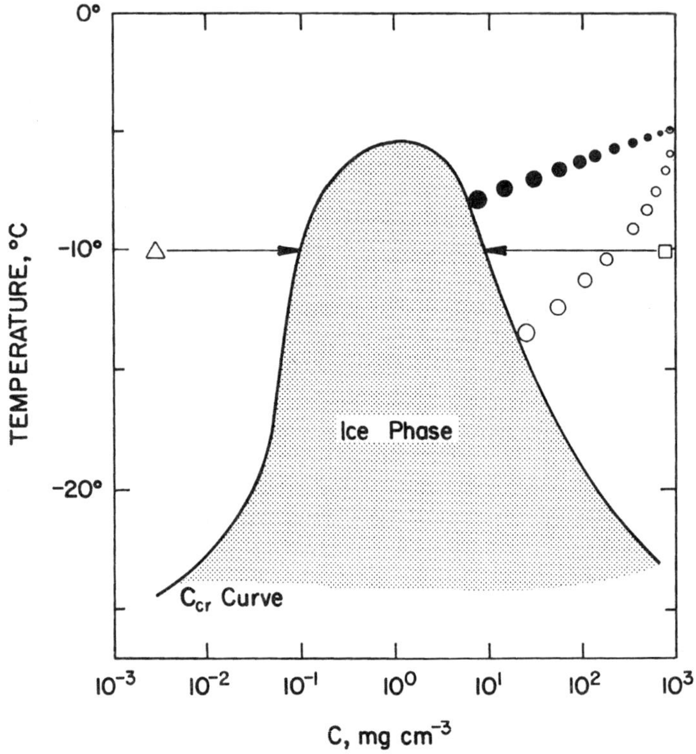

FIGURE 5.7 Hypothetical curve based on experimental data showing temperature of freezing of droplets. Critical concentration curve, C_{cr}: △, evaporating, and □, growing droplets, ●, at $S_w = 10\%$, ○, at $S_w = 0.3\%$.

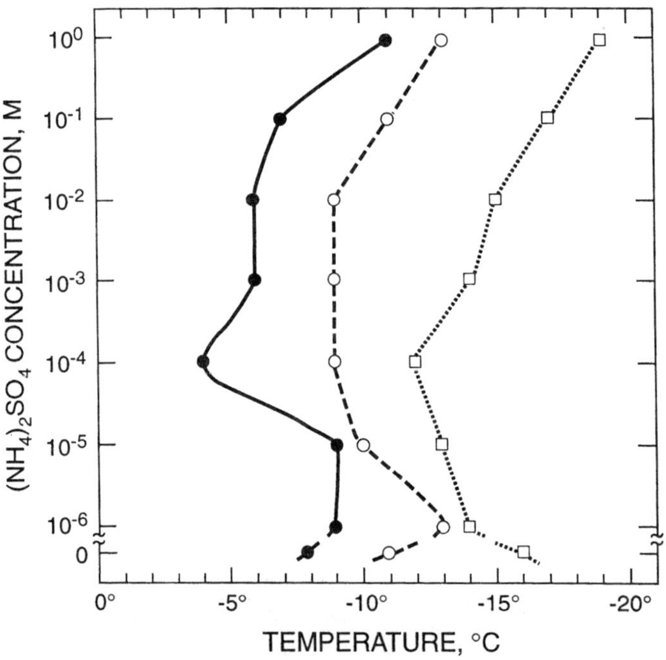

FIGURE 5.8 Maximum freezing temperatures are a function of ammonium sulfate concentration. (●, marine aerosol particles; ○, continental aerosol particles; □, no particles).

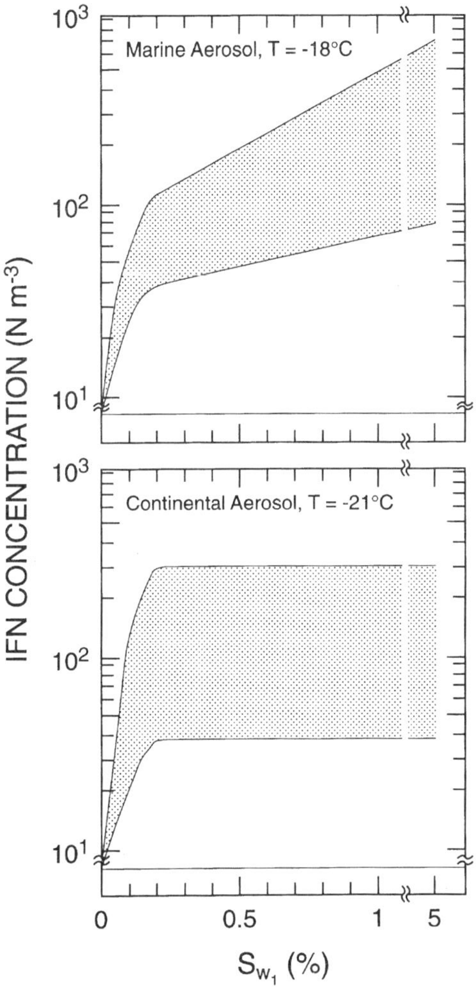

FIGURE 5.9 Cumulative concentrations of IFN of marine and continental origin as a function of water vapor supersaturation, S_w%.

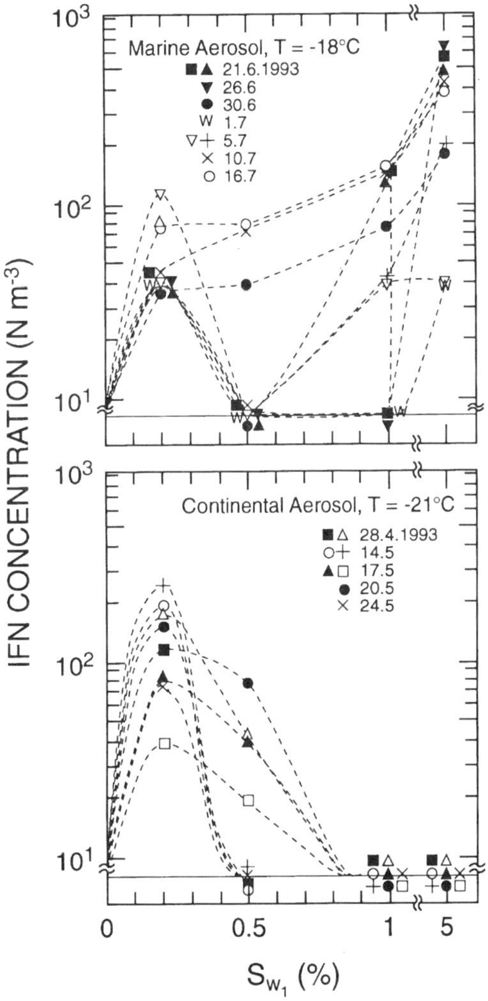

FIGURE 5.10 Differential concentrations of IFN of marine and continental origin as a function of water vapor supersaturation, S_w%.

will take place at a critical temperature. The critical temperature of ice nucleation is the maximum temperature at which the L → S phase transition takes place. For drop freezing, there are three temperature maxima: one is for a pure solution of ammonium sulfate in water, and the second and third are for the solution in contact with aerosol particles of continental and marine origins. The ~4 μm diameter droplet must therefore cross the altitude corresponding to one of the critical temperatures to freeze. If the temperature is higher than the critical temperature when the growing droplet reaches ~4 μm in diameter, then it will continue to grow by condensation in an updraft and freeze later at some lower temperature corresponding to a freezing temperature of a more dilute solution. The diameter of a droplet formed on a 0.3 μm diameter ammonium sulfate particle (2.5×10^{-4} g) is 17 μm for 10^{-4} *M* solute concentration. This cloud droplet size is a better candidate for freezing than the 4 μm diameter droplet because it will reach this diameter at a lower temperature within a cloud.

All three concentrations of solute vs. temperature of ice nucleation curves (see Figure 5.8) are more or less parallel with the maximum for the L → S phase transition occurring at a concentration of 10^{-4} M. If particles of marine or continental origin would be solely responsible for the nucleation of ice, then the temperatures of drop freezing should be the same as the temperature of ice nucleation of the individual particles. This is not the case, however, and the temperature of ice nucleation is governed by the concentration of the solute to a large degree. Particles seem to change temperature in an orderly manner. The internal structure of liquid water is not uniform. Water has local regions in its interior of hydrogen-bound clusters. In the presence of particles, ions present in the solution must somehow either increase the number or size of ice-like clusters and move the ice nucleation temperature upward. Ammonium and sulfate ions are larger than H^+ and OH^- ions, and they probably form their own network of ions, thus "squeezing" or "caging" water molecules and maybe stabilizing hydrogen-bound clusters. All this must take place on the surface of hydrosol particles. Particles of marine origin always contain organic matter. Particles of continental origin, on the other hand, contain everything that is picked up by wind from the surface of the ground. The ice nucleating ability of particles of continental origin is associated with the presence of clays and other minerals. Adsorption of hydrogen and hydroxyl ions — and maybe hydrogen-bounded clusters — and of ammonium and sulfate ions is different for organic and inorganic surfaces, and this may account for the difference in temperatures of ice nucleation for the two different classes of IFN surfaces.

The population of IFN generally increases with decreasing temperature. This was observed all over the world. The variations were up to a factor of ten at any given temperature. Using data of Bigg and Stevenson,[28] it is possible to derive an expression for the IFN concentration (C_{IFN}, m^{-3}, the median number concentration) as a function of temperature (T°C):

$$C_{IFN} = 0.036 \times 10^{-0.222T} \tag{5.2}$$

The C_{IFN} is equal to 10 m^{-3} and 1000 m^{-3} for the temperatures of –11°C and –20°C, respectively. Equation 5.2 suggests that the IFN concentration increases tenfold for every 4.5°C temperature decrease. It should be emphasized that this relation exists for the identical method of detection of IFN under investigation. Different techniques will yield different concentrations of IFN because of different modes of activation of aerosol particles in different chambers.

There are exceptions. It was found, for example, that the aerosol particles of marine origin existing in the equatorial region of the Pacific Ocean act as IFN that are independent of S_w and temperature (see Figure 5.11). The mode of ice nucleation was condensation-followed-by-freezing. Concentrations of 100 m^{-3} active at a temperature of –3.3°C and of 3×10^4 m^{-3} active at and below –4°C were located over the South Equatorial Current. These concentrations were patchy and by no means represent the IFN concentration over the Pacific Ocean. Concentrations of IFN collected in the coastal region of the Pacific Ocean increased with increasing S_w; but for each water vapor supersaturation, they were independent of temperature (Figure 5.12). The data are scattered due to the different times of day of sampling the aerosol particles from the same air mass for each of the temperature curves. There are C_{IFN} vs. T curves that exhibit a concentration plateau for some temperature ranges.[29–31] It was assumed that the part of the curve showing C_{IFN} independent of T was due to the presence of aerosol particles of marine origin.

It has been shown up until recently that CCN are the source of IFN active by freezing; but this is not the general case. In storms, it was found that the changes in concentrations of IFN do not follow the curves showing the concentrations of CCN vs. time; it looks, as a matter of fact, like these two curves are completely independent of each other. The IFN concentration vs. time curves are parallel to the wind speed curves, indicating that the source of IFN is aerosol particles lifted from the ground surface by winds. It was also found that IFN storms exist; the concentration of IFN rises, quite often two orders of magnitude and lasts sometimes for a few hours (see Figures 5.13 and 5.14).

The size distribution of aerosol particles lifted from the surface of land depends on the condition of the land surface (either wet or dry, covered with grass or open soil) and wind speed. Particles

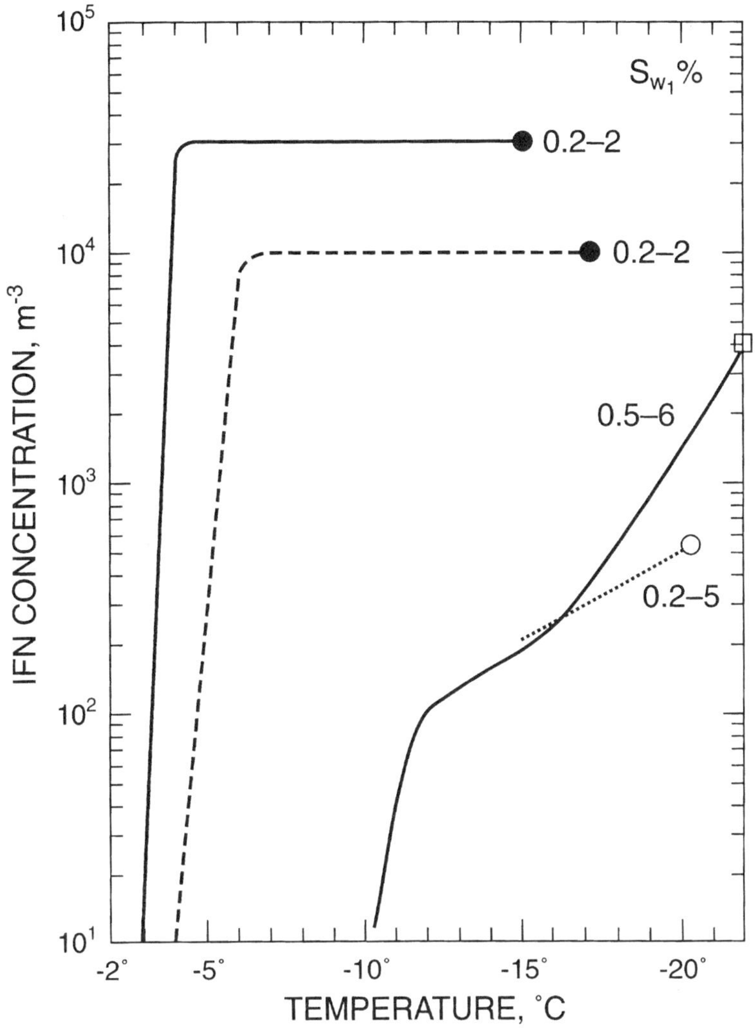

FIGURE 5.11 Concentrations of IFN active by condensation-followed-by-freezing as a function of temperature for different ranges of S_w% (continental aerosol particles: □, South Africa; ○, United States marine aerosol particles; ●, Pacific Ocean equatorial region).

found in updrafts were generally up to 250 μm in diameter (Table 5.1); particles larger than that size were found in severe storms. Aerosol particles therefore extend size distribution of cloud droplets before their full size distribution can be developed inside a cloud (given in Table 5.1). The large aerosol particles accrete cloud droplets; they become wet practically from the time they enter the cloud, and continue to grow and form large droplets or raindrops. The liquid phase of such droplets consists of one large hydrosol particle, and sometimes of large (several hundred) numbers of submicrometer-sized and a few micrometer-sized particles. The smaller hydrosol particles are shed from the surfaces of larger particles (see Figure 5.3).

Results from laboratory experiments have shown that the parent large particles nucleate ice at temperatures higher than the temperatures of ice nucleation of the shed particles (see Figure 5.15). The large hydrosol particles freeze large drops within clouds, and the frozen drops serve as transparent hailstone embryos.[32–38] The dimensions of the largest soil particle located centrally in a transparent embryo was 265 × 750 μm.

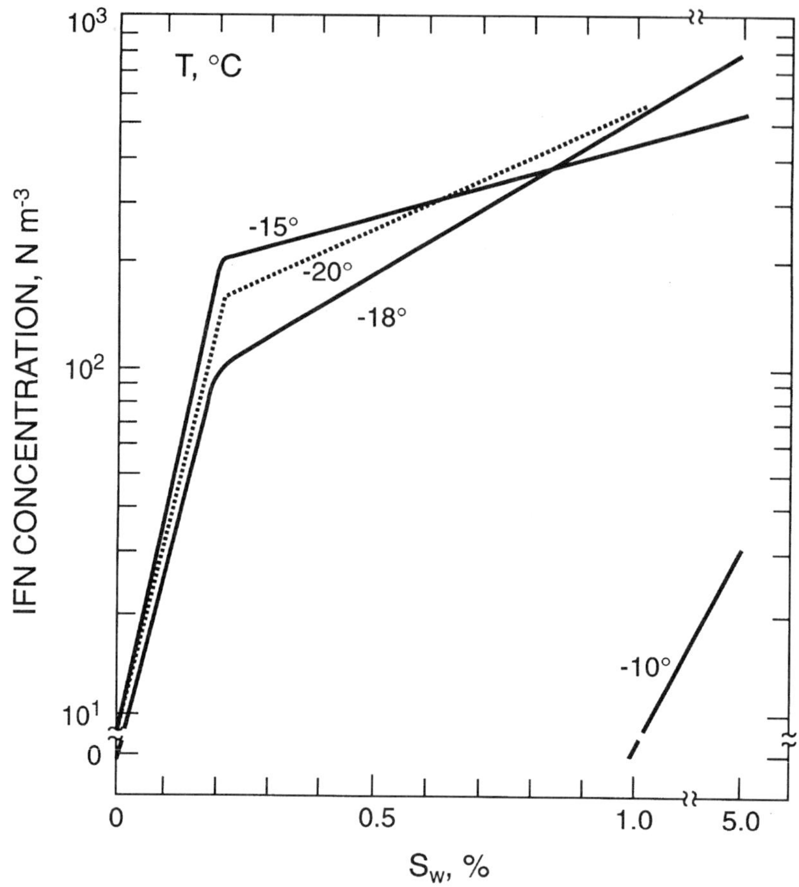

FIGURE 5.12 Concentrations of IFN from a Pacific Coastal region as a function of S_w% for different temperatures.

There are many hydrosol particles in a droplet or drop. One of the particles will start the L → S phase transition. This particle will nucleate ice at the highest temperature; the rest of the hydrosol particles may be capable of nucleating ice at temperatures infinitesimally lower than the one that started the transition. Formation of ice nucleating sites on the surfaces of hydrosol particles is time dependent.[39,40] At the time of contact of the surface of a particle with liquid water, the surface is exposed to a highly turbulent water layer. The dissolution of water-soluble chemical compounds and the shedding of submicrometer and micrometer hydrosol particles starts from the moment a particle is wetted. Heat of immersion (i.e., the sum of different heats associated with immersion, whether positive or negative) is released during the time of wetting of aerosol particles. Microscale turbulent diffusion moves the dissolved solids and the shed hydrosol particles away from the surface of the parent particle. In a few minutes, the system (hydrosol particles and solution) returns to thermal equilibrium and the process of shedding small particles ceases. During the period of particle shedding and dissolution of water-soluble chemical compounds, high rates of ice nucleation are observed. The shed particles, and the parent particle when resuspended again in supercooled water, did not exhibit enhanced ice nucleation rates and nucleated ice at lower temperatures. It is clear, therefore, that refreezing of collected precipitation cannot produce ice nucleation temperature spectra of aerosol particles ingested by a storm. Aerosol particles undergo chemical and physical changes when they become hydrosol particles.

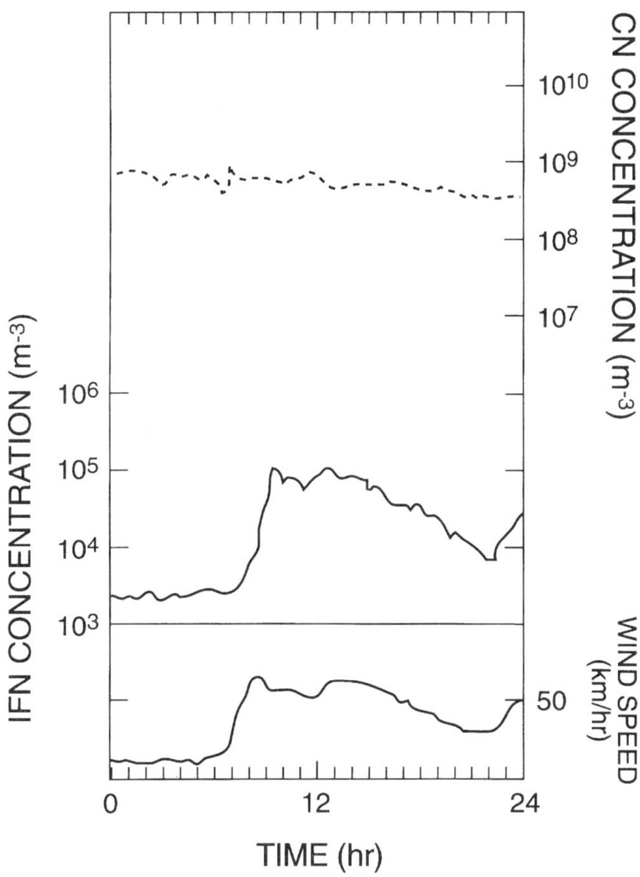

FIGURE 5.13 An example of changes in CN and IFN (–21°C) concentrations and of wind speed during a 24-hour period.

Nucleation of Ice During Collision of an Aerosol Particle with Supercooled Water Drop: Contact Nuclei

A cloud consists of cloud droplets and "dry" aerosol particles. Cloud droplets grow rapidly from the time of the V → L phase transition taking place at a cloud base. Aerosol particles that acted as CCN are removed from the aerosol population. Aerosol particles larger than about 40 μm in diameter act as an accretion center for cloud droplets from the time they enter a cloud and are removed from the aerosol population. The temperature at the cloud base is often not low enough for these particles to act as IFN at the time of first contact with liquid droplets. They act as IFN through freezing after they form larger droplets that are lifted by an updraft into lower temperatures at higher altitudes. The remaining aerosol particles may collide with supercooled drops. They may bounce off the surface of a supercooled drop (water non-wettable particles), be captured by a supercooled drop on contact (wettable particles), or penetrate through the surface of a supercooled drop and be captured. Particles that are transferred into the interior of drops become hydrosol particles and will nucleate ice only through freezing. The relationship[41] between the impact velocity v(cm^{-1}), its component normal to the surface of a drop required for penetration of an impacting non-wettable aerosol particle (V_{PN} cm sec^{-1}), water surface tension, T_W (dynes cm^{-1}), and water non-wettable aerosol particle diameter (d, μm), and its density (ρ, g cm^{-3}), is:

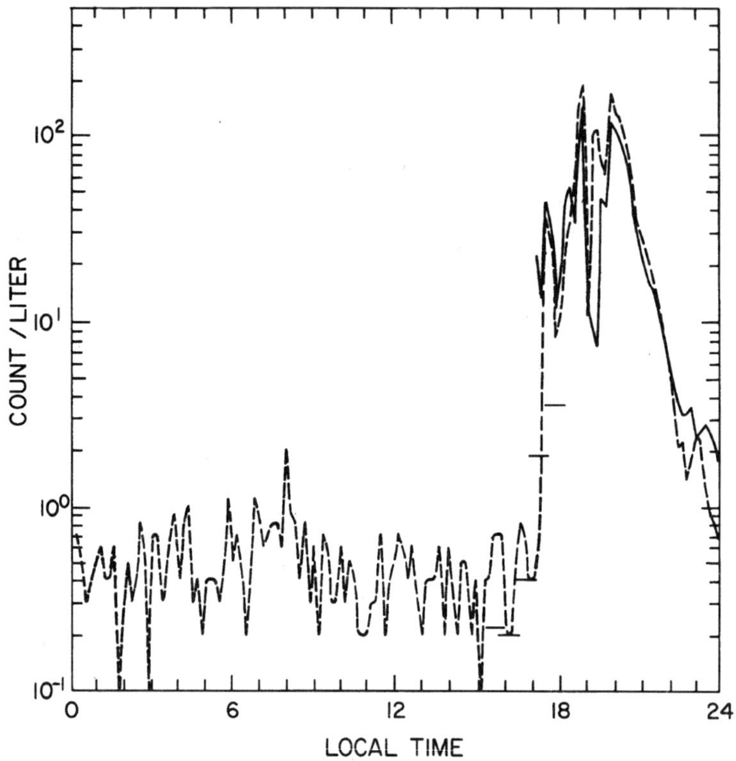

FIGURE 5.14 An example of an IFN storm (dashed line: data from NCAR ice nucleus counter, and solid line: data from the filter technique).

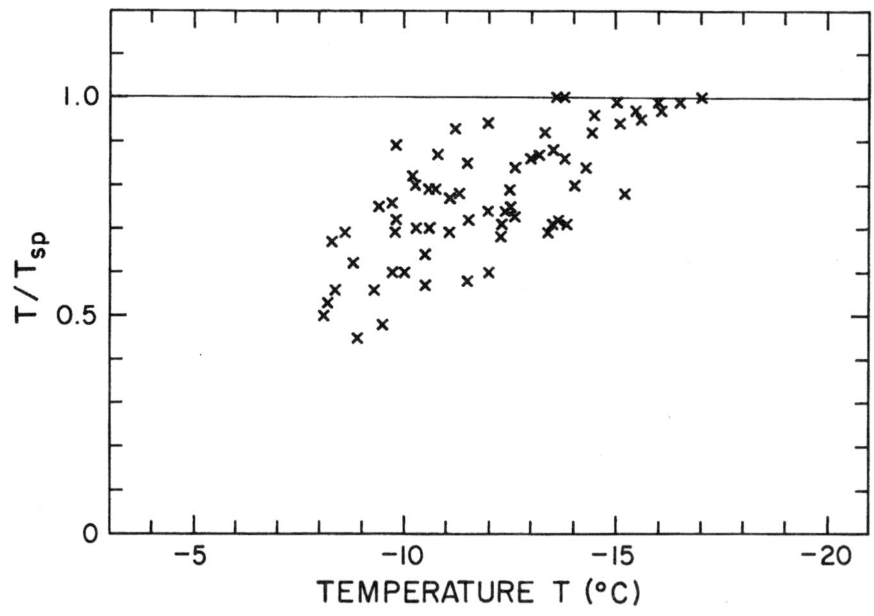

FIGURE 5.15 The ratios of ice nucleation temperatures of parent particles (T) to the shed particles (T_{sp}).

TABLE 5.1
Average Concentrations of Different-Sized Cloud Droplets and Aerosol Particles in a Severe Storm (Number of Particles per Cubic Meter per Given Size Interval)

	Particle diameter (μm)															
	0.01–0.1	**0.1–0.5**	**0.5–5**	**5–10**	**10–15**	**15–20**	**20–25**	**25–30**	**30–40**	**40–50**	**50–60**	**60–70**	**70–100**	**100–150**	**150–200**	**200–250**
Cloud droplets			10^6	1.5×10^7	3×10^7	2.5×10^7	1.5×10^7	5×10^6	5×10^6	3×10^6	2×10^6	10^2	10			
Aerosol	10^9	10^7	10^6	10^5	2×10^4	10^4	10^4	10^4	10^4	10^4	10^4	10^4	9000	500	500	50

$$\frac{v^2 - V_{PN}^2}{\left(v^2 \cos^2 \alpha - V_{PN}^2\right)^{1/2}} \leq \frac{DC}{m} \tag{5.3}$$

and

$$V_{PN} = \left(\frac{8T_w}{\rho d}\right)^{1/2} \tag{5.4}$$

where α is the angle of incidence (°), D is the raindrop diameter (mm), m is mass of an aerosol particle (g), and C is equal to $3\pi\mu_w d$.

In the Stoke's laws, μ_w is the viscosity of water (g cm^{-1} sec^{-1}).[41] The angle of incidence was determined experimentally by Rosinski for different geometrical objects.[42–46] The minimum diameter of aerosol particles transferred into the interior of a 1 mm diameter water drop falling at its terminal velocity of 4 m s^{-1} is 15 µm; for a 3 mm drop and 8 m s^{-1} velocity, the diameter is 3 µm. All these particles and larger may be nucleating ice by freezing. Aerosol particles collide with a spherical drop at different angles. In a vacuum, aerosol particle trajectories are straight lines (see Figure 5.16). Particle trajectories deviate from the streamlines of airflow. Particles following trajectories 1 and 2 collide with the surface of the spherical sector confined between angles α_2 and α_1. A tangent particle trajectory defines the zone of the sphere downstream from the circle of intersection AA′ (corresponding to angle γ). This zone is free of particles because particles deposited on the circle AA′ block deposition downstream. The number of particles collected at time t on the surface of a spherical sector positioned at $\bar{\alpha}$ is:

$$N_{\bar{\alpha},t} = \left(\frac{D}{d}\right)^2 \left[1 - \exp\left(-\frac{\pi d^2 F t}{2}\right)\right] \sin\beta(\bar{\alpha}) \sin\alpha \Big|_{\alpha_1}^{\alpha_2} \tag{5.5}$$

and at saturation:

$$N_{\bar{\alpha},t\to\infty} = \left(\frac{D}{d}\right)^2 \sin\beta(\bar{\alpha}) \sin\alpha \Big|_{\alpha_1}^{\alpha_2} \tag{5.6}$$

where $\bar{\alpha}$ is the mean of α_2 and α_1, and where $(\alpha_2 - \alpha_1)$ is very small. The equations permit calculation of an angle of approach of an aerosol particle during aerodynamic capture. N is the number of particles per square centimeter.

An example of the angle at which aerosol particles are captured by a sphere is given in Figure 5.17. Aerosol particles are also collected on the lee side of a sphere; results of experiments with 2.8 µm diameter particles have shown that higher number of particles were collected at lower air velocities. Every collision following deflection of a particle or its capture can start the L → S phase transition. A question remains, however: Is the angle at which an aerosol particle colliding with a liquid surface of a drop or droplet important?

Some of the aerosol particles may start the L → S phase transition during penetration through the skin of a drop. In this case, they can nucleate ice through mechanical disturbance of the surface of a supercooled drop; to do this, they may or may not possess ice nucleating sites active at the temperature of a supercooled drop. Smaller aerosol particles than the minimum size will collide with drops, but they will not penetrate through the surface of the drops. Particles may be captured through aerodynamic capture or, if aerosol particles are in the submicron diameter size range,

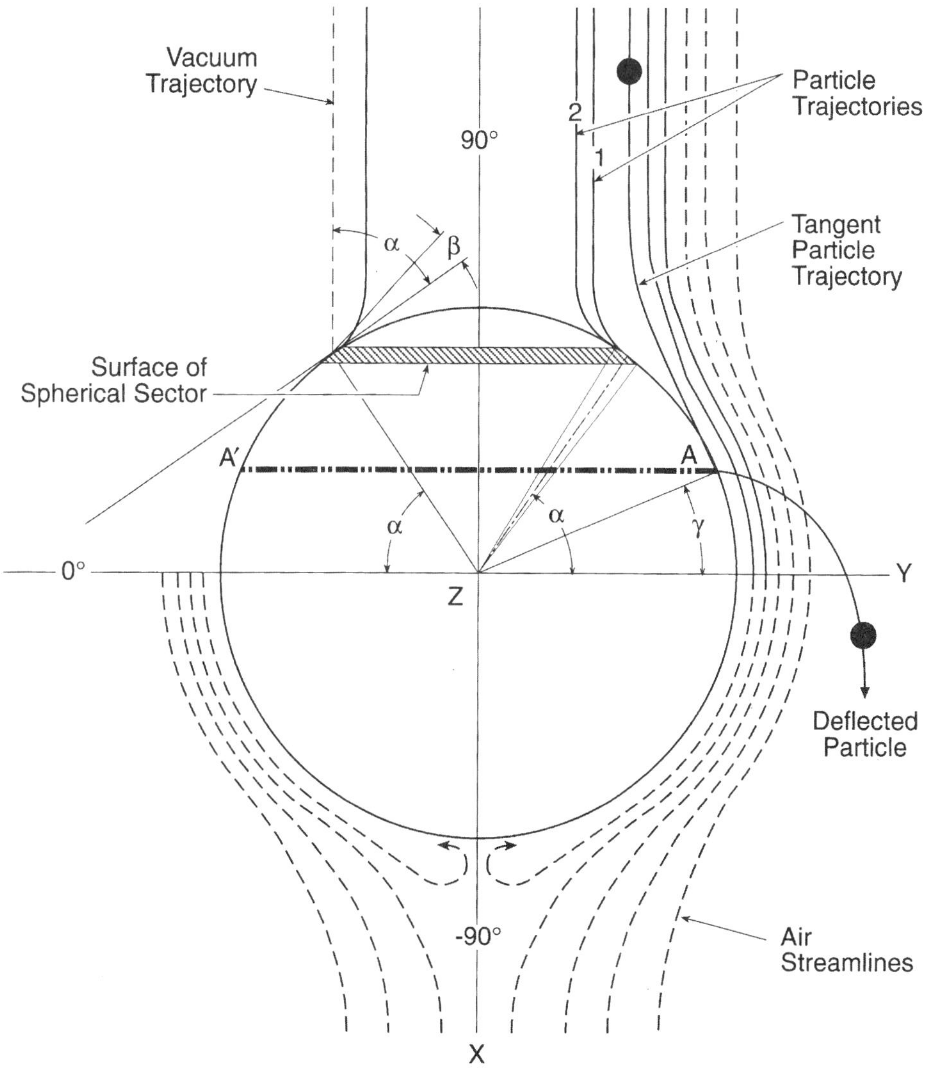

FIGURE 5.16 Schematic drawing showing aerodynamic capture of an aerosol particle.

through Brownian motion. Pure Brownian motion can be modified by thermophoresis. Electrostatic forces can also play an important role in transferring aerosol particles toward the surface of a droplet. Diffusiophoresis must be taken into account. Nucleation of ice by contact can be subdivided into three subgroups using time as a variable. The first one might be nucleation on contact during a collision and subsequent bouncing-off of an aerosol particle. This will be most pronounced when dealing with particles that are hydrophobic. The time of nucleation is a fraction of a second. The second one is during collision and capture of an aerosol particle; in this case, nucleation takes place at the time of collision or between the collision and extremely short residence time of an aerosol particle on the surface of a drop. The third category is when an aerosol particle is captured on or in the surface of a drop and spends some time floating before nucleating ice; this is called the delayed on-surface ice nucleation. Pitter and Pruppacher,[47] in an elegant experiment using homogeneous sources of aerosol particles, have shown that aerosol particles present on the surface of supercooled drops nucleated at temperatures higher than when they were present inside drops as hydrosol particles. Particles of montmorilonite and koalinite nucleated ice at –3°C and –5°C when

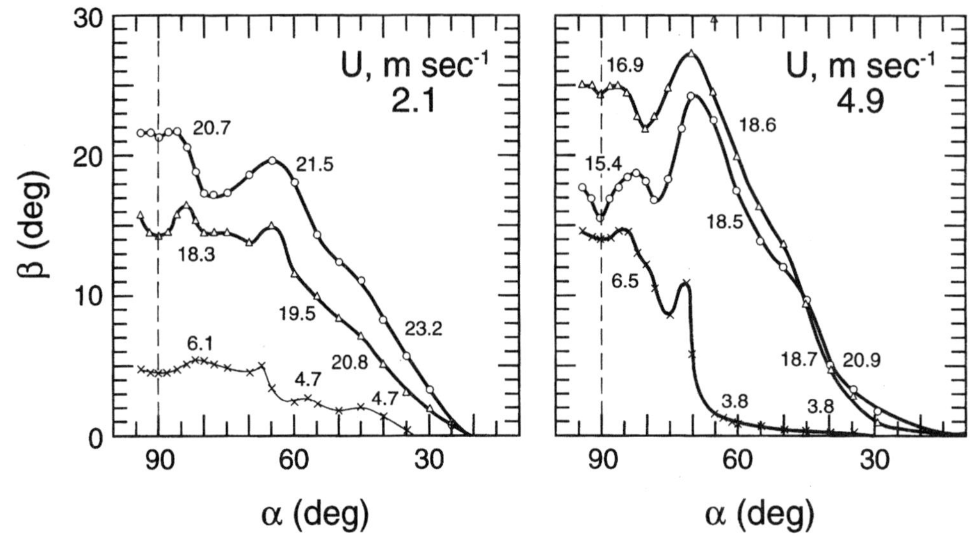

FIGURE 5.17 Angle (β) at which aerosol particles were captured by a sphere at position defined by angle α for different sized particles (d, μm) and two different air flow velocities U, (m s^{-1}).

present on the surface of supercooled drops (nucleation by contact), but the temperatures of ice nucleation were –12°C and –14°C for particles suspended in water and acting as hydrosol particles (nucleation by freezing). The above temperatures are the highest temperatures and there are temperature distributions connected with each experiment.

Ice nucleation by contact starts on or in the surface of a drop. It has not been possible, however, to identify a particle that nucleated ice because it could act alone or maybe in a cluster of particles floating on the surface of a drop. Experimental evidence indicates that the fraction of aerosol particles nucleating ice by contact at the time of collision increases with increasing size of colliding particles and with decreasing temperature.[39,40,48] The results of these experiments are given in Figure 5.18 for different natural aerosol particles derived from two different soils. The diameter range of the aerosol particles was 5 to 40 μm. If these "dust" particles (soil-derived aerosol particles) are lifted by an updraft and subject to scavenging by cloud droplets or drops, they will produce ice particles through delayed on-surface ice nucleation as a function of temperature, aerosol particle size, and time of residence of particles on the surface of droplets.

The number of ice crystals formed in a cloud parcel at time t by delayed on-surface ice nucleation is given by:

$$\int_0^{\infty} drN_p(r)\int_0^t dsI(N_d,r,s)\exp\left(-\int_0^s I(N_d,r,w)dw\right)\max_{s\le w\le t} P\{r,\alpha(w),l[s,t;\alpha(w)]\} \qquad (5.7)$$

where $N_p(r)$ is the aerosol particle number density (number cm^{-3}); $I(N_d, r, t)$ is an expected rate at which soil particles of radius r collide with cloud droplets (number s^{-1}); and $P(r, t, s)$ is a probability of a captured aerosol particle of radius r at temperature T nucleating ice by delay time s (dimensionless).

As mentioned above, the fraction of aerosol particles nucleating ice at the time of collision increases with increasing size of colliding particles and decreasing temperature. It is very small or even equal to zero for temperatures above –10°C for natural aerosol particles. Extrapolation of data into the submicron size range indicates that the fraction is extremely small, even for low temperatures.

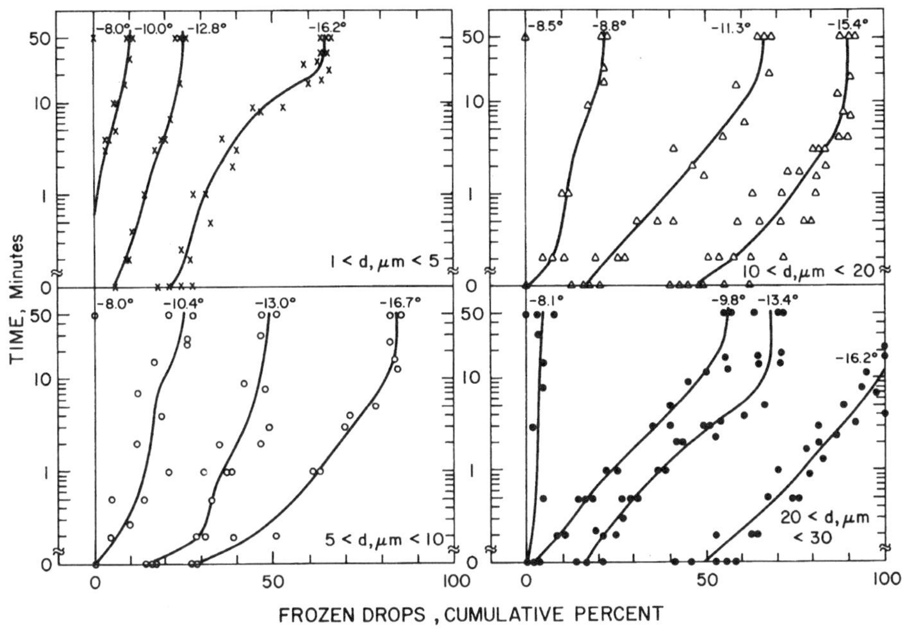

FIGURE 5.18 Delayed on-surface ice nucleation as a function of temperature and size of aerosol particles.

Delayed on-surface ice nucleation should therefore be considered as an important process for primary ice particle formation in clouds. Older clouds should have larger concentrations of ice particles because of the longer residence time available for aerosol particles captured by droplets to nucleate ice. At temperatures of ~ –10°C, 10 to 20 μm diameter aerosol particles or larger must be present to have nucleation of ice by contact. Captured aerosol particles start to nucleate ice with time; smaller particles will require longer residence times and lower temperatures, and larger particles require shorter times and higher temperatures. At temperatures of –15°C, the ice particle concentration during the lifetime of a cumulus cloud should reach ~10^4 m^{-3} (it is clear that this concentration will depend on the ice nucleating temperature spectrum of aerosol particles ingested by a cloud, the size distribution of aerosol particles, temperature, size distribution of cloud droplets, and the updraft velocity). However, estimated ice concentrations are far below the observed ones in cumulus clouds. Blyth and Latham,[49] studying development of ice in New Mexican summertime cumulus clouds, found ice particle concentrations of up to 1.3×10^6 m^{-3}. These large concentrations are far greater than concentrations of IFN active at –15°C present in that part of the country. IFN concentrations of up to 10^4 m^{-3} were observed at a temperature of –16°C on some occasions.[10,11]

The number of aerosol particles in the micron and submicron diameter size range colliding with and captured by a drop due to aerodynamic capture is very low. Aerosol particles in these size ranges are in constant random motion due to collisions with gas molecules. The steady-state flux of aerosol particles of radius r to a surface of a droplet of radius R is given by:

$$4\pi D_p(r) R N_p(r) dr \Psi_w(r), \text{ number cm}^{-2} s^{-1} \tag{5.8}$$

where $N_p(r)$ is the aerosol particle number density at some distance from a drop surface (number cm^{-3} cm^{-1}); $D_p(r)$ is the diffusion coefficient for aerosol particles in air (cm^2 s^{-1}); and $\Psi_w(r)$ is the factor by which the pure Brownian transport is modified in order to include phoretic effects. This factor is:

$$\Psi_w(r) = \frac{\exp(-\beta)}{1-\exp(-\beta)} \tag{5.9}$$

where

$$\beta = 8.1\times10^{14} r^2 \left(\frac{T_w}{T_\infty}\right)\exp\left(3.8\times10^4 r\right) - -3.4\times10^{18} r^2\left(1-7.2\times10^3 r\right)\left(\rho_\infty-\rho_w\right) \tag{5.10}$$

The first term in Equation 5.9 describes the thermophoretic effects, while the second term describes the diffusiophoretic effects. T_w and T_∞(°C), respectively, are the water droplet temperature and the mean temperature of the cloud parcel, and ρ_w and ρ_∞ are the water vapor density at a droplet surface and the mean water vapor density in a cloud parcel (g cm^{-3}), respectively. The upper limit of applicability of this equation is $d \leq 1$ μm.

An example of the rate of aerosol particle capture (cm^{-3} s^{-1}) is given in Figure 5.19. The factor β depends on the thermodynamic state of the surrounding cloud parcel. In-cloud conditions used in this example are $T = -15$°C, $N_{ice} = 0.1N$ droplets, and cloud water content = 4 gm^{-3}. Three (5, 10, and 15 m s^{-1}) updraft and three (–5, –10, and –15 m s^{-1}) downdraft velocities were used in calculations. Capture of aerosol particles by evaporating droplets in the downdraft is larger for all downdraft velocities than in an updraft; droplets evaporate not only due to the presence of ice, but also due to the downward transport of an air (cloud) parcel. This latter effect is the predominant cause of enhanced particle capture (collision), and the collision rate is proportional to the downdraft velocity. All collision rates in the downdraft regime are higher than the one predicted by pure Brownian theory. In the downdraft at 0 m s^{-1}, cloud droplets evaporate in the presence of the ice phase only. At 10 m s^{-1} updraft, the rate of evaporation of droplets is compensated by the rate of condensation of water vapor on droplets; these two practically cancel one another and the modification of the pure Brownian theory is eliminated. At this and higher updraft velocities, the thermophoretic effect is practically eliminated and nucleation of ice by contact ceases.

Slinn and Hales[50] were the first to point out that ice nucleation in clouds can take place on submicron particles by means of thermophoresis. The fraction of aerosol particles nucleating ice is a function of the particle diameter. The smaller the size of the particle, the smaller the fraction of particles starting the phase transition. Below 1 μm in diameter, this fraction is reduced by 4 or 5 orders of magnitude. Consequently, the large number of collisions is counteracted by small numbers of particles capable of nucleating ice. However, this last statement might be incorrect. Different size fractions of aerosol particles nucleating ice were determined for different temperatures using supercooled drops that were neither condensing nor evaporating. For a water drop in equilibrium with temperature, the number of condensing and evaporating water molecules are equal to each other in any time interval. During the thermophoretic collision of a submicron-diameter aerosol particle with a water droplet, water molecules are leaving the surface; the configuration of water molecules in the evaporating surface may be different from that in the surface at rest and, consequently, the fraction of aerosol particles nucleating ice may be different. It should be pointed out that the captured submicron aerosol particles will not float separately on the surface of a drop, but they will form clusters. Some of these may be capable of nucleating ice.

Aerosol particles in the phoretic size range, when moving toward an evaporating drop, must cross a layer of high water vapor supersaturation adjacent to the surface of an evaporating drop. If the particle can nucleate ice through the vapor → ice phase transition, then it will be coated with at least a molecular layer of ice. Nucleation of ice in an evaporating drop will take place, not by an aerosol particle, but by an ice-coated particle. In natural clouds, this process will be restricted

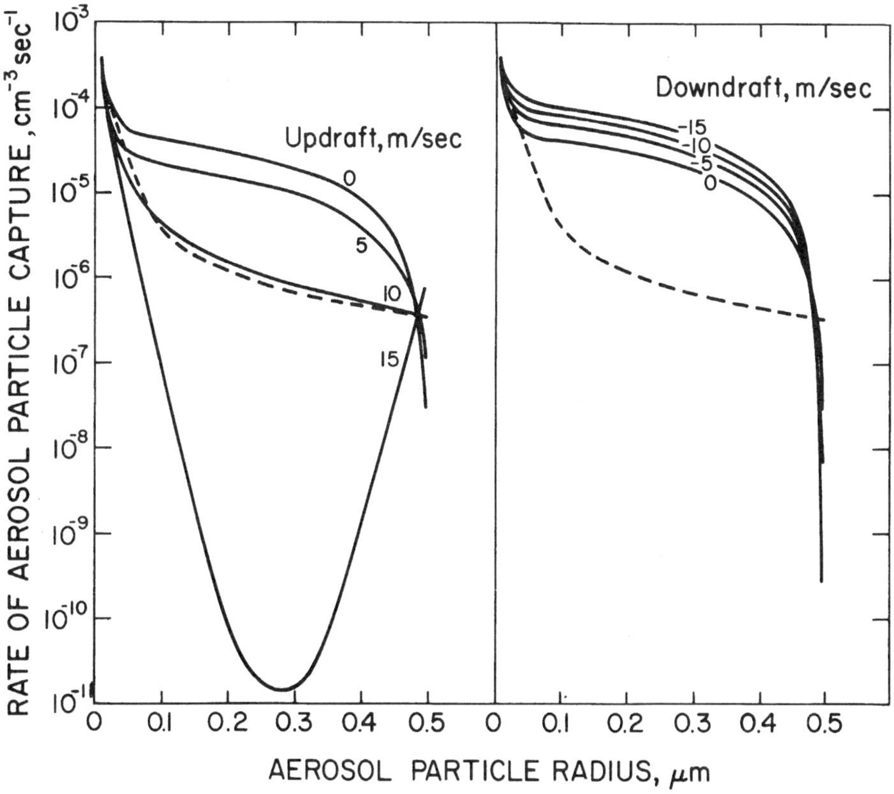

FIGURE 5.19 The rate of capture of aerosol particles as a function of their size (in-cloud conditions are given in text).

to lower temperatures (maybe below –20°C). In the presence of silver iodide particles, ice nucleation should take place at temperatures just below 0°C.

There is also an internal temperature gradient within an evaporating droplet. Hydrosol particles moving in that gradient will collide with the internal surface of an evaporating droplet. Again, the configuration of water molecules in that surface (the surface facing the interior of a drop) may be different from that on the evaporating surface. The possibility of the nucleation of ice by internal collisions was suggested by Weickmann.[51]

Experiments have been performed to study capture of submicron- and micron-sized aerosol particles by evaporating and condensing water drops.[17,44,45,52] Hydrophilic and hydrophobic particles were used to see if there was a difference in capture efficiency of particles with different wettabilities. The experiment was not designed to duplicate processes taking place in clouds. The results are given in Figure 5.20. Cooling of a drop was provided by a thermoelectric element of a single-stage bismuth telluride p-n junction. Heat of condensation of water vapor was removed by excessive cooling, which was necessary to keep the temperature of a condensing drop below the temperature of the chamber for the entire duration of the experiment. The results shown in the right side of Figure 5.20 therefore represent the capture of different-sized aerosol particles by three mechanisms acting simultaneously: Brownian diffusion, thermophoresis, and diffusiophoresis. For particles of 0.05 μm diameter, Brownian diffusion modified by thermophoresis is the most effective collision and capture mechanism. Capture of aerosol particles decreased with increasing size of particles, and capture increased for all aerosol particle sizes with increasing rate of condensation of water vapor (in this experiment, it corresponds to increased cooling). The fastest increase was for 0.05 μm, followed by 0.37 μm and 1.9 μm diameter particles. Capture of hydrophilic aerosol particles

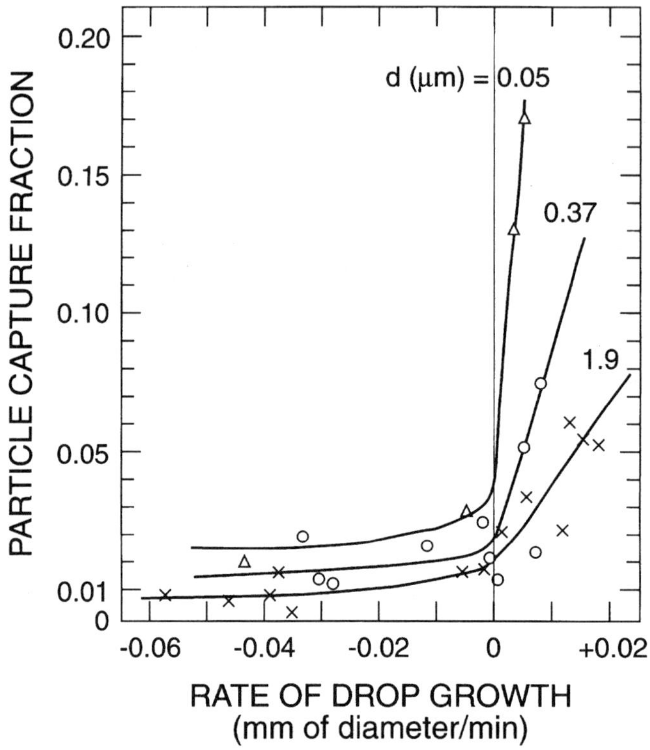

FIGURE 5.20 Aerosol particle capture by an evaporating and condensing water drop.

was found to be up to four times larger than that of hydrophobic particles. The number of collisions was the same for both types of particles (number cm^{-2} s^{-1}). Hydrophobic particles are therefore better candidates for the delayed-on-surface ice nucleation because of their indefinite residence time on the surface of a drop. In a real cloud, evaporation of droplets takes place in a downdraft, in the presence of ice particles or during dilution with outside drier air; condensation of water vapor takes place in an updraft. Phoretic forces are an important mechanism in capturing aerosol particles by cloud particles and, consequently, in contact nucleation of ice.[48]

Ice Nucleation from the Vapor Phase: Sorption Nuclei

Aerosol particles present in the rising parcel of a cloud are exposed to water vapor present at different supersaturations with respect to ice. The degree of supersaturation in a cloud depends on the updraft velocity and the concentration of cloud condensation nuclei.[53] The most active CCN will produce the first droplets, which will continue to grow by further condensation and coagulation with adjacent droplets. A droplet or a drop may freeze through freezing or contact nucleation. At the time of initiation of the liquid → solid (ice) phase transition, the temperature of a drop is the same as the temperature of the surroundings. From time zero, the heat of the phase transition of freezing is released and the temperature of a freezing drop is higher than that of the surrounding cloud parcel; the temperature of a freezing drop is 0°C during the entire process of freezing. Heat and water vapor are released into the surrounding air in proportion to the size (diameter) of the freezing drop. Due to nonequilibrium thermal conditions, a region of supersaturated water vapor surrounds a freezing drop. The released water vapor condenses on water droplets, ice crystals, and aerosol particles present in the vicinity of a freezing drop. Water vapor supersaturation in natural clouds reaches about 3 and 1% in marine and continental atmospheres, respectively, for a 10 m s^{-1} updraft velocity; at 1 m s^{-1} updraft, the S_w is 0.8% and 0.3%, respectively. For these updraft

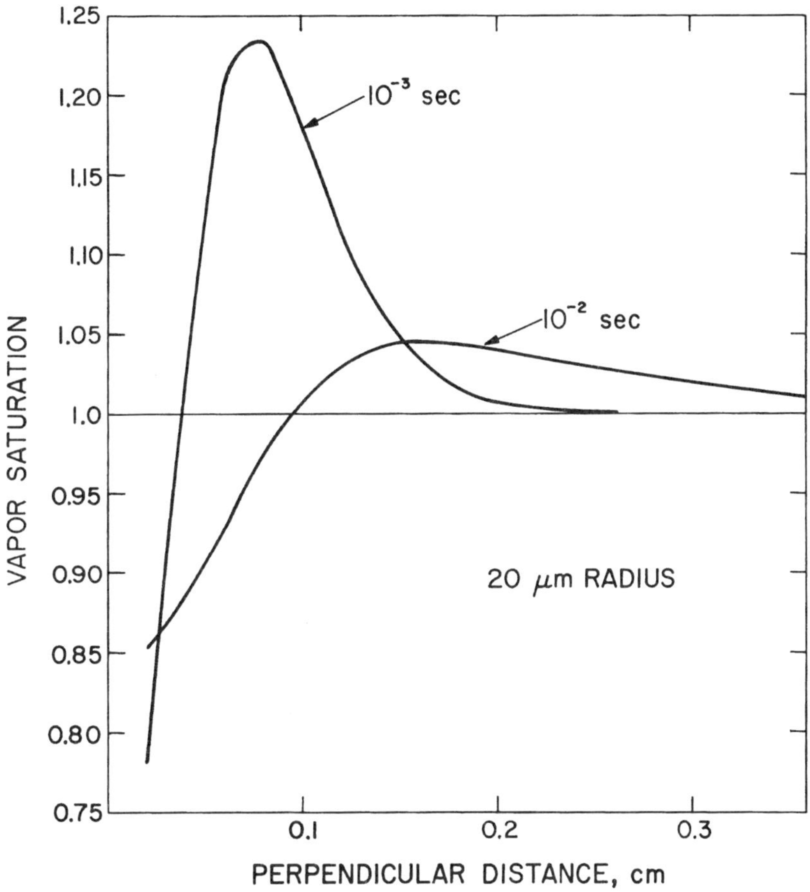

FIGURE 5.21 Water vapor supersaturation around a freezing droplet (d = 40 μm).

velocities, the concentration of cloud droplets is about 95 and 60 cm^{-3} for marine, and 925 and 515 cm^{-3} for continental clouds, respectively. Supersaturation of water vapor around a freezing drop, on the other hand, reaches values in excess of 20%. Supersaturation lasts for a fraction of a second around freezing cloud droplets, but it persists for the entire time of freezing of large water drops, which may take over one minute (Figure 5.21). Water vapor supersaturation in an ascending cloud is distributed more or less uniformly above the cloud base, but the supersaturation present around the freezing drops is spread non-uniformly in time and space within a storm. Weickmann gave the name "Rosinski-Nix effect" to the formation of ice crystals on aerosol particles at water vapor supersaturations present around a freezing drop (see "Acknowledgments").

Wegener[4] suggested the existence of natural aerosol particles that absorb water vapor on their surfaces and grow ice crystals directly from vapor. The existence of such particles was shown by Findeisen[54] (in 1938), who grew ice crystals at water vapor supersaturation with respect to ice but below liquid water saturation. Roberts and Hallett,[55] using different minerals, found the threshold temperature to be –19°C and the minimum supersaturation with respect to ice about 20%. Rosinski et al.,[56,57] using different-sized soil particles, have shown that ice nucleation in the vicinity of a freezing drop or when exposed to a controlled supersaturation in a dynamic chamber depends on the size of particles, the nature of particles, and the temperature (Figure 5.22). For particles larger than 40 μm in diameter, the temperature of ice nucleation was –16.8°C and independent of size. The temperature of ice nucleation on particles below 40 μm diameter increased with increasing size; at 15 μm diameter, it was below –20°C. On the left side of the demarcation line in Figure

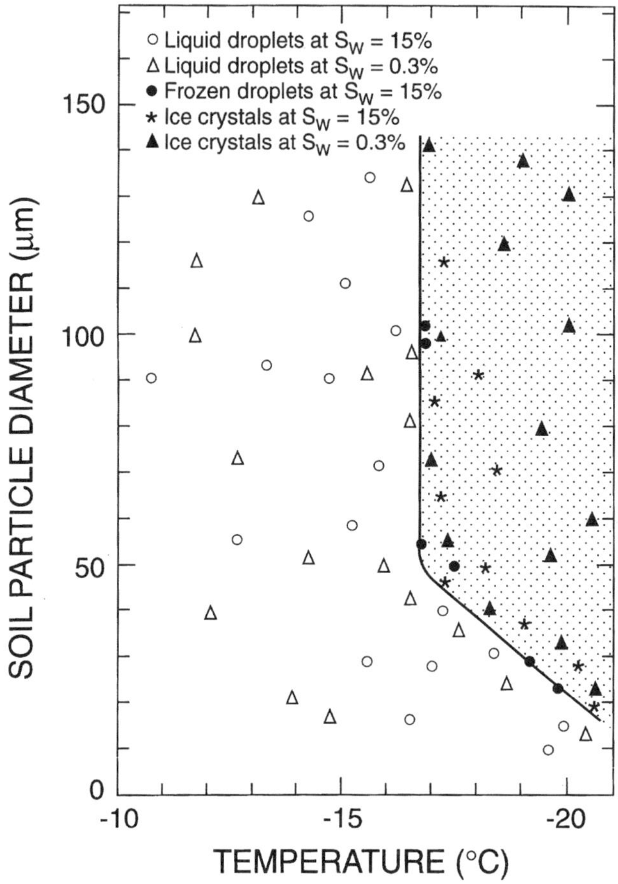

FIGURE 5.22 Formation of liquid and solid (ice) phases on different-sized aerosol particles.

5.22, soil particles acted as condensation nuclei; only water droplets were observed down to the temperature of –21°C. In the water vapor → solid (ice) phase transition zone around a freezing drop, formation of liquid droplets was observed on some occasions. Consequently, one obtains ice nucleation through two mechanisms: condensation-followed-by-freezing, and sorption. In storms, particles larger than 40 μm in diameter will always be "wet" through collisions with cloud droplets and, therefore, they will not nucleate ice through sorption. Small aerosol particles will require low temperatures to form ice directly from water vapor, even in the vicinity of a freezing drop. The soil particles used do not represent the entire aerosol population, but it is probably safe to state that the Rosinski-Nix effect produces insignificant numbers of ice crystals in a natural cloud.[58]

The distribution of water vapor is not uniform around a freezing drop. This is shown in Figure 5.23 for the phase transition on silver iodide particles. It should also be noted that on some occasions, there was a delay of ice particle formation — with a maximum of a few seconds. Water vapor → solid phase transition took place through ice nucleation by sorption only; liquid droplets and subsequently freezing droplets were not observed. The threshold temperature was –9.8°C ± 0.1°C.

Dessens[59] and Gagin[60] have shown that the IFN concentration increased with relative humidity at a constant temperature. Huffman[61] plotted normalized data of IFN concentrations vs. water vapor supersaturation over ice for natural aerosol particles and for laboratory-prepared particles of silver iodide. The IFN concentrations, N_{IFN}, were found to be independent of temperature and obey a power law:

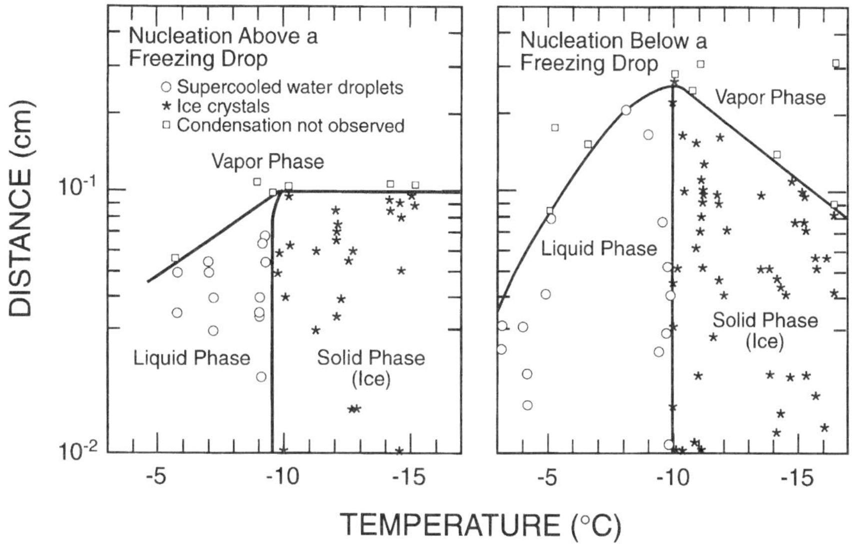

FIGURE 5.23 Water vapor → solid (ice) phase transitions around a freezing drop.

$$N_{IFN} = CS_i^{\alpha} \tag{5.11}$$

where S_i is the water vapor supersaturation over ice (in percent), and C and α are constants (Figure 5.24). The constant α was found to be equal to about 3 for aerosol particles present over Colorado and Wyoming, and about 8 in the vicinity of St. Louis, MO. The relation is of the same form as to the one expressing concentration of CCN vs. water vapor supersaturation.

The nature of aerosol particles is governed by their origin and their life history in the atmosphere. The concentration of IFN is strongly dependent on temperature. Experiments performed at different temperatures and water vapor supersaturations for a number of aerosol particles of different origins[31,62] have shown that:

1. There is a certain minimum water vapor supersaturation over ice below which aerosol particles cannot nucleate ice by sorption.
2. This minimum S_i increases with decreasing temperature.
3. There are two different slopes (two α values) for the aerosol particles under investigation: the first one is for data from the minimum S_i up to water vapor saturation, $S_{i,\min} < S_i < (S_w = 0\%)$, and the second one is for data points lying above $S_w = 0\%$.
4. Slopes of curves for $S_i < (S_w = 0\%)$ are steeper than for $S_i > (S_w = 0\%)$; the first curve is for ice nucleation by sorption, and the second is for ice nucleation by condensation-followed-by-freezing. Both are a function of temperature (Figure 5.25).
5. The concentration of sorption IFN generally increases with decreasing temperature at a constant water vapor supersaturation with respect to ice, S_i.
6. The concentration of sorption IFN generally increases with increasing S_i at a constant temperature.
7. The concentration of aerosol particles supplying IFN differs for different sources of particles. It was discovered that there are aerosol particles that nucleate ice independently of changes in water vapor supersaturation with respect to ice at a constant temperature and with temperature at a constant S_i. This takes place for a certain range of S_i and of T.

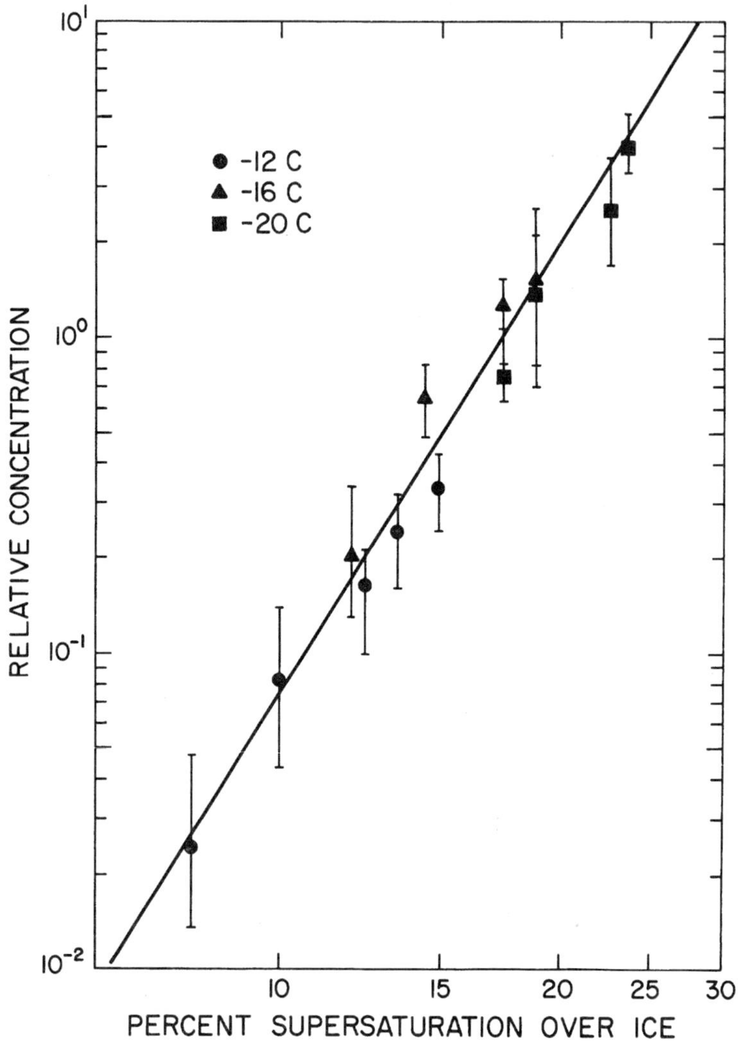

FIGURE 5.24 Relative concentration of IFN as a function of water vapor supersaturation over ice.

Changes in concentrations of IFN on some occasions are shown in Figures 5.26 and 5.27. Curve C in Figure 5.26 represents the most common relation between concentration of IFN and temperature, that is, the IFN concentration generally increases with decreasing temperature. The data were obtained using simultaneously the Bigg-Warner chamber, the NCAR continuous IFN counter, and a dynamic diffusion chamber. However, in the presence of storms and overcast skies, it was found that on some occasions the curve contains a minimum in the IFN concentration–temperature curve (curve B). Theoretically, the C_{IFN}–T curve should look like curve A. Simply, the C_{IFN} cannot have a minimum at some intermediate temperature. Aerosol particles nucleating ice at temperature 1 should nucleate ice at a lower temperature 2. In the absence of particles nucleating ice between these two temperatures, 1 and 2, the curve should be a straight line between the two temperatures; the C_{IFN}–T curve is a cumulative curve. But data obtained from the three instruments have shown the existence of a minimum. It is assumed at present that the nature of the aerosol particles is primarily responsible for the existence of a minimum in the C_{IFN}–T curve. Anomalous behavior was also observed in data representing IFN concentration as a function of water vapor supersaturation with respect to ice (Figure 5.27). Again, curves are cumulative and it is impossible

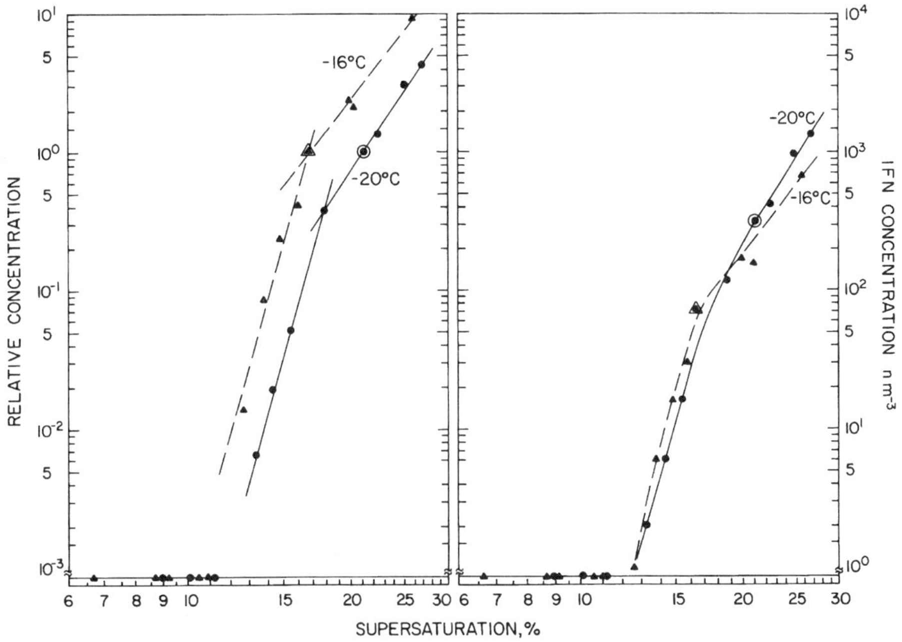

FIGURE 5.25 Relative and actual concentrations of ice-forming nuclei vs. water vapor supersaturation with respect to ice.

to explain the existence of a minimum in the C_{IFN}–S_i curves at constant temperature; with increasing S_i at constant temperature, the C_{IFN} should monotonically increase. It should be noted here that the minimum in the C_{IFN}–T curve appeared between temperatures of –16°C and –20°C, and that the minima in the C_{IFN}–S_i curves were just below S_i equal to S_w = 0% (water vapor saturation) for the temperature range between –12°C and –14°C; curves for temperatures above and below this range have shown normal increases with increasing S_i and all had inflection points at S_w = 0% (two α values).

A plot of IFN concentration, water vapor supersaturation over ice at water vapor saturation, and the number of collisions of water molecules per second per unit area of an aerosol particle as a function of temperature is given in Figure 5.28. The N_{IFN} and S_i curves are similar and, consequently, N_{IFN}–T or N_{IFN}–S_i curves should be similar. The decreasing number of collisions with decreasing temperature and increasing IFN concentration indicate that the IFN concentration is primarily a function of temperature. The number of collisions at –5°C at S_w = 0% (S_i = 5.03%) is 5.1×10^{12} μm^{-2} s^{-1}; at a temperature of –19°C and S_w = 0% (S_i = 20.36%), it is 1.7×10^{12} μm^{-2} s^{-1}. Despite the three times lower collision rate at –19°C vs. –5°C, the IFN concentration increases about a thousand times. The number of collisions at S_i = 7% and T = –19°C is 1.5×10^{12} μm^{-2} s^{-1}; the ratio of the number of collisions at S_w = 0% to the one at S_i = 7% is 1.1. The C_{IFN} is about 100 times higher at S_i = 20% than at S_i = 13% for IFN active by sorption at T = –19°C. The ratio of corresponding collisions is 1.06.

In a cloud, parcel water droplets will evaporate when S_w is below 0%. If at the same time, S_i is higher than 0%, then any aerosol particle capable of acting as sorption IFN may produce an ice crystal. Residues of evaporated droplets could act as sorption IFN. They start to nucleate ice at –14°C and S_i = 14.5%. The highest concentration found in clouds over the Pacific Northwest was 7×10^3 m^{-3}. Usually, the concentration has reached 400 m^{-3} at –20°C. When ice crystals were sublimed off the residues, they lost their ability to nucleate ice by sorption. On two occasions, however, they nucleated ice by sorption at –5°C and –6°C.[22] One of the sources of sorption IFN in the atmosphere consists of aerosol particles that are the residues of evaporated cloud droplets.

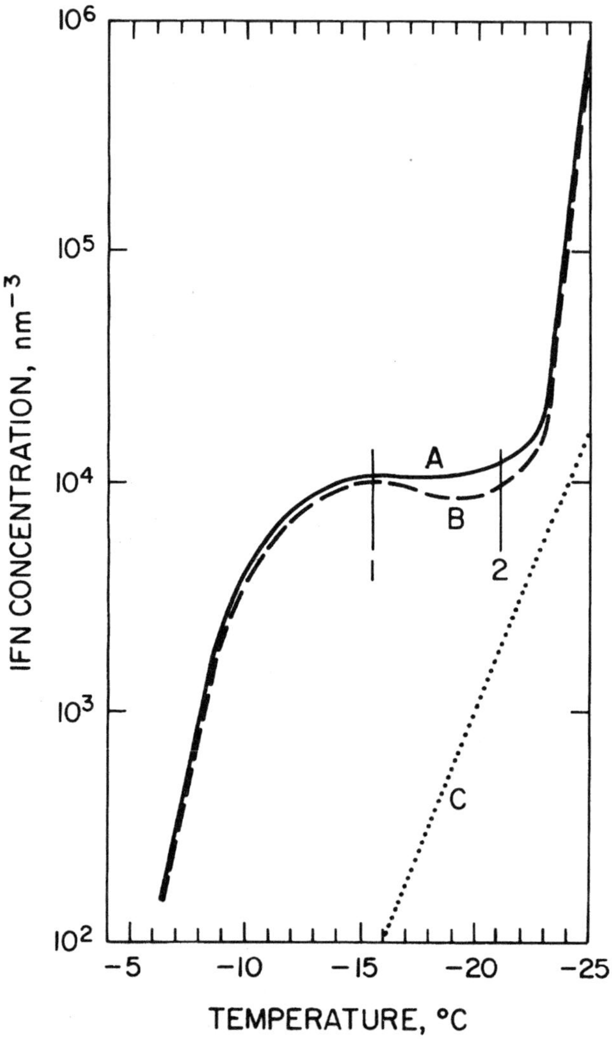

FIGURE 5.26 Concentration of ice-forming nuclei vs. temperature.

Their response to the temperature and water vapor saturation appears to depend on the origin and life history of aerosol particles that acted as cloud condensation nuclei.[22,30,31] The temperature in a cloud decreases with increasing altitude. It has always been believed that a certain temperature (e.g., –15°C) is uniform at a certain altitude. It was shown, however, that there can be temperature variations at a constant altitude of as much as 5°C over short distances.[63] Nucleation of ice, therefore, can proceed at a given altitude at different temperatures. Once ice is formed on an IFN active at –20°C, it will continue to grow when transferred a short distance into the region of temperature –15°C. The concentration of ice crystals, therefore, at a certain altitude is a mosaic of concentrations of ice crystals formed on IFN active at the various temperatures present at that altitude.

TEMPERATURE OF ICE NUCLEATION AS A FUNCTION OF THE SIZE OF AEROSOL PARTICLES

It is impossible to examine an aerosol particle and predict its ice nucleating ability. Consequently, an aerosol particle must be caught in the act of ice formation and then examined in the laboratory.

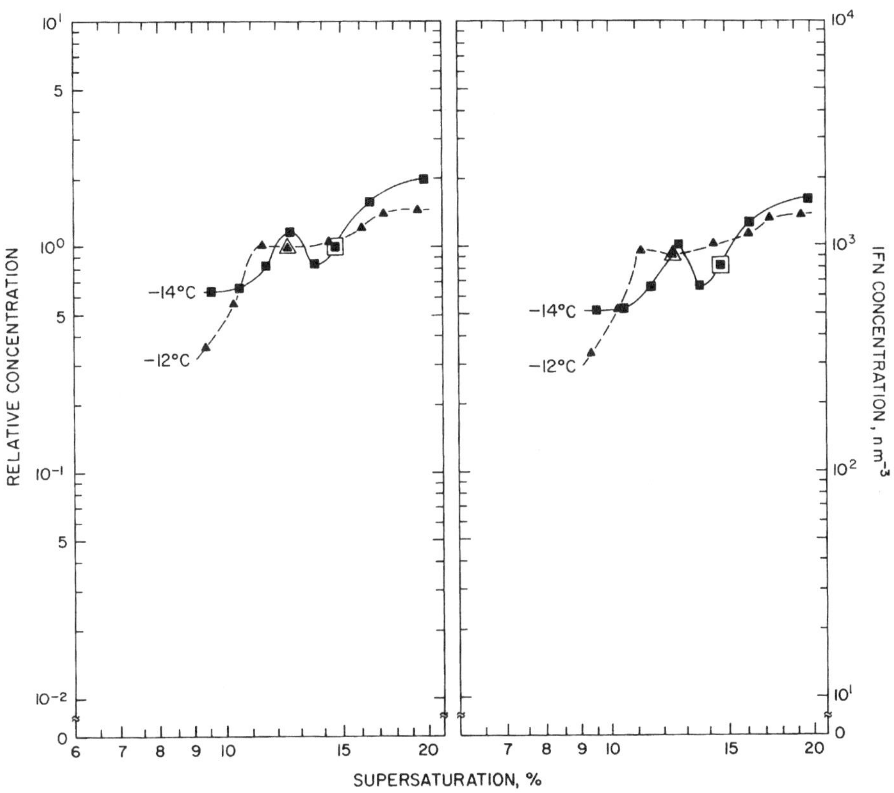

FIGURE 5.27 Relative and actual concentrations of ice-forming nuclei vs. water vapor supersaturation with respect to ice.

It can be safely assumed that a centrally located particle in a, for example, hexagonal ice crystal nucleated that crystal. In nature, there are many aerosol particles present in an ice crystal; they were being scavenged during the lifetime of the ice particle. The early work of Soulage,[64,65] Kumai,[66–68] and Isono[69,70] has shown that the minimum aerosol particle diameter is 0.1 μm. The upper limit was 80 μm. The maximum frequency varied from 0.1 μm to 1.0 μm to 5 to 15 μm in diameter. The temperature of ice nucleation of these particles was above –20°C. In 1962, Rosinski decided to build a chamber in which single aerosol particles acting as IFN could be separated into single ice crystals. The chamber, known later as the NCAR continuous ice nucleus counter, was operated at controlled temperatures.[71] Some results are summarized in Figure 5.29. If there is any relation between the size of aerosol particles and temperature of ice nucleation, it is hardly noticeable. The lower the temperature, the smaller the size of aerosol particles acting as IFN; the particle size here is the size present in the maximum frequency in the size distribution. Rosinski classified separated IFN into two groups: the first group consisted of inorganic particles and the second of organic particles. Inorganic particles were always larger than organic ones, but the difference was small. On one occasion, liquid droplets in the size range of 0.017 μm to 0.1 μm in diameter were separated as IFN. Their size distribution was associated with two maxima: one at 0.035 μm and another at 0.054 μm. These particles are most likely terpenes released from pine trees by needles. Melting points of some terpenes are around –75°C and, consequently, they are always in a form of liquid droplets in the atmosphere.

The dependence of temperature on the size of soil particles nucleating ice through freezing as hydrosol particles is shown in Figure 5.29. Peaks in temperature curves indicate the presence of specific sources of freezing nuclei associated with certain particle size ranges. For each temperature,

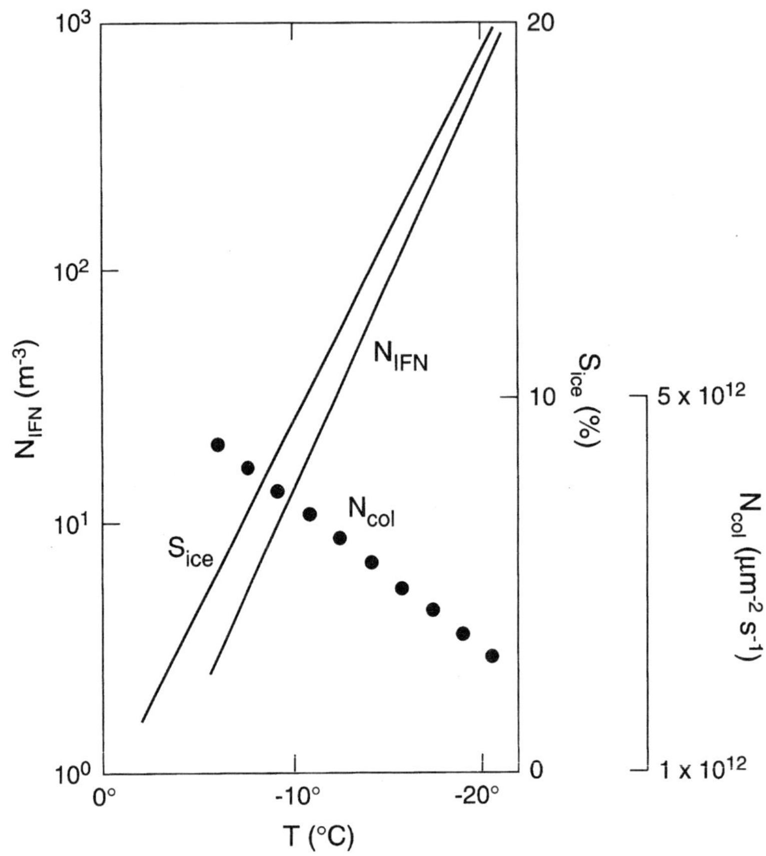

FIGURE 5.28 Concentration of IFN, N_{IFN} water vapor supersaturation with respect to ice, S_{ice}, ad the rate of water vapor molecules colliding with unit area of a particle, N_{Col}.

the fraction of IFN is increasing with increasing size of hydrosol particles. A similar relation holds for aerosol particles acting through condensation-followed-by-freezing. It can be seen that at lower temperatures, the fraction of particles nucleating ice is practically independent of particle diameter larger than about 100 μm. At S_w = 0.6%, therefore, there is always an ice nucleating site on the surface of such large particles. Ice nucleating ability is different for different soils, but the general relation seems to be valid worldwide. A similar relation also exists for aerosol particles acting by contact and sorption.

Soil particles are generally subdivided mechanically into smaller particles. The smallest inorganic particles seem to be around 0.1 μm; they can nucleate ice at low temperatures only (e.g., –20°C). Organic particles, on the other hand, can nucleate ice when they are as small as 0.017 μm (17 Å). And they can be produced in such a small size in the atmosphere through condensation of organic vapors, evaporation of droplets, and ejection of terpenes from pine needles in an electrostatic field.[72] There is no relation between the temperature of ice nucleation and the size of the Ångstrom-sized particles. Droplets of phloroglucinol show a concentration peak around 1000 Å particle diameter at temperatures of –20°C; this peak moved to 750 Å at –17°C. Clearly, the temperature of ice nucleation moved in the opposite direction.[72] Silver iodide shows two peaks in the activity curve. The first one is at 30 Å and the second one between 200 and 500 Å particle diameter. Both peaks exist for the temperature range between –10°C and –18°C.[74,75] It is clear that ice nucleation on the Ångstrom-size particles depends on the configuration of molecules and ions in the surface

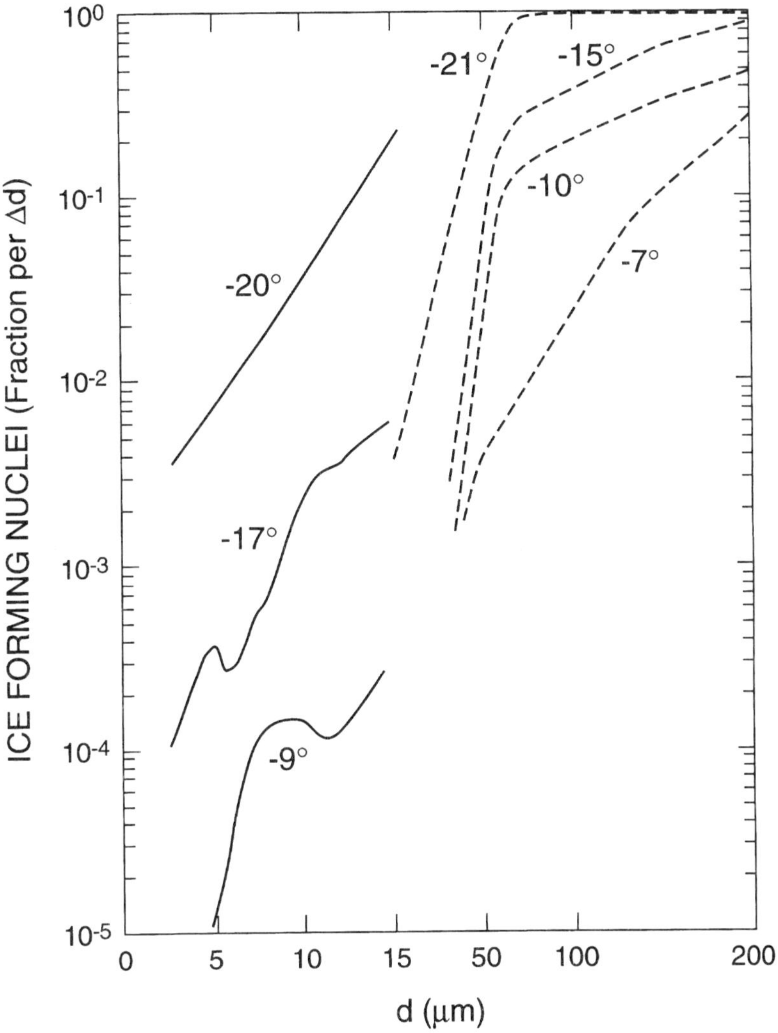

FIGURE 5.29 Fractions of hydrosol particles nucleating ice by freezing (solid lines) and of aerosol particles nucleating ice by condensation-followed-by-freezing (S_w = 0.6%) (dotted lines) for different temperatures.

and not on the diameter of the particle. The lifetime of these very small aerosol particles in the atmosphere is short. They coagulate with larger particles and probably lose their ice nucleating ability.

NATURE OF ICE-FORMING NUCLEI PRESENT IN THE ATMOSPHERE

There are two major sources of IFN: aerosol particles generated from the surface of the solid Earth, and from the oceans.

Particles generated from the Earth consist of inorganic particles, mostly soil, and organic particles, mostly decomposed vegetation (Table 5.2). In reality, they are all mixed particles. In eastern Colorado, on average, 15% of particles were classified as organic and 85% as inorganic particles. Using electron diffraction patterns, it was possible to identify the mineralogical composition of aerosol particles and of separated aerosol particles that act as IFN (Table 5.3). These particles were separated in the NCAR ice nucleus counter; approximately 3700 particles were

TABLE 5.2
Size of Natural Ice Forming Nuclei

Size of IFN (Size Range)	d; μm (Maximum Frequency)	Temperature (°C)	Location	Observer, Date
0.5–8	3	> –20°C	Japan	Kumai, 1961
1–80	5–15	> –20°C	France	Souilage, 1953
0.2–8		> –21°C	Michigan	Kumai, 1961
0.1–7			Japan	Isono, et al., 1959, 1960
0.1–.13	0.5–1		Greenland	Kumai and Francis, 1960
0.5–7 I*	2–3	–12°C	Colorado	Rosinski, et al., 1976
0.5–5 O*				
0.5–10 I	2–3	–12°C	Montana	Rosinski, et al., 1976
1–5 O				
1.5–10 I	2–5	–15°C	Montana	Rosinski, et al., 1976
1–5 O				
0.2–7 I	2–3	–19°C	Colorado	Rosinski, et al., 1976
0.5–3 O				Rosinski, et al., 1976
0.5–10 I	2	–20°C	Montana	Rosinski, et al., 1976
1–5 O				Rosinski, et al., 1976
0.1–15 I	1	–20°C	Colorado	Rosinski, et al., 1980
0.1–13 O	0.5–1.5			
0.017–0.1 O	0.035, 0.054	–20°C	Colorado	Rosinski, et al., 1980

*I, inorganic; O, organic particles.

TABLE 5.3
Aerosol Particles Separated as Ice-Forming Nuclei (July 1975)

Chemical Compounds	Aerosol Particles (%)	IFN at T (°C) –12°C	–19°C
		(%)	
Montmorillonite	35	34	33
Feldspar	24	41	28
Illite	17	23	38
Miscellaneous	8	0	0
Organic	16	2	1
Mixed particles containing:	Number (m^{-3})		
AgI	0	3	7
NaCl	2	4	5
CuO_x	1	3	2
IFN concentration (nm^{-3})		19	100

Note: Atmospheric conditions: daytime windy, night calm.

separated and 750 were analyzed. Montmorillonite, feldspar, and illite were present in the aerosol particles and acted as IFN. Organic particles practically did not nucleate ice over grassland. Consequently, they cannot contribute to any significant extent to the IFN population as mixed particles, that is, as mineral/organic mixed particles. Some of the mixed particles contained silver

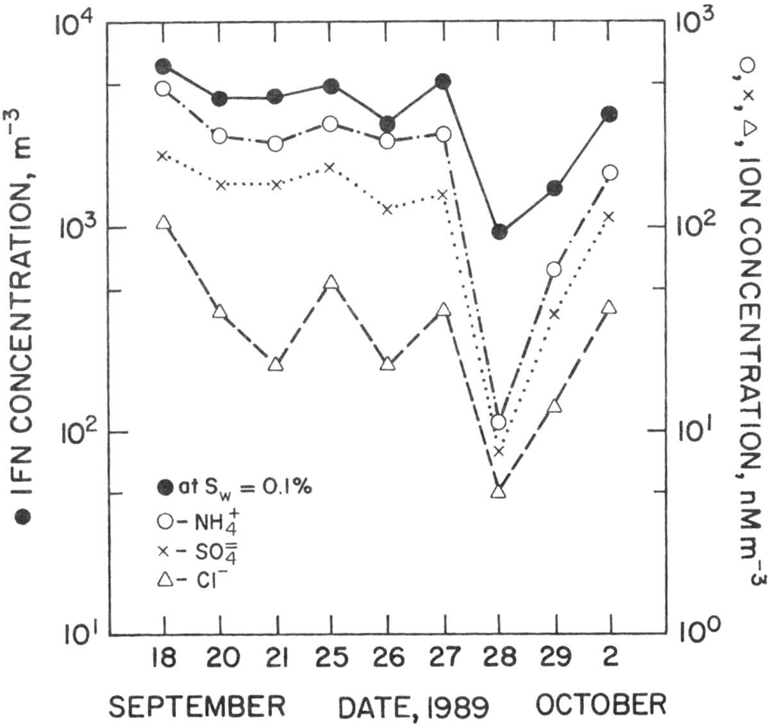

FIGURE 5.30 Concentrations of IFN active by condensation-followed-by-freezing ($S_w = 0.1\%$, $T = -19.7°C$) and of different ions from September 18 to October 2, 1989, in Italy.

iodide; they were separated as IFN. Silver iodide was left behind on grasses and soil long after experimental programs involving seeing clouds were terminated. Copper compounds are always present in trace quantities and act as IFN. Sodium chloride acts as IFN through condensation-followed-by-freezing.

Earlier studies[66–68,76,77] have shown the presence of montmorillonite, illite, feldspar, kaolinite, and other minerals active in nature as IFN. Local sources of IFN exist all over the world. Isono and co-workers[69,70] have shown that very high concentrations of IFN reported over Tokyo were associated with yellow loess dust that originated in China. Langer et al. (1974; in Fullerton et al., 1975) have shown that there is an increase in IFN concentration during volcanic activity. It was not, however, associated directly with volcanic eruption. It can be speculated that oxidation of sulfur compounds to sulfates causes the increase in IFN concentration. Rosinski et al.[8,21,31,77] have shown that sulfates, probably ammonium sulfate, are an integral part of the ice-forming nuclei population in marine atmospheres (see Figure 5.30); also, aerosol particles of marine origin in the presence of sulfates nucleate ice at higher temperatures than the temperatures of ice nucleation by aerosol particles acting alone. Mason and colleagues[79,80] examined 35 minerals for their ice-nucleating ability; ten were active at temperatures higher than –10°C.

In addition to the IFN released from the crust of the Earth, there are sporadic natural and anthropagenic releases of aerosol particles that contain IFN. Forest fires and wood and coal burning produce very small numbers of IFN. Smoke from cane fires contains very low IFN concentrations, but the ashes left behind when dispersed by the wind were an order of magnitude richer in IFN. The higher activity of ash is attributed to the presence of copper compounds.[81] Metal smelters and steel furnaces release, on occasion, substantial concentrations of IFN; their contribution to the global population of IFN is insignificant because the sources of IFN are localized. Over some parts of the

U.S., however, they can substantially increase the population of IFN and even cause weather modification.[82]

Very little is known about the contribution of organic particles to the formation of ice in the atmosphere. Terpenes and plant oils in the form of very small droplets, or their vapors condensed on aerosol particles, can act as IFN. As mentioned earlier, Fish[72] has shown that in the presence of an electrostatic charge, spruce trees release terpenes directly into the atmosphere in the form of liquid aerosol particles. Experiments with different terpenes were performed in the laboratory and in the open atmosphere. The temperatures of ice nucleation of terpene vapors condensed on sand were: –10°C for α-terpineol[82]; Linalool, a simpler acyclic terpene, nucleated ice at –6.4°C. Its solution in carbon tetrachloride nucleated ice over a wide range of concentrations. At concentrations down to ~80% (v/v), it nucleated ice over a temperature range of –6.4°C to –8.7°C; at concentrations of 1%, it nucleated –16°C. Water drops placed on the surface of carbon tetrachloride freeze at –19.2°C.[84]

Ice nucleation on the surfaces of liquids is different than on the surfaces of solids. Water molecules adsorbed on a surface of a crystal will move around until they are forced to settle on a site created by positions of ions in a crystal lattice. Ions in a crystal lattice do not move; only water molecules travel over the surface looking for sites with larger binding energy for water dipoles. Heterogeneous nucleation of ice at the liquid–liquid interface takes place where two liquids are in contact; molecules of both liquids are in constant motion and their dipole moments influence each other. Terpene molecules are large and can orient water molecules that are organized in ice-like clusters.[83] The configuration of the constituent link dipole moments in a terpene or other organic molecule is a decisive factor for the existence of a resultant dipole moment of an organic molecule. Rosinski[121] proposed that the spatial distribution of vectors corresponding to the constituent link dipole moments should be considered in studying ice nucleation at the liquid–liquid interface. Results of experiments with two trichlorobenzenes can serve as an example. In these experiments, chlorobenzenes were dissolved in carbontetrachloride, a nonpolar liquid in which ionization is absent (water is a polar, partially ionized liquid). All water drops placed on a solution of 1,2,4-trichlorobenzene were frozen in the temperature range between –19°C and –20°C on carbontetrachloride. This temperature range is actually the range in which water drops freeze on pure carbontetrachloride; 1,2,4-trichlorobenezene did not contribute, obviously, to the initiation of the L → S phase transition. 1,3,5-trichlorobenzene started ice nucleation at –8°C and ended at –25°C. A group of drops freezing between –22°C and –25°C should be frozen at –20°C on carbontetrachloride. Is it possible to suggest that the symmetrical trichlorobenzene can produce two different orientations of molecules at the liquid–liquid interface? The first one promotes the L → S phase transition, and the second one stabilizes the water structure and prevents formation of the ice-like clusters. The symmetrical 1,3,5-trichlorobenzene does not possess a resultant dipole moment. It is clear now that the hetergeneous nucleation of ice does not have to take place at the solid–liquid (supercooled water) interface, but it can be initiated at the liquid (terpene)–liquid (supercooled water) interface. Orientation of dipole moments of both liquids should be responsible for the L → S phase transition. Terpenes with OH groups are active as IFN at the highest temperatures; the OH group serves as a link in the attachment of water molecules. The oxygen-bearing groups appear to act in the following order: alcohol, ketone, oxide, and aldehyde.

All available terpenes were tested for their ability to act as CCN; none of them initiated formation of liquid drops (V → L phase transition) for S_w up to 3%. Terpene droplets can act as IFN through contact or a delayed-on-surface ice nucleation. The freezing temperature of water drops on, for example, the surface of tung oil is therefore dependent on the life history of the surface of the oil. Experiments were performed in which water drops were frozen on the surface of tung oil, the frozen drops were melted at temperatures between 15°C and 20°C, and then refrozen; this process was repeated up to five times. It was discovered that all the drops would freeze at a single temperature that was higher than the temperature of freezing freshly used drops. This implies that the freezing drops left an ice-like imprint in the liquid surface of tung oil (Table 5.4). Terpenes were tested for this memory effect, and it was found that it does not exist for liquids. The only

TABLE 5.4
Temperatures of Freezing Water Drops at the Water–Tung Oil Interface (T_i°C)

Exp. 1 (10 drops)			Exp. 2 (10 drops)		Exp. 3 (4 drops)		
Fresh	Refreezing		Fresh	Refreezing	Fresh	Refreezing	
Drops	First	Second	Drops	First through Fifth	Drops	First	Second & Third
−20.4 ±1.4	−21.5 ±0.9	−17.6 ±0.1	−14.1 ±2.5	−13.2 ±0.0	−23.1 ±1.5	−14.3 ±0.3	−13.2 ±0.0

explanation is that imprints are left in the semi-solid surface of tung oil and that they survived repeated melting and refreezing. But the most important property of these imprints is that they all must be identical to initiate the L → S phase transition at the same temperature. Orientation of molecules in the surface of tung oil must be introduced by ice during the freezing of drops. Only ice can leave a spot with memory for ice nucleation. Fresh water drops placed on an oil surface froze at temperatures corresponding to the freezing temperatures of fresh drops.[85]

Schnell and Vali[86,87] found that decaying vegetation produces about 10^8 IFN per gram of decaying leaves; they were active by freezing at temperatures as warm as −8°C. There is no known mechanism that could aerosolize decomposed leaves and other plant litter. Schaefer[88] studied microscopic flora and fauna and reported that these aerosol particles can contribute to ice formation in the atmosphere below temperatures of −30°C. Soulage[65] found that some kinds of spore or yeast in the size range of 5 to 20 μm nucleates ice at −20°C; its concentration was 10^3 m^{-3} of air. They were present under some specific meteorological conditions and in the presence of very high humidity. Maki et al.[89] isolated a bacteria from decaying leaves that was capable of freezing water in the temperature range −1.8°C to −3.8°C. It appears that nucleation was associated with intact bacteria cells. Broken cells did not nucleate ice. Concentration of bacteria equal to 10^6 cm^{-3} was necessary to initiate ice nucleation at higher temperatures. The concentration of bacteria in unpolluted atmospheres is very low and their contribution to the formation of ice is probably negligible.

Particles generated from the surface of an ocean consist of water droplets and evaporated droplets. Evaporated droplets may serve as CCN. In addition to the water-soluble part of aerosol particles, there is always an insoluble part that consists of organic matter. The water-insoluble part may act as IFN. Brier and Kline[90] showed that ocean water can generate very large concentrations of aerosol particles active as IFN. This was observed with general storminess and widespread precipitation. They suggested that strong winds generated aerosol particles that were carried over land with air masses. These aerosol particles contained unknown substances that acted as IFN. Schnell[91,92] found that ocean plankton can produce up to 10^7 IFN active per gram of plankton through freezing at $T = -10$°C. Bands of aerosol particles active as IFN were discovered over the ocean by Bigg.[93] This may suggest that ocean plankton is a source of IFN and that aerosol particles are released from the ocean surface by bursting bubbles. The presence of bands indicates the existence of local sources and that IFN do not survive long travel. The decay of IFN may be biological (drying of cells) or physical (settling of aerosol particles).

Aerosol particles collected over the Pacific Ocean were examined for the ability to nucleate ice by sorption, freezing, and condensation-followed-by-freezing.[78] No particles nucleating ice by sorption were found for the temperature range between −5°C and −17°C and $S_w < 0\%$. Particles collected over the western coast of North America were in the 6- to 8-μm diameter size range. Very small concentrations of particles below 0.5 μm diameter nucleated ice at −12°C; these particles were present over the South Pacific Ocean. Concentrations of IFN active by condensation-followed-by-freezing were 100 m^{-3} at −3.3°C, and 3×10^4 m^{-3} at −4°C; there was no increase in IFN concentration with lowering of the temperature down to −14°C. The IFN concentration was independent of temperature in the −4°C to −14°C range. The fraction of aerosol particles nucleating ice at and below –4°C was found to be around 10^{-3}. This is a very large fraction; usually, this fraction is 10^{-6} or even lower. These aerosol particles were found over the South Equatorial Current only. They were associated with biological activity in that current and they seem to consist mostly of organic matter; aerosol particles nucleating ice were less that 0.5 μm in diameter.

In air masses over the equatorial region of the Pacific Ocean between 7°N and 5°S latitude and 110°W to 142°W longitude, the IFN active by sorption were again absent. It therefore seems that the Pacific Ocean does not produce sorption IFN. Aerosol particles below 0.1 μm diameter nucleated ice at –4°C by freezing; the initial (the highest) temperature for ice nucleation on particles larger than 0.3 μm diameter was –12°C. The concentration of IFN active by condensation-followed-by-freezing was patchy over the ocean; the maximum concentration was 4.5×10^4 m^{-3} active at –4°C. Again, the concentration of IFN active by condensation-followed-by-freezing was independent of temperature over the range from –4°C to –17°C. It therefore seems that the presence of IFN particles acting independently of temperature indicates the presence of aerosol particles of marine origin. The size range of aerosol particles was between 0.1 μm and 0.3 μm in diameter. They were all, or practically all, hydrophobic and consisted of chemical compounds or a mixture of compounds that vaporized completely in vacuum. These compounds were therefore neither proteins nor bacteria, but were definitely produced by living organisms in the upwelling regions of the ocean where biological activity takes place. The sulfate ion was found to be an integral part of the ice-nucleating aerosol particles. IFN present over the Pacific Ocean can be classified as a physical/chemical system consisting of a hydrophobic surface with absorbed sulfate ion clusters. The critical concentration of sulfate ions necessary to initiate the L → S phase transition during ice nucleation by condensation-followed-by-freezing is different for different hydrophobic surfaces.[21]

Dimethyl sulfide (DMS) in the ocean water is a precursor for SO_4^{-2} ions in the atmosphere. The relation between DMS, its oxidation intermediate chemical compounds leading to the formation of sulfate ion, and the formation of aerosol particles active as IFN is very complex. The flux of DMS from the ocean water into the atmosphere is proportional to the concentration of DMS in ocean water. The relation between the concentrations of DMS and IFN is shown in Figure 5.31[94]; it supports the findings that SO_4^{-2} ions play a major role in L → S phase transitions taking place in the marine atmosphere.

In addition to the two main sources producing pure marine and continental aerosol particles, there is a thrid one that consists of a mixture of the two types of aerosol particles. One might think that by mixing the two types of aerosol particles, one would create a new aerosol particle population behaving as a mixture: aerosol particles of marine origin would supply IFN as if they were alone, and aerosol particles of continental origin would do the same.[29] In reality, the continental-marine air masses contain mostly modified marine and continental aerosol particles.[95] IFN active by sorption ($S_w < 0\%$) and by condensation-followed-by-freezing at S_w up to 30% were not present in the mixed continental-marine atmosphere for the temperature range between –8°C and –20°C. Aerosol particles that normally would nucleate ice were coated with aerosol particles in the Aitken particle size range. Their surfaces probably absorbed gases produced in the ocean and over land. The coating of continental aerosol particles consisted of a mixture of sulfates and nitrates. This should actually produce IFN, but the coating reacts chemically with some of the chemical compounds present in the surface of particles of continental origin, producing new chemical compounds that do not nucleate ice. Removal of the coating restores the ice nucleating ability. This was observed when marine air masses crossed the Florida peninsula and were mixed with a highly polluted continental air mass. IFN active by freezing were nucleating ice at a temperature of –5.2°C. The Aitken nuclei and absorbed gases that deactivate IFN by sorption may actually create new surfaces through chemical reactions when exposed to large dilution within water droplets during freezing. The distribution of IFN active at different temperatures along a flight path is shown in Figure 5.32. It is quite clear that the average IFN concentration along a flight path cannot represent any physical reality; it will be simply meaningless. Trajectories of air masses must be known to deduce the origin of aerosol particles. Sampling of aerosol particles must show their distribution in time and space.

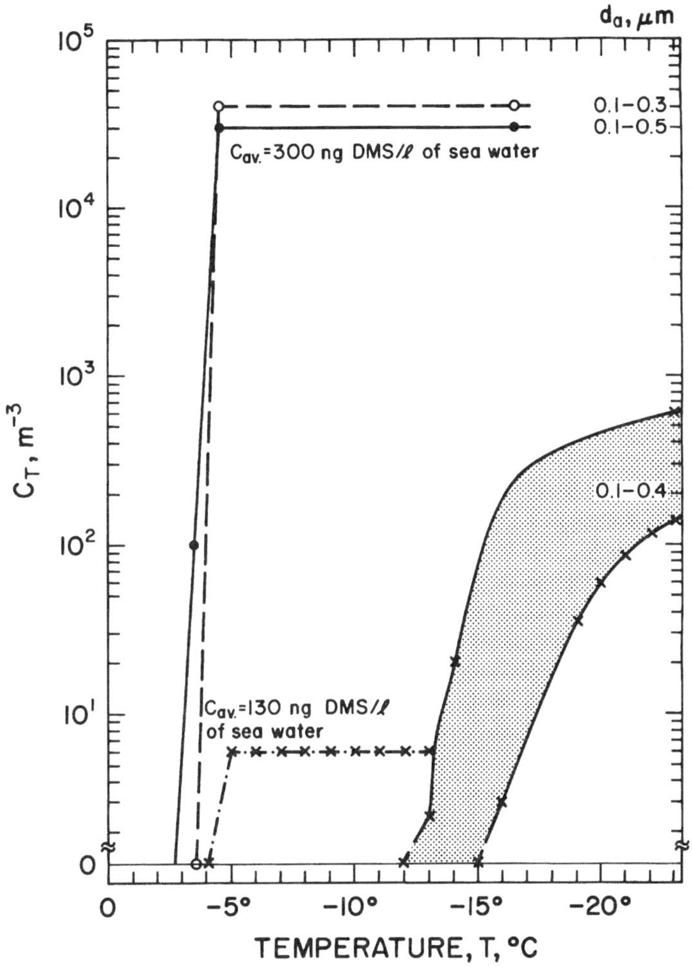

FIGURE 5.31 Concentrations of IFN of marine (○, ●, X–X) and continental (grey area) as a function of temperature and DMS concentration in sea water.

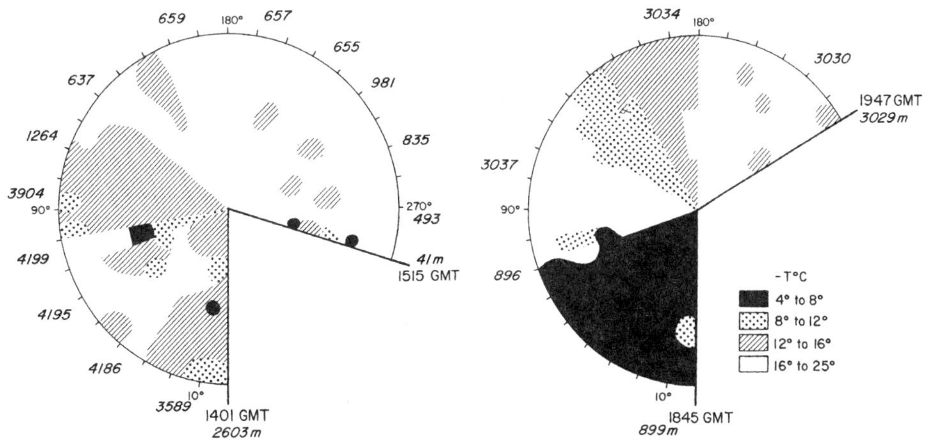

FIGURE 5.32 Temperature spectra of ice-forming nuclei present at different altitudes along a flight path.

Average concentrations of IFN (present in surface air) active at temperatures of –20°C in the northern and southern hemispheres were 2600 m^{-3} and 2090 m^{-3} (from mixing chambers) and 300 m^{-3} and 80 m^{-3} (from expansion chambers), respectively. The average concentration over oceans was 100 m^{-3}.[96] The average concentrations over deserts and semi-arid regions were higher than that over foothill regions and seas.[19]

RADIONUCLIDES AS ICE-FORMING NUCLEI

Man-made and cosmogenic radionuclides are present in the atmosphere. The number of radioactive atoms that undergo radioactive decay, N_i, is equal to the difference between the number of radioactive atoms at some arbitrarily chosen time $t = o$ (n_o) and the number of radioactive atoms present at time t (n_t). N_i, therefore, is given by:

$$N_i = n_o - n_t = n_o\left(1 - e^{\gamma t}\right) \tag{5.12}$$

where γ is the radioactive decay constant of the specific radioactive atom and is equal to 0.69 divided by the half-life of that specific radionuclide. If every disintegration could produce an ice particle, then for n_o = 100, the number of ice crystals after 10 min and 60 min would be 17 and 67 due to ^{38}Cl, and 1 and 5 due to ^{24}Na, respectively. This of course should take place in a cloud parcel. Not every disintegration produces a freezing event and thus the number of ice crystals will be much smaller than calculated. Concentrations of ^{24}Na and ^{38}Cl, in cloud water on one occasion in 1969, were 1.2×10^{-3} dpm ml^{-1} and 3.6×10^{-2} dpm ml^{-1}. If all cosmogenic radionuclides would attach themselves to cloud droplets, then their concentration would be ~12 dpm ml^{-1}. If all gaseous cosmogenic radionuclides (^{41}Ar, ^{15}O, ^{14}O, ^{13}N) had attached to cloud droplets, then the total radioactivity would be ~10^3 dpm ml^{-1}. The concentration of radionuclides produced by nuclear weapons (e.g., ^{90}Sr) is diminishing with time. Concentration of short-lived cosmogenic radionuclides is 3 orders of magnitude higher at an altitude of 18 km than at the ground level. The concentrations at 9.4 km altitude were 50, 40, and 4 dpm per standard cubic meter of air for man-made, cosmogenic, and radon daughters, respectively.

It was found that β^- and γ radiation alters the temperature of the L → S phase transition. In the case of ^{24}Na in water solution, the temperature of the phase transition was increased by 9.3°C. An increase of 5.6°C was recorded for ^{38}Cl. The temperature increase was proportional to the concentration of radioactive species. The increase of the ice nucleation temperature was influenced by β^- radiation; the role of γ radiation is unknown.[97] γ-rays are usually accompanied by secondary electrons during their passage through air.

It is possible to use silver iodide with different radioactive atoms, for example, $Ag^{131}I$, ^{110}AgI, or $^{110}Ag^{131}I$. In the case of $Ag^{131}I$, the transformation is:

$$Ag^{131}I \rightarrow \left[Ag^{131}Xe + \beta\right] \rightarrow Ag^{+} + ^{131}Xe \tag{5.13}$$

The radioactive $Ag^{131}I$ was a more effective ice-nucleating chemical compound than the non-radioactive one. But what nucleated the ice: freezing, condensation-followed-by-freezing, or contact? β^- is released during the decay of every radioactive atom; it can cause L → S phase transition. The Ag^+ ion is situated in the surface of an $Ag^{131}I$. The Ag^+AgI lattice may be the actual spot where ice nucleation takes place. This view is supported by the nucleation of ice taking place at higher temperatures in the presence of high concentrations of hydrogen ions.[98]

The number of ice crystals produced by different radionuclides in clouds is probably very low. Nevertheless, they may be the first ice crystal that will start to modify the colloidal structure of a cloud.

ICE-FORMING NUCLEI AND CLIMATE

Freezing temperatures of drops containing ammonium sulfate depend on the concentration of the salt. There exists a maximum temperature of freezing at a certain salt concentration. For pure water drops placed on a clean filter, on a filter coated with aerosol particles of continental and marine origins, the freezing temperatures were –16°C, –11°C, and –8°C, respectively; these temperatures are the maximum and they belong to the first drops frozen during continuous cooling of the filter. In the case of an ammonium sulfate solution, the temperatures were –12°C, –9°C, and –4°C, respectively; the corresponding concentrations were 10^{-4}, 10^{-4} to 10^{-2}, and 10^{-4} *M*. Cloud droplets present in clouds located in the northern and southern hemispheres, north and south of about 25° latitude, will start to freeze at –4°C in the marine and at –9°C in the continental atmosphere. The albedos of clouds containing ice particles are higher than of similar clouds containing water droplets only. The increasing concentration of sulfate in the atmosphere will increase the albedos of clouds; it will also expand the belt of clouds containing ice north in the southern and south in the northern hemisphere. This in turn will increase daytime global cooling.[8]

FORMATION OF ICE IN CLOUDS

Phase transitions, either vapor → solid (ice) or liquid → solid (ice), take place in clouds on aerosol particles that can act as ice-forming nuclei. All mechanisms are working within a cloud, producing ice particles during the lifetime of the cloud. In a mixed cloud, droplets evaporate in the vicinity of ice crystals. Some of the droplets may freeze before evaporating completely; they will produce new ice particles. Other droplets will evaporate completely, leaving "dry" aerosol particles; new aerosol particles may or may not act as CCN or may act as ice-forming nuclei. These aerosol particles are actually cloud condensation nuclei that were modified chemically during their residence as cloud droplets within a cloud. The chemical reactions take place within the liquid phase when CCN form a droplet. After evaporation of water, the new "dry" particle may or may not act as CCN, but may act as IFN.[10,11] An example of new chemical compounds formed in the residue of an evaporated droplet is given in Figure 5.33. The residue contained sodium, calcium, and sulfur. The most probable chemical compounds are sulfates of sodium and calcium. The most probable chemical compounds in CCN were ammonium sulfate, and sodium and calcium chlorides. Droplets formed on CCN were collected, the water solution was filtered, and drops of filtered solution were evaporated. The residues (Figure 5.33) contained sodium, calcium, and sulfur. Evaporation of droplets can also take place when a liquid water droplet present in a parcel of a cloud is exposed to drier air during mixing with outside air (Figure 5.34).[30] The concentration of new aerosol particles active as IFN at temperatures of –4.9°C and –15.9°C were 367 m^{-3} and 3000 m^{-3}, respectively. These aerosol particles did not lose IFN activity when exposed to a repeated evaporation/condensation cycle. The population of new IFN active by sorption reached a concentration of 6400 m^{-3} at –4.4°C and 7000 m^{-3} at –20°C. These concentrations were found in air parcels present in Italy. In the state of Washington, the concentrations were from 0 m^{-3} at high temperatures down to –13°C, and up to 2000 m^{-3} for the temperature range between –14°C and –20°C; these aerosol particles acted as condensation-followed-by-freezing IFN. IFN active by sorption reached concentrations of 7000 m^{-3} at $T = -17$°C and $S_{ice} = 0.2\%$. Not all aerosol parcels contained particles that participated in the chain process: aerosol particle [CCN] → cloud droplet → "dry" residue left after evaporation → new aerosol particle [IFN] → ice particle. It seems that clouds containing aerosol particles of marine origin and/or sulfates have the capability of forming new aerosol particles that act as IFN. Those new IFN sometimes survive repeated condensation/evaporation cycles, but there are aerosol particles that, after the first cycle, are completely deactivated. This process of ice formation in clouds is therefore only applicable to some air masses. Icing at the edges of clouds may be due to this mechanism. The primary formation of ice particles takes place on ice-forming nuclei; secondary

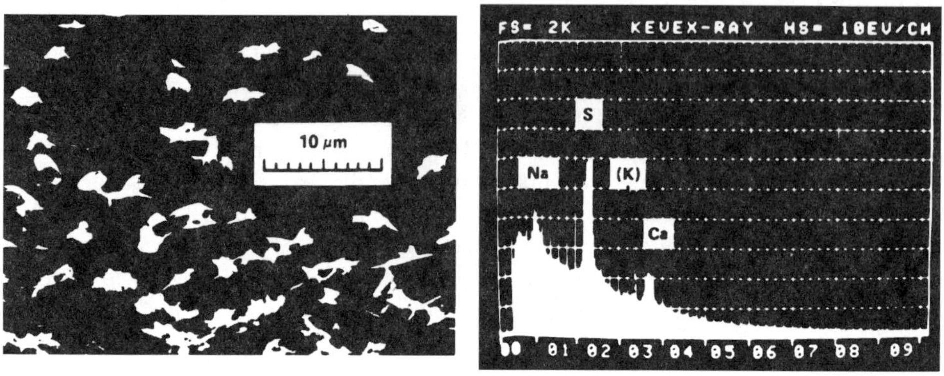

FIGURE 5.33 Chemical compounds present in the residues of evaporated droplets (new aerosol particles).

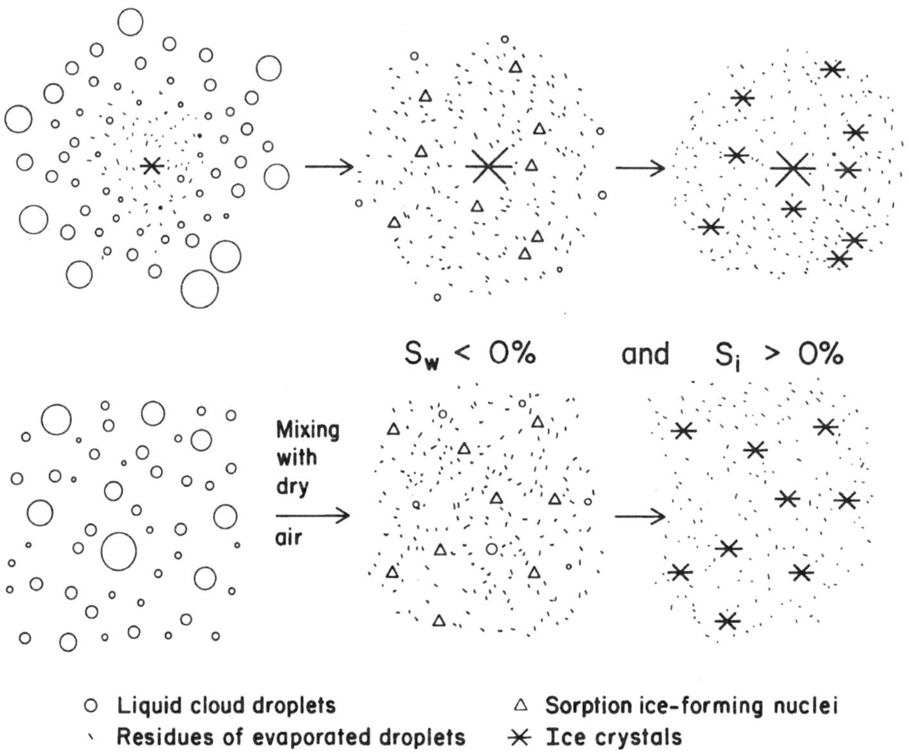

FIGURE 5.34 Formation of ice crystals through evaporation of cloud droplets.

ice particle production takes place when the number concentration of ice particles exceeds the concentration of IFN.

An example of a secondary ice particle production mechanism is the Hallett-Mossop[99] process. In their experiment, a metal rod rotated in the chamber at –4.5°C. In the presence of supercooled water droplets, a large number of ice particles were formed. In natural clouds, Blyth and Latham[49] showed that, in the presence of both supercooled drops of 0.5 mm diameter in concentration 10^4 m^{-3} and ice crystals, a new generation of ice crystals was produced with concentrations of up to 1.3×10^6 m^{-3}. These secondary ice particles did not form on IFN. The first process of ice particle formation is time dependent; the first ice particle forms and then evaporation of droplets takes place, producing some aerosol particles that can act as IFN and some frozen drops. This is a chain reaction, starting very slowly and moving faster, probably in 15 to 30 minutes. In the Hallett-Mossop process, ice particles are formed in a very short time.

Nikandrov[100], in his publictions that went unnoticed in the U.S., has shown that during the near collision between an ice particle and a supercooled water drop, the drop will freeze and a number of ice particles will form. Water vapor from the evaporating droplet will move toward an ice crystal or aerosol particle acting as IFN by sorption. The vapor stream, when reaching the surface of an ice particle, will freeze; and the freezing process will propagate formation of ice toward the evaporating surface. Upon reaching the surface, ice will touch the drop and the drop will freeze. The heat released from a freezing drop is distributed non-uniformly around the freezing drop, and the drop will move abruptly and break the thread made of ice. That frozen thread of condensed water vapor will break into many fragments of ice. These fragments will grow in a parcel of a cloud as long as there is water vapor supersaturation with respect to ice. In Nikandrov's experiment, the distance between an evaporating drop and an ice surface was about 50 μm or larger. The higher the temperature, the higher the water vapor pressure. At –5°C, $P_{supercooled\ liquid\ water}$ = 3.163, P_{ice} = 3.013, and ΔP = 0.15 mmHg. At temperatures between –4°C and –6°C, elementary needles should be formed and they should break easily into fragments. It should be noted here that the ice fragment should be free from any aerosol particles, and snowflakes they produce should be without a centrally located foreign nucleus. Such "empty" snowflakes are indeed found in nature.

Nikandrov's work preceded Hallett's experiment by 20 years; his contribution to this ice multiplication mechanism should be recognized and therefore this author proposes to name it the Nikandrov-Hallett-Mossop process.

FREEZING OF WATER DROPS

Cloud droplets form on cloud condensation nuclei; subsequently, in a liquid cloud, they grow by condensation and by collisions with other droplets present within the cloud. During severe rainstorms and thunderstorms, large soil particles are ingested by a storm and these large particles accrete cloud droplets. These large aerosol particles will produce single, large raindrops. Direct observation of the aerosol particle content of the raindrops has revealed that larger particles (100 to 200 μm diameter) produce smaller raindrops (1 to 3 mm diameter); smaller particles (40 to 50 μm) produced larger raindrops (4 to 6 mm). This relation was explained by calculating the residence time of a growing raindrop. Raindrops growing on smaller aerosol particles resided longer inside an updraft and grew larger. Large aerosol particles accreted cloud droplets rapidly and fell out of an updraft after a brief residence time in a cloud. Large drops contained one large aerosol particle per drop (80%).[35,101] Large frozen raindrops provided embryos for hailstones.[32,37–39] All the scavenged aerosol particles will act as IFN at different temperatures; each individual particle is associated with a specific temperature of ice nucleation, and that temperature will depend on the mode of ice nucleation. Is it therefore possible to take a sample of precipitation and deduce what processes took place in a cloud that lead to ice formation? Or, is it possible to take a sample of aerosol particles and deduce their contribution to ice and raindrop formation in a storm? In an attempt to deduce the concentration of IFN active by freezing in precipitation, Vali[102] derived an equation expressing a cumulative IFN spectrum $K(T)$ as a function of temperature T:

$$K(T) = \frac{\ln N_o - \ln N_t}{V} \left(\mathrm{cm}^{-3}\right) \qquad (5.14)$$

where N_o is the total number of drops present on filter, N_t is the number of drops still unfrozen at temperature T, V is the volume of a drop, and $K(T)$ is the cumulative IFN concentration of nuclei active at all temperatures higher than T.

The drop freezing technique gives the freezing nucleus content present in a sample of precipitation. These IFN active by freezing are hydrosol particles that were originally present in the atmosphere as aerosol particles. The number of hydrosol particles differs appreciably from the number of aerosol particles transferred into the precipitation.[5–7,14,103–105] Some of the aerosol particles take an active part in the formation of precipitation and others are scavenged by different mechanisms, discussed previously. The first group will be called k_1 and the second k_2. Large (d > 40 μm) aerosol particles shed smaller hydrosol particles when transferred into a precipitation; these smaller hydrosol particles will belong to the k_3 category. When droplets evaporate, especially at low temperatures, the dissolved chemical compounds may precipitate within a supercooled droplet, or new water-insoluble chemical compounds may form in chemical reactions between ions; water-insoluble compounds are actually compounds that are slightly soluble in water; this is category k_4. The concentration of freezing nuclei present in a precipitation is equal to the sum of four k values at some temperature T; $k(T)$ is given by:

$$k(T) = \left(k_{1.1} + k_{1.2} + k_{1.3}\right) + \left(k_{2.1} + k_{2.2} + k_{2.3} + k_{2.4} + k_{2.5}\right) + k_3 + \left(k_{4.1} + k_{4.2}\right) \tag{5.15}$$

where

- $k_{1.1}$ is derived from hydrosol particles that act as freezing nuclei within a cloud
- $k_{1.2}$ represents aerosol particles that nucleate ice by contact
- $k_{1.3}$ represents aerosol particles that nucleated ice through condensation-followed-by-freezing
- $k_{2.1}$ are aerosol particles that were transferred into precipitation elements by aerodynamic capture
- $k_{2.2}$ by Brownian motion
- $k_{2.3}$ by diffusiophoresis
- $k_{2.4}$ by thermophoresis
- $k_{2.5}$ by electrostatic forces
- k_3 are hydrosol particles shed from larger aerosol particles after they are transferred into the liquid phase of a cloud
- $k_{4.1}$ are hydrosol particles formed in precipitation elements during cooling
- $k_{4.2}$ are hydrosol particles formed in precipitation elements by chemical reactions, forming new water-insoluble chemical compounds

All aerosol particles in the k_1 category can nucleate ice in three different ways, at three different temperatures. The largest difference in temperature is between nucleation by contact and freezing. In a cloud, therefore, aerosol particles will nucleate ice at temperatures higher than that determined from the drop freezing technique. Aerosol particles that are scavenged by precipitation elements do not contribute to ice formation in clouds, but they do contribute to IFN population determined by the drop freezing technique. Also, dissolved chemical compounds are distributed non-uniformly within cloud particles. Thus, the chemical-physical processes taking place in clouds are not reproduced during the freezing of drops. Therefore, the freezing drop technique cannot give information about phase transitions taking place in clouds. This also applies to IFN concentrations and temperatures of ice nucleation.

Another drop freezing technique consists of sampling aerosol particles using a filter and then placing water drops on the filter. Instead of a filter, one can use impactor plates or other substrate. Each water drop will be in contact with a certain number of collected aerosol particles. If a large soil particle sheds small particles, then all the shed particles will be confined to that particular drop and they will not be distributed within the precipitation water. It is true that concentrations of water-soluble compounds will be much lower than in cloud droplets, but they will be higher than in a precipitation sample. Contact nucleation can be assessed by “raining small, fog-like droplets”onto the substrate. Experiments with single soil particles have shown that one can produce up to 1000 IFN active by freezing when they are placed into 1 ml of water. When a drop is placed on one soil particle, it will freeze at the temperature corresponding to the temperature of ice nucleation of that particle or one of the shed particles; the particle nucleating ice at the highest temperature will cause the drop to freeze. The pH of rain water depends on the soluble part of scavenged aerosol particles and water soluble gases. In the semi-acid region of Colorado, the pH is constant at 7.30 in the first part of a storm and drops to 7.00 in the second part. In Italy, the pH oscillated between 6.4 and 5.6, indicating the presence of dissolved CO_2, SO_2, and NO_x.[106]

EXTRATERRESTRIAL PARTICLES AND PRECIPITATION

Findeisen[54] suggested the presence of particles of extraterrestrial origin active as IFN in the atmosphere. He proposed this as a possible explanation of the discrepancy between concentrations of IFN and ice particles in clouds. Extraterrestrial particles originate through the fragmentation of meteors during entry into the Earth's atmosphere. Bowen, in a series of publications,[96,107,108] suggested the existence of a relationship between meteor streams and precipitation. The latter involved statistically derived maximum rainfall days. He also suggested the existence of a time lag of 28 ± 2 days between the maximum activity within meteor showers and a rainfall anomaly. Mason[80] summarized Bowen's position as follows:

> Bowen also states that, in general, a peak in the rainfall curve is produced not only by one or two very heavy falls on that day, nor by an extraordinarily large number of rainy days on that date, but by a greater than normal number of rainy days on which the rainfall was consistently higher than average.

Rosinski, in the early 1960s in a letter to Bowen, pointed out that even if the rainfall anomalies could be real, they cannot be associated with meteor streams. If the association was true, then similar number concentrations of magnetic spherules (particles of extraterrestrial origin) and their size distribution should be similar for every meteor shower. But, for example, the entry velocity of meteors in the Perseids stream is 60 km s^{-1}, meteors vaporize completely, and what is left are coagulation products of meteor vapors (Rosinski secondary particles).[108] They are in the Ångstrom diameter size range, between 5Å and 150Å. The settling time of secondary particles is years, and they cannot reach the lower atmosphere in 28 days. Their existence was documented by Dzienis and Kopcewicz[110] and Kopcewicz.[111] They consist of agglomerates of ultra-fine particles and they are composed of Fe_2O_3. The low-velocity (~30 km s^{-1}) meteor streams can supply extraterrestrial particles that can survive entry into the atmosphere. Because of different entry velocities associated with different meteor streams, it is impossible to have a constant time lag of 28 ± 2 days between rainfall anomalies and days of maximum meteor activity.

Particles of extraterrestrial origins are present in the atmosphere. Consequently, there must exist a source for these particles. If sporadic meteors are the source, then the extraterrestrial particles should be distributed at random in the atmosphere in time and space. But if a meteor stream is the source, then particles should be found every year at the same time in the atmosphere. To pin down the specific source, sampling of extraterrestrial particles must be performed at different latitudes over the northern and southern hemispheres. Remote places should be used to minimize industrial contamination. If similar number concentration peaks of magnetic spherules (extraterrestrial particles) appear simultaneously at most of the sampling stations, then the contamination from local sources can be discounted and the existence of an extraterrestrial source of particles established.

Not all rainfall anomalies are significant. As a matter of fact, most of them are insignificant, and some of them may not even exist.[111] A rainfall anomaly in October was chosen as a first sampling period. The reason for this choice was that there are no meteor showers in September 28 days prior to the rainfall peak, and consequently one should be unable to find extraterrestrial particles. The results of sampling in 1967, 1969, and 1971 are shown in Figure 5.35. It is clear

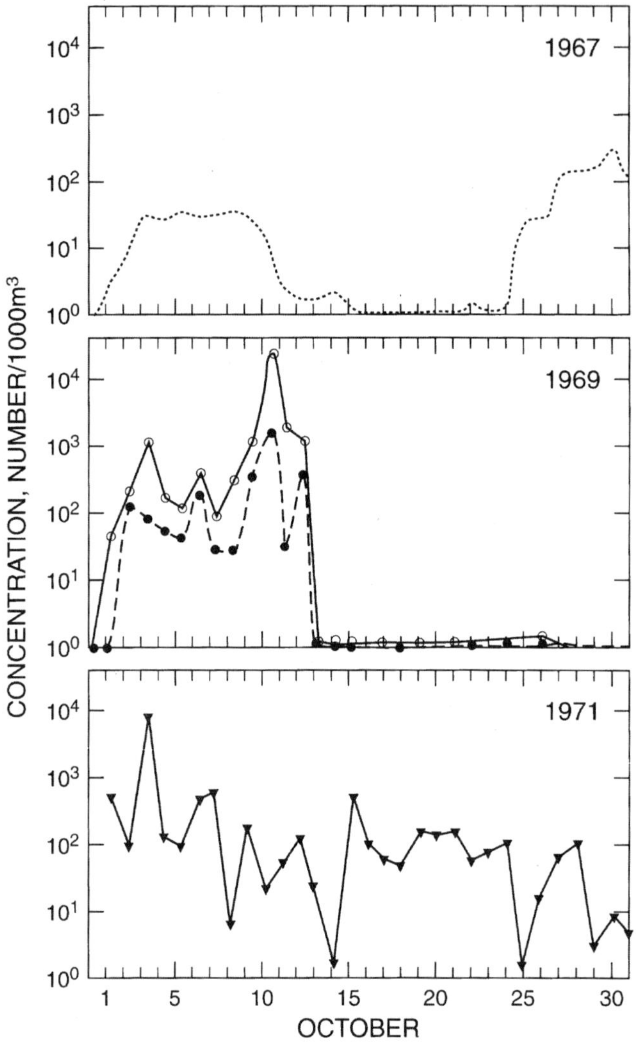

FIGURE 5.35 Concentrations of 5–20 μm diameter magnetic spherules in the Southern Hemisphere (▼,…, 1°S, 36°E; ○, 26°S, 27°E; ●, 43°S, 172°E).

that the source of magnetic spherules exists in space and that the Earth intercepts it every year in October. The source is unknown, but it consists of low entry velocity micrometeorites. In addition to a flux of magnetic spherules (up to 500 m^2 day^{-1}), particles consisting of silicates were found in large concentrations (5×10^4 m^2 day^{-1}). Ice crystals from cirrus clouds subsequently seed storms below, thus producing a rainfall anomaly. Subsequently, the rainfall anomaly in October was classified as a significant peak in the southern hemisphere.[113] All the data indicate that the presence of extraterrestrial particles is real and therefore this author would like to suggest the existence in October of a singular Findeisen-Bowen meteor shower–rainfall anomaly relation. More research needs to be performed to firmly establish the existence of this relation.

Sampling of magnetic spherules was performed at ground level. These particles have to settle through the entire atmosphere. Their sizes vary between 1 and 30 μm in diameter; the density range is between 1.33 g cm^{-3} and 6.45 g cm^{-3}.[5-7,114] Particles with surfaces that can be oxidized will react with ozone molecules when crossing the ozone layer. The surface of magnetic spherules consists of

$FeO_{1.3}$; the subsurface consists of $FeO_{1.1}$. Rosinski[115] presented evidence showing that the ozone hole in October may be a natural phenomenon. The basic chemical reactions are discussed below.

Oxygen atoms must be present to produce ozone molecules; the chemical reactions are

$$O_2 + h\nu = O + O \tag{5.16}$$

and

$$O_2 + O = O_3 \tag{5.17}$$

The destruction of ozone molecules is shown below:

$$O_3 + O = O_2 + O_2 \tag{5.18}$$

and

$$O_3 + Cl = ClO + O_2. \tag{5.19}$$

Chlorine atoms can be substituted by OH groups or NO. The final result of all these reactions is the production of oxygen molecules. These oxygen molecules dissociate into oxygen atoms via reaction 5.16. In this way, there is a constant supply of oxygen atoms to produce ozone molecules, which are decomposed by, for example, chlorine atoms.

The chemical reactions between iron (Fe) and ozone or oxygen are extremely slow, as long as the iron is in a bulk state. When the iron is in a finely divided state (particle diameters less than 100 μm), it becomes pyrophoric. Such iron burns in oxygen and should burn even more readily in ozone. Magnetic spherules are all iron oxides, with other metals present. The reaction is

$$Fe + O_3 = FeO + O_2 \tag{5.20}$$

$$2Fe + O_2 = 2FeO \tag{5.21}$$

and

$$Fe + O + FeO \tag{5.22}$$

The destruction of ozone molecules leaves oxygen molecules. They can reform ozone molecules via reactions 5.16 and 5.17. In the case of reaction with iron, all oxygen atoms are removed from the atmosphere and the ozone hole is formed. There is no oxygen available to form ozone in the path of a magnetic spherule in the ozone layer. The boundaries of the ozone hole should be similar to those of a meteor stream.

The chemical formula of FeO is $FeO_{0.90-0.95}$. The final product of oxidation is $FeO_{1.5}$, found in ultrafine magnetic particles. Different metals were found in the surface of magnetic spherules (e.g., Ni, Cr, etc.). The possibility exists, therefore, that ozone molecules can be destroyed by a catalytic reaction:

$$O_3 + [A] = [A] + O + O_2 \tag{5.23}$$

where A is the surface of a magnetic spherule.

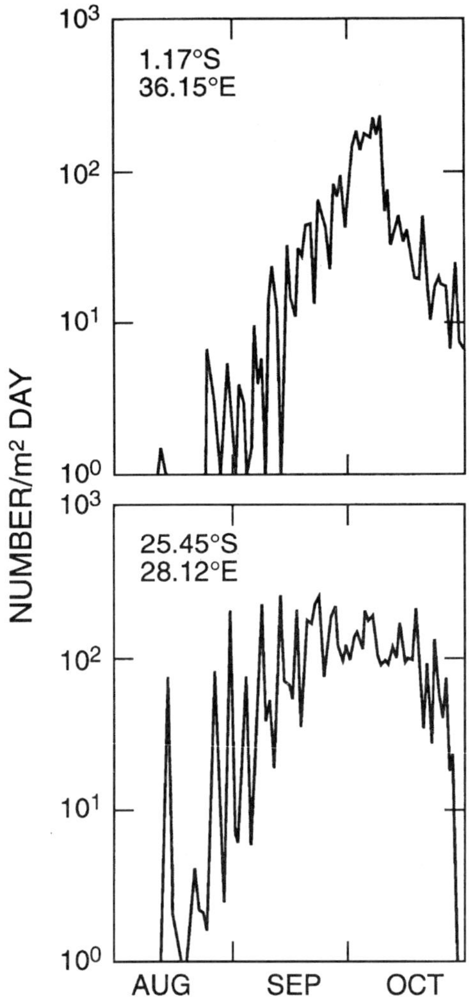

FIGURE 5.36 Flux by number of magnetic spheroids in the 5 to 20 μ size range at 70 km altitude necessary to reproduce experimental results from October 1971.

The total number of iron atoms required to destroy all the ozone molecules via reaction 5.20 is 4.86×10^{18} m^{-3}. To destroy all the oxygen molecules produced in reaction 5.20, the number of Fe atoms needed is 9.72×10^{18} m^{-3}. The total number of Fe atoms required to destroy O_3 and O_2 molecules is about 1.5×10^{19} m^{-3}. Ozone destruction was about 10%. The magnetic spherules should be of about 12 μm in diameter to supply enough Fe atoms for chemical reation 5.20. The magnetic spherules collected were between 2 μm and 25 μm in diameter. It seems, therefore, that there is sufficient concentration of magnetic spherules to produce the ozone hole in October.

The surface of a pyrophoric particle produced by reduction of iron oxide by, for example, hydrogen is different from that of a magnetic spherule. Both surfaces react with oxygen atoms and/or molecules and produce oxides of iron. The surface of a magnetic spherule falling into the ozone layer is partially oxidized. Ozone molecules can be destroyed via collisions with iron (Equation 5.20) or with iron oxides.

$$O_3 + FeO_x = FeO_{x+1} + O_2 \tag{5.24}$$

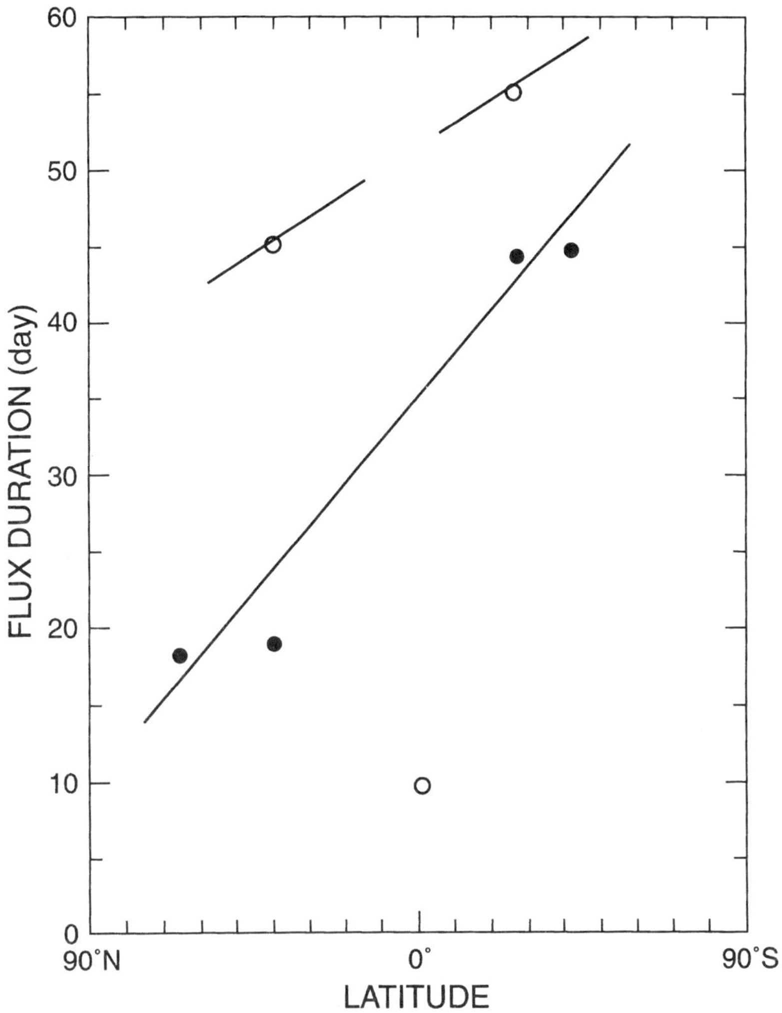

FIGURE 5.37 The duration of fluxes of magnetic spheroids vs. latitude for 1969 (●) and 1971 (○). Flux of magnetic spheroids is equal to or greater than 10 m^{-2}.

and

$$Fe_nO_x + O = Fe_mO_{x+1} \tag{5.25}$$

where n and m are the number of Fe atoms sharing electrons with oxygen.

The destruction of ozone could be between 5% and 20% during the 10-day crossing of the flux of magnetic spherules through the ozone layer in the second part of September and October.[17,26,27] An example of a flux of magnetic spheroids entering the Earth's atmosphere at a 70-km altitude is given in Figure 5.36; the duration of the flux vs. latitude (Figure 5.37) indicates the presence of the southern (unknown) meteor stream. Extraterrestrial particles consisting of magnetic spherules, slightly magnetic and non-magnetic particles must settle down to the altitude of cirrus clouds, where they may produce the cirrus cloud (ice) particles. If the aerosol particles that are responsible for the formation of the ozone hole in September/October are the same that later produce the singular rainfall anomaly in October (if such a phenomenon does exist), then the Earth must intercept the meteor stream every year. In that case, the ozone hole is produced by intercepting a higher concentration of particles containing iron within a meteor stream.

An ozone hole exists also in the arctic region in April. Its magnitude is much smaller than that of the October ozone hole present in the southern hemisphere. Both ozone holes are produced, according to Rosinski, by the magnetic spherules falling into the Earth's atmosphere after the Earth crosses the same meteor stream 6 months apart on the other side of the sun. The difference in the size of the ozone holes is due to the differences present in concentrations of micrometeorites in different parts of the meteor stream. The ozone hole is a natural phenomenon.

ACKNOWLEDGMENTS

Special thanks go to Dr. Claudio Tomasi and to Prof. Franco Prodi at ISAO (formerly FISBAT).

The term "Rosinski-Nix effect" used in this article was introduced for the first time by Dr. H. Weickmann in the Welcoming Address at the *Third International Workshop on Ice Nucleus Measurements*, 1976.

My appreciation goes to Martine Bunting of NCAR for preparing this manuscript, and to C.T. Nagamoto and G. Morgan for their help.

Research described in the author's publications published prior to the year 1982 was supported by the National Center for Atmospheric Research (NCAR).

During 1982, NCAR terminated the author's research. The author acknowledges the courtesy of Dr. Farn Pwu Parungo of the National Oceanic and Atmospheric Administration (NOAA) Boulder, CO. Dr. Parungo furnished her laboratory, thus making it possible for the author to continue his research.

The author would like to acknowledge the use of ships of NOAA and the People's Republic of China, and ground-sampling stations in Italy and the Republic of South Africa.

Acknowledgments to scientists who helped each specific program are listed in pertinent publications.

DEDICATION

To the memory of my wife, Barbara, who lent me all her support in my work.

REFERENCES

1. Dorsey, N.E., Properties of Ordinary Water Substance, Reinhold, New York, 1940.
2. Coulier, P.J., *J. Pharm. Chiem., Paris*, 22, 165, 1875.
3. Aitken, J., *Trans. R. Soc. Edinburgh*, 30, 337, 1880.
4. Wegener, A., *Thermodynamik der Atmosphäre*, Barth, Leipzig, 1911.
5. Rosinski, J., Nagamoto, C.T., and Kerrigan, T.C., *J. Atmos. Terr. Phys.*, 35, 95, 1973.
6. Rosinski, J., Nagamoto, C.T., Kerrigan, T.C., and Langer, G., *J. Atmos. Sci.*, 30, 644, 1973.
7. Rosinski, J., Langer, G., Nagamoto, C.T., Thomas, C.W., Young, J.A., and Wogman, N.A., *J. Appl. Meteor*, 12, 1303, 1973.
8. Rosinski, J. and Nagamoto, C.T., *Acta Oceanologica Sinica*, 8, 498, 1994.
9. Junge, C.G., *Arch. Met. Geophys. Biokdim.*, 5, 44, 1952.
10. Rosinski, J., Nagamoto, C.T., and Gandrud, B.W., *Meteorol. Rdsch.*, 34, 47, 1981.
11. Rosinski, J., Morgan, G., Weickmann, P., Baird, J., Lecinski, A., Murr, L.E., and Emch, K., *Meteorol. Rdsch.*, 34, 77, 1981.
12. Nguyen, B.C., Bosang, B., and Gaudry, A., *J. Geophys. Res.*, 88, 10903, 1983.
13. Charlson, R.J., Lovelock, J.E., Andreae, M.O., and Warren, S.G., *Nature*, 326, 655, 1987.

14. Rosinski, J. and Nagamoto, C.T., *J. Colloid Interface Sci.*, 40, 116, 1972.
15. Twomey, S., *Atmospheric Aerosols*, Elsevier Sci. Publ., 1977.
16. Rosinski, J. and Langer, G., *J. Aerosol Sci.*, 5, 373, 1974.
17. Rosinski, J., *Adv. Colloid Interface Sci.*, 10, 315, 1979.
18. Langer, G., *Proc. Int. Conf. on Cloud Physics*, Toronto, 34, 1968.
19. Khorguani, V.G., Myekonkye, G.M., and Stepanon, G.B., *Proc. 9th Int. Conf. on Atmos. Aerosols, Condensation and Ice Nuclei*, Galway, Ireland, 1977.
20. Sano, I., Fujitaini, Y., and Uzu, Y., *J. Meteorol. Sci. Japan*, 38, 195, 1960.
21. Rosinski, J., Haagenson, P.L., Nagamoto, C.T., and Parungo, F., *J. Aerosol Sci.*, 18, 291, 1987.
22. Rosinski, J., *Atmos. Res.*, 26, 509, 1991.
23. Dinger, J.E., Howell, H.B., and Wojciechowski, T.A., J. Atmos. Sci., 27, 791, 1970.
24. Parungo, F.P., Nagamoto, C.T., Rosinski, J., and Haagenson, P.L., *J. Atmos. Chem.*, 4, 199, 1976.
25. Rosinski, J., Nagamoto, C.T., Gandrud, B.W., and Parungo, F., *J. Aerosol Sci.*, 15, 709, 1984.
26. Rosinski, J., *Atmos. Res.*, 38, 351, 1995.
27. Rosinski, J., *Il Nuovo Cimento*, 19C, 417, 1997.
28. Bigg, E.K. and Stevenson, C., *J. Rech. Atmos.*, 4, 41, 1970.
29. Rosinski, J. and Morgan, G.M., *J. Aerosol Sci.*, 19, 531, 1988.
30. Rosinski, J. and Morgan, G., *J. Aerosol Sci.*, 22, 123, 1991.
31. Rosinski, J., Haagenson, P.L., Nagamoto, C.T., and Morgan, G., *Acta. Meteorologica Sinica*, 5, 497, 1991.
32. Rosinski, J., Knight, C.A., Nagamoto, C.T., Morgan, B.M., Knight, N.C., and Prodi, F., *J. Atmos. Sci.*, 36, 882, 1979.
33. Rosinski, J., *J. Atmos. Terr. Phys.*, 29, 1201, 1967.
34. Rosinski, J., *J. Appl. Meteor*, 6, 1062, 1967; *J. Appl. Meteor*, 6, 1066, 1967.
35. Rosinski, J. and Kerrigan, T.C., *J. Atmos. Sci.*, 26, 695, 1969. (Corrigendum: *J. Atmos. Sci.*, 27, 178, 1979.)
36. Rosinski, J., Browning, K.A., Langer, G., and Nagamoto, C.T., *J. Atmos. Sci.*, 33, 530, 1976.
37. Rosinski, J., Langer, G., Nagamoto, C.T., Bayard, M.C., and Parungo, F.P., *J. Rech. Atmos.*, 10, 201, 1976.
38. Rosinski, J., Langer, G., Nagamoto, C.T., and Bayard, M.C., *J. Rech. Atmos.*, 10, 243, 1976.
39. Rosinski, J. and Nagamoto, C.T., *J. Aerosol Sci.*, 7, 1, 1976.
40. Rosinski, J. and Nagamoto, C.T., *J. Aerosol Sci.*, 7, 479, 1976.
41. Pemberton, C.S., *Aerodynamic Capture of Particles*, Pergamon Press, New York, 1960.
42. Rosinski, J., Nagamoto, C., and Ungar, A., *Kolloid-Z*, 164, 26, 1959.
43. Rosinski, J. and Nagamoto, C., *Kolloid-Z*, 175, 29, 1961.
44. Rosinski, J., Stockham, J., and Pierrard, J., *Kolloid-Z*, 190, 126, 1963.
45. Rosinski, J., Pierrard, J., and Nagamoto, C., *Kolloid-Z*, 192, 101, 1963.
46. Rosinski, J. and Church, T., *Powder Tech.*, 1, 272, 1968.
47. Pitter, R.L. and Pruppacher, H.R., *Quart. J. Roy. Meteor. Soc.*, 99, 540, 1973.
48. Rosinski, J. and Kerrigan, T.C., *J. Rech. Atmos.*, 11, 77, 1977.
49. Blyth, A.M. and Latham, J., *Quart. J. Roy. Meteor. Soc.*, 119, 91, 1993.
50. Slinn, W.G. and Hales, J.M., *J. Atmos. Sci.*, 28, 1465, 1971.
51. Weickmann, H., private communication, 1979.
52. McCully, C.R., Fisher, M., Langer, G., Rosinski, J., Glaess, H., and Werle, D., *Ind. Eng. Chem.*, 48, 1512, 1956.
53. Squires, P., *Australian J. Sci. Res.*, 5, 473, 1952.
54. Findeisen, W., *Met. Z.*, 55, 121, 1938.
55. Roberts, P. and Hallett, J., *Quart. J. Roy. Meteor. Soc.*, 16, 53, 1968.
56. Rosinski, J., Nagamoto, C.T., and Bayard, M., *J. Atmos. Terr. Phys.*, 37, 1231, 1975.
57. Rosinski, J., Nagamoto, C.T., and Kerrigan, T.C., *J. Rech. Atmos.*, 9, 107, 1975.
58. Rosinski, J. and Kerrigan, C.T., *Z. Agnew. Math. Phys.*, 23, 288, 1972.
59. Dessens, J., *Proc. Int. Conf. on Cloud Physics*, Toronto, 773, 1968.
60. Gagin, A., *J. Rech. Atmos.*, 6, 175, 1972.
61. Huffman, P.J., *J. Appl. Meteor*, 12, 1080, 1973.
62. Rosinski, J. and Lecinski, A., *J. Aerosol Sci.*, 14, 49, 1983.

63. Haman, K.E. and Malinowski, S.P., *Atmos. Res.*, 41, 161–175.
64. Soulage, G., *Ann. Geophys.*, 23, 103 and 167, 1953.
65. Soulage, G., *Nubila*, 4, 143, 1961.
66. Kumai, M., *J. Meteor.*, 18, 139, 1961.
67. Kumai, M. and Francis, K.E., *J. Atmos. Sci.*, 19, 474, 1962.
68. Kumai, M., *J. Atmos. Sci.*, 33, 833, 1976.
69. Isono, K.M., Komgoyashi, M., and Ono, A., *J. Meteor. Soc. Japan*, 37, 211, 1959.
70. Isono, K. and Ikebe, Y., *J. Meteor. Soc. Japan*, 37, 211, 1960.
71. Langer, G., Rosinski, J., and Edwards, E.P., *J. Appl. Meteor.*, 6, 114, 1967.
72. Fish, B.R., *Science*, M5, 1239, 1972.
73. Langer, G., Cooper, G., Nagamoto, C.T., and Rosinski, J., *J. Appl. Meteor.*, 17, 1039, 1978.
74. Rosinski, J., Cooper, G., and Kerrigan, T.C., *J. Phys. Chem.*, 84, 1464, 1980.
75. Rosinski, J., Morgan, G., Nagamoto, C.T., Langer, G., Yamate, G., and Parungo, F., *Meteorol. Tdsch.*, 33, 97, 1980.
76. Aufm Kampe, H.J., Weickmann, H.K., and Kedesdy, H.H., *J. Meteor.*, 9, 374, 1952.
77. Rucklidge, J., *J. Atmos. Sci.*, 22, 301, 1965.
78. Rosinski, J., Haagenson, P.L., Nagamoto, C.T., and Parungo, F., *J. Aerosol Sci.*, 17, 23, 1986.
79. Mason, B.J. and Maybank, J., *Quart. J. Roy. Meteor. Soc.*, 84, 235, 1958.
80. Mason, B.J., *Quart. J. Roy. Meteor. Soc.*, 86, 552, 1960.
81. Pueschel, R.F. and Langer, G., *J. Appl. Meteor.*, 12, 549, 1973.
82. Changnon, S.A., *AMS Bull.*, 49, 4, 1968.
83. Rosinski, J. and Parungo, F., *J. Appl. Meteor.*, 5, 119, 1966.
84. Rosinski, J., *J. Phys. Chem.*, 84, 1829, 1980.
85. Rosinski, J. and Lecinski, A., *J. Phys. Chem.*, 85, 2993, 1981.
86. Schnell, R.C. and Vali, G., 236, 163, 1972.
87. Schnell, R.C. and Vali, G., 246, 212, 1983.
88. Schaefer, V.J., *Icas. Rep. #28 Project Cirrus*, General Electric Co., 1950.
89. Maki, L.R., Galyan, E.L., Chien, M.C., and Caldwell, D.R., *Microbiology*, 28, 456, 1974.
90. Brier, G.W. and Kline, D.B., *Science*, 130, 717, 1959.
91. Schnell, R.C., *Tellus*, 27, 321, 1975.
92. Schnell, R.C., *J. Atmos. Sci.*, 34, 1299, 1977.
93. Bigg, E.K., *J. Atmos. Sci.*, 30, 1157, 1973.
94. Rosinski, J., Haagenson, P.L., Nagamoto, C.T., Quintana, B., Parungo, F., and Hoyt, D.S., *J. Aerosol Sci.*, 19, 539, 1988.
95. Nagamoto, C.T., Rosinski, J., Haagenson, P.L., Smak, A.M., and Parungo, F., *J. Aerosol Sci.*, 15, 147, 1984.
96. Bowen, E.G., *Nubila*, 4, 7, 1961.
97. Rosinski, J., Langer, G., and Nagamoto, C.T., *J. Appl. Meteor.*, 11, 405, 1972.
98. Rosinski, J., Nagamoto, C.T., and Young, J.A., *J. Rech. Atmos.*, 6, 693, 1972.
99. Hallett, J. and Mossop, S.C., *Nature*, 249, 26, 1974.
100. Nikandrov, V. Ya., *Tr. Gl. Geofiz. Dbs.*, 31, 93, 1956.
101. Rosinski, J., *J. Rech. Atmos.*, 7, 221, 1973.
102. Vali, G., *J. Atmos. Sci.*, 22, 402, 1971.
103. Rosinski, J. and Nagamoto, C.T., *J. Rech. Atmos.*, 6, 469, 1972.
104. Rosinski, J., Nagamoto, C.T., Kerrigan, T.C., and Langer, G., *J. Atmos. Sci.*, 31, 1459, 1974.
105. Rosinski, J., Langer, G., Nagamoto, C.T., and Bogard, J., *Proceedings, Atmospheric Surface Exchange of Particulate and Gaseous Pollutant — 1974 Symposium*, Richland, Washington, September 4-6, 1974.
106. Rosinski, J., Mancini, G., Nagamoto, C.T., and Prodi, F., *Acta. Meteorologica Sinica*, 12, 382, 1998.
107. Bowen, E.G., *Aust. J. Phys.*, 6, 490, 1953.
108. Bowen, E.G., *J. Met.*, 13, 142, 1956.
109. Rosinski, J. and Snow, R.H., *J. Meteor.*, 18, 736, 1961.
110. Dzienis, B. and Kopcewicz, B., 1973.
111. Kopcewicz, B., 1978.
112. Swinbank, W.C., *Aust. J. Phys.*, 7, 354, 1954.

113. Gobinathan, R. and Ramasamy, P., *Mausam*, 34, 101, 1983.
114. Nagamoto, C.T. and Rosinski, J., *J. Atmos. Terr. Phys.*, 33, 1559, 1971.
115. Rosinski, J., *Il Nuovo Cimento*, 19, 217, 1996.
116. Woodcock, A.H., *J. Meteor.*, 10, 362, 1953.

OTHER RELEVANT PUBLICATIONS

117. Kerrigan, T.C. and Rosinski, J., *Proc. Precipitation Scavenging–1974 Symposium*, Champaign, IL, October 14-16, 1974.
118. Kerrigan, T.C. and Rosinski, J., *Z. Angew. Math. Phys.*, 33, 513, 1982.
119. Kopcewicz, B., Nagamoto, C.T., Parungo, F., Harris, J., Miller, J., Sievering, H., and Rosinski, J., *Atmos. Res.*, 26, 245, 1991.
120. Langer, G., Morgan, G., Nagamoto, C.T., Solak, M., and Rosinski, J., *J. Atmos. Sci.*, 38, 2484, 1979.
121. Langer, G., Rosinski, J., and Bernsen, S., *J. Atmos. Sci.*, 20, 557, 1963.
122. Nagamoto, C.T., Rosinski, J., and Langer, G., *J. Appl. Meteor.*, 6, 1123, 1967.
123. Nagamoto, C.T., Rosinski, J., and Zhou, M.Y., *Atmos. Res.*, 1994.
124. Parungo, F., Nagamoto, C.T., Rosinski, J., and Haagenson, P.L., *J. Aerosol Sci.*, 17, 23, 1986.
125. Parungo, F., Rosinski, J., Wu, M.L.C., Nagamoto, C.T., Zhou, M., and Zhang, N., *Acta Oceanologica Sinica*, 12, 521, 1993.
126. Parungo, F.P., Nagamoto, C.T., Madel, R., Rosinski, J., and Haagenson, P.L., *J. Aerosol Sci.*, 18, 277, 1987.
127. Rosinski, J. and Langer, G., *Powder Tech.*, 1, 167, 1967.
128. Rosinski, J. and Lieberman, A., *Appl. Sci. Res. Sec. A*, 6, 92, 1956.
129. Rosinski, J. and Nagamoto, C.T., *Kolloid-Z*, 204, 111, 1965.
130. Rosinski, J. and Pierrard, J., *J. Atmos. Terr. Phys.*, 24, 1017, 1962.
131. Rosinski, J. and Pierrard, J., *J. Atmos. Terr. Phys.*, 26, 51, 1964.
132. Rosinski, J. and Stockham, J., *Bull. Am. Meteor. Soc.*, 42, 688, 1961.
133. Rosinski, J., Glaess, H.E., and McCully, C.R., *Annal. Chem.*, 28, 486, 1956.
134. Rosinski, J., *J. Appl. Meteor*, 481, 1966.
135. Rosinski, J., *J. Atmos. Terr. Phys.*, 32, 805, 1970.
136. Rosinski, J., *J. Atmos. Terr. Phys.*, 34, 487, 1972.
137. Rosinski, J., Kopcewicz, B., and Sandoval, N., *J. Aerosol Sci.*, 21, 87, 1990.
138. Rosinski, J., Langer, G., and Bleck, R., *J. Atmos. Sci.*, 26, 289, 1969.
139. Rosinski, J., Langer, G., Nagamoto, C.T., Kerrigan, T.C., and Prodi, F., *J. Atmos. Sci.*, 28, 391, 1971.
140. Rosinski, J., Nagamoto, C.T., and Langer, G., *J. Aerosol Sci.*, 10, 555, 1979.
141. Rosinski, J., Nagamoto, C.T., Langer, G., and Parungo, F.P., *J. Geophys. Res.*, 75, 2961, 1970.
142. Rosinski, J., National Center for Atmospheric Research, Boulder, Colorado, Tech. Note STR-71, 63 pp., 1971.
143. Rosinski, J., *Rev. Geophys. Space Phys.*, 12, 129, 1974.
144. Rosinski, J., Werle, D., and Nagamoto, C., *J. Colloid. Sci.*, 17, 703, 1962.

6 Reversible Chemical Reactions in Aerosols

Mark Z. Jacobson

CONTENTS

INTRODUCTION

Aerosols in the atmosphere affect air quality, meteorology, and climate in several ways. Submicron-sized aerosols (smaller than 1 μm in diameter) affect human health by directly penetrating to the deepest part of human lungs. Aerosols between 0.2 and 1.0 μm in diameter that contain sulfate, nitrate, and organic carbon, scatter light efficiently. Aerosols smaller than 1.0 μm that contain elemental carbon, absorb efficiently. Aerosol absorption and scattering are important because they affect radiative fluxes and, therefore, air temperatures and climate. Aerosols also serve as sites on which chemical reactions take place and as sinks in which some gas-phase species are removed from the atmosphere.

The change in size and composition of an aerosol depends on several processes, including nucleation, emissions, coagulation, condensation, dissolution, reversible chemical reactions, irreversible chemical reactions, sedimentation, dry deposition, and advection. In this chapter, dissolution and reversible chemical reactions are discussed. These processes are important for determining the ionic, solid, and liquid water content of aerosols.

0-87371-829-1/00/$0.00+$.50

DEFINITIONS

Dissolution is a process that occurs when a gas, suspended over a particle surface, adsorbs to and dissolves in liquid on the surface. The liquid in which the gas dissolves is a *solvent.* A solvent makes up the bulk of a solution, and in atmospheric particles, liquid water is most often the solvent. In some cases, such as when sulfuric acid combines with water to form particles, the concentration of sulfuric acid exceeds the concentration of liquid water, and sulfuric acid may be the solvent. Here, liquid water is assumed to be the solvent in all cases.

A species, such as a gas or solid, that dissolves in solution is a *solute*. Together, solute and solvent make up a *solution*, which is a homogeneous mixture of substances that can be separated into individual components upon a change of state (e.g., freezing). A solution may contain many solutes. Suspended material (e.g., solids) may also be mixed throughout a solution. Such material is not considered part of a solution.

The ability of the gas to dissolve in water depends on the solubility of the gas in water. *Solubility* is the maximum amount of a gas that can dissolve in a given amount of solvent at a given temperature. Solutions usually contain solute other than the dissolved gas. The solubility of a gas depends strongly on the quantity of the other solutes because such solutes affect the thermodynamic activity of the dissolved gas in solution. Thermodynamic activity is discussed shortly. If water is saturated with a dissolved gas, and if the solubility of the gas changes due to a change in composition of the solution, the dissolved gas can *evaporate* from the solution to the gas phase. Alternatively, dissociation products of the dissolved gas can combine with other components in solution and *precipitate* as solids.

In solution, dissolved gases can dissociate and react chemically. *Dissociation* of a dissolved molecule is the process by which the molecule breaks into simpler components, namely ions. This process can be described by *reversible chemical reactions*, also called *chemical equilibrium reactions* or *thermodynamic equilibrium reactions.* Such reactions are reversible, and their rates in the forward and backward directions are generally fast. Dissociated ions and undissociated molecules can further react reversibly or irreversibly with other ions or undissociated molecules in solution. *Irreversible chemical reactions* act only in the forward direction and are described by first-order ordinary differential equations. When they occur in solution, irreversible reactions are called *aqueous reactions.*

EQUILIBRIUM EQUATIONS AND RELATIONS

Reversible chemical reactions describe dissolution, dissociation, and precipitation processes. In this section, different types of equilibrium equations are discussed and rate expressions, including temperature dependence, are derived.

EQUILIBRIUM EQUATIONS

An *equilibrium equation* describes a reversible chemical reaction. A typical equation has the form

$$\nu_D D + \nu_E E + \ldots \Leftrightarrow \nu_A A + \nu_B B + \ldots, \tag{6.1}$$

where D, E, A, and B are species and the ν's are dimensionless *stoichiometric coefficients* or number of moles per species divided by the smallest number of moles of any reactant or product in the reaction. Each reaction must conserve mass. Thus,

$$\sum_i k_i \nu_i m_i = 0, \tag{6.2}$$

where m_i is the molecular weight of each species and k_i = +1 for products and −1 for reactants. The reactants and/or products of an equilibrium equation can be solids, liquids, ions, or gases.

Reversible dissolution reactions have the form

$$AB(g) \Leftrightarrow AB(aq), \tag{6.3}$$

where (g) indicates a gas and (aq) indicates that the species is dissolved in solution. In this equation, the gas phase and dissolved (solution) phase of species AB are assumed to be in equilibrium with each other at the gas–solution interface. Thus, the number of molecules of AB transferring from the gas to the solution equals the number of molecules transferring in the reverse direction. In the atmosphere, gas–solution interfaces occur at the air–ocean, air–cloud drop, and air–aerosol interfaces. Examples of *dissolution reactions that occur at these interfaces* are

$$HCl(g) \Leftrightarrow HCl(aq) \tag{6.4}$$

$$HNO_3(g) \Leftrightarrow HNO_3(aq) \tag{6.5}$$

$$CO_2(g) \Leftrightarrow CO_2(aq) \tag{6.6}$$

$$NH_3(g) \Leftrightarrow NH_3(aq) \tag{6.7}$$

The reaction

$$H_2SO_4(g) \Leftrightarrow H_2SO_4(aq) \tag{6.8}$$

is also a reversible dissolution reaction. In equilibrium, almost all sulfuric acid is partitioned to the aqueous phase; thus, the relation is rarely used. Instead, sulfuric acid transfer to the aqueous phase is treated as a diffusion-limited condensational growth process.

Once dissolved in solution, the species on the right sides of Equations 6.4 to 6.8 often dissociate into ions. Substances that undergo partial or complete dissociation in solution are *electrolytes*. The degree of dissociation of an electrolyte depends on the acidity of solution, the strength of the electrolyte, the concentrations of other ions in solution, the temperature, and other conditions.

The *acidity* of a solution is a measure of the concentration of *hydrogen ions* (*protons* or H^+ ions) in solution. Acidity is measured in terms of *pH*, defined as

$$pH = -\log_{10}\left[H^+\right], \tag{6.9}$$

where $[H^+]$ is the *molarity* of H^+ (moles H^+ L^{-1} solution). The more acidic the solution, the higher the molarity of protons and the lower the pH. Protons in solution are donated by acids that dissolve. Examples of such acids are H_2CO_3(aq), HCl(aq), HNO_3(aq), and H_2SO_4(aq). The abilities of acids to dissociate into protons and anions vary. HCl(aq), HNO_3(aq), and H_2SO_4(aq) dissociate readily, while H_2CO_3(aq) does not. Thus, the former species are *strong acids* and the latter species is a *weak acid*. Because all acids are electrolytes, a strong acid is a *strong electrolyte* (e.g., it dissociates significantly) and a weak acid is a *weak electrolyte*. Hydrochloric acid is a strong acid and strong electrolyte in water because it almost always dissociates completely by the reaction

$$HCl(aq) \Leftrightarrow H^+ + Cl^-. \tag{6.10}$$

Sulfuric acid is also a strong acid and strong electrolyte and dissociates to bisulfate by

$$H_2SO_4(aq) \Leftrightarrow H^+ + HSO_4^-. \tag{6.11}$$

While HCl(aq) dissociates significantly at a pH above –6, H_2SO_4(aq) dissociates significantly at a pH above –3. Another strong acid, *nitric acid*, dissociates significantly at a pH above –1. Nitric acid dissociates to nitrate by

$$HNO_3(aq) \Leftrightarrow H^+ + NO_3^-. \tag{6.12}$$

Bisulfate is also a strong acid and electrolyte because it dissociates significantly at a pH above about +2. Bisulfate dissociation to *sulfate* is given by

$$HSO_4^- \Leftrightarrow H^+ + SO_4^{2-}. \tag{6.13}$$

Carbon dioxide is a weak acid and electrolyte because it dissociates significantly at a pH above only +6. Carbon dioxide converts to *carbonic acid* and dissociates to bicarbonate by

$$CO_2(aq) + H_2O(aq) \Leftrightarrow H_2CO_3(aq) \Leftrightarrow H^+ + HCO_3^-. \tag{6.14}$$

Dissociation of *bicarbonate* occurs at a pH above +10. This reaction is

$$HCO_3^- \Leftrightarrow H^+ + CO_3^{2-}. \tag{6.15}$$

While acids provide hydrogen ions, *bases* provide *hydroxide ions* (OH^-). Such ions react with hydrogen ions to form neutral water via

$$H_2O(aq) \Leftrightarrow H^+ + OH^-. \tag{6.16}$$

An important base in the atmosphere is ammonia. *Ammonia* reacts with water to form *ammonium* and the hydroxide ion by

$$NH_3(aq) + H_2O(aq) \Leftrightarrow NH_4^+ + OH^-. \tag{6.17}$$

Since some strong electrolytes, such as HCl(aq) and HNO_3(aq), dissociate completely in atmospheric particles, the undissociated forms of these species are sometimes ignored in equilibrium models. Instead, gas-ion equilibrium equations replace the combination of gas-liquid, liquid-ion equations. For example, the equations

$$HCl(g) \Leftrightarrow H^+ + Cl^- \tag{6.18}$$

can replace Equations 6.4 and 6.10. Similarly,

$$HNO_3(g) \Leftrightarrow H^+ + NO_3^- \tag{6.19}$$

can replace Equations 6.5 and 6.12.

Once in solution, ions can precipitate to form *solid electrolytes* if conditions are right. Alternatively, existing solid electrolytes can dissociate into ions if the particle water content increases sufficiently. Examples of *solid precipitation/dissociation reactions* for ammonium-containing electrolytes include

$$NH_4Cl(s) \Leftrightarrow NH_4^+ + Cl^- \tag{6.20}$$

$$NH_4NO_3(s) \Leftrightarrow NH_4^+ + NO_3^- \tag{6.21}$$

$$(NH_4)_2SO_4(s) \Leftrightarrow 2NH_4^+ + SO_4^{2-}. \tag{6.22}$$

Examples of such reactions for sodium-containing electrolytes are

$$NaCl(s) \Leftrightarrow Na^+ + Cl^- \tag{6.23}$$

$$NaNO_3(s) \Leftrightarrow Na^+ + NO_3^- \tag{6.24}$$

$$Na_2SO_4(s) \Leftrightarrow 2Na^+ + SO_4^{2-}. \tag{6.25}$$

If the relative humidity is sufficiently low, a gas can react chemically with another adsorbed gas on a particle surface to form a solid. Such reactions can be simulated with *gas-solid equilibrium reactions*, such as

$$NH_4Cl(s) \Leftrightarrow NH_3(g) + HCl(g) \tag{6.26}$$

$$NH_4NO_3(s) \Leftrightarrow NH_3(g) + HNO_3(g). \tag{6.27}$$

In sum, equilibrium relationships usually describe aqueous-ion, ion-ion, ion-solid, gas-solid, or gas-ion reversible reactions. Relationships can be written for other interactions as well. Table 6.1 shows several equilibrium relationships of atmospheric importance.

Equilibrium Relations and Constants

Species concentrations in a reversible reaction, such as Equation 6.1, are interrelated by

$$\frac{\{A\}^{\nu_A}\{B\}^{\nu_B}\ldots}{\{D\}^{\nu_D}\{E\}^{\nu_E}\ldots} = K_{eq}(T), \tag{6.28}$$

where $K_{eq}(T)$ is a temperature-dependent equilibrium coefficient and {A}..., etc., are *thermodynamic activities*. Thermodynamic activities measure the effective concentration or intensity of the substance. The activity of a substance differs, depending on whether the substance is in the gas, undissociated aqueous, ionic, or solid phases. The activity of a gas is its *saturation vapor pressure* (atm). Thus,

$$\{A(g)\} = p_{s,A}. \tag{6.29}$$

TABLE 6.1
Equilibrium Reactions, Coefficients, and Coefficient Units

No.	Reaction		A	B	C	Units	Ref.[a]
1	$HNO_3(g)$	$\Leftrightarrow HNO_3(aq)$	2.10×10^5			mol kg^{-1} atm^{-1}	D
2	$NH_3(g)$	$\Leftrightarrow NH_3(aq)$	5.76×10^1	13.79	-5.39	mol kg^{-1} atm^{-1}	A
3	$CO_2(g)$	$\Leftrightarrow CO_2(aq)$	3.41×10^{-2}	8.19	-28.93	mol kg^{-1} atm^{-1}	A
4	$CO_2(aq) + H_2O(aq)$	$\Leftrightarrow H^+ + HCO_3^-$	4.30×10^{-7}	−3.08	31.81	mol kg^{-1}	A
5	$NH_3(aq) + H_2O(aq)$	$\Leftrightarrow NH^{4+} + OH^-$	1.81×10^{-5}	−1.50	26.92	mol kg^{-1}	A
6	$HNO_3(aq)$	$\Leftrightarrow H^+ + NO_3^-$	1.20×10^1	29.17	16.83	mol kg^{-1}	N
7	$HCl(aq)$	$\Leftrightarrow H^+ + Cl^-$	1.72×10^6	23.15		mol kg^{-1}	O
8	$H_2O(aq)$	$\Leftrightarrow H^+ + OH^-$	1.01×10^{-14}	−22.52	26.92	mol kg^{-1}	A
9	$H_2SO_4(aq)$	$\Leftrightarrow H^+ + HSO_4^-$	1.00×10^3			mol kg^{-1}	R
10	HSO_4^-	$\Leftrightarrow H^+ + SO_4^{2-}$	1.02×10^{-2}	8.85	25.14	mol kg^{-1}	A
11	HCO_3^-	$\Leftrightarrow H^+ + CO_3^{2-}$	4.68×10^{-11}	−5.99	38.84	mol kg^{-1}	A
12	$HNO_3(g)$	$\Leftrightarrow H^+ + NO_3^-$	2.51×10^6	29.17	16.83	mol^2 kg^{-2} atm^{-1}	A
13	$HCl\ (g)$	$\Leftrightarrow H^+ + Cl^-$	1.97×10^6	30.19	19.91	mol^2 kg^{-2} atm^{-1}	A
14	$NH_3(g) + H^+$	$\Leftrightarrow NH_4^+$	1.03×10^{11}	34.81	−5.39	atm^{-1}	A
15	$NH_3(g) + HNO_3(g)$	$\Leftrightarrow NH_4^+ + NO_3^-$	2.58×10^{17}	64.02	11.44	mol^2 kg^{-2} atm^{-2}	A
16	$NH_3(g) + HCl(g)$	$\Leftrightarrow NH_4^+ + Cl^-$	2.03×10^{17}	65.05	14.51	mol^2 kg^{-2} atm^{-2}	A
17	$NH_4NO_3(s)$	$\Leftrightarrow NH_4^+ + NO_3^-$	1.49×10^1	−10.40	17.56	mol^2 kg^{-2}	A
18	$NH_4Cl(s)$	$\Leftrightarrow NH_4^+ + Cl^-$	1.96×10^1	−6.13	16.92	mol^2 kg^{-2}	A
19	$NH_4HSO_4(s)$	$\Leftrightarrow NH_4^+ + HSO_4^-$	1.38×10^2	−2.87	15.83	mol^2 kg^{-2}	A
20	$(NH_4)_2SO_4(s)$	$\Leftrightarrow 2\ NH_4^+ + SO_4^{2-}$	1.82	−2.65	38.57	mol^3 kg^{-3}	A
21	$(NH_4)_3H(SO_4)_2(s)$	$\Leftrightarrow 3\ NH_4^+ + HSO_4 + SO_4^{2-}$	2.93×10^1	−5.19	54.40	mol^5 kg^{-5}	A
22	$NaNO_3(s)$	$\Leftrightarrow Na^+ + NO_3^-$	1.20×10^1	−8.22	16.01	mol^2 kg^{-2}	A
23	$NaCl(s)$	$\Leftrightarrow Na^+ + Cl^-$	3.61×10^1	−1.61	16.90	mol^2 kg^{-2}	A
24	$NaHSO_4(s)$	$\Leftrightarrow Na^+ + HSO_4^-$	2.84×10^2	−1.91	14.75	mol^2 kg^{-2}	A
25	$Na_2SO_4(s)$	$\Leftrightarrow 2\ Na^+ + SO_4^{2-}$	4.80×10^{-1}	0.98	39.50	mol^3 kg^{-3}	A

Note: The equilibrium coefficient reads,

$$K_{eq}(T) = \mathrm{A}\exp\left\{\mathrm{B}\left(\frac{T_0}{T} - 1\right) + \mathrm{C}\left(1 - \frac{T_0}{T} + \ln\frac{T_0}{T}\right)\right\},$$

where $T_0 = 298.15\mathrm{K}$ and the remaining terms are defined in Equation 6.45.

[a] A: Derived from data in Reference 21; D: From Reference 22; N: Derived from a combination of other rate coefficients in the table; O, R: From Reference 23. With permission.

The activity of an ion in solution or an undissociated electrolyte is its *molality* ($\mathbf{m}_A$) (moles solute kg^{-1} solvent) multiplied by its *activity coefficient* (γ) (unitless). Thus,

$$\{A^+\} = \mathbf{m}_{A^+}\gamma_{A^+} \text{ and} \tag{6.30}$$

$$\{A(\mathrm{aq})\} = \mathbf{m}_A\gamma_A, \tag{6.31}$$

respectively. An *activity coefficient* accounts for the deviation from ideal behavior of a solution. It is a dimensionless parameter by which the molality of a species in solution is multiplied to give the species' thermodynamic activity. In an ideal, infinitely dilute solution, the activity coefficient of a species is unity. In a nonideal, concentrated solution, activity coefficients may be greater than or less than unity. Debye and Huckel showed that, in relatively dilute solutions, where ions are far apart, the deviation of molality from thermodynamic activity is caused by Coulombic (electric) forces of attraction and repulsion. At high concentrations, ions are closer together, and ion-ion interactions affect activity coefficients more significantly than do Coulombic forces.

The activity of liquid water in an atmospheric particle is the ambient relative humidity (fraction). Thus,

$$\{H_2O(aq)\} = a_w = f_r, \tag{6.32}$$

where a_w denotes the activity of water and f_r is the relative humidity, expressed as a fraction. Finally, solids are not in solution, and their concentrations do not affect the molalities or activity coefficients of electrolytes in solution. Thus, the activity of any solid is unity; that is,

$$\{A(s)\} = 1. \tag{6.33}$$

Equation 6.28 is derived by minimizing the Gibbs free-energy change of a system. The Gibbs free-energy change per mole (ΔG) (J mole^{-1}) is a measure of the maximum amount of useful work per mole that may be obtained from a change in enthalpy or entropy in the system. The relationship between the Gibbs free-energy change and the composition of a chemical system is

$$\Delta G = \sum_i k_i \nu_i \mu_i, \tag{6.34}$$

where μ_i is the chemical potential of the species (J mole^{-1}) and k_i = +1 for products and –1 for reactants. *Chemical potential* is a measure of the intensity of a chemical substance and is a function of temperature and pressure. It is really a measure of the change in free energy per change in moles of a substance, or the partial molar free energy. The chemical potential is

$$\mu_i = \mu_i^{\circ} + R^* T \ln\{a_i\}, \tag{6.35}$$

where μ_i^o is the chemical potential at a reference temperature of 298.15K, and $\{a_i\}$ is the thermodynamic activity of species i. The chemical potential can be substituted into Equation 6.34 to give

$$\Delta G = \sum_i \left(k_i \nu_i \mu_i^{\circ}\right) + R^* T \sum_i \left(k_i \nu_i \ln\{a_i\}\right). \tag{6.36}$$

Rewriting this equation yields

$$\Delta G = \Delta G^{\circ} + R^* T \ln \prod_i \{a_i\}^{k_i \nu_i}, \tag{6.37}$$

where

$$\Delta G^{\circ} = \sum_i \left(k_i \nu_i \mu_i^{\circ}\right) \tag{6.38}$$

is the *standard molal Gibbs free energy of formation* (J mole[-1]) for the reaction. Equilibrium occurs when $\Delta G = 0$ at constant temperature and pressure. Under such conditions, Equation 6.37 becomes

$$\exp\left[-\Delta G^{\circ}/\left(R^{*}T\right)\right] = \prod_{i}\{a_i\}^{\nu_i}. \tag{6.39}$$

The left side of Equation 6.39 is the *equilibrium coefficient*. Thus,

$$K_{eq}(T) = \exp\left[-\Delta G^{\circ}/\left(R^{*}T\right)\right]. \tag{6.40}$$

Substituting Equation 6.40 into Equation 6.39 and expanding the product term gives

$$K_{eq}(T) = \prod_{i}\{a_i\}^{k_i\nu_i} = \frac{\{A\}^{\nu_A}\{B\}^{\nu_B}\ldots}{\{D\}^{\nu_D}\{E\}^{\nu_E}\ldots} \tag{6.41}$$

which is the relationship shown in Equation 6.28.

TEMPERATURE DEPENDENCE OF THE EQUILIBRIUM COEFFICIENT

The temperature dependence of the equilibrium coefficient is calculated by solving the *Van't Hoff equation*,

$$\frac{\mathrm{d}\ln K_{eq}(T)}{\mathrm{d}T} = \frac{\Delta H_T^{\circ}}{R^{*}T^{2}}, \tag{6.42}$$

where ΔH_T^o *is* the change in total enthalpy (J mole[-1]) of the reaction. The *change in enthalpy* can be approximated by

$$\Delta H_T^{\circ} \approx \Delta H_{T_o}^{\circ} + \Delta c_p^{\circ}\left(T - T_o\right) \tag{6.43}$$

when the *standard change in molal heat capacity of the reaction* ($\Delta c_P{}^o$) (J mole[-1] K[-1]) does not depend on temperature. In this equation, $\Delta H_{T_o}^{\circ}$ is the standard enthalpy change in the reaction (J mole[-1]) at temperature $T_o = 298.15$K. Combining Equations 6.42 and 6.43 and writing the result in integral form gives

$$\int_{T_o}^{T} \mathrm{d}\ln K_{eq}(T) = \int_{T_o}^{T}\left[\frac{\Delta H_{T_o}^{\circ} + \Delta c_p^{\circ}\left(T - T_o\right)}{R^{*}T^{2}}\right]\mathrm{d}T. \tag{6.44}$$

Integrating this equation yields the temperature-dependent equilibrium coefficient expression

$$K_{eq}(T) = K_{eq}\left(T_o\right)\exp\left[-\frac{\Delta H_{T_o}^{\circ}}{R^{*}T_o}\left(\frac{T_o}{T} - 1\right) - \frac{\Delta c_p^{\circ}}{R^{*}}\left(1 - \frac{T_o}{T} + \ln\left(\frac{T_o}{T}\right)\right)\right] \tag{6.45}$$

where $K_{eq}(T_o)$ is the equilibrium coefficient at temperature, T_0. Values of $\Delta H^{\circ}_{T_0}$ and $\Delta c_p{}^o$ are measured experimentally. Table 6.1 shows temperature-dependent parameters for several equilibrium reactions.

FORMS OF EQUILIBRIUM COEFFICIENT EQUATIONS

Each reaction in Table 6.1 can be written in terms of thermodynamic activities and an equilibrium coefficient. For example, an equilibrium coefficient equation for the reaction

$$HNO_3(g) \Leftrightarrow HNO_3(aq) \tag{6.46}$$

is

$$\frac{\{HNO_3(aq)\}}{\{HNO_3(g)\}} = \frac{\mathbf{m}_{HNO_3(aq)}\gamma_{HNO_3(aq)}}{p_{s,HNO_3(g)}} = K_{eq}(T), \tag{6.47}$$

where $p_{s,HNO_3(g)}$ is the saturation vapor pressure of nitric acid (atm), $\mathbf{m}_{HNO_3(aq)}$ is the molality of nitric acid in solution (moles kg^{-1}), and $\gamma_{HNO_3(aq)}$ is the activity coefficient of dissolved, undissociated nitric acid (unitless). The equilibrium coefficient has units of (moles kg^{-1} atm^{-1}).

When the equilibrium coefficient relates the saturation vapor pressure of a gas to the molality (or molarity) of the dissolved gas in a dilute solution, the coefficient is called a *Henry's constant*. Henry's constants (moles kg^{-1} atm^{-1}), like other equilibrium coefficients, are temperature and solvent dependent. *Henry's law* states that, for a dilute solution, the pressure exerted by a gas at the gas–liquid interface is proportional to the molality of the dissolved gas in solution. For a dilute solution, $\gamma_{HNO_3(aq)} = 1$, and Equation 6.47 obeys Henry's law.

A dissociation equation has the form

$$HNO_3(aq) \Leftrightarrow H^+ + NO_3{}^-. \tag{6.48}$$

The equilibrium coefficient expression for this reaction is

$$\frac{\{H^+\}\{NO_3^-\}}{\{HNO_3(aq)\}} = \frac{\mathbf{m}_{H^+}\gamma_{H^+}\mathbf{m}_{NO_3^-}\gamma_{NO_3^-}}{\mathbf{m}_{HNO_3(aq)}\gamma_{HNO_3(aq)}} = \frac{\mathbf{m}_{H^+}\mathbf{m}_{NO_3^-}\gamma^2_{H^+,NO_3^-}}{\mathbf{m}_{HNO_3(aq)}\gamma_{HNO_3(aq)}} \tag{6.49}$$

where the equilibrium coefficient has units of (moles kg^{-1}).

In Equation 6.49, the activity coefficients are determined by considering a mixture of all dissociated and undissociated electrolytes in solution. Thus, the coefficients are termed *mixed activity coefficients*. More specifically, γ_{H^+} and $\gamma_{NO_3^-}$ are *single-ion mixed activity coefficients*, and γ_{H^+,NO_3^-} is a *mean (geometric mean) mixed activity coefficient*. When H^+, and $NO_3{}^-$ are alone in solution, γ_{H^+} and $\gamma_{NO_3^-}$ are *single-ion binary activity coefficients*, and γ_{H^+,NO_3^-} is a *mean (geometric mean) binary activity coefficient*. Activity coefficients for single ions are difficult to measure because single ions cannot be isolated from a solution. Single-ion activity coefficients are easier to estimate mathematically. Mean binary activity coefficients are measured in the laboratory. Mean mixed activity coefficients can be estimated from mean binary activity coefficient data through a mixing rule.

A geometric mean activity coefficient is related to a single-ion activity coefficient by

$$\gamma_{\pm} = \left(\gamma_{+}^{\nu_+}\gamma_{-}^{\nu_-}\right)^{1/(\nu_+ + \nu_-)}, \tag{6.50}$$

where $\gamma_{\pm}$ is the mean activity coefficient, γ_+ and γ_- are the activity coefficients of the single cation and anion, respectively, and ν_+ and ν_- are the stoichiometric coefficients of the cation and anion, respectively. In Equation 6.48, $\nu_+ = 1$ and $\nu_- = 1$.

Raising both sides of Equation 6.50 to the power $\nu_+ + \nu_-$ gives

$$\gamma_{\pm}^{(\nu_+ + \nu_-)} = \gamma_{+}^{\nu_+}\gamma_{-}^{\nu_-}, \tag{6.51}$$

which is form of the mean activity coefficient used in Equation 6.49.

When $\nu_+ = 1$ and $\nu_- = 1$, the electrolyte is *univalent*. When $\nu_+ > 1$ or $\nu_- > 1$, the electrolyte is *multivalent*. When $\nu_+ = \nu_-$ for a dissociated electrolyte, the electrolyte is *symmetric*; otherwise, it is *nonsymmetric*. In all cases, a dissociation reaction must satisfy the charge balance requirement

$$z_+\nu_+ + z_-\nu_- = 0, \tag{6.52}$$

where z_+ is the positive charge on the cation and z_- is the negative charge on the anion.

MEAN BINARY ACTIVITY COEFFICIENTS

The mean binary activity coefficient of an electrolyte, which is primarily a function of molality and temperature, can be determined from measurements or estimated from theory. Measurements of binary activity coefficients for several species at 298.15K are available. Parameterizations have also been developed to predict the mean binary activity coefficients. One parameterization is *Pitzer's method*,[1,2] which estimates the mean binary activity coefficient of an electrolyte at 298.15K with

$$\ln\gamma_{12b}^{0} = Z_1Z_2f^{\gamma} + \mathbf{m}_{12}\frac{2\nu_1\nu_2}{\nu_1+\nu_2}B_{12}^{\gamma} + \mathbf{m}_{12}^{2}\frac{2(\nu_1\nu_2)^{3/2}}{\nu_1+\nu_2}C_{12}^{\gamma} \tag{6.53}$$

where γ^0_{12b} is the mean binary activity coefficient of electrolyte 1-2 (cation 1 plus anion 2) at the reference temperature (298.15K), Z_1 and Z_2 are the absolute value of the charges of cation 1 and anion 2, respectively, $\mathbf{m}_{12}$ is the molality of electrolyte dissolved in solution, and ν_1 and ν_2 are the stoichiometric coefficients of the dissociated ions (assumed positive here). In addition,

$$f^{\gamma} = -0.392\left[\frac{I^{1/2}}{1+1.2I^{1/2}} + \frac{2}{1.2}\ln\left(1+1.2I^{1/2}\right)\right] \tag{6.54}$$

$$B_{12}^{\gamma} = 2\beta_{12}^{(1)} + \frac{2\beta_{12}^{(2)}}{4I}\left[1 - e^{-2I^{1/2}}\left(1+2I^{1/2}-2I\right)\right], \tag{6.55}$$

TABLE 6.2
Pitzer Parameters for Three Electrolytes

Electrolyte	$\beta_{12}^{(1)}$	$\beta_{12}^{(2)}$	$C_{12}^{(\gamma)}$
HCl	0.17750	0.2945	0.0012
HNO_3	0.1119	0.3206	0.0015
NH_4NO_3	–0.0154	0.112	–0.000045

Source: From Reference 13. With permission..

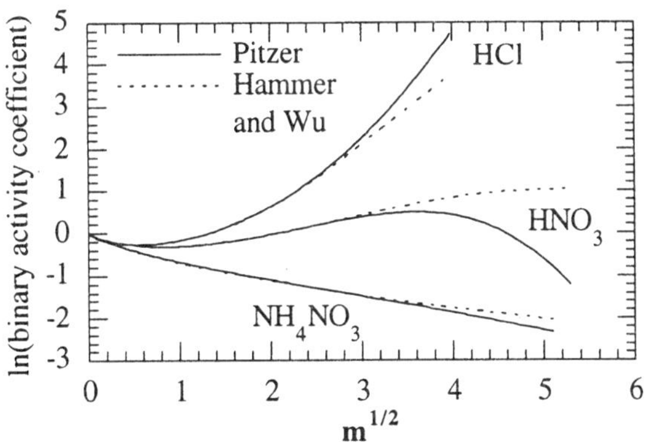

FIGURE 6.1 Comparison of binary activity coefficient data measured by Hamer and Wu[3] to those computed using Pitzer's method.[3] (From Reference 18. With permission.)

where I is the ionic strength of the solution (moles kg^{-1}). The *ionic strength* is a measure of the interionic effects resulting from attraction and repulsion among ions and is given by

$$I = \frac{1}{2}\left[\sum_{i=1}^{N_C} \mathbf{m}_{2i-1} Z_{2i-1}^2 + \sum_{i=1}^{N_A} \mathbf{m}_{2i} Z_{2i}^2\right] \tag{6.56}$$

In this equation, N_C is the number of different cations, N_A is the number of different anions in solution, odd-numbered subscripts refer to cations, and even-numbered subscripts refer to anions. In the case of one electrolyte, such as HCl(aq) alone in solution, $N_C = 1$ and $N_A = 1$. The quantites, $\beta_{12}^{(1)}$, $\beta_{12}^{(2)}$, and $C_{12}^{(\gamma)}$ are empirical parameters derived from measurements. Pitzer parameters for three electrolytes are shown in Table 6.2.

While Pitzer's method accurately predicts mean binary activity coefficients at 298.15K from physical principles, its limitation is that the coefficients are typically valid up to about 6 molal (m) only. Figure 6.1 shows a comparison of activity coefficients predicted by Pitzer's method to those measured by Hamer and Wu.[3] The measured data are accurate to higher molalities.

Whether molality-dependent mean binary activity coefficients at 298.15K are determined from measurements or theory, they can be parameterized with a *polynomial fit* of the form

$$\ln \gamma_{12b}^{0} = B_0 + B_1 \mathbf{m}_{12}^{1/2} + B_2 \mathbf{m}_{12} + B_3 \mathbf{m}_{12}^{3/2} + \ldots. \tag{6.57}$$

where B_0, B_1,... are fitting coefficients. Coefficients for several electrolytes are given by Jacobson et al.[4] Polynomial fits are used to simplify and speed up the use of binary activity coefficient data in computer programs.

TEMPERATURE DEPENDENCE OF MEAN BINARY ACTIVITY COEFFICIENTS

The temperature dependence of solute mean binary activity coefficients can be derived[5] from thermodynamic principles as

$$\ln \gamma_{12b}(T) = \ln \gamma^0_{12b} + \frac{T_L}{(\nu_1 + \nu_2)R^* T_0}\left[\phi_L + \mathbf{m}\frac{\partial \phi_L}{\partial \mathbf{m}}\right] + \frac{T_C}{(\nu_1 + \nu_2)R^*}\left[\phi_{c_p} + \mathbf{m}\frac{\partial \phi_{c_p}}{\partial \mathbf{m}} - \phi^{\text{o}}_{c_p}\right], \tag{6.58}$$

where $\gamma_{12b}(T)$ is the binary activity coefficient of electrolyte 1-2 at temperature T, T_0 is the reference temperature (298.15K), R^* is the gas constant (J mole^{-1} K^{-1}), ϕ_L is the relative apparent molal enthalpy (J mole^{-1}) of the species at molality **m** (with subscript, 12, omitted), ϕ_L is the apparent molal heat capacity (J mole^{-1} K^{-1}) at molality **m**, and $\phi^{\text{o}}_{c_p}$ is the apparent molal heat capacity at infinite dilution. Further,

$$T_L = \frac{T_0}{T} - 1 \text{ and} \tag{6.59}$$

$$T_C = 1 + \ln\left(\frac{T_0}{T}\right) - \frac{T_0}{T} \tag{6.60}$$

are temperature-dependent parameters.

The relative apparent molal enthalpy equals the negative of the heat of dilution, ΔH_D. With heat of dilution and apparent molal heat capacity data, polynomials of the form

$$\phi_L = U_1\mathbf{m}^{1/2} + U_2\mathbf{m} + U_3\mathbf{m}^{3/2} + \ldots \tag{6.61}$$

and

$$\phi_{c_p} = \phi^{\text{o}}_{c_p} + V_1\mathbf{m}^{1/2} + V_2\mathbf{m} + V_3\mathbf{m}^{3/2} + \ldots \tag{6.62}$$

can be constructed. Apparent relative molal enthalpy is defined as $\phi_L = \phi_H - \phi^0_H$, where ϕ_H is the *apparent molal enthalpy* and ϕ^0_H is the *apparent molal enthalpy at infinite dilution,* which occurs when $\mathbf{m} = 0$. Equations 6.58, 6.61, and 6.61 can be combined to give temperature-dependent, mean binary activity coefficient polynomials of the form

$$\ln \gamma_{12b}(T) = F_0 + F_1\mathbf{m}^{1/2} + F_2\mathbf{m} + F_3\mathbf{m}^{3/2} + \ldots, \tag{6.63}$$

where $F_0 = B_0$ and

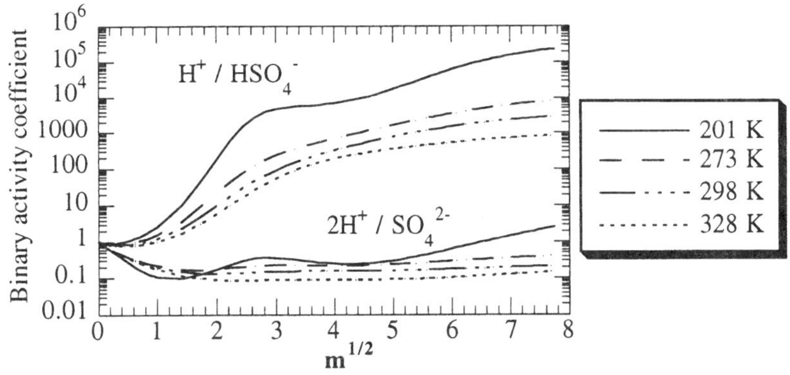

FIGURE 6.2 Binary activity coefficients of sulfate and bisulfate, each alone in solution. Results are valid for 0 to 40 **m** total H_2SO_4. (From Reference 18. With permission.)

$$F_j = B_j + G_j T_L + H_j T_C \tag{6.64}$$

for each additional term, beginning with $j = 1$. In Equation 6.64,

$$G_j = \frac{0.5(j+2)U_j}{(\nu_1 + \nu_2)R^* T_0} \quad \text{and} \tag{6.65}$$

$$H_j = \frac{0.5(j+2)V_j}{(\nu_1 + \nu_2)R^*}. \tag{6.66}$$

With sufficient data, many temperature- and molality-dependent mean binary activity coefficients can be written in terms of Equations 6.63 and 6.64. Jacobson et al.[4] list *B*, *G*, and *H* values for 10 electrolytes and the range of validity for data.

Determining the temperature-dependent binary activity coefficients of bisulfate and sulfate is more difficult. They can be found by combining equations from the model of Clegg and Brimblecombe[6] with Equations 6.72 and 6.73 of Stelson et al.[7] in a Newton-Raphson iteration. Figure 6.2 shows results tabulated for the temperature range <201 to 328K and the molality range 0 to 40 m.

MEAN MIXED ACTIVITY COEFFICIENTS

The mean binary activity coefficients described by Equations 6.57, 6.58, and 6.63 were defined under the assumption that an electrolyte dissociated alone in a solution. In atmospheric particles, several electrolytes usually coexist in solution. For example, dissolved sulfuric acid, nitric acid, hydrochloric acid, ammonia, and sodium chloride often exist together. In such cases, activity coefficients are approximated with an empirical mixing rule that accounts for interactions among ions. One such rule is *Bromley's method*,[8] which gives the *activity coefficient* of electrolyte 1-2 in a mixture as

$$\log_{10} \gamma_{12m}(T) = -A_\gamma \frac{Z_1 Z_2 I_{\mathbf{m}}^{1/2}}{1 + I_{\mathbf{m}}^{1/2}} + \frac{Z_1 Z_2}{Z_1 + Z_2}\left[\frac{W_1}{Z_1} + \frac{W_2}{Z_2}\right], \tag{6.67}$$

where A_γ is the *Debye-Huckel parameter* (0.392 at 298K), Z_1 and Z_2 are the absolute values of charge of cation 1 and anion 2, respectively, I_m is the total ionic strength of the mixture, and W_1 and W_2 are functions of all electrolytes in solution. W_1 and W_2 are

$$W_1 = Y_{21}\left[\log\gamma_{12b}(T) + A_\gamma \frac{Z_1Z_2I_{\mathrm{m}}^{1/2}}{1+I_{\mathrm{m}}^{1/2}}\right] + Y_{41}\left[\log\gamma_{14b}(T) + A_\gamma \frac{Z_1Z_4I_{\mathrm{m}}^{1/2}}{1+I_{\mathrm{m}}^{1/2}}\right] + \ldots + \tag{6.68}$$

$$W_2 = X_{12}\left[\log\gamma_{12b}(T) + A_\gamma \frac{Z_1Z_2I_{\mathrm{m}}^{1/2}}{1+I_{\mathrm{m}}^{1/2}}\right] + X_{32}\left[\log\gamma_{32b}(T) + A_\gamma \frac{Z_3Z_2I_{\mathrm{m}}^{1/2}}{1+I_{\mathrm{m}}^{1/2}}\right] + \ldots + \tag{6.69}$$

where

$$Y_{21} = \left(\frac{Z_1+Z_2}{2}\right)^2 \frac{\mathbf{m}_{2,\mathrm{m}}}{I_{\mathrm{m}}}, \tag{6.70}$$

$$X_{12} = \left(\frac{Z_1+Z_2}{2}\right)^2 \frac{\mathbf{m}_{1,\mathrm{m}}}{I_{\mathrm{m}}}. \tag{6.71}$$

Similar expressions are written for X_{32}, X_{52} ..., etc. Y_{41}, Y_{61}..., etc. In these equations, $\gamma_{12b}(T)$, $\gamma_{14b}(T)$, $\gamma_{32b}(T)$..., etc. are temperature-dependent mean binary activity coefficients, and odd-numbered subscripts refer to cations while even-numbered subscripts refer to anions. For example, $\mathbf{m}_{1,m}$ and, $\mathbf{m}_{2,m}$ are molalities in the mixture of a cation and anion, respectively.

THE WATER EQUATION

Interaction between solvent and solute in solution is *solvation.* An example of solvation is when a solvent bonds to a cation, anion, or nonelectrolyte (such as sucrose) in solution. When the solvent is liquid water, the bonding is *hydration.* During hydration of a cation, the lone pair of electrons on the oxygen atom of a water molecule bonds to the cation-end of the dipole. During hydration of an anion, the water molecule attaches to the anion-end of the dipole via hydrogen bonding. Several water molecules can hydrate to each ion.

When liquid water molecules bond to ions in solution, water vapor condenses to maintain saturation over the solution surface, increasing the liquid water content. Liquid water content is a unique function of electrolyte molality and sub-100% relative humidity. As the relative humidity increases up to 100%, hydration increases the aerosol liquid water content. The liquid water content also increases with increasing solute molality in solution. Above 100% relative humidity, particles grow rapidly by condensation. When particles are large and dilute, the volume of water added to them by hydration is small compared to the volume of water already present. Thus, hydration does not affect water content much when the relative humidity exceeds 100%.

At ambient relative humidities below 100%, an important aspect of modeling aerosols is determining their liquid water content as a function of electrolyte concentration. A convenient parameterization of aerosol liquid water content is the Zdanovskii-Stokes-Robinson (*ZSR*) equation.[9] The equation can be applied to electrolytes or nonelectrolytes. The simplest form of the equation, for two species x and y is

TABLE 6.3
Demonstration of the ZSR Equation Prediction Accuracy for a Sucrose (Species *a*) — Mannitol (Species *b*) Mixture at Two Different Water Activities

$\mathbf{m}_{x,a}$	$\mathbf{M}_{y,a}$	$\mathbf{m}_{x,m}$	$\mathbf{m}_{y,m}$	$\frac{\mathbf{m}_{x,m}}{\mathbf{m}_{x,a}}+\frac{\mathbf{m}_{y,m}}{\mathbf{m}_{y,a}}$
0.7751	0.8197	0.6227	0.1604	0.9990
0.9393	1.0046	0.1900	0.8014	1.0000

Source: From Stokes, R.H. and Robinson, R.A., *J. Phys. Chem.*, 70, 2126, 1966. With permission.

$$\frac{\mathbf{m}_{x,m}}{\mathbf{m}_{x,a}}+\frac{\mathbf{m}_{y,m}}{\mathbf{m}_{y,a}}=1, \tag{6.72}$$

where, $\mathbf{m}_{x,a}$ and $\mathbf{m}_{y,a}$ are the molalities of x and y, alone in solution at a given water activity, while $\mathbf{m}_{x,m}$ and $\mathbf{m}_{y,m}$ are the molalities of x and y, when mixed together, at the same water activity. *Water activity* is redefined as

$$\left\{\mathrm{H_2O(aq)}\right\}=a_w=f_r=\frac{p_{s,c,\mathrm{H_2O}}}{p_{s,d,\mathrm{H_2O}}}, \tag{6.73}$$

where $p_{s,c,\mathrm{H_2O}}$ is the saturation vapor pressure of water over a pure (dilute) liquid water surface, and $p_{s,c,\mathrm{H_2O}}$ is the saturation vapor pressure of water over a liquid water solution containing solute. The latter term is always smaller than the former term because, when a solute hydrates, it binds liquid water, requiring vapor to condense to replace the hydrated liquid water, reducing the vapor-phase concentration of water. In the dilute solution case, the vapor-phase concentration of water is not reduced.

The mixed and binary molalities in Equations 6.72 differ from each other because, in a mixture, the quantity and type of ions differ from in a binary solution; thus, a different quantity of water is hydrated in each case. Table 6.3 gives mixed and binary molalities of sucrose and mannitol alone and mixed together in water. The table also shows that, when the molalities are applied to Equation 6.72, the equation is satisfied.

Equation 6.72 can be generalized for a mixture with any number of components by

$$\sum_k \frac{\mathbf{m}_{k,m}}{\mathbf{m}_{k,a}}=1, \tag{6.74}$$

where the summation is over all solutes in solution, $\mathbf{m}_{k,m}$ is the molality of solute k in a solution containing all solutes at the ambient water activity (moles kg^{-1}), and $\mathbf{m}_{k,a}$ is the molality of solute k as if it were alone in solution at the ambient water activity (moles kg^{-1}). For atmospheric aerosols, this equation is rewritten as

$$c_w = \frac{1000}{\mathbf{m}_v} \sum_k \frac{c_{k,m}}{\mathbf{m}_{k,a}}, \tag{6.75}$$

where c_w is the liquid water content of particles in units of mole concentration (moles $H_2O(aq)$ cm^{-3} air), m_v is the molecular weight of water (g $mole^{-1}$), $c_{k,m}$ is the mole concentration (moles cm^{-3} air) of solute k in a solution containing all solutes at the ambient water activity, and 1000 converts g to kg.

Experimental data for water activity as a function of binary electrolyte molality are available (e.g., see References 1, 10, and 11). Such data can also be fit to polynomial expressions of the form

$$\mathbf{m}_a^{1/2} = Y_0 + Y_1 a_w + Y_2 a_w^2 + Y_3 a_w^3 + ..., \tag{6.76}$$

where a_w is the water activity (relative humidity expressed as a fraction), $\mathbf{m}_a$ is the molality of an electrolyte alone in solution, and the Y values are polynomial coefficients. Jacobson et al.[4] list sets of Y values for 12 electrolytes.

In comparison to the temperature dependence of binary solute activity coefficients, the temperature dependence of binary water activity coefficients under ambient surface conditions is relatively small. The temperature dependence of water activity can be rewritten from Harned and Owen[5] as

$$\ln a_w(T) = \ln a_w^0 - \frac{m_v \mathbf{m}^2}{1000R^*}\left[\frac{T_L}{T_0}\frac{\partial \phi_L}{\partial \mathbf{m}} + T_C \frac{\partial \phi_{cP}}{\partial \mathbf{m}}\right]. \tag{6.77}$$

If the water activity at the reference temperature is expressed as

$$\ln a_w^0 = A_0 + A_1\mathbf{m}^{1/2} + A_2\mathbf{m} + A_3\mathbf{m}^{3/2} + ... + ..., \tag{6.78}$$

then Equations 6.77, 6.78, 6.61, and 6.62 can be combined to form

$$\ln a_w(T) = A_0 + A_1\mathbf{m}^{1/2} + A_2\mathbf{m} + E_3\mathbf{m}^{3/2} + E_4\mathbf{m}^2 ...., \tag{6.79}$$

where

$$E_l = A_l - \frac{0.5(l-2)m_v}{1000R^*}\left[\frac{T_L}{T_0}U_{l-2} + T_C V_{l-2}\right] \tag{6.80}$$

for each l greater than 2. Equation 6.79 shows that temperature affects the water-activity polynomial beginning only in the fourth term. In Equation 6.79, temperature affected the solute activity beginning with the second term of the polynomial. These equations indicate that the effect of temperature on water activity is usually less than that on solute activity. At high molalities (above 10 **m**) and at ambient surface temperatures (273–310K), temperature affects water activity only slightly. For example, at 16 **m**, HCl gives binary water activities of 0.09 at T = 273K and 0.11 at 310K. At lower molalities, temperature has even less of an effect.

In an atmospheric model containing mixed aerosols, the water equation is rearranged from Equation 6.74 to

$$c_w = \frac{1000}{m_v} \sum_{i=1}^{N_C} \left(\sum_{j=1}^{N_A} \frac{c_{i,j,m}}{\mathbf{m}_{i,j,a}} \right), \tag{6.81}$$

where binary molalities of species alone in solution ($\mathbf{m}_a$) are obtained from Equation 6.76 at the given relative humidity. In this equation, i,j is an electrolyte pair (where the odd/even subscripts used previously are ignored), and c is the hypothetical mole concentration of the pair when mixed in solution with all other components. In a model, hypothetical mole concentrations of electrolyte pairs are not usually known; instead, mole concentrations of individual ions are. Thus, individual ions must be combined into electrolyte pairs for Equation 6.81 to be solved.

METHOD OF SOLVING EQUILIBRIUM EQUATIONS

Equilibrium equations, activity coefficient equations, and the water equation are often solved together in an atmospheric model to estimate particle composition, including liquid water content. One method of solving these equations is with a Newton-Raphson iteration (e.g., see Reference 12). Other methods are the bisectional-Newton method (e.g., see References 13 and 14) and a method[15,16] that minimizes free energy. These methods require iteration and are mass and charge conserving.

Another method used to solve equilibrium problems is a mass flux iteration (MFI) method.[4,17-19] This method can converge thousands of equilibrium equations simultaneously, cannot produce negative concentrations, and is mass- and charge-conserving at all times. The only constraints are that the equilibrium equations must be mass- and charge-conserving, and the system must start in charge balance. For example, the equation $HNO_3(aq) = H^+ + NO_3^-$ conserves mass and charge. The charge balance constraint allows initial charges to be distributed among all dissociated ions, but the initial sum, over all species, of charge multiplied by molality must equal zero. The simplest way to initialize charge is to set all ion molalities to zero. Initial mass in the system can be distributed arbitrarily, subject to the charge balance constraint. If the total nitrate in the system is known to be, say, 20 $\mu g\ m^{-3}$, the nitrate can initially be distributed in any proportion among $HNO_3(aq)$, NO_3^-, $NH_4NO_3(s)$, etc.

The MFI method requires the solution of one equilibrium equation at a time by iteration. A system of equations is solved by iterating all equations many times. Suppose a system consists of a single aerosol size bin and 15 equations representing the equilibrium chemistry within that bin. At the start, the first equation is iterated. When the first equation converges, the updated and other initial concentrations are used as inputs into the second equation. This continues until the last equation has converged. At that point, the first equation is no longer converged, because the concentrations used in it have changed. The iteration sequence must be repeated over all equations several times until the concentrations no longer change upon more iteration.

Equilibrium among multiple particle size bins and the gas phase is solved in a similar manner. Suppose a system consists of several size bins, equations per bin, and gases that equilibrate with dissolved molecules in each bin. Each gas' saturation vapor pressure over a particle surface is assumed to equal the gas' partial pressure, which is a single value. In reality, the saturation vapor pressure differs over every particle surface. In order to account for variations in saturation vapor pressure over particle surfaces, nonequilibrium gas-aerosol transfer equations must solved.

Gas-particle equilibrium over multiple size bins is solved by iterating each equilibrium equation, including gas-solution equations, starting with the first size bin. Updated gas concentrations from the first bin affect the equilibrium distribution in subsequent bins. After the last size bin has been iterated, the sequence is repeated in reverse order (to speed convergence), from the last to first size bin. The marches back and forth among size bins continue until gas and aerosol concentrations do not change upon more iteration.

To demonstrate the solution to one equilibrium equation, an example where two gases equilibrate with two ions is shown. The sample equation has the form of Equation 6.1, with two gases on the left side of the equation. The first step is to calculate Q_d and Q_n, the smallest ratio (MIN) of mole concentration to moles among species appearing in the denominator and numerator, respectively, of Equation 6.28. Thus,

$$Q_d = \text{MIN}\left\{\frac{C_{D,1}}{\nu_D}, \frac{C_{E,1}}{\nu_E}\right\} \tag{6.82}$$

$$Q_n = \text{MIN}\left\{\frac{c_{A,1}}{\nu_A}, \frac{c_{B,1}}{\nu_B}\right\}, \tag{6.83}$$

where the subscript "1" refers to initial concentration. Initial concentrations can be selected arbitrarily with the requirement that mole concentrations (moles cm^{-3}) of all individual species in a mole-balance group must sum up to the total moles in the group. If an equilibrium equation contains a solid, each solid's concentration is included in Equation 6.82 or Equation 6.83.

Second, two parameters are initialized as $z_1 = 0.5(Q_d + Q_n)$ and $\Delta x_1 = Q_d - z_1$. The iteration begins by adding the mass flux factor (Δx, which may be positive or negative) to each mole concentration in the numerator, or subtracting it from each mole concentration in the denominator of the equilibrium equation. Thus,

$$c_{A,l+1} = c_{A,l} + \nu_A \Delta x_l, \qquad c_{B,l+1} = c_{B,l} + \nu_B \Delta x_l, \tag{6.84}$$

$$C_{D,l+1} = C_{D,l} - \nu_D \Delta x_l, \qquad C_{E,l+1} = C_{E,l} - \nu_E \Delta x_l, \tag{6.85}$$

respectively. Starting with Equation 6.84, iteration numbers are referred to by subscripts l and l + 1. If the equation contain solids, then the change in each solid's concentration is calculated with Equation 6.84 or 6.85 (solid, aqueous, and ionic mole concentrations are all identified with a c). The above equations show that, during each iteration, mass and charge are transferred either from reactants to products or vice versa. This transfer continues until $\Delta x = 0$. Thus, the scheme conserves mass and charge each iteration.

Third, the ratio of activities is compared to the equilibrium coefficient. The ratio is

$$F = \left[\frac{\left(\mathbf{m}_{A,l+1}\right)^{\nu_A}\left(\mathbf{m}_{B,l+1}\right)^{\nu_B}\left(\gamma_{AB,l+1}\right)^{\nu_A+\nu_B}}{\left(p_{D,l+1}\right)^{\nu_D}\left(p_{E,l+1}\right)^{\nu_E}}\right] \Big/ K_{eq}(T). \tag{6.86}$$

To perform this calculation, mole concentrations are converted to units of either molality or atmospheres. In the case of solids, the activities are unity; thus, none appears in Equation 6.86. Further, mean mixed activity coefficients ($e.g., \gamma_{AB,l+1}$) are updated before each iteration sequence. They converge after all iteration sequences are complete. Finally, the liquid water content (c_w) is updated either during or before each iteration sequence.

The fourth step in the process is to recalculate z for the next iteration. Thus,

$$z_{l+1} = 0.5 z_l \tag{6.87}$$

Finally, convergence is checked with the *convergence criterion*:

$$F = \begin{cases} > 1 & \rightarrow \quad \Delta x_{l+1} = -z_{l+1} \\ < 1 & \rightarrow \quad \Delta x_{l+1} = +z_{l+1} \\ = 1 & \rightarrow \quad \text{Convergence} \end{cases}. \tag{6.88}$$

Each nonconvergence, Δx is updated, the iteration number is advanced, and the code returns to (84). Ultimately, all molalities converge to positive numbers.

SOLID FORMATION AND DELIQUESCENCE RELATIVE HUMIDITY

Insoluble solids can form within a particle by precipitation or on its surface by chemical reaction. *Precipitation* is defined as the formation of an insoluble compound from solution and can be simulated as a reversible equilibrium process, such as

$$NH_4NO_3(s) \Leftrightarrow NH_4^+ + NO_3^-, \tag{6.89}$$

where the equilibrium coefficient for this reaction is called the *solubility product*. A solid precipitates from solution when the product of its reactant ion concentrations and mean activity coefficient exceeds its solubility product. In other words, precipitation occurs when

$$\mathbf{m}_{NH_4^+}\mathbf{m}_{NO_3^-}\gamma^2_{NH_4^+,NO_3^-} > K_{eq}(T). \tag{6.90}$$

Similarly, gas deposition and solid-forming reaction on a surface can be simulated with a reaction such as

$$NH_4NO_3(s) \Leftrightarrow NH_3(g) + HNO_3(g). \tag{6.91}$$

In this case, the solid can form when one gas adsorbs to a surface and the other gas collides and reacts with the adsorbed gas. Alternatively, both gases can adsorb to a surface and then diffuse on the surface until a collision and reaction occur. In Equation 6.91, a solid is assumed to form on the surface when

$$p_{s,NH_3}p_{s,HNO_3} > K_{eq}(T). \tag{6.92}$$

In either of the above two cases, solid formation is accounted for with the MFI equilibrium solution method, described above. When a solid forms, F from Equation 6.88 converges to 1.0 and solid, ion, and/or gas concentrations are updated with Equations 6.84 and 6.85. When a solid does not form, F does not converge, and Equations 6.84 and 6.85 predict no net change in concentrations.

The process by which an initially dry particle lowers its saturation vapor pressure and takes up liquid water is *deliquescence*. If a particle consists of an initially solid electrolyte at a given relative humidity, and the relative humidity increases, the electrolyte does not take on liquid water by hydration until the *deliquescence relative humidity* (DRH) is reached. At the DRH, water rapidly hydrates with the electrolyte, dissolving the solid, and increasing the liquid water content of the particle. Above the DRH, the solid phase no longer exists, and the particle takes up additional liquid water to maintain equilibrium.

TABLE 6.4
DRHs and CRHs of Several Electrolytes at 298K

Electrolyte	DRH (percent)	CRH (percent)	Electrolyte	DRH (percent)	CRH (percent)
NaCl	75.28[a]	47[c]	$(NH_4)_2SO_4$	79.97[a]	37–40[b]
Na_2SO_4	84.2[b]	57–59[b]	NH_4HSO_4	40.0[b]	0.05–22[b]
$NaHSO_4$	52.0[d]	<0.05[d]	NH_4NO_3	61.83[a]	25–32[d]
$NaNO_3$	74.5[d]	0.05-30[b]	$(NH_4)_3H(SO4)_2$	69[b]	35–44[b]
NH_4Cl	77.1[a]	47[e]	KCl	84.26[a]	62[c]

[a] From Reference 24. With permission.
[b] From Reference 20. With permission.
[c] From Reference 25. With permission.
[d] From Reference 26. With permission.
[e] From Reference 10. With permission.

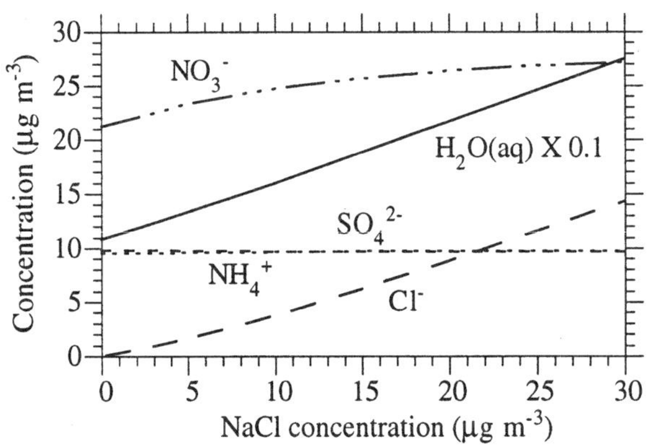

FIGURE 6.3 Aerosol composition vs. NaCl concentration when the relative humidity was 90%. Other initial conditions were $H_2SO_4(aq)$ = 10 μg m^{-3}-air, HCl(g) = 0 μg m^{-3}, $NH_3(g)$ = 10 μg m^{-3}, $HNO_3(g)$ = 30 μg m^{-3}, and T = 298K. NaCl dissolves and dissociates completely at this relative humidity. (From Reference 4. With permission.)

If a particle consists of an initially aqueous electrolyte, and the relative humidity decreases below the DRH, water evaporates, but dissolved ions in solution do not necessarily *precipitate* (crystallize) immediately. Instead, the solution is supersaturated and remains so until solid nucleation occurs. The relative humidity at which nucleation occurs and an initially aqueous electrolyte becomes crystalline is the *crystallization relative humidity* (CRH). The CRH is always less than or equal to the DRH. Table 6.4 shows the DRHs and CRHs of several electrolytes at 298K. Some electrolytes, such as NH_3, HNO_3, HCl, and H_2SO_4, do not have a solid phase at room temperature. These substances, therefore, do not have a DRH or a CRH. In a mixed solution, the DRH of a solid in equilibrium with the solution is lower than the DRH of the solid alone.[16,20]

EQUILIBRIUM SOLVER RESULTS

Graphical results from the equilibrium solution method discussed in the section "Method of Solving Equilibrium Equations" are shown here for two cases. Figure 6.3 shows the change in composition of a bulk particle solution as a function of sodium chloride mole concentration. The figure shows

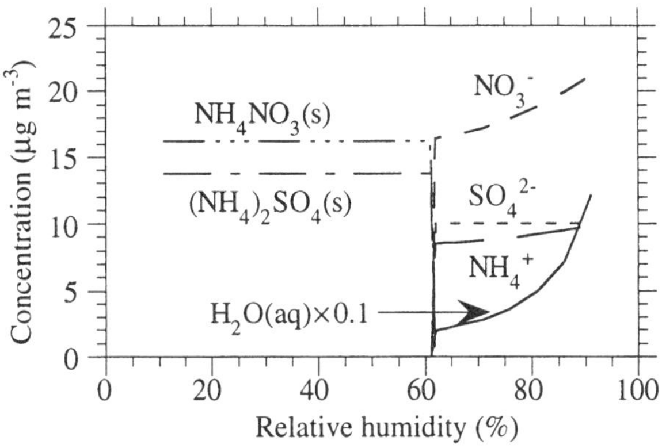

FIGURE 6.4 Aerosol composition vs. relative humidity. Initial conditions were $H_2SO_4(aq) = 10\ \mu g\ m^{-3}$, $HCl(g) = 0\ \mu g\ m^{-3}$, $NH_3(g) = 10\ \mu g\ m^{-3}$, $HNO_3(g) = 30\ \mu g\ m^{-3}$, and T = 298K. (From Reference 4. With permission.)

that an increase in sodium chloride caused water to condense and hydrate, increasing the rate of dissolution and dissociation of nitric acid.

Figure 6.4 shows a model simulation of the change in aerosol composition as a function of relative humidity. As humidity decreased down from 100%, water, chlorine, nitrate, and ammonium decreased steadily. At about 62% relative humidity, which is near the DRH of ammonium nitrate, both ammonium nitrate and ammonium sulfate precipitated. Ammonium sulfate did not precipitate at relative humidities of 62 to 80%, although its DRH is about 80% because it was undersaturated at those humidities. When the relative humidity is decreasing, ammonium sulfate can remain in the aqueous phase until the relative humidity reaches 40% (Table 6.4).

SUMMARY

In this chapter, chemical equilibrium equations were discussed. When equilibrium equations are solved in a model, mean mixed activity coefficient and water content equations are also needed. A method of calculating the water content and expressions for temperature-dependent mean binary activity coefficients were given. Mean mixed activity coefficients were calculated from mean binary activity coefficients with a mixing rule. A method of iterating equilibrium, activity coefficient, and water content equations together was also given.

REFERENCES AND FURTHER READING

1. Pitzer, K.S. and Mayorga, G., Thermodynamics of electrolytes. II. Activity and osmotic coefficients for strong electrolytes with one or both ions univalent, *J. Phys. Chem.*, 77, 2300, 1973.
2. Hamer, W.J. and Wu, Y.-C., Osmotic coefficients and mean activity coefficients of uni-univalent electrolytes in water at 25°C, *J. Phys. Chem. Ref. Data,* 1, 1047, 1972.
3. Pitzer, K.S., Ion interaction approach: theory and data correlation, *Activity Coefficients in Electrolyte Solutions,* 2nd ed., edited by Pitzer K.S., CRC Press, Boca Raton, FL, 1991.
4. Jacobson, M.Z., Tabazadeh, A., and Turco, R.P., Simulating equilibrium within aerosols and nonequilibrium between gases and aerosols, *J. Geophys. Res.*, 101, 9079, 1996.
5. Harned, H.S. and Owen, B.B., *The Physical Chemistry of Electrolyte Solutions*, Chap., 8, Reinhold, New York, 1958.
6. Clegg, S.L. and Brimblecombe, P., Application of a multicomponent thermodynamic model to activities and thermal properties of 0–40 mol kg^{-1} aqueous sulfuric acid from <200K to 328K, *J. Chem. Eng. Data* 40, 43, 1995.

7. Stelson, A.W., Bassett, M.E., and Seinfeld, J.H., Thermodynamic equilibrium properties of aqueous solutions of nitrate, sulfate and ammonium, *Chemistry of Particles, Fogs and Rain,* edited by Durham, J.L., Ann Arbor Publication, Ann Arbor, MI, 1984.
8. Bromley, L.A., Thermodynamic properties of strong electrolytes in aqueous solutions, *AIChEJ* 19, 313, 1973.
9. Stokes, R.H. and Robinson, R.A., Interactions in aqueous nonelectrolyte solutions.I. Solute-solvent equilibria, *J. Phys. Chem.,* 70, 2126, 1966.
10. Cohen, M.D., Flagan, R.C., and Seinfeld, J.H., Studies of concentrated electrolyte solutions using the electrodynamic balance., 1, Water activities for single-electrolyte solutions, *J. Phys. Chem.*, 91, 4563, 1987.
11. Cohen, M.D., Flagan, R.C., and Seinfeld, J.H, Studies of concentrated electrolyte solutions using the electrodynamic balance. 2. Water activities for mixed-electrolyte solutions, *J. Phys. Chem.*, 91, 4575, 1987.
12. Press, W.H., Flannery, B.P., Teukolsky, S.A., and Vetterling, W.T., *Numerical Recipes: The Art of Scientific Computing*, Cambridge University Press, Cambridge, 1992.
13. Pilinis, C. and Seinfeld, J.H., Continued development of a general equilibrium model for inorganic multicomponent atmospheric aerosols, *Atmos. Environ.*, 21, 2453, 1987.
14. Kim, Y.P., Seinfeld, J.H., and Saxena, P., Atmospheric gas-aerosol equilibrium I. Thermodynamic model, *Aerosol Sci. Technol.*, 19, 157, 1993.
15. Bassett, M.E. and Seinfeld, J.H., Atmospheric equilibrium model of sulfate and nitrate aerosol, *Atmos. Environ.,* 17, 2237, 1983.
16. Wexler, A.S. and Seinfeld, J.H., The distribution of ammonium salts among a size and composition dispersed aerosol, *Atmos. Environ.*, 24A, 1231, 1990.
17. Jacobson, M.Z., Developing, Coupling, and Applying a Gas, Aerosol, Transport, and Radiation Model to Study Urban and Regional Air Pollution, Ph.D. thesis, Dept. of Atmospheric Sciences, University of California, Los Angeles, 1994.
18. Jacobson, M.Z., *Fundamentals of Atmospheric Modeling*, Cambridge University Press, New York, 656, 1999.
19. Villars, D.S., A method of successive approximations for computing combustion equilibria on a high speed digital computer, *J. Phys. Chem.,* 63, 521, 1958.
20. Tang, I.N. and Munkelwitz, H.R., Composition and temperature dependence of the deliquescence properties of hygroscopic aerosols, *Atmos. Environ.*, 27A, 467, 1993.
21. Wagman, D.D., Evans, W.H., Parker, V.B., Schumm, R.H., Halow, I., Bailey, S.M., Churney, K.L., and Nuttall, R.L., The NBS tables of chemical thermodynamic properties: selected values for inorganic and C_1 and C_2 organic substances in SI units, *J. Phys. Chem. Ref. Data,* 11, Suppl., 2, 1982.
22. Schwartz, S.E., Gas- and aqueous-phase chemistry of HO_2 in liquid water clouds, *J. Geophys. Res.*, 89, 589, 1984.
23. Perrin, D.D., *Ionization Constants of Inorganic Acids and Bases in Aqueous Solution*, 2nd ed., Pergamon, New York, 1982.
24. Robinson, R.A. and Stokes, R.H., *Electrolyte Solutions,* Academic Press, New York, 1955.
25. Tang, I.N., Thermodynamic and optical properties of mixed-salt aerosols of atmospheric importance, *J. Geophys. Res.,* 102, 1883, 1997.
26. Tang, I.N., Chemical and size effects of hygroscopic aerosols on light scattering coefficients, *J. Geophys. Res.,* 101, 245, 1996.

Part II

Laboratory Studies

7 LAMMA and Raman Study of Oxidation States of Chromium in Aerosols: Application to Industrial Hygiene

A. Hachimi, E. Poitevin, G. Krier, and J.F. Muller

CONTENTS

INTRODUCTION

Since its discovery in 1797, chromium has been increasingly used in industry. Its effects on human health have been gradually understood. The two most important sights of professional pathology, due to chromium and to its derivatives, are represented by "allergogene" action and the cancer-producing power of some products. Indeed, hexavalent chromium derivatives cause dermatosis of contact, bronchitic asthma, perforation of the nasal septum, and bronchiopulmonary cancers.[1,2]

0-87371-829-1/00/$0.00+$.50

Experimental studies have shown that chromates and bichromates are able to induce *in vitro* and *in vivo* cancers in animals.[3-5] Epidemiologic survey have shown that the lungs represent a target organ of hexavalent derivatives of chromium. However, only a few experimental studies exist that have allowed for the thorough study of toxicity mechanisms. In welding and steel working, high concentrations of fumes and gas are emitted, which contain chromium in either the hexavalent or trivalent form; thus, health problems in relation to the presence of chromium in dust fumes can occur. Therefore, methodology application on valency determination of chromium in environmental dusts was desirable.

In the literature, one finds two kinds of techniques for the determination of chromium valency in aerosols of industrial origin:

- Chemical spectrophotometric analysis techniques, also called wet chemical techniques
- Physical and chemical techniques using direct measurement and spectroscopic techniques

Wet chemical techniques, such as colorimetry, ion exchange resins, luminescence, and atomic absorption,[6-8] have the ability to determine both chromium identification and correct proportioning in its numerous oxides forms. A main drawback concerns aerosol sampling because the sample can be physically altered (e.g., by extraction, dissolution, electrochemical reactions during lixiviation operation, etc.) with the possibility of induced chemical changes, matrix effects, and differential solubility artifacts.[9]

Spectroscopic techniques such as Raman, X-ray diffraction, FTIR,[10,11] XPS,[12-14] and XRF[15,16] allow *in situ* characterization of solid aerosols, but there are many instrumental limitations related to detection, sensitivity, sampling, and data interpretation.

Among the spectroscopic techniques recently described in the literature to identify *in situ* element oxidation rates, mass spectrometry presents new insight on this problem, especially laser microprobe LAMMA.

One can apply the valency determination method for chromium to a complex matrix: on the one hand, arc welding fumes on stainless steel (MMA/SS) and, on the other hand, dust from the steel industry. Thus, one can investigate the determination of major oxidation rates of chromium derivatives contained in the dust that is directly inhaled by workers.

STUDY OF CHROMIUM VALENCY IN POLYPHASIC DUST

The method proposed here involves the determination of chromium valency, in aerosols less than and greater than 10 μm in diameter, emitted by welding and in aerosols, 0.4 to 10 μm in diameter from various steel works. This method has been improved upon by precise sampling with granulometric discrimination of aerosols (using cyclone and Andersen impactors) and targeting of fume emission sites at steel-making locations. Microprobe Raman techniques, and ESCA and SEM, have been used as complementary techniques of confirmation.

MATERIALS AND METHODS

SAMPLING

Welding dust was collected after a welding operation on stainless sheet steel (MMA/SS) of type 304L 18-10 (lower carbon) with rods AROSTA — basic rutile electrode 304. Dusts pass in a cyclone impactor by suction, drawing up a flow of 1.71 l min^{-1} for 46 min. Dusts are finally collected onto a nitrocellulose filter of 0.2 μm porosity and 32-mm diameter. Sampling selection of dust was fixed to 10 μm coating element of the rods and trimming dusts (metallic projectiles) also have been collected in the experimentation workshop for LAMMA analysis.

Dust sampling was performed at a site rich in fumes containing large amounts of chromium using two kinds of impactors:

- A normal Andersen cascade impactor with an air flow rate of 28 l min^{-1} for discrimination of aerosols in air flux by inertial impaction relative to their mean aerodynamic diameter. Size fractionation in nine stages (>9 μm; 9 to 5.8 μm; 5.8 to 4.7 μm; 4.7 to 3.3 μm; 3.3 to 2.1 μm; 2.1 to 1.1 μm; 1.1 to 0.7 μm; 0.7 to 0.4 μm; and <0.4 μm) gave information about the distribution of chromium compounds in relation to aerosol size.
- Dusts were collected by impaction on aluminium filters during a period of 12 to 24 h; a portable cascade impactor consisting of five stages, with aerosols size ranging from 0.4 to 9.8 μm and air flow rate of 1.7 l min^{-1} carried by workers during an 8-h shift. Dusts were collected on nitrocellulose or polyethylene terephthalate filters.

Analytical Techniques

SEM Analysis

Element analysis was achieved using a JEOL 840 electron microscope coupled to a dispersive energy spectrometer. An electron beam current of 200 to 300 pA and accelerating voltage of 15 kV were used.

Raman Microprobe

MicroRaman analysis of aerosols collected on the portable cascade impactor was performed on a standard DILOR XY apparatus equipped with an argon ion laser and multichannel detector (1024 diodes). Excitation wavelength was 514.5 nm using a power of 75 mW to avoid fluorescence emission that could mask Raman diffusion. Spectral resolution was 4 cm^{-1}, and integration time varied from 1 to 15 s.

The spectrometer was coupled to an optical microscope (Olympus), permitting a spatial resolution of 2 μm. Spectra were obtained in reflective mode and presented in arbitrary units vs. wavenumbers (cm^{-1}).

X-ray Photoelectronic Spectroscopy (XPS)

The used apparatus from Leybold Heraeus™ had a resolution of 1 eV. The radiation source comes from the Kα radiation of aluminum. The deposited dusts are directly mounted on the sampling support. An interval of 3 eV, on average, has been recorded for the characteristic lines of C(1*s*), K(2*p*), and K(2*s*), O(1*s*), and F(1*s*). This interval is due to the charge effects from fibrous dusts and the nitrocellulose filter.

Laser Microprobe Mass Analysis (LAMMA)

LAMMA was developed for localization and determination of elements in various samples — either conducting or isolating ones. An interesting feature of this technique is its ability to characterize the molecular composition of inorganic substances. Moreover, it allows for elementary analysis without the traditional separation step.

Element detection limits are 10^{-15} to 10^{-19} g. This sensitivity allows for LAMMA analysis of aerosols in biological and environmental studies.[17,18]

The study of ionized clusters in both positive and negative modes is correlated with the sample chemical composition (e.g., SO^-, SO_2^-, SO_3^-, SO_4^-, $NaSO_3^-$, $NaSO_4^-$). Major ions obtained by laser ionization are representative of sulfate and sodium thiosulfates.[19] The information is useful to complete data obtained by other techniques. Additionally, a link between morphological properties of particles can be established by sample observation under visible light.

Technological progress (laser, configuration of ionization chamber) has permitted increased use of this method, and has allowed for the analysis of organic and inorganic matter ((nitro)PAH desorption) and matrix identification.[20,21] The laser microprobe already has a privileged place and ongoing instrumentation progress will make certain its success.

Impacted dusts were extracted from filters by superficial scraping and set by simple pressure on a microscopic grid coated with a formvar film.

Comparisons could be made with constant instrumental parameters, and the LAMMA apparatus had the following features:

- Wavelength: 266 nm
- Pulse width: 12 ns
- Laser focus: 2–3 μm
- Energy on sample: 1–3.5 μJ
- TOF voltage: ±3000 V
- Extraction lens voltage: ±1000 V
- Reflection voltage: ±790 V
- Cathode voltage: ±6000 V

All spectra were recorded on Nicolet 4094C, a recorder connected to an Apple Macintosh II CX computer with a 40-Mbyte system. Spectra were calibrated and linearized following mass spectrometry conventions.

ANALYSIS OF DUSTS FROM WELDING FUMES

Welding fumes can contain large amounts of chromium compounds with varying concentrations that depend on the welding process and the rod composition. A systematic study was achieved on these different elements for the following:

- Coating of the rod (AROSTA, rutile basis electrode type 304 L)
- Trimming dusts (metallic projectiles)
- Welding aerosols less than 10 μm in diameter that have metallic microspherical form and are collected in the impactor after welding operation
- Welding aerosols more than 10 μm in diameter that are microfibers collected onto nitrocellulose filters

QUALITATIVE ANALYSIS

We have investigated fingerprint spectral analysis of the different elements cited above. LAMMA spectra are presented in Figures 7.1 to 7.5. General observations of these spectra lead to to the following remarks:

- The constitutive elements of the coatings of the rod, such as Na, K, Ca, F, Cl, Mn, P, and Si (in aluminosilicate form), are present in all the aerosols.
- Barium, which is present in its oxide and fluoride forms and comes from coatings of the rod, is simply present in the microspherical aerosols (>10 μm).
- Trimming compounds, likely microspheres, contain metals (Cr, Al, Ti, and Ni) that are representative of the elemental composition of the rod and stainless steel sheet-metal.
- Fibrous dusts (<10 μm) with a basic nature (fluorides and chloride compounds) likely contain an alumino- and ferrosilicate matrix in which metallic elements are scattered.
- Microspheres contain a lot of sulfate compounds, like trimming compounds, characterized by SO_x^- clusters in negative ionization mode, whereas microfibrous aerosols do not.

TABLE 7.1
Elemental Analysis of Welding Dust by LAMMA

Element	Rod	Coating	Trimming	Dust (>10 μm)	Dust (<10 μm)
Si	+++	+++	++	+++	+++
Mn	+	+	+	+	++
Cr	++	–	++	+++	++
Ni	++	–	+	+	++
Al	–	++	++	++	++
O	–	++	++	++	+++
F	–	++	+++	+++	+++
Na	–	+++	++	++	+++
Cl	–	++	++	++	++
K	–	+++	+++	+++	+++
Ca	–	+++	+	+++	+
Fe	–	++	+++	++	+
S	–	+	+	++	–
Ti	–	–	++	+++	+
Mg	–	–	+	+	+
Ba	–	++	–	+++	–
Type of Cluster Ions Present on Positive and Negative Modes					
Al_xO_y		+++	++	+++	+
Si_xO_y		+++	++	+++	+++
PO_x		++	+++	+	+++
Fe_xO_y		+	++	++	+
$AlSi_xO_y$		+++	++	+	++
CaF		+	+	+++	+
SO_x		+	++	++	–
$FeSi_xO_y$		+	+	–	–
CrO_x		–	++	++	++
BaOH		++	–	++	–
TiO		–	+	+++	+
BaF		–	–	+	–
K_xF_y		+	–	–	+++
Na_xCl_y		–	–	–	++
CaCl		–	–	–	+
K_xCl_y		–	–	–	+
$Al(Na_xCl_x)$		–	–	–	++
$Na(K_xF_x)$		–	–	–	++
$K(Na_xCl_x)$		–	–	–	++
Na_2OH		–	–	–	+
K_2OH		–	–	–	+

Note: Absent: –; Present: +; Majority: ++; Abundant: +++

- The two different spectra of fibrous aerosols in positive mode show the heterogeneity of the matrix, one can obtain either spectra with metallic nature of spectra with alumino- and ferrosilicates matrix.
- Chromium and nickel are systematically present in positive spectra (Cr^+ and Ni^+) in both kinds of aerosols.

The entire LAMMA analysis is presented in Table 7.1 where elemental ions and combinations present in plasma are displayed.

TABLE 7.2
Ratio Calculation of CrO_2^-/CrO_3^- of Welding Dust

Type of Dust	CrO_2^-/CrO_3^-	N Spectra	Mode	Wavelength (nm)
Fibrous	0.218 ± 0,12	85	Individual	225.7
Dust <10 μm	0.155	70	Accumulation	225.7
	0.228	200	Accumulation	286.5
	0.181	100	Accumulation	286.5
		Total: 455		
Weighted mean	0.205			
Standard deviation	0.03			
Variance	0.0008			
Microsphericals	0.754 ± 0.194	24	Individual	225.7
Dust >10 μm	0.673	100	Accumulation	225.7
	0.543	100	Accumulation	225.7
	0.56	25	Accumulation	225.7
	0.551	25	Accumulation	225.7
	0.629	50	Accumulation	225.7
Weighted mean	0.614	Total: 324		
Standard deviation	0.205			
Variance	0.042			

Calculation of Chromium Amount with Different Stoechiometry

Oxidation state studies were performed on 85 and 24 individual spectra of fibrous and microspherical aerosols, respectively. Investigation of 455 and 325 accumulated spectra allow for the determination of the major chromium oxidation state in the fibrous and microspherical dusts, respectively. All results are summarized in Table 7.2. The table displays the CrO_2^-/CrO_3^- ratios of different individual and mean spectra.

Fibrous Aerosols: (<10 μm)

Spectral analysis of fibrous aerosols has shown that chromium was only present in its elementary form Cr^+ in positive ions, while CrO_2^-, CrO_3^- clusters (and sometimes CrO_4^-) were present in negative mode. Energy variation and length variation do not interfere on spectra and intensity ratios of negative clusters.

Results are presented in Table 7.3 with analysis details of individual spectra. The methodology was applied on fibrous aerosols with no ambiguity and it confers to chromium an oxidation degree of VI (93% of individual spectra have a chromium VI fingerprint). The salt character of fibers and a value of CrO_2^-, CrO_3^- that corresponds to an anhydrous chromium salt could show chromium in chromate form. In fact, this is in agreement with other results[22,23] on similar type dusts, where chromium has been identified in sodium or potassium form.

However, LAMMA does not detect the type of cluster $K_xCrO_y^+$, $CrxO_y^-$ ($x = 2$; $y = 4, 5, 6$); this is due to either the lower sensitivity or other major recombinations in plasma. Therefore, the assumption of chromium VI presence in the chromate form must be verified by other analytical techniques.

Microspherical Aerosols (>10 μm)

After analysis, the 24 individual spectra of studied microspheres (20 μm in diameter) gave the following results:

TABLE 7.3
% Chromium VI Calculated in Welding Dust

Dust <10 µm	**CrO_2/CrO_3 < 0.3**	**With CrO_4**	**Without SO_x**	**% of Chromium**
Spectrum calc.				
Spectrum number	79	12	85	79 Cr VI
	CrO_2/CrO_3 > 0.3	With CrO_4		6 Cr III
Spectrum number	6	3		93% Cr VI
				7% Cr III
Accumulation	R = CrO_2/CrO_3	Valency of majority of chromium		
455 spectra	0.205	Value of R confer on chromium VI valency		
		More than 90% chromium		
Dust >10 µm	**R > 0.3**	**With CrO_4**	**With SO_x**	**24 Cr III**
Spectrum calc.				
Spectrum number	24	24	24	100% Cr III
Accumulation	R	Valency of majority of chromium		
324 spectra	0.614	Value of R confer on chromium III valency		
		Mixture of chromium III and VI		

- SO_x^- ions clusters and CrO^- chromium are systematically present (100% of spectra), in addition to CrO_2^- and CrO_3^- ions clusters.
- Cluster intensity of sulfur does not interfere with the CrO_2^-/CrO_3^- ratio, which is about 0.8 for the aerosol (arithmetic average of 0.75). A CrO_2^-/CrO_3^- ratio of 0.75 might correspond either to hydrated or anhydrous chromium salt. The lack of CrO_4^- ion and simultaneous presence of CrO^- and SO_x^- for 100% of spectra would reveal a chromium presence as anhydrous chromium sulfate, thus in the III form.

One cannot conclude definitively about the real state of chromium in microspheres because the $Cr_2O_y^-$ series has not been detected. Nevertheless, two other observations from complementary works would show that chromium is in its III form and in the salt and oxide mixture form:

- The CrO_2^-/CrO_3^- ratio varies according to studied microsphere (from 10 to 40 µm in diameter, with value between 0.5 and 4, which corresponds to chromium III and is constant for a particular microsphere).
- All microspheres have the same composition and spectral fingerprint (systematic presence of CrO^-, SO_x^-, and lack of CrO_4^-).

These two results support the hypothesis of a mixture containing chromium III salts (sulfate compounds, for instance) and chromium III oxide in variable amounts depending on microspheres. Variation of the CrO_2^-/CrO_3^- ratio from 0.5 to 4 could mean that microspherical aerosols contain an increasing amount of chromium III oxide, depending on microspheres.

Complementary Analysis

Raman analysis was only effective on fibrous aerosols with a minimal energy to avoid absorption or destruction of dusts. Microspheres study is revealed to be unfeasable because of the large absorption of the aerosols, even with minimal energy application.

Results of Raman analysis are shown in Figure 7.1. Two specific bands of the CrO_4^{2-} ion are obtained at 849 and 901 cm^{-1} and given hexavalent chromium in salt form in fibrous dust.

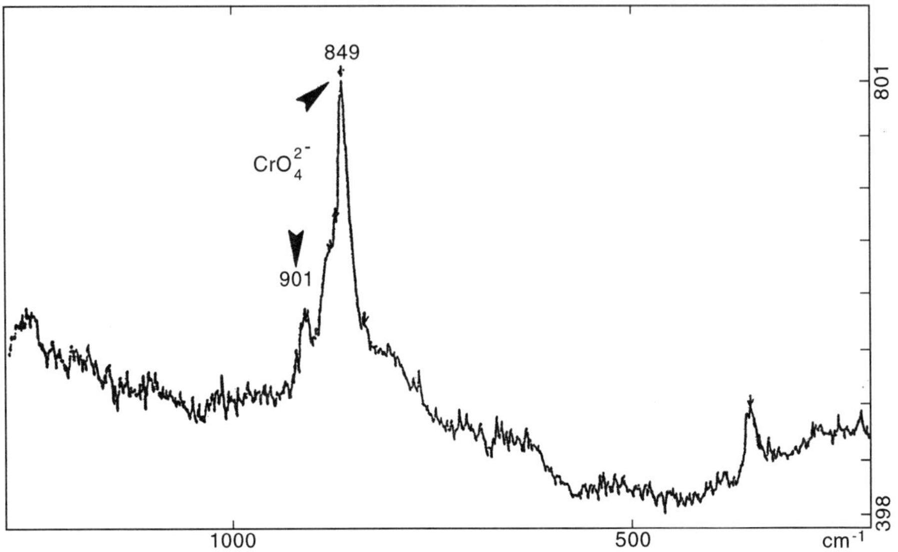

FIGURE 7.1 Raman spectrum of dust particles smaller than 10 μm.

XPS analysis has likely been performed only on fibrous aerosols because of the deficient amount of microspheres (>10 μm) for effective detection. Major elements already detected by LAMMA analysis are found, and Cr(2*p*) and Cr(3*p*) bands for chromium and Ni(2*p*) and Ni(3*p*) bands for nickel are observed. The presence of the Cr(2*p*) band at 579.4 eV after correction for the binding energy indicates that chromium is present in its VI oxidation form, whereas no band is detected at 675.4 eV, which is specific for chromium III. So, the band only at 579.4 eV reveals the presence of chromium exclusively in its hexavalent form. Nickel is present in an oxidized form (a 4-eV shift from the metal binding energy value) with two bands at 858.1 and 876.8 eV for Ni(2*p*) and Ni(3*p*), respectively.

The LAMMA microprobe turns out to be a sufficiently sensitive and rapid technique for the determination of the oxidation state of chromium included in complex matrices like aerosols emitted from industrial environments. However, it appears from this work that dust sampling (i.e., the conditions of granulometric and morphological selection) and high-risk site targeting are important factors to consider in improving the methodology. That is the reason why we have established a sampling strategy that is more refined for the case of the steel industry, which is an important source of dusts containing non-oxidative products.

DETERMINATION OF CHROMIUM VALENCY IN AEROSOLS LESS THAN 10 μM IN SIZE EMITTED AS DUST FROM THE STEEL INDUSTRY

This study deals with the determination of chromium valency in aerosols less than 10 μm in diameter emitted as dust from the steel industry, and with the development of a strategy for controlling the hazards these aerosols represent.

Within the context of industrial hygiene, there is great interest in the way the distribution of trivalent and hexavalent states of chromium differ among the various dust-emitting sites. Dust sampling by cascade impactors at a given site does not necessarily reflect the conditions of inhalation of the aerosols by a person working at that site. Sampling can be improved by fixing a portable cascade impactor to a person in the course of his/her normal activities in an area where chromium-rich dusts are found.

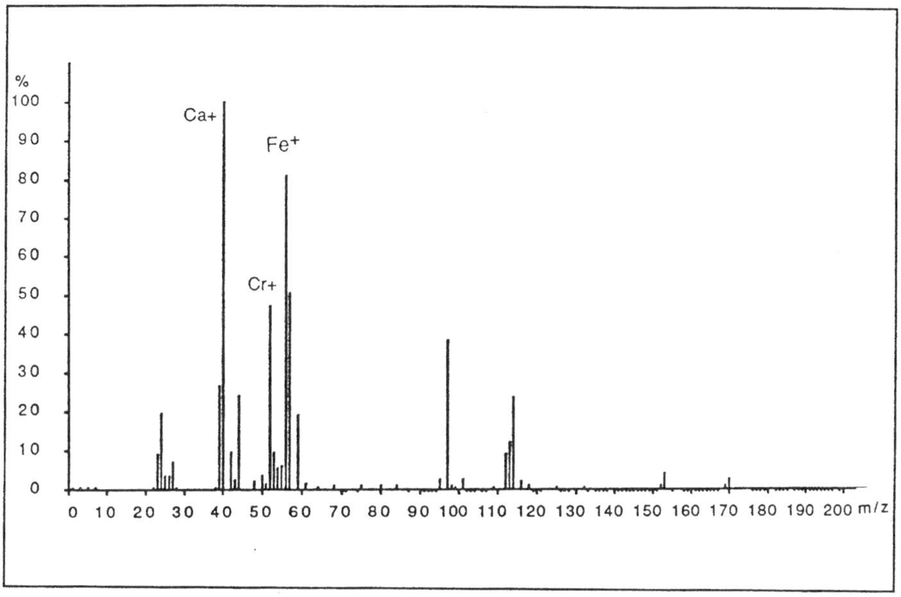

FIGURE 7.2 LAMMA spectrum of dust particles smaller than 3.3 μm.

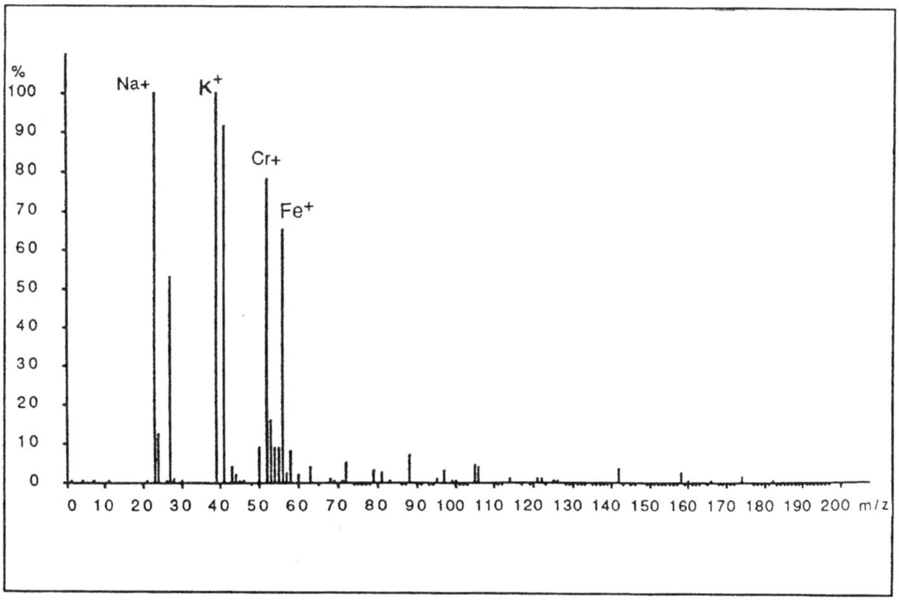

FIGURE 7.3 LAMMA spectrum of dust particles smaller than 0.4 μm.

ELEMENTAL ANALYSIS OF DUST COLLECTED FROM THE SITE

The LAMMA analysis of aerosols in size-fractionated dust samples shows that small aerosols mainly consist of potassium and sodium, with relatively little calcium (Figure 7.2), whereas the large dust aerosols are mainly comprised of calcium (Figure 7.3).

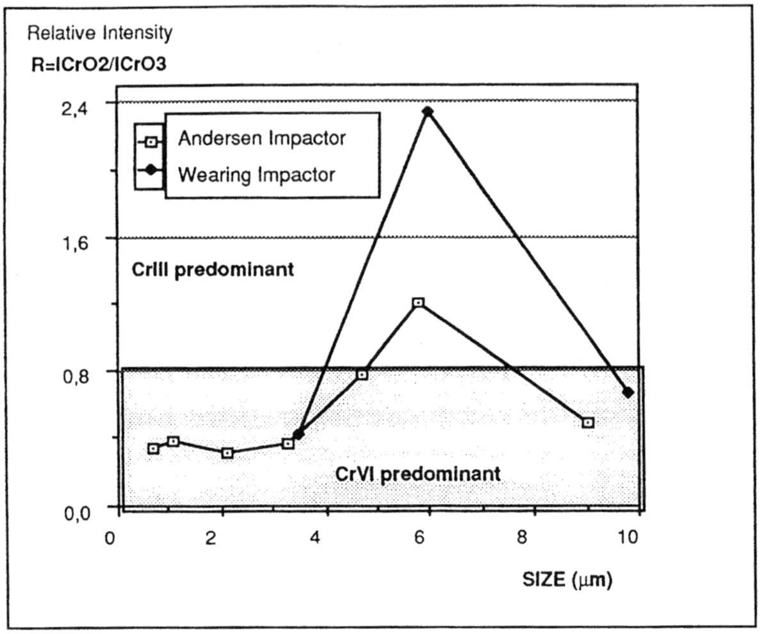

FIGURE 7.4 Comparison of chromium valency between dusts collected in an Andersen cascade impactor and that collected with a portable cascade impactor in relation to the particles.

Study of the Valency of Chromium in Dusts Collected by a Portable Impactor and by a Normal Andersen Impactor at the Same Site

If a comparison is made between analyses of dusts collected on the two kinds of impactors, the same majority distribution of chromium valency is found for both aerosol sizes collected on the portable impactor and those collected on the Andersen impactor at the same site (Figure 7.4).

This means that chromium is mainly present in the hexavalent form in the smallest and largest aerosols from both impactors. In fact, chromium is exclusively hexavalent for aerosols smaller than 3.5 μm and larger than 6 μm and trivalent for intermediate sizes. It is noteworthy that in the context of industrial hygiene, dust as emitted from the site has the same characteristics as that sampled on a worker's portable impactor.

A more detailed study of the relative ratio of chromium III/VI from all filters of the Andersen cascade impactor revealed that, for the entire dust sample (Figure 7.5), 30% of chromium is trivalent and 70% is hexavalent.

Furthermore, the ratio of chromium VI increases as aerosol size decreases and more than 60% of hexavalent chromium is present in the smallest sizes (<3.3 μm).

In the context of occupational health, the similarity of distributions of the valency of chromium as collected by the two impactors at the same site indicates an accurate simulation of the inhalation by a worker of fumes emitted.

It is remarkable that aerosols larger than 8 μm contain such a large amount of chromium VI. We predict that they are composed of a particular matrix that we believe is deserving of further study.

Chromium Analysis of Large Dust Aerosols (8 to 10 μm) Collected with the Portable Impactor

The aerosols of greatest size collected in the cascade impactor carried by a worker have been studied by SEM, LAMMA, and Raman microprobes.

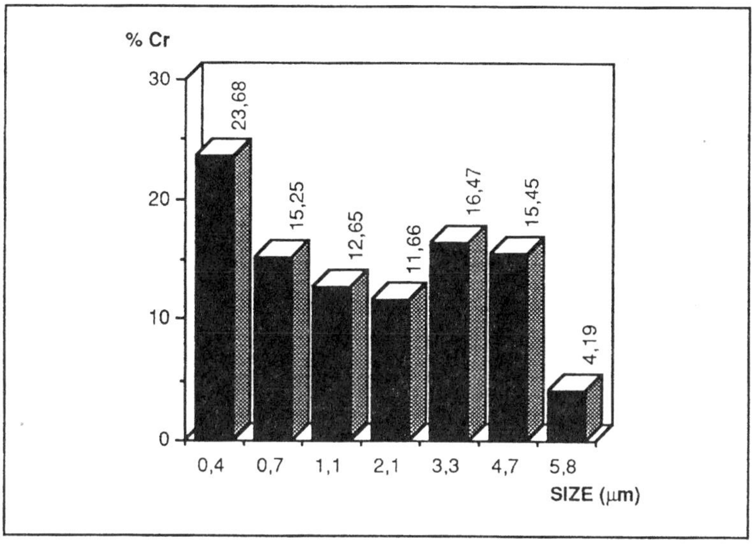

FIGURE 7.5 Relative chromium concentration in relation to particle size in dust collected with an Andersen impactor.

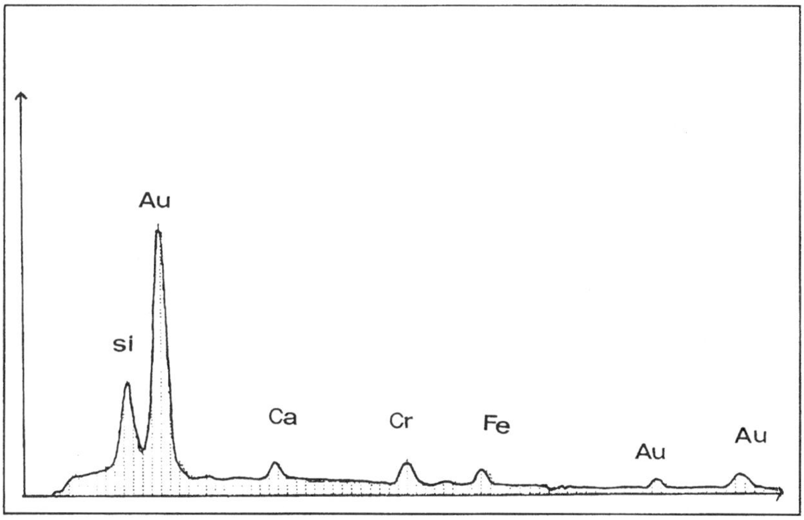

FIGURE 7.6 SEM spectrum of dust particle (8 to 10 μm) collected with portable cascade impactor.

SEM microscopy permits elemental analyses of aerosols, whereas the two other analytical techniques, which are complementary, enable detection and identification of chromium compounds in aerosols. Thus, SEM analysis has shown that the dust contains many of the following elements: silicon, calcium, iron, and chromium (Figure 7.6).

LAMMA analyses of these aerosols confirmed the presence of chromium characterized by systematic formation of $Ca_xCrO_y^-$ ions (m/z 108, 124, 140, 148, 180, and 196) in the positive mode (Figure 7.7). In the negative mode, the cluster class $Cr_xO_y^-$ (m/z 84, 100, 116, 168, 184, and 200) are always present, with systematic formation of CrO_4^- being characteristic for the solvation of the chromium compound clusters in a hydrated or oxygen-rich environment.[24] This hypothesis is consistent with an observation ratio of cluster CrO_2^-/CrO_3^- intensities less than 0.8, and CrO_4^-/CrO_3^-

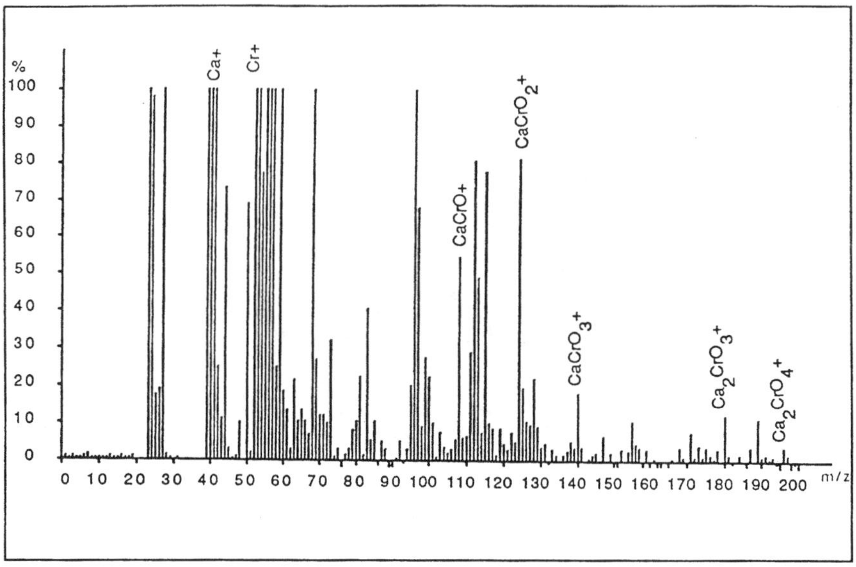

FIGURE 7.7 LAMMA spectrum of dust particles (8 to 10 μm) collected with portable cascade impactor (positive ionization mode).

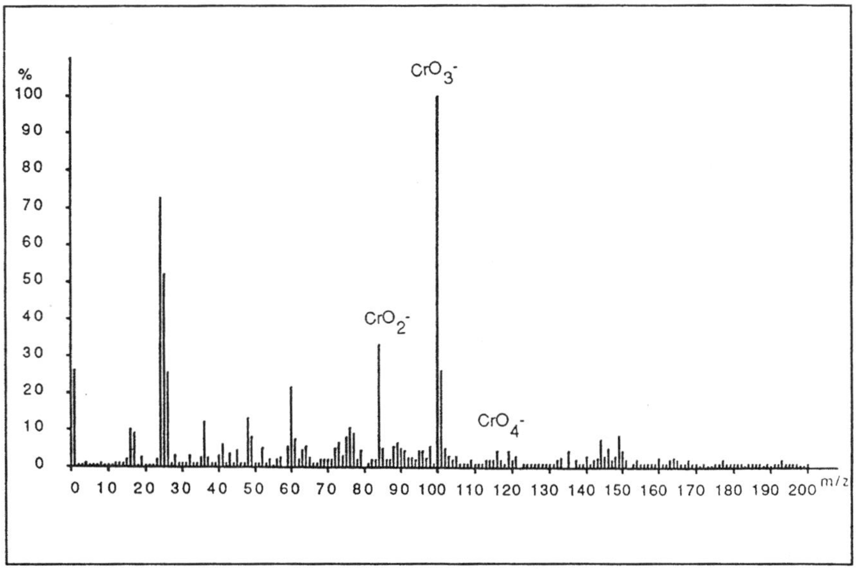

FIGURE 7.8 LAMMA spectrum of dust particles (8 to 10 μm) collected with portable cascade impactor (negative ionization mode).

intensities less than 0.1 (Figure 7.8). The investigations by LAMMA in both modes of ionization suggest hexavalent chromium is present in the form of calcium chromate.

Raman measurements permit the observation of the four modes of vibration of CrO_4^{2-} that belong to T_d symmetry. This last one is highly reduced by the presence of others ions in the crystalline structure. However, in the case of hexavalent chromates, anion vibrations are hardly

TABLE 7.4
Attribution of Raman Vibration Bands for Different Standard Compounds and for Dust Collected with a Portable Cascade Impactor

Sample		Wavenumber (cm^{-1})
$PbCrO_4$		839 (M)
		378 (vW)
Na_2CrO_4, xH_2O		3340, 3250 (vH2O)
		938 (M), 923 (M), 891 (H), 853 (vH), 810 (M)
		438 (W), 350 (W)
K_2CrO_4		906 (M), 877 (M), 869 (H), 853 (H)
		394 (vW), 389 (W), 348 (M)
$CaCrO_4$		905 (M), 879 (H)
		465 (vW), 383 (W), 302 (W)
Dust particles	A	905 (M), 880 (H)
(8–10 μm)	B	900 (M), 875 (H)
	C	900 (M), 875 (H), 855 (M), 833 (H), 707 (W)
	D	908 (H), 875 (M), 757 (W), 650 (W)
	E	926 (M), 855 (H)

Note: Intensity: vH = very high, H = high, M = medium, W = weak, vW = very weak.

disrupted.[25] Fundamental stretching vibrations of Cr-O (ν_1 and ν_2) are intense and localized between 950 and 800 cm^{-1}, whereas deformation vibration intensities are weak and localized between 420 and 370 cm^{-1} (ν_2 and ν_4).[26-30]

Campbell[30] had defined two types of chromates from vibration spectra:

1. "Type I chromate" spectra previously described (Na-, Ca-, Ba-, Sr-, Rb-, Li-, Ag-, and Pb-CrO_4)
2. "Type II dichromate: spectra, which are differentiated by the presence of bands in the range 800 to 700 cm^{-1}, similar to those of the ion $Cr_2O_7^{2-}$ (Fe-, Ni-, Zn-, Cu-, Co-, Al-, and Cd-Cro_4)

Raman results of standard chromium compounds and dust aerosols (8 to 10 μm) collected on the portable impactor are summarized in Table 7.4 and in Figures 7.9 and 7.10. The presence of calcium chromate implied by the SEM and LAMMA analyses is confirmed by this study: the Raman spectra of dust from sample A is identical to the $CaCrO_4$ reference spectrum. The spectrum of dust sample B, however, shows a shift of vibration bands toward low wave-numbers. This shift could be explained by modifications in the environment of chromate ion or in its crystalline mode. A similar spectrum was also obtained from dust sample C and is possibly an unidentified compound belonging to type II dichromate (band at 707 cm^{-1}). The Raman spectrum obtained for dust sample D can probably be compared to the reference sodium chromate spectrum. The Raman study of all dust samples collected at the given site using the portable cascade impactor confirms the hypothesis of the presence of hexavalent chromium — essentially as calcium chromate species — and permits us to expect either sodium chromate or type II chromates.

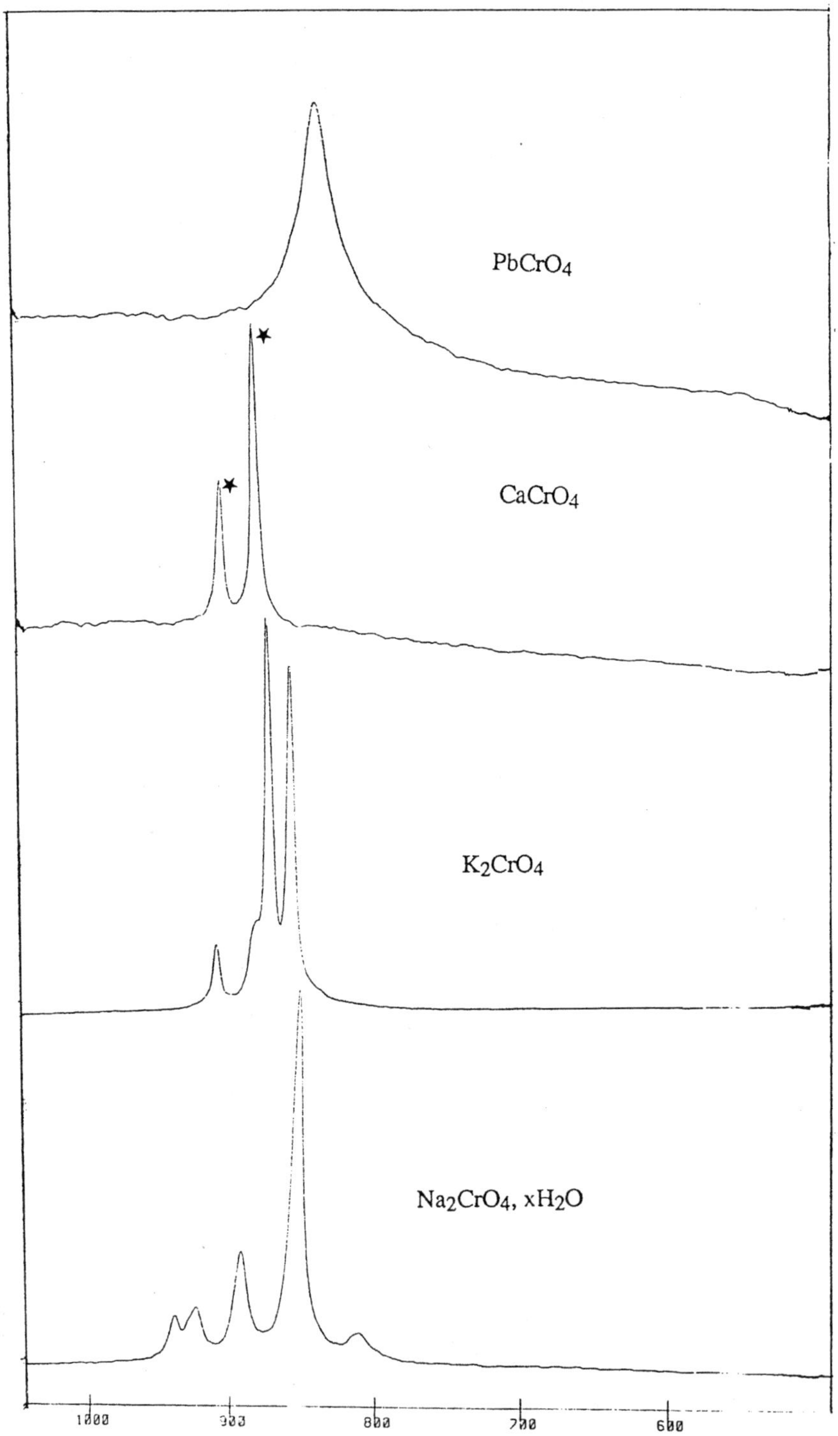

FIGURE 7.9 Raman reference spectra of type I chromates (for band positions, see Table 7.4).

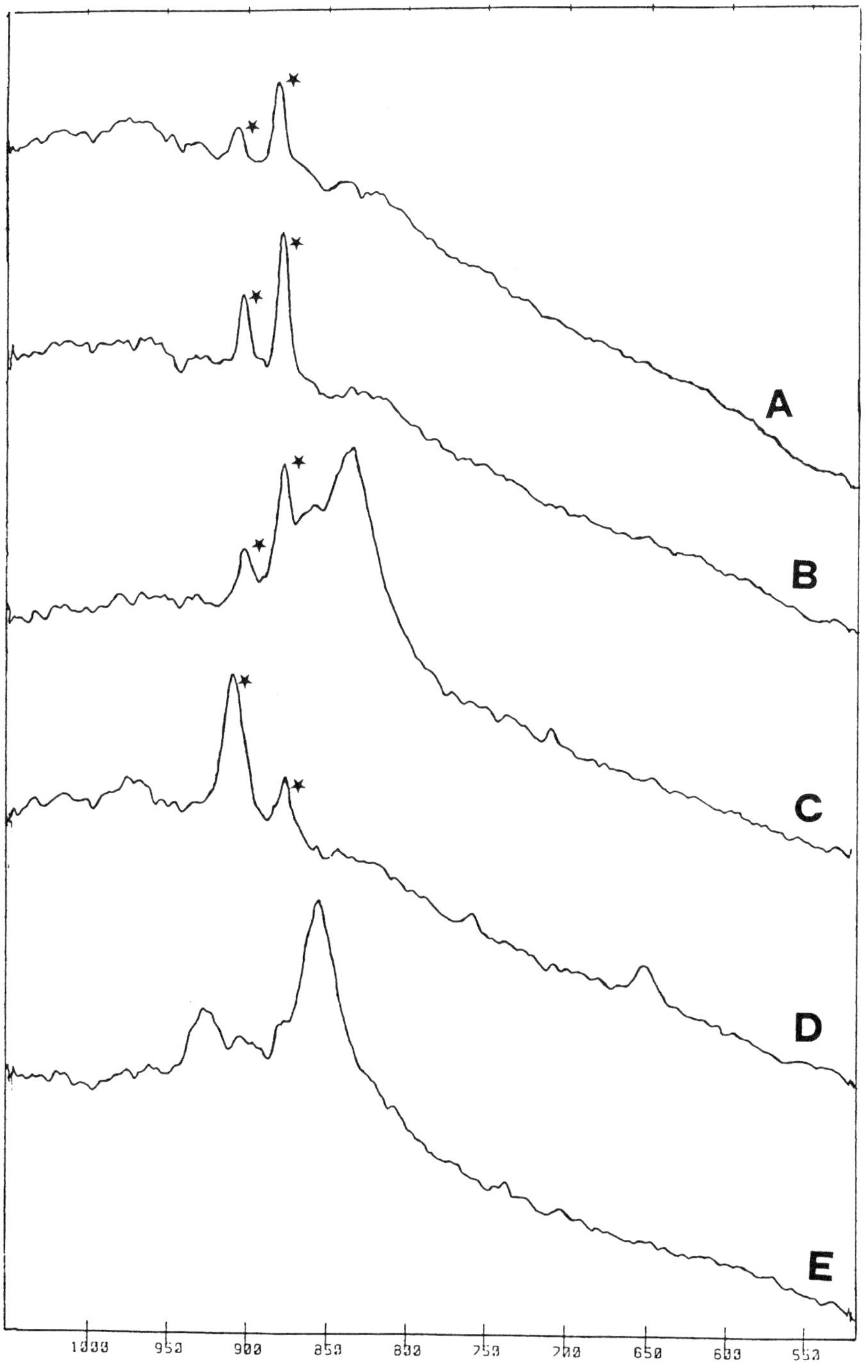

FIGURE 7.10 Raman specta of dust particles (8 to 10 μm) collected with portable cascade impactor (for band positions, see Table 7.1).

CONCLUSION

Systematic LAMMA analyses of oxidation states of chromium on aerosols less than 10 μm in diameter emitted in fumes from steelworks have proved that the valency of this metal varies according to the size of the aerosols. The smallest aerosols (<1.1 μm) and the largest ones (>5.8 μm) contain chromium in the hexavalent state, whereas dusts of intermediary sizes mainly have trivalent chromium.

Sampling of dust with a portable cascade impactor carried by a person working in an environment containing chromium fumes has shown a composition similar to that found for dust sampled with a fixed Andersen impactor.

Detailed and complementary characterization by three *in situ* analytical techniques of inhaled aerosols that do not penetrate the lungs (8–10 μm) have proved that hexavalent chromium is present as calcium chromate in this kind of aerosol.

The valency and type of chromium appear to correlate with the overall composition of the aerosol: indeed, small aerosols (<1.1 μm) are basically composed of potassium and sodium matrices, whereas large one (>6 μm) are mainly calcium. The presence of calcium chromates in dust aerosols 8 to 10 μm collected by a portable impactor carried by a worker at the same site confirms this fact. The small aerosols contain hexavalent chromium forms with the possible presence of sodium and/or potassium chromates.

This study shows the capabilities of microprobe LAMMA for *in situ* characterization of aerosols of different sizes (0.4–10 μm). The analytical speed and high sensitivity of chromium detection, combined with a well-chosen dust sampling protocol, favor the LAMMA technique for this type of study.

However, other complementary *in situ* techniques — the microRaman technique in particular are desirable for the verification of chromium oxidation state and of the nature chromium compounds present in aerosols.

REFERENCES

1. Haguenoer, J.M., Lefrançois, H., Mercier, J.F., and Boniface, B., Socieáté Française de Toxicologie, Toulouse, 1981.
2. Davies, J.M., *Br. J. Ind. Med.*, 41, 158, 1984.
3. Sen, P. and Costa, M., *Carcinogenesis,* 7, 1527, 1986.
4. Sugiyama, M., Wang, X.W., and Costa, M., *Cancer Res.,* 46, 4547, 1986.
5. Levy, L.S., Martin, P.A., and Bidstrup, P.L., *Br. J. Ind. Med.,* 43, 243, 1986.
6. Bemst, A.V. and Stern, R.M., *Welding in the World,* 21, 10, 1983.
7. Naranjit, D., Thomassen, Y., and Van Loon, J.G., *Anal. Chim. Acta,* 110, 307, 1979.
8. Korber, D. and Fiban, M., *Anal. Chem.,* 10, 13, 1981.
9. Cox, X.B., Linton, R.W., and Butler, F.E., *Environ. Sci. Technol.,* 19, 345, 1985.
10. Kentgen, G.A., *Anal. Chem.,* 56, 69 R, 1984.
11. Kimura, S., Kobayashi, M., Godai, T., and Minato, S., *AWS, 60th Annual Meeting*, Welding Research, Supplement, 1979, 1955.
12. Delamar, M., *Analusis,* 16, 419, 1988.
13. Inni, M.E., Gustafsson, T.E., Koponen, M., and Kalliomaki, P.L., *J. Aerosol Sci.,* 15, 57, 1984.
14. Lautner, G.M., Carler, J.C., and Konzen, R.B., *Am. Ind. Hyg. Assoc.,* 39, 651, 1978.
15. Schroeder, W.H., Dobson, M., Kane, D.M., and Johnson, N.D., *J.A.P.C.A.,* 37(12), 67, 1987.
16. Arber, J.M. and Urch, D.S., *Analyst.,* 113, 779, 1988.
17. Fischmeister, H.F., *Z. Anal. Chem.,* 332, 421, 1988.
18. Wieser, P., Wurster, R., and Seiler, H., *Atmos. Environ.,* 14, 485, 1980.
19. Dennemont, J., Jaccard, J., and Landry, J.C., *Int. J. Environ. Anal. Chem.,* 21, 115, 1985.
20. Delmas, S. and Muller, J.F., *Analusis,* 20, 165, 1992.
21. Hachimi, A., Krier, G., Poitevin, E., Muller, J.F., Schweigert, M.C., Klein, F., and Sowa, L., submitted.

22. Kimura, S., Kobayashi, M., Godai, T., and Minati, S., *Colloqium on Welding and Health,* Estoril, 1980, 1.
23. Delamar, M., *Analusis.,* 16, 419, 1988.
24. Hachimi, A., Millon, E., Poitevin, E., and Muller, J.F., *Analusis,* 21, 11, 1993.
25. Doyle, W.P. and Eddy, P., *Spectrochim. Acta,* 23A, 1903, 1967.
26. Bensted, J., *Naturwissenschaften,* 63, 193, 1976.
27. Farmer, V.C., *The Infrared Spectra of Minerals,* Mineralogical Society, London, 1974, 539.
28. Griffith, W.P., Advances in the Raman and infrared spectroscopy of minerals, *Spectroscopy of Inorganic-based Materials,* Clark, R.J.H. and Hester, R.E., Eds., John Wiley & Sons, New York, 1987, 119.
29. Karr, C., *Infrared and Raman Spectroscopy of Lunar and Terrestrial Minerals,* Academic Press, New York, 1975, 375.
30. Campbell, J.A., *Spectrochim. Acta,* 21, 1333, 1956.

8 Chemical Characterization of Aerosol Particles by Laser Raman Spectroscopy*

K. Hang Fung and Ignatius N. Tang

CONTENTS

INTRODUCTION

The importance of aerosol particles in many branches of science, such as atmospheric chemistry, combustion, interfacial science, and material processing, has been steadily growing during the past decades. One of the unique properties of these particles is the very high surface-to-volume ratios, thus making them readily serve as centers for gas-phase condensation and heterogeneous reactions. These particles must be characterized by size, shape, physical state, and chemical composition. Traditionally, optical elastic scattering has been applied to obtain the physical properties of these particle (e.g., particle size, size distribution, and particle density). These physical properties are particularly important in atmospheric science as they govern the distribution and transport of atmospheric aerosols.

The chemical characterization of airborne particles has always been tedious and difficult. It involves many steps in the process, namely, sample collection, species and/or size separation, and chemical analysis. There is a great need for non-invasive methods for *in situ* chemical analysis of suspended single particles. For bulk samples, Raman scattering fluorescence emission, and infrared absorption are the most common spectroscopic techniques. While fluorescence spectroscopy is extremely sensitive in terms of detection limit,[1] it lacks the spectral specificity required for chemical

* This research was performed under the auspices of the U.S. Department of Energy under Contract No. DE-AC02-76CH00016.

speciation. Furthermore, this technique can only be used for materials that fluoresce in the visible region and, therefore, is quite limited as an analytical tool for general application. Infrared spectroscopy has successfully been applied to chemical characterization of the organic and inorganic species in size-segregated aerosol samples collected on impactor plates.[2] Deposited single particles can also be analyzed by infrared microscopy.[3] On the other hand, although Arnold and co-workers[4-7] have obtained infrared spectra of levitated single aqueous droplets, the infrared absorption of the species is not directly measured in the experiment. Instead, the Mie scattering from the droplet is monitored and the size change due to evaporation as a result of infrared absorption is detected. The experiment is interesting but rather involved. It is difficult to adapt this technique to routine particle analysis because it requires the particle to be spherical in shape and to change size by evaporation during infrared absorption.

Despite the inherent low scattering cross-section of the spontaneous Raman scattering process, Raman spectroscopy has been used rather successfully in particle analysis. In contrast to fluorescence emission and infrared absorption techniques, Raman scattering can be applied to optically opaque, irregular-shaped samples. It is also ideally suited for microscopic samples as well. Moreover, it delivers rich vibrational molecular information that is comparable to infrared spectroscopy for identification purposes. The use of the Raman microprobe is a well-established method for analyzing samples collected on a substrate. Early work in this research area was led by Rosasco and co-workers.[8-12] Aerosol particles were collected on a filter substrate at first. Then the sample was illuminated by a high-power laser. Various type of compounds, such as inorganic minerals and carbonaceous materials, were analyzed by this technique. Adar and co-workers[13,14] have subsequently developed a highly automated micro/macroRaman spectrometer. The sensitivity and signal-to-noise ratio of the instrument are high enough to enable a spatial resolution of one micron.

However, there was still a lack of suitable measurement techniques for *in situ* chemical characterization of a levitated particle containing only about 10^{12} molecules. Thurn and Kiefer,[15,16] in an effort to develop a microprobe technique for suspended particles, have obtained Raman spectra of optically levitated glass particles. The optical levitation of a particle was first demonstrated by Ashkin and Dziedzic.[17] This is, in essence, a turning point for the application of Raman spectroscopy in aerosol research.[18-25] Raman spectroscopy of aerosol particles has several interesting properties that are of special interest to aerosol science. The morphology-dependent optical resonances that occur in the Mie scattering of dielectric spheres can interact with the Raman scattered photons. This interaction leads to two physical processes. At the low energy field regime, the simple Mie resonance can interfere and sometimes mask the Raman frequencies.[26] The overall inelastic scattered signal can be viewed as a linear summation of the spontaneous Raman scattering and the morphology-dependent Mie resonance. The Mie interference diminishes for larger spheres, as the resonance peaks become lower in amplitude and higher in numbers per spectral bandwidth. At the high energy regime, stimulated Raman emissions can be generated.[27-29] The Mie resonance peaks provide a high Q-factor for the Raman scattered photons to amplify coherently, and the intensity of the stimulated Raman peaks depend exponentially on the Q-factor of each Mie resonance peak. The stimulated Raman scattering is a nonlinear process, whose intensity is given by

$$I_{sr} = I_s \exp\left(g_s I_i z\right), \tag{8.1}$$

where I_s is the spontaneous Raman intensity, g_s is the gain factor, I_i is the incident laser intensity, and z is interaction path length. Mie resonances thus affect the stimulated Raman in two ways. First, the pump path for the laser through the interaction volume is lengthened, typically from the physical size of the particle of a few microns to several meters. The second effect is on the gain factor of the stimulated Raman scattering.[57-59] This gain factor is proportional to the number density of the Raman active species that are present in the particle. The effective number depends again on the particular Mie resonance peak. Despite the nonlinearity of the intensity in stimulated Raman scattering, some quantitative measurements have been carried out with streams of solution droplets, containing nitrates, sulfates, and phosphates.[28-30]

Resonance Raman scattering is another area of much interest to aerosol characterization. The resonance Raman effect arises when the incident laser frequency is chosen to approach or fall within an absorption band. There are several features that set the resonance Raman scattering technique apart from the spontaneous Raman scattering technique. The most important feature is it capability to probe extremely low concentration samples. However, due to absorption of the incident photons, the sample medium is no longer transparent, resulting in unwanted effects such as fluorescence and heating. In the condensed phase, fluorescence is much reduced by quenching and thus may not constitute an overwhelming problem as it would in the gas phase. Nevertheless, the heating effect is still formidable and this requires special sample-handling techniques for bulk media,[31] as well as aerosol particles.[32,33]

This chapter reviews the recent advances in the chemical and physical characterization of suspended single particles by laser Raman spectroscopy. Many of the current experiments outfitted with the state-of-the-art instrumentation are described. Various types of experimental set-ups for aerosol laser Raman spectroscopy are discussed in detail. The detection limits and the analytical applications of the spontaneous Raman and resonance Raman scattering are described and discussed at length. The limitations and future expectations of the Raman techniques in the field of aerosol research are also given.

EXPERIMENTAL TECHNIQUES

A variety of experimental set-ups with different lasers, particle containment chambers, and optical detectors have been used to measure Raman scattering from aerosol particles. It is best to divide the methodologies into two categories. One is the single-particle suspension method and the other is the monodisperse particle stream. These two sampling methods are most frequently used in Raman scattering experiments today.

Although commercial Raman microprobe systems are readily available, many of the aerosol Raman experiments are based on the needs of individual experiments. As a result, only the monochromator and detector components are obtained directly from commercial suppliers without any modifications. In general, an aerosol Raman experiment is designed with specific analytical purpose and the apparatus is built on a modular design basis for maximum flexibility.

LASER SOURCES

Currently, there is a wide range of commercially available lasers suitable for aerosol Raman scattering experiments. For spontaneous Raman scattering, the most frequently used continuous wave (CW) laser is the argon-ion laser. The argon-ion laser typically provides a line-tunable source in the visible and the near-ultraviolet regions. The wavelengths and their relative powers are tabulated in Table 8.1. The argon-ion laser is chosen for aerosol Raman experiments because it has several high-powered laser lines in the blue and green regions of the visible spectrum. Raman emission from these excitation lines fall within the maximum sensitivity region of most optical detectors. Even molecules with very large Raman frequency shifts, such as the OH band in a water molecule (3200 cm^{-1}), can be covered with these optical detectors. In contrast, a krypton-ion laser has nearly as high single-line output powers as the argon-ion laser; however, it has its high-power output lines in the red region (i.e., at 6470.88 Å and 6764.42 Å). Consequently, the typical Raman shifted symmetric vibrational bands for the inorganic and OH groups would appear near 7000 Å and 8200 Å, respectively, making the krypton laser less desirable. Moreover, the Raman scattering cross-section increases with frequency. Therefore, the blue region in the visible is spectrally most suitable for Raman excitation. For stimulated Raman scattering experiments, the most widely used laser for excitation is the solid-state YAG pulsed laser. The second harmonic line of the YAG laser

TABLE 8.1
Spectral Characteristics of Commonly Used Lasers

Argon-ion laser lines:	
Wavelength (Å)	Relative Intensity
3511.12	0.01
3637.78	0.01
4545.05	0.07
4579.34	0.18
4657.89	0.07
4726.85	0.10
4764.86	0.36
4879.86	0.93
4965.07	0.28
5017.16	0.18
5145.31	1.00
Krypton-ion laser lines:	
Wavelength (Å)	Relative Intensity
5208.31	0.14
5308.65	0.40
5681.88	0.20
6470.88	1.00
6764.42	0.24

at 5320 Å produces a stable and high-power output that is well-suited for stimulated Raman scattering. The third harmonic line is less frequently used than the 5320 Å line. The reason for its low popularity is twofold: (1) this line is higher in photon energy, and thus, increases the probability of multiphoton ionization, and (2) Rayleigh scattering presents some technical problems because the availability of optical filters for the ultraviolet region is still quite limited.

Sample Generation and Illumination

The most important consideration for sample containment and illumination is the efficiency of the optical elements involved. The physical dimensions of the particle containment chamber and the vibrating orifice particle generator are usually the determining factors for how the laser beam should be focused when only one laser beam is considered as the sole source for illumination, the minimum focal spot size of the beam for a diffraction-limited beam waist can be easily calculated.[34] The spot diameter is given by

$$d_{1/e} = (\pi\lambda/4)(f/D_{1/e}), \tag{8.2}$$

where $d_{1/e}$, $D_{1/e}$, λ, and f are the spot diameter, laser beam diameter, laser wavelength, and focal length, respectively. For a typical argon-ion laser with $D_{1/e}$ = 2 mm, at 4880 Å and 10 to 15 cm focal length, the spot diameter is between 20 to 30 μm. Thus, in the laboratory, suspended particles in the 15-μm diameter range can be easily illuminated by this beam. On the other hand, the pulsed YAG laser generates a laser beam with diameter equal to about 9 mm in the second harmonic. Therefore, the corresponding spot size is about 5 to 7 μm.

The most commonly used single-particle containment technique is the quadrupole electrodynamic suspension. A schematic diagram is shown in Figure 8.1. It consists of two dc endcaps and

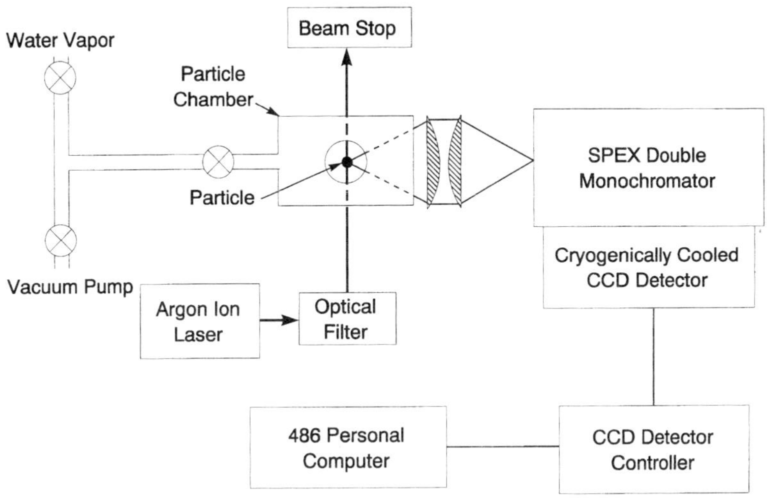

FIGURE 8.1 Schematic diagram of the experimental set-up for single-particle Raman spectroscopy.

an ac ring electrode. The dc field balances the particle against the gravitational force and the ac field maintains the particle at the center of the cell. A detailed description of the principles is given by Frickel et al.[35] Since the introduction of this quadrupole electrodynamic cell concept, there have been several modifications and variations of this design. Davis et al.[24,25] and Ray et al.[36] have used two ac ring electrodes with a dc offset over a glass tube to maximize the collection angle for Raman scattering and fluorescence experiments. Arnold et al.[6] have used a spherical void design to maximize the light collection efficiency.

In resonance Raman and stimulated Raman experiments, particles no longer suspended in electrodynamic cells. Instead, a stream of droplets are continuously generated by the Berglund-Liu vibrating orifice particle generator.[37] This piezoelectric vibrating orifice is made commercially available by TSI (Minneapolis, MN). The feed mechanism in the commercial model consists of a solution reservoir and a syringe pump. The flow rate is found to be uneven when highly monodisperse particles are desired. Snow et al.[27] and Lin et al.[38] showed that the reservoir can be pressurized by a compressed inert gas such as nitrogen to maintain a steady liquid flow, thus eliminating the use of the syringe pump. In addition, a high throughput, submicron-pore size solution filter can greatly enhance the stability of particle generation.

Collection Optics, Spectrometers, and Detectors

The collection optics and spectrometer should always be considered together in aerosol particle Raman scattering experiments. The size of the scattering source is very often the physical diameter of the particle that is imaged onto the entrance slit of the spectrometer. There are two aspects critical for the collection optics that are very important; namely, the magnification of the image and the desired resolution of the Raman spectrum. Assume that the *f*-numbers of the collection optics and the spectrometer are f_1 and f_2, respectively. Then, the magnification of the particle image with 100% transmission at the entrance slit would be

$$M = f_2 / f_1 . \tag{8.3}$$

However, the slit width, which limits the spectrometer resolution, must be set to at least a size of Md in order to transmit the entire particle image (d is the diameter of the particle). Therefore, the larger the particle, the lower the resolution one can obtain for a given dispersion of the spectrometer.

On the other hand, the best approach for high resolution in Raman scattering experiments is to use a spectrometer with high dispersion, which requires the use of both large grating and/or high groove density. This is because the product, *Md*, is fixed and the resolution of the spectrometer can only be increased by increasing the resolution of the grating.

In practice, Raman experiments require photon-counting techniques that yield the minimum noise level. Although some experiments are still carried out with photomultipliers, most recent experiments are carried out with more efficient detectors, such as the intensified photodiode and charged-coupled device (CCD) array detectors. These modern detectors offer an array approximately 25 mm long. The spatial resolution at the image field is in the vicinity of 22 to 25 μm. Considering the fact that the entrance slit of a typical spectrometer is normally set between 100 and 150 μm to accommodate the image of the aerosol particle, these array detectors thus serve the purpose very effectively. The spectral ranges of these detectors are comparable to those of photomultipliers; they can reach from 250 nm in the ultraviolet to 1100 nm in the infrared. A personal computer is currently a necessity for online control of both the spectrometer and the array detector, as well as for data acquisition and analysis.

There is a major difference between intensified array and non-intensified array detectors. The intensifier resembles a photomultiplier and therefore has intrinsic dark counts. The addition of dark counts due to the intensifier limits the exposure time for the array detector. However, the intensifier can be gated, or turned on momentarily in a pulsed laser experiment; hence, the dark counts are substantially reduced. Furthermore, the CCD detector can be cryogenically cooled to the point where the dark count is nearly zero. Therefore, the CCD detectors are extremely well-suited for very low signal level experiments. The CCD detectors have one intrinsic problem: namely, being subject to cosmic ray interference. As a result, the spectra obtained from long-time exposure of CCD arrays always contain numerous random high-intensity spikes due to cosmic rays. These spikes are typically one to two channels in width and can be numerically removed by software routines.

CURRENT ADVANCES IN CHEMICAL ANALYSES OF AEROSOL PARTICLES

The application of laser Raman spectroscopy in the field of aerosol research has steadily grown during the past decade. Although the work published in the literature covers a vast array of topics, it is helpful to categorize them into three general areas that hold special interests for aerosol researchers. These three areas are: (1) physical and chemical characterization of aerosol particles, (2) quantitative analyses by Raman spectroscopy, and (3) the development of resonance Raman spectroscopy for aerosol particles.

Characterization and Identification of Aerosol Particles

Aerosol particles of inorganic salts in the crystalline state usually exhibit characteristic Raman frequency shifts with a very narrow bandwidth; whereas, in solution, the corresponding Raman frequency shifts are slightly displaced and the peaks are broadened by molecular motion.[21,39,40] A typical example is shown in Figure 8.2, where the Raman spectra taken of a sodium nitrate ($NaNO_3$) particle (a) as a solution droplet, (b) during phase transformation from liquid solution to solid state, and (c) as a crystalline particle, clearly show the changes in the molecular vibrational band features for the same particle in different physical states. The observed Raman shifts at 1051 cm^{-1} for the free nitrate ion (NO_3^-) in aqueous solution droplets and at 1067 cm^{-1} for $NaNO_3$ crystalline particles are in good agreement with the literature data obtained for bulk samples. The measured linewidth for the droplet is typically 6 cm^{-1}, compared with only 2 cm^{-1} for the solid particle. Thus, the

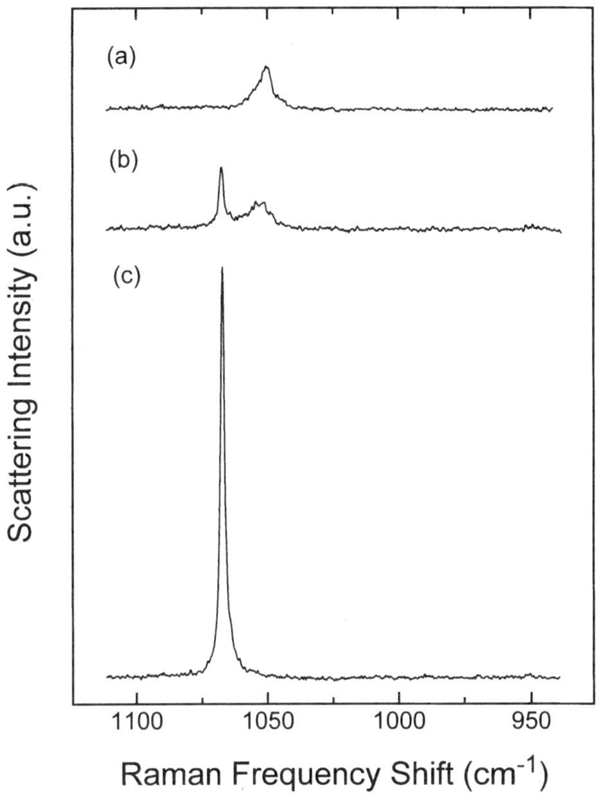

FIGURE 8.2 Raman spectra of an $NaNO_3$ solution droplet undergoing phase transformation to form a crystalline particle.

Raman shifts, combined with the large difference in the linewidth between the solid and liquid states, provide a viable means for particle characterization.

Many inorganic salts in the crystalline form can exist either as anhydrous salts or as hydrated salts containing one or more water molecules of crystallization, depending on the chemical nature and the crystallization conditions. Ammonium sulfate is a common constituent of atmospheric aerosols and it always exists in the anhydrous form. In bulk solutions, sodium sulfate crystallizes below 35°C to form the stable hydrated solid, $Na_2SO_4 \cdot 10H_2O$. Some inorganic salts may have more than one stable hydrated form. Chang and Irish[41] have reported Raman and infrared studies of hexa-, tetra-, and dihydrates of crystalline magnesium nitrate. The latter two hydrates are formed from partial dehydration of the hexahydrate under vacuum at 30 to 40°C. However, given the temperature extremes that can be attained in the atmosphere, most inorganic salts are not expected to exist in more than two different crystalline forms in atmospheric aerosols. For example, magnesium nitrate has two stable hydrated states that are expected to be present in ambient aerosols. At temperatures below –20°C, it exists as $Mg(NO_3)_2 \cdot 9H_2O$; and above –8°C, it exists as $Mg(NO_3)_2 \cdot 6H_2O$. These two hydrates may coexist at temperatures between –20 and –8°C. The anhydrous state and other hydrates of magnesium nitrate can only be prepared under conditions that are not encountered in the atmospheric environment.

In order to identify the hydrated or anhydrous forms present in an aerosol particle, it is necessary to have band resolutions better than a few wavenumbers (cm^{-1}). Table 8.2 gives a list of Raman frequencies for several common nitrates and sulfates. The proximity of these Raman vibrations clearly illustrates the need for high-resolution spectrometers for aerosol particle analyses. For

TABLE 8.2
Summary of Raman Frequencies (cm^{-1}) Observed for Inorganic Salt Particles

Nitrates		Sulfates			Phosphates		
$LiNO_3$	1070	$Li_2SO_4 \cdot H_2O$		1008	Na_2HPO_4		935
$LiNO_3 \cdot 3H_2O$	1056	Na_2SO_4		996	$(NH_4)_2HPO_4$		913
$NaNO_3$	1067	$Na_2SO_4 \cdot 10H_2O$		992	$NH_4H_2PO_4$		913
KNO_3	1053	K_2SO_4		983			
NH_4NO_3	1050	$(NH_4)_2SO_4$		975			
$Mg(NO_3)_2$	1064	$MgSO_4 \cdot 7H_2O$		983			
$Mg(NO_3)_2 \cdot 6H_2O$	1059						
$Ca(NO_3)_2 \cdot 4H_2O$	1050				**Chromates**		
$Sr(NO_3)_2$	1056				Na_2CrO_4		851
$Ba(NO_3)_2$	1047				K_2CrO_4		852
$Pb(NO_3)_2$	1047						
		Solution Droplets			**Mixed Salts**		
		NO_3^-		1048	$Na_2SO_4 \cdot NaNO_3$	996	1063
		SO_4^{2-}		980	$(NH_4)_2SO_4 \cdot NH_4NO_3$	975	1043
		HSO_4^-	892	1048	NH_4HSO_4	860	1025
					$(NH_4)_3HSO_4$	960	1065

example, the presence of anhydrous sodium sulfate (Na_2SO_4) or the hydrated form ($Na_2SO_4 \cdot 10H_2O$) in aerosol particles can only be confirmed with a minimum resolution of ±1 cm^{-1}, which is needed to identify the corresponding Raman frequencies of 996 cm^{-1} and 992 cm^{-1}, respectively.

Aerosol particles composed of inorganic salts such as chlorides, sulfates, and nitrates are hygroscopic and exhibit the properties of deliquescence and efflorescence in humid air. These aerosols play an important role in many atmospheric processes that affect local air quality, visibility degradation, as well as global climate. The hydration behavior, the oxidation and catalytic capabilities for trace gases, and the optical and radiative properties of the ambient aerosol all depend crucially on the chemical and physical states in which these microparticles exist. The existence of hygroscopic aerosol particles as metastable aqueous droplets at high supersaturation has routinely been observed in the laboratory[42-44] and verified in the ambient atmosphere.[45] Because of the high degree of supersaturation at which a solution droplet solidifies, a metastable amorphous state often results. The formation of such state is not predicted from bulk-phase thermodynamics and, in some cases, the resulting metastable state is entirely unknown heretofore.[46] Figure 8.3 shows the hydration behavior of the $Sr(NO_3)_2$ particle, where the particle mass change resulting from water vapor condensation or evaporation is expressed in moles H_2O per mole solute and plotted as a function of relative humidity (%RH). A crystalline anhydrous particle, whose Raman spectrum shown in Figure 8.4b, displays a narrow peak at 1058 cm^{-1} and a shoulder at 1055 cm^{-1}, was first subjected to increasing RH (filled circles). The solid particle was seen to deliquesce at 83% RH when it spontaneously gained weight by water vapor condensation and transformed into a solution droplet containing about 13 moles H_2O/ moles solute. Further growth of the droplet, as RH was again increased, was in complete agreement with the curve computed from bulk solution data.[47] As RH was reduced, the droplet started to lose weight by evaporation (open circles). It remained a supersaturated metastable solution droplet far below the deliquescence point until it abruptly transformed into an amorphous solid particle at ~60% RH. The particle retained some water even in vacuum. The Raman spectrum of such a particle is shown in Figure 8.4d, displaying a broad band at 1053 cm^{-1}, in sharp contrast to those of the anhydrous particle and the bulk solution (Figure 8.4c). In most cases, an amorphous solid particle would continuously absorb a very small amount of water upon increasing RH until they deliquesced at 69% RH. Once in solution, the particle

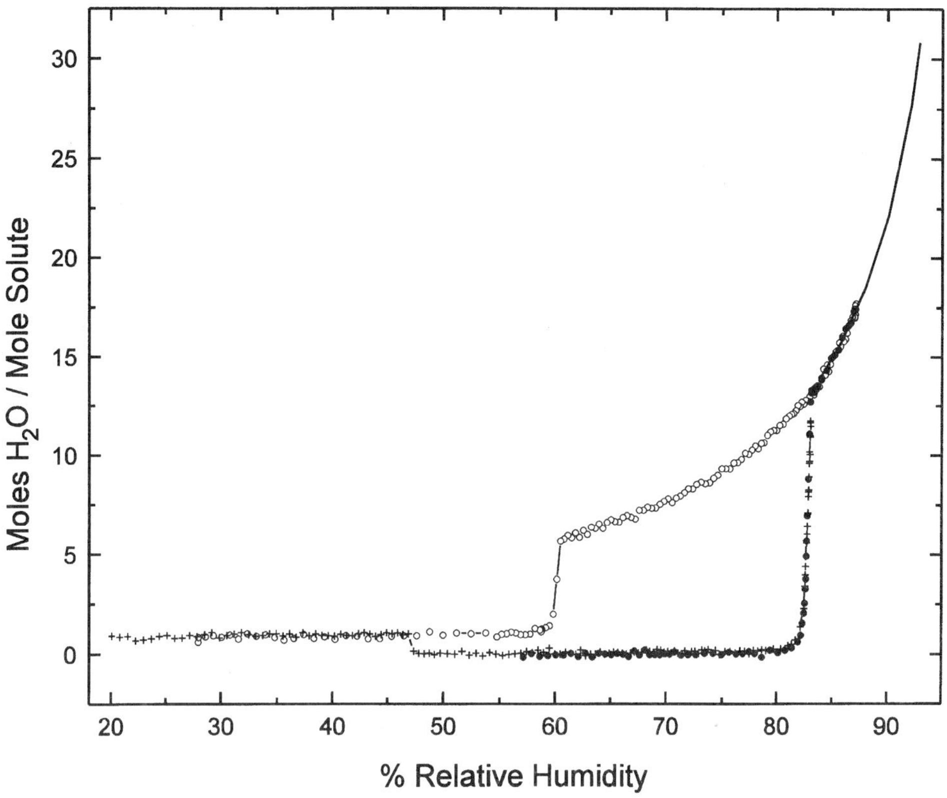

FIGURE 8.3 Growth and evaporation of a suspended $Sr(NO_3)_2$ particle in a humid environment: (a) particle growth #1 (•); (b) particle growth #2 (+); (c) droplet evaporation (o); and, (d) literature data (solid line).

would behave like a typical solution droplet. In the special case shown in Figure 8.3, however, the particle (crosses) was observed to have transformed first into an anhydrous particle during increasing RH and the deliquesced at 83% RH, indicating that the amorphous solid particle was metastable with respect to the anhydrous state. The Raman spectrum of the hydrated $Sr(NO_3)_2 \cdot 4H_2O$ is shown in Figure 8.4a for comparison. This hydrated form of strontium nitrate is the one that exists in bulk samples, but is not found in particles.

Other nitrate systems such as calcium nitrate and magnesium nitrate also show the formation of amorphous state upon recrystallization of solution droplets. Typically, the water content of these amorphous particles increases slightly with increasing relative humidity. They have a distinctive deliquescence point that is lower than that of their respective crystalline counterparts. In addition to these nitrate systems, metastable states are observed in several bisulfate systems. Figure 8.5b shows a Raman spectrum of ammonium bisulfate, NH_4HSO_4, in bulk samples. The strongest bisulfate bands are centered at 1013 and at 1041 cm^{-1}. However, the ammonium bisulfate particle shows a completely different spectrum, as shown in Figure 8.5a. The strongest band is no longer split, but centers at 1021 cm^{-1}. All the other spectral features are simpler and slightly shifted as well. It has been proposed[48] that the bisulfate has two different structures in the crystalline form. As a result, a splitting occurs at the bisulfate vibration bands. When a bisulfate solution droplet recrystallizes at high supersaturation, it is likely that, due to kinetic constraints, only one of the two proposed structures emerges to form the crystalline phase, yielding a Raman spectrum with less vibration bands.

Ambient aerosols are far from being a single-component system. In fact, the chemical composition of atmospheric aerosols is highly complex and may vary considerably with time and location.

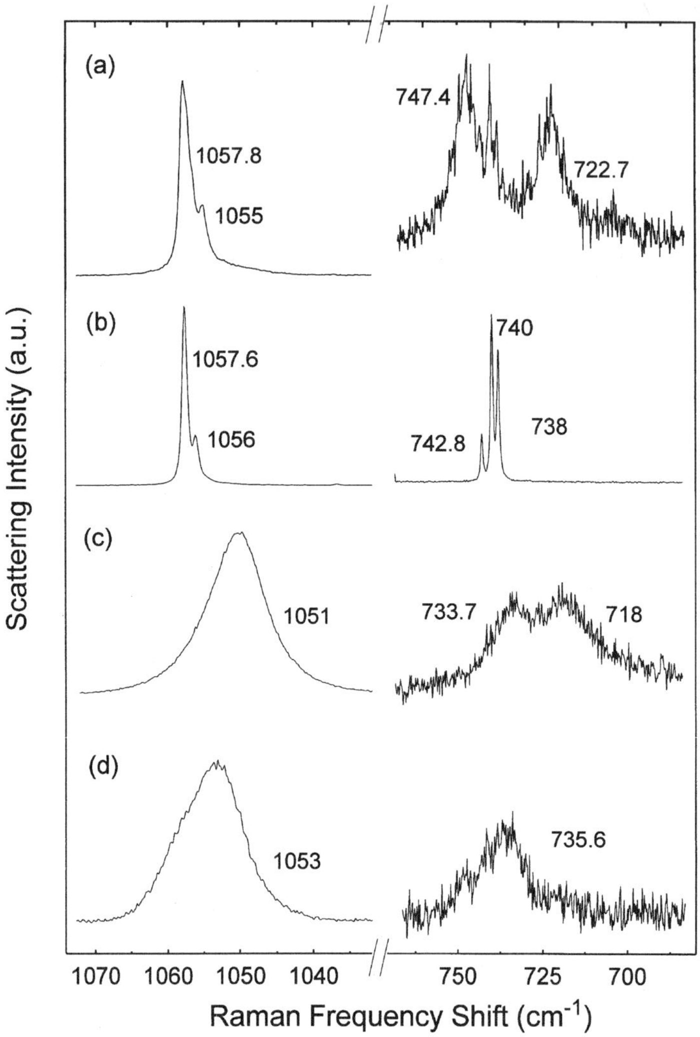

FIGURE 8.4 Raman spectra of $Sr(NO_3)_2$ in different physical and chemical states: (a) crystalline $Sr(NO_3)_2 \cdot 4H_2O$; (b) anhydrous $Sr(NO_3)_2$; (c) bulk $Sr(NO_3)_2$ aqueous solution; and d) $Sr(NO_3)_2$ particle.

In solution droplets, the presence of different cations does not appreciably affect the vibration frequencies of the anions that are being monitored by the Raman spectroscopic technique; therefore, the free ions (such as nitrate and sulfate ions) exhibit their characteristic Raman shifts, for all practical purposes, irrespective of the different kinds of cations present in the droplet. However, when a droplet containing multicomponent electrolytes transforms into a solid particle under low humidity conditions, the chemistry and kinetics of the system will operate to govern the outcome of crystallization process.

Thus, for non-interacting systems, the droplet will simply solidify to contain salt mixtures that make up the composition of the original dry-salt particle. For these particles, the composition can be determined from the relative peak intensities and the Raman cross-sections of the respective components. Figure 8.6a shows a Raman spectrum of a potassium nitrate and potassium sulfate solution droplet, indicating only SO_4^{2-} at 980 cm^{-1} and NO_3^- at 1049 cm^{-1} without any information about the cation. The Raman spectrum of the recrystallized solid particle is shown in Figure 8.6b, where the peaks reveal the characteristic Raman shifts of K_2SO_4 at 983 cm^{-1} and KNO_3 at 1053 cm^{-1}.

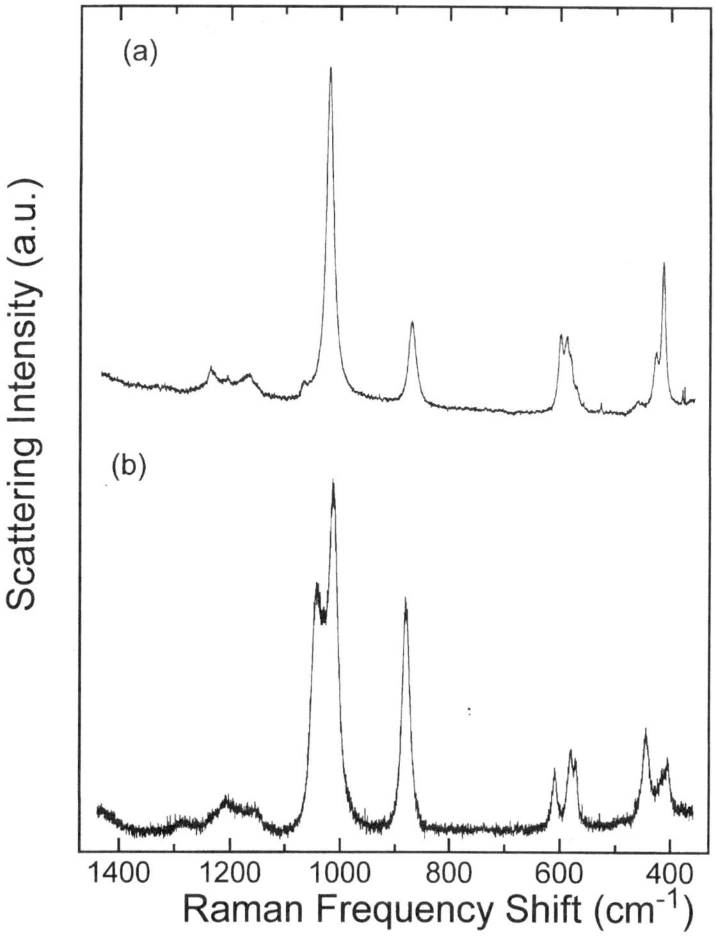

FIGURE 8.5 Raman spectra of NH_4HSO_4 in (a) a particle, and (b) in bulk phase.

Note that the band broadening effect in the droplet is quite apparent as compared to the crystalline particle.

However, many inorganic salts upon crystallization from its aqueous solution are known to form mixed salts that are stable stoichiometric compounds. Mixed salts have been shown to be present in ambient aerosols and in laboratory-generated aerosols. The Raman lines of mixed salts may be very different from those of the pure component salts, or they may represent a slight displacement that only becomes apparent with ultra-high spectral resolution. For example, in the crystallization of a solution droplet containing sodium and ammonium cations and sulfate and nitrate anions,[39] the solid particle may contain salts of all possible combinations, namely, NH_4NO_3 $(NH_4)_2SO_4$, $NaNO_3$, and Na_2SO_4, which have strong symmetric Raman bands at 1050 cm^{-1}, 975 cm^{-1}, 1067 cm^{-1} and 996 cm^{-1}, respectively (see Table 8.2). In addition, mixed salts can also form.[49] For example, in the solid particle formed from a solution droplet containing Na_2SO_4 and $NaNO_3$ (molar ratio 1:4), the Raman spectrum shown in Figure 8.7a reveals the presence of not only the pure components at 1067 cm^{-1} and 996 cm^{-1}, but a new band at 1063 cm^{-1}, which is attributed to the presence of mixed salt $NaNO_3 \cdot Na_2SO_4 \cdot H_2O$. Similarly, Figure 8.7b shows the Raman spectrum of a solid particle containing a 1:4 mixture of $NaNO_3$ and NH_4NO_3, where a new Raman band observed at 1053 cm^{-1} is attributed to the formation of the mixed salt $2NH_4NO_3 \cdot NaNO_3$. The formation of the mixed crystal in an aerosol particle is largely governed by the kinetic

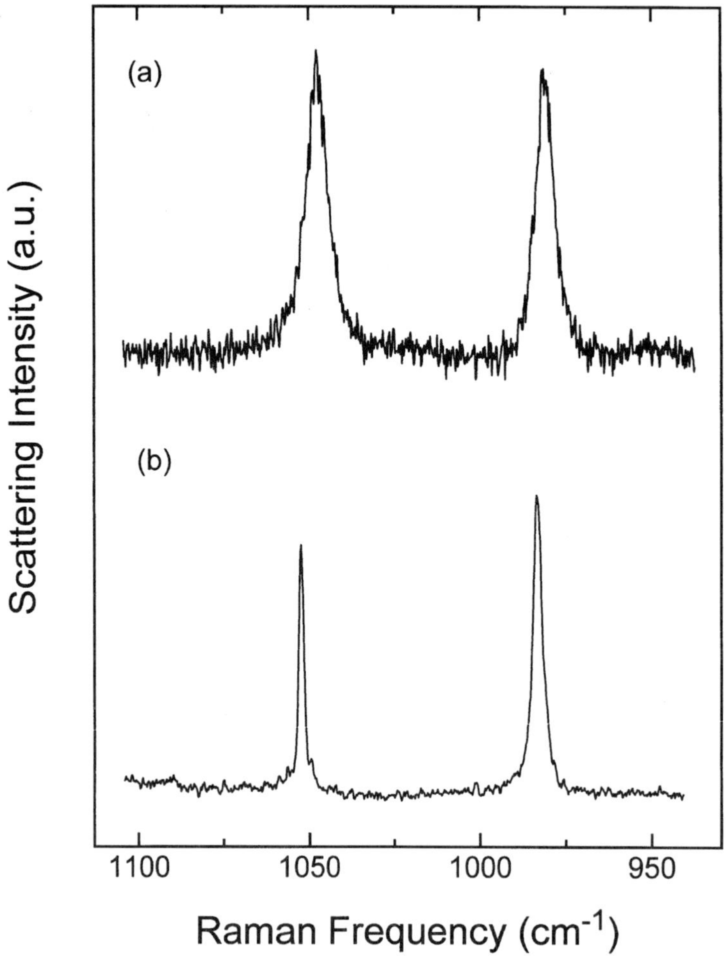

FIGURE 8.6 Raman spectra of a solution droplet containing K_2SO_4 and KNO_3 (a) before and (b) after crystallization.

conditions at crystallization. For droplets of identical composition, the outcome of the mixed crystals is not always the same.

Quantitative Analyses

There are several aspects in considering the use of spontaneous Raman scattering as a quantitative measuring technique for aerosol particles. In principle, the Raman scattering intensity, I_s is proportional to the total number of Raman active scattering molecules or centers, $l\sigma\rho$, and the intensity of the excitation source, I_i:

$$I_s = I_i l\sigma\rho, \tag{8.4}$$

where l is the interaction length, σ is the Raman scattering cross-section, and ρ is the density. However, for aerosol particles, these parameters are extremely difficult to measure in practice. As the size of the particle changes, the number of Raman active scattering molecules or centers will be different, and the overlap between the laser beam and the particle can also vary. Moreover, for

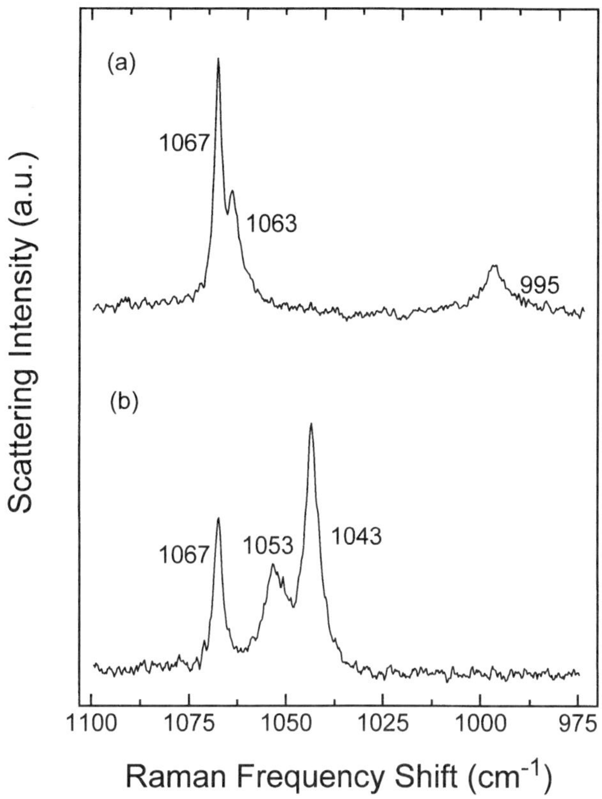

FIGURE 8.7 Raman spectra of multicomponent salt particles showing the presence of mixed salts: (a) 4:1 mixture of $NaNO_3$ and Na_2SO_4, and (b) 1:4 mixture of $NaNO_3$ and NH_4NO_3.

solution droplets, the morphology dependent Mie resonances can interfere with and modify the overall Raman scattering intensity. Therefore, it is helpful to have an internal standard for aerosol Raman intensity measurement. This internal standard can easily eliminate the particle size variation and the fluctuation in the intensity of the excitation source. In laboratory studies, many non-interacting Raman active species can be added to the samples of interest. For ambient aerosols, water is often a dominant component and can be used as an intensity reference.[50] On a positive note, aerosol particles are physically thin samples. Typically, they are only a few micrometers in diameter. Thus, problems arising from optical diffusiveness as encountered[51] in bulk samples have less effect on aerosol particles.

An example of quantitative measurement is illustrated with the system of ammonium sulfate and sodium sulfate solid mixtures.[49] A Raman spectrum of an aerosol particle composed of $(NH_4)_2SO_4$ and Na_2SO_4 is shown in Figure 8.8. This spectrum represents an exposure of 10 seconds, producing a signal intensity about 6000 counts/s. The peak shape is entirely Lorentian. The symmetric vibrational bands of the two sulfate groups show a small overlap. To account for the proper integrated peak-area signal, the spectrum is computer-resolved and best-fitted with a set of optimal values of peak position and width by a numerical routine. The optimization algorithm follows the nonlinear least-squares method outlined by Marquardt.[52] In Figure 8.9, a plot of the scattering intensity ratio against the molar mixing ratio of Na_2SO_4 to $(NH_4)_2SO_4$ is shown. The linearity of this plot is very good. The slope of the line, which represents the relative Raman cross-section ratio of Na_2SO_4 to $(NH_4)_2SO_4$ in this case, is found to be 0.65 ± 0.01 by liner regression analysis. Experimental data points, in general, represent the average results from at least three

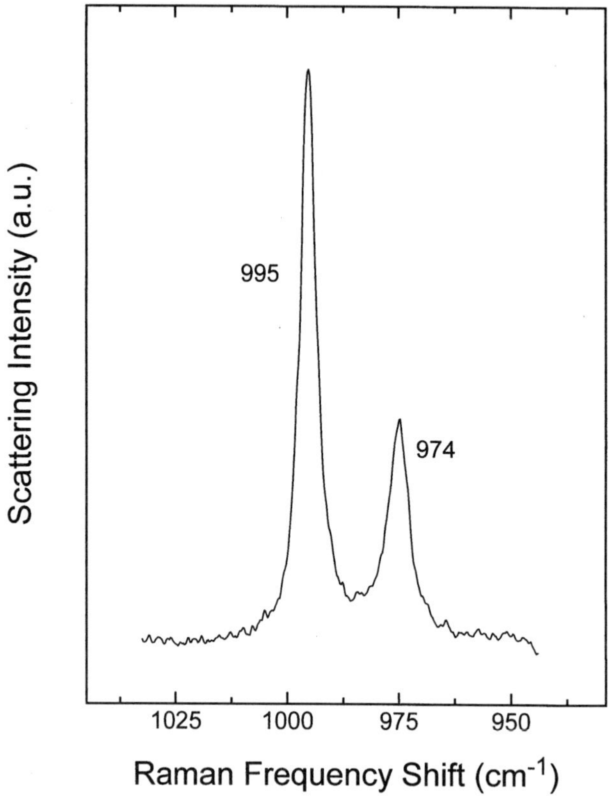

FIGURE 8.8 Raman spectra of a suspended $(NH_4)_2SO_4$ + Na_2SO_4 (1:4) particle.

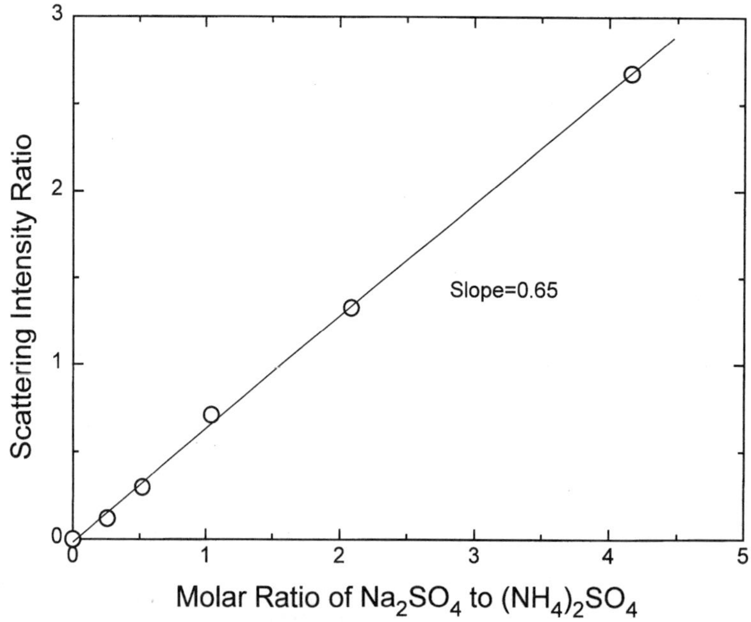

FIGURE 8.9 Dependence of relative Raman intensity on molar ratio of Na_2SO_4 to $(NH_4)_2SO_4$ in particles.

different aerosol particles; this is to ensure the even distribution of samples. The line width of both sulfate peaks shows a small increase, when compared to that of the pure component form. The slight broadening of the Raman peaks indicates the presence of the solid mixture of the two sulfates.

The quantitative Raman analysis for microdroplets[53,54] needs special attention. The morphology-dependent Mie resonances can affect the over-all Raman scattering intensity. The incident intensity is given by

$$I_i = 1/2(e/\mu)^{1/2}|E_1|^2, \tag{8.5}$$

where e is the electric inductive capacity and μ is the magnetic inductive capacity. $|E_1|^2$ is the internal electric field strength due to the incident beam. According to Mie theory, this internal field for spherical particles is different from that for bulk samples. The Raman scattering intensity is linearly proportional to the intensity of the incident radiation:

$$I_s = \text{constant} \times I_i. \tag{8.6}$$

Therefore, the morphology-dependent resonances directly modify the Raman emission from a spherical droplet. In order to compensate for this effect, the best approach is to use an internal standard to correct for this input Mie resonance effect. The Raman scattered photons are also subject to the Mie resonance condition. This output resonance effect can be seen to produce superimposed components on the spontaneous Raman signals.[15,18]

Resonance Raman Spectroscopy

As mentioned earlier, the resonance Raman effect arises when the incident laser frequency is tuned to the absorption band of the species of interest. The absorption spectra of the aqueous solutions of sodium dichromate, sodium chromate, potassium permanganate, and *p*-NDMA (*p*-nitrosodimethylaniline) are shown in Figure 8.10.[33] In this example, the excited state of both dichromate and chromate lie outside the range of the wavelengths available in the argon-ion excitation laser. Therefore, the resonance effects can be interpreted as pre-resonance Raman. The absorption band of the *p*-NDMA and the permanganate solutions provides a better overlap with the laser coverage. Thus, they can be considered in the resonance Raman regime. However, the permanganate may be governed by some of the post-resonance effects, as the excitation energy is higher than the maximum of the absorption band.

Due to the pre-resonance Raman effect, the dichromate and chromate ions were found to have cross-sections only about 12 and 10 times larger than that of the nitrate ion, respectively. Here in this study of aerosol particles, the nitrate ion was used as the internal standard, enabling the measurement of relative Raman cross-sections. The permanganate solution shows dominantly post-resonance effects, as the laser energy lies beyond the absorption maximum. A detailed study of this wavelength dependence has been made by Kiefer and Bernstein[31,55] with bulk solution samples. In droplets, the permanganate ion was found to have its Raman cross-section about 300 times larger than that of the nitrate ion. For *p*-NDMA, there are two strong Raman bands at 1164 cm^{-1} and 1613 cm^{-1}, which are the phenyl-nitroso deformation and symmetric benzene ring-stretching vibrations, respectively. Figure 8.11 shows the Raman spectrum of a solution droplet containing potassium nitrate (0.02 *M*), potassium sulfate (0.02 *M*), and *p*-NDMA (10^{-5} *M*). At the 4880 Å excitation wavelength, the measured enhancement for *p*-NDMA with respect to nitrate or sulfate is 3×10^4. The detection limit in this example is of the order of 10^{-7} *M* for *p*-NDMA.[30]

As is well known, the Mie theory precisely describes the light scattering from spherical particles. The Mie scattering function can be strongly influenced by the imaginary, or the absorption part of

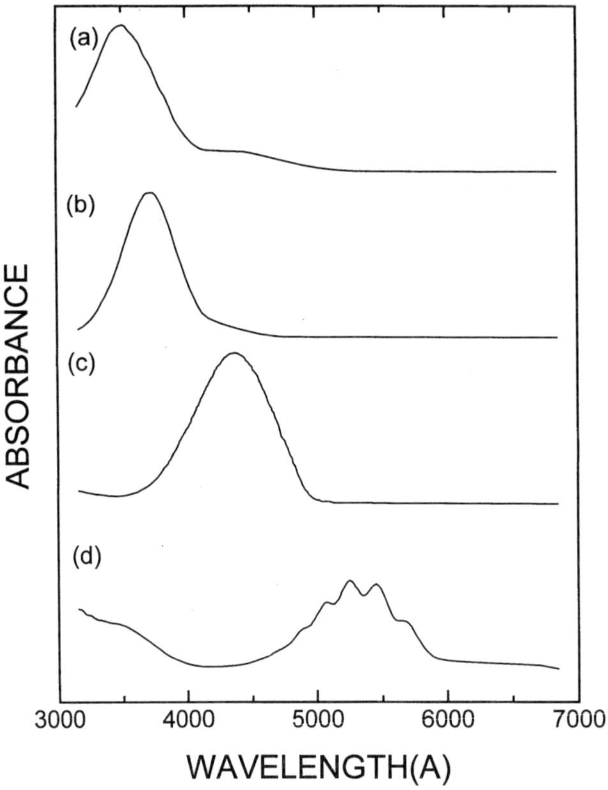

FIGURE 8.10 Absorption spectra of aqueous solutions of (a) sodium dichromate, (b) sodium chromate, (c) *p*-NDMA, and (d) potassium permanganate.

the index of refraction. As this imaginary part increases, both the angular scattering intensity distribution and the size scattering intensity, distribution become more monotonic. Effectively, the morphology-dependent peaks are softened by the absorption component and the scattering function approaches the absorption limit of the scattering center. Kerker[56] has given a detailed discussion as well as graphic illustration of the effects of the imaginary part on the scattering function. Besides the effects of the imaginary part on the Mie scattering function, the variation in the droplet size can also affect the Mie resonances. For example, a 45-μm droplet would have very dense morphology-dependent resonance peaks. Typically, the change in the droplet diameter is about 0.26 μm between adjacent resonance peaks. The large light collection angle (approximately 60°) used in the Raman scattering experiment further reduces this 0.26-μm spacing to 0.12 μm. Meanwhile, the Mie resonance peak width is also broadened, from 0.05 μm to 0.02 μm, by the large light collection angle. Therefore, an estimate of less than 0.1 μm or 0.2% variation in the droplet diameter would sufficiently smooth out most of the Mie resonance features. The absence of the Mie elastic scattering features in the spectra can be attributed to the two factors mentioned above. Even in the event of highly monodisperse droplets, this unique property of resonance Raman spectroscopy can be used to dampen the Mie resonance peaks. Hence, a more meaningful quantitative measurement can be obtained.

Another unique feature in the resonance Raman scattering is the occurrence of a long progression of overtones. From the point of molecular spectroscopy, these overtones allow the determination of anharmonicity in the molecular vibration. Such observation was obtained on solid potassium chromate by Kiefer and Bernstein.[55] A total of ten harmonics of the internal stretching mode, ν_1,

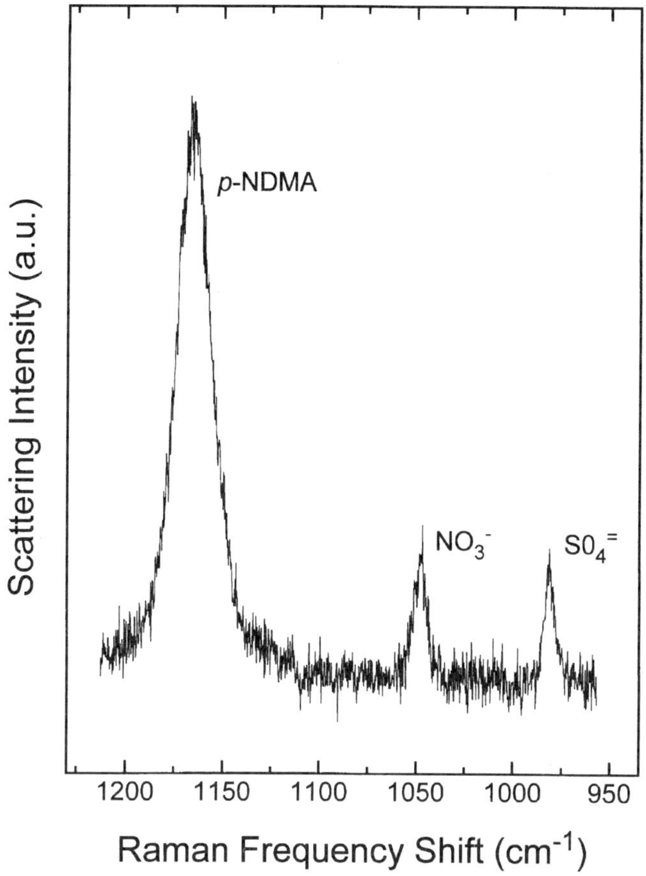

FIGURE 8.11 Raman spectrum of a solution droplet containing potassium nitrate, potassium sulfate, and *p*-NDMA.

at 853 cm^{-1} was observed. In addition, only total symmetric vibrations have such characteristics. It was observed that some B symmetry vibrations were absent in the resonance Raman spectrum. In the case of solution droplets, overtones for the permanganate ion were also observed.[32] The progression was limited to a few overtones due to the lack of sensitivity. The anharmonicity obtained from the solution droplet is in good agreement with the one derived from bulk samples.

SUMMARY AND FUTURE DEVELOPMENT

As laser Raman spectroscopy of aerosol particles is only in its infancy, new developments leading toward higher sensitivity and better selectivity for chemical characterization are anticipated. Extraction of information from Mie scattering-affected Raman bands is of particular importance to microdroplet analysis. During the past decade, there have been many advances made in optical instrumentation development, the generation and containment of aerosol particles, and other spectroscopic analytical techniques as well. However, particle Raman spectroscopy has almost become a standard laboratory technique for microparticle research. Current Raman scattering techniques largely focus on the aerosol particle as a whole. Resonance enhancement techniques would allow the investigation of the surface layer coverage of the aerosol particles by selectivity tuning the excitation wavelength to the absorption bands of the species of interest. Since many chemical and physical processes are occurring at the gas–particle interface, an ultimate challenge is to probe and study these surface layers. The resonance Raman technique may emerge as an important tool in this respect.

REFERENCES

1. Barnes, M.D., Whitten, W.B., and Ramsey, J.M., *Anal. Chem.*, 67, 418A, 1995.
2. Allen, D.T., Palen, E.J., Haimov, M.I., Hering, S.V., and Young, J.R., *Aerosol Sci. Technol.*, 21, 325, 1994.
3. Allen, D.T. and Palen, E.J., *J. Aerosol Sci.*, 20, 441, 1989.
4. Arnold, S. and Pluchino, A.B., *Appl. Optics,* 21, 4194, 1982.
5. Arnold, S., Murphy, E.K., and Sageev, G., *Appl. Optics,* 24, 1048, 1985.
6. Arnold, S., Neuman, M., and Pluchino, A.B., *Optics Lett.*, 9, 4, 1984.
7. Sageev-Grader, G., Arnold, S., Flagan, R.C., and Seinfeld, J.H., *J. Chem. Phys.,* 86, 5897, 1987.
8. Rosasco, G.J., Etz, E.S., and Cassatt, W.A., *Appl. Spectrosc.,* 29, 396, 1975.
9. Rosasco, G.J., Roedder, E.R., and Simmons, J.H., *Science,* 190, 557, 1975.
10. Rosasco, G.J. and Blaha, J.J., *Appl. Spectrosc.,* 34, 140, 1980.
11. Blaha, J.J., Rosasco, G.J., and Etz, E.S., *Appl. Spectrosc.,* 32, 292, 1978.
12. Blaha, J.J. and Rosasco, G.J., *Anal. Chem.,* 50, 892, 1978.
13. Grayzel, R., LeClerq, M., Adar, F., Lerner, J., Hutt, M., and Diem, M., *Microbeam Analysis,* Armstrong, J.T., Ed., San Francisco Press, San Francisco, 1985.
14. Adar, F., *ACS Symposium Series,* 295, 230, 1986.
15. Thurn, R. and Kiefer, W., *Appl. Spectrosc.,* 38, 78, 1984.
16. Thurn, R. and Kiefer, W., *Appl. Optics.,* 24, 1515, 1985.
17. Ashkin, A. and Dziedzic, J.M., *Phys. Rev. Lett.,* 38, 1351, 1977.
18. Lettieri, T.R. and Preston, R.E., *Optics Comm.,* 54, 349, 1985.
19. Preston, R.E., Lettieri, T.R., and Semerjian, H.G., *ACS Langmuir J.,* 1, 365, 1985.
20. Schrader, B., *Physical and Chemical Characterization of Individual Airborne Particles* K.R. Spurny, Ed., Chapt. 19. Halsted, New York, 1986.
21. Fung, K.H. and Tang, I.N., *Appl. Optics,* 27, 206, 1988.
22. Schweiger, G., *Particle Charact.,* 4, 67, 1987.
23. Schweiger, G., *J. Aerosol Sci.,* 21, 483, 1990.
24. Davis, E.J. and Buehler, M.F., *Mater. Res. Soc. Bull.,* 15, 26, 1990.
25. Davis, E.J., Buehler, M.F., and Ward, T.L., *Rev. Sci. Instrum.,* 61, 1281, 1990.
26. Chew, et al. (1976).
27. Snow, J.B., Qian, S.X., and Chang, R.K., *Opt. Lett.,* 10, 37, 1985.
28. Eickmans, J.H., Qian, S.X., and Chang, R.K., *Part. Charact.,* 4, 85, 1987.
29. Serpengüzel, A., Chen, G. and Chang, R.K., *Part. Sci. Technol.,* 8, 197, 1990.
30. Fung, K.H., Imre, D.G., and Tang, I.N., *J. Aerosol Sci.,* 25, 479, 1994.
31. Kiefer, W. and Bernstein, H.J., *Appl. Spectrosc.,* 25, 609, 1971.
32. Fung, K.H. and Tang, I.N., *J. Aerosol Sci.,* 23, 301, 1992.
33. Fung, K.H. and Tang, I.N., *Appl. Spectrosc.,* 46, 159, 1992.
34. Strong, J., *Concepts of Classical Optics,* Freeman, San Francisco, 1958.
35. Frickel, R.H., Schaffer, R.H., and Stamatoff, J.B., *Chamber for the Electrodynamic Containment of Charger Aerosol Particles,* National Technical Information Service Report No. AD/A 056236, U.S. Department of Commerce, Springfield, VA, 1978.
36. Ray, A.K., Souyri, A., Davis, E.J., and Allen, T.M., *Appl. Optics,* 30, 3974, 1991.
37. Berglund, R.N. and Liu, B.Y.H., *Environ. Sci. Technol.,* 7, 147, 1973.
38. Lin, H.-B., Eversole, J.D., and Campillo, A.J., *Rev. Sci. Instrum.,* 61, 1018, 1990.
39. Fung, K.H. and Tang, I.N., *J. Colloid Interface Sci.,* 130, 219, 1989.
40. Tang, I.N. and Fung, K.H., *J. Aerosol Sci.,* 20, 609, 1989.
41. Chang, G.T. and Irish, D.E., *Can. J. Chem.,* 85, 995, 1973.
42. Orr, C., Hurd, F.K., and Corbett, W.J., *J. Colloid Sci.,* 13, 472, 1958.
43. Tang, I.N., Munkelwitz, H.R., and Davis, J.G., *J. Aerosol Sci.,* 8, 149, 1958.

44. Richardson, C.B. and Spann, J.F., *J. Aerosol Sci.,* 15, 563, 1984.
45. Rood, M.J., Shaw, M.A., Larson, T.V., and Covert, D.S., *Nature* 337, 537, 1989.
46. Tang, I.N., Fung, K.H., Imre, D.G., and Munkelwitz, H.R., *Aerosol Sci. Technol.,* 23, 443, 1995.
47. Robinson, R.A., and Stokes, R.H., *Electrolyte Solutions,* 2nd Ed., Butterworth, London, 1970.
48. Payan, F. and Haser, R., *Acta Cryst.* B32, 1875, 1976.
49. Fung, K.H. and Tang, I.N., *Appl. Spectrosc.,* 45, 734, 1991.
50. Chang, C.K., Flagan, R.C., and Seinfeld, J.H., presented at the *1990 American Association for Aerosol Research Annual Meeting,* Philadelphia, PA, 1990.
51. Blomer, F. and Moser, H., *Z. Angew. Phys.,* 5, 302, 1969.
52. Marquardt, D.W., *J. Soc. Industr. Appl. Math.,* 11, 431, 1963.
53. Buehler, M.F., Allen, T.M., and Davis, E.J., *J. Coll. Interface Sci.,* 146, 79, 1991.
54. Buehler, M.F. and Davis, E.J., *Colloids and Surfaces: A Physicochemical and Engineering Aspects,* 79, 137, 1993.
55. Kiefer, W. and Bernstein, H.J., *Molec. Phys.,* 23, 835, 1972.
56. Kerker, M., *The Scattering of Light and Other Electromagnetic Radiation,* Academic Press, New York, 1969.

OTHER RELEVANT PUBLICATIONS

57. Kwok, A.S. and Chang, R.K., *Optics & Photonics News,* 4, 34, 1993.
58. Lin, H.-B., Eversole, J.D., and Campillo, A.J., *Opt. Lett.,* 17, 828, 1992.
59. Mazumder, M.D., Schaschek, K., Chang, R.K., and Gillespie, J.B., *submitted to Optics Letters,* 1995.

9 Novel Applications of the Electrodynamic Levitater for the Study of Aerosol Chemical Processes

Glenn O. Rubel

CONTENTS

INTRODUCTION

By virtue of their name, aerosols are involved in a wide variety of chemical and physical processes that are operative at the gas/liquid–solid interface. A complete list of these processes would probably fill this page and it is not the intent of this chapter to be a comprehensive review of aerosol physics and chemistry. Indeed, this chapter focuses on the implementation of a specific experimental method — single-particle electrodynamic levitation (SPEL) — to study a narrow class of aerosol rate processes, including water condensation-evaporation, heterogeneous reactions, monolayer resistance to evaporation and reaction, droplet microencapsulation, and gas adsorption onto solid particulates. For a more complete discussion of the development and application of SPEL, see the review by Davis.[1] The selection of scientific problems discussed in this chapter was governed by three primary factors: areas where data gaps exist; areas of scientific controversy; and areas where single-particle analysis could substitute for bulk analysis methods. As an example of the use of SPEL to fill an existing data gap, we cite the study on the evaporation of multicomponent oil solutions.[2,3] Questions on the ideality and the validity of correlative relationships between molecular weight and hydrocarbon vapor pressure were addressed using single-particle levitation.

Perhaps more intriguing are the opportunities to resolve scientific controversies that arise from studies employing bulk and/or aerosol measurement methods. One such controversy concerned the heterogeneous reaction between acid aerosols and base gases. Robbins and Cadle[4] and Huntzicker et al.[5] both measured the reaction rate between sulfuric acid aerosol and ammonia gas. However, because the researchers were employing discrete time analysis methodologies, they were unable to detect the transition from a surface-phase reaction to a gas-phase diffusion-controlled reaction. In contrast, SPEL levitates single droplets, which permits a continuous measurement of the droplet-gas reaction. As a result, the transition from surface-phase to gas-phase diffusion-controlled reaction could be detected using SPEL.

In addition to the advantages of continuous measurement, another powerful feature of SPEL is its capability to measure rate processes for single micrometer-sized particles. Because measurement time increases with increasing particle size for surface area-dependent processes, the SPEL measurement time is significantly smaller than most bulk method measurement times. This translates into a cost-saving feature that makes SPEL an attractive concept for future development as a standard test measurement method. This is exemplified by the case where water vapor isotherms for carbon were measured using SPEL in one-tenth the time required by standard bulk measurement methods.[6] This chapter begins with a brief discussion on the operation of single-particle electrodynamic levitation (SPEL).

SPEL DESIGN AND OPERATION

Millikan[7] was the first to use electrical levitation to study the properties of single particles. Using two parallel plates with opposite applied potentials to establish an electric field, he suspended individual, charged oil droplets by counterbalancing the droplet's weight against the static electric field. By measuring the droplet fall rate with and without the electric field, Millikan was able to determine the elementary charge of an electron for which he was subsequently awarded a Nobel Prize in physics. While the Millikan cell was used successfully by many researchers, the cell was disadvantaged by the fact that the droplet was stabilized in one direction only. Because horizontal stabilization did not exist, the droplet tended to drift laterally, causing significant measurement problems.

To circumvent this deficiency in the Millikan cell, Straubel[8,9] and Wuerker et al.[10] demonstrated that an electrically charged droplet can be localized in three dimensions by applying an oscillating potential across an electrode configuration with a specific geometry. The electrode configuration must be such that the electric field intensity increases linearly with distance from the field origin. Because of particle drag, the particle oscillation will lag the electric field oscillation and the droplet will experience a net time-averaged central force that drives the particle to the field origin. This is the principle for particle stablization in SPEL.

Figure 9.1 shows a schematic of the SPEL apparatus. The chamber shown is referred to as the bihyperboloidal balance because the cross-section of the electrode is described by two hyperbolae. Other electrode geometries exist[1] that produce fields that result in particle levitation, but we consider only the bihyperboloidal chamber in this chapter. The electrode surfaces are described by

$$2z^2 - r^2 = C_{\pm}, \tag{9.1}$$

where z and r are axial and radial coordinates, respectively, and $C_{\pm}$ are constants, a positive constant for the two-sheet hyperboloid forming the top and bottom of the chamber, and a negative constant for the one-sheet hyperboloid forming the sides of the chamber. These electrodes produce an electric field strength that is zero at the origin and has the described linearity. The top and bottom hyperboloidal sheets were separated from the central, single-sheet hyperboloid by two horizontal Teflon rings. The alternating voltage was applied to the one-sheet bihyperboloidal electrode. The

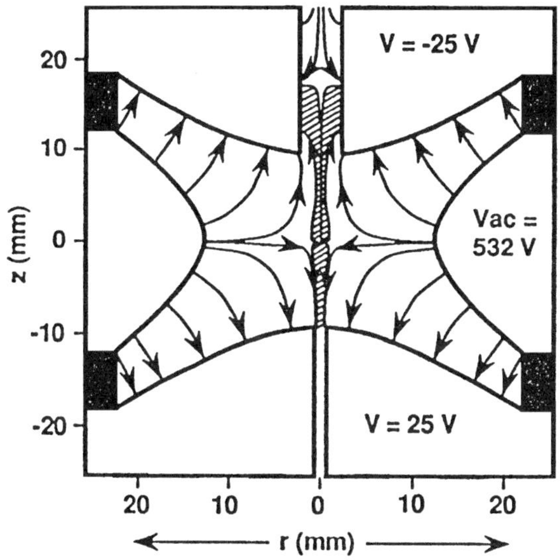

FIGURE 9.1 Schematic of SPEL apparatus with electric field lines shown.

direct current voltage was applied symmetrically across the top and bottom electrodes using halving resistors, which were connected to the common ground of the alternating voltage. The symmetry of the dc field resulted in an analytic solution for the particle motion in the electrodynamic field. The dc voltage was varied between 0 and 200 volts, and the ac voltage was varied between 0 and 1000 volts.

The stability of particle motion was shown to depend on two parameters: a particle drag coefficient K_D and an electric field intensity coefficient E. Analyzing the stability of the solution for the equation of motion of a charged particle in the bihyperboloidal field, Frickel et al.[11] showed that the particle motion could be described by stability zones mapped out in K_D-E space. Figure 9.2 shows such a stability map where the particle drag is defined as $K_D = 18\eta/\rho\omega d^2$, where η is air viscosity, ρ is the air density, ω is the field frequency, and d is the particle diameter; the electric field strength parameter E is defined as $E = CqV/\omega^2 m$, where C is a geometric constant, q is the particle charge, V is the oscillating field potential, and m is the particle mass. Thus, at constant field frequency and air properties, the complete stability map can be traversed by independently varying the particle charge-to-mass ratio and the field potential. An interesting feature of Figure 9.2 is the condition that for a given drag coefficient, one can pass from a stable zone to an unstable zone and back to a stable zone by increasing the field potential. In principle, particle "sorting" can be accomplished by setting the field potential so that only a specific particle charge-to-mass ratio leads to a stable configuration.

Particle Detection and Measurement

Two principal particle measurement methods are used with SPEL: optical and gravimetric. A 1-mW He-Ne laser is used to illuminate the particle and scattered radiation is detected at 90° and 35° from the forward direction. Radiation at 35° is detected with a split photodiode that is used to monitor the vertical position of the particle. If the particle is not centered in the chamber, the diode generates an error signal that is converted to a correction voltage by the use of a proportional-integral-derivative controller. The proportional derivative section gives a quick response for the particle position adjustment and it also damps particle oscillation, whereas the integral section offsets changes in mass, charge, or external forces. This electro-optical feedback system allows for automatic monitoring of the particle weight-balancing potential and is best-suited for spherical

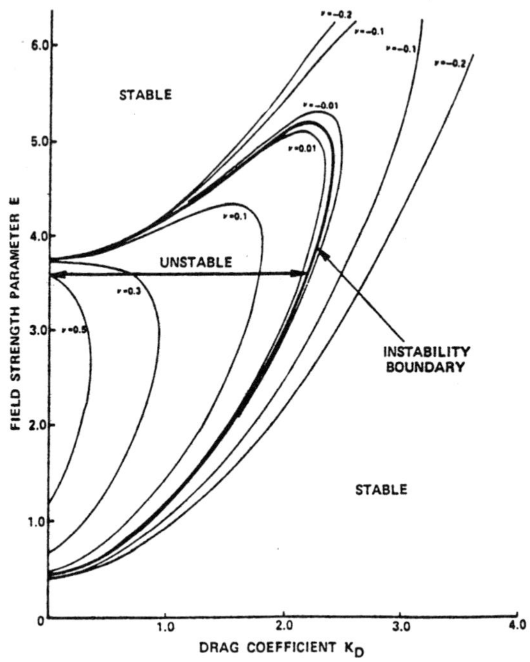

FIGURE 9.2 Stability phase space for a charged particle in SPEL.

particles such as liquid droplets. Radiation at 90° is detected with a photomultiplier (Products for Research) and is recorded on a y-t recorder. At 90°, light scattering resonances are detectable and particle size changes can be inferred from resonance spacing. The particle is also back-illuminated with a white light source, which permits manual control of the position of the particle in the chamber.

The static field potential is adjusted until the particle oscillation ceases. At this point, the particle is at rest at the center of the chamber. When the particle is centered, the particle weight is exactly balanced by the static electric field and the following condition applies:

$$mg = qC_{DC}V_{DC}. \tag{9.2}$$

When this condition is valid, relative particle masses are determined from the relative static voltages. It is this relation that is used extensively in the aerosol studies discussed in this chapter. The particle was observed through a telemicroscope equipped with a 35-mm objective by back-illuminating the droplet with white light. The telemicroscope was attached to a Sony camera, which permitted viewing of the particle on a 13-in monitor for ease of positioning. The telemicroscope was also equipped with a scanning graticule that permitted *in situ* particle size measurement within ±1 micrometer.

Particle Generation

Charged particles were generated using two methods of dissemination: electrospray and contact charging. Electrospray was used to disseminate liquid droplets and colloidal particles that formed solid particles by flash distillation. Conductive solid particles were generated using a dry process referred to as contact charging. Contact charging involved bringing a potential field in the vicinity of the conductive powder that was placed in contact with ground. The conductive powder attained a net charge opposite to the field polarity. Aspiration of the powder resulted in a dry charged solid aerosol. Contact charging proved important when it was necessary to generate an uncontaminated particle for vapor sorption studies, such as for single-particle isotherm studies.

For the electrospray dissemination method, the liquid was placed in a capillary tube connected to a dc high-voltage power supply. By raising the voltage to a value between 4000 and 7000 volts, a spray of charged droplets was generated. The spray was directed toward the levitation cell for particle trapping.

Environmental Control

The composition of the gas entering the levitation cell is controlled by a series of compressed gas sources regulated by flowmeter controls. The water humidity is established by passing compressed air through water bubblers and dessicants and by controlling the relative flow rate through the two chambers. Dew points are measured with an EG&G hygrometer. Other condensible/reactive gases are introduced by a bleed-in line. Vapor concentrations are measured with a Varian 3000 FID gas chromatograph.

SPEL APPLICATION STUDIES

Multicomponent Oil Droplet Studies

The study of the evaporation of multicomponent oils is important for several reasons: predicting lubricant oil lifetimes, modeling the obscuration performance of oil smokes, and understanding the combustion of oil droplets, to name a few. Davis and Ray[12] showed that the evaporation of single droplets could be measured using electrical levitation. Studying the evaporation of single-component droplets, they were able to determine the gas-phase diffusion coefficients of the condensible species from the droplet evaporation rate.

The isothermal, diffusion-controlled evaporation of a multicomponent liquid droplet can be represented by the flux rate

$$\dot{n}(m,t) = \frac{-2\pi d D(m) P\big(\chi(m,t),d\big)}{KT}, \tag{9.3}$$

where $n(m,t)$ is the molecular number of mass m, d is the droplet diameter, $D(m)$ is the diffusion coefficient of mass m, K is Boltzmann constant, and $P(\chi(m,t),d)$ is the droplet vapor pressure of component m characterized by mole fraction χ. It is assumed that the gas-phase partial pressure of component m is zero. The droplet vapor pressure of component m is alternately described by the relation

$$P(m,t) = \gamma(m)\chi(m,t)P^{\circ}(m), \tag{9.4}$$

where $\gamma(m)$ is the activity coefficient of species m, and $P^{\circ}(m)$ is the saturation vapor pressure of species m. Subsituting Equation 9.4 into Equation 9.3 and taking the first mass moment, the flux rate becomes

$$\frac{d}{dt}\int_{m_1}^{m_2} mn(m,t)dm = -2\pi d\int_{m_1}^{m_2} mD(m)\gamma(m)\chi(m,t)P^{\circ}(m)dm. \tag{9.5}$$

It was shown[2] earlier that for realistic mass ranges, the product $mD(m)$ is a slowly varying function and can be approximated by a constant value. In addition, if we further consider cases where the droplet diameter does not change appreciably, then the r.h.s of Equation 9.5 is proportional to the total droplet vapor pressure and can be written as

$$P_T = \frac{-KT}{2\pi d\,\overline{mD(m)}}\dot{M}(t), \tag{9.6}$$

where $mD(m)$ is a predetermined constant.

Figures 9.3A and B show the evaporation data and droplet vapor pressure, respectively, for the following multicomponent hydrocarbons: distilled 100 pale oil at 40°C (△), undistilled 100 pale oil at 40°C (○), No.2 diesel fuel at 25°C (▽), and No.2 diesel fuel at 40°C (□). The droplet mass is measured from the levitation voltage and the optically determined droplet size. The droplet density is assumed constant during evaporation. The slope of the evaporation curve is determined graphically using a chord-area method. Substitution of the mass decay rate and the instantaneous droplet diameter into Equation 9.6 gives the total droplet vapor pressure as a function of percent mass evaporated M_E. Figure 9.3B shows the dependence of natural log of the total pressure on the percent mass evaporated and reveals that these hydrocarbons can be characterized by the same functional relationship. Subsequently, Rubel[3] measured the evaporation rate of binary oil droplets composed of dioctyl and dibutyl phthalate. Assuming ideal solution theory in the evaporation model, comparison of experiment and theory was excellent.

Hygroscopic Growth Studies

The equilibrium growth of hygroscopic particles has generated considerable interest in the atmospheric sciences community due to the effect of humidity on air visibility. Previous attempts to measure the uptake of water vapor by hygroscopic particles have focused on multi-particle analysis such as electrical mobility analysis. Orr et al.[13] measured the water gain/loss of hygroscopic salt particles using humidity-dependent electrical mobility measurements. The growth of particles of NaCl, $(NH_4)H_2SO_4$, $CaCl_2$, AgI, PbI_2, and KCl was measured and the deliquescence humidity was determined for each of the salts. Analogous single-particle measurements were conducted by Twomey,[14] who measured the thermodynamic growth of atmospheric salt particles by suspending the particles on spider webs. While this method was successful for most of the salts studied, sodium carbonate measurements deviated from predicted values. The influence of the spider web on particle growth is uncertain and for this reason a nonintrusive method like electrical levitation would be advantageous.

To study the thermodynamic growth of hygroscopic particles, the electrodynamic balance is equipped with a humidity-controlled gas flow that is monitored continuously with a dew point hygrometer. The humidity is precisely controlled by mixing two flow sources: a water-dessicated flow and a water-saturated flow. By controlling the relative flow rates of the two flows, it was possible to vary the ambient dew point from –30°C to 22°C. As with the oil studies, relative mass changes are determined from relative voltage changes. Droplet diameters are measured using a scanning graticule in combination with a telemicroscope.

Figure 9.4 shows a comparison between the volume increase of a phosphoric acid solution droplet with increasing relative humidity as measured with SPEL and as predicted from a regression equation developed from the water activity data of Mellor.[15] Deviations are generally less than 5%. Demonstration of the water activity measurement capabilities of the SPEL was crucial in conducting kinetic studies where data analysis required knowledge of the thermodynamic "tracking" characteristics of hygroscopic particles. Thus, as the relative humidity of the levitater is changed, the chemical composition of the solution droplet can be accurately predicted.

The capability to measure the hygroscopic growth of particles was applied to several developmental problems in the U.S. Army, including the use of phosphorus smoke as a visible and infrared

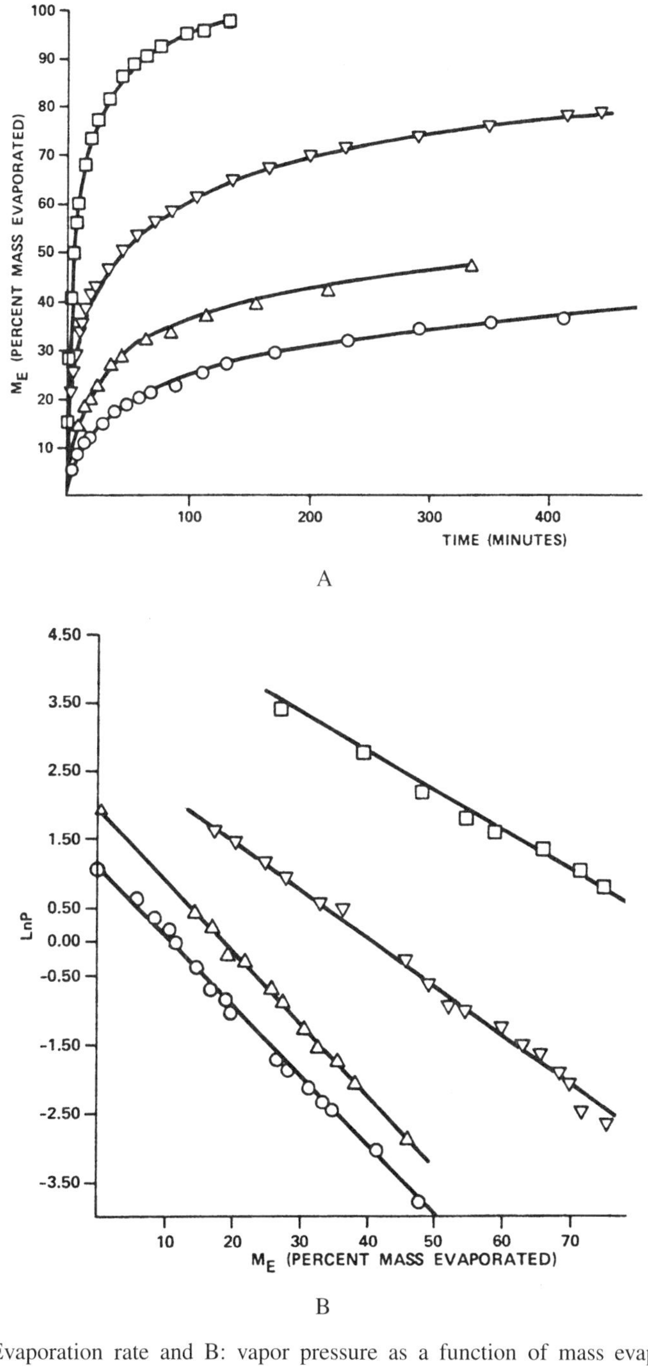

FIGURE 9.3 A: Evaporation rate and B: vapor pressure as a function of mass evaporated of different multicomponent hydrocarbons.

obscuring smoke. While infrared transmission measurements revealed that the infrared spectra of phosphorus smoke is different from that of an orthophosphoric acid aerosol, the SPEL showed that the hygroscopic characteristics of the two aerosols were equivalent.[16] This finding significantly simplified smoke modeling efforts by permitting the use of orthosphosphoric acid/water activity data to predict phosphorus smoke hygroscopic growth.

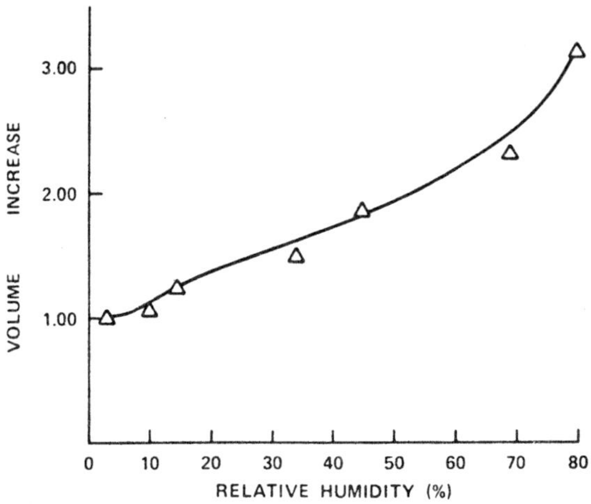

FIGURE 9.4 Comparison of volume increase of H_3PO_4 droplet measured by SPEL (△) and predicted by Mellor.[15]

Droplet Kinetics with Monolayers

The evaporation of water through insoluble monolayers has been studied extensively.[17-19] It was shown that evaporation resistance increased with surface pressure π, or inversely the area per molecule σ, and with the monolayer chain length. Using water troughs with controllable surface areas, these researchers performed a detailed analysis of the π-σ isotherms and discovered characteristic "kinks" in the isotherms. They attributed the "kinks" to impurities that were introduced through the administration of monolayer-solvent solutions necessary for the uniform spreading of the monolayer over the water surface. It was our intent to: (1) determine if the "kinks" are present in the single droplet evaporation data, and (2) determine the value of the water accommodation coefficient for various monolayer surface coverages.

Surfactant vapor is introduced to the levitater using a carrier flow that is passed over a heated quantity of hexadecanol. A phosphoric acid droplet is stabilized in the levitater and hexadecanol is adsorbed onto the droplet surface. Humidification and dehumidification of the levitater causes the droplet to alternately evaporate and condense with a concomitant compression and expansion of the monolayer. Employing mass transfer theory for the water and hexadecanol, the monolayer surface coverage is predicted, as well as the water accommodation coefficient, the fraction of water molecules that intercept the droplet surface that are retained.[20]

Figures 9.5A and B show the mass of an evaporating and condensing phosphoric acid solution droplet, respectively, in the presence of hexadecanol vapor. In the case of evaporation (Figure 9.5A), the droplet is initially covered with a partial monolayer of hexadecanol, and evaporation is initially rapid with water accommodation coefficients on the order of 10^{-3}. After approximately 6 s, the droplet evaporation slows dramatically and the water accommodation coefficient decreases by almost an order of magnitude to 10^{-4}. The dramatic decrease in the droplet evaporation rate is associated with the formation of a critical coverage of hexadecanol monolayer, which represents the transition between the liquid condensed and solid monolayer.

The growth cycle (Figure 9.5B) reveals a similar "kink" in the condensation kinetics with now a dramatic increase in the accommodation coefficient as the hexadecanol monolayer transitions between the solid and liquid condensed monolayer. The reversibilty of the "kink" kinetics contradicts the argument of Archer and La Mer,[18] who hypothesized from surface pressure-area measurements that the sharp change in the evaporation rate was due to surface impurities being "squeezed out" of the monolayer during monolayer compression. However, the "kink" in the growth kinetics could not be explained by such a impurity hypothesis.

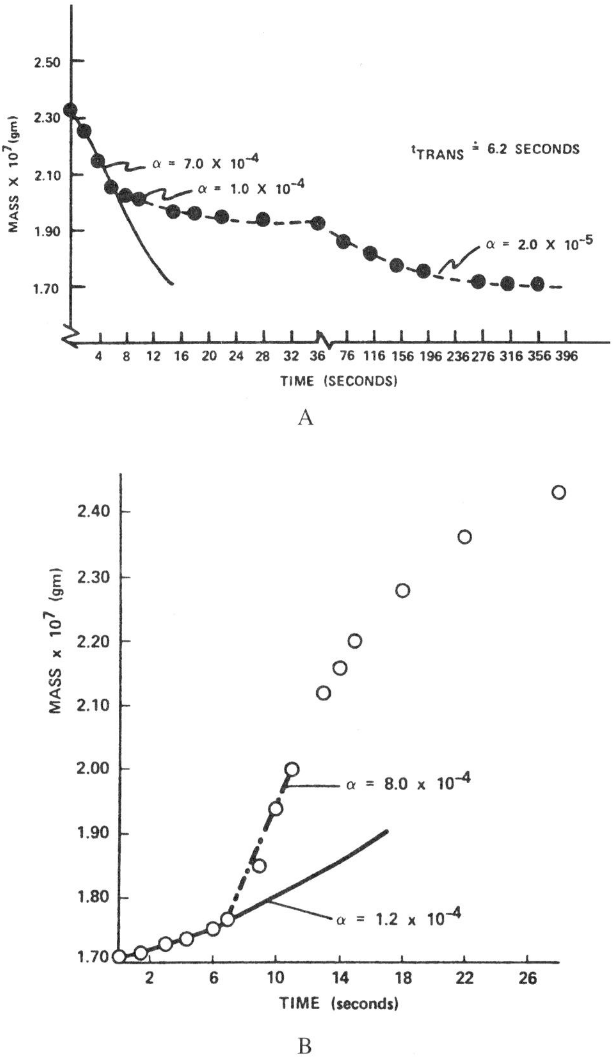

FIGURE 9.5 A: Evaporation and B: growth of a monolayer-coated H_3PO_4 droplet.

Heterogeneous Reactions

One of the earlier investigations of the heterogeneous reaction between droplets and reactive gases was conducted by Robbins and Cadle,[4] who measured the reaction between sulfuric acid droplets and ammonia gas. They observed that surface-phase reaction models with constant reaction coefficients did not correctly predict the reaction rate, and that the experimental rate was significantly smaller than the "best-fit" reaction model. Furthermore, the droplet reaction rate was unchanged when the nitrogen carrier gas was mixed with helium gas, leaving the researchers to conclude that gas-phase diffusion-controlled processes were not rate-limiting. To shed light on the mechanisms that control the heterogeneous reaction between the acid droplet and a reactive gas, single-particle levitation was used to continuously monitor the droplet reaction dynamics. Continuous monitoring of an isolated droplet permits the identification of discontinuous changes in the reaction rate, which are difficult to identify with discrete time analyses such as that of Robbins and Cadle.

In this study, single phosphoric acid droplets, varying in diameter from 42 to 72 μm, were levitated in the cell at varying ammonia gas partial pressures (from 115 to 1000 dyne cm^{-2}). Because the droplet weight is balanced by the force of the electric field, the ratio of the levitating voltages is equal to the ratio of droplet masses. Weight changes are solely due to the addition of ammonia molecules by heterogeneous reaction because the acid is nonvolatile. The extent of reaction ξ, the ratio of accreted ammonia molecules to the initial number of acid molecules, can be expressed in terms of the levitation voltages as

$$\xi = \frac{M_p\left(V(t)/V(0)-1\right)}{M_A f + M_p(f-1)}, \tag{9.7}$$

where $V(t)$ is the levitation voltage at time t, $M_{P,A}$ are the molecular weights of phosphoric acid and ammonia, respectively, and f is the initial acid weight fraction of the droplet.

Figure 9.6 depicts the reaction dynamics for different-sized phosphoric acid droplets at constant ammonia gas partial pressure. For all cases, for the first second of reaction, a rapid increase in the extent of reaction occurs. The maximum extent of reaction achieved during the initial growth phase is relatively insensitive to particle size. Subsequently, a slower growth rate dominates, the rate of which increases as the particle size decreases. During this latter growth phase, the maximum extent of reaction increases with increasing particle size. Remarkably, for the 42-μm diameter droplet, the second growth phase is completely inhibited. Telemicroscopic examinations indicated that the droplet was encapsulated by a thin transparent shell of a glass ammonium phosphate.[21] At a later time, a sharp transition in the reaction dynamics occurred. Microscopic examination showed that particle crystallization occurred at this transition rate, and it was concluded that gas-phase diffusion was no longer rate-limiting but that internal particle diffusion became rate-limiting. These conclusions were confirmed by model analysis by Rubel and Gentry,[21] who showed that the heterogeneous reaction history of the droplet could be modeled as a sequential reaction set given by surface phase, gas-phase diffusion-controlled, and finally internal particle diffusion. The onset of internal particle diffusion-controlled reactions was initiated by the surface crystallization of ammonium phosphate. Rubel and Gentry[22] developed a model for the time-dependent surface concentration of ammonium phosphate resulting from the heterogeneous reaction of ammonia gas with phosphoric acid. It was shown that for all acid droplets, independent of particle size, the surface phosphate concentration was the same value at the time of particle crystallization.

As a corollary to studies involving the effects of monolayers on water condensation and evaporation from aqueous droplets, a study was conducted to examine the effect of the same monolayers on ammonia gas accommodation at the droplet surface.[16] In this study, phosphoric acid droplets initially covered with a hexadecanol monolayer were immersed in ammonia gas. Vis-a-vis the reaction dynamics studies discussed earlier, extents of reaction were determined as a function of time for various states of the monolayer. The results of the study showed that for both the solid and liquid monolayer, the ammonia accommodation coefficient was an order of magnitude smaller than the water accommmodation coeffcient.

Interestingly, during droplet reaction, the ammonia and water accommodation coefficients decreased with increasing extent of reaction. One possible explananation is that the monolayer contracts, that is, the area per molecule decreases during droplet reaction. Monolayer contraction results in greater monolayer cohesion and thus a greater free energy barrier for monolayer permeation. Monolayer contraction with decreasing substrate acidity has also been demonstrated by Langmuir.[23]

Droplet Microencapsulation

One of the earlier attempts to encapsulate aerosols inside impermeable films involved reacting droplets of phosphoric acid with 1,3-butadiene gas to form polymer films at the droplet surface.[4] Rubel[22] showed that it was possible to encapsulate phosphoric acid droplets in ammonium phosphate

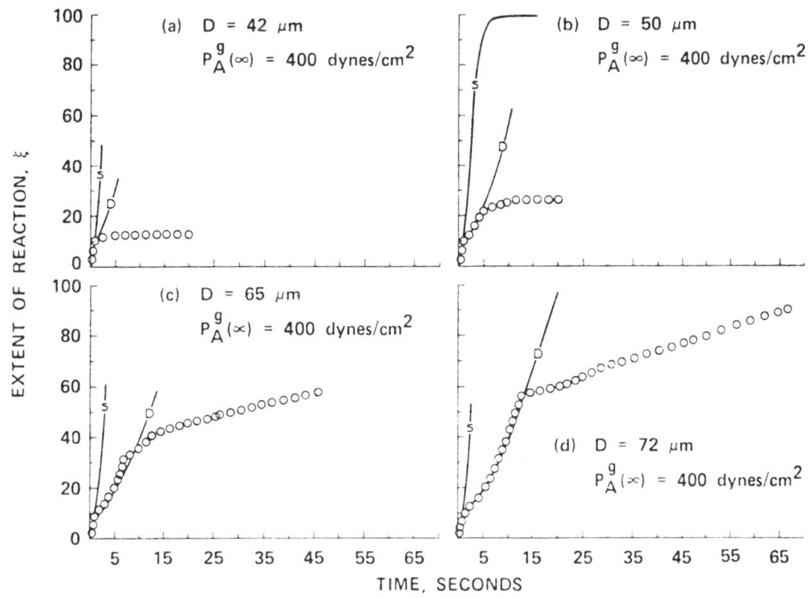

FIGURE 9.6 Reaction rate of a phosphoric acid droplet immersed in ammonia gas as a function of particle size.

by immersing the droplet in ammonia gas. It was shown that the shell porosity is a function of the ammonia gas pressure, decreasing as the ammonia pressure is increased. A different problem is posed by the encapsulation of volatile compounds such as low molecular weight hydrocarbons. For these hydrocarbons, the film formation process will depend on the evaporation rate of the volatiles. Durand-Keklikian and Partch[24] investigated the microencapsulation of dodecane and diesel fuel droplets by doping the droplets with metal alkoxides and then reacting the droplets with water vapor. They investigated the effects of alkoxide concentration, hydrocarbon type, temperature, liquid aerosol flow rate, and length of reaction tube on the morphology of the film. No attempt was made to determine the reaction dynamics.

To develop a more complete understanding of the microencapsulation dynamics, single-particle levitation was used to investigate the microencapsulation of titanium ethoxide-doped dodecane droplets. The relationship between time of film formation, droplet size, and reactant concentrations was investigated. By monitoring the droplet evaporation rate, it was possible to infer the qualitative porosity of the film. The inferred porosities are compared to scanning electron micrographs.

Figure 9.7a shows the time-dependent levitation voltage for a 77-μm evaporating dodecane droplet doped with titanium ethoxide at a concentration of 5% by weight. Initially, the droplet is spherical and free from surface films, and the evaporation rate is essentially that of pure dodecane. After more than 2 minutes, the film is first observed. The film is detected as an irregularity in the normally spherical droplet surface. Interestingly, droplet evaporation continues unaffected by the film formation. Electron photomicrographs show that the droplet surface is encrusted with a nonuniform film with significant defect structure. It is the porosity of the shell that leads to a permeable characteristic and fails to inhibit dodecane evaporation.

To investigate the effect of water partial pressure on the film formation process, the dew point temperature was increased. Figure 9.7b shows a similar evaporation event, except the ambient dew point is increased to 20°C. Now, the onset of film formation occurs at 1.83 minutes, approximately 3/4 the time required at 11°C dew point. To investigate the effect of particle size on the onset of film formation, the conditions in Figure 9.7a are duplicated, except the droplet size is reduced to 46 μm. Now, the onset of film formation is at 0.83 minutes. Again, the film has no observable

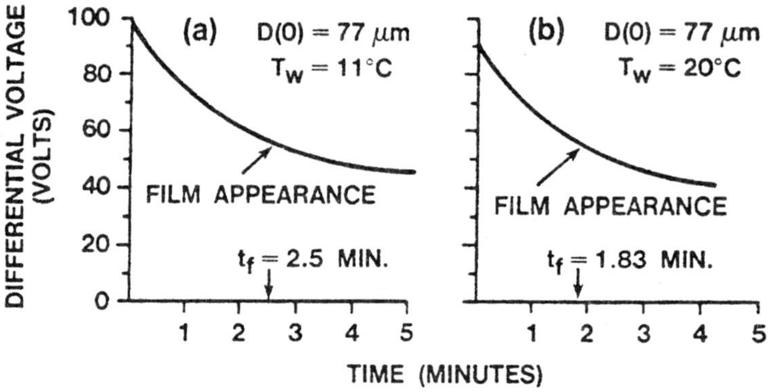

FIGURE 9.7 Levitation voltage of an evaporating dodecane droplet doped with titanium ethoxide.

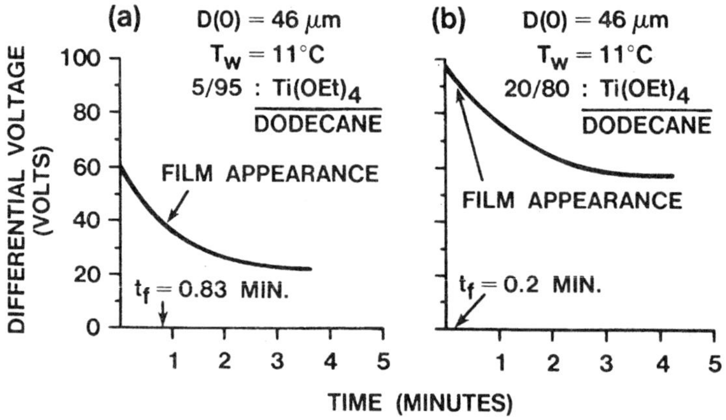

FIGURE 9.8 Levitation voltage of an evaporating dodecane droplet doped with titanium ethoxide.

effect on the evaporation rate. Finally, the reactant concentration was increased to 20% by weight: all other conditions are set to match those in Figure 9.8a. Now, the film forms rapidly at 0.2 minutes. No observable effect on the droplet evaporation rate was observed. When the ethoxide concentration was increased to 50% by weight, encapsulation halted droplet evaporation.

Several observations can be made from the results shown in Figures 9.7 and 9.8. First, because the time for film formation depends on the initial reactant concentration, it is concluded that the dynamics are not gas-phase diffusion-controlled. In fact, one can derive an expression relating the time for film formation to the water partial pressure under the assumption that the reaction is gas-phase diffusion-controlled. Then, for two distinct water partial pressures, the respective times for film formation (equivalent product concentrations) are related by

$$\frac{P_W(1)}{P_W(2)} = \frac{1-\left(1+kt_2/d_0^2\right)^{-3/2}}{1-\left(1+kt_1/d_0^2\right)^{-3/2}}, \tag{9.8}$$

where d_o is the initial diameter and k is the evaporation rate parameter defined as

$$k = -\frac{8v_d D_d P_d}{RT}. \tag{9.9}$$

Here, v_d is the molar volume, D_d is the gas-phase diffusion coefficient, and P_d is the saturation vapor pressure of dodecane. Using typical values for these parameters as given by the *CRC Handbook of Chemistry and Physics*,[27] the relative partial pressure calculated from Equation 9.8 corresponding to the times of film formation as shown in Figures 9.7 and 9.8 is 0.73. This value exceeds the experimental value of 0.56, a discrepancy that is beyond the range of experimental error. While the water partial pressure dependence cannot be explained in terms of gas-phase diffusion-controlled reaction, there clearly exists a strong dependence on the water partial pressure. One possibility is that the overall reaction is controlled by a liquid-phase reaction that depends on the droplet reactant concentration. The concentration of water in the dodecane and titaniun ethoxide mixture could depend on the water partial pressure.

Gas Adsorption onto Solid Particles

Microporous carbons are used extensively by industry and the military in filter systems for air purification. Because water vapor can affect the adsorption of hazardous vapors onto the carbon adsorbents, considerable research has been conducted on the water isotherms of porous carbons.[25] A common experimental procedure for isotherm analysis is the gravimetric method where water vapor adsorption is determined from the weight change of the adsorbent beds. The range of techniques varies from the large macroscopic studies where adsorption tubes containing grams of carbon adsorbent are used, to the smaller microscopic studies involving electrobalances where milligram quantities are required. The measurement time increases dramatically with the quantity of carbon used in the study, varying from days for the tube studies to hours for the electrobalance.

It is object of this study to describe a new method for isotherm determination that reduces the measurement time and accurately reflects the adsorption capacity of the adsorbent bed. Single-particle levitation is used to measure the water isotherms for the well-characterized material silica gel and adsorbent carbons.

Particle charging was accomplished in the following manner. For the dielectric silica gel, the silica granules were first crushed to a fine powder and then mixed with methyl alcohol to a slurry consistency. The colloidal mixture was drawn into a pipette with an orifice diameter of 1 mm. An 8-KV charge was applied to the mixture through an immersed wire, producing a fine spray of charged particles. More than one particle was levitated in the trap and a glass rod was used to eliminate all particles except one. For the activated carbons, induction charging was used to charge solid microparticles of carbon. Because carbon is electrically conductive, the carbon particles attain a net charge in the presence of an electric field if the carbon powder is grounded. Both BPL and ASC carbon were used in this study.

Figure 9.9 compares the water loading of silica gel as reported by Davison Chemical[26] and that measured by the single-particle approach. The water loading, expressed in terms of grams of water to grams of adsorbent, is expressed in terms of the levitation voltages as

$$\text{Water loading} = 1 - \frac{V_D}{V}, \tag{9.10}$$

where V_D is the levitation voltage of the dry particle. The isotherm generated by SPEL represents five separate runs averaged together. However, the reproducibility is such that the error bars are comparable to the size of the data points. Although the SPEL data slightly underestimate the values reported by Davison Chemical, the functional dependence of the water loading on relative humidity is accurately reflected in the single-particle data. As an aside, it was found that the silica gel isotherm was independent of the number of times the powder was crushed. This is perhaps not surprising because most of the water adsorption is taking place in micropores, which are unaffected by the crushing process.

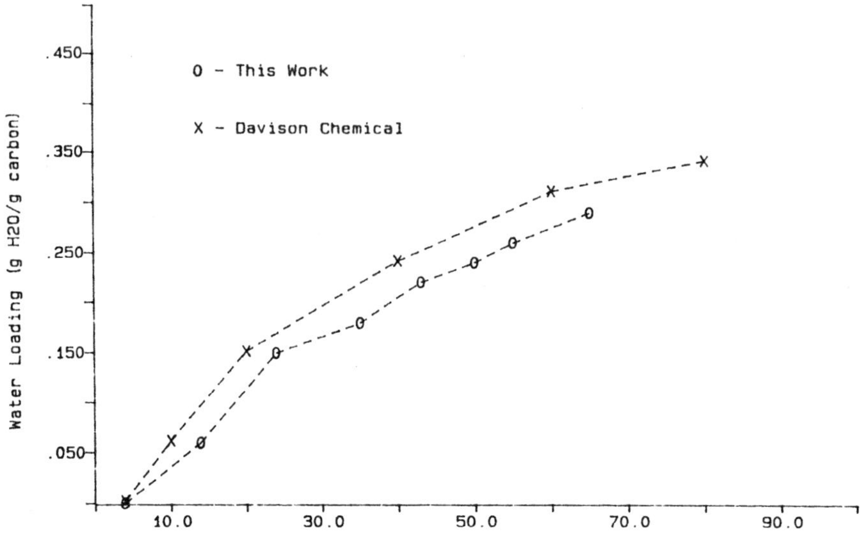

FIGURE 9.9 Comparison of silica gel water isotherms as measured by SPEL (0) and reported by Davison Chemical (×).

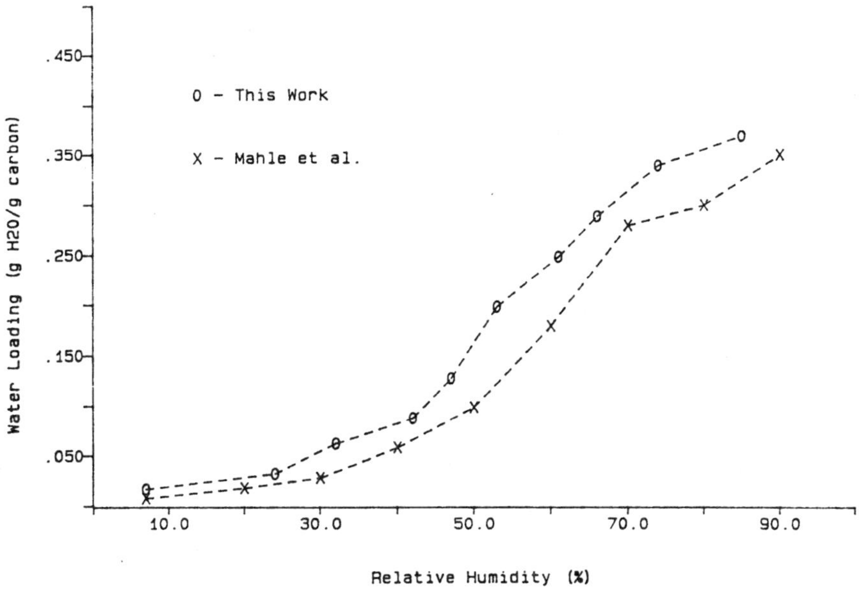

FIGURE 9.10 Comparison of carbon water isotherms as measured by SPEL (0) and as reported by Mahle and Friday (×). (From Reference 28. With permission.)

Mahle and Friday[28] measured the water adsorption charcateristics of ASC and BPL carbon using the tube adsorption methodology. Figure 9.10 shows a comparison of the water adsorption for ASC carbon as measured with SPEL and with tube adsorption. Other than a small overestimation of the data, the SPEL isotherm follows closely the data of Mahle and Friday. An interesting difference in the water vapor desorption branch of the ASC carbon for the SPEL and tube methodologies does exist. The two methods give almost identical results from 100 to 50% relative humidity; however, below 50%, the SPEL desorption branch exceeds the tube data. Mahle and Friday[28] argued that the non-closure of the water isotherm loop was due to impregnants in the carbon, such as copper, silver, and chromium salts, that react with water vapor once adsorbed. Then, the difference in the water isotherms could be due to different amounts of impregnants in the carbon.

CONCLUSIONS

This chapter reviewed a narrow class cf studies conducted by this author to investigate novel applications of the single-particle levitater. Specifically, the following aerosol chemical and physical processes were discussed: water condensation onto aqueous droplets, heterogeneous reactions, monolayer resistance to evaporation and reactions, droplet microencapsulation and gas adsorption onto solid particles. The results of this study showed that the single-particle levitater is a valuable tool for the study of microparticle dynamics.

REFERENCES

1. Davis, E.J., *Aerosol Sci. and Tech.*, 2, 121, 1983.
2. Rubel, G.O., *J. Colloid Int. Sci.*, 81, 188, 1981.
3. Rubel, G.O., *J. Colloid Int. Sci.*, 85, 549, 1982.
4. Robbins, R.C., Thomas, J., and Cadle, R., *J. Colloid Int. Sci.*, 18, 483, 1963.
5. Huntzicker, J.J., Cary, R.A., and Ling, C., *Environ. Sci. Tech.*, 14, 819, 1980.
6. Rubel, G.O., *Carbon*, 30, 1007, 1992.
7. Millikan, R.A., *Phys. Rev.*, 15, 545, 1920.
8. Straubel, H., *Z. Elektrochem.*, 60, 1033, 1956.
9. Straubel, H., *Dechema. Monogr.*, 32, 153, 1959.
10. Wuerker, R.F., Shelton, H., and Langmuir, I., *Appl. Phys.*, 30, 342, 1959.
11. Frickel, R.H., Shaffer, R.E., and Stamatoff, J.B., Rpt. No. ARCSL-TR-77041, Chemical Systems Laboratory, APG, MD., 1978.
12. Davis, E.J. and Ray, A.K., *Chem. Phys.*, 67, 414, 1977.
13. Orr, C., Hurd, F.K., and Corbett, W.J., *J. Colloid Int. Sci.*, 13, 472, 1958.
14. Twomey, S., *J. Meteor.*, 335, 1954.
15. Mellor, A.J., *Mellor's Comprehensive Treatise on Inorganic and Theoretical Chemistry*, Vol. VIII, Suppl. III, Phosphorus, Wiley Interscience, NY, 1971, 1900.
16. Rubel, G.O. and Gentry, J.W., *J. Aerosol Sci.*, 16, 571, 1985.
17. Langmuir, I. and Schaffer, V.J., *J. Franklin Inst.*, 235, 119, 1943.
18. Archer, B.J. and La Mer, V.K., *J. Phys. Chem.*, 59, 200, 1954.
19. La Mer, V.K., Healy, T.W., and Alymore, L.A.G., *J. Colloid Int. Sci.*, 19, 673, 1964.
20. Rubel, G.O. and Gentry, J.W., *J. Phys. Chem.*, 88, 3142, 1984.
21. Rubel, G.O. and Gentry, J.W., *J. Aerosol Sci.*, 15, 661, 1984.
22. Rubel, G.O. and Gentry, J.W., *J. Aerosol Sci.*, 18, 23, 1987.
23. Langmuir, I., *Proc. Roy. Soc.*, 39, 1848, 1917.
24. Durand-Keklikian, L. and Partch, R.E., *J. Aerosol Sci.*, 19, 511, 1988.
25. Dubinin, M.M. and Serpinski, V.V., *Dokl. Akad. Nauk. SSR.*, 99, 1035, 1954.
26. Davison Chemical, IC-16-782, Ind. Chem. Dept., Baltimore, MD, 1985.
27. Weast, R.C., Ed., *CRC Handbook of Chemistry and Physics*, 62nd ed., CRC Press LLC, Boca Raton, FL, 1981.
28. Mahle, J.J and Friday, D.K., CRDEC-TR-018, U.S. Army Chemical Research, Development and Engineering Center, A.P.G., MD, 1988.

10 Radioactive Labeling in Experimental Aerosol Research

Kvetoslav R. Spurny

CONTENTS

INTRODUCTION

From an experimental point of view, in studying physical, chemical, and biological behavior and effects of solid and liquid aerodisperse systems under atmospheric and laboratory conditions, the application of radioactive labeling procedures is very useful.

Radioactive aerosols and radioactively labeled aerosols have existed in nature probably as long as our planet has existed. In 1995, Renoux published an overview of the history of the natural atmospheric radioactivity. The discovery of the rare radioactive gas **radon** is attributed to Pierre

0-87371-829-1/00/$0.00+$.50

and Marie Curie in 1898 and Dorn in 1900. Thoron (^{220}Rn) was discovered by Rutherford and Owens in 1899–1900, and actinon (^{219}Rn) by Debierne and Geisel about the same time.

The first scientist to find radioactive aerosols was Marie Curie in 1905.[2] She studied the influence of gravitational field on the decay products of radon. Radon's radiotoxicity was first studied in France in 1904 by Bouchard and Balthazard, and in 1924 it was hypothesized that the great mortality observed in uranium mines of Schmeeberg in Germany and Joachimsthal in Czechoslovakia was due to radon. In 1939, Read and Mottram found that radioactive aerosols are biologically more effective than radon itself.[2,3] Elster and Geitel were the first to see (in 1901) that radioactivity is present in the atmosphere.[1]

Since World War II, radioactive aerosols have become well known, and the object of increasing studies and use. Their physical properties and effects started to be intensively studied in the 1950s. Wilkening estimated the size distribution of the natural radioactive aerosols in the atmosphere in 1952 and, in 1959, Jacobi found that more than 50% of the natural atmospheric radioactivity is deposited on aerosol particles smaller than 0.2 μm. The first theory of small particle labeling by radioactive ions was developed by Bricard in 1949.

The exploitation of radioactive labeling in aerosol research also dates back to the 1950s. Nevertheless, a very fast development started about ten years later.[3] Since that time, basic theoretical investigations have led to a complex description of the nuclear methods applied in physical and chemical research.[4]

What is the difference between a radioactive aerosol and a radioactively labeled aerosol? There may be no precisely definable difference. From a historical point of view, all radioactively labeled aerosols in the atmosphere and space are called radioactive aerosols. But from a radiochemical point of view, for aerosols used in the laboratory conditions, it is best to use the expression "radioactively labeled aerosols." This means that only some aerosol particles are radioactive, and that only a portion of each particle is in fact radioactive. In contrast, the expression "radioactive aerosol" means that all particles are radioactive, and that each particle consists predominantly of radioactive species.

LABORATORY METHODS OF PREPARING RADIOACTIVELY LABELED AEROSOLS

Different methods can be used in the preparation of radioactively labeled aerosols under laboratory conditions. The most important labeling methods for practical and laboratory purposes are listed in Table 10.1. Neutron activation of the aerosol itself (method 1) is not very suitable or economical, and therefore will not be discussed here. The most suitable methods are those in which the aerosol is first prepared with the desired properties, and the particles are then labeled by condensing a radioactive substance on their surfaces (methods 2 and 3). Another convenient method for preparing radioactively labeled aerosols involves condensation or dispersion of radioactive substances (method 4). The processes of preparing radioactive nuclei (method 5) and preparing condensation aerosols can be combined.

LABELING BY MEANS OF DECAY PRODUCTS OF RADON AND THORON

This method is used very often and is similar to the natural radioactive labeling of fine aerosol particles in the atmosphere. Through a diffusion process, the natural aerosols are labeled by means of radon and thoron decay products.[3] The relative distribution of the activity on particles of different sizes was first described by Lassen in 1965.[5] This distribution function was constructed assuming the validity of Junges's distribution of natural aerosols,[6] including the condition of coagulation (see Table 10.1). Wire screen diffusion batteries have been found to be the most suitable method for measuring the activity size distribution of radon and thoron progeny.[7-9]

TABLE 10.1
Some Methods for Preparing Radioactively Labeled Aerosols

1. Preparation by means of *neutron activation* of aerosols in a nuclear pile or other neutron source (not very suitable).
2. Labeling by means of *decay products* of radon and thoron. Relative distribution of activity on particles of different sizes (L. Lassen):

$$A(r)\,dr = \Phi(r)\,N(r)\,dr$$

3. Labeling by means of *radioactive gases* (Rn, Tn, Xe, etc.) in high-frequency discharge at low pressure.
4. Preparation by means of *radioactively labeled elements and compounds* (condensation aerosols, disperse aerosols, and plasma aerosols).
5. Preparation by means of *radioactively labeled condensation nuclei.*

The short-lived decay products of ^{222}Rn and ^{220}Rn are formed initially in an atomic, positively charged state that rapidly combines with submicron (mainly with nano-sized) aerosols. The resulting ultra-fine aerosols consist of a complex mixture of charged and neutral particles. Under normal conditions, the average electrical charge of the ^{222}Rn and ^{222}Rn progeny atmosphere is substantially less than one elementary unit. The electrical charge distribution is mostly symmetrical.[10-13]

The decay-product method of labeling is relatively easy to use in the laboratory. An artificial inactive aerosol is passed through a cylinder filled with radon or thoron. When aerosol particles remain in this atmosphere for a sufficient length of time, they become alpha-radioactive. It should be noted, however, that if concentrations of thoron greater than about 1 μCi (27 kBq)/liter are used, aerosols may be produced by radiolytic reactions with impurities in the air. these may also become labeled with ThB and confuse the picture.[3]

Ultra-fine Aerosols by Radiolysis

It has been reported for many years that condensation nuclei can be produced by ionizing radiation. For example, radiolysis following the decay of ^{222}Rn results in the production of ultra-fine aerosols. Recent studies were able to improve the measurement of activity size distribution of these ultra-fine particles produced by radon and its daughters. It has been found that the activity that was conventionally referred to as the "unattached" fraction is actually an ultra-fine particle aerosol from water molecule radiolysis with a size range of 0.5 nm to 3 nm.[14] Oxidizable species such as SO_2 react promptly with hydroxyl radicals and form a condensed phase. These molecules coagulate and become ultra-fine particles. The size distribution of these ultra-fine particles can be shifted upward with the increase of SO_2 concentrations.

Further investigation[13] showed that ^{218}Po formed — during radon decay in well-controlled composition atmospheres (e.g., N_2) — clusters in the size range between 0.7 nm and 2.0 nm. Figure 10.1 shows the diagram of such a ^{218}Po cluster generation system. The size of the produced clusters could be efficiently measured by means of a SMEC (*spectrometre de mobilite electrique circulaire*) device.

The clusters formed in the radiolysis of radon include progeny particles and nonradioactive particles. In more recent investigations, the activity size distributions of ^{212}Pb- and ^{212}Bi-borne nanometer particles were produced and measured. When thoron gas enters the spherical chamber (Figure 10.2), it soon decays to ^{212}Pb and can be oxidized. Since most ^{212}Pb ions have positive charge, they attract polar molecules and form clusters. The cluster sizes measured by means of a diffusion battery (DB) were less than 2 nm.[9]

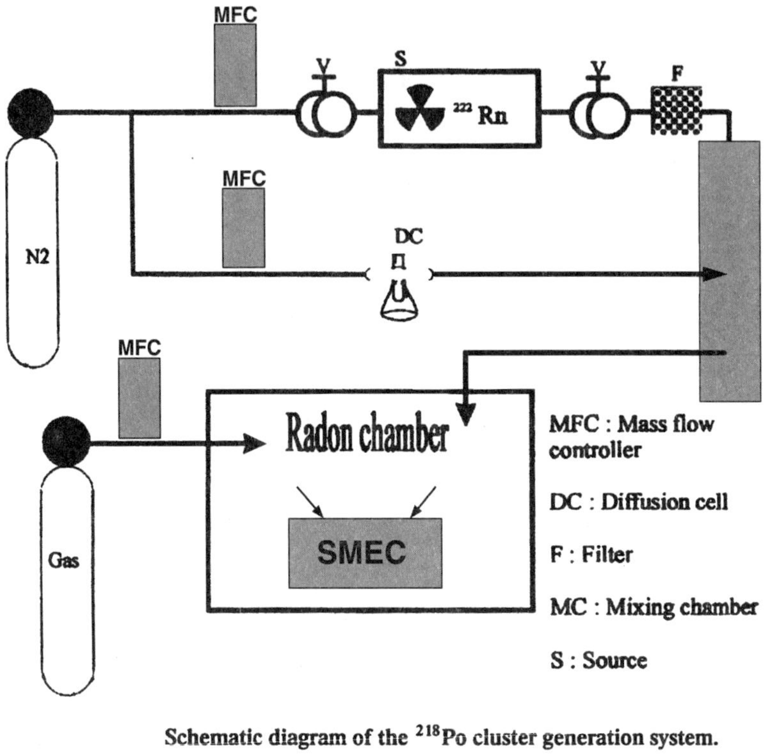

FIGURE 10.1 Schematic diagram showing the system for the generation of ^{218}Po cluster aerosols. (From Mesbah, B., Fitzgerald, B., Hopke, P.K., and Pouprix, M., *Aerosol Sci. Technol.*, 27, 381-393, 1997. With permission.)

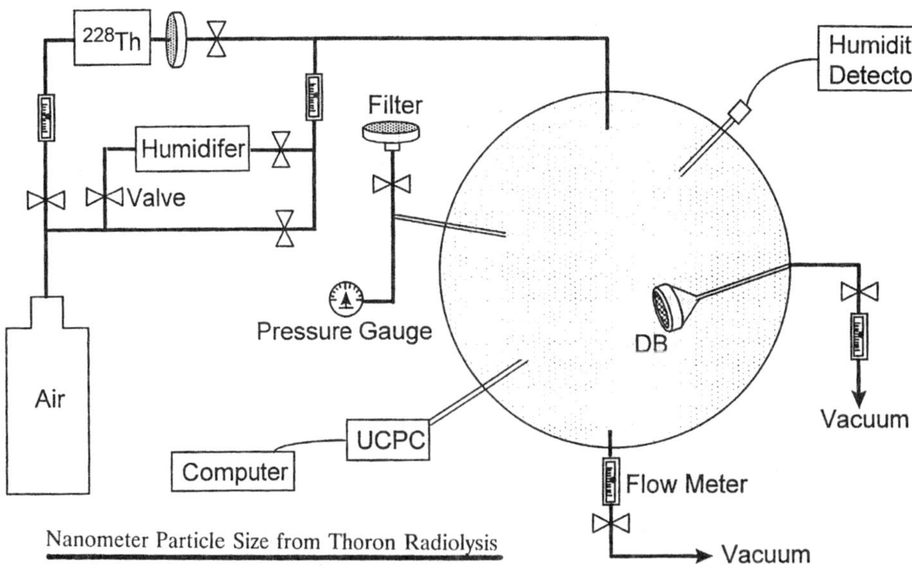

FIGURE 10.2 Experimental setup for the generation and measurement of nanometer-sized ^{212}Pb- and ^{212}Bi-borne particles. (From Chen, T.R., Tung, C.J., and Cheng, Y.S., *Aerosol Sci. Technol.*, 28, 173-181, 1998. With permission.)

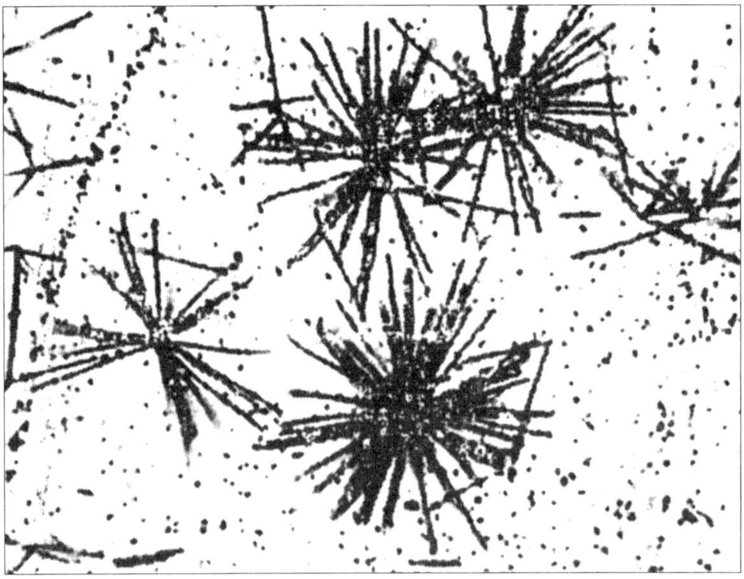

FIGURE 10.3 The tracks of alpha-rays from single aerosol particles.

Labeling by Means of Radioactive Gases

This method consists of exposing an aerosol or an aerosol sample to a high-frequency discharge at low pressure in a mixture of radon, krypton, or xenon, etc., and air. The atoms of a radioactive gas, ionized and accelerated in the electric discharge, penetrate and are retained on the surface of the aerosol particles. The method of labeling has two attractive features. First, the position of individual particles can be determined by autoradiography. When radon is used for labeling and radiography is carried out with nuclear emulsion, individual particles show up in the radiogram as stars consisting of the tracks of alpha-rays (Figure 10.3). The frequency of the tracks in each star is an indication of the particle's size. Second, the action of a suitable gaseous medium (a chemical surface reaction) on the aerosol particles can release the radioactive gas from the aerosol sample. This feature provides the possibility of chemically identifying individual particles in the aerosol sample.

An aerosol can be activated directly in a suspended state, independent of its chemical composition, in a stream of gas. By repeated measurements of aerosol activity, the aerosol concentration can be measured continuously. Labeling with decay products of radon is most suitable for these purposes, because the radon is not used in gaseous form and is attached to the surface of the solid substances. Radon can be firmly fixed, for example, on the inner wall of a glass tube with the aid of an electric discharge at low pressure.[15,16] The radon is retained near the surface and a large proportion of the RaA atoms originating from the decay are ejected by recoil into the gas inside the tube. Because of their low energy, these atoms traverse a small distance, roughly 0.1 mm; and if the air is free of aerosols, they quickly diffuse back to the surface of the tube, where they are retained. If the air contains aerosol particles, however, some of the RaA atoms are retained by the aerosol; and the retention of RaA atoms increases with increasing concentration of the aerosol.

An instrument for continuous measurements of inactive aerosol concentration, based on this principle, was built and described by Jech in 1963.[16] The function of the instrument is shown schematically in Figure 10.4. The aerosol sample in air flows at a speed of roughly 0.25 1/min through the activating tube (A), which contains 5 to 10 mCi (185 to 370 MBq) radon. The activated aerosol emerges from the tube, is filtered by a Millipore filter (F), and its activity is measured differentially. The activity of the filter was continuously measured by a Geiger-Müller counter (in its proportional region); and the counts were integrated and registered by the ratemeter (Rm) and

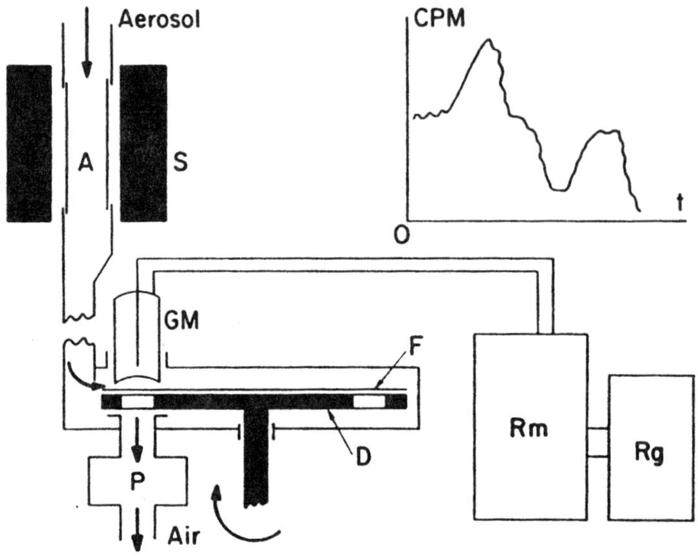

FIGURE 10.4 Schematic diagram of apparatus for continuous recording of aerosol concentration, and an example (inset) of a recording that shows aerosol concentration in unfiltered and filtered air from the laboratory. A = activating tube; F = Millipore® filter; GM = Geiger–Müller tube; Rm = ratemeter; Rg = recorder; S = lead shield; D = revolving metal disk; P = pump; CPM = counts per minute, t = time.

recorder (Rg). The relative amount of RaA atoms retained by individual particles was dependent on the size of the particles as well as on the numerical concentration of the aerosol. Therefore, the instrument had to be calibrated for an aerosol of given dispersity.[17]

Labeling by Means of Radioactively Labeled Elements and Compounds

In this case, there are two principle possibilities: (1) preparation of dispersed aerosols by spraying or nebulizing solutions or powders; and (2) preparation of condensation aerosols by spontaneous vapor condensation, or vapor condensation in the presence of radioactive condensation nuclei.

The first method has some disadvantages: the possibility of contamination is great; the consumption of radioactive material is large; and the aerosol particles show little specific radioactivity. Nevertheless, it was used in some cases to great effect in the 1960s and thereafter.[18]

However, the second possibility — the use of condensation methods — provides highly dispersed aerosols, approximately monodisperse, and the particles show a high specific radioactivity. Through the nucleation process, the particle size and the aerosol concentration can be changed by changing the supersaturation of the vapor. From nucleation theory, it is known that the particle concentration for a given time is an exponential function of the supersaturation of a vapor. This supersaturation is controlled in practice by changing the evaporation temperature of the substance and the flow rate of dilution gas. When all conditions are constant, the concentration can be calibrated and the particle size determined as a function of evaporation temperature and gas flow, the particle size being measured with an electron microscope, diffusion battery, etc.[19]

The chemical elements and compounds for preparing condensation aerosols have to be stable; they should not decompose on heating. Evaporation is often accompanied by oxidation, so that the aerosol being prepared becomes oxidized. Tables 10.2 and 10.3 describe the elements and inorganic and organic compounds that are suitable for preparing condensation aerosols and which are easy to label with different radioisotopes. Table 10.4 shows more detail concerning some radioactively labeled inorganic condensation aerosols that were described and used in laboratory experiments in the 1960s.[19-27]

TABLE 10.2
Inorganic Material Suitable for Preparing Radioactively Labeled Condensation Aerosols

Element or Compound	Melting Point (°C)	Temperature (°C) at Vapor Pressure 10^{-5} mmHg
Hg	–38.9	126.2
H_2SO_4	10.5	145.8
Ga	30	1349
$H_4P_2O_7$	61	—
Se	217	356
Re_2O_7	296	215.5
Tl	303.5	825
SeO_2	340	157
Te	452	520
AgCl	455	912
BeI_2	488	283
$PbCl_2$	501	547
LiF	547	1047
AgI	552	820
CsI	621	738
CsBr	636	748
CsCl	646	744
NaI	651	767
V_2O_5	690	—
NaCl	800	865
Ag	961	767
Au	1063	1083
Mn	1244	717
Be	1284	942
Si	1410	1024
Ni	1455	1157
Co	1478	1249
Fe	1535	1094
V	1710	—
Pt	1774	1606
Cr	1900	907
SrO	2430	2068
Mo	2622	1923
Os	2697	2101
Ta	2996	2407
W	3382	2554

Equipment and Procedures

Different kinds of equipment can be used for spontaneous condensation under constant conditions. Three of them have proven to be very suitable for the generation of highly dispersed radiolabeled condensation aerosols. Such model aerosols make it possible to measure more rapidly and sensitively numerous processes in the mechanics of aerosols (e.g., coagulation, phase transformation, filtration, deposition, etc.).

TABLE 10.3
Organic Compounds Suitable for Preparing Condensation Aerosols (Radioactive Labeling by Means of Radioactive Condensation Nuclei)

Compound	Formula	Melting Point (°C)	Temperature (°C) at Vapor Pressure 1 mmHg
Dichloro-1-naphthylsilane	$C_{10}H_8C_{12}Si$	—	106.2
Trethylene glycol	$C_6H1_4O_4$	—	114.0
Tetraethylene glycol	$C_8H_{18}O_5$	—	153.9
Nitroglycerine	$C_3H_5N_3O_9$	11.0	127.0
Capric acid	$C_{10}H_{20}O_2$	31.5	125.0
Palmitic acid	$C_{16}H_{32}O_2$	64.0	153.6
Diacetamide	$C_4H_7NO_2$	78.5	70.0
Glutaric acid	$C_5H_8O_4$	97.5	155.5
Acridine	$C_{13}H_9N$	110.5	129.4
Resorcinol	$C_6H_6O_2$	110.7	108.4
Sebacic acid	$C_{10}H_{18}O_4$	133.0	—
Adipic acid	$C_6H_{10}O_4$	152.0	159.5
Hydroquinone	$C_6H_6O_2$	170.3	132.4
Benzanthrone	$C_{17}H_{10}O$	174.0	225.0
Hexachlorobenzene	C_6C_{16}	230.0	114.4
Dioctyl sebacate	$C_{26}H_{50}O_4$	—	—
Dioctyl-phthalate	$C_{18}H_{30}O_4$	—	—
Dibutyl-phthalate	$C_{16}H_{22}O_4$	—	—

Furnace Generators

The apparatus for the generation of condensation aerosols by sublimation of the solid phase or by evaporation of the liquid phase or inorganic substances consists of an electric furnace in which the substance under study is heated to an adjustable and controlled temperature. The dry gas passing through the furnace at an adjustable flow rate is enriched with the vapor or aerosol particles of the same substance used. After passing through the furnace, the gas with aerosol particles is led into a condenser and then into a homogenizer. Several types of furnace, each specially designed for an individual aerosol or aerosol group, proved suitable.[19-27]

A longitudinal furnace (Figure 10.5) was employed to prepare sodium chloride aerosols.[19] A ceramic tube was heated by two electric coils. In the first part of this tube, radiolabeled NaCl [^{24}Na, 10 to 500 mCi (370 MBq to 18.5 GBq)] was heated to the desired temperature in a porcelain boat. The second part of the tube was heated to a temperature about 10% higher than that in the first part. A vertical furnace (Figure 10.6) was employed to prepare silver iodide aerosols. The furnace consisted of two halves that were heated by electric, ceramic heating elements with a power output of 800 W. The gas entered the space over the substance (AgI) in the middle of a sealed silica tube. The vapor and the aerosols of silver iodide were drawn off from the upper part of the furnace. The yellow powder of AgI was added through a wider tube into a platinum crucible placed on the bottom of the tube. The AgI can be radiolabeled by ^{131}I or by ^{110}Ag.

For substances with a low melting point or high vapor pressure (e.g., Se, SeO_2, H_2SO_4, $H_4P_2O_7$), an apparatus with a double glass orifice was found suitable (Figure 10.7). Here, a vapor was condensed in a gas stream. After going through the double orifice (2), the vapor and cold clean gas were combined in the mixing reservoir (5). The produced aerosols were radiolabeled by ^{35}S, ^{75}Se, and ^{32}P.[19]

TABLE 10.4
Radioactively Labeled Inorganic Condensation Aerosols

Compound or Element[a]	Temperature Range (°C)	Range of Particle Radii (μm)	Radioactive Isotopes (half-life)
Pt-oxides, mo, s	600–1300	5×10^{-3}–3×10^{-2}	^{197}Pt (18 h) ^{199}Au (3 d)
Ag, s	600–1300	2×10^{-2}–2×10^{-1}	^{110}Ag (249 d)
Au, s	700–1200	10^{-2}–10^{-1}	^{198}Au (2.7 d)
WO_3, N_2, mo, s	900–1200	2×10^{-2}–8×10^{-2}	^{185}W (73 d)
NaCl, mo, s	400–1100	3×10^{-3}–10^{-1}	^{24}Na (15 h) ^{22}Na (2.6 y)
V_2O_5, s	400–950	5×10^{-2}–1.5×10^{-1}	^{50}V (6 × 1014 y)
Se, N_2, mo, s	150–300	3×10^{-2}–3×10^{-1}	^{75}Se (27 d)
Te, N_2, mo, s	200–300	10^{-2}–5×10^{-2}	^{127}Te (105 d)
$Re(Re_2O_7)$, s	100–350	4×10^{-3}–3×10^{-2}	High spec. act. ^{186}Re (90 h) 188Re (17 h)
AgI, N_2, mo, s	200–600	5×10^{-2}–3×10^{-1}	^{131}J (8 d) ^{110}Ag (249 d)
$H_4P_2O_7$, s	150–300	10^{-2}–10^{-1}	^{32}P (14 d)
H_2SO_4, l	50–200	10^{-1}–10°	^{35}S (87 d)
Hg, N_2, l	50–200	$5 \times 1^{0-1}$–10^{-1}	^{203}Hg (48 d)
AgCl, s, N_2	350–1000	10^{-2}–7×10^{-1}	^{110}Ag (249 d)
$Fe(Fe_2O_3)$, s, air	450–800	2×10^{-3}–1.5×10^{-2}	^{55}Fe (2.5 y)
$GaCl_3$, s, N_2, air	60–200	0.4–2.5	^{67}Ga (78 h)
S, s, He	50–200	10–$1^{-5} \times 10^{-1}$	^{35}S (87.6 d)
Tl, s, N_2	300–700	2×10^{-3}–5×10^{-3}	^{204}Tl (3.9 y)

[a] mo = monodisperse aerosol; s = solid; l = liquid.

Wire Generators

Aerosol generators in which metal wires can be evaporated have also proved very suitable. This type of condensation aerosol generator produces a constant concentration of aerosol particles and constant particle sizes; these are reproducible. Furnace generators require a few hours before they work stably. On the other hand, wire generators, 10 minutes after they are turned on, produce constant particle sizes. The preparation of radioactively labeled aerosols from platinum wire and nickel-chromium wire were reported in the middle of the 1960s.[28,29]

The principle of such a generator is shown in Figure 10.8. Clean, dry, and preheated air (G) flows across a platinum wire, which is heated electrically. The produced aerosols can be labeled by ^{197}Pt and ^{199}Au. Similarly, other types of metal wires have been found suitable, such as Re (^{186}Re, ^{188}Re), Au (^{198}Au), etc.[22]

Another apparatus that can be used for preparing radiolabeled aerosols by wire evaporation is a "plasma" aerosol generator.[30] The principle of this method is shown in Figure 10.9. The tungsten or platinum wire (W) is exploded using energy stored in a bank of condensers (about 30 J). Such wire explosions are possible in atmospheres of various gases.[31]

Sintering Metal Generators

Highly dispersed silver aerosols have found useful applications in various physical, chemical, and biological investigations. Generation procedures for this metallic aerosol have been reported in several publications since the middle 1970s.[32]

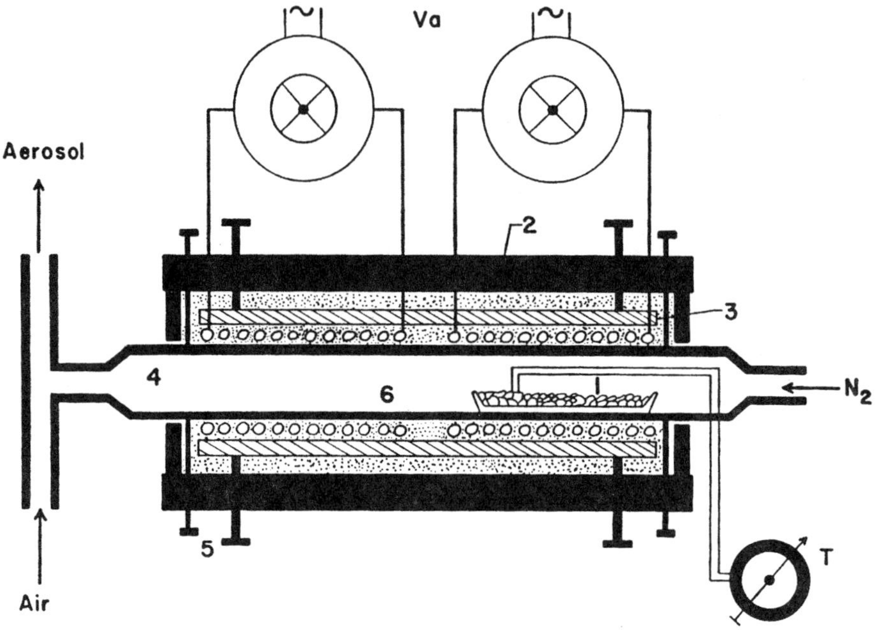

FIGURE 10.5 A longitudinal furnace for preparing inorganic condensation aerosols. 1 = boat containing porous ceramic and an inorganic substance; 2 = metal shield; 3 = ceramic and asbestos shield; 4 = reheater; 5 = adjusting screws; 6 = quartz tube; N_2 = nitrogen; T = thermometer; Va = Variac.

Sutugin et al.[33-36] have developed a fundamental theoretical basis for the nucleation of metal and metal oxide molecules, and their results were later exploited for practical aerosol preparation.

The Ag-aerosol can be easily radiolabeled. At the end of the 1970s, Spurny developed and described a generator for highly dispersed (Ag + ^{110}Ag) aerosols.[27] In this generator, disks of sintered silver particles (produced as "silver membrane filters" by Flotronics, U.S.) were used as the initial material (Figure 10.10). The disks were labeled with ^{110}Ag by neutron activation.

A schematic and a photograph of the apparatus for preparation of condensation aerosols of radiolabeled silver are shown in Figure 10.11. The silver filter disk (Ag) was heated by electric current. Nitrogen, helium, or argon was used as the inert gas. Approximately monodisperse radiolabeled aerosols of silver were prepared at furnace temperatures between 400 and 1000°C. (Ag melting point is 960.8°C.) In the temperature range below the melting point (400 to 950°C), very fine aerosols could be obtained at concentrations between 10 ng l^{-1} and 5 μg l^{-1}. Mean particle diameters ranged between about 2 nm and 6 nm. At temperatures above the melting point (sintered silver was maintained in a porcelain boat), the particle sizes increased rapidly up to over 0.1 μm (Figure 10.12). This aerosol reacts easily with gases and vapors, such as O_2, H_2S, Cl_2, Br_2, I_2, etc.

Labeling by Means of Radiolabeled Condensation Nuclei

The condensation methods described can be used to prepare aerosols of relatively small particle size; for example, those smaller than 0.5 μm in diameter. However, these methods are not convenient for labeling organic aerosols because oxygen has no usable radioisotopes, and carbon and hydrogen yield soft radiation. In such cases, preparation by means of radiolabeled condensation nuclei should be considered.

A combination of two kinds of aerosol generators is useful for the preparation of liquid organic aerosols labeled by radioactive nuclei. It is composed of a furnace generator for preparing radioactive condensation nuclei, and a modified Sinclair-LaMer generator (Figure 10.13). An organic

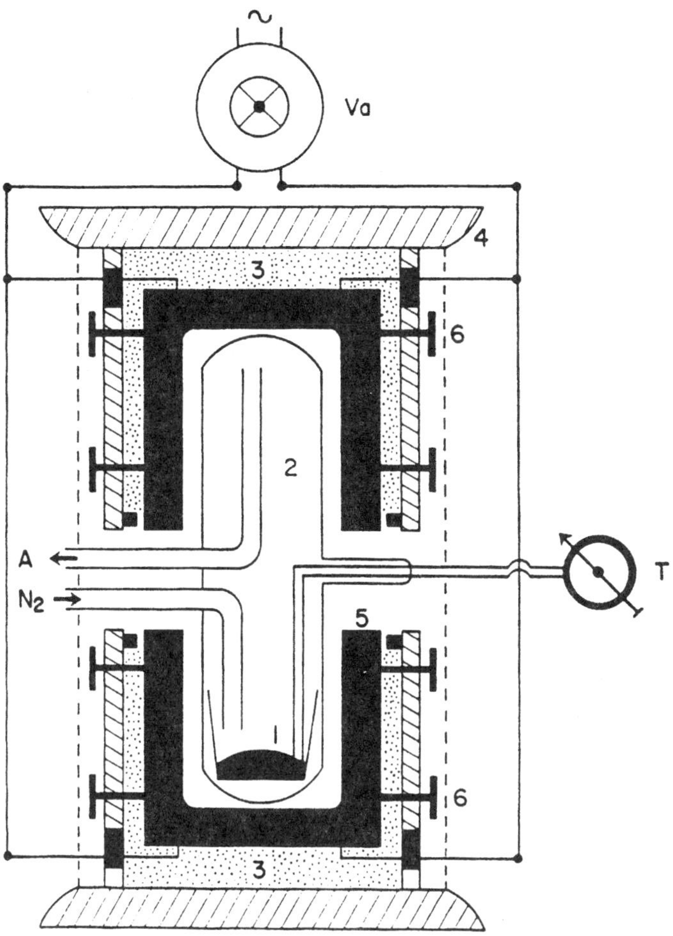

FIGURE 10.6 A vertical furnace for preparing inorganic condensation aerosols. 1 = platinum or gold dish containing an inorganic substance; 2 = quartz tube; 3 = shields; 4 = metal cover; 5 = heating bodies; 6 = metal network; T = thermometer; Va = Variac; A = aerosol; N_2 = nitrogen.

compound (e.g., dioctyl-phthalate, dioctyl-sebacate, etc.) is absorbed on the surface of silica gel (S). The organic vapor diffuses through the inner orifice (I) after heating. The condensation nuclei pass through the diluting space (D2) and reach the outer orifice (O) radioactively labeled. Under suitable conditions, organic vapor condensed on these nuclei. The degree of supersaturation in the mixer D1 depends on the nature of the liquid, the velocity of both gas streams, the temperature of the vaporizer, and the concentration of nuclei.[37] Good results were obtained with radiolabeled nuclei of NaCl, Se, and $H_4P_2O_7$.[20]

Another useful generator for preparing radiolabeled organic aerosols was the apparatus described by Prodi in 1970.[38] A collision generator disperses very diluted solutions of inorganic substances (e.g., NaCl) (see Figure 10.14). After drying (C) fine condensation nuclei are introduced into a thermostated bubbler (D) containing melted carnauba was. The outcoming particles act as condensation nuclei and, as a result, a solid monodisperse wax aerosol is formed. This procedure also works very well using radiolabeled nuclei. Good results were obtained by nuclei of $^{24}NaCl$ and $(NH_4)_6{}^{99}Mo_7O_{24} \cdot 4H_2O$ (see Figure 10.15).[25]

Solid nuclei of $^{99}MoO_3$ were also successful.[26] The ammonium paramolybdate decomposes by heating, and fine molybdenum trioxide is produced (Figure 10.16).

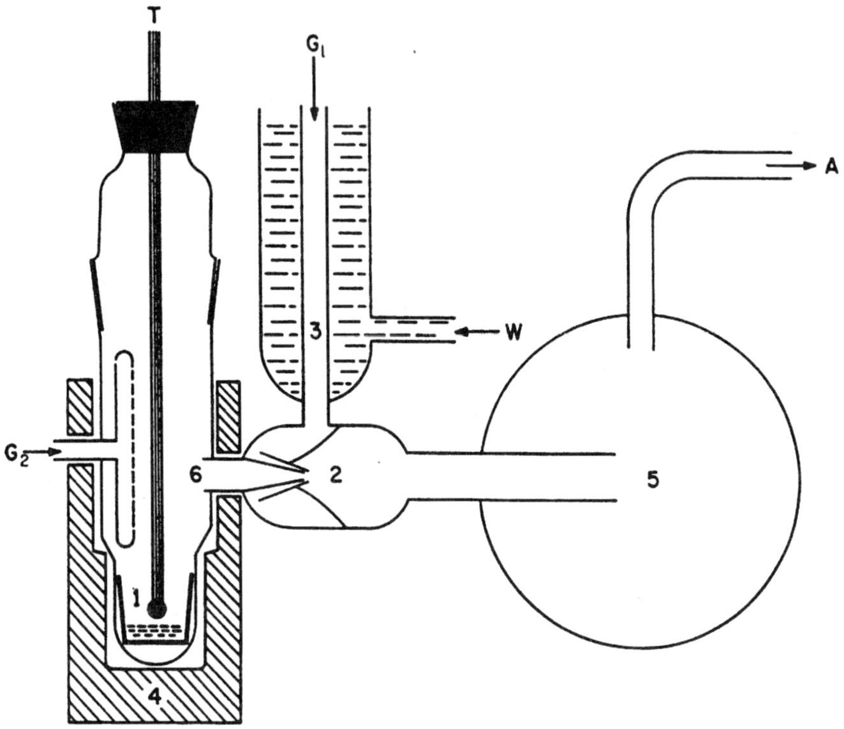

FIGURE 10.7 Evaporation equipment with double orifice. 1 = dish containing an inorganic substance; 2 = double orifice; 3 = path of cold gas stream (G1); 4 = furnace; 5 = mixing reservoir; 6 = glass evaporator; G2 = gas supply; T = thermometer; W = water for cooling; A = aerosol.

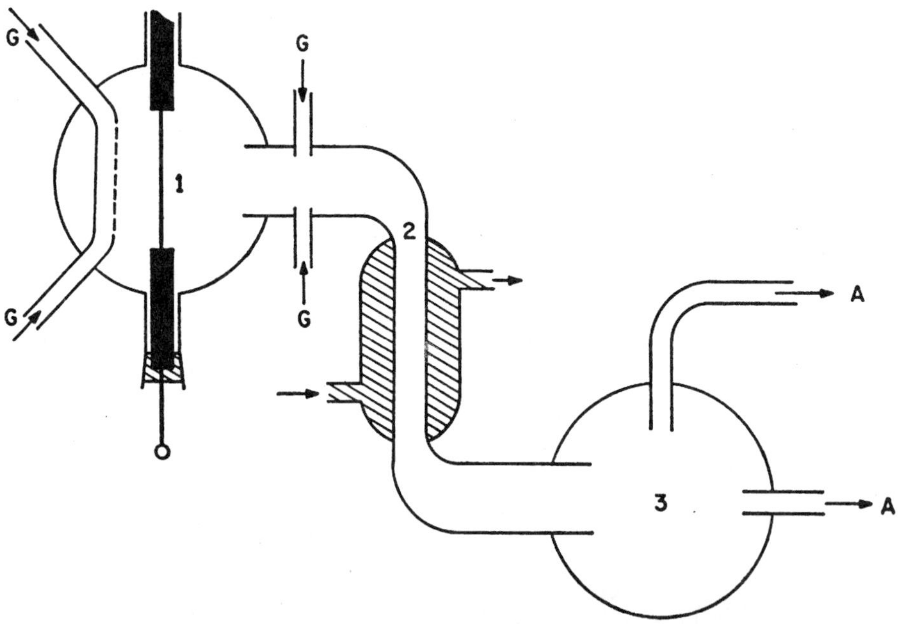

FIGURE 10.8 Platinum wire aerosol generator. 1 = platinum wire in a ceramic insulator; 2 = cooling section; 3 = glass sphere for diluting; G = gas; A = aerosol.

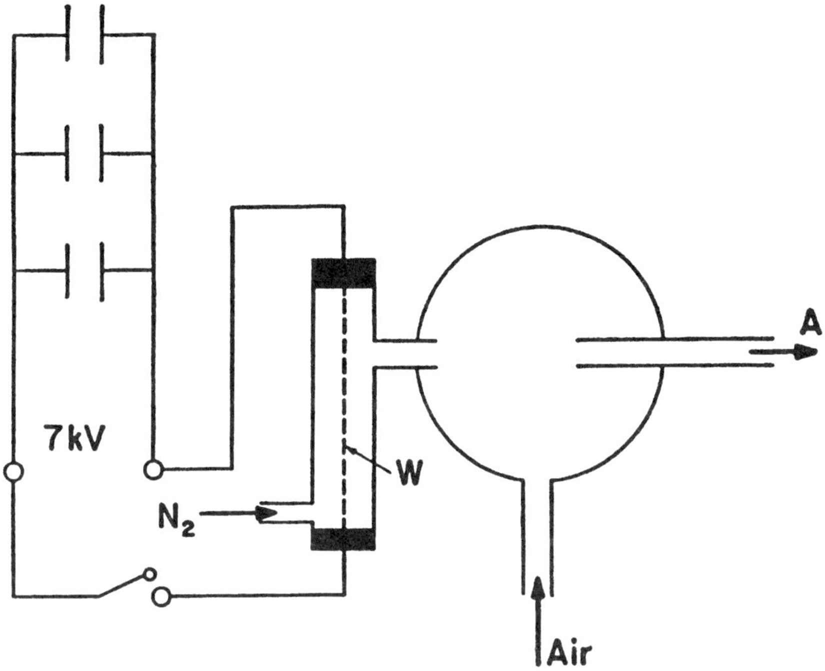

FIGURE 10.9 Schematic of a plasma aerosol generator. W = wire (e.g., a tungsten wire about 0.05 mm in diam. and 2 mm long); N_2 = nitrogen; A = aerosol.

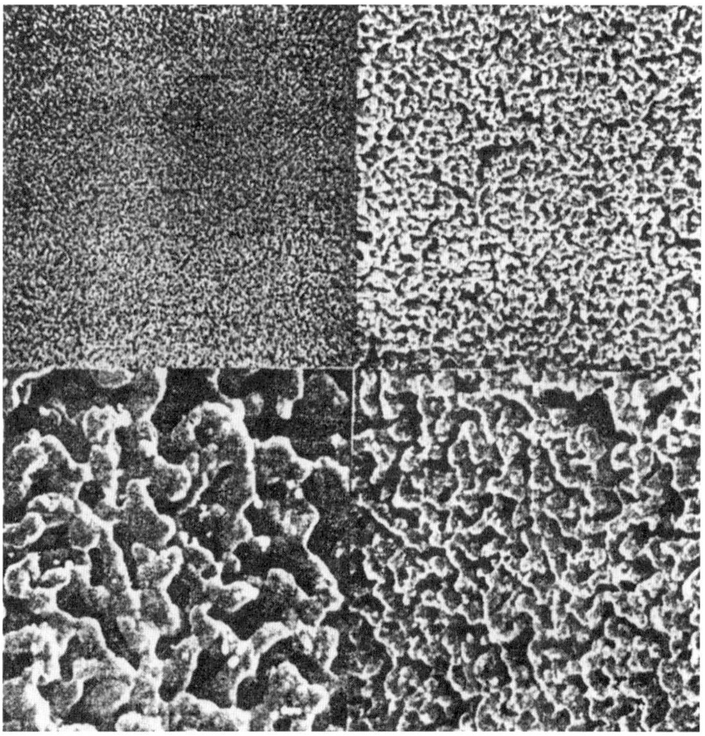

FIGURE 10.10 Scanning electron micrographs of the surface of a silver membrane filter (pore diameter = 0.2 μm; different magnifications).

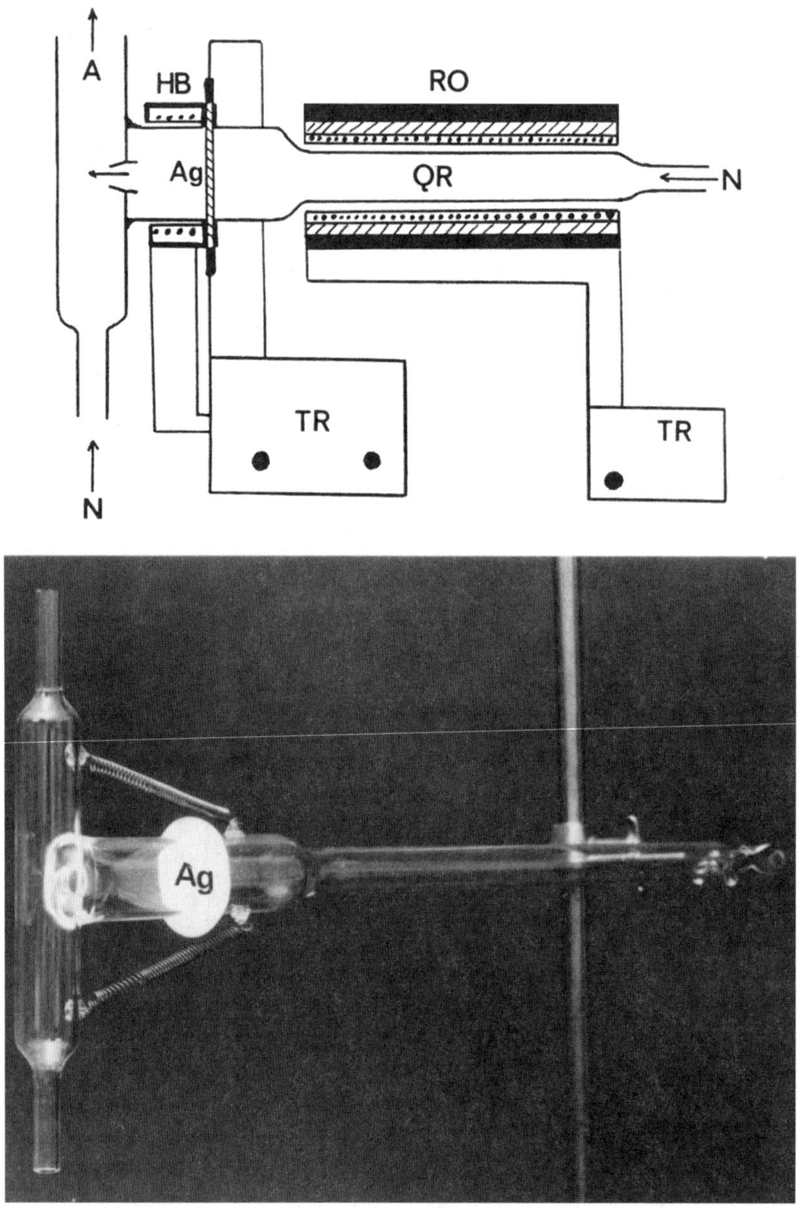

FIGURE 10.11 Equipment for the preparation of silver aerosols. A = aerosol; Ag = silver filter; N = nitrogen; HB = reheater; RO = tubular furnace; QR = quartz tube; TR = transformer.

Radioactively Labeled Carbonaceous Aerosols

Carbonaceous (black carbon or soot) aerosols in polluted atmospheres are totally respirable, with particle sizes less than 1 μm. They carry many, mostly organic, toxic substances (e.g., PAH, nitro-PAH, etc.).

Radiolabeling is a very useful methodology for physico-chemical studies as well as animal toxicological investigations of the production, behavior, and health effects of this group of aerosols. Combustion processes are the most important source of production of carbonaceous aerosols in the atmospheric environment.

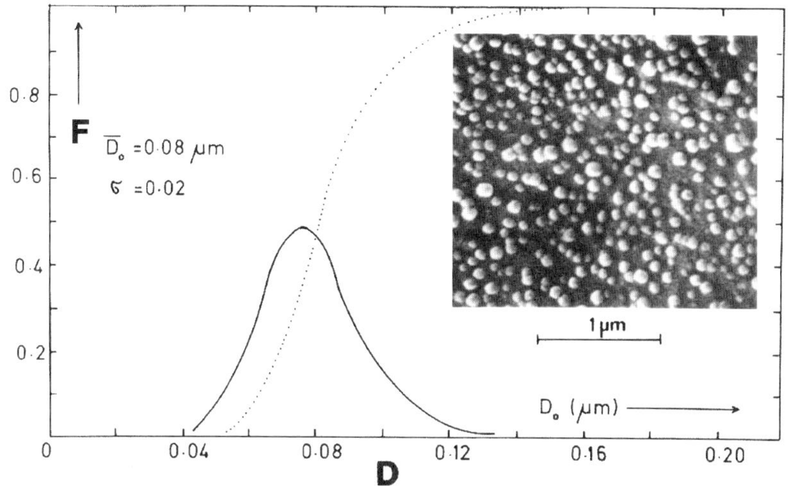

FIGURE 10.12 Scanning electron micrograph and size distribution curves of silver aerosol particles (particle diameter D and particle number frequency F) prepared at furnace temperature of 980°C.

Therefore, it is also reasonable to use the combustion of well defined fuels for generating model carbonaceous aerosols under laboratory conditions and to apply them in physical, chemical, and animal toxicological studies. For radioactive labeling, ^{14}C is the most suitable radioisotope.

Very fine dispersed carbonaceous aerosols can be produced in the laboratory by an incomplete combustion of acetylene or acetylene + benzene.[39,40] A mixture of acetylene, benzene, and oxygen, introduced into a small special burner, is additionally and continuously labeled by a radioactive gas or vapor (e.g., by ^{14}C-acetylene, ^{14}C-benzene, etc.). The system for very sensitive dosage of the radiolabeled benzene vapor is schematically shown in Figure 10.17. The radiolabeled benzene is available in glass vials (Ra-S). After opening, the vial is heated (heater D, regulation transformer RT, and thermometer T) so that the rate of diffusion of the radiolabeled benzene vapor through the glass fritt (F) into a chamber (M) can be well controlled. The original gas mixture (B) is labeled by the ^{14}C-benzene (Ra-B) by this procedure, and is then introduced into the burner. The fine carbonaceous aerosol thus produced is labeled by ^{14}C (Figure 10.18) and has specific activities in the range 1 to 10 μCi (37 to 370 kBq)/mg. Such a model carbonaceous aerosol can be loaded with different PAH (non-active or radioactive) and used in physico-chemical as well as toxicological studies.[39] By studying the behavior of a benzo(a)pyrene (BaP) aerosol itself and in combination with soot particles, different resublimation characteristic curves can be observed (Figure 10.19).

Radiolabeled BaP-aerosol and a mixture of "soot" + BaP were sampled on silver membrane or glass fiber filters. Then, the filter sample was gradually heated and the released radioactivity was measured. The BaP alone could be completely resublimated very quickly. The BaP "coated" on soot particles sublimated much more slowly, at higher temperatures, and incompletely. After being heated at temperatures over 200°C, about 10% of the BaP was still bound on the soot sample.

Similar useful labeling procedures can be applied directly in a diesel engine. Radiolabeled diesel exhaust is then produced and can be used for physical, chemical, and mainly for inhalation toxicological studies on animals.[41] A single-cylinder diesel engine was used to burn diesel fuel containing trace amounts of ^{14}C-labeled hexadecane, dotriacontane, benzene, phenanthrene, or benzo(a)pyrene. Greater than 98% of the ^{14}C in all additives was converted to volatile materials upon combustion. It has been found that aromatic additives labeled carbon particles more efficiently than aliphatic additives.

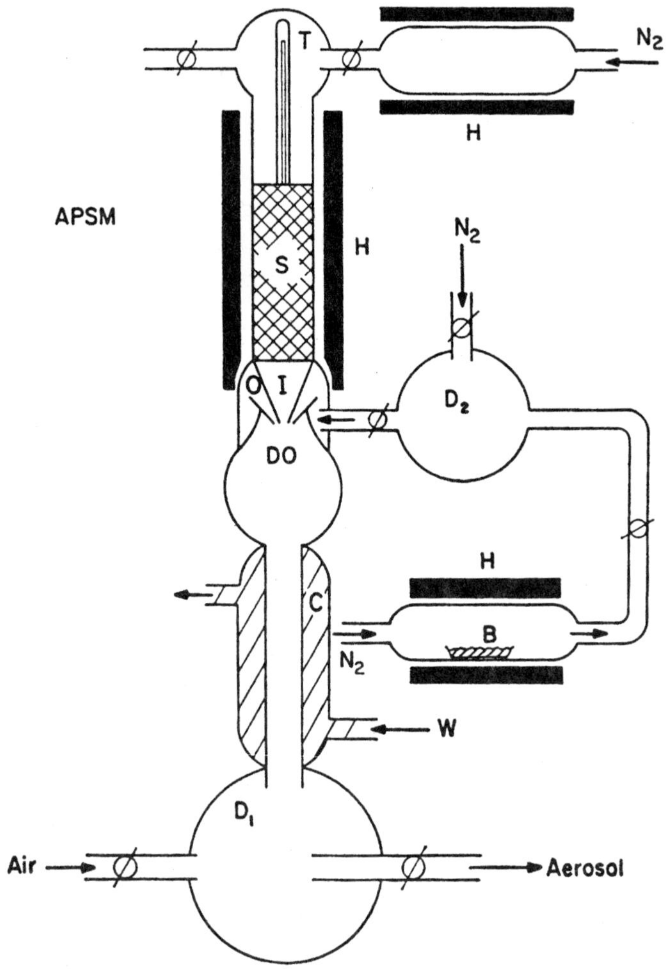

FIGURE 10.13 Schematic of a condensation aerosol generator: aerosol particle size multiplier (APSM). B = boat containing nucleus material; N2 = nitrogen; S = silica gel; DO = double glass orifice; I = inner orifice; O = outer orifice; D1, D2 = glass spheres for diluting; H = heaters; T = thermometer; C = cooler; W = water supply for cooler.

Radioactively Labeled Fibrous Mineral Aerosols

Fibrous mineral aerosols belong in the group of aerosols consisting of nonspherical particles. The particle shape, size, and chemical composition are parameters characterizing the physical, chemical, as well as toxic effects of any fibrous aerosol.[42-51]

The procedure of radioactive labeling is therefore of basic importance in physico-chemical and toxicological studies in this field. The fibrous mineral aerosols are produced by dispersing natural (mainly asbestos) and man-made (glass, ceramic, etc.) mineral fibers. These are mostly silicate fibers containing several elements, which are often characteristic for different types of fibers, and can be radiolabeled, for example, by irradiation with thermal neutrons in a reactor (by fluxes of about $2.10^{14}/cm^2$ s).

Particle size — fiber diameter and fiber length — is involved in the dynamic behavior and in the toxic health effects. Before labeling, the fibrous powders to be used should be well classified with respect to fiber diameter and fiber length.[46] Such quasi-monodispersed powders (Figure 10.20) can be irradiated for a period of 7 days.[42-45]

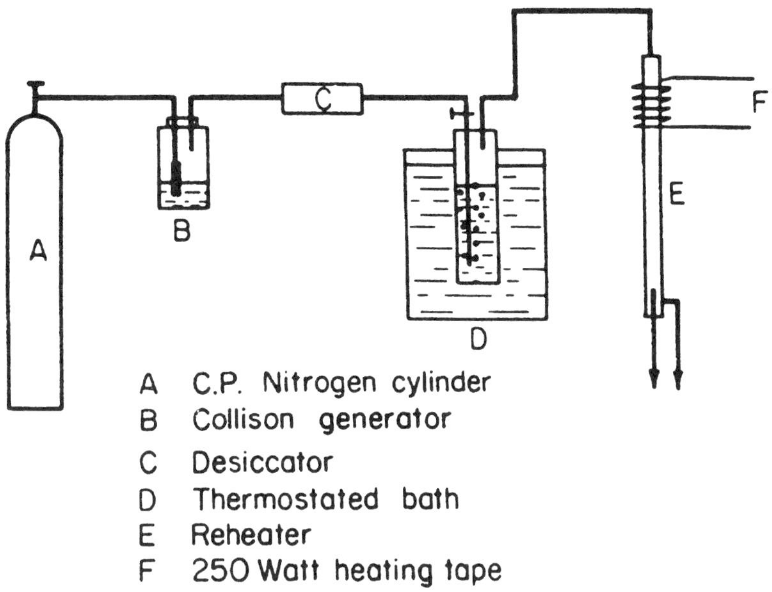

FIGURE 10.14 Schematic diagram of the "Prodi"-generator.

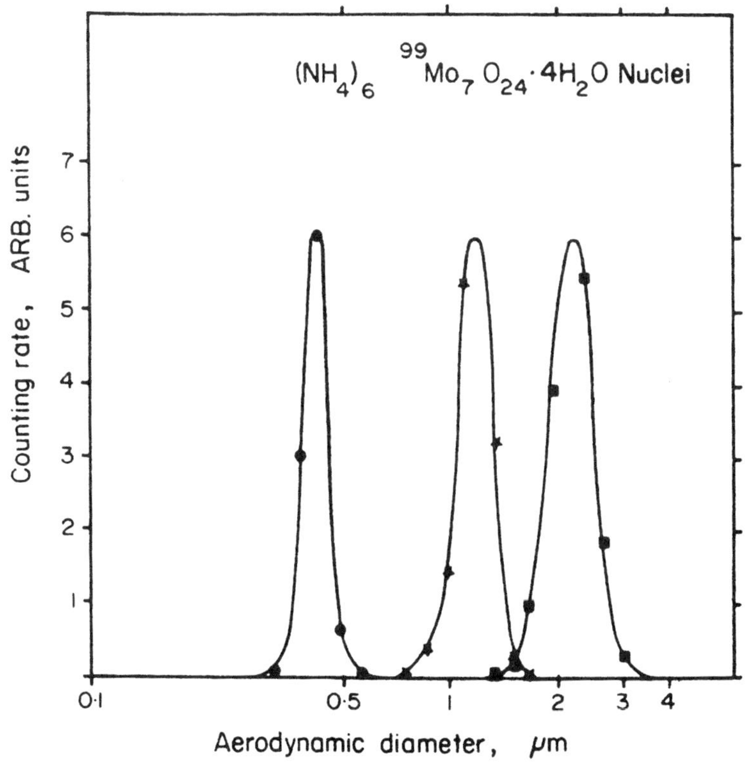

FIGURE 10.15 Activity size distribution spectra with ammonium paramolybdate nuclei obtained at three different temperatures.

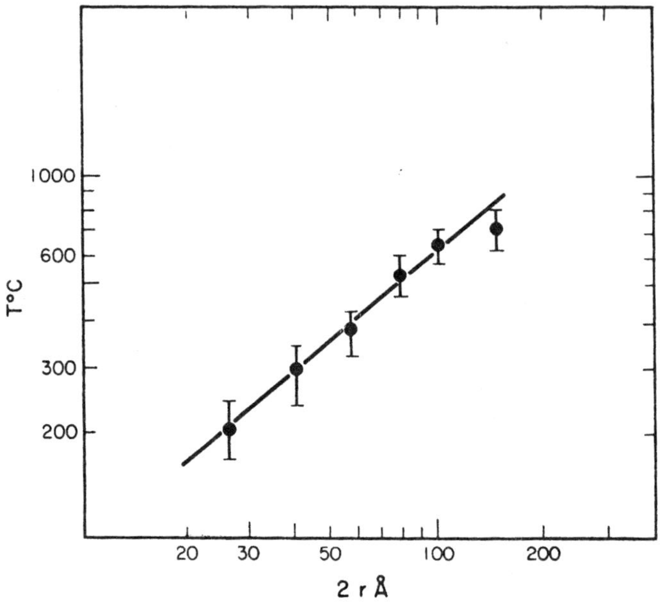

FIGURE 10.16 Particle size (r = particle radius in Ångstroms; 1 Å = 10^{-8} cm) as a function of sublimation temperature.

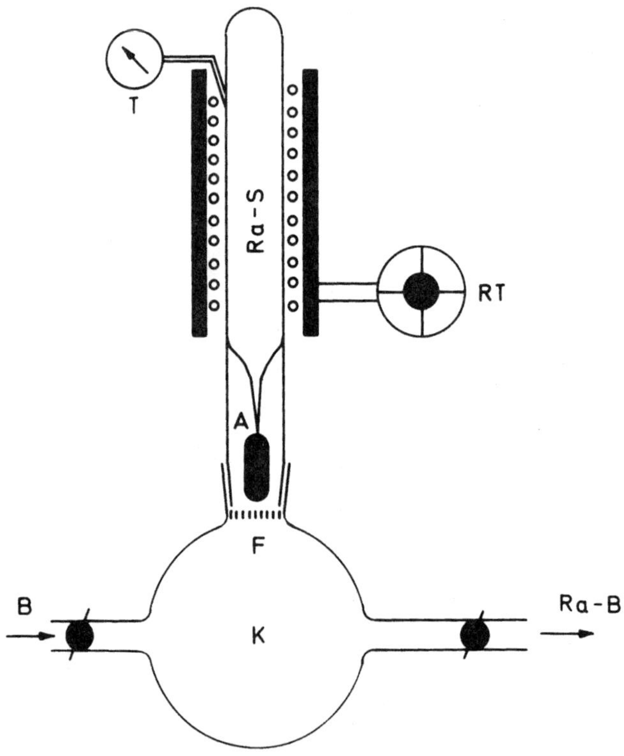

FIGURE 10.17 Schematic diagram of the equipment for the dosage of the radiolabeled benzene vapor.

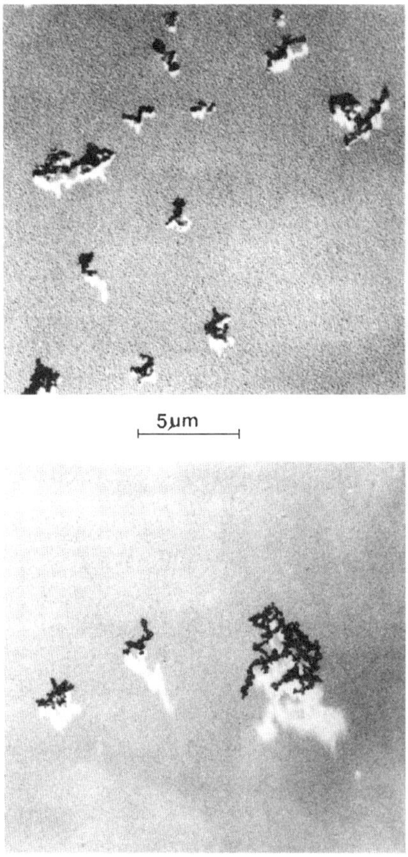

FIGURE 10.18 Transmission electron micrographs of carbon-black particles produced by burning the acetylene and benzene.

The gamma-ray spectra of neutron-irradiated fibrous samples exhibit the same range of radionuclides. In the Table 10.5, the principal gamma-emitting products of neutron-irradiated chrysotile asbestos are listed. The principal, relatively long-lived activation products ^{46}Sc, ^{51}Cr, ^{59}Fe, and ^{60}Co are induced in the (neutron, gamma) reaction on the corresponding stable element.

Similar trace and impurity elements, which are shown in Table 10.5, are present in other asbestos minerals (e.g., in amphiboles) and also in the products of man-made mineral fibers. ^{51}Cr and ^{59}Fe are therefore the most important radio-tracers of mineral fibers. The classified and radiolabeled fibrous probes are then used for aerosol generation. A vibrating bed aerosol generator[47] can be then used to obtain a reproducible cloud of radiolabeled fibrous aerosols with desirable size distributions.

Neutron activation is the principal method used for radiolabeling on mineral fibers. Nevertheless, a few other tcchniques are also mentioned in the literature.[48-51] Tewson et al. have successfully used the radioisotope ^{68}Ga, with a half-life of 68 min.[48] The tracing was very effective and specific activities of about 1 μCi (37 kBq)/mg were obtained.

Turnok et al. labeled the mineral fibers using T_3O. The labeling was realized at 300°C and a pressure of 2000 atm. For biological application, labeling by ^{99c}Tc was also useful.[48,49]

RADIOACTIVE LABELING OF SAMPLING FILTERS

Like aerosols, filters, which are used for aerosol sampling, can also be radioactively labeled. By means of such filters, the mass of particles separated on the filter surface can be measured quickly,

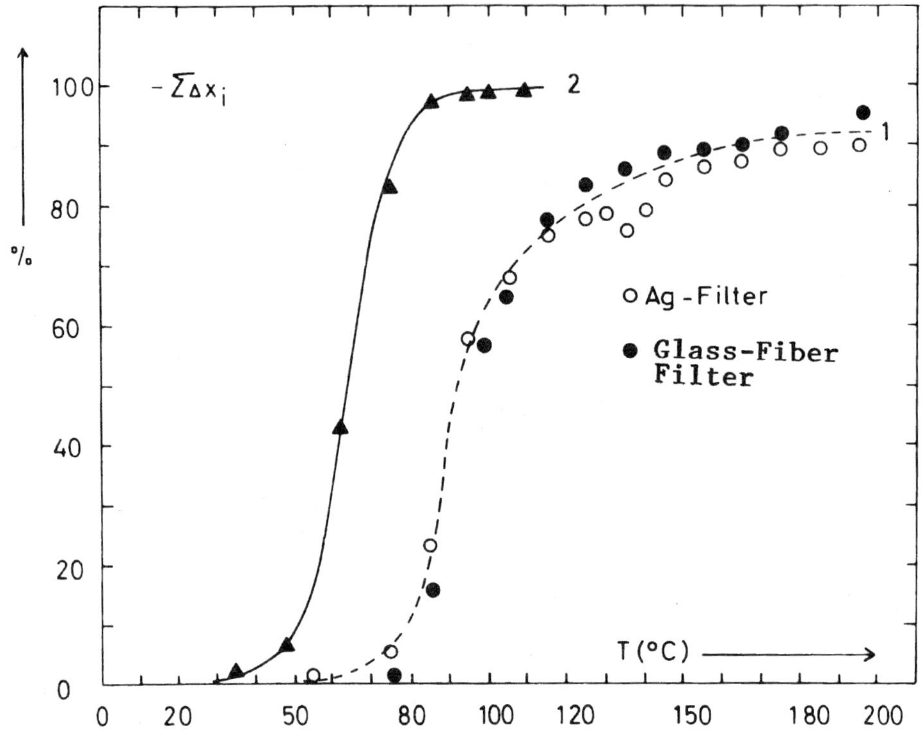

FIGURE 10.19 Resublimation of BaP and BaB + soot from filter probes (1 = soot + BaP; 2 = BaP only).

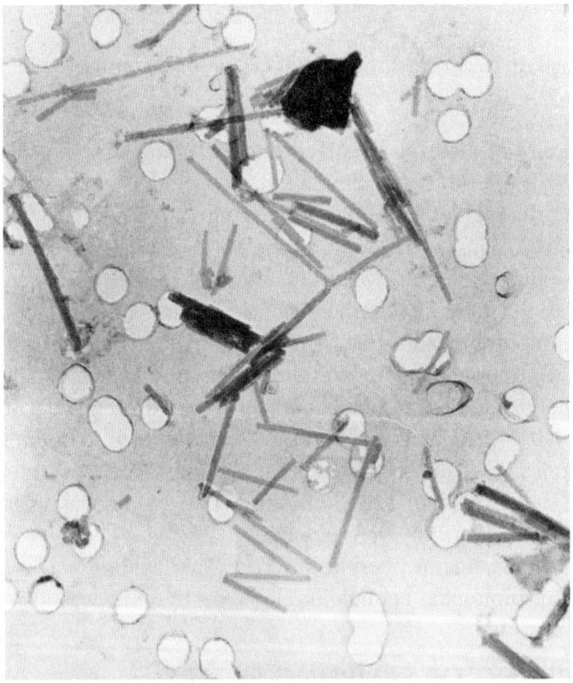

FIGURE 10.20 Transmission electron micrograph showing size-selected, fine chrysotile fibers sampled on the surface of a Nuclepore filter (pore diamater = 0.2 μm).

TABLE 10.5
Principal γ-emitting Activation Products in Neutron-Irradiated Chrysotile with Half-lives > 10 days

Activation Product	Production Process	Half-life	E_γ (MeV) (>10% abundance)
^{46}Sc	$^{45}Sc(n,\gamma)^{46}Sc$	84.0 d	0.89 (100%)
			1.12 (100%)
^{51}Cr	$^{50}Cr(n,\gamma)^{51}Cr$	27.8 d	0.32 (8%)
^{54}Mn	$^{55}Mn(n,2n)^{54}Mn$	312.0 d	0.84 (100%)
^{58}Co	$^{58}Ni(n,p)^{58}Co$	71.0 d	0.81 (99%)
^{59}Fe	$^{58}Fe(n,\gamma)^{59}Fe$	45.0 d	1.10 (56%)
			1.29 (44%)
^{60}Co	$^{59}Co(n,\gamma)^{60}Co$	5.27 y	1.17 (100%)
			1.33 (100%)

Source: From Morgan, A. and Talbot, R.J., *Ann. Occup. Hyg.*, 41, 269-279, 1997. With permission.

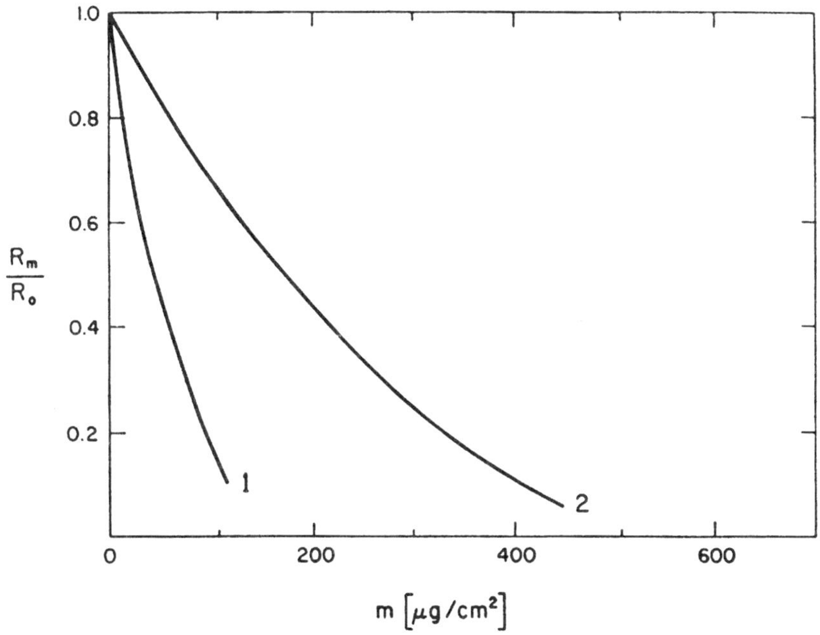

FIGURE 10.21 Relative decrease (R_m/R_o) of the amount of recoil atoms after deposition of m μg cm^{-2} of ammonium chloride aerosols (1) and coal dust (2).

sensitively, and online. In the 1960s, membrane filters labeled with ^{63}Ni (half-life = 92 y) were produced. They were used for sampling and measurement of industrial dusts in the workplace.[52] labeling an analytical filter with decay products of radon and thoron is sensitive.[53] Figure 10.21 shows the results of a Millipore filter that was radioactively labeled with ThC''.[51] The labeling was realized in the following way. The decay products of thoron were collected on the surface of a platinum wire by a negative 600-V potential. Then, the wire was heated to 1200°C, and the platinum aerosol with decay products of thoron was collected on the surface of the Millipore filter.

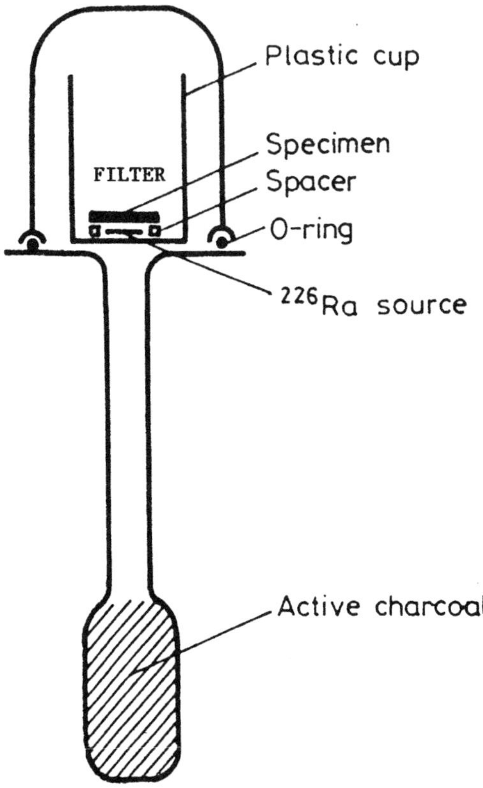

FIGURE 10.22 Apparatus used for labeling in vacuum. (From Jech, C., *Staub*, 20, 75-81, 1960. With permission.)

The adsorption of recoil atoms of ThC'' is measured after aerosol sampling. Radioactive recoil atoms emitted by alpha-decay with an energy on the order of 100 keV penetrate a layer only a fraction of a micron thick. By measuring the relative number of ThC'' recoil atoms, collected layers of particles as thin as 10 $\mu g/cm^2$ can be measured. Figure 10.21 shows that absorption of recoil atoms is a function of not only the area density of the aerosol layer, but also (because of the discontinuous character of that layer) depends on the size of aerosol particles of which the layer is composed. The ThC'' recoil atoms are measured by placing an aluminum foil about 1 mm from the filter surface in a vacuum. After 15 min, a steady state is reached between collection of ThC'' recoil atoms and their decay. The foil is then removed and its radioactivity determined as a function of time.

The principle of another labeling technique is shown in the Figure 10.22. The exposure can be carried out in a small, O-ring-sealed glass chamber that is evacuated.[54,55] ^{226}Ra source is used; it emits recoil atoms of ^{222}Rn and also recoil atoms of ^{218}Po and ^{214}Pb. Immediately after the end of the implantation period, the activity of the implanted specimen (filter) is dominated by the activity of injected decay products (^{218}Po, ^{214}Pb, ^{214}Bi). At 4 hr after the end of the implantation period, the activity in the specimen decreases to the level of ^{222}Rn that is in equilibrium with its decay products (Figure 10.23).

RADIOACTIVE AEROSOL LABELING IN ANIMAL INHALATION TOXICOLOGY AND MEDICAL RESEARCH

Radioactively labeled aerosols are useful for basic and applied research in inhalation toxicology studies using laboratory animals, as well as in basic and applied medical research and diagnostics.[56-59]

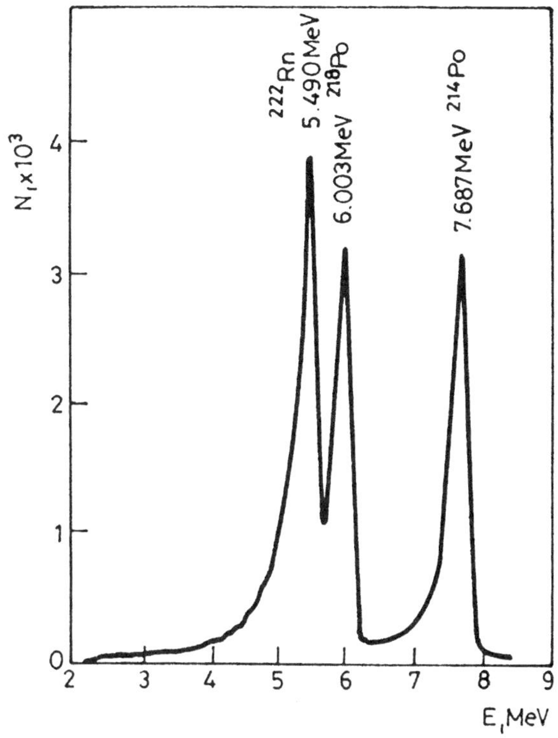

FIGURE 10.23 Spectrum of alpha-particles emitted from an equilibrated recoil labeled specimen. (From Jech, C., *Staub*, 20, 75-81, 1960. With permission.)

Animal Inhalation Toxicology

Assessment of potential inhalation hazards from particulate air pollutants requires information on the deposition, retention, and translocation of toxic aerosols after inhalation exposure. Radioactively labeled aerosols are very useful for such studies and investigations.

Radioactively labeled aerosols provide improved accuracy and speed of measurements. Radiolabeling can further provide a determination sensitivity in parts per billion or smaller concentrations.

Generation Techniques

One of the simplest techniques for generating radiolabeled aerosols for animal inhalation is the nebulization of solutions containing dissolved salts. For example, CsCl with radioactive labels such as ^{137}Cs, ^{137}Ba, ^{144}Ce, ^{144}Pr, ^{90}Sr, ^{90}Y, ^{95}Zr, ^{140}Ba, and ^{140}La is a very useful model.[56] The resultant radiolabeled aerosols are the residue particles after solvent evaporation. Another extremely useful technique is the gamma-labeling (^{51}Cr and ^{160}Yb) of actinide oxides.

Insoluble aerosols can be provided by nebulization of solution suspensions with subsequent heat treatment. Among such belong $^{144}CeCl_3$, $^{60}Co_3O_4$ from $^{60}Co(NO_3)^2$, $^{95}ZrO_2$ from ^{95}Zr-oxalate, etc.[60]

A useful method is the radioactive labeling of standard particles such as polystyrene latex. These particles with standardized size can be labeled in liquid suspensions by, for example, ^{51}Cr or ^{99m}Tc.[61,62] Another source of insoluble radiolabeled particles is aluminosilicates (e.g., montmorillonite clay $Si_8A_{14}O_2(OH)_4$). Radioactive labeling can also be realized by cation exchange methods.[63] Radioisotropes such as ^{90}Y, ^{90}Sr, ^{140}Ba, ^{140}La, ^{67}Ca, ^{68}Zn, etc., are used for such labeling. Several metal oxide aerosols proposed by the already-mentioned condensation procedures are

TABLE 10.6
Physical Characteristics of Some Gamma-Emitting Radionuclides Used for Aerosol Labeling

Nuclide	Physical Half-life	Principal Gamma-Ray Energy	Beta-rays	Production Method
^{11}C	20 min	511keV	Yes	Cyclotron
^{18}F	1.8 h	511 keV	Yes	Cyclotron
^{24}Na	15.0 h	1.37, 2.75 MeV	Yes	Reactor
^{51}Cr	27.0 d	323 keV	No	Reactor
^{77}Br	58.0 h	239, 521 keV	Yes	Cyclotron
^{82}Br	36.0 h	560, 780 keV	Yes	Reactor
^{99m}Tc	6.0 h	140 keV	No	Reactor
^{111}In	2.8 d	171, 245 keV	No	Cyclotron
^{113m}In	1.7 h	393 keV	No	Reactor
^{123}I	13.0 h	160 keV	No	Cyclotron
^{131}I	8.0 d	360 keV	Yes	Reactor
^{198}Au	2.7 d	412 keV	Yes	Reactor

Source: From Newman, S.P., *Aerosols and the Lung,* Clarke, S.W. and Pavia, D., Eds., Butterworths, London, 1984, 71-91. With permission.

representative of fine and ultra-fine aerosol. Useful oxide radiolabeled aerosols include $^{239}PuO_2$, $^{144}CeO_2$, $^{90}Y_2O_3$, $^{67}Ga_2O_3$, $^{60}Co_3O_4$, and $^{140}La_2O_3$.

Radioactively Labeled Model Aerosols in Human Medicine

Studies with radiolabeled aerosols produce reliable data on the deposition, retention, and clearance of inhaled particles. Such techniques are accurate, sensitive, usually rapid to perform, and generally involve little discomfort for volunteer subjects or patients.[57]

Choice of Particles and RadioLabel

The choice of particle and label for radiolabeled aerosol studies in human medicine depends partly on the objectives of the particular investigation being undertaken. Solid particles that are insoluble in body fluids have several advantages. They can be used, for example, for the measurement of mucociliary clearance, lung retention, alveolar deposition, etc. For this reason, the radiolabeled polystyrene latex aerosols are of great interest.

Useful techniques employing radiolabeled aerosols are largely confined to those radionuclides that emit gamma-rays. The gamma-ray energy should be high enough ($\geq$ 100 keV) for penetration through tissue, but low enough ($\leq$ 300 keV) to minimize the amount of radiation shielding required. The half-life of the radionuclide should be as short as possible, but must of course be long enough to enable the study to be performed.

Leaching and dissociation of the label from the particles should be minimal, and high specific activities should be obtainable.[57] Examples of some useful radionuclides are summarized in Table 10.6. Another field is the radiolabeling of different organic compounds with a higher molecular weight. Labeling by the radionuclide ^{99m}Tc is widely used (see Table 10.7). A variety of solid radiolabeled aerosols used in medical studies can be produced by means of spinning disk generator methods.[57] Some examples are illustrated in Table 10.8.

TABLE 10.7
Nebulization of Radiolabeled Solutions

Material Nebulized	Label	Type of Nebulizer	Object of Study
Pertechnetate	^{99m}Tc	Ultrasonic, jet	Deposition measurement
Pertechnetate	^{99m}Tc	Jet	Ventilation scanning
Phytate	^{99m}Tc	Ultrasonic	Development of prototype nebulizer
Albumin	^{113m}In	Jet	Ventilation scanning
Pertechnetate	^{99m}Tc	Ultrasonic	Measurement of pulmonary clearance
Phytate	^{99m}Tc		
Chloride	^{111}In		
DTPA	^{111}In		
DTPA	^{99m}Tc	Jet	Measurement of alveolar epithelial permeability
Antipyrene	^{123}I		
DTPA	^{99m}Tc	Jet	Ventilation scanning
Gluconate	^{99m}Tc		

Source: From Newman, S.P., *Aerosols and the Lung*, Clarke, S.W. and Pavia, D., Eds., Butterworths, London, 1984, 71-91. With permission.

TABLE 10.8
Radioaerosols Generated by Spinning Disk Method

Aerosol Material	Radiolabel
Polystyrene	^{51}Cr
	^{99m}Tc
	^{82}Br, ^{131}I
	^{11}C*
Teflon	^{99m}Tc
	^{18}F*
	^{51}Cr
^{111}In	
	^{198}Au
Iron oxide	^{7}Be, ^{51}Cr, ^{59}Fe, ^{141}Ce, ^{234}Th, ^{198}Au, ^{99m}Tc
	^{99m}Tc
Lucite	^{7}Be, ^{51}Cr, ^{59}Fe, ^{95}Zr, ^{95}Nb

*Following cyclotron irradiation.

Source: From Newman, S.P., *Aerosols and the Lung*, Clarke, S.W. and Pavia, D., Eds., *Butterworths,* London, 1984, 71-91. With permission.

Radiolabeled Aerosols for Ventilating Imaging

The lung ventilation inquiry is one of the most important test methods in nuclear human medicine. The alveolar deposition of model aerosols is measured using an analyzing gamma camera and suitable aerosols, such as DTPA (diethylenetriamine pentaacetic acid) radiolabeled with ^{99m}Tc or ^{81m}Kr. In Figure 10.24, examples of such applications are shown.[64]

RADIOACTIVE LABELING OF ATMOSPHERIC AEROSOLS

Radioactively labeled, finely dispersed aerosols are produced in the atmosphere — partly in a natural way and partly emitted into the atmosphere from several anthropogenic sources.[65-79] They represent, in some situations, a non-negligible health risk for the general population. Nevertheless, they are also "useful" tracers that enable important studies dealing with global tropospheric and stratospheric transport processes, with chemical processes in clouds, precipitations, etc.[65] Three main sources contribute to the production and input of radiolabeled aerosols in the atmosphere.

LABELING BY DECAY PRODUCTS OF RADON AND THORON

The Earth's crust contains the radioactive nuclides ^{238}U, ^{235}U, and ^{232}Th, which by decay produce isotopes of the noble gases radon, thoron, and actinon. After formation in the ground, radon and thoron diffuse into the atmosphere. Direct emanation from soils and rocks to the atmosphere is therefore the major source of atmospheric radon. Groundwater is the second most significant source. Typical atmospheric concentrations of radon range from about 0.1 to 0.4 pCi/l (4 to 15 Bq/m^3). When ^{222}Rn decays, the atom of freshly formed ^{218}Po may quickly attach to a particle of the ambient aerosol. Most of the radon progeny attach to ambient particles, the magnitude of the fraction depending on the number of airborne particles available. The radioactivity is attached mostly to a very fine particle fraction, lying between particle diameters of 0.015 and 0.5 μm.[65-67]

LABELING BY COSMIC RADIATION

Cosmic radiation constantly produces a number of radioisotopes in the atmosphere by nuclear reactions with atmospheric gases. The most important radioisotopes produced by cosmic radiation include ^{3}H, ^{7}Be, ^{10}Be, ^{14}C, ^{22}Na, ^{32}Si, ^{32}P, ^{33}P, and ^{35}S. The majority of these isotopes are produced by secondary, low-energy neutrons. They are formed as very fine particulates. Their concentrations in the troposphere lie in the range of 10^{-9} to 10^{-2} Bq/ml.[65]

LABELING BY ARTIFICIAL RADIOACTIVITY

Artificial atmospheric radioactivity is produced by emissions from anthropogenic sources. Atomic bomb tests, nuclear power plants, as well as their accidents, and the use of radionuclides in industrial and medical processes are the major emission sources of artificial radiolabeled atmospheric aerosols.

Fission Products

The global distribution of fission products from atomic bomb tests has been the object of considerable public and scientific interest, mainly during the period of time between 1950 and 1960. These tests took place in the atmosphere at different test locations as, for example, in the U.S., U.K., U.S.S.R., Christmas Island, etc. In nuclear tests, most radioactive isotopes were produced by neutron fission of ^{235}U, ^{238}U, and ^{239}Pu. The most important labeling was produced by radioisotopes ^{137}Cs, ^{50}Sr, ^{144}Ce, ^{95}Zr, ^{89}Sr, and ^{140}Ba. From the early studies,[68-70] it became apparent that the atmosphere acted as a worldwide reservoir for the accumulation of fission products. The radioactivity was bound on particles in the size range between 0.01 and 0.1 μm. The finest radiolabeled aerosols reached the troposphere and stratosphere. In the troposphere, these very fine particles soon became involved in the processes of water vapor condensation, cloud evaporation, aerosol coagulation, etc. Precipitation accounted, at that time, for 80 to 90% of the fallout deposition of fission-product-labeled aerosols.

Industrial Sources

Some industrial processes use radioactive materials for various specialized applications. Aerosols are released during manufacturing and handling of such materials and products.[66]

Nuclear Power Plants

The use of nuclear power for electrical energy production contributes, under conventional conditions, very little to atmospheric contamination by radiolabeled aerosols. Nevertheless, the release can be very significant in the case of a nuclear reactor accident.

In the accident at the Three Mile Island nuclear generating station in 1979, a large portion of the reactor core underwent a meltdown. Fortunately, the release of radiolabeled aerosols was relatively small.

Unlike the accident at Three Mile Island, the April 26, 1986, accident at the Chernobyl nuclear generating station in the Ukraine did involve very high and dangerous release of radiolabeled aerosols. An explosion and fire in the reactor core dispersed radiolabeled aerosols into the nearby environment (Figure 10.25) as well as into the global atmsophere.[71-77]

Chernobyl Aerosol Characterization

The Chernobyl accident resulted in the discharge from the damaged reactor of about 50 million Curies of different radioactive isotopes, in addition to the larger amounts of noble gases (^{85}Kr and ^{133}Xe). The total core inventory of the exploded unit 4 of the Chernobyl Atomic Energy Station was at the level of 1000 MCi ($3.7 \cdot 10^{19}$ Bq) before the accident.[71,72]

The total release of radionuclides was estimated by the International Atomic Energy Agency in Vienna in 1996 (Table 10.9). The atmospheric transport of the released radiolabeled aerosol cloud was predestined by the existing weather and wind parameters (Figure 10.26). Subsequent atmospheric transport to European countries is illustrated in Table 10.10.

The particle-size distributions of the "Chernobyl Aerosol" were measured in different countries. They depended on several meteorological and geographical parameters. The labeling with different radioisotopes was also particle-size dependent. Not only small particles (e.g., less than 5 μm in diameter), but also relatively large radiolabeled particles (> 20 μm) were transported hundreds of kilometers from the Chernobyl plant.[72]

The Chernobyl accident had, of course, several very negative impacts on the total global environment,[73] including the health of the general population,[75] ecological systems,[76] etc. On the other hand, the measurements of the transported radiolabeled aerosols made it possible to develop, improve, and verify different useful mathematical models to characterize the global atmospheric particle transport.[77]

RADIOLABELED ATMOSPHERIC AEROSOLS AND RADIATION SMOG

The presence of large, highly concentrated clouds of radiolabeled aerosols can initiate several radiochemical reactions in the surrounding polluted atmosphere.[78,79] This was also the case during and after the Chernobyl accident. The most important radiochemical process is the formation of radiolytic products. The adsorbed radiation initiates the formation of ions and free radicals (Figure 10.27), such as O^{2-}, OH^+, OH^- and H_3O^+. These products are designated in the literature as "atmospheric radiation smog."[79] Such free radicals can furthermore play important roles in the formation process of finely dispersed, secondary aerosols and thereby enhance their toxic potential.

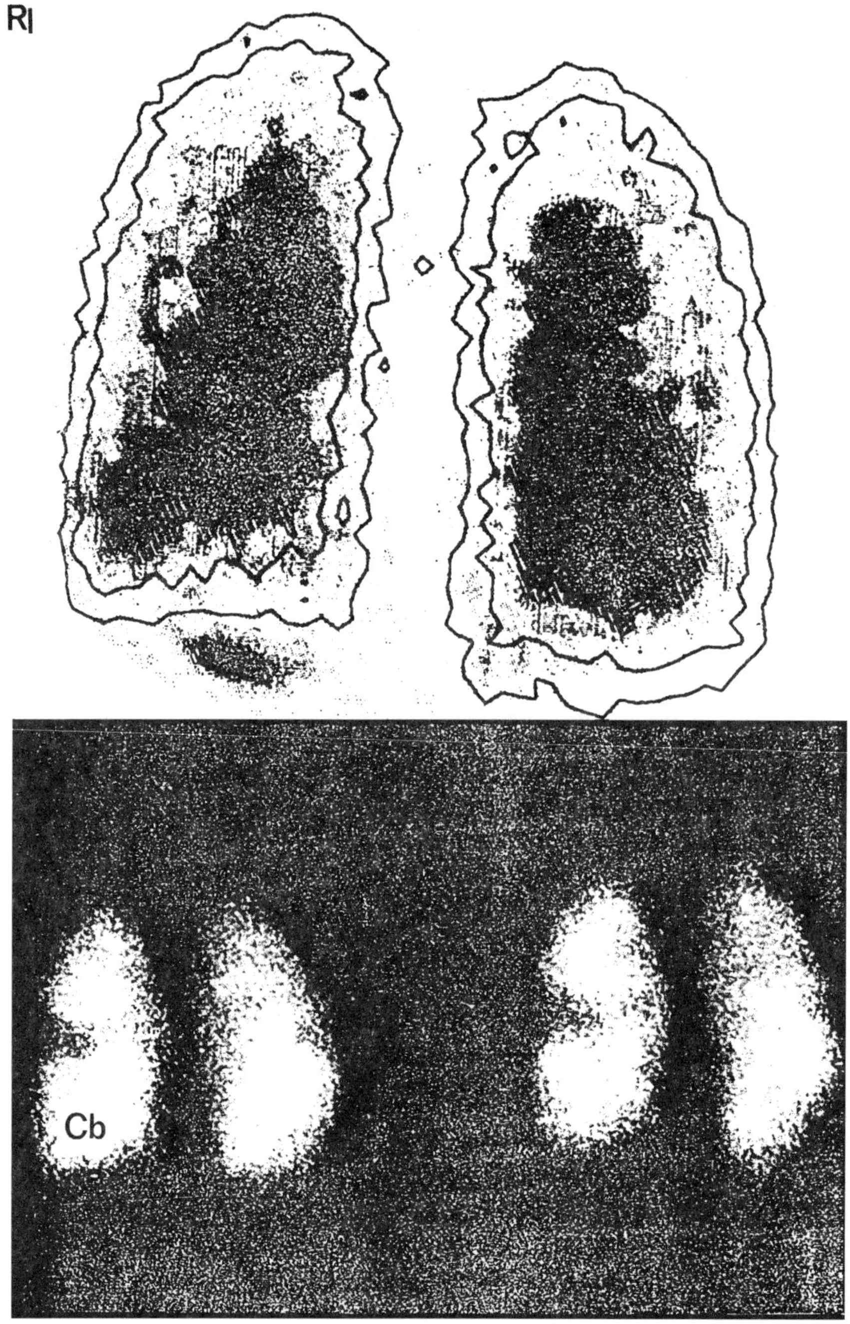

FIGURE 10.24 Radioaerosol lung penetration. Top: Aerosol image aligned with ^{81m}K contours (RI). Bottom: Patient with chronic bronchitis (Cb): aerosol image with ^{99m}Tc-DTPA (left) and with ^{81m}K (right). (From Newman, S.P., Aerosols and the Lung, Clarke, S.W. and Pavia, D., Eds., *Butterworths,* London, 1984, 71-91. With permission.)

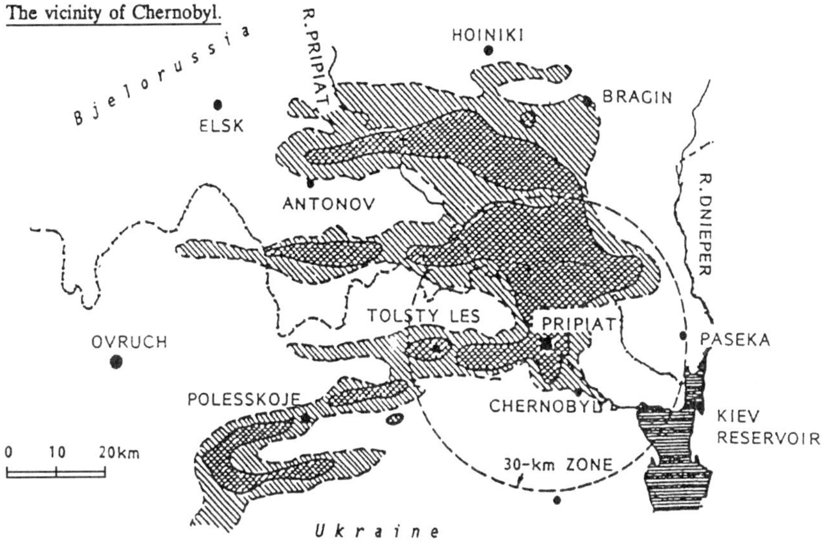

FIGURE 10.25 The vicinity of Chernobyl. The hatched areas belong to the heavy radioactive-contaminated zones. (From Medvedev, Z.A., *The Environmentalist,* 7, 201-209, 1987. With permission.)

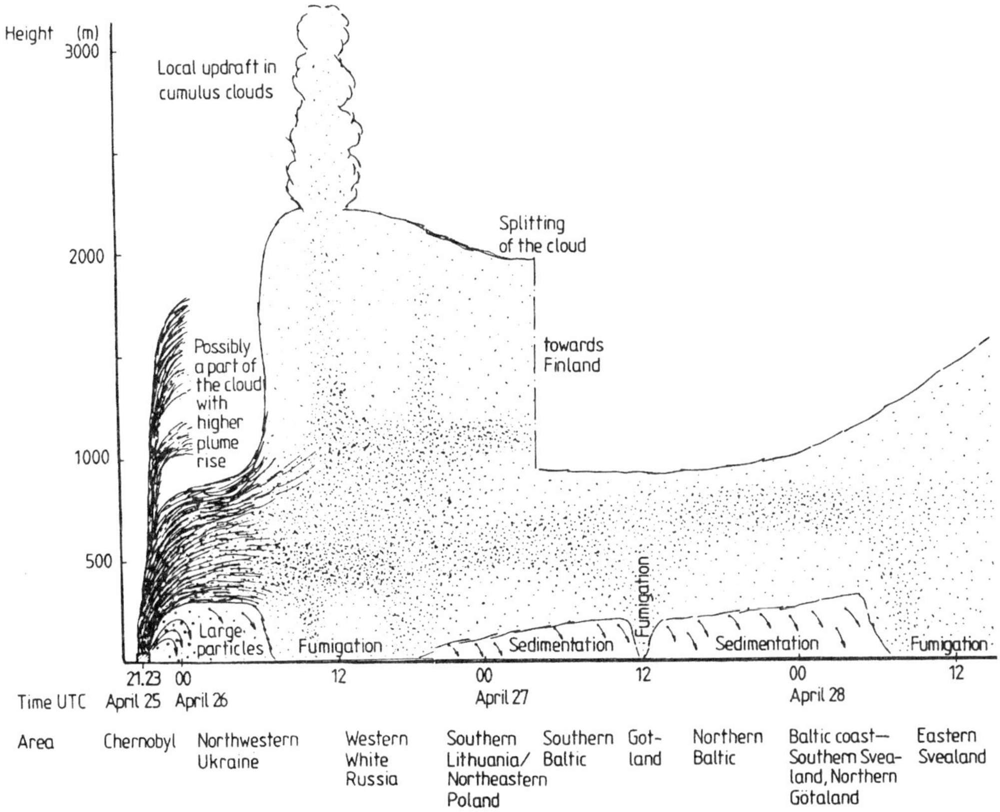

FIGURE 10.26 Qualitative picture of the transport and dilution of the initial cloud of radionuclides from the Chernobyl reactor. (From Persson, C., Rodhe, H., and deGeer, L.E., *Ambio,* 16, 20–31, 1987. With permission.)

TABLE 10.9
Core Inventories and Total Released from Reactor No. 4, Chernobyl Nuclear Power Plant

Element	Half-life (d)	Inventory (Bq)	Percentage Released
^{85}Kr	3930.0	3.3×10^{16}	~100.0
^{133}Xe	5.27	1.7×10^{18}	~100.0
^{131}I	8.05	1.3×10^{18}	20.0
^{132}Te	3.25	3.2×10^{17}	15.0
^{134}Cs	750.0	1.9×10^{17}	10.0
^{137}Cs	1.1×10^{4}	2.9×10^{17}	13.0
^{99}Mo	2.8	4.8×10^{18}	2.3
^{95}Zr	65.5	4.4×10^{18}	3.2
^{103}Ru	39.5	4.1×10^{18}	2.9
^{106}Ru	368.0	2.0×10^{18}	2.9
^{140}Ba	12.8	2.9×10^{18}	5.6
^{141}Ce	32.5	4.4×10^{18}	2.3
^{144}Ce	284.0	3.2×10^{18}	2.8
^{89}Sr	53.0	2.0×10^{18}	4.0
^{90}Sr	1.02×10^{4}	2.0×10^{17}	4.0
^{239}Np	2.35	1.4×10^{17}	3.0
^{238}Pu	3.15×10^{4}	1.0×10^{15}	3.0
^{239}Pu	8.9×10^{6}	8.5×10^{14}	3.0
^{240}Pu	2.4×10^{6}	1.2×10^{15}	3.0
^{241}Pu	800.0	1.7×10^{17}	3.0
^{242}Cm	164.0	2.6×10^{16}	3.0

Source: From Medvedez, Z.A., *The Environmentalist,* 7, 201-209, 1987. With permission.

TABLE 10.10
Regions in Europe Affected by Emissions During Different Days

Period	Day of Emission	Day of Arrival	Target Zone
1	April 26	April 27–30	Baltic states, Scandinavia, Finland
2	April 27	April 28–May 2	Eastern central Europe, southern Germany, Italy, Yugoslavia
3	April 28–29	April 28–May 2	Ukraine and eastward
4	April 29–30	May 1–4	Romania, Bulgaria, Balkan
5	May 1–4	May 2–7	Black Sea, Turkey
6	May 5	May 6–9	Central Europe, Scandinavia, Finland

Source: From Pöllänen, R., Valkama, I., and Toivonen, H., *Atm. Environ.,* 31, 3575-3590, 1997. With permission.

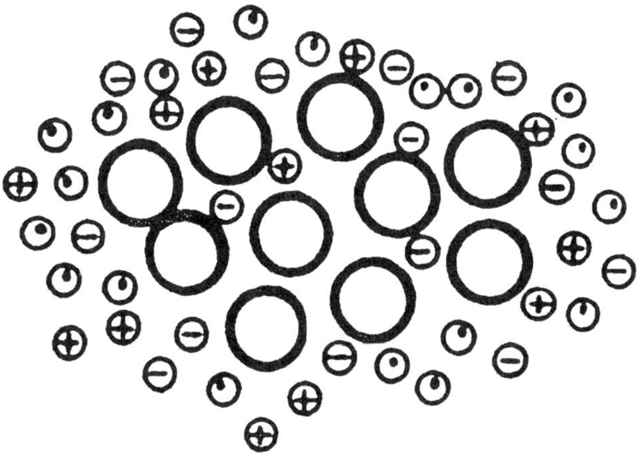

FIGURE 10.27 Schematic picture showing the cloud of radiolabeled aerosol particles and the formation of radiolytic products (ions and radicals). (From Spurny, K., *Atm. Protection* (*Ochrana Ovzdusi*), June 1988, 156-158. With permission.)

REFERENCES

1. Renoux, A., La radioactivité atmospherique naturelle. Son histoire-ses decouvertes. *Actes des 11 emes Journees d'Etudes sur les Aerosols,* COFERA, Paris, 1995, 94-102.
2. Behounek, F., Research on aerosols, *Proc. First Natl. Conf. on Aerosols,* Czechoslovak Acad. Sci. Prague, 1965, 331-334.
3. Spurny, K.R. and Lodge, P.J., Radioactively labeled aerosols, *Atm. Environment,* 2, 429, 1968.
4. Gosman, A. and Jech, C., *Nuclear Methods in Chemical Research,* Academia Publishers, Prague, 1989.
5. Lassen, L., The distribution of natural radioactivity on different particle sizes in atmospheric aerosols, *Proc. First. Natl. Conf. on Aerosols,* Czechoslovak Academy of Sci., Prague, 1965, 345-354.
6. Junge, Ch.E., *Air Chemistry and Radioactivity,* Academic Press, New York, 1963.
7. Morawska, L. and Jamriska, M., Determination of the activity size distribution of radon progeny, *Aerosol Sci. Technol.,* 26, 459, 1997.
8. Schery, S.D., Wasiolek, P.T., Nemetz, B.N. et al., Relaxed eddy accumulator for flux measurement of nanometer-size particles, *Aerosol Sci. Technol.,* 28, 159, 1998.
9. Chen, T.R., Tung, C.J., and Cheng, Y.S., Nanometer particle size and concentration from thoron radiolysis, *Aerosol Sci. Technol.,* 28, 173, 1998.
10. Hopke, P.K., Ed., Radon and its decay products, *ACS Symposium Ser. 331, Am. Chem. Soc.,* Washington, D.C., 1987.
11. Bigu, J., Some observations on the electrical characteristics of radioactive (222Rn progeny) and nonradioactive aerosols, *Health Phys.,* 58, 21, 1990.
12. Bigu, J., Electrical charge characteristics of long-lived radioactive dust, *Health Phys.,* 58, 341, 1990.
13. Meshbah, B., Fitzgerald, B., Hopke, P.K., and Pouprix, M., A new technique to measure the mobility size of ultrafine radioactive particles, *Aerosol Sci. Technol.,* 27, 381, 1997.
14. Chu, K.D., Hopke, P.K., Knutson, E.Q., Tu, K.W., and Holub, R.F., Induction of ultrafine aerosols by radon radiolysis, P.K. Hopke, Ed., Radon and its decay products, ACS Symposium Ser., 331, *Am. Chem. Soc.,* Washington, D.C., 1987.
15. Jech, C., Untersuchung von abgeschiedenen Aerosolproben mittels radioaktiver Oberflächenmar-keirung, *Staub,* 22, 40, 1962.
16. Jech, C., Nachweis von submikronischen Aerosolen mittels kontinuierlicher Rückstossmarkierung mit den Radon-Zerfallsprodukten, *Staub,* 23, 55, 1963.
17. Jech, C., Utilization of radioisotopes and their radiation in aerosology, *Proc. First Natl. Conf. on Aerosols,* Czechoslovak Acad. Sci. Prague, 1965, 413-424.

18. Booker, D.V., Chamberlain, A.C., Rundo, J., Muir, D.C.F., and Thomson, M.F., Elmination of 5 μm particles from the human lung, *Nature,* London, 215, 30, 1967.
19. Spurny, K. and Hampl, V., Preparation of radioactive labeled condensation aerosols. I. Aerosols of sodium chloride, silver iodide and sulfuric acid, *Coll. Czech. Chem. Commun.,* 30, 507, 1965.
20. Spurny, K. and Hampl, V., Preparation of radioactive labeled condensation aerosols. II. Aerosols of mercury, platinum, selenium and pyrophosphoric acid, *Coll. Czech. Chem. Commun.,* 32, 4190, 1967.
21. Spurny, K. and Kubalek, J. Preparation of radioactively labeled condensation aerosols. III. Equipment for filter efficiency measurement by means of radioactivly labeled aerosols, *Coll. Czech. Chem. Commun.,* 36, 2362, 1971.
22. Spurny, K. and Lodge, P.J. Preparation of radioactively labeled condensation aerosols. IV. Aerosols of gold, rhenium oxide, silver, tellurium and vanadium oxide, *Coll. Czech. Chem. Commun.,* 36, 3358, 1971.
23. Spurny, K. and Lodge, P.J. Preparation of radioactively labeled condensation aerosols. V. Aerosols of gallium chloride, iron oxide, sulfur and thallium oxide, *Am. Ind. Hyg. Assoc. J.,* 33, 431, 1972.
24. Spurny, K. and Lodge, P.J., Herstellung hochdisperser Modellaerosole für Staubforschung and Filterprüfung, *Staub-Reinhalt. Luft,* 33, 166, 1973.
25. Prodi, V. and Spurny, K.R., Radioactively labeled monodisperse aerosols, *J. Aerosol Sci.,* 7, 43, 1976.
26. Spurny, K.R. and Prodi, V., On the production of radioactive molybdenum trioxide ultrafine aerosols, *J. Aerosol Sci.,* 7, 101, 1976.
27. Spurny, K.R., Opiela, H., Weiss, G., and Lodge, J.P., On the preparation of highly dispersed radioactively labeled condensation aerosols of silver and silver compounds, *Atm. Environ.,* 14, 871, 1980.
28. Megaw, W.J. and Wiffen, R.D. The nature of condensation nuclei. *Proc. First Natl. Conf. on Aerosols,* Czechoslovak Acad. Sci. Prague, 1965, 511-524.
29. Goldsmith, P. and May, F.G., The chemical nature of the chromium oxide aerosols produced by heated 80:20 nickel-chromium alloy, *Br. Corrosion J.,* 1, 323, 1966.
30. Karioris, F.G. and Fish, B.R. An exploding wire aerosol generator, *J. Colloid Sci.,* 17, 155, 1962.
31. Jech, C., Research on methods of preparing radioactively labeled aerosols for use in chemistry and biology, Technical Report No. 412/CF, IAEA, Vienna, 1967.
32. Porstendörfer, J. and Mercer, T.T., Experimentelle Untersuchungen zum Anlagerungsprozess von Atomen and Ionen im Partikelgrößenbereich 0.1 μm, *Staub-Reinhalt. Luft,* 38, 49, 1978.
33. Sutugin, A.G., Fuchs, N.A., and Kotsev, A.I., Formation of condensation aerosols under rapidly changing environmental conditions, *J. Aerosol Sci.,* 2, 361, 1971.
34. Sutugin, A.G., Lushnikov, A.A., and Chernyaeva, G.A., Bimodal size distribution in highly dispersed aerosols, *J. Aerosol Sci.,* 4, 295, 1973.
35. Zagainov, V.A., Sutugin, A.G. et al., Nucleation of silver vapours on molecular complex mixtures, *Dokl. Acad. Sci. U.S.S.R.,* 238, 1377, 1978.
36. Zagainov, V.A., Sutugin, A.G. et al., On the sticking probability of molecular clusters to solid surfaces, *J. Aerosol Sci.,*7, 389, 1976.
37. Fuchs, N.A. and Sutugin, A.G., Generation and use of monodisperse aerosols, C.N. Davies, Ed., *Aerosol Science,* Academic Press, London, 1966, 1-30.
38. Prodi, V., A condensation aerosol generator for solid monodisperse aerosols, T.T. Mercer, P.E. Morrow, and W. Söber, Eds., *Assessment of Airborne Particles,* C.C. Thomas Publisher, Springfield, IL, 1972, 169-181.
39. Spurny, K.R., Baumert, H.P., Weiss, G., and Opiela, H., Investigation on the carcinogenic burden by air pollution in man: Formation of combined carbon black and benzo(a)pyrene aerosol, *Zbl. Bakt. Hyg. I. Abt. Orig. B.,* 165, 139, 1977.
40. Spurny, K.R. and Baumert, H.P., Formation of a combined benzo(a)pyrene and carbon black aerosol, *Carcinogenesis*, Vol. 3: *Polynuclear Aromatic Hydrocarbons.* P.W. Jones and R.I. Fruedenthal, Eds., Raven Press, New York, 1978, 217-229.
41. Dutcher, J.S., Sun, J.D., Lopez, J.A. et al., Generation and characterization of radiolabeled diesel exhaust, *Am. Ind. Hyg. Assoc. J.,* 45, 491, 1984.
42. Morgan, A., Holmes, A., and Gold, C., Study of the solubility of constituents of chrysotile asbestos *in vivo* using radioactive tracer techniques, *Environ. Res.,* 4, 558, 1971.
43. Morgan, A. and Timbrell, V., The use of neutron activation analysis to determine the composition of blended samples of asbestos, *Int. J. Appl. Radiat. Isotopes,* 22, 745, 1971.

44. Spurny, K.R., Schörmann, J., Weiss, G., and Opiela, H., On the problem of liquids and gases contaminated by fibers when using filters containing chrysotile, *Sci. Total Environ.*, 19, 143, 1981.
45. Morgan, A. and Talbot, R.J., Acid leaching studies of neutron-irradiated chrysotile asbestos, *Ann. Occup. Hyg.*, 41, 269, 1997.
46. Spurny, K.R., Fiber generation and length classification, Willeke, K., Ed., *Generation of Aerosols and Facilities for Exposure Experiments,* Ann Arbor Sci., Ann Arbor, MI, 1980, 257-298.
47. Spurny, K.R., Vibrating bed aerosol generator and its applications, *Staub-Reinhalt. Luft,* 41, 330, 1981.
48. Tewson, T.J., Pransechini, M.P., Scheule, R.K., and Holian, A., Preparation of radiolabeled bioactive asbestos fibers, *Appl. Radiat. Isotop.*, 42, 499, 1991.
49. Turnock, A.C., Brycks, S., and Bertalanffy, F.D., The synthesis of tritium labeled asbestos for uses in biological research, *Environ. Res.*, 4, 86, 1971.
50. Man, S.F.P., Lee, T.K., Gibnet, R.T.N. et al., Canine tracheal mucus transport of particulate pollutants, *Arch. Environ. Health,* 35, 283, 1980.
51. Chan, H.K. and Gonda, I., Preparation of radiolabeled materials for studies of deposition of fibers in the human respiratory tract, *J. Aerosol Med.*, 6, 241, 1993.
52. Spurny, K. and Kubie, G., Herstellung und Anwendung von mit radioaktiven 63-Ni-Isotop markierten Membranfilter, *Coll. Czech. Chem. Commun.*, 26, 1991, 1961.
53. Jech, C., Die Ausnutzung des Alpha-Rückstosses zur Untersuchung der Eindringtiefe radioaktiver aerosole in Filterschichten, *Staub,* 20, 75, 1960.
54. Jech, C., Radioactive recoil used to study marker atom movement during gaseous anodization of silver, *Radiochem. Radioanal. Lett.*, 29, 55, 1977.
55. Jech, C. and Gosman, A., A simple technique for surface labeling of solids with 222Rn and its short-lived decay products, *J. Radioanal. Nucl. Chem. Lett.*, 164, 221, 1992.
56. Newton, G.J., Kanapilly, G.M., Boecker, B.B., and Raabe, O.G., Radioactive labeling of aerosols. Generation methods and characteristics. Willeke, K., Ed., *Generation of Aerosols and Facilities for Exposure Experiments,* Ann Arbor Sci., Ann Arbor, MI, 1980, 339-425.
57. Newman, S.P., Production of radioaerosols, Clarke, S.W. and Pavia, D., Eds., *Aerosols and the Lung,* Butterworths, London, 1984, 71-91.
58. Matthys, H. and Köhler, D., Pulmonary deposition of aerosols by different mechanical devices, *Respiration,* 48, 269, 1985.
59. Phipps, P., Borham, P., Gonda, I., Bailey, D., Bautovich, G., and Anderson, S., A rapid method for the evaluation of diagnostic radioaerosol delivery system, *Eur. J. Nucl. Med.*, 13, 183, 1987.
60. Kanapilly, G.M., Raabe, O.G., and Newton, G.J., A new method for the generation of aerosols of insoluble particles, *Aerosol Sci. Technol.*, 1, 313, 1970.
61. Szende, G. and Udvarheli, K., Production and labeling of monodisperse latexes, *Int. J. Appl. Radiat. Isot.*, 26, 53, 1975.
62. Hass, P.J., Lee, P.S., and Lourenco, R.V., Tagging of iron oxide particles with ^{99m}Tc, *J. Nucl. Med.*, 17, 122, 1976.
63. Ginnell, W.S. and Simon, G.P., Preparation of tagged spherical clay particles, *Nucleonics,* 2, 49, 1953.
64. Clarke, S.W. and Pavia, D., Eds., *Aerosols and the Lung: Clinical and Experimental Aspects,* Butterworths, London, 1984.
65. Junge, C.E., *Air Chemistry and Radioactivity,* Academic Press, London, 1963.
66. Hoover, M.D. and Newton, G.J., Radioactive aerosols, Willeke, K. and Baron, P., Eds., *Aerosol Measurement,* Van Nostrand Reinhold, New York, 1993, 768-798.
67. Cohen, B.S., Radon and its short-lived decay product aerosols, Willeke, K. and Baron, P., Eds., *Aerosol Measurement,* Van Nostrand Reinhold, New York, 1993, 799-815.
68. Libby, W.F., Radioactive strontium fallout, *Proc. Natl. Acad. Sci. U.S.A.*, 42, 365, 1956.
69. Libby, W.F., Radioactive fallout and radioactive strontium, *Science,* 123, 657-660, 1956.
70. Libby, W.F., Radioactive fallout particularly from the Russian series, *Proc. Natl. Acad. Sci. U.S.A.*, 45, 959, 1959.
71. Medvedev, Z.A., The environmental impact of Chernobyl in the Soviet Union, *The Environmentalist,* 7, 201, 1987.
72. Persson, C., Rodhe, H., and deGeer, L.E., The Chernobyl accident — a meteorological analysis of how radionuclides reached and deposited in Sweden, *Ambio,* 16, 20, 1987.
73. Lieser, K.H., Ed., Radionuclides from the Chernobyl accident, *Radiochim. Acta,* 41, 131, 1987.

74. Santschi, P.H., Bollhalder, S., Farrenkothen, K. et al., Chernobyl radionuclides in the environment. Tracers for the tight coupling of atmospheric, terrestrial, and aquatic geochemical processes, *Environ. Sci. Technol.,* 22, 510, 1988.
75. Anspaugh, L.R., Catlin, R.J., and Goldman, M., The global impact of the Chernobyl reactor accident, *Science,* 242, 1513, 1988.
76. Levi, H.W., Radioactive deposition in Europe after the Chernobyl accident and its long-term consequences, *Ecological Res.,* 6, 201, 1990.
77. Pöllänen, R., Valkama, I., and Toivonen, H., Transport of radioactive particles from the Chernobyl accident, *Atm. Environ.,* 31, 3575, 1997.
78. Spurny, K., *Atmospheric Ionization,* Academia Publishers, Prague, 1985.
79. Spurny, K., Radiation Smog, *Atm. Protection (Ochrana Ovzdusi),* Prague, June 1988, 156-158.

Part III

Aerosol Synthetic Chemistry

11 Synthesis and Online Characterization of Zirconia Powder Produced by Atomization

Claude Landron

CONTENTS

INTRODUCTION

New manufacturing techniques for ceramic processing have been carefully explored as means of producing powders with a variety of well-defined characteristics. The quality of the final product can be significantly improved by the choice of an efficient production process. The control of the shrinkage depends drastically on the size and on the microporosity of the particles. It is well-known that the formation of hollow spheres has a negative effect on densifying and sintering. Another important problem related to the particle shape is illustrated by the fact that strongly agglomerated powders give difficulties in sintering conventional powders. Atomization is an alternative procedure that presents many advantages to produce ceramic.[1] It has been used to elaborate high-Tc superconductor powders.[2] As a matter of fact, the synthesis of powders by the spray pyrolysis of the precursor solution can be considered a containerless experiment. Thus, chemical segregation of cations is reduced to a single microsized drop, thereby ensuring good stoichiometry control. Among the techniques for preparation of aerosols (such as condensation methods, dispersion methods, electrical atomization, vibrating orifice technique), we have selected ultrasonic atomization, which is suitable for controlling the distribution of the powder particle size.

Although aqueous solutions occupy an important place in material chemistry, relatively few studies have dealt with structural investigations of the liquid state, for which detailed, accurate information related to atomic distance and coordination number would be useful. It has been shown[3] that a liquid zirconium precursor used at high concentration reproduces an environment similar to that of crystallized compounds. In recent years, new methods for the characterization at the micrometer scale of small liquid particles which play an increasingly important role in many fields of material science, has been developed with high spatial and temporal resolution. EXAFS

0-87371-829-1/00/$0.00+$.50

(Extended X-ray Absorption Fine Structure) is presently the best *in situ* technique that can be used to detect differences on the Angstrom scale in precursor transformation during ceramic synthesis. This spectrometric technique gives accurate information concerning the environment of the cations and the presence of anions in the coordination sphere, which has an effect on the structure of the final product. This chapter describes the modification of the environment of Zr^{4+} cations in samples selected during zirconia synthesis by an atomization route. An *in situ* structural approach is developed in association with zirconia (ZrO_2) powder processing.

X-RAY ABSORPTION

Interest in local structure determination has recently stimulated the growth of EXAFS studies.[4] The principal advantage of this spectrometry is that it is particularly suitable for the study of disordered, glassy, or vitreous materials.[4] EXAFS is based on the measurement of the attenuation of X-ray beam propagation in a condensed medium. Attenuation results from three principal processes: pair production, scattering, and photoelectric absorption. In the energy range of the X-rays used in EXAFS (1 to 20 keV), photoelectric absorption dominates the attenuation process. The atomic distribution of a selected atom is obtained by Fourier transforming the EXAFS signal as a function of incident photon wavenumber. This oscillating signal is related to the variation of the absorption cross-section for the photoexcitation of an electron from a deep core state to a continuum state. The oscillations of the absorption spectrum above an absorption edge of one particular type of atom is related to the final state of the excited atom. They result from the interferences between the wave function of an outgoing photoelectron that is emitted during the photoemission process and the part of this wavefunction that is reflected by the neighboring atoms. Because of the shortness of its mean free path, the photoelectron can be considered a local probe for structure determination. The oscillations essentially depend on the number, the distance, the disorder, and the type of nearest neighbor of the excited atom. From theoretical considerations, the absorption function $\chi(k) = (\mu - \mu_0)/\mu_0$ for a given shell i is given by

$$\chi(k) = \Sigma\left(\left[N_i/kR_i^2\right]f_i(k)B(k)\right) \tag{11.1}$$

where

$$B(k) = \exp\left(-2k^2\sigma_i^2\right)\exp\left(-2R_i\right)\sin\left\{2kR_i + \phi_i(k)\right\} \tag{11.2}$$

and where

- k is the wave vector of the photoelectron
- R_i is the distance between the absorbing atoms and the atoms of the shell i
- $f_i(k)$ is the back-scattered amplitude of the shell i
- $\phi_i(k)$ is the phase shift related to the atoms of the shell i
- σ_i results from the variation of the R_i containing information related to static and dynamic disorder
- λ is the mean free path of the photoelectron

EXAFS depends only on the local structure by the fact that this length is short, thus resulting in information that is particularly useful in the case of amorphous materials.

The radial distribution functions around an excited atom are obtained by extracting[5] the modulus of the complex Fourier transform of $k^n\chi(k)$:

$$TF(R_i) = \pi^{1/2}\int_{k_{min}}^{k_{max}} \exp\left(2ikR_i\right)\chi(k)k^n\,dk \tag{11.3}$$

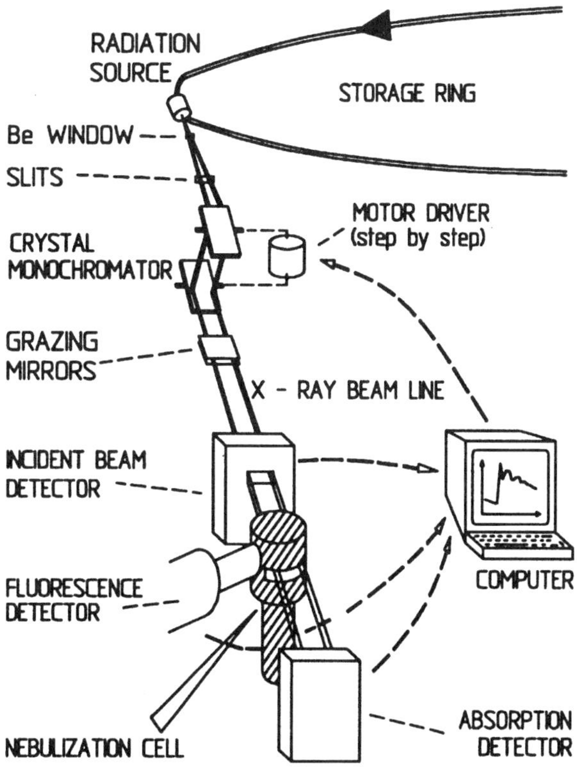

FIGURE 11.1 Outline of the experimental set-up used for the X-ray absorption measurements of an aerosol of micrometer-sized droplets of solutions. The apparatus is composed of an ultrasonic atomizer producing the mist from a 5-cm high geyser generated from the solution. The spray chamber is designed to promote a good mixing and transport of the aerosol toward the analysis cell where the droplets are irradiated in various situations. A closed circuit permits one to keep a constant level in the solution and a stable density of particles in front of the X-ray source.

SAMPLE ELABORATION

The starting solution was prepared by dissolving reagent-grade zirconium nitrate salts in distilled water to yield a 0.5 N solution. The experimental set-up used for *in situ* X-ray absorption measurements of precursors during ceramic synthesis comprises an ultrasonic nebulizer coupled with a detection chamber[6] as shown in Figure 11.1 and Figure 11.2. Note two important advantages of the ultrasonic nebulizer: (1) the aerosol is practically monodispersed, and (2) both the aerosol production rate and the carrier gas flow rate can be independently varied.

The micrometer-sized spheres arc circulated in the X-ray irradiation cell for XAS experiments. The spray chamber is designed to promote good mixing and transport of the aerosol toward the analysis cell where the droplets are irradiated.

EXAFS EXPERIMENTS

Spectroscopic measurements were performed at the synchrotron radiation facilities of LURE (Orsay, France). The spectra were recorded above the Zr K-edge in the energy range 179000 to 19000 eV. The X-ray beam provided by the 1.85-GeV storage ring of DCI was monochromatized by a double Bragg reflection (311) of two parallel Si crystals. The energy resolution was estimated to be 2 eV. The maximum intensity of the positron beam was about 300 mA. The experimental device designed

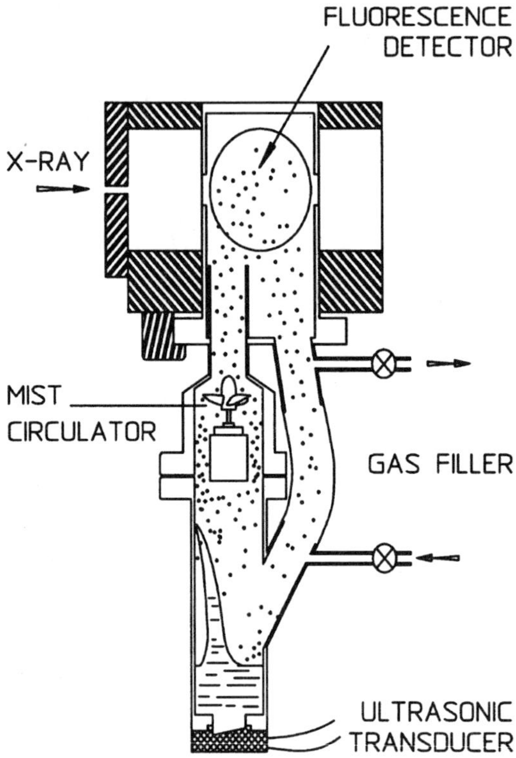

FIGURE 11.2 Schematic illustration of the device used for *in situ* EXAFS measurements on nebulized solutions. This cell was developed for an adaptation to the EXAFS IV station of the DCI storage ring at LURE (Orsay, France).

for *in situ* EXAFS analysis of aerosols is placed after the first ionization chamber of the X-ray beam line. *In situ* absorption measurements of nebulized solutions were thus performed using the fluorescence detection mode, which is more efficient than the transmission mode. The conditions for the absorption measurements of nebulized solutions were similar to those of trace analysis in very dilute samples. The EXAFS measurement of samples produced by atomization and dried at different temperatures were performed, post mortem, in the transmission mode, on powders. They are deposited on a series of tapes in order to obtain a sample thickness, x, which in optimal conditions is determined by $\mu x = 1$, where μ is the absorption coefficient of the powder.

RESULTS AND DISCUSSION

EXAFS data were treated using the program developed by Bonnin,[7] which follows the suggestions of the report on the International Workshop on Standards and Criteria in XAS. A Victoreen-type baseline of the pre-edge was removed, and the oscillations were extracted from the background approximated by a cubic spline function and then Fourier transformed as seen in Figure 11.3. The structural parameters were determined through a fitting procedure using the theoretical amplitude and phase function.[8] Some unknown parameters were evaluated from EXAFS analysis of monoclinic zirconia used as a standard and which has a well-known crystalline structure.[9] We have preferred this structural form of zirconia among its various polymorphs (cubic,[10] tetragonal,[11] or amorphous zirconium oxide[12]). Stabilized cubic or tetragonal forms of zirconia are avoided as standards because yttrium atoms used in substitution to stabilize these structures introduce oxygen vacancies that alter the coordination shell of the cations.

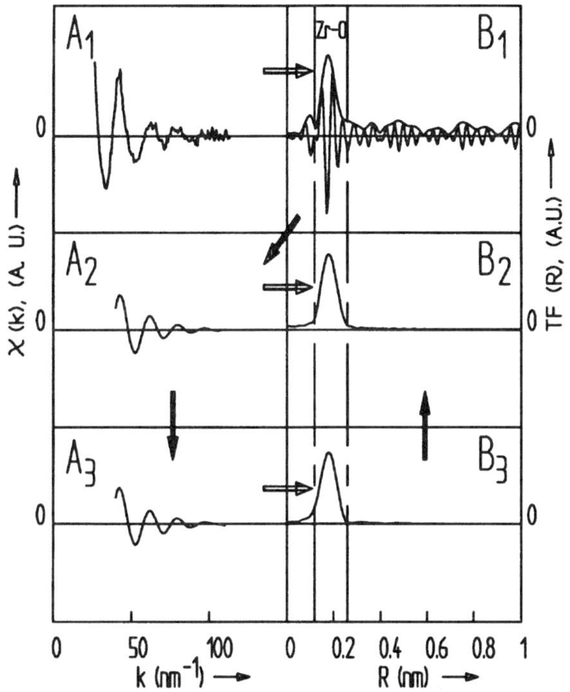

FIGURE 11.3 Zirconium K-edge EXAFS of an aerosol of a 0.5 N aqueous solution of zirconium nitrate. A_1: $\chi_1(k)$ function extracted from experimental EXAFS data; B_1: Fourier transform $TF_1(R)$ of the $k^3\chi_1(k)$ function magnitude: solid line; imaginary part: dotted line); A_2: Fourier filtered experimental oscillations $\chi_2(k)$; B_2: Fourier transform $TF_2(R)$ of the filtered oscillations $k^3\chi_2(k)$; A_3: Least-squares fit $\chi_3(k)$ of the filtered spectrum; and B_3: Fourier transforms $TF_3(R)$ of the function $k^3\chi_3(k)$.

Our structural results concerning an aerosol of a 0.5 N aqueous solution of zirconium nitrate presented are in quantitative accordance with the mean values of the bond lengths obtained by X-ray powder diffraction data.[13] They are calculated from the EXAFS spectra of Figure 11.3. The coordination sphere around the zirconium atoms contains four OH groups, two water molecules, and two oxygen atoms belonging to one bidentate nitrate ligand. We find a similar environment for zirconium in the atomized solution of zirconium nitrate. Note that previous published data[3] are also in agreement; these authors have given d(Zr-O) = 0.222 nm, which is the average of those values found by EXAFS.

We have noted that the structure of the precursor depends on its state in the starting solution. The anion complexing power has an effect on the formation of regularly shaped powders by avoiding hollow sphere formation. The high degree of polymerization of the sulfate precursor solution of hydrolyzed Zr(IV) species is related to the viscosity of the liquid. By heating the powder at different temperatures, we have obtained precursors at different polymerization states. Another fact is able to explain the morphology differences observed according to the precursor salt used: a comparison of thermal decomposition by thermal gravimetric analysis of powders from $ZrOCl_2 \cdot 8H_2O$ and $ZrSO_4$ clearly shows that HCl continuously leaves the spheres, while the SO_2 remains up to the crystallization of zirconia.

Figure 11.4 shows the EXAFS spectra of the samples produced by atomization and dried at different temperatures. For the amorphous samples elaborated between 200°C and 650°C, no significant changes in the Zr-O distances are observed before crystallization and d(Zr-O) is close to 0.221 nm. The coordination number in the amorphous samples seems to increase with the temperature of elaboration. The treatment of the spectrum of the sample elaborated at 800°C shows

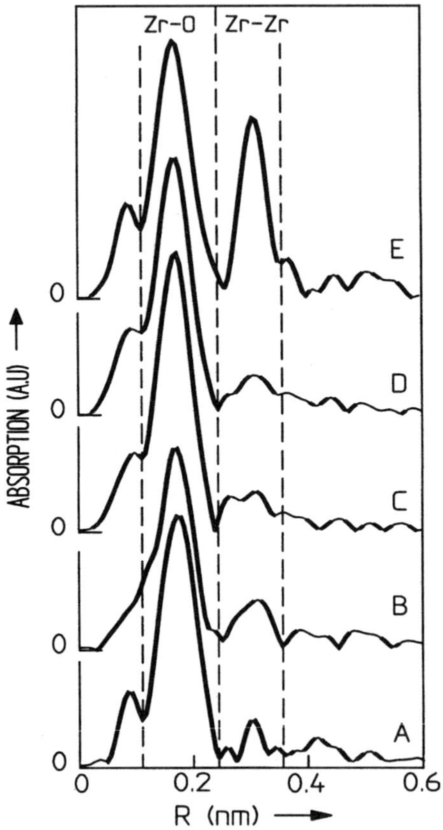

FIGURE 11.4 Comparison of the partial radial distribution function around a zirconium atom of a nebulized solution (0.1 *M*) of zirconium sulfate heated at different temperatures: (A) 200°C, (B) 350°C, (C) 500°C, (D) 650°C, and (E) 800°C.

that the structure of this sample is close to that of the monoclinic zirconia, in which Zr atoms are surrounded by four oxygen atoms at a distance d = 0.211 nm and three atoms by distance d = 0.230 nm. We can conclude that there is no significant modification of the zirconium environment during this first stage of drying. The variation of the coordination number can be related to the growth of crystallites and by the modification of the chains $[Zr(OH)_2(SO_4)_2(H_2O)_4]^n$ linked by water molecules. We have pointed out that the strongly concentrated solution exhibits an order around the zirconium atoms that can be compared to the order around zirconium in a crystallized zirconium sulfate. The local structure in the solution is built from chains of dodecahedrons formed by zirconium atoms that are eightfold oxygen coordinated. The present analysis gives a Zr-O average distance of 0.224 nm, which is not far from the sum of the ionic radii of O^{2-} and $^{VIII}Zr^{4+}$. The final zirconia is phase pure, as indicated by XRD.

CONCLUSION

Zirconia powders have been synthesized with well-defined characteristics such as hardness, and resistance useful for high temperature technical applications. The elaboration process is based on atomizing a solution containing the precursors. A good conditioning of the aerosol is necessary to improve our strategy of reducing the defects that can profoundly affect the sintering. The application of the ultrasonic nebulization technique to the synthesis of zirconia leads to materials composed of homogeneous and narrow size distributed powders with the required particle morphology and size distribution suitable for sintering. Structural study by X-ray absorption spectroscopy has shown that the different stages of the application of heat treatment appear to control the alteration of particle morphology. Absorption experiments have evidenced that crystallite growth in the individual particles is strongly related to the local structural development around the Zr^{4+} cations. One can characterize the zirconium environment during a heating process between its precursor liquid state and its final monoclinic ZrO_2 form. Interesting results has been derived about the correlation between the complexing power of an active anion and the final particle morphology, that is an important research challenge in ceramic science.

REFERENCES

1. Dubois, B., Ruffier, D., and Odier, P., Preparation of fine, spherical yttria-stabilized zirconia by the spray pyrolysis method, *J. Am. Ceram. Soc.*, 72, 713, 1989.
2. Chadda, S., Ward, T.L., Carim, A., Kodas, T.T., Ott, K., and Kroeger, D., Synthesis of $Yba_2Cu_3O_7$-Y and $Yba_2Cu_4O_8$ by aerosol decomposition, *J. Aerosol Sci.*, 22, 601, 1991.
3. McWhan, D.B. and Lundgren, G., The crystal structures of some zirconium hydroxide salts, *Acta Cryst.*, 16, A36, 1963.
4. Koningsberger, D.C., General principle of analysis, *X-Ray Absorption*, J.D. Winefordner, Ed., Chap. 6, John Wiley & Sons, New York, 1989, 243.
5. Teo, B.K., EXAFS: basic principles and data analysis, *Inorg. Chem. Conc.*, Vol. 9, Springer-Verlag, Berlin, 1986, 26.
6. Landron, C., Odier, P., and Bazin, D., *In situ* XAS of aerosol systems: application to the structural study of a zirconia precursor, *Europhysics Lett.*, 21, 859, 1993.
7. Landron, C., Ruffier, D., Dubois, B., Odier, P., Bonnin, D., and Dexpert, H., Etude par EXAFS du rôle des précurseurs dans la synthése de ZrO_2-Y2O_3, *Phys. Stat. Sol.*, 121, 359, 1990.
8. McKale, A.G., Veal, B.W., Paulikas, A.P., Chan, S.K., and Knapp, G.S., Generalized Ramsauer-Townsend effect in extended XAS, *Phys. Rev. B*, 38, 10919, 1988.
9. Catlow, C.R.A., Chadwick, A.V., Greaves, G.N., and Moroney, L.M., EXAFS study of yttria stabilized zirconia, *J. Am. Ceram. Soc.*, 69, 272, 1986.
10. Clearfield, A., Structural aspects of zirconium chemistry, *Rev. Pure and Appl. Chem.* 14, 91, 1964.
11. Berthet, P., Berton, J., and Revecolevschi, A., Etude par EXAFS de l'environnement du zirconium dans des materiaux prepares ... partir d'alcoxydes metalliques, *J. Phys.*, 47, C8-729, 1987.
12. Livage, J., Doi, K., and Mazieres, C., Nature and thermal evolution of amorphous hydrated zirconium oxide, *J. Am. Ceram. Soc.*, 51, 349, 1968.
13. Bénard, P., Loüer, M., and Loßer, D., Crystal structure determination of $Zr(OH)^2(NO_3) \cdot 4.7H_2)$ from X-ray powder diffraction data, *J. Solid State Chem.*, 94, 27, 1991.

12 Recent Developments in the Structural Investigation of Aerosols by Synchrotron Radiation: Application to Ceramic Processing

Claude Landron

CONTENTS

INTRODUCTION

In the last few years, the use of fine powder with controlled characteristics has become increasingly interesting for industrial applications such as reinforcement of material, surface coating, and technical ceramics production.[1] Microsized, monodisperse, and homogeneous powders result in optimized densification rates during the preparation of advanced ceramics and exhibit higher performances than the same compositions produced by traditional techniques.[2] The flexibility of the spray-pyrolysis technique offers manufacturers the opportunity of powder production with controlled physical and chemical characteristics. Atomization eliminates the reactions due to the container; this process is very useful for production of materials with improved performances as high temperature and oxidation resistance. In order to increase the quality of the final products, spray-dryers have been designed. They benefit from the unusual properties of aerosols resulting from the small size of the particles. Spray-pyrolysis is a powerful tool for containerless production of micro-sized particles.

0-87371-829-1/00/$0.00+$.50

An important point in powder processing is structural control during particle formation.[3] Developments in online structural analysis at the atomic level are required for production of high-performance materials. The aim of this study is to improve our knowledge of the mechanisms that govern the structural evolution of materials at high temperature. In conventional processes, material interactions with containers drastically limit the physical and chemical studies of high-temperature processes. Among available spectroscopic techniques, XAS (X-ray absorption spectroscopy) yields atomic distribution with considerable precision. This spectroscopy, which probes the atomic environment of crystalline or disordered materials, has been used to elucidate some problems concerning the evolution of the precursor structure with temperature during powder processing. The experimental device in this study is based on the production of small droplets from a mother solution that is nebulized by an ultrasonic aerosol generator. The particles are rapidly dried, chemically treated, and analyzed in a convenient cell for XAS analysis.

The chemistry of the evolution of the precursor system during processing plays an important role in the resulting properties of zirconia, emphasizing the importance of the local study around a cation in the intermediate phases of the transformation of the precursor toward the final product.[4] In the first place, the predominant species are more or less polymeric in the precursor solution according to the complexing power of the solvent (cationic and anionic).[5] Afterward, during the drying stage, salt precipitation and solvent evaporation determine the degree of crystallization and the gas departure. Last, the pyrolysis conditions settle the precursor alteration characteristics into polymorph phases. Even the chemistry of the early steps is important (i.e., in the liquid),[6] and requires attention. The correlation between the complexing power of an anion and the final particle morphology is also an important research challenge in ceramic science.[7] The principal difficulty in this type of study arises from the amorphous nature of the aerosol under transformation.

This work is intended to serve engineers and scientists concerned with the production and the processing control of fine powder involving the development of an X-ray absorption spectrometer for *in situ* analysis of aerosol at the atomic scale. We will describe the state of *in situ* X-ray absorption studies in aerosol analysis. From this technique, one can learn a lot about the behavior of this particular state of matter. These experiments are very useful because the modification of reactive or volatile species due to temperature increases can rapidly modify the equilibrium between the aerosol particles and the surrounding gas.

INDUSTRIAL APPLICATION OF SPRAY-DRYING TO FINE POWDERS

In the past years, a wide variety of techniques of preparation of aerosols has emerged. These techniques depend on the requirements of the final material. The gaining popularity of the spray-drying route for ceramics production results from the fact that this chemical processing is an inexpensive way to make ultra-fine, homogeneous powders with a high surface area and a broad range of chemical composition. The techniques used for aerosol production are based on one of the following methods: (1) condensation of droplets from a supersaturated vapor, (2) gas-phase chemical reaction induced by heating in a laser beam, or (3) dispersion of a liquid or fine particles into a gas. Laboratory devices for the preparation of aerosols based on liquid dispersion can be classified as rotating disk sprayers, compressed air nebulizers, vibrating orifice generators, electrical atomizers, and ultrasonic nebulizers. The problem in subdividing a liquid into micro-sized particles involves adding sufficient energy to the liquid to overcome the bounding forces and to form a droplet surface.[8] The choice of an ultrasonic nebulizer for XAS analysis of aerosols is mainly justified by its ability to form highly dense aerosols of monodisperse particle size with a typical standard deviation less than 30%. The perfectly controlled diameter of the droplets results from the selection of the frequency of the vibrator and the concentration of the starting solution. In this type of mist generator, the mechanism of droplet production results from the formation and collapse of cavities generated by an intense beam of high-frequency ultrasonic waves propagating in the starting solution. Addition of 1% sodium polyelectrolyte is used as a powerful dispersing and surfactant agent in order to decrease the surface tension in accordance with a good spray aptitude. Inorganic salt precursors are usually required as starting solutions for preparing the powders by nebulization.

EXPERIMENTAL PROCEDURE

The details of the device for ultrasonic spray generation and the procedure for the preparation of the starting mother solution has been partially described in previous papers.[9,10] The experimental system consists of two main parts: a laminar flow aerosol generator and an X-ray analysis cell. The micrometer powder delivery comprises a wet aerosol generation system associated with its power supply, a furnace, and a particle dryer. The active part of the ultrasonic nebulizer is composed of a piezoelectric transducer designed by RBI (Meylan, France) driven by a power supply operating at 0.85 MHz. The mist of micrometer-sized droplets is generated from a 5-cm high geyser and injected into a helium gas stream. The diameter of the droplets is given by the relation: $d = [\pi\sigma_t/4\rho f^2]^{1/3}$, where σ_t and ρ are, respectively, the surface tension and the density of the solution, f is the frequency of the transducer, and $d = 2$ μm in our experiments. A narrow size distribution of the droplets is obtained by avoiding turbulence flow in the reactor. Under high gas flow, some spheres coalesce in the aerosol, giving larger drop size than expected. The tendency of the aerosol particles to agglomerate was very low using our experimental conditions. The spray chamber is designed to promote good mixing and transport of the aerosol toward the analysis cell where the droplets are irradiated in various situations. We use a closed-circuit liquid loop to keep a constant height level of the solution above the transducer and a stable density of particles in front of the X-ray source. The mist can be thermally or chemically treated; the rate of particle transformation is rapid because of the large surface/volume ratio of a spherical droplet, which improves thermal and chemical exchanges. Calcination rates are optimized in order to obtain non-porous spherical particles. The electrically heated desolvation furnace is followed by a cooling condenser that removes the water vapor from the carrier gas. An additional drying of the aerosol is carried out in a device consisting of two concentric cylinders, with silica gel in the volume between a porous screen and the external cylinder. The resultant dry aerosol is transported to the analysis chamber. The jet is crossed at 90° by the X-ray beam, as represented in the Figure 12.1.

The behavior of the particle suspension is of considerable importance for *in situ* analysis results. When a particle falls under gravity in a viscous fluid, it is acted upon by the following forces: gravitational force, buoyant force, drag force and, if the particle is small, interaction with individual fluid molecules. A comparison of Brownian motion and gravitational settling is given in Table 12.1. The path lengths due to the Brownian motion and the path lengths resulting from the gravity effect have been calculated for spherical particles of monoclinic zirconia in air at room temperature and at 76 cmHg pressure, using the formula expressed by Fuchs[11] and including the correction of Cunningham. We deduce that for our nebulizer which utilizes a piezoelectric ceramic operating at 0.8 MHz, the laminar flow settling conditions are fulfilled and Brownian motion can be considered negligible.

This special device was also used to study dried and heated aerosols. The rate of particle transformation is rapid because of the large surface/volume ratio of the droplets that furthers exchanges. The drying stage with a progressive thermal gradient is necessary for obtaining spherical particles because a rapid evaporation of the surface of the particles forms a crust that prevents the gas from being expelled. The spheres are composed of crystallites with sizes in the range 10 to 20 nm. The particle size of dried particles depends on the concentration of the salt in the solution. The dependency of the diameter d_s of a dried sphere vs. the diameter d of a droplet and the salt concentration C is given by: $d_s = d(C/\rho)^{1/3}$, where ρ is the density of the particles. However, for this type of experiment, EXAFS (extended X-ray absorption fine structure) measurements have

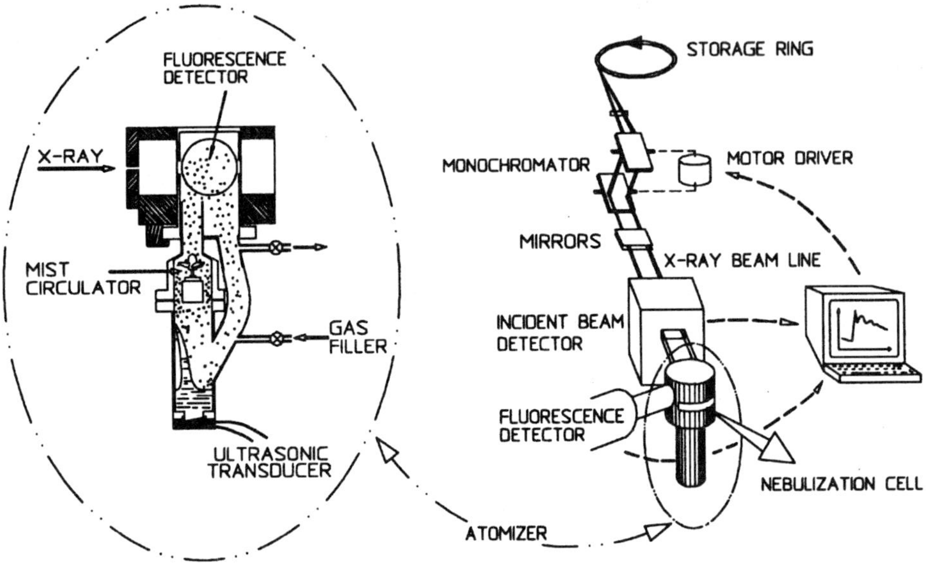

FIGURE 12.1 Schematic illustration of the experimental set-up used for *in situ* EXAFS measurements on atomized solutions. The cell was developed for an adaptation to the EXAFS IV station of the DCI storage ring at LURE (Orsay, France). The transducer levitates liquid droplets surrounded by a continuously flowing helium gas for periods of time up to days. We observe that small cavities are generated in the liquid. The collapse of the bubbles induces a pressure gradient in the vicinity of the liquid surface. The forces are strong enough to produces drops that break away from the liquid surface at the knots of the ultrasonic vibration. Aerosol circulates in the analysis chamber, where it is irradiated by the X-ray beam.

TABLE 12.1
Comparison of the Calculated Path Lengths P_g and P_b Traveled per Second vs. Particle Diameter Due to Gravitational and Brownian Motions, Respectively

Particle Diameter (μm)	P_g (μm)	P_b (μm)
0.1	4.81	29.4
0.25	17.5	14.2
1.0	193.5	5.91
2.5	1112.0	3.58
10.0	4309	1.75

Note: The density of a monoclinic zirconia particle is 5560 kg m^{-3}. For sufficiently small particles (d < 0.25 μm), their motion results mainly from the random molecular bombardment in the gas.

been performed in a post-mortem manner. Our first results suggest that the mechanism of grain formation involving eventually hollow particles is governed by solvent evaporation at low temperature, followed by densification to porous spheres when the elaboration temperature becomes sufficiently high. XAS studies of the heated samples show that the porosity of the ceramic particles is very sensitive to the environment of the cations.

X-RAY ABSORPTION ANALYSIS

Finding a technique for studying the structure of disordered materials has plagued chemists for decades. It is now possible to obtain intense sources of continuous electromagnetic radiation in the X-ray range produced in synchrotron radiation storage rings. These photon lines have stimulated new techniques such as X-ray absorption spectroscopy (XAS), which are especially useful for studies requiring a knowledge of the local atomic environment in materials, even in the absence of long range order.[12] The observed fine structures of absorption spectra result from the local diffraction of an excited photoelectron, giving information on the coordination sphere and on the thermal and static disorder at the atomic level. The possibility of studying disordered, amorphous, or liquid precursors presents an important advantage of this technique over the more commonly used X-ray diffraction technique. Surprisingly, although X-ray absorption has been heavily used in various structural studies, pushing the limits of experimental parameters such as temperature, pressure, and dilution, until recently it has not been used to any great extent in aerosol science.

In the current view of EXAFS, the normalized oscillations: $\chi(k) = (\mu(k) - \mu_0(k))/\mu_0(k)$ of the absorption result from the interferences between the waves associated with the outgoing photoelectron and the back-scattered part of this wave on the neighboring atoms, where $\mu(k)$ and $\mu_0(k)$ are the absorption coefficient of an atom normalized to the background absorption and k is the photoelectron wave vector. The oscillations for a given shell i depend essentially on the number N_i, the distance R_i, the disorder σ_i, and the type of nearest neighbour of the excited atom. From theoretical considerations, the absorption function $\chi(k)$ is given by the relation:

$$\chi(k) = \Sigma\left[N_i/kR_i^2\right]f_i(k)\exp\left(-2k^2\sigma_i^2 - 2R_i/\lambda\right)\sin\left\{2kR_i + \phi_i(k)\right\} \tag{12.1}$$

where:

$f_i(k)$ is the back-scattered amplitude of the shell i
$\phi_i(k)$ is the phase shift related to the atoms of the shell i
λ is the mean free path of the photoelectron

The radial distribution functions around an excited atom are obtained by extracting the modules of the complex Fourier transform of $k^n\chi(k)$[13]

$$\mathrm{TP}\left(R_i\right) = \pi^{1/2}\int_{k_{min}}^{k_{max}}\exp\left(2ikR_i\right)\chi(k)k^n dk \tag{12.2}$$

The intensity of the scattered light depends essentially on the dimensionless parameter $P_s = d/\lambda_r$, where d is the diameter of the particle and λ_r is the wavelength of the incident light. For the droplets used in the present experiments, the condition $P_s > 1$ is fulfilled for the X-ray interaction; then, for X-ray irradiation, the aerosol particles are Rayleigh scatterer centres and the scattered radiation distribution is given in terms of the Rayleigh scattering cross-section.

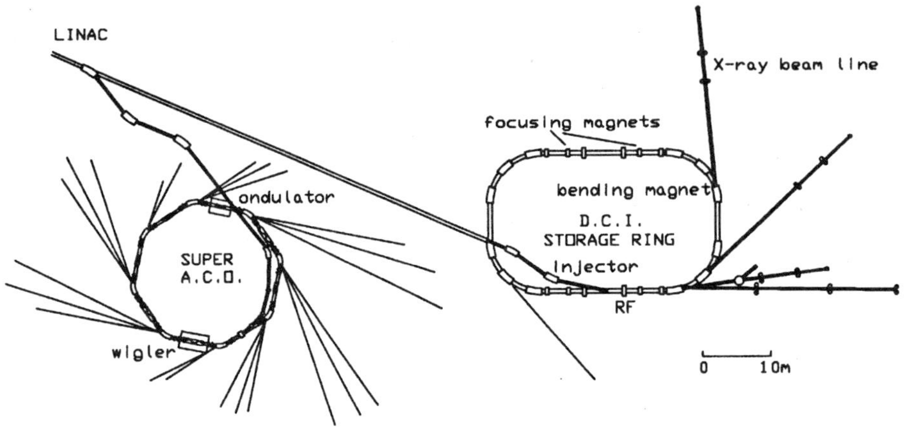

FIGURE 12.2 Schematic layout of the storage rings of LURE, where the EXAFS experiments were performed. A relativistic positron beam is forced to circulate in a curved path around a storage ring. It emits a narrow X-ray beam tangential to the positron orbit.

The Scanning electron microscope (SEM; Cambridge Stereoscan 100) has been used for the observation of the final powder, and X-ray diffraction (XRD) has been used for the identification of intermediate crystallized phases. Experimental data have been recorded on dried and annealed powders using a Philips PW 1729 diffractometer outfitted with a potential X-ray generator operating at 40 kV, 30 mA, with a Cu target.

The absorption experiments were carried out at Orsay (France). The synchrotron radiation was emitted from the 1.85-GeV DCI storage ring of the LURE facility represented in Figure 12.2. It was monochromatized by the Bragg reflections (111) of two parallel, Si crystals. The crystals were adjusted slightly off parallel inducing a reduction of harmonics. The EXAFS signals were recorded above the Zr K-edge in the energy range 17900 to 18900 eV using the EXAFS IV spectrometer. The positron intensity was 300 mA at the beginning of the run with a lifetime of about 110 h. The ionization chambers were filled with pure argon at a pressure that optimized the signal. The system must first be calibrated for aerosol size and density, and thermally stabilized.

Data were recorded at intervals of 2 eV over the EXAFS region. The energy resolution was estimated to be about 2 eV. The measurements were recorded under the same experimental conditions in order to minimize statistical errors. The experimental device designed for *in situ* XAS analysis of aerosols is placed after the first ionization chamber of the X-ray beam line. The nebulization cell is first filled with very clean helium gas; it is then introduced into the nebulization chamber and carries the particles through the cell analysis. The experimental conditions for the absorption measurements of nebulized solutions are similar to those of trace analysis in very dilute samples. *In situ* absorption measurements of nebulized solutions were thus performed using the fluorescence detection mode, which is more efficient than the transmission one. The XAS measurement of samples produced at different temperatures by nebulization were performed, post mortem, in the transmission mode, on powders. They are deposited on a series of tapes in order to obtain a sample thickness x which, under optimal conditions, is determined by $\mu x = 1$. The absorption of helium is negligible, and the flow of particles has been directly measured in the analysis chamber in order to ensure the constancy of the particle density during signal measurement.

Data Treatment

EXAFS spectra give structural information associated with the radial distribution function that can be directly obtained from a treatment of the measured data. A description of the various atomic shells is evidenced by Fourier transforming the oscillations extracted from experimental signals.

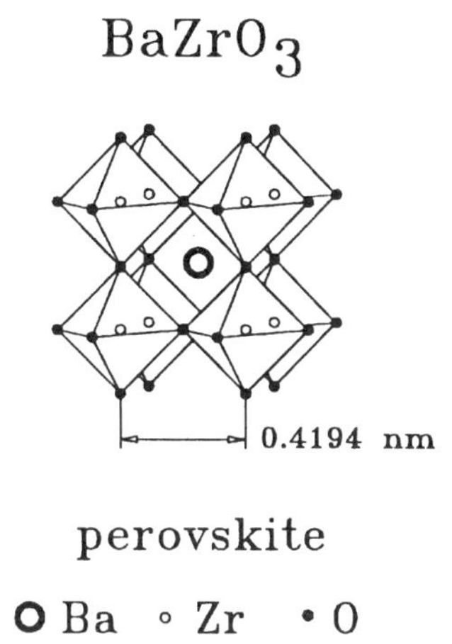

FIGURE 12.3 Schematic diagram of the ideal perovskite structure of barium zirconate oxide with cubic symmetry. Each zirconium cation has six oxygen ions in octahedral coordination, and each zirconium atom is at the corner of a cube.

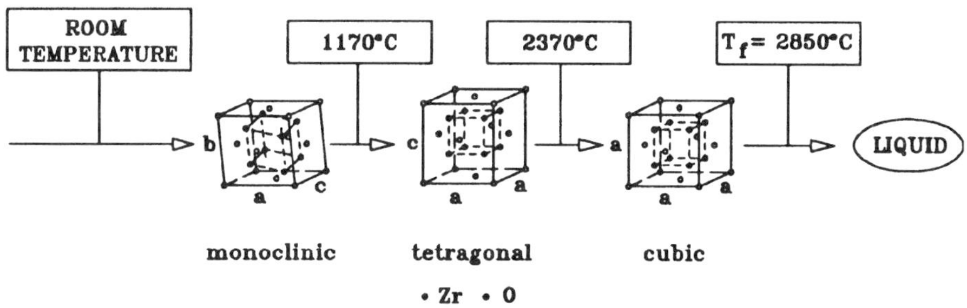

FIGURE 12.4 Polymorph structure of zirconia. The parameters for the following phases are: 1. Monoclinic: a_m = 0.5156 nm, b_m = 0.5191 nm, c_m = 0.5354 nm; 2. Tetragonal: a_t = 0.5094 nm, c_t = 0.5177 nm; 3. Cubic: a_c = 0.5124 nm.

Experimental data were treated using the procedure recommended in the report of the International Workshop on Standards and Criteria in XAFS. All spectra were subjected to the same mathematical treatment, the XAS signal above the Zr K-edge was analyzed in terms of the known theory of X-ray absorption developed by Teo[13]; that is,

A) Background subtraction
B) Fourier transform, using a Hanning window for apodization of the k^3-weighted XAS experimental spectra.
C) Inverse Fourier transform of the peak of interest to obtain distance and coordination number. A least-squares fitting procedure for the oxygen coordination shell determination was employed with theoretical values of phases and amplitudes calculated by McKale et al.[14]

EXAFS RESULTS

References

First, we chose references where zirconium is sixfold, sevenfold, and eightfold coordinated: barium zirconate oxide (Figure 12.3), monoclinic zirconia, and tetragonal zirconia (Figure 12.4).

The barium zirconate oxide is a double oxide belonging to the ideal perovskite type with cubic symmetry (space group P_{m3m}).[15] Zirconia presents three polymorphs[16]: a monoclinic phase (space group P_2/c), a tetragonal phase (space group $P4_2/nmc$), and a cubic phase (space group F_{m3m}).

TABLE 12.2
Structural Parameters Obtained by Least-Squares Fitting the EXAFS Spectra for the Oxygen Coordination Shell Around a Zirconium Atom in a Nebulized Solution and Dried Powders of Zirconium Nitrate

Zirconium Nitrate	Nebul. Sol.	T = 100°C	T = 300°C	T = 450°C	T = 550°C
d(Zr-O) (nm)	0.221	0.223	0.218	0.219	0.219
N(O)	8	7.7	6.8	7.0	6.6
σ (nm)	0.0065	0.0078	0.0072	0.0085	0.0070

Note: A double shell is assumed. Bond length *d*(Zr-O), coordination number N(O), and Debye-Waller factor σ.

Zirconia Precursors: Nitrate Route

Table 12.2 and Figure 12.5A describe the first coordination shell of zirconium cations for nebulized samples and nitrate compositions studied at various temperatures. In contrast with data on solids, information concerning second neighbors of liquid solutions is lost because of the disorder and damping to which EXAFS data are very sensitive. For samples elaborated at 300°C by the nitrate route, Zr atoms are surrounded by eight oxygen atoms at an average distance d = 0.218 nm. This result is coherent with structural data of McWhan and Lundgren.[17] It is possible that the higher dilution used in this latter case (0.1 mole l^{-1}) may modify the structure in averaging it. However, it is concluded that there is no significant alteration of the zirconium environment during this first stage of drying. These modifications do not significantly affect the first shell. This fact is probably explained by the formation of linked chains $[Zr(OH)_2(NO_3)_2(H_2O)_2^+]_n$ with nitrate groups and free water molecules located between the chains.

Zirconia Precursors: Chloride Route

Muha et al.[18] have determined the structure of the complex in a zirconyl oxychloride solution. They found that the zirconyl ion is a tetramer $(Zr_4(OH)_8(H_2O)_{16})^{8+}$.

EXAFS results show that four zirconium ions are located at the corners of a slightly distorted square and are linked together above and below the plane of the square. XAS data given in Table 12.3 show that the eight oxygen atoms of the zirconium coordination sphere are in accordance with XRD studies. A decrease of the distance *d*(Zr-O) vs. temperature is observed in Figure 12.5B; it can be explained by the replacement of Zr-[O H_2] bonds with *d*(Zr-O) = 0.227 nm by shorter bonds Zr-[O H], with *d*(Zr-O) = O.214 nm.

Zirconia Precursors: Sulfate Route

In Figure 12.6, we have compared the radial distribution functions of the nebulized precursor and the final zirconia. The Fourier transformed EXAFS data associated with the nebulized solution exhibits only features related to the nearest neighbor in the environment of the zirconium atom. For the amorphous samples produced at 200°C from the sulfate route, no significant changes in

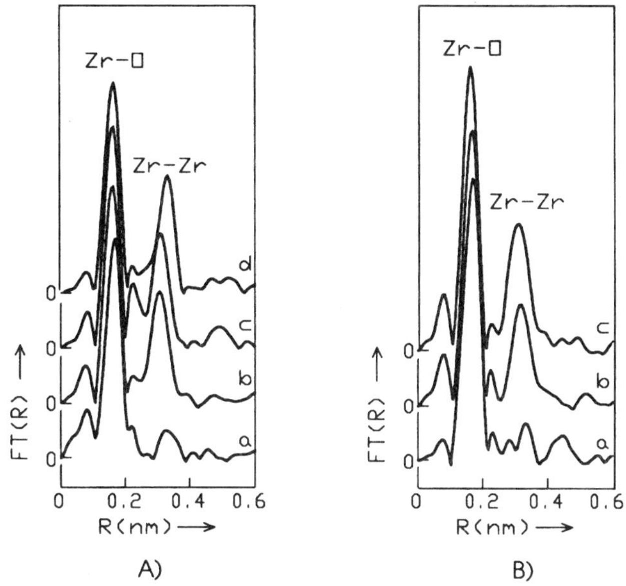

FIGURE 12.5 Comparison of the module of the Fourier transform TF(R) of the $\chi(k)$ function recorded at the Zr K-edge on the following samples. A. Powder prepared by the nitrate route at: (a) T = 100°C, (b) T = 300°C, (c) T = 550°C, and (d) T = 650°C. B. Powder prepared by the oxychloride route at: (a) T = 150°C, (b):T = 300°C, and (c) T = 400°C. The distributions functions are uncorrected for the phase shift of the Zr-O pair.

TABLE 12.3
Structural Parameters Obtained by Least-Squares Fitting the EXAFS Spectra for the Oxygen Coordination Shell Around a Zirconium Atom in a Nebulized Solution and Dried Powders of Zirconyl Chloride

Zirconyl Chloride	Nebul. Sol.	T = 150°C	T = 300°C	T = 400°C
d(Zr-O) (nm)	0.223	0.222	0.220	0.219
N(O)	8	7.7	7.7	7
σ (nm)	0.0075	0.0105	0.0098	0.0091

Note: A double shell is assumed. Bond length *d*(Zr-O), coordination number N(O), and Debye-Waller factor σ.

the Zr-O distances are observed (Table 12.4 and Figure 12.7B) before crystallisation, and *d*(Zr-O) is close to 0.211 nm. The coordination number in the amorphous samples seems to increase with the temperature of fabrication. We can conclude that there is no significant modification of the zirconium environment during this first stage of drying. The variation of the coordination number can be related to the growth of crystallites and by the modification of the chains $[Zr(OH)_2(SO_4)_2(H_2O)_4]_n$ linked by water molecules. We have pointed out that the strongly concentrated solution exhibits an order around the zirconium atoms that can be compared to the order around zirconium in a crystallized zirconium sulfate. The local structure in the solution is built from chains of dodecahedrons formed by zirconium atoms that are eightfold oxygen-coordinated.

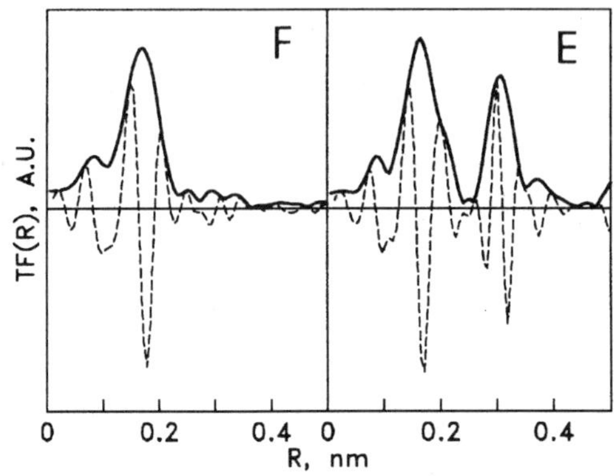

FIGURE 12.6 Comparison of the module (full line) and imaginary part (dotted line) of the Fourier transforms corresponding to the k^3-weighted EXAFS spectra of a nebulized zirconium sulfate aqueous solution (0.5 N), [curve F] and monoclinic zirconia produced from the sulfate precursor. The partial radial distribution functions TF(R) is Uncorrected for the Phase shift of the Zr-O pair (Zr K-edge)-curve E-.

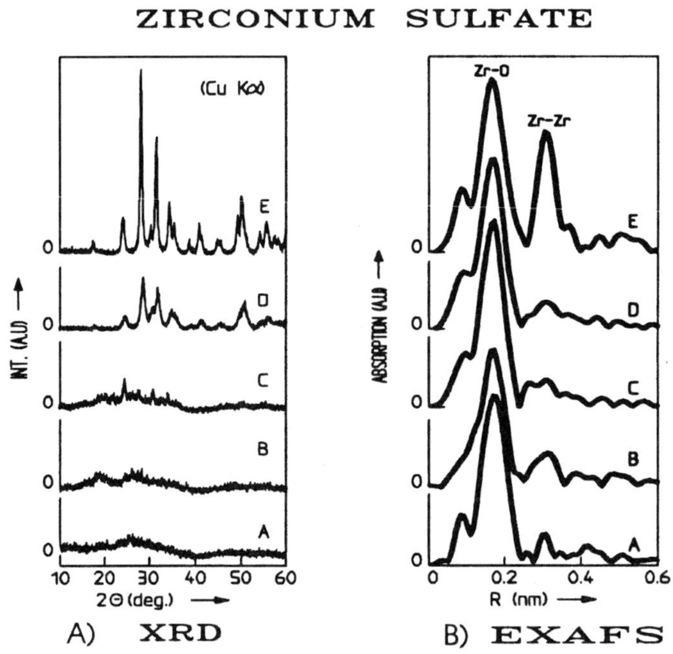

FIGURE 12.7 (A) Comparison of the X-ray diffraction pattern of dried powders produced from an aqueous solution (0.5 N) precursor of zirconium sulfate via atomization after convenient heating at: (A) 200°C, (B) 350°C, (C) 500°C, (D) 650°C, and (E) 800°C. (B) Comparison of the module of the Fourier transform TF(R) of the $\chi(k)$ function recorded at the Zr K-edge on dried powders produced from an aqueous solution (0.5 N) precursor of zirconium sulfate via atomization after convenient heating at (A) 200°C, (B) 350°C, (C) 500°C, (D) 650°C, and (E) 800°C. TF(R) is uncorrected for the phase shift of the Zr-O pair.

The present analysis gives a Zr-O average distance of 0.224 nm, which is not far from the sum of the ionic radii of Zr^{4+} eightfold coordinated and O^{2-}.

The treatment of data collected on crystallized powders shows that, for oxygen-coordinated zirconium, the results obtained for the theoretical and experimental parameters were compatible. The results of the treatment of the XRD and the EXAFS data of samples A to E is illustrated are

TABLE 12.4
Structural Parameters Obtained by Least-Squares Fitting the EXAFS Spectra for the Oxygen Coordination Shell Around a Zirconium Atom in a Nebulized Solution and Dried Powders of Zirconium Sulfate

Zirconium Sulfate	Nebul. Sol.	T = 200°C	T = 500°C	T = 650°C	T = 800°C
d(Zr-O) (nm)	0.216	0.221	0.221	0.221	0.219
N(O)	7.5	5.5	6.4	6.5	7
σ (nm)	0.0090	0.0092	0.0101	0.101	0.00097

Note: A double shell is assumed. Bond length *d*(Zr-O), coordination number N(O), and Debye-Waller factor σ.

Figure 12.7A and B respectively; they are in agreement with previous studies.[19-22] The first peak of the radial distribution function is correlated with the cation-ligand bonds, while the second peak corresponds to the diffusion from cation-cation first neighbors. The fit confirms that the powders E (i.e., annealed at 800°C) exhibit a monoclinic-like structure. Especially, the measured cation-oxygen distance *d*(Zr-O) of the crystallized sample reflects the description of the zirconium environment in monoclinic zirconia given in the literature.[22-24] The structural behavior of the amorphous samples (A to D) is characterized by a relatively low intensity of the Zr-Zr peak in the radial distribution function, while this intensity is equal to zero in the sulfate solutions studied *in situ*.

This is clearly shown in Figure 12.6 where the Fourier transform of the nebulized liquid precursor and final monoclinic zirconia powder are compared. The part of the curve corresponding to the Zr-Zr bonds is drastically different. Contrarly to the case of the crystallized material (E), only nearest neighbors (Zr-O) can be quantitatively described by the present EXAFS data processing. The lack of a second peak in the radial distribution function of the nebulized product implies that mainly monomeric species or short chains exist in the solution at the experimental conditions used.

Important information that one can obtain from EXAFS treatment are the coordination number (CN) and the cation-ligand bond distance for the first layer. As one can see in Figure 12.7B, both increase in the intermediate phases (A to D) after a significant decrease with respect to the precursor solution. The weak value observed for CN in the intermediate phases illustrates the consequences of small dimension and edge effects in crystallites on EXAFS results. Previous TEM analyses[25] have described the internal microstructure of micrometer spheres of zirconia obtained by nebulization of a nitrate precursor. They have produced evidence that, within a certain range of temperature, both amorphous and crystallized grains coexist in the aerosol (but in separate spheres). Crystallized grains increase in size with increasing temperature in the 2- to 10-nm range. EXAFS treatment seems to point out that the coordination numbers and distances of the amorphous powders are significantly shifted to weaker values for amorphous samples produced at lower temperatures. Our approach for tackling this problem consists of interpreting the reduction of the coordination number as a consequence of the smaller size (nanometric) of the primary particles, which then presents an increased concentration of atoms located at the grain surface with respect to the bulk. Similarly, the Zr-O distances are shorter than in the bulk. Both effects appear to be a result of the granularity (nanometric scale) of the studied samples.

The analyses of the various experiments lead to conclusions in agreement with other types of spectrometry, but they give detailed information on the local arrangement, which cannot be obtained with the same accuracy by another method. We show, for highly concentrated solutions, the existence of polymeric chains rather than monomeric complexes. We have noted that the structure of the precursor depends on its state in the starting solution. The anion-complexing power has an effect on the formation of regularly shaped powders by avoiding hollow sphere formation. The high degree of polymerization of the sulfate precursor solution of hydrolyzed Zr(IV) species is related to the viscosity of the liquid. By heating the powder at different temperatures, we have obtained precursors in different polymerization states. Another fact can explain the morphology differences observed according to the precursor salt used: a comparison of thermal decomposition by thermal gravimetric analysis of powders from $ZrOCl_2 \cdot 8H_2O$ and $ZrSO_4$ clearly shows that HCl continuously leaves the spheres, while the SO_2 remains up until the crystallization of zirconia.

CONCLUSION

The synthesis by nebulization of multicomponent oxides is a very attractive process. In comparison with conventional techniques, containerless thermal treatment of a spray introduces greater control over the size, homogeneity, dispersivity, and morphology of the particles. We have explored spray-pyrolysis routes for producing fine ceramic precursor particles, well-suited for sintering. The effort was focused on the investigation of the structure alteration during processing. *In situ* X-ray absorption spectroscopy is an important contribution to the knowledge of the evolution of an atom environment during the production of fine powder by the sol-gel process. This study has demonstrated the feasibility of using an ultrasonic transducer as a wave source for a stable mist generator in order to perform XAS measurements. This local probe has evidenced the modification of the radial distribution function around Zr atoms during zirconia synthesis. The appropriate modification of the solute species, the transition (even temporary) by a hydroxide network, avoids the formation of coarse porosity. XAS has shown that high purity, chemical homogeneity, particle size, and sphericity — all required for improving ceramic production — are related to the structural information during synthesis from the liquid phase.

REFERENCES

1. Masters, K., *Am. Ceram. Soc. Bull.*, 73, 63, 1994.
2. Messing, G.L., Zhang, S.C., and Jayanthi, G.V., *J. Am. Ceram. Soc.*, 76, 2707, 1993.
3. Arnold, S., Spectroscopy of single levitated micron sized particles, *Optical Effects Associated with Small Particles*, S. Ramaseshan, Ed., Word Publishing Co., Singapore, 1988.
4. Tosan, J.L., Durand, B., Roubin, M., Chassagneux, F., and Bertin, F., *J. Non-Cryst. Solids,* 168, 22, 1994.
5. Livage, J., Henry, M., and Sanchez, C., *Solid State Chem.,* 18, 259, 1989.
6. Clearfield, A., *Rev. Pure and Appl. Chem.*, 14, 91, 1964.
7. Dubois, B., Ruffier, D., and Odier, P., *J. Am. Ceram. Soc.*, 72, 713, 1989; Dubois, B., Ruffier, D., and Odier P., Homogeneous and fine Y_2O_3 stabilized zirconia powders prepared by a spray pyrolysis method, *Ceramic Powder Processing Science,* H. Hausner, G. L. Messing, and S. Hirano, Eds., Deutsche Keramische Gesellschaft, Koln, Germany, 1988, 229.
8. Landron, C., Ruffier, D., Dubois, B., Odier, P., Bonnin, D., and Dexpert, H., *Phys. Status Solidi.*, 121, 359, 1990.
9. Landron, C., Ruffier, D.,.Odier, P, Bazin, D., and Dexpert, H., *J. Phys.*, III 1, 1971, 1991.
10. Landron, C., Odier P., and Bazin, D., *Europhysics Lett.*, 21, 859, 1993.
11. Fuchs, N.A., *The Mechanics of Aerosol,* Pergamon, New York, 1964.
12. Koningsberger, D.C., *X-Ray Absorption,* Wiley Interscience Publ., New York, 1988.
13. Teo, B.K., EXAFS: basic principles and data analysis, *Inorg. Chem. Conc.*, Vol. 9, Springer-Verlag, Berlin, 1986.
14. Mac Kale A.G., Veal B.W., Paulikas A.P., Chan S.K. and Knapp G.S., *J. Am. Ceram. Soc.,* 110, 3763, 1988.
15. Megaw, H.D., *Proc. Phys. Soc.*, 58, 133, 1946.

16. Garvie, R.C., Zirconium dioxide and some of its binary systems, *High Temperature Oxides,* A.M. Alper, Ed., Academic Press, New York, 1970, 117.
17. McWhan, D.B. and Lungren, G., *Acta Cryst.,* A19, 26., 1963.
18. Muha, G.M. and Vaughan, P.A., *J. Chem. Phys.*, 33, 194, 1960.
19. Slamovich, E.B. and Lange, F.F., *Mat. Res. Symp. Proc.,* 121, 257, 1988.
20. Okasaka, K., Nasu, H., and Kamia, K., *J. Non-Cryst. Solids,* 136, 103, 1991.
21. Zeng, Y.W., Fagherazzi, G., Pinna, F., Riello, P., and Signoretto, M., *J. Non-Cryst. Solids,* 155, 259, 1993.
22. Smith, D.K. and Newkirk, H.W., *Acta Cryst.,* 983, 1965.
23. Catlow, C.R.A., Chadwick, A.V., Greaves, G.N., and Moroney L.M., *J. Am. Ceram. Soc.,* 69, 272, 1986.
24. Li, P., Chen, J.W., and Penner-Hahn, J.E., *Phys. Rev. B,* 48, 10063, 1993.
25. Odier, P., Dubois, B., Clinard, C., Stroumbos, H., and Monod, Ph., Ceramic *Transaction, Ceramic Powder Science III,* G.L. Messing, S. Hirano, and H. Hausner, Eds., The American Society, Columbus OH, 1990.

13 Fundamentals and Performance of the MCVD Aerosol Process

Vlastimil Matějec, Ivan Kašík, and Miroslav Chomát

CONTENTS

INTRODUCTION

Since the Derjagin-Waldmann theory of thermophoretic movement of aerosol particles was developed, many practical applications of this effect have been successfully accomplished and realized. In the last 20 years, techniques employing the thermophoretic effect have also been widely used for the fabrication of preforms for drawing silica optical fibers. One can say that the techniques based on chemical vapor deposition (CVD) methods have brought about progress in the fabrication of silica optical fibers with low attenuation, high bandwidth, and excellent mechanical properties, and have been used as a transmission medium in the near-infrared spectral region.

CVD methods can be divided into two classes. Methods of the first class are used primarily for the fabrication of semiconductor devices. Typically, a low-pressure and low-concentration vapor stream of organometallic compounds or halides reacts at or near a heated substrate surface, resulting in a uniform, defect-free deposition on the substrate. The reaction can be heterogeneous (occurring on the surface) or homogeneous (occurring in the vapor phase). Owing to small differences between the substrate and reaction temperatures, thermophoresis evidently has little impact on the transport of reactants. Plasma chemical vapor deposition,[1] developed at Philips in 1979 for production of silica optical fibers, belongs to this class of CVD methods.

0-87371-829-1/00/$0.00+$.50

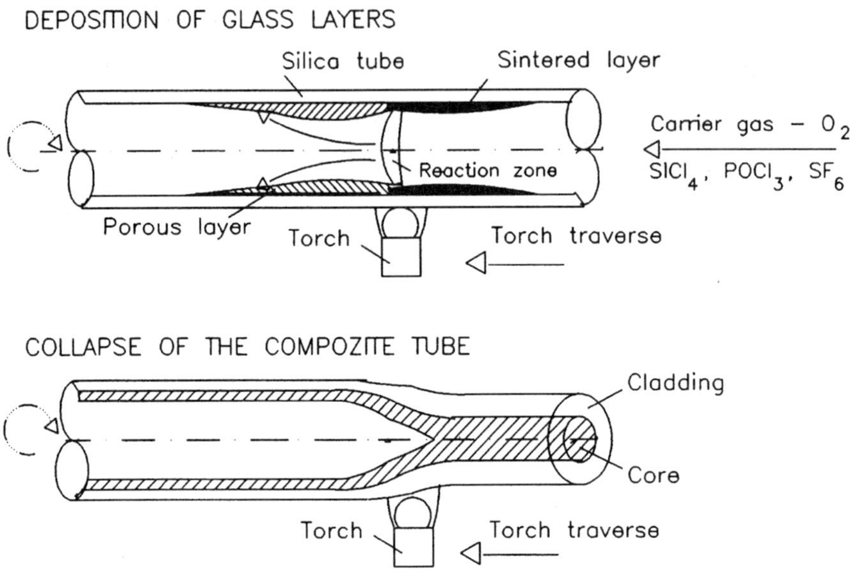

FIGURE 13.1 Set-up of the MCVD process.

The second class of CVD, most commonly used for the fabrication of bulk glass products, relies on a thermally activated, homogeneous reaction of a starting mixture of metal halides with oxygen and in some cases also with hydrogen. Amorphous particles nucleate and grow in the reaction zone. They form an aerosol system with gaseous reactants and are subsequently deposited on a cooled substrate or target owing to the thermophoretic force created by differences between the temperatures of the particles and the substrate. Depending on the temperature, the deposited particles can be fused to form a transparent glass object or collected as a porous preform. This application typically uses higher concentrations of reactants and is carried out at ambient pressure. As a result, the deposition rate can be more than an order of magnitude higher than in the first case. The majority of techniques for manufacturing preforms for drawing silica optical fibers (e.g., outside vapor deposition,[1] vapor axial deposition,[1] and modified vapor chemical deposition (MCVD)) belong to this second class of CVD methods.

This chapter reviews the MCVD technique, focusing on studies of the processes and mechanisms involved, as well as on their employment for controlled fabrication of typical fiber-optic structures. It updates some review reported by Nagel et al.[2,3] or Grigor'yanc et al.[80] and attempts to compare results achieved in this area in former Czechoslovakia, the GDR, and the U.S.S.R. with those in western Europe and the U.S.

DESCRIPTION OF THE MCVD PROCESS

The modified chemical vapor deposition (MCVD) method, developed by MacChesney et al.[4] from AT&T Bell Laboratories, has become one of major manufacturing processes for the fabrication of high-quality preforms for drawing silica optical fibers. It is practiced by many companies in the U.S., Europe, and Japan.

General features of CVD methods can also be found in the MCVD method. In a standard experimental set-up (Figure 13.1), a gaseous mixture of raw materials, oxygen, or other carrier gases is injected into a reactor — a rotating silica tube, typically 15 to 30 mm in diameter, which is heated by an exterior torch. The reactant gases entering the tube at relatively low temperatures, are heated as soon as they approach the hot zone of the traversing torch. Amorphous oxide products

PREPARATION OF GASEOUS MIXTURE OF RAW MATERIALS	DEPOSITION OF GLASS LAYERS	TUBE COLLAPSE
t<70°C	Chemical reaction t>1300°C Nucleation and growth of particles t>1300°C Particle deposition $T_{WALL} < T_{GAS}$ Consolidation at 1500-1800°C	1900-2100°C

FIGURE 13.2 Basic stages of the MCVD process.

are formed, mainly by a homogeneous gas-phase oxidation of raw materials which occurs wherever the gas temperature reaches the value of about 1200°C.

Amorphous oxide particles, with sizes typically between 0.15 and 1 μm suspended into hot gases, flow through the section of the tube downstream of the hot zone, where the wall is at a lower temperature than the gas. Here, part of the particles deposit thermophoretically due to the radial temperature gradients in the tube. Further downstream, the gas and wall temperatures equilibrate and the deposition ceases. The torch traversing slowly (10 to 30 cm min^{-1}) in the same direction as the interior gas flow consolidates the deposited porous layer into a transparent glass layer. A modification of the MCVD technique has been reported, in which particles are deposited without consolidation when the torch moves in the direction opposite to the interior gas flow.[31,50,65] The torch is traversing repeatedly in order to build up — layer by layer — the desired structure of the preform; typically, 30 to 100 layers are deposited. The composition of the layers can be varied by changing the flow rates of raw materials. Following deposition, the composite tube is collapsed into a solid rod — the preform.

FUNDAMENTALS OF THE MCVD METHOD

Since the first report on the MCVD technique in 1974, several studies have focused on understanding fundamental physics and chemistry of the process. In the following paragraphs, results of experimental and theoretical studies of the MCVD process will be discussed on the basis of distinguishing several process stages as depicted in the diagram in Figure 13.2.

PREPARATION OF GASEOUS MIXTURE OF RAW MATERIALS

The MCVD method usually utilizes raw materials in the gaseous state. For this purpose, chemical reagents are mostly entrained in a gas stream either by passing carrier gases such as O_2, Ar, He, N_2 through liquid halides with reasonably high vapor pressures at room temperatures (e.g., $SiCl_4$, $GeCl_4$, BBr_3, $POCl_3$, $C_2Cl_3F_3$), or by using gaseous raw materials such as SiF_4, CCl_2F_2, SF_6, CF_4, or BCl_3.[3]

There are only a few articles dealing with passing carrier gases through liquid halides in the MCVD process. Their results are summarized in Table 13.1, where μ(EX) is the saturation coefficient of the halide EX defined by Equation 13.1 V_{ox} is the oxygen flow rate, t_t is the temperature in the thermostat, t_l is the temperature of a liquid measured directly in the bubbler, and h denotes the liquid level.

TABLE 13.1
Results on Efficiency of O_2 Saturation by Halides

Sandoz et al.[5]:
$\mu(SiCl_4) = 1.108 - 5.98.10^{-4}\ V_{ax}$
$t_t = 35°C,\ 77 < V_{ox} < 190\ cm^3\ min^{-1}$
Choc[58]:
$\mu(SiCl_4) = 0.98\ \mu(POCl_3) = 0.94$
$\mu(GeCl_4) = 0.93\ \mu(BBr_3) = 0.97$
$15 < t_t < 35°C,\ 50 < h < 190$ mm, $V_{ox} < 500\ cm^3\ min^{-1}$

TABLE 13.2
Results on the Chemical Kinetics of the MCVD Method

	$k_1 \times 10^{-14}$ [s^{-1}]	$k_2 \times 10^{-15}$ [dm^3 mol^{-1} s^{-1}]	$k_3 \times 10^{-12}$ [s^{-1}]	E [kJ mol^{-1}]
$SiCl_4$	1.7 [8]	30 [8]	—	402 [8]
	—	—	800 [7]	410 [7]
	5.4 [6]	17 [6]	700 [6]	418 [6]
$SiBr_4$	—	—	0.5 [7]	280 [7]
$POCl_3$	—	—	0.3 [7]7	259 [7]
$GeCl_4$	—	—	$0.02 = F(C_{Ge})$	264 [7]

$$V_{EX} = V_{ox}\frac{P^o_{EX}\mu(EX)}{P_c - P^o_{EX}\mu(EX)} \tag{13.1}$$

In by Equation 13.1, V_{EX} is the flow rate of the evaporated halide, and P_c and P^0_{EX} are the overall gas pressure and vapor pressure of the halide EX, respectively.

On the basis of Table 13.1, one can conclude that the control of the flow rates of carrier gases, liquid temperatures, and levels are of key importance for achieving the desired characteristics of fibers.

Chemical Reactions in the MCVD Process

Reaction Kinetics

Chemistry of the MCVD process plays a key role in achieving the desired properties of the final fiber product. One can expect the MCVD reactions, in essence the burning of halides, to be described by chain mechanisms. However, there are only a few articles dealing with the kinetics of the MCVD oxidation and, moreover, they use just simple reaction kinetics. The results of these studies carried out in the temperature range from 800 to 1300°C can be expressed by Equations 13.2 or 13.3 and are summarized in Table 13.2.

$$-\frac{d[EX]}{d\tau} = \left(k_1 + k_2[O_2]\right)[EX]\exp\left(-\frac{E}{RT}\right) \tag{13.2}$$

$$-\frac{d[EX]}{d\tau} = k_3[EX]\exp\left(-\frac{E}{RT}\right) \tag{13.3}$$

In Equations 13.2 and 13.3 and Table 13.2, [EX] denotes the molar concentration of the halide EX, C_{Ge} is the input concentration of $GeCl_4$, E is the activation energy, T is the temperature and k_1,k_2,k_3 are the preexponential factors, respectively. Equation 13.3 holds for large excess of O_2 over halide EX.

On the basis of Table 13.2, one can conclude that oxidation of $SiCl_4$, $SiBr_4$, or $POCl_3$ under the MCVD conditions can be described by the first-order kinetic Equation 13.3. For $GeCl_4$, French et al.[7] found the rate constant k_3 to decrease with increasing C_{Ge}. The kinetics of the oxidation of BCl_3 did not fit Equation 13.3 at all. Investigating simultaneous oxidation of $SiCl_4$ and $GeCl_4$, they determined that in this case, $SiCl_4$ reacted with a lower activation energy of about 250 kJ mol^{-1}.

Grigor′yanc et al.[77] found that at temperatures below 1250°C, when only the heterogeneous reaction on the tube wall took place, the activation energies were equal to 130 kJ mol^{-1} for SiO_2 and 105 kJ mol^{-1} for SiO_2-P_2O_5 deposition. At higher temperatures, when the homogeneous reaction was observed, they determined a value of the activation energy of 480 ± 8 kJ mol^{-1} for SiO_2 deposition.

Morphology of Solid Products of the MCVD Eeactions

Papers dealing with size growing of solid oxide particles in the hot zone have shown the particles to be formed with sizes from 0.1 to about 1 μm and with the morphology dependent on the composition of the gaseous phase. Tanaka and Kato,[6] investigating solid products from oxidation of $SiCl_4$, showed these particles to be amorphous and nearly spherical, with diameters from 0.15 to 1 μm. The size distribution became narrower and the mean particle size smaller if the oxygen concentration decreased or the $SiCl_4$ concentration increased. They attributed these effects to different dependencies of the nucleation and growth rates on the composition of the gaseous phase.

Kleinert et al.[46] investigating solid products of simultaneous $SiCl_4$ and $GeCl_4$ oxidation at 2023K, found increasing size of the formed particles with GeO_2 content, namely, 0.1 μm for 19 mol.% and 1 ± 0.3 μm for 65 to 84 mol.% GeO_2.

In the case of gas-phase oxidation of $TiCl_4$, George et al.[9] attributed the further particle growth beyond a certain range of particle sizes primarily to Brownian coagulation. Walker et al.,[10] supposing that a similar mechanism took place in the MCVD process, estimated the time needed for particle growth from 1 nm to 0.1 μm to be about 0.01 s.

Thermodynamical Constraints in the MCVD Process

The results summarized above allow us to make the conclusion that, under standard MCVD operating conditions (i.e., the temperature is above 1400°C and the residence time in the hot zone is about 0.1 s), the rates of the MCVD reactions are high enough that thermodynamic equilibrium can be attained. Several papers investigating this equilibrium have determined thermodynamic limits in the case of doping SiO_2 with GeO_2, F, or P_2O_5.

Systems Containing GeO_2

Germanium dioxide is the most frequently used dopant in the fabrication of cores of silica optical fibers. It increases the refractive index of silica glass. This effect is usually described by Equation 13.4,

$$\Delta n = F_D x_D, \tag{13.4}$$

where Δn, F_D, and x_D are the change of the refractive index of silica induced by doping, the experimentally determined factor, and the molar fraction of the particular dopant in silica glass, respectively. For GeO_2, values of F_D from 0.13 to 0.14 have been determined.[11,12,57]

In general, $SiCl_4$ is completely oxidized during the MCVD process, while $GeCl_4$ oxidation is strongly affected by the unfavorable thermodynamic equilibrium

$$GeCl_4(g) + O_2(g) \Leftrightarrow GeO_2(gl) + 2Cl_2(g) \tag{13.5}$$

described by the apparent equilibrium constant K_a (Equation 13.6):

$$K_a = \frac{K}{\gamma} = \frac{X_{GeO_2} P^2_{Cl_2}}{P_{GeCl_4} P_{O_2}} \tag{13.6}$$

In Equation 13.6, P_i are the normalized partial pressures of gaseous species related to the standard pressure 101.3 kPa, X_i denotes the mole fraction of the particular species in the solid, and γ is the activity coefficient pertinent to GeO_2.

Kleinert et al.[45] examined the equilibrium (13.5) under MCVD conditions. They found 100% reaction extent for $SiCl_4$ oxidation and determined the amount of unreacted $GeCl_4$ to be controlled by the input oxygen concentration; and in the case of simultaneous oxidations of $GeCl_4$ and $SiCl_4$, also by the input $SiCl_4$ concentration. On the basis of thermodynamic calculations, they predicted GeO_2 to be incorporated in a continuous glass phase and not in a liquid phase.

In 1984, Kleinert et al.[46] reported results of an infrared (IR) spectroscopical study of the oxidation of gaseous mixtures of $SiCl_4$ and $GeCl_4$ at 2023K. They observed the absorption band of the Si-O-Ge structural unit at 660 cm^{-1} in the spectrum of particles, confirming the existence of a continuous $Si_{1-x}Ge_xO_2$ phase. Measuring effluent concentrations of $GeCl_4$ from the MCVD process, they determined the value of K_a in Equation 13.6 to be equal to about 0.14. Later on, they calculated nearly the same value of 0.15 from the GeO_2 concentrations measured on sintered soot.[47]

Virtually at the same time as Kleinert, Wood et al.[13] examined the GeO_2 incorporation using IR analysis of effluent gases. They found full oxidation of $SiCl_4$ above 1800K and the minimum in the effluent content of $GeCl_4$ at about 1800K, and explained these effects by the transition from a regime that is rate-limited to a regime that is equilibrium-limited. In the same paper, they showed the equilibrium (13.5) to be shifted toward the reaction products by substituting a part of $SiCl_4$ with $SiBr_4$.

In their further paper, Wood et al.[14] found the GeO_2 deposition efficiency to be strongly dependent on the hot zone temperature when no P_2O_5 was present. Measuring the GeO_2 content in the glass layers, they found the value of K_a in Equation 13.6 to be equal to 0.1 to 0.14, depending on the concentration of P_2O_5. They explained these facts by the increase in the particle size due to viscous sintering accelerated by P_2O_5, which decreased GeO_2 diffusion.

Continuing their investigations on GeO_2 incorporation, Wood et al.[15] observed a small impact of reactions 13.7 and 13.8 on the final GeO_2 content in the glass layers; and from this content, they determined the value of the constant K_a to be equal to 0.115 at 1700K.

$$Cl_2(g) \Leftrightarrow 2Cl(g) \tag{13.7}$$

$$GeO_2(s) \Leftrightarrow GeO(g) + \frac{1}{2}O_2(g) \tag{13.8}$$

The efficiency of GeO_2 incorporation in the MCVD process was also investigated by Choc.[58] He determined the value of this parameter, defined as the ratio of the measured and stoichiometric molar fractions of GeO_2 in glass, to be equal to 0.33 to 0.38. The experimentally determined dependence of GeO_2 concentration in glass on flow rates of the raw materials could be satisfactorily explained using the value of K_a in Equation 13.6 equal to 0.125[60,66,72] (Figure 13.3).

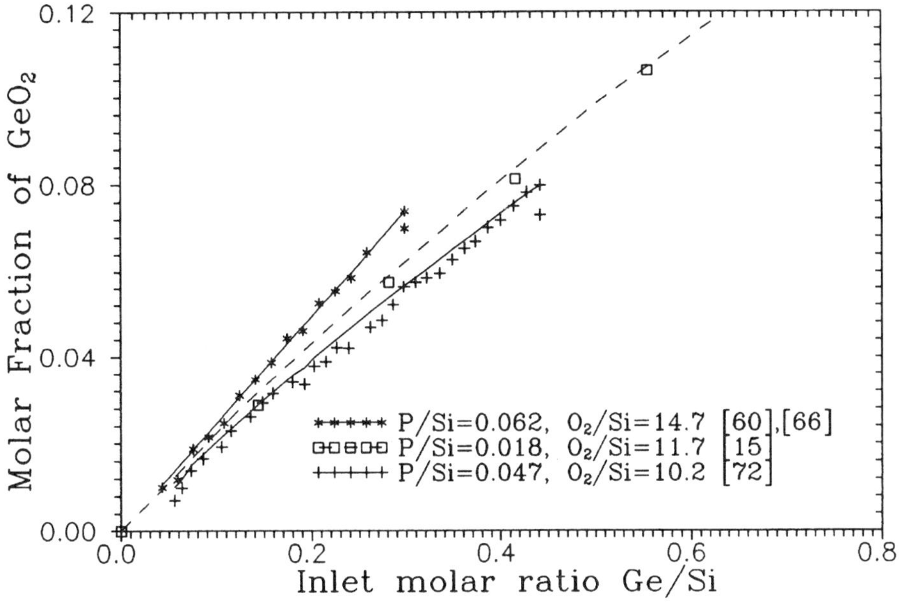

FIGURE 13.3 Dependence of the GeO_2 concentration in glass on flow rates of raw materials.

The problem of germanium incorporation in the MCVD process has been treated theoretically by McAfee et al.[16] in a system initially containing $SiCl_4$, $GeCl_4$, and oxygen. Supposing the presence of 30 gaseous species in the equilibrium gaseous phase and a nearly ideal solution for the GeO_2-SiO_2 glass system, they calculated that the maximum of GeO_2 concentration in the glass occurred near 1800K. Fitting this model to the measured GeO_2 concentrations in glass layers prepared by the MCVD process, they further determined that the value of Gibbs free energy for GeO_2(l) equalled –724 kJ/mol^{-1}.[17]

Some of the results on incorporation of GeO_2 in glass layers in the MCVD process are summarized in Figure 13.3. They allow us to draw a conclusion about the possibility of describing this process using the value of the equilibrium constant (Equation 13.6) in the range of from 0.1 to 0.13. Moreover, a relative independence of this value of experimental conditions shows the GeO_2 incorporation into glass layers to be controlled both by the chemical reactions and by the transport and sintering processes.

Systems with P_2O_5 and B_2O_3

Boron trioxide and phosphorus pentoxide are also used as dopants in the MCVD fabrication of silica optical fibers. They decrease the viscosity of silica and therefore the sintering temperature of soot. Phosphorus pentoxide increases the refractive index of silica by the factor F_D in Equation 13.4 with the value of from 0.06 to 0.1.[48,57] On the other hand, B_2O_3 decreases this refractive index by the factor F_D equal to –0.03.[11] It is also used for inducing optical birefringence in preforms owing to its high thermal expansion coefficient of about $10^{-5}K^{-1}$.

In the preparation of P_2O_5 by oxidation of $POCl_3$, only negligible amounts of $POCl_3$ were found in the effluent gases at temperatures above 1400°C.[14,15] This fact was also confirmed by the determined deposition efficiency of P_2O_5 in the range of 0.95 to 1.[48]

Similar to the case of GeO_2 incorporation, there is the effect of soot sintering on the P_2O_5 content in the glass layers. Kleinert et al.[48] determined the maximum content of about 8 mol.% P_2O_5 in the glass layers, even when they found about 25 mol.% P_2O_5 at 1973K in soot. This effect was explained by reevaporation of P_2O_5 during soot sintering. Reevaporation of P_2O_5 or B_2O_3 was

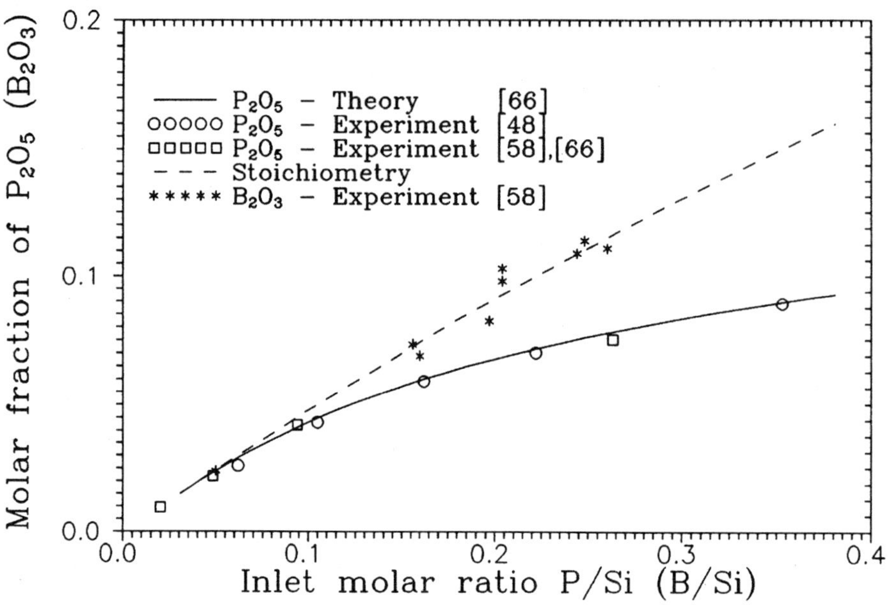

FIGURE 13.4 Dependence of P_2O_5 and B_2O_3 concentrations in glass on flow rates of raw materials.

also observed by Botvinkin et al.,[79] who found free P_2O_5 in soot of SiO_2 doped with P_2O_5 and low amounts of free B_2O_3 in soot of SiO_2 doped with B_2O_3.

The incorporation of phosphorus into germanium-doped silica glass was theoretically studied by McAfee et al.[18] Supposing the presence of 41 compounds in the equilibrium and estimating the value of Gibbs free energy of $PO_{2.5}(l)$, they determined that phosphorus concentrations in the glass rapidly decreased with temperature above 1800K.

The process of doping silica with P_2O_5 or B_2O_3 in silica glass during the MCVD process was also investigated by Choc et al.,[58,61] who determined the value of the deposition efficiency of about 0.97 for P_2O_5 concentrations in glass up to 3 mol.% and of about 0.99 for B_2O_3 concentrations up to 11 mol.%. These investigations were continued by Kašík et al.,[66] who observed the tendency of P_2O_5 content in the glass to saturation at about 7 mol.%. They attributed this effect to the equilibrium

$$P_4O_{10}(g) \Leftrightarrow 4PO_{2.5}(gl), \quad K = 0.29 \tag{13.9}$$

An essentially complete incorporation of B_2O_3 into silica glass was confirmed on the basis of thermodynamic calculations by Botvinkin,[79] Proft,[59] and Matějec.[60]

The results on P_2O_5 and B_2O_3 incorporation, which are summarized in Figure 13.4, allow us to draw a conclusion on incomplete incorporation of P_2O_5 into silica glass in the MCVD method, especially if its concentration in the glass exceeds 5 mol.%. These effects are controlled by evaporation of P_2O_5 during soot sintering.

Systems Containing Fluorine

Fluorine is considered a very important dopant because it, similar to B_2O_3, decreases the refractive index of silica by the factor F_D equal to –0.14.[11] Under MCVD conditions, fluorine incorporates into the glass with low efficiency. Moreover, the amount of the deposited glass can decrease owing to SiF_4 formation.

Fluorine incorporation into silica glass under MCVD conditions was studied by Walker et al.[19] for SiF_4 as a fluorine source. From the composition of the effluent gases during fluorine doping,

TABLE 13.3
Comparison of Theoretically and Experimentally Determined Content of F in Glass for Raw Materials SF_6 and R113

Raw Material	Inlet Molar Ratio			Content of F (wt.%)	
	F/Si	P/Si	O_2/Si	Theory	Experiment
SF_6	0.632	0.0445	9.8	0.48	0.5 (62)
	1.446	0.0434	9.5	0.61	0.6 (62)
$C_2Cl_3F_3$	0.065	0.0266	5.4	0.32	0.25 (66)
	1.796	0	5.3	0.76	0.72 (66)
	0.45	0.02	7.2	0.50	0.47 (50)
	0.90	0	7.0	0.62	0.66 (50)

they determined a value of 0.47 for the equilibrium constant for the reaction 13.10 at 2000K. They attributed the decreased amount of the deposited glass to the reactions 13.7 and 13.10.

$$3SiF_4(g) + SiO_2(s) + 2Cl_2(g) \Leftrightarrow 4SiF_3Cl(g) + O_2(g) \tag{13.10}$$

Moreover, they found that the decrease of the glass refractive index with respect to silica was proportional to $P(SiF_4)^{0.25}$, where $P(SiF_4)$ is the partial pressure of SiF_4 in the hot zone. They theoretically explained this dependence by the equilibrium 13.11

$$SiF_4(g) + 3SiO_2(s) \Leftrightarrow SiO_{1.5}F(s), \tag{13.11}$$

where $SiO_{1.5}F$ represented a silica tetrahedron in the glass with one fluorine atom and three bridging oxygen atoms. Evidence of the existence of Si-F bonds in the fluorine-doped glass layers prepared by the MCVD process was given by Cocito et al.[20] who found the Si-F vibration band at 10.58 μm in the Raman spectrum.

Kirchhof et al.[49] examined the incorporation of fluorine into silica using $SiCl_4$, $C_2Cl_3F_3$ (R113), and oxygen. Measuring the composition of effluent gases from the MCVD process and comparing it with that one from thermodynamic analysis, they identified 19 fluorine-containing gaseous species. They determined no solid phase above a molar ratio of R113/$SiCl_4$ > ≈1.2. Similar results were obtained by Grigor'yanc et al.,[80] who found no deposition above a molar ratio R113/$SiCl_4$ > ≈1.

Continuing their study on fluorine incorporation, Kirchhof et al.[50] found the fluorine content in glass to be practically independent of both the deposition temperature between 1650 and 1900°C and the total flow rate between 500 and 1600 sccm. They showed the fluorine content in the glass to be determined by the equilibrium 13.11, with the equilibrium constant equal to 6.10^{-6}.

The thermodynamic approach was also used for the explanation of the results on fluorine incorporation into silica glasses using $SiCl_4$, $POCl_3$, SF_6, or R113 as raw materials by Sedlář et al.[62] and Kašík et al.[66] They showed that experimentally determined fluorine concentrations in glass could be well fitted with the model based on the value of the equilibrium constant for reaction 13.11 equal to 5.10^{-6} (Table 13.3).

The relationship between the inlet gaseous stream composition, the refractive index difference, and the thickness of the deposited layer was derived by Aulitto et al.[21] for $SiCl_4$, SF_6, and O_2 raw materials.

TABLE 13.4
Essential Chemical Equilibria in the MCVD Process

Reaction	K
$P_4O_{10}(g) = 4\ PO_{2.4}(gl)$	$K \approx 0.3$
$GeCl_4\ (g) + O_2(g) = GeO_2\ (gl) + 2Cl_2(g)$	$K \approx 0.1–0.13$
$SiF_4(g) + 3SiO_2(gl) = 4SiO_{1.5}F(gl)$	$K \approx 5.10^{-6}$
$3SiF_4(g) + SiO_2(gl) + 2Cl_2(g) = 4SiF_3Cl(g) + O_2(g)$	$K \approx 0.5$

The reaction kinetic approach was used by Marschal,[22] who supposed that $SiCl_4$ oxidation and reaction 3.12

$$2SiCl_3F(g) + \frac{3}{2}O_2(g) \rightarrow 2SiO_{1.5}F(gl) + 3Cl_2(g) \tag{13.12}$$

were rate-limiting steps described by the first-order rate equations, with activation energies of 251 to 268 kJ mol^{-1}. He reported good agreement between the experimentally and theoretically determined glass amounts and fluorine concentrations in the glass prepared by the reactions of $SiCl_4$ with CCl_2F_2 and oxygen. Using this approach for the reactions of $SiCl_4$, R113 with O_2, Kirchhof et al.[50] found it did not account for the variation of fluorine content in the glass due to changes of the inlet $SiCl_4$ concentration.

Formation of OH Groups in the MCVD Process

In contrast to previous dopants, the incorporation of OH groups into silica glass during the MCVD process should be prevented. It is well-known that OH groups in glass give rise to absorption bands at 0.95, 1.24, and 1.39 μm, which are the overtones and combination vibrations of the O-H fundamental stretching vibration at 2.73 μm.[3,11] In addition, small amounts of P_2O_5 were found to cause an additional overtone in the 1.5 to 1.6 μm region. All these bands increase fiber losses at the wavelengths of 0.85, 1.3, and 1.55 μm, which are important operating wavelengths for telecommunications applications.

Wood[23] showed that OH group incorporation into glass layers from hydrogen-containing compounds was controlled by the equilibria 13.13 and 13.14

$$2HCl(g) + \frac{1}{2}O_2(g) \Leftrightarrow H_2O(g) + Cl_2(g), \tag{13.13}$$

$$H_2O(g) + [Si-O-Si](s) \Leftrightarrow 2[Si-OH](s) \tag{13.14}$$

The information given above enables us to conclude that the chemical thermodynamic approach provides us with a powerful means to explain the composition of the glass layers prepared by the MCVD process. Important values of equilibrium constants are shown in Table 13.4

TRANSPORT PROCESSES IN THE MCVD METHOD

The basic mechanism of transport processes in the MCVD method is thermophoresis. The first model on thermophoretic deposition of small particles in a laminar incompressible tube flow was reported by Walker et al.[24] Supposing temperature-independent gas properties they wrote governing equations for the velocity profile (Equation 13.15), energy (Equation 13.16) and mass transfer (17). These equations were solved with boundary conditions (Equation 13.18) and (Equation 13.19).

$$\mathrm{v}_r = \mathrm{v}_\varphi = 0, \quad \mathrm{v}_z = \frac{2Q}{\pi R^2}\left(1 - r^2\right), \tag{13.15}$$

$$\left(1 - r^2\right)\frac{\partial T}{\partial z} = \frac{1}{Pe}\Delta T, \tag{13.16}$$

$$\left(1 - r^2\right)\frac{\partial \phi}{\partial z} = \frac{\pi R D}{2Q}\Delta\phi + \frac{\pi R}{2Q}K\nu\nabla\left(\phi\frac{\nabla T}{T}\right), \tag{13.17}$$

$$r \le 1; \quad \begin{matrix} z = 0; & \phi = 1, T = T_{\max} \\ z > 0; & \phi, T\text{-finite} \end{matrix}, \tag{13.18}$$

$$r = 1; \quad \begin{matrix} z \le zp; & T = T_{\max} - \dfrac{T_{\max} - T_{\min}}{zp} \\ z > zp; & T = T_{\min} \end{matrix}. \tag{13.19}$$

In Equation 13.15 to 13.19, $Pe = 2Q/(\pi\alpha R)$ is the thermal Peclet number, ν is the kinematic viscosity of the gas phase, and Q, R, T, α, D, and Φ are the volumetric flow rate, the tube radius, the temperature, the thermal diffusivity, the diffusivity, and the particle concentration, respectively. The radial and axial coordinates r and z are normalized with respect to R. The thermophoretic coefficient K in Equation 13.17 was estimated from Equation 13.20

$$K = k_T \frac{1 + C_1 \dfrac{\lambda}{R_p}\dfrac{k_p}{k_g}}{1 + \dfrac{k_p}{2k_g} + C_1 \dfrac{\lambda}{R_p}\dfrac{k_p}{k_g}}. \tag{13.20}$$

In Equation 13.20, $c_l = 2.17$, $k_T \approx 1.1$, and λ, R_p, k_p, and k_g are the mean free path length, the radius of the particle, and the thermal conductivities of the particle and gas, respectively. For micron-sized and submicron-sized particles, the values of K were estimated to vary from 0.5 to 1.1.

Results obtained by Walker et al.[24] can be expressed via the overall thermophoretic efficiency $E(z)$, defined as a fraction of the particles deposited on the wall within a distance z from the beginning of the deposition zone. For a step change in the wall temperature ($zp \to 0$), $E(z)$ can be given by Equation 13.21:

$$\frac{\theta^*}{\Pr K\phi_0} E\left(\frac{z}{Pe}\right) \approx \begin{cases} 4.07\left(\dfrac{z}{Pe}\right)^{2/3} & z < 0.002Pe \\ 1 - \exp\left(-7.317\dfrac{z}{Pe}\right) & z > 0.2Pe \end{cases} \tag{13.21}$$

In this equation, $Pr = \nu/\alpha$, $\theta^* = T_{\min}/(T_{\max} - T_{\min})$, and ϕ_0 is the apparent concentration of the particles at the tube wall. It was determined from the boundary layer approximation for low values of z and expressed by Equation 13.22.

$$\phi_0 \xrightarrow{\Pr K \to 1} \frac{T_{\min}}{T_{\max}} \tag{13.22}$$

For $zp = 0.3$, Walker et al.[24] found lower values of $E(z)$ than for $zp = 0$ and the same limiting value of the efficiency E_T given by Equation 13.23

$$E_T = \Pr K\phi_0\theta^* \tag{13.23}$$

In their further paper, Walker et al.[10] tested this model on experimental data from the deposition of pure silica layers. They found good agreement (maximum relative error of 5%) between the values of E_T from the experiments and those obtained by solving the equation set 3.15 to 3.17 for the measured temperature profiles of the wall. Analyzing the calculated temperature fields and the particle trajectories inside the tube, they identified two transport regimes.

Under standard MCVD conditions (i.e., in the particle-transport limited regime), E_T was found to be only a function of the reaction temperature T_r and the temperature T_e at which the gas and walls temperatures equilibrate downstream of the torch. In this case, the approximate relation (13.24) holds:

$$E_T = 0.8\left(1 - \frac{T_e}{T_r}\right) \tag{13.24}$$

At high flow rates in the tube, the reaction-limited regime was encountered in which the particles were formed inside a limited volume near the inner tube wall, wherever the gas temperature exceeded the reaction temperature. In this case, the deposition efficiency was extremely sensitive to the temperature in the hot zone, its length, and the gas flow rate.

Simpkins et al.[25] concluded theoretically that thermophoresis was the dominant mechanism of mass transfer in the MCVD process. They solved the equation set 13.15 to 13.16 with a wall temperature distribution given by a sinusoidal pulse superimposed on $T_{\min}$ and showed the temperature gradient downstream of the pulse to decrease exponentially in the axial direction. Experimentally, in both cases, they found nearly the same exponential decay for the temperatures and the amount of the deposited particles downstream of the torch.

Weinberg[26] showed that for a general wall temperature profile $T(r = 1, z)$ and $PrK \to 1$, the particle concentration at the tube wall is given by Equation 13.25, which is a generalization of Equation 13.22. He used Equations 13.25 and 13.26 for the calculation of $E(z)$.

$$\phi(r = R, z) = \frac{T(r = 1, z)}{T_{\max}} \tag{13.25}$$

$$E(z_s) = -4K\Pr\int_0^{z_s} \frac{\phi(r = 1, z)}{T(r = 1, z)}\left(\frac{\partial T(rz)}{\partial r}\right)_{r=R} dz \tag{13.26}$$

Doupovec et al.[63] investigated the thermophoretic deposition in an axially non-symmetrical MCVD process (i.e., for an angularly and axially dependent wall temperature profile). Using the approximate Pohlhausen method for the integration of Equation 13.16, and Equation 13.25 as the solution of Equation 13.17, they derived an approximate analytical expression for the temperature and concentration profiles. They showed that in this case, E_T is an angularly dependent parameter. Experimentally, they proved this result by depositing layers with non-circular cross-sections using a non-rotating substrate tube, a round burner, two linear heaters, and two linear coolers placed downstream of the burner and traversing at the same velocity as the burner.

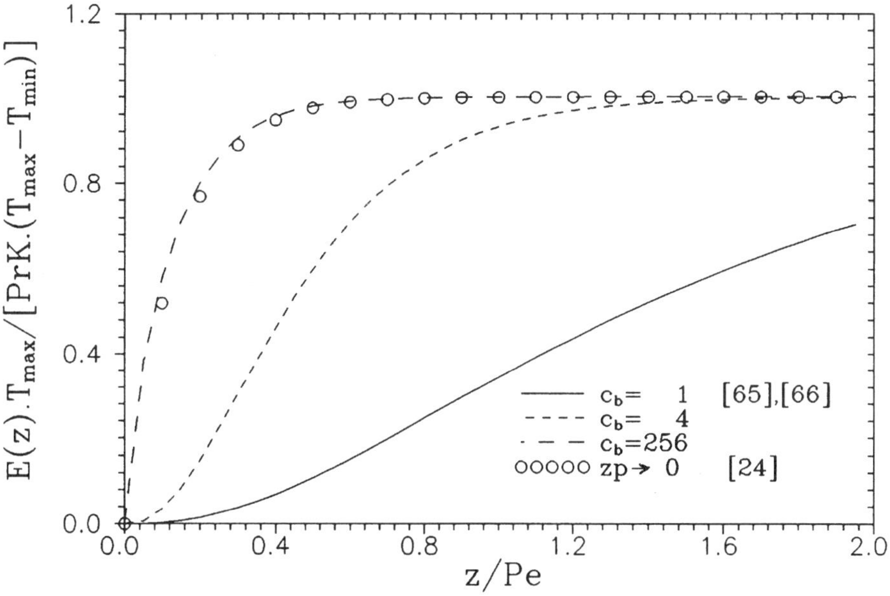

FIGURE 13.5 The dependence of the overal thermophoretic efficiency on the axial coordinate for three wall temperature profiles.

$$T(r=1,z) = T_{\text{min}} + \left(T_{\text{max}} - T_{\text{min}}\right)\left(2\exp\left(-C_b z\right) - \exp\left(-2C_b z\right)\right) \tag{13.27}$$

Kašík et al.[65,66] used the approximate Pohlhausen method for the integration of Equation 13.16 with the wall temperature profile given by Equation 13.27, and Equations 13.25 and 13.26 for the calculation of the overall thermophoretic efficiency. In Equation 13.27, c_b is a parameter that was adjusted from the measured wall temperature profile. An example of the results of this calculation and comparison with the results of Walker et al.[24] can be seen in Figure 13.5.

In addition to the above discussed papers, several authors have investigated the dependence of the overall thermophoretic efficiency and deposition rates in the MCVD process on experimental conditions. Sandoz et al.[5] observed a drop of E_T either with increasing the flow rates of reactants or with decreasing the torch speed for deposition rates below 0.23 g min^{-1}. This drop correlated with the measured changes of the temperature T_e.

Koel[1] showed that deposition rates of about 1 g min^{-1} could be attained if T_e was decreased by water cooling. Sandoz et al.,[27] using quenching of the tube with nitrogen vaporized from a liquid source, increased E_T for SiO_2 nearly to 95% at deposition rates of about 0.1 g min^{-1}. This value decreased with increasing tube diameter, which was attributed to the reaction-limited regime.

Walker et al.[10] showed the tendency of the reaction-limited regime to decrease using a torch with a broad hot zone, addition of He to raw materials, or by lowering the oxygen content in the reaction mixture. They achieved a deposition rate of 2.3 g min^{-1}. The effect of He addition on increasing the amount of the deposited GeO_2-SiO_2 glass was also observed by Nordvik et al.[28]

Choc[58] determined a value for E_T of about 0.68 for the deposition of P_2O_5-SiO_2 or B_2O_3-SiO_2 layers at the deposition rates of about 0.31 g min^{-1}, and about 0.42 for the deposition of P_2O_5-GeO_2-SiO_2 layers at rates of about 0.12 g min^{-1}. Continuing this study, Kašík et al.[65] showed the possibility of increasing E_T up to 0.70 by cooling the tube with a stream of air. In the same paper, they studied the deposition process in a circular flow induced by a helix placed near the starting point of the deposition zone. They found that deposition in a circular flow could be described by a power function given by Equation 13.22 with the exponent lower than 2/3.

On the basis of the above discussion, one can conclude that only part of particles formed in MCVD chemical reactions is deposited on the tube wall.

CONSOLIDATION OF THE DEPOSITED PARTICULATE LAYERS

The next step in the MCVD process, the consolidation of the flocculent deposit on the tube wall into a vitreous layer, is very important for the preparation of layers with a low content of light-scattering centers and low concentration variations across the layer.

Results on the first extensive experimental and theoretical study of the consolidation process in the MCVD process for SiO_2, GeO_2-SiO_2, B_2O_3-SiO_2 and B_2O_3-GeO_2-SiO_2 flocculent layers were published by Walker et al.[29] They showed that consolidation in the MCVD process was controlled by a viscous sintering process with the rate dependent on the capillary number C expressed by Equation 13.28,

$$C = \mu l_0 \frac{(1-\varepsilon)^{1/3}}{\sigma t_s}, \tag{13.28}$$

where μ, l_0, ε, and σ are the glass viscosity, the size of the initial void regions, the initial void fraction, and the surface tension, respectively, and t_s is the sintering time. They also developed a theoretical model of the sintering process based on energy equations for the particulate layer and the gas.

In the same paper, they explained the bubble formation during consolidation by supposing the existence of a critical temperature T or the composition x controlling whether the gas flux was in or out a closed pore. This mechanism is summarized in Equation 13.29,

$$\begin{gathered} P_v < P_s \text{ pore shrinking, } P_v > P_s \text{ bubble formation,} \\ P_v = P(T, x); \quad P_s = \frac{2\sigma}{R} \end{gathered} \tag{13.29}$$

where P_v, P_s, and R are the vapor pressure of the most volatile component, the pressure due to surface tension σ, and the pore radius, respectively.

Several papers have dealt with the influence of the composition of gaseous and solid phases during the sintering of a flocculent layer on the dopant concentration in glass. Wood et al.[14,15] experimentally found higher GeO_2 contents in silica layers doped with GeO_2 and P_2O_5 compared to that doped only with GeO_2. They attributed this effect to higher sintering rates of P_2O_5 co-doped particles which decreased the time for GeO_2 diffusion.

DiGiovanni et al.[30] also experimentally studied P_2O_5, GeO_2, and fluorine incorporation during the sintering of porous layers in various atmospheres. Their results are summarized in Table 13.5. Similar effects were also observed by Kirchhof et al.[50] at the sintering of silica porous layers doped with fluorine, using R113. In a companion paper, DiGiovanni et al.[31] reported a theoretical model on P_2O_5 incorporation into a porous silica layer. The model was based on the energy equation in the layer, the diffusion equation for dopant in the voids with a source of dopants, the equilibrium relation for the solubility of P_2O_5 in SiO_2, the sintering equation, and the diffusion equation for the dopant concentration in solid. Solving this equation set, they found that the modulation in P_2O_5 concentration across the sintered layer was primarily created during the layer sintering. The temperature dependence of the reaction kinetics for P_2O_5 formation was found to be of secondary importance.

TABLE 13.5
Effect of Sintering Atmosphere on Dopant Content in Glass

Sintering Atmosphere	Refractive Index Difference Relative to SiO_2			
	SiO_2	GeO_2-SiO_2	P_2O_5-SiO_2	F-SiO_2
Oxygen	—	0.002	0.0001	0.0003
O_2 + $GeCl_4$	0.001	0.011	—	—
O_2 + $POCl_3$	0.0017	—	0.0028	—
O_2 + CCl_2F_2	–0.001	—	—	–0.001

Input molar fractions: $SiCl_4$ = 0.22; $POCl_3$ = 0.004; $GeCl_4$ = 0.016; CCl_2F_2 = 0.003; O2 = 0.22; He = 0.55.

COLLAPSE OF THE TUBE WITH DEPOSITED LAYERS

The collapse of the silica tube with deposited inner layers converts the viscous composite tube to a solid rod — to the preform. This process involves a slow viscous flow of the glass tube, driven by surface tension and differential pressures on the inner and outer tube surfaces. Lewis[32] treated the radial collapse of a viscous circular composite tube consisting of two concentrically arranged circular layers characterized by uniform viscosities. Solving the equations for the radial motion of a viscous incompressible fluid, he found an approximate equation (Equation 13.30) to hold for the collapse time t_c.

$$t_c \approx \frac{2\mu_s}{\Delta P}\left[\frac{\nu}{(1+\varepsilon)}\ln\frac{\alpha_0(1+\varepsilon)+\varepsilon}{\alpha(1+\varepsilon)+\varepsilon}+(1-\nu)\ln\frac{\gamma_0}{\gamma}-\ln\frac{\beta_0}{\beta}\right]$$

$$\varepsilon=\frac{\sigma}{b_c\Delta P},\quad \alpha=\frac{a}{b_c},\quad \beta=\frac{b}{c_c},\quad \gamma=\frac{c}{b_c},\quad \nu=\frac{\mu_g}{\mu_s} \tag{13.30}$$

In Equation 13.30, a is the inner radius, b is the outer radius of the composite tube, c determines the interface between the inner and outer layers, ΔP, μ_g, and μ_s are the difference between the pressures loaded on the outer and inner tube surfaces, the viscosity of the inner and the outer concentric layers, respectively.

Kirchhof[51] derived the collapse equation for a homogeneous tube in the form

$$v_r=\frac{\Delta P+\sigma\left(\frac{1}{a}+\frac{1}{b}\right)}{2\eta r\left(\frac{1}{b^2}-\frac{1}{a^2}\right)}, \tag{13.31}$$

where v_r, and r are the fluid velocity and the radius coordinate, respectively. Integrating Equation 13.31 for time-dependent glass viscosity owing to the torch traverse, he derived the relation[52]

$$(a_2-a_1)R^*=\left(\frac{\Delta P}{S^*}-2\sigma\right)\frac{\Delta z_T}{2\eta v_T}$$

$$R^*=1-\frac{a_2+a_1}{b_2+b_1},\quad S^*=\left(\frac{b_1a_1}{b_1+a_1}+\frac{b_2a_2}{b_2+a_2}\right)^{-1}. \tag{13.32}$$

In Equation 13.32, Δz_T is the width of the axial temperature profile, v_T is the torch velocity, and the subscripts 1 and 2 denote the values of a geometrical parameter before and after the torch pass. On the basis of experiments and Equation 13.33, Kirchhof determined the value of surface tension of silica to be equal to 0.4 Nm^{-1}. In their further paper, Kirchhof et al.[53] used collapsing experiments and Equation 13.33 for the determination of the temperature dependence of the silica viscosity and for the estimation of temperature gradients in the wall of the substrate tube.

Geyling et al.[33] extended the modeling by Lewis to the case of two-dimensional collapsing the tube with departures from the nominally circular shape. Solving the equation set for the stationary viscous flow of the glass in the radial and azimuthal directions for initially sinusoidal disturbances of the circular tube shape with a magnitude of about 10 μm, they obtained the dependence of the preform elipticity on the processing variables. Their results for the homogeneous tube can be expressed approximately by Equation 13.33. Similar results were obtained for the composite tube.

$$\varepsilon_b = 6.7\times10^{-4}\exp\left(4.394\Delta P\frac{b}{\sigma}\right)\left(\frac{t}{b}\right)^{-3.211} \tag{13.33}$$

In Equation 13.33, $t = b - a$ and holds for $a = 0.95$ cm, $b = 1.3$ cm, the deposit viscosity equal to 200 Pa·s, and the tube viscosity equal to 10^4 Pa·s.

Yarin et al.[64] presented a model of a two-dimensional viscous collapse of composite tubes with arbitrary cross-sections describing the glass flow by the biharmonic equation for the stream function in each glass layer. They applied this model to the modeling of the preparation of preforms with dumb-bell cores.

Diffusion Processes in Glass Layers

During the collapse of the composite tube, diffusion processes occur that decrease the dopant concentrations in the glass and smooth the modulations on the concentration profiles. Kirchhof et al.[54] investigating the diffusion in B_2O_3-SiO_2 glass layers found that:

$$D(T,X) = 0.001\exp\left(-1000\frac{60-95X+250X^2}{T}\right) \tag{13.34}$$

$$X < 0.12, \quad 2140\text{K} < T < 2450\text{K}$$

In their following paper on the diffusion of GeO_2 in thin GeO_2-SiO_2 or GeO_2-B_2O_3-SiO_2 glass layers at 2400K, they estimated the diffusion coefficient of GeO_2 to be of the order of 10^{-13} m^2s^{-1}.[55] DiGiovani et al.[30] reported values of 1.1×10^{-15} and 1.7×10^{-16} m^2s^{-1} for P and Ge, respectively.

Another important effect is the diffusion of OH groups from the substrate tube into the deposited glass layers. Kirchhoff et al.,[56] summarizing published results for OH group diffusion in silica, showed that the values of the preexponential factor and the activation energy ranged from 2.5×10^{-9} to 2.7×10^{-11} m^2s^{-1} and from 72.4 to 125.6 kJ mol^{-1}, respectively.

PERFORMANCE OF THE MCVD METHOD FOR THE FABRICATION OF OPTICAL FIBERS

In the fabrication of optical fibers by the MCVD method, a lightguiding structure is built up layer by layer in the preform and then reproduced in the fiber. The design of this structure, which follows from physical theory,[34] consists of the specification of properties of materials and their spatial arrangement in the fiber in order to achieve specified physical effects. In the simplest case, the lightguiding structure consists of a core of refractive index n_l surrounded by a cladding glass of a lower refractive index n_2. Owing to a finite core diameter (up to several tens of μm), there are a finite number of optical modes in which the light can propagate in the core. Single-mode, few-mode, and multimode fiber structures can be distinguished.[34]

The choice of materials with specified properties and the development of procedures for achieving their spatial arrangement are two basic aspects of fiber preparation. The first aspect, the choice of materials, is related to the glass composition. The optical losses, refractive index, and fluorescence are the most important properties depending on the glass composition.

The attenuation coefficient α, representing the optical losses of the fiber, is usually given by Equation 13.35

$$\alpha = \frac{A}{\lambda^4} + B_a(\lambda) + C_w(\lambda) \tag{13.35}$$

In Equation 13.35, the first term represents Rayleigh scattering losses. Values of the scattering coefficient A increase with dopant concentration and range from 0.5 to about 2 dB km^{-1} μm^{-4}.[2]

The second term is related to losses attributed to absorption transitions. In particular, absorption transitions related to OH group vibrations can be detrimental because, for every ppm of OH groups, losses of 48 dB km^{-1} at 1.39 μm, of 2.5 dB km^{-1} at 1.25 μm, or of 1.2 dB km^{-1} at 0.95 μm are added.[3,11] There are two primary sources of OH groups in the MCVD process, namely hydrogenic impurities in raw materials and the thermal diffusion from substrate tubes, because customary silica tubes contain from 3 to 1000 ppm OH groups.[3,78]

In agreement with thermodynamic considerations, several approaches have been used for reducing the OH content. They include photochemical purification of raw materials,[3,67] addition of Cl_2 at the collapse, or lowering the partial pressure of O_2 using inert gases.[3,80] Deposition of the cladding with a low content of OH groups as a barrier or exchanging the OH groups in the tube by deuteration[3,80] has been used to prevent the diffusion of OH from the tube into the fiber core. The first approach, relying on increasing the deposition length for OH groups, is mainly used in practice, especially in the fabrication of single-mode fibers.[80]

The third term in Equation 13.35 represents optical losses due to the waveguiding structure. It is known that the attenuation increases due to waveguide imperfections such as bubbles and core imperfections.[34,78] The formation of these inhomogeneities can be prevented by controlling the sintering process. As the mode can propagate in the fiber only below a certain wavelength limit (the so called cut-off wavelength), the optical losses are also increased, owing to poor light confinement near this wavelength.

Applying the procedures above, standard single-mode and multimode fibers with low attenuation values have been prepared.[35,68,80] Examples of the attenuation spectra can be seen from Figure 13.6.

Fiber Amplifiers and Lasers

While in standard optical fibers it is important to decrease the attenuation as much as possible, in fibers for fiber lasers and amplifiers the effort is concentrated on fabricating fibers with a controlled spectral attenuation and amplified stimulated emission (ASE). These fibers are used in optical communication systems, both as amplifiers for signal amplification and as lasers generating optical solitons. The amplified stimulated emission is observed in silica fibers doped with rare-earth ions (RE^{3+}) such as Er^{3+} (1.55 μm), Nd^{3+} (1.06 μm), and Yb^{3+} (1.06 μm).[36] The effect of clustering RE^{3+} ions in high- silica glasses, which decreases the amplification, can be prevented by co-doping silica with Al_2O_3 and P_2O_5.[37,38] In order to induce ASE, these fibers are pumped at suitable wavelengths. For pumping, the absorption bands of RE^{3+} ions can be used. In some cases, additional ions, called sensitizers, are used, shifting the pumping wavelength to a region where commercial pump sources can be used. Fibers doped with Er^{3+} and sensitized with Yb^{3+} are a typical example of a fiber pumped at 1.06 μm (Yb^{3+}) and emitting at 1.55 μm (Er^{3+}).[38,69]

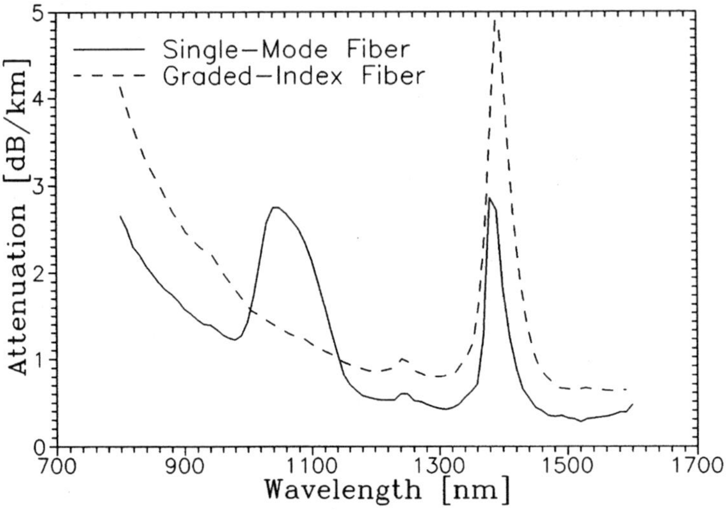

FIGURE 13.6 Attenuation spectra of the standard single-mode and graded-index fibers. (From Reference 68. With permission.)

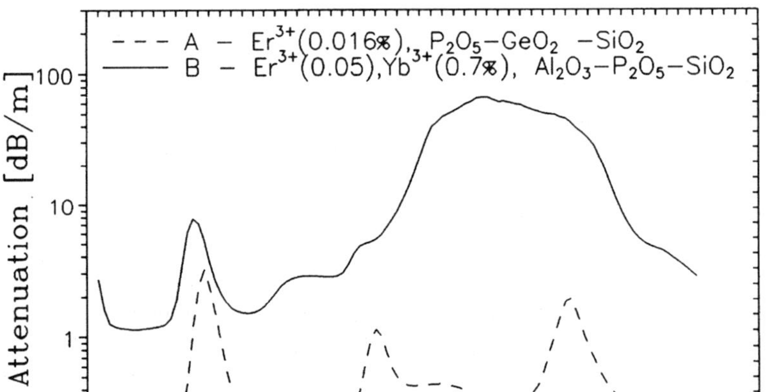

FIGURE 3.7 Attenuation spectra of silica optical fibers doped with Er^{3+} and/or TB^{3+} ions. (From References 69–71. With permission.)

These fiber structures have stimulated the use of solid raw materials with very low vapor pressures at normal or elevated temperatures. In-line evaporation of solid halides (e.g., $NdCl_3$, $ErCl_3$) directly in a chamber placed in the substrate silica tube was employed for the extension of the MCVD method. A typical absorption spectrum of Er^{3+} in a P_2O_5-GeO_2-SiO_2 glass core can be seen in Figure 13.7, curve A.[70,71]

Solutions of solid raw materials were also employed in the MCVD method. In one of such extensions, called the solution-doping method,[37,73] a porous layer deposited on the inner wall of a silica tube is soaked with the solution of $AlCl_3$, $ErCl_3$, and/or other dopants. After solvent evaporation, the porous layer is consolidated to a transparent layer. An example of the absorption spectrum of the fiber with the core of P_2O_5-Al_2O_3-SiO_2 glass doped with Yb^{3+} and Er^{3+} is shown in Figure

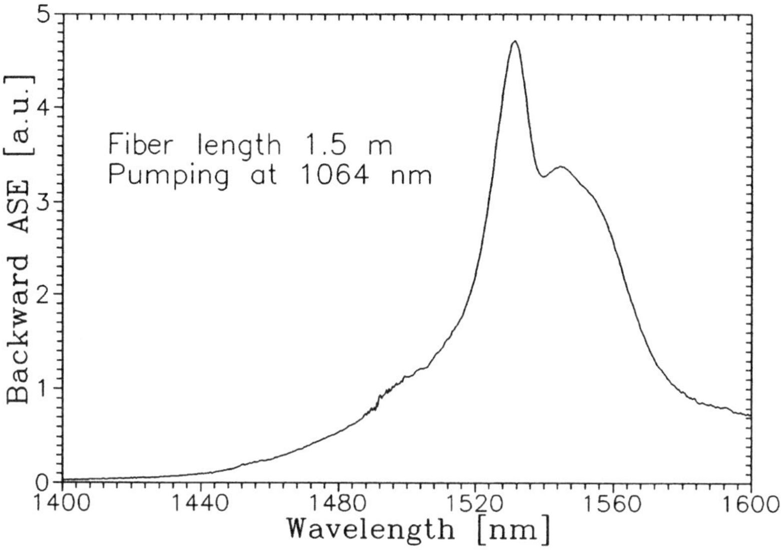

FIGURE 3.8 The spectrum of the backward ASE of the fiber pumped at 1064 nm. (From Reference 69. With permission.)

13.7, curve B.[69] The spectrum of the backward ASE of this fiber pumped at 1.064 μm can be seen from Figure 13.8.[69]

These two extensions of the MCVD method have been used for doping silica core with various rare-earth elements,[37-39] MgO.[70] Fibers doped with Er^{3+} have been used as in-line amplifiers in submarine telecommunication cable.

The second aspect of the fabrication of the lightguiding structure — the arrangement of glass layers into a proper refractive-index profile — influences the signal broadening and the polarization properties of the wave propagating in the fiber. The refractive-index profiles of single-mode and multimode fibers have been theoretically optimized in order to minimize signal broadening. In multimode fibers, a graded-index profile resulted from these calculations.

For building-up this refractive-index profile during the MCVD preparation of graded-index P_2O_5-GeO_2-SiO_2 glass cores, a physico-chemical model has been developed.[72] This model is based on the calculation of the composition of glass layers using equilibrium constants summarized in Table 13.4. It uses Equation 13.4 and the values of F_D for the estimation of the refractive index of the layer, and Equation 13.25 for the determination of the amount of the deposited material. An example of the results is shown in Figure 13.9. On this basis, graded-index fibers with a mean transmission bandwidth Δf(3dB) ≈ 700 MHz.km at 0.85 μm (maximum values of about 1.5 GHz.km) and the spectral attenuation shown in Figure 13.7 have been fabricated.[68] These results are comparable with the best reported results.[35]

In single-mode fibers, the pulse broadening is attributed to chromatic dispersion D_c, which is a function of the material dispersion D_m due to the wavelength dependence of the glass refractive index, and to waveguide dispersion D_w, which is determined by the refractive-index profile.[34,74,80] Standard single-mode fibers are produced with the value of D_c equal to ±3.5 ps nm^{-1} km^{-1} at 1.3 μm. An example of the attenuation spectrum of fibers with cores of GeO_2-SiO_2 glass and optical claddings of F-P_2O_5-SiO_2 glass, which has been prepared in Czechoslovakia, is shown in Figure 13.6.[40]

Fibers with specific refractive-index profiles have been developed in which the zero value of D_c was achieved at 1.55 μm (dispersion shifted fibers) or over the range of from 1.3 μm to 1.55 μm (dispersion compensated fibers).[3,35]

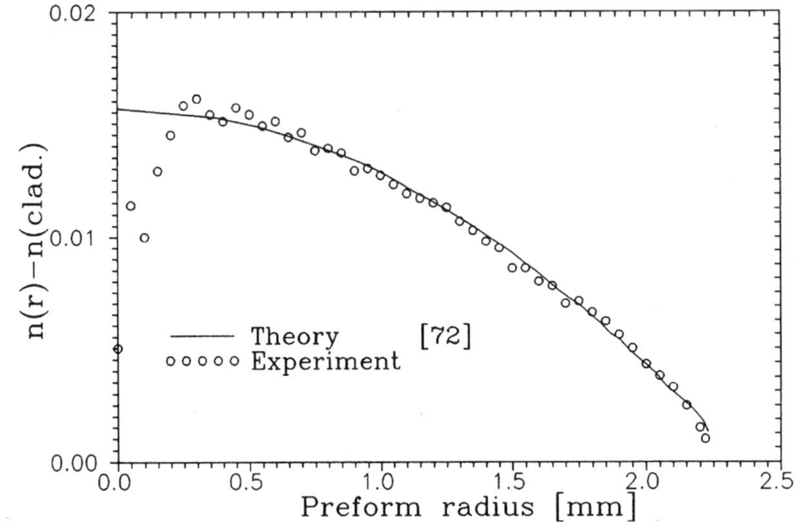

FIGURE 13.9 Comparison of measured and calulated graded-index profile.

Polarization Maintaining Fibers

Polarization maintaining (PM) fibers play an important role in coherent optical communications systems, in fiber gyroscopes, and in some other sensors. During the fabrication of PM fiber structure, optical birefringence in the fiber core is created. Although birefringent layered structures with circular symmetry have been theoretically proposed,[42] the realized structures always exhibited an angular dependence of the refractive index. Birefringence has been achieved either by using elliptical cores (geometrical birefringence) or induced due to photoelastic effect by angularly dependent thermal stresses created by doping silica with B_2O_3 which possesses a high thermal expansion coefficient (stress birefringence).[41,75,81] As the geometrical birefringence is rather low (about 10^{-5}), usually a combination of the both mechanisms is used.

There are several methods for the preparation of birefringent fiber structures. One method consists of drilling two holes in a circularly symmetric MCVD preform with the core doped with GeO_2, and inserting there stress rods doped with B_2O_3. In this way, “panda” fibers have been fabricated.[41,75] A similar method was applied by Grigor'yanc et al.[82] to the preparation of preforms with elliptical stress claddings. For the modeling of the last stage of this process, they developed a mathematical model based on viscous flow of a composite rod with non-circular borders driven by surface tension.

Birch et al.[43] developed a method for the fabrication of "bow-tie" fibers based on non-circular gas-phase etching of silica glass layers by fluorine compounds. This method was extended by Matějec et al.,[76] who used thermal screens for creating an angularly dependent temperature field in the tube. Both these modifications are shown in Figure 13.10 and an example of the resulting preform profile is shown in Figure 13.11. Using these methods, fibers with optical birefringence of about 2×10^{-5} have been prepared.[43,76]

Several other methods have been developed for the fabrication of polarization maintaining fibers; for example, the collapse of the tube under vacuum[44] or the deposition of the layers with angularly dependent thickness and the collapse of this tube.[63,64]

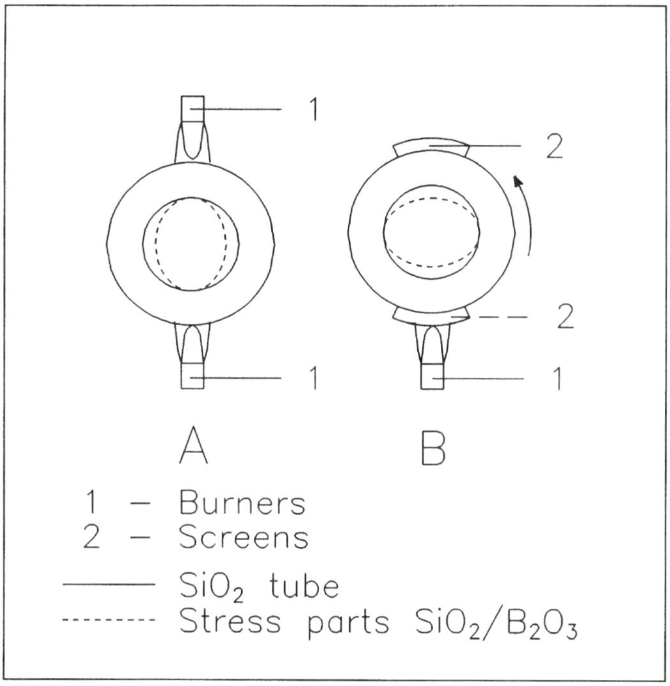

FIGURE 13.10 Gas-phase etching methods. (A, from Reference 43 and B, from Reference 76. With permission.)

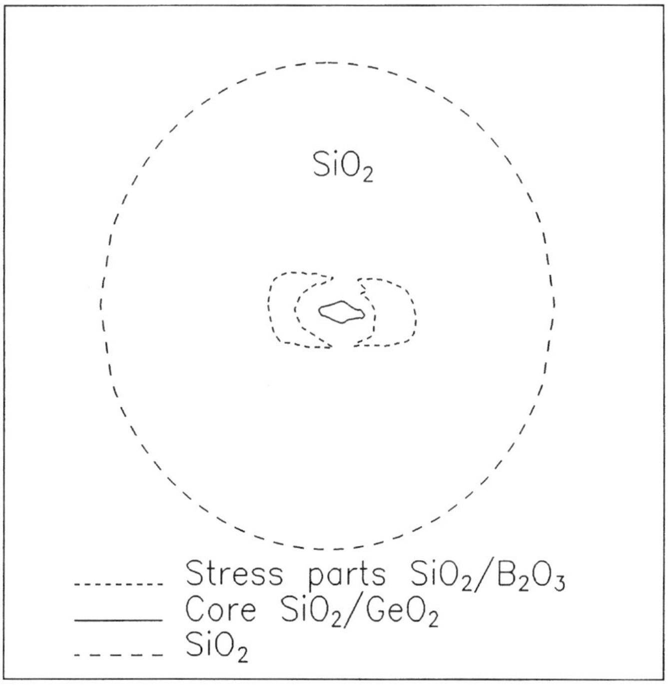

FIGURE 13.11 A structure of a "bow-tie" PM fiber.

CONCLUSIONS

Since their development, the MCVD methods have proved their capabilities in the production of silica optical fibers for optical telecommunications. Now, they are also employed for the fabrication of special fibers for fiber lasers, fiber sensors, investigations of nonlinear optical effects, or for laser power delivery in medical applications. Much effort has focused on the investigation of this method and fabrication of optical fibers in former Czechoslovakia, the GDR, and the U.S.S.R. Some of these laboratories continue their activities in the preparation of optical fibers for fiber lasers and sensors.

REFERENCES

1. Koel, G.J., Proc. *8th Eur. Conf. Optical Commun.* (*ECOC*), Cannes, France, 1982, 1.
2. Nagel, S.R., MacChesney, J.B., and Walker, K.L., *IEEE J. Quant. Electr.,* QE-18(4), 459, 1982.
3. Nagel, S.R., MacChesney, J.B., and Walker, K.L., *Optical Fiber Communications I.*, Ed. Tingye Li, Academic Press, Orlando, FL, 1985, 1.
4. MacChesney, J.B., O'Connor, P.B., DiMarcello, F.V., Simpson, J.R., and Lazay, P.D., Proc. 10*th Int. Congr. Glass*, Kyoto, Japan, 6-40–6-44, 1974.
5. Sandoz, F., Hamel, Ph., and Piffareti J., Proc. 9*th ECOC*, Geneva, Switzerland, 25–28, 1983.
6. Tanaka, J., and Kato, A., *Yogya-Kyokai-Shi,* 81(5), 179, 1972.
7. French, W.G., Pace, L.J., and Foertmayer, V.A., *J. Phys. Chem.* 82(20), 2191, 1978.
8. Powers, D.R., *J. Am. Ceram. Soc.* 61(7-8), 295, 1978.
9. George, A.P., Murley, R.D., and Place E.R., *Symp. Faraday Soc.*, 63, 1973.
10. Walker, K.L., Geyling, F.T., and Nagel, S.R., *J. Am. Ceram. Soc.,* 63(9-10), 552, 1980.
11. Bagley, B.G., Kurkjian, C.R., Mitchel, J.W., Peterson, G.E., and Tynes, R.E., *Optical Fiber Telecommunications*, Eds., S.E. Miller and A.G. Chynoweth, Academic Press, 167, 1979.
12. Fleming, J.W., *Appl. Opt.,* 23, 4486, 1984.
13. Wood, D.L., Walker, K.L., Simpson, J.R., MacChesney, J.B., Nash, D.L., and Angueira P., *Proc. 7th ECOC*, Copenhagen, Denmark, 1981, 1.2.1.
14. Wood, D.L., Walker, K.L., Simpson, J.R., and MacChesney, J.B., *Opt. Fiber Commun. Tech. Dig.,* TUCC4, 10, 1982.
15. Wood, D.L., Walker, K.L, MacChesney, J.B., Simpson, J.R., and Csencsits, R., *J. Lightwave Technol.,* LT-5(2), 277, 1987.
16. McAfee, K.B., Hozack, R.S., and Laudise, R.A., *J. Lightwave Technol.,* LT-1(4), 555, 1983.
17. McAfee, K.B., Walker, K.L., Laudise, R.A., and Hozack, R.S., *J. Am. Ceram. Soc.,* 67(6), 420, 1984.
18. McAfee, K.B., Gay, D.M., Walker, K.L., and Hozack, R.S., *J. Am. Ceram. Soc.,* 68(6), 359, 1985.
19. Walker, K.L., Csencsits, R., and Wood, D.L., *Opt. Fiber Commun. Tech. Dig.,* TUA7, 36, 1983.
20. Cocito, G., Cognolato, L., Modone, E., and Parisi, G., *J. Non-Cryst. Solids,* 93, 296, 1987.
21. Aulitto, V., Zuccala, A., and Madone, E., *J. Opt. Commun.* 11(2), 65, 1990.
22. Marshall, A., Irven, J., Boag, N., and Brooke, B.N., *Proc. 10th ECOC*, Stuttgart, Germany, 1984, 76.
23. Wood, D.L., and Shirk, J.S., *J. Am. Ceram. Soc.,* 64(6), 325, 1981.
24. Walker, K.L., Homsy, G.M., and Geyling, F.T., *J. Colloid Interface Sci.,* 69(1), 138, 1979.
25. Simpkins, P.G., Greenberg-Kosinski, S., and McChesney J.B., *J. Appl. Phys.,* 50(9), 5676, 1979.
26. Weinberg, M.C., *J. Am. Ceram. Soc.,* 65(2), 81, 1982.
27. Sandoz, F., Hamel, Ph., and Pifaretti J., *Proc. 10th ECOC*, Stuttgart, Germany, 1984, 298.
28. Hordvik, A., and Eriksrud, M., *Proc. 9th ECOC*, Geneva, Switzerland, 1983, 369.
29. Walker, K.L., Harveey, J.W., Geyling, F.T., and Nagel, S.R., *J. Am. Ceram. Soc.,* 63(1-2), 96, 1980.
30. DiGiovanni, D.J., Morse, T.F., and Cipolla, J.W., *J. Lightwave Technol.,* LT-7(12), 1967, 1989.
31. DiGiovanni, D.J., Morse, T.F., and Cipolla, J.W., *J. Am. Ceram. Soc.,* 71(11), 914, 1988.
32. Lewis, J.A., *J. Fluid Mech.,* 81(1), 129, 1977.
33. Geyling, F.T., Walker, K.L, and Csencsits, R., *J. Appl. Mech.,* 50(7), 303, 1983.
34. Marcuse, D., Gloge, D., and Marcatili, E.A.J., *Optical Fiber Telecommunications*, S.E. Miller and A.G. Chynoweth, Eds., Academic Press, 37, 1979.

35. Lilly, C.J., *Proc. 8th ECOC*, Cannes, France, 1982, 17.
36. Payne, D.N., and Reekie L., *Proc. 14th ECOC*, Brighton, U.K., 1988, 49.
37. Poole, S.B., *Proc. 14th ECOC*, Brighton, U.K., 1988, 433.
38. Towsend, J.E., Barnes, W.L., and Jedrzejewski, K.P., *Electr. Lett.*, 27(21), 1958, 1991.
39. Poole, S.B., Payne, D.N., and Ferman, M.E., *Electr. Lett.*, 21(17), 736, 1985.
40. Desurvire, E., *Physics Today* 28, 1994.
41. Payne, D.N., Barlow, A.J., and Ramskov Hansen, J.T., *IEEE J. Quantum Electr.*, QE-18, 477, 1982.
42. Fujji, Y., *Appl. Optics*, 25, 1061, 1986.
43. Birch, R.D., Payne, D.N., and Varnham, M.P., *Electr. Lett.*, 18(24), 1036, 1982.
44. Kutsuyama, T., Matsumura, H., and Suganuma, T., *J. Lightwave Technol.*, LT-2(5),634, 1984.
45. Kleinert, P., Schmidt, D., Kirchhof, J., and Funke A., *Kristall und Technik*, 15(9), K 85-K90, 1980.
46. Kleinert, P., Schmidt, D., Kirchhof, J., Laukner, H.J., and Knappe B., *Z. Anorg. Allg. Chem.*, 508, 176, 1984.
47. Kleinert, P., Kirchhof, J., and Schmidt, D., *Proc. 5th Internat. School of Coherent Optics, I.*, Jena, Germany, 1984, 42.
48. Kleinert, P., Kirchhof, J., Schmidt, D., and Knappe B., *Proc. 5th Internat. School of Coherent Optics II.*, Jena, Germany, 1984, 54.
49. Kirchhof, J., Kleinert, P., Unger, S., and Funke, A., *Cryst. Res. Technol.*, 21(11), 1437, 1986.
50. Kirchhof, J., Unger, S., Knappe, B., Kleinert, P., and Funke, A., *Cryst. Res. Technol.*, 22(4), 495, 1987.
51. Kirchhof, J., *Phys. Stat. Sol. (a)*, 60, K127, 1980.
52. Kirchhof, J., *Cryst. Res. Technol.*, 20(5), 705, 1985.
53. Kirchhof, J., and Funke, A., *Cryst. Res. Technol.* 21(6), 763, 1986.
54. Kirchhof, J., Kleinert, P., Knappe, B., and Müller, H.R., *Proc. 5th Internat. School of Coherent Optics II.*, Jena, Germany, 1984, 7.
55. Kirchhof, J., Kleinert, P., Funke, A., and Müller, H.R., *Cryst. Res. Technol.*, 22(6), K105, 1987.
56. Kirchhof, J., Kleinert, P., Radloff, W., and Below, E., *Phys. Stat. Sol. (a)*, 101, 391, 1987.
57. Pospíšlová, M., Veselý, J., Choc, Z., and Havránek, V., Factors of the refractive index of SiO_2-P_2O_5-GeO_2 glasses fabricated by the MCVD method, (in Czech), *Čs. ŭcasopis pro fyziku*, 36, 43, 1986.
58. Choc, Z., Deposition Efficiencies and Rates in the MCVD Fabrication of Optical Fiber (in Czech), Ph.D. thesis, Inst. of Chemical Technology Praha, 1987.
59. Proft, B., Investigation of Systems Containing $SiCl_4$, $GeCl_4$, $POCl_3$ and BBr_3 (in Czech), M.Sc. thesis, Inst. of Chemical Technology, Praha, 1985.
60. Matějec, V. and Choc, Z., Physico-chemical model of the MCVD process (in Czech), *Proc. Optické Komunikace* 1986, Ed. C. Anderle, Praha, 1986, 54.
61. Choc, Z. and Götz, J., Study of basic parameters of the MCVD fabrication of silica optical fibers (in Czech), *Sklář a Keramik*, 39(3), 67, 1989.
62. Sedlář, M., and Matějec, V., Physico-chemical model of fabrication of silica optical fibers doped with fluorine (in Czech), *Proc. Optické Komunikace 1988*, Ed. C. Anderle, Praha, 1988, 131.
63. Doupovec, J. and Yarin, A.L., *J. Lightwave Technol.*, LT-9(6), 695, 1991.
64. Yarin, A.L., Bernát, V., Doupovec, J., and Mikloš, P., *J. Lightwave Technol.*, LT-11(2), 199, 1993.
65. Kašík, I., and Matějec, V., *J. Aerosol Sci.*, 26(3), 399, 1995.
66. Kašík, I., personal communication.
67. Ležal, D., and Doležal, J., Fabrication and chemical analysis of $SiCl_4$ (in Czech), *Proc. Optické Komunikace 1986*, Ed. C. Anderle, Praha, 1986, 42.
68. Götz, J., Choc, Z., Matějec, V., Kuncová, G., and Pospíšlova, M., Technology of graded-index optical fibers (in Czech), *Proc. Optické komunikace 1988*, Ed. C. Anderle, Praha, 1988, 99.
69. Kašík, I., Matějec, V., Pospíšlova, M., Kaňka, J., and Hora, J., *Proc. EOS Int. Conf. Photonics' 95*, Praha, to be published (1995).
70. Matějec, V., Sedlář, M., and Götz, J., *Proc. Int. Conf. Solid State Chemistry*, Pardubice, Czech Republic, 24, 1989.
71. Matějec, V., Sedlář, M., and Poláková S., Fabrication of silica optical fibers doped with Nd^{3+} or Er^{3+} (in Czech), *Proc. Optické Komunikace* 1990, Ed. C. Anderle, Praha, 1990, 73.
72. Matějec, V., Choc, Z., Sysala, O., Novotná N., Pospíšilová, M., and Götz, J., Optimization of the refractive-index profile of graded-index fibers employing the model of the MCVD process (in Czech), *Proc. Optické Komunikace 1988*, Ed. C. Anderle, Praha, 1988, 104.

73. Sysala, O., Kašík, I., and Spejtková, I., *Ceramics-Silikáty,* 35(4), 363, 1991.
74. Vichr, R., Karásek, M., Choc, Z., and Götz, J., *Ceramics-Silikáty,* 36(1), 7, 1992.
75. Sašek, L. and Sochor, V., Polarization single-mode optical fibres (in Czech), *Ceramics-Silikáty,* 34(3), 257, 1990.
76. Matějec, V., Sašek, L., Götz, J., Ivanov, G.A., Koreněva, N.A., and Grigor′yanc, V.V., *Proc. SPIE 1513 Glasses for Optoelectronics II,* Paper 19, 1991.
77. Grigor'yanc, V.V., and Guljajev, Ju.V., Optical-fiber communication systems (in Russian), *Proc. Problemy Sovremennoj Radiotechniki i Elektroniki,* Nauka, Moskva, USSR, 1980, 192.
78. Grigor′yanc, V.V., Zigunskaja, A.V., Ivanov, G.A., Koreneva, N.A., Čamarovskij, Ju.K., and Šemet, V.V., Influence of properties of substrate tubes on attenuation of optical fibers (in Russian), *Radiotechnika,* 37(4), 25, 1982.
79. Botvinkin, M.I. and Ivanov, G.A., Some aspects of fabrication of optical-fiber preforms by the MCVD method (in Russian), *Proc. Polučenije i analiz čistych věščest* Gorkij, USSR, 1984, 16.
80. Grigor′yanc, V.V., Ivanov, G.A., and Čamarovskij, Ju.K., Single-mode fibers (in Russian), *Itogi Nauki i Techniki,* I, 67, 1988.
81. Grigor′yanc, V.V., Zalogin, A.N., Ivanov, G.A., Isaev, V.A., Kozel, S.M., Listvin, V.N., and Čamarovskij, Ju.K., Polarization effects in optical fibers with elliptical cladding, *Kvantovaja Elektronika,* 13(10), 2080, 1986.
82. Grigor′yanc, V.V., Entov, V.M., Ivanov, G.A., Čamarovskij, Ju.K., and Yarin, A.L., Fabrication of preforms for optical fibers with non-circular core (in Russian), *Dokl. Akad. Nauk S.S.S.R.,* 305(4), 855, 1989.

Part IV

Aerosols and Buildings

14 Aerosol Particles Deposited on Building Stone

Anders G. Nord

CONTENTS

INTRODUCTION

It is well-known that industrialized society has created a manifold of air pollutants in the form of gaseous compounds, aerosol particles, and other particulate matter. They constitute a large problem, and their effect is clearly visible in polluted districts as dirty house facades and deteriorating building stone. Still worse is their harmful effect on vegetation, animals, and human beings. Fortunately, recent decades have witnessed a steadily growing interest in this important field, with the aim to quantify the pollutants and help suggest appropriate countermeasures.

The traditional way to penetrate the problem and analyze airborne particles is to collect the latter on some kind of filter at carefully controlled monitoring stations, followed by conscientious chemical and physical analyses. This chapter, however, describes how deposits on building stone have instead been used as the source for extensive studies. Thus far, more than 2000 samples of building stone have been collected in Sweden and other European countries and analyzed at the Conservation Institute of National Antiquities in Stockholm (cf. References 1–8). A few similar investigations have been carried out at other institutions. Although the material collected in this way is inhomogeneous in terms of composition, climate, age, variable atmospheric scavening, etc., the large number of samples constitute a valuable source for detailed investigations.

The chapter decribes sampling, relevant analytical techniques, particles found on building facades, and their possible origin. Selected results are summarized, and examples of damaging effects to stone materials are given. Finally, some concluding remarks and aspects on future developments are given.

PARTICULATE MATTER AND ORIGIN

Not even prehistoric ages offered a completely clean air free from pollutants. There has always been emission of mineral particles from eroded rocks, ash particles and sulfur compounds from volcanoes, soot and smoke from forest fires, marine salt particles, constituents from biological putrefaction, and biological microparticles like pollen grains, spores, etc. In ancient Rome, many

0-87371-829-1/00/$0.00+$.50

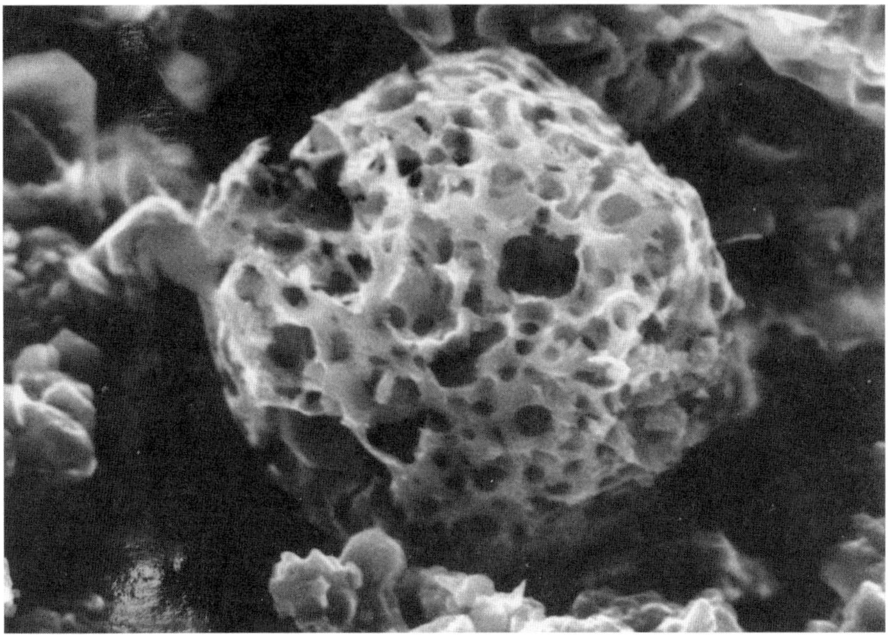

FIGURE 14.1 Eroded spheroidal particle found on the facade of the Kreuzkirche in Dresden, Germany. The particle is mainly composed of Ca, Al, Si, C, and O. Scanning electron microscope photograph at a magnification of 10,000X.

intellectual leaders complained of high concentrations of smoke and soot from house-heating, glassworks, and other industries.[9] In parts of England, air pollution problems have been documented since the 13th century. It was noticed early on how heating with coal and wood caused unaesthetic black layers on building stone, and possibly also affected human health.[9] Modern industrialization has, of course, made the situation a thousandfold worse. This section describes various aerosol particles found in polluted air today, and their main sources. To avoid confusion, it must be stressed that the gaseous pollutants create the greatest problems in terms of materials deterioration and danger to human health. Examples of such compounds are $SO_2/SO_3/H_2SO_4$, NO_x, HCl, organic solvents, organic acids, etc.

Among the solid pollutants, it is clear that soot particles for hundreds of years have been, and still are, a major problem. Soot is generated from incomplete combustion of carbon-rich fuels such as oil, coal, wood, peat, garbage, etc., and is produced in power plants and from house heating, industries, garbage combustion, car traffic, etc. Soot may not only be an outdoor problem; the continuous burning of wax candles in churches is known to have ruined many valuable textiles and mural paintings.

In addition to soot, the burning of fuel generates a number of other pollutants — poisonous gases as well as aerosols. Typical aerosols produced in this way are inorganic fly-ash particles from fossil fuel. They are spheroidal and about 0.01 mm in diameter. A characteristic particle is shown in Figure 14.1. The burning of fossil fuel also produces many organic constituents, including carcinogenic polycyclic aromatic hydrocarbons, usually abbreviated PAH.[10-13] Some details are given in the following sections. Many fuels, in particular oil, contain low amounts of metals like iron and vanadium;[14,15] when burned, their oxides are spread into the atmosphere as aerosol particles.

If we scrutinize car traffic, we see that it creates gaseous pollutants, unburned organic solvents, compounds of lead and other heavy metals, soot, and organic substances. However, the traffic is also responsible for the wear of car-tire rubber, asphalt pavements, brake shoes, engines, etc. Furthermore, soot, compounds of heavy metals, and other pollutants are emitted from industry.

Typical "aerosol metals" are iron, nickel, chromium, lead, copper, and zinc, along with many others. The former tradition in eastern Europe of locating steelworks or melting furnaces in the center of the town has indeed turned out to be disasterous.

Compounds rich in chlorine or phosphorus are emitted in large amounts from industry and from the burning of garbage; although less harmful, marine salt particles must not be forgotten. Modern agriculture and sewage systems create immense quantities of compounds rich in nitrogen, phosphorus, and sulfur. Various ammonium sulfates have been identified in the air along the Swedish west coast.[16] Natural erosion of rocks and emission from building industries are responsible for the spread of large amounts of particulate minerals like quartz, feldspar, calcite, gypsum, etc. Accordingly, it is obvious that a major part of the aerosol particles circulating in polluted areas can be attributed to the industrialized community.

ANALYTICAL TECHNIQUES

The analytical techniques reviewed here are basically those that have been used at the Conservation Institute of National Antiquities for quantification and surface analysis of small stone samples.[17] The samples most suitable for chemical analysis are, of course, pieces of stone taken from the surface or drilled cores; however, often it has only been possible to obtain a powdered sample scratched off the stone surface. Gaseous pollutants like sulfur dioxide may have penetrated deep into a porous stone, but such mechanisms are not the subject of the present chapter and have been descibed elsewhere.[2,5,8,17]

The study of aerosol deposits on a piece of stone had better start with the optical microscope to give a general survey of the sample. The next logical step is to apply a thin layer of conducting material (Au, Pd, C) in a sputtering device, followed by analysis with a scanning electron microscope equipped with a unit for energy-dispersive X-ray microanalysis (SEM/EDS). (An electron microprobe may give still better results.) A tiny particle (less than 0.01 mm) may thus be analyzed and photographed (see Figure 14.2). All elements except those with the lowest atomic numbers, somewhat different for various instruments, can be quantitatively determined. However, by means of a special or so-called window-less EDS detector crystal, the number of elements that can be detected — although not quantitatively determined — can be extended. Accordingly, particles of asphalt can often be identified from their carbon and sulfur contents. The characteristics of car-tire rubber are low amounts of sulfur and (usually) zinc in a matrix of carbon. Larger amounts of carbon on a stone surface can be quantitatively determined with a specific carbon analyzing instrument.

Metals and metal-rich particles are easy to identify and analyze. An iron particle is shown in Figure 14.3. It should also be emphasized that an "overall scan" of a stone surface sample in SEM/EDS gives a semiquantitative measure of the ubiquitous elements chlorine, sulfur, and phosphorus. Mineral particles, however, are seldom possible to distinguish from the stone surface, but are instead easily perceivable on other materials (e.g., on a bronze statue or a metal window- sill). Still better results are of course obtained from the study of particles collected on specially designed filters. A typical quartz particle in high magnification is shown in Figure 14.4.

To get a general survey of a sample composition, XRF (X-ray fluorescence), SIMS (secondary ion mass spectrometry), or PIXE (proton-induced X-ray emission analysis) may be used. The latter two methods, however, are sophisticated and expensive techniques. Neutron activation analysis (NAA) has been used in special cases to determine extremely low concentrations of the manifold of elements found on a stone surface (e.g., Reference 18). An analytical technique suitable for extremely thin surface layers is ESCA (electron spectroscopy for chemical analysis), also abbreviated XPS (X-ray photoelectron spectroscopy). In this technique, the sample is irradiated by monochromatic X-rays in ultra-high vacuum. The outer 50-Å layer gives a well-defined electron energy spectrum reflecting the elements present, and also gives a clue to their chemical bonds. A typical ESCA spectrum is shown in Figure 14.5.

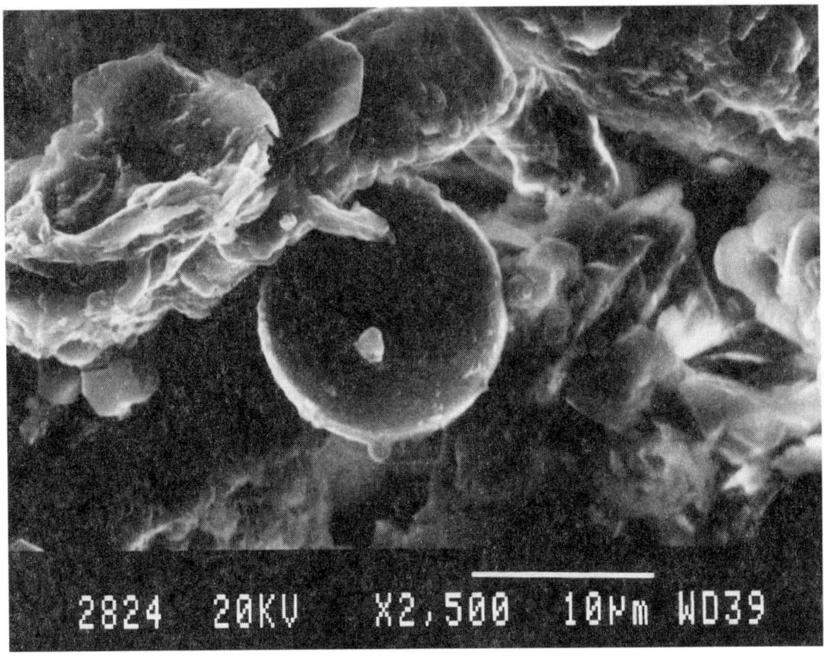

FIGURE 14.2 Non-eroded spherical particle formed from the burning of coal. It contains Si, Al, C, and O. (Sample is from the Torré Asinelli tower in Bologna, Italy.)

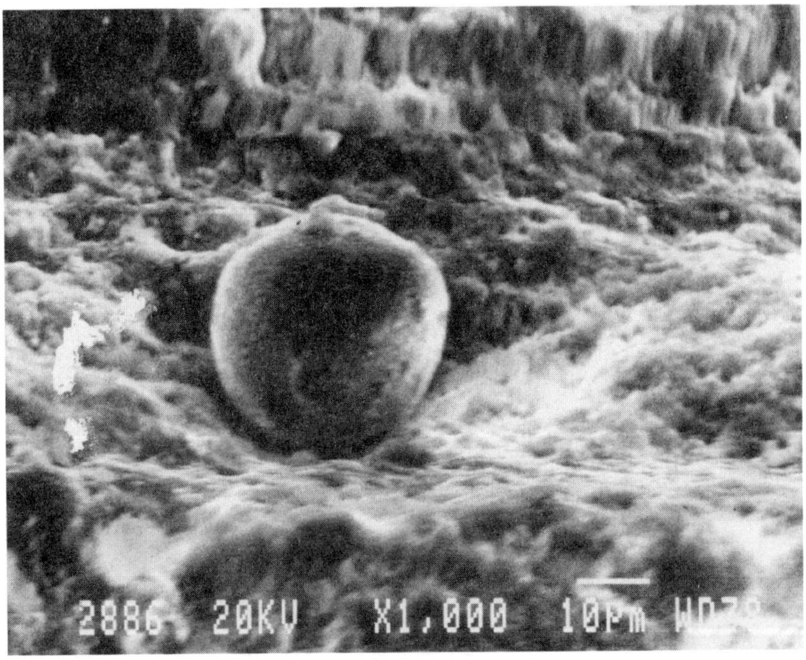

FIGURE 14.3 A spheroidal iron-rich particle observed on building stone near Huta Warszawa, Poland.

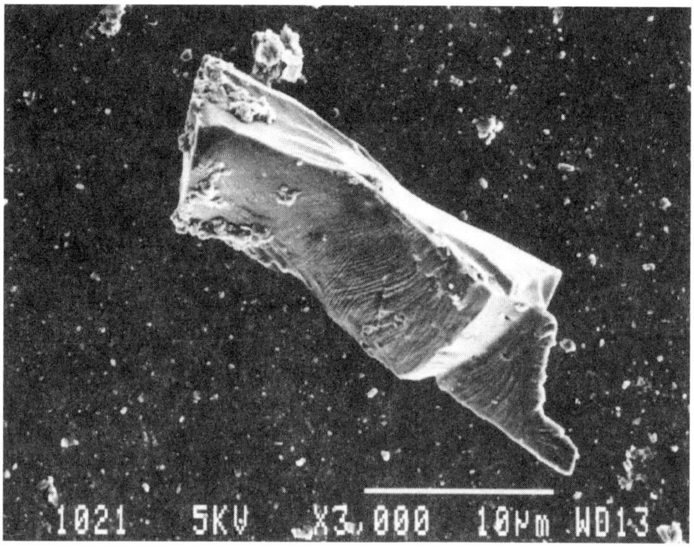

FIGURE 14.4 A typical quartz particle viewed in a scanning electron microscope at high magnification.

Heavy metals in low concentration can be determined from a water solution obtained by reacting the suface deposits with concentrated acids, followed by analysis with AAS (atomic absorption spectroscopy) or ICP (induced-coupled plasma spectroscopy). An iron-rich particle can also be identified by SEM/EDS. However, the iron compounds found on a stone surface are usually so badly crystallized that even if at least 1 mg of a component is available, X-ray diffraction techniques cannot be used for an exact identification. Mössbauer spectroscopy is instead a possible technique (see Reference 6).

Analysis of organic constituents creates a more difficult problem than for inorganic compounds. The existence of organic substances may be established by thermoanalytical methods like DTA or DSC, in which phase transitions and decomposition of the substances can be registered. With only one component or a limited number of compounds present, spectroscopic methods like FTIR (Fourier-transform infrared spectroscopy) or LRS (laser-Raman spectroscopy) can be used. However, due to interfering overlap of peaks, these "fingerprint" techniques are not suitable for the "mess" of organic and biological components often found on a contaminated stone surface. The best technique is to first separate and then analyze the mixture of organic substances by means of GC-MS. The sample is treated with a powerful organic solvent like cyclohexane to extract most organic components. These are separated in a gas chromatograph (GC) interfaced with a mass spectrometer (MS). The NBS/MS Peak Index database is preferably utilized for identification of the mass fragments. Another useful instrument for analysis in a similar way, based on separation and identification of dissolved components, is HPLC (high-performance liquid chromatography). The latter technique, however, is less accurate than GM-MS.

SOME RESULTS

In many city centers, the concentration of SO_2, NO_x, and soot particles are continuously being measured at stationary monitoring stations. Typical values for Sweden today are 10 to 20 µg soot per cubic meter in city centers, and values as low as <1 µg m^{-3} in countryside districts in the northern part of the country. Corresponding values in polluted cities in continental Europe are usually much higher. For example, values around 100 µg m^{-3} have not been uncommon in cities like London (see Reference 9). It is clear that the concentration of airborne soot particles is reflected

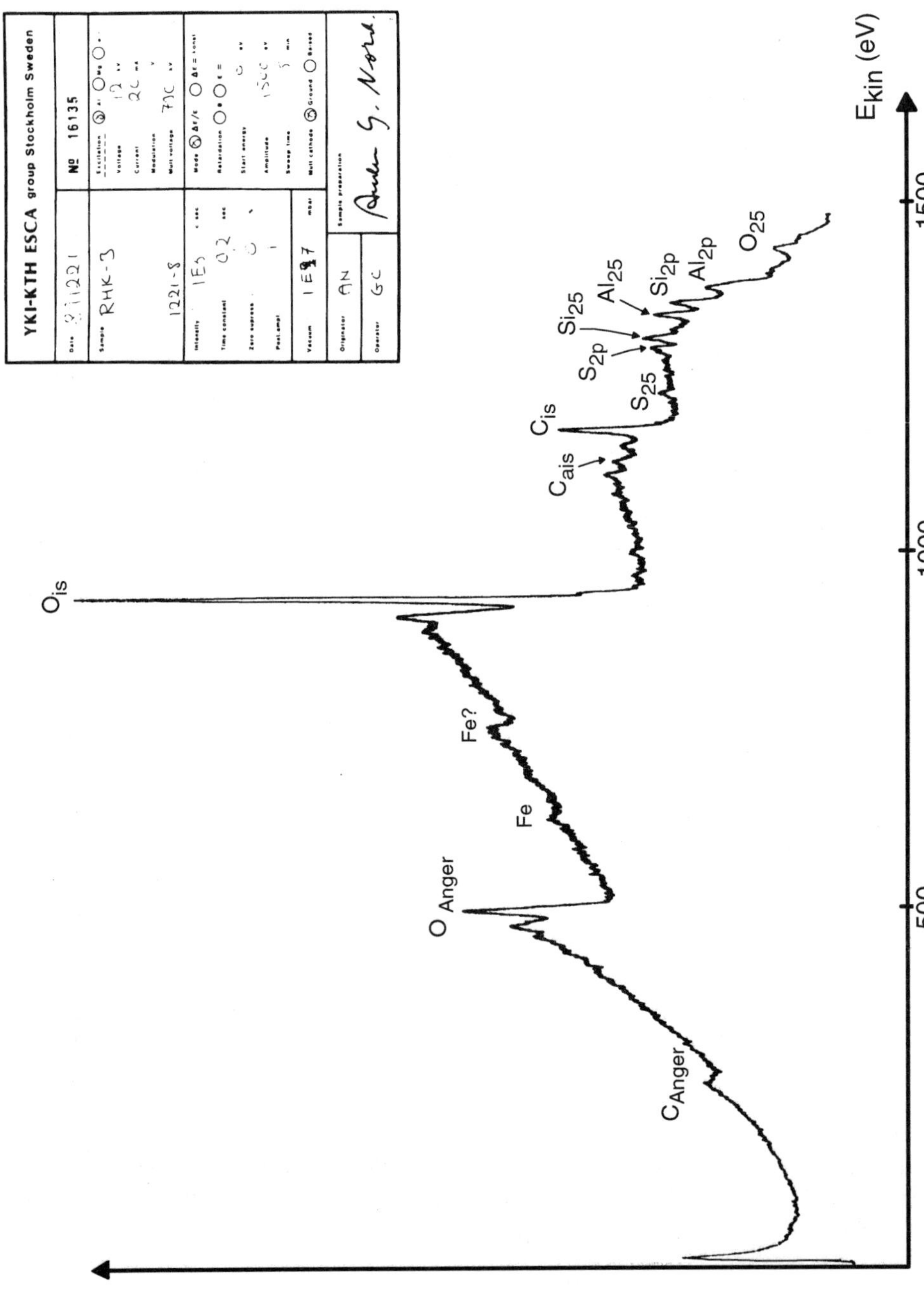

FIGURE 14.5 ESCA spectrum of a sample from the medieval Riddarholm church in central Stockholm, Sweden. (From: Nord, A.G. and Ericsson, T., *Studies in Conserv.*, 38, 25, 1993.)

in deposits on building stone. This was found early on by Gillberg et al. in Sweden.[19] In a more recent paper, Nord et al.[8] have analyzed 48 samples of non-calcitic stone for carbon (i.e., mainly soot) by means of a Carlo-Erba analyzing instrument. Twelve samples from the Swedish countryside contained 0.16 to 0.26 mg carbon per square centimeter (average: 0.20), to be compared with 0.88 to 2.74 mg cm^{-2} (average: 1.07) for 16 samples from Swedish city samples. The study in question also included the analysis of samples of building stone from cities like Copenhagen, London, Glasgow, Paris, Bordeaux, Dresden, Katowice, Krakow, and Budapest. Typical values for carbon (soot) were in the range 0.7 to 3.0 mg cm^{-2}. However, the samples from southwestern Poland were exceptional with significantly higher values, around 4 to 6 mg cm^{-2}, and with an observed maximum of 7.66 mg cm^{-2} (i.e., 77 g m^{-2}) (0.8 m above street level at Mariacka church, central Katowice). The latter value corresponds to a coffee cup full of soot at every square meter! The conspicuous amount of soot is certainly noticeable as unaesthetic black layers on many buildings in towns like Katowice, Chorzów, Zabrze, Bytom, Gliwice, etc.[3]

When a stone surface sample from a polluted area is viewed in a scanning electron microscope at a magnification around 2000 to 4000X, it is not unusual to observe small eroded particles, around 0.01 mm in size, rich in calcium, aluminium, silicon, oxygen, and (sometimes) carbon, cf. Figure 14.1. These are formed upon the burning of oil, and have been described by Del Monte et al.[20,21] (see also Wik and Renberg[22-24]). Burning of coal, on the other hand, mainly generates un-eroded, solid particles of similar size and chemical composition. Car traffic is responsible for emission of eroded particles, namely from diesel engines. The traffic also creates small particles of rubber (from car tires) and asphalt (from road pavements). Chemically, car tires are made of natural or synthetic rubber, vulcanized with 3 to 4 wt% sulfur using zinc oxide as a catalyst.

The wear of car tires is considerable. If every city, car tire is supposed to lose 1 kg in weight every year, the annual emission of rubber for 250,000 four-wheeled cars will amount to as much as $250{,}000 \times 4 \times 1$ kg = 1,000,000 kg = 1000 tons, an impressive figure indeed! The wear of asphalt is also imposing. Asphalt mainly consists of high-molecular aromatic hydrocarbons, with molecular weights in the region 1000 to 100,000 a.m.u. and a C:H ratio around 0.8 to 0.9. In addition, there are low (1 to 4 wt%) concentrations of sulfur, nitrogen, and oxygen, and mineral grains containing sodium, potassium, calcium, aluminium, silicon, oxygen, etc. Asphalt pavements are worn by traffic, especially during wintertime in northern European countries when many cars are equipped with studded tires. Dirty building facades in central Stockholm contain a total amount of around 5 g m^{-2} of rubber and asphalt particles, in exceptional cases up to 20 g m^{-2}.

Traffic, house-heating, power plants, etc. pollute the atmosphere with a number of other organic particles, which partly are deposited on building stone. (More volatile constituents like aliphatic hydrocarbons, benzene, and xylene from unburnt fuel are also emitted from car traffic in large amounts, but generally these are not adsorbed on building stone). In a study by Nord and Ericsson,[6] a sample was taken from the medieval Riddarholm church in central Stockholm, situated close to a motor-way with heavy traffic (100,000 vehicles per day). As many as 93 different organic compounds were identified by GC-MS. The predominant group, amounting to 0.5 to 10 µg g^{-1} sample, were linear and branched hydrocarbons (C_{12}–C_{30}), esters of fatty acids, and some alkenes (C_{11}–C_{18}). A second group consisted of alcohols, aldehydes, ketones, and phenol derivatives, usually <0.1 µg g^{-1} sample. A third group, including polycyclic aromatic hydrocarbons (PAH) in very low concentrations, was also observed. As many as 25 different PAH compounds, which have been found on city samples, are listed in Table 14.1. Many of these have been reported in non-selective deposits.[10,25,26] Also, 15 PAH-N compounds were found by Nord and Ericsson,[6] such as azafluorene, acridine, nitropyrene, nitroperylene, etc. It is indeed alarming that a carcinogeneous compound like benzopyrene (see Figure 14.6) has been adsorbed on the facades of European city buildings. It is noticeable that in a reference sample from countryside Sweden (Forshem, Västergötland, south Sweden), only 17 different organic compounds were found.

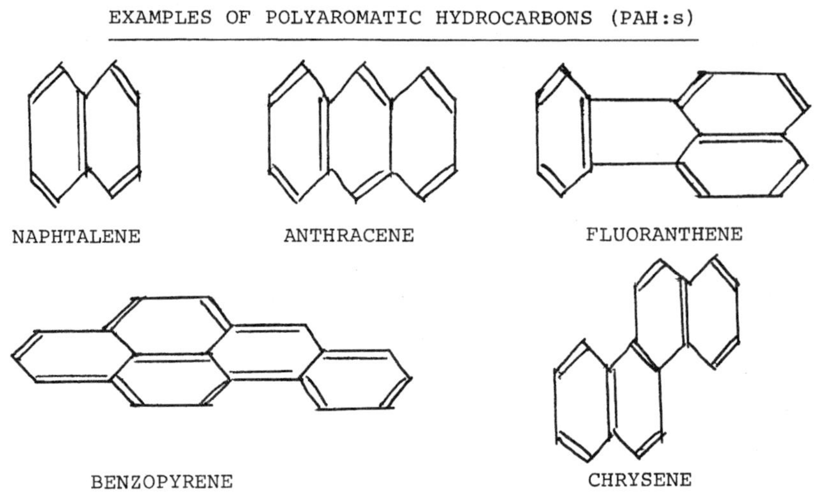

FIGURE 14.6 Structural formulas of some polycyclic aromatic hydrocarbons that have been found on the facades of European city buildings.

TABLE 14.1
Polycyclic Aromatic Hydrocarbons (PAH) Identified by GC-MS on Stone Samples from Stockholm, Katowice, and Brussels

Compound	M (a.m.u.)	Empirical Formula
Naphtalene	128	$C_{10}H_8$
Acenaphthylene	152	$C_{12}H_8$
Acenaphthene	154	$C_{12}H_{10}$
Fluorene	166	$C_{13}H_{10}$
Phenanthrene	178	$C_{14}H_{10}$
Anthracene	178	$C_{14}H_{10}$
Fluoranthene	202	$C_{16}H_{10}$
Pyrene	202	$C_{16}H_{10}$
Benzo[a]fluorene	216	$C_{17}H_{12}$
Benzo[b]fluorene	216	$C_{17}H_{12}$
Benzo[a]anthracene	228	$C_{18}H_{12}$
Chrysene	228	$C_{18}H_{12}$
Triphenylene	228	$C_{18}H_{12}$
Benzo[b]fluoranthene	252	$C_{20}H_{12}$
Benzo[j]fluoranthene	252	$C_{20}H_{12}$
Benzo[k]fluoranthene	252	$C_{20}H_{12}$
Benzo[a]pyrene	252	$C_{20}H_{12}$
Benzo[e]pyrene	252	$C_{20}H_{12}$
Perylene	252	$C_{20}H_{12}$
Indeno-[1,2,3,c,d]pyrene	276	$C_{22}H_{12}$
Benzo[g,h,i]-perylene	276	$C_{22}H_{12}$
Anthanthrene	276	$C_{22}H_{12}$
Dibenzo[a,c]anthracene	278	$C_{22}H_{14}$
Dibenzo[a,h]anthracene	278	$C_{22}H_{14}$
Coronene	300	$C_{24}H_{12}$

The organic constituents clearly indicate an origin from fossil fuels, mainly petroleum products (cf. References 10 to 12). Polyaromatic hydrocarbons are supposed to be products of incomplete combustion of gasoline or diesel fuel.[27] Results by Saiz-Jimenez[13] on the deposition of airborne organic pollutants on gypsum crusts confirm the above-mentioned results. The latter study includes stone samples from Mechelen (Belgium), Dublin (Ireland), and Sevilla (Spain).

It is well-known that small particles containing heavy metals are spread into the atmosphere from car traffic, industries, steelworks, etc. However, the combustion of fossil fuels significantly increases the number of various metal particles. For example, petroleum oil contains low amounts of a lot of heavy metals, for instance vanadium (~ 30 to 60 ppm) and iron (~ 1000 to 3000 ppm; see Reference 15). Accordingly, it can be estimated that the annual fuelling of 1,000,000 m^3 in a large city adds about 3000 tons of iron particles to the atmosphere. The traffic further increases the fuel-based metal emissions, but also produces substantial emissions from the wear of engines and in particular from the wear of brake shoes. The latter were previously made of asbestos; but for environmental reasons, they now contain iron powder in a polymer matrix. In a large city with heavy traffic, the annual wear of car brake shoes may amount to and even exceed 100 tons. We have analyzed heavy metals on building stone by means of SEM/EDS and ICP (cf. References 7 and 17). Iron is present in large amounts in dirty black layers, often covering the surface with about 10 to 20 g iron m^{-2}. However, part of this iron originates from weathered minerals in the stone, which have been dissolved by acid rain, whereupon iron has been enriched at the surface and precipitated there as (mainly) iron oxide hydroxides.[6,8] Therefore, the surface iron concentration does not unequivocally reflect the atmospheric concentration of iron aerosol particles. The iron compounds, in combination with soot and grime, create a thin black layer on the stone surface.

City building stone usually contain 0.01 to 0.03 mg cm^{-2} of vanadium; nickel and chromium around 0.01 to 0.10 mg cm^{-2}, and lead 0.05 to 0.20 mg cm^{-2}.[7] For countryside samples, the surface concentration of vanadium, nickel, and chromium seldom exceeds 0.003 mg cm^{-2}. The spread of lead into the atmosphere has decreased significantly during the last decade since the introduction of unleaded car fuel. Of course, industries and metallurgical melting furnaces may locally emit a lot of metal particles such as those mentioned above and, in addition, copper, zinc, manganese, cadmium, mercury, etc. They may be detected on building stone but arc more easily detected in soil or vegetation.[28]

In our analysis of building stone samples, two further elements have come out as ubiquitous: chlorine and phosphorus. In a study of 627 samples from all over Europe, the maximum surface concentration of chlorine was as large as 0.92 wt% (average for Venice, Italy). The concentration is substantial in all marine surroundings; but in cities far from the coast, the chlorine contents are usually enhanced due to emission from industries, burning of garbage, etc. Chlorine as observed on building stone is basically bound in inorganic chlorides. These are readily soluble in water and may thus be dissolved and rinsed away by rain. This fact suggests a constant additional contribution from the atmosphere. Many other salt minerals like chlorides, carbonates, sulfates, nitrates, etc. have been identified on building stone (e.g., see Reference 29).

The main origins of phosphorus in the environment is fertilizers, combustion of fuel and garbage, sanitary drainage, and industries. Some of this phosphorus is spread into the atmosphere, and partly deposited on building facades (typically) as inorganic phosphates. Average surface concentrations of phosphorus are in the range 0.2 to 0.4 wt%, with increased values noticed in agricultural districts as well as in large cities (0.4 to 0.8 wt%). Of course, many other inorganic particles circulate in the atmosphere. Among these are minerals like quartz, calcite, feldspar, gypsum, etc., originating from the natural erosion of rocks, building industries, traffic, and so forth. However, for obvious reasons, these are not easily detected on building stone but should rather be collected and analyzed on some kind of filter.

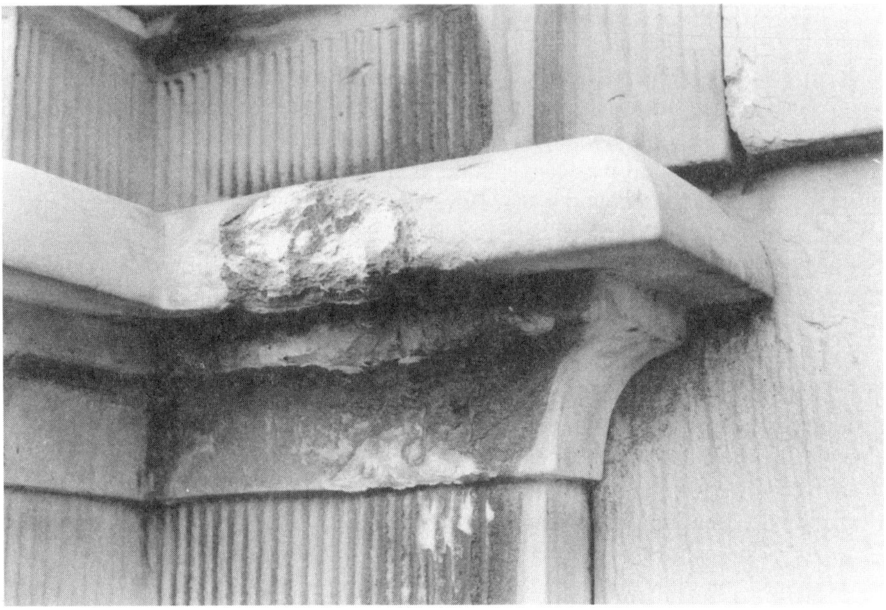

FIGURE 14.7 Soot and grime enriched at a rain-sheltered part of a sculptural decoration on The Royal Palace, central Stockholm.

FINAL REMARKS

From the data presented here, one can see that numerous different particles of aerosol type are continuously emitted into the atmosphere, in principle due to the modern community. The emitted particles are often harmful to people's health, and the ultimate aim of every society must be to reduce the emissions to an absolute minimum. The particles also have negative effects on many materials, such as stone. There is a positive correlation between air pollution level and the deposits observed on building stone. A clearly visible and striking effect of atmospheric aerosol pollutants is the occurrence of dirty house facades and dirty sculptural decorations in city centers and other polluted areas (see Figure 14.7). Grey or black layers are formed from soot and grime, iron oxide hydroxides, organic constituents, etc., sometimes, there are also remains of biological origin.

It must be emphasized, however, that aerosol particles usually play a minor role in stone-weathering, and that deterioration of stone is mainly caused by weather and wind in combination with gaseous pollutants like sulfur dioxide. However, it has been shown by many authors that soot and transition metal compounds may catalyze the oxidation of sulfur dioxide into sulfur trioxide and thus accelerate the decay of calcareous stone (see, e.g., References 21 and 30 to 34). In this way, aerosol particles can accelerate the stone weathering. An example of decaying calcitic sandstone is shown in Figure 14.8. It should finally be noted that although highly toxic substances like benzopyrene are found on building stone today, they do not affect the stone material in any way.

Much work has already been done in industrialized countries to reduce harmful emissions of gaseous and particulate pollutants. This has been accomplished by means of catalytic exhaust emission control, industrial emission filtering, more effective combustion techniques for fossil fuels, etc. Despite of these endeavors, there is an urgent need of further research in this field to minimize all kinds of emissions and damages. This work includes a number of scientific disciplines: chemistry, physics, biology, meteorology, toxicology, medicine, engineering, social science, etc. The investigation and analysis of aerosol particles is an important part of this work, which is likely to continue for several decades.

FIGURE 14.8 Part of the Royal Carolean Burial Chapel in central Stockholm, close to a motorway. The photograph shows seriously deteriorated calcitic sandstone.

REFERENCES

1. Nord, A.G., *Konserv. Tekn. Studier* Vol. 2, Central Board of National Antiquities, Stockholm, (in Swedish with English summary), 1990, 1.
2. Nord, A.G., *Geol. Fören. Stockholm Förhandl.*, 117, 43, 1995.
3. Nord, A.G. and Svärdh, A., *Internal Report from a journey to Poland*, Conservation Institute of National Antiquities, Stockholm, (in English), 1991, 1.
4. Nord, A.G. and Tronner, K., *Proc. 7th Int. Congr. Deter. Conserv. Stone*, Lisbon 1992, Vol. 1, 217, 1992.
5. Nord, A.G. and Tronner, K., *Water, Air and Soil Poll.*, 85, 2719, 1995.
6. Nord, A.G. and Ericsson, T., *Studies in Conserv.*, 38, 25, 1993.
7. Nord, A.G., Svärdh, A., and Tronner, K., *Atmos. Environ.*, 28, 2615, 1994.
8. Nord, A,G., Tronner, K., and Säfström, A., *Geol. Fören. Stockholm Förhandl.*, 116, 105, 1994.

9. Brimblecombe, P. and Rodhe, H., *Durabil. Build. Mater.*, 5, 291, 1988.
10. Yu, M.L. and Hites, R.A., *Anal. Chem.,* 53, 951, 1981.
11. Simoneit, B.R.T., *Int. J. Environ. Anal. Chem.*, 22, 203, 1985.
12. Simoneit, B.R.T., *Int. J. Environ. Anal. Chem.*, 23, 207, 1986.
13. Saiz-Jimenez, C., *Atmos. Environ.,* 27B, 77, 1993.
14. Leonardi, A., Burtscher, H., and Siegmann, H.C., *Atmos. Environ.,* 26A, 3287, 1992.
15. Sabbioni, C. and Zappia, G., *Atmos. Environ.,* 27A, 1331, 1993.
16. Brosset, C., Andreasson, K., and Ferm, M., *Atmos. Environ.*, 9, 631, 1975.
17. Nord, A.G. and Tronner, K., *Stone Weathering — Air Pollution Effects Evidenced by Chemical analysis. Konserv. Tekn. Studier* Vol. 4, Central Board of National Antiquities, Stockholm, in English, 1991, 1.
18. Tomza, U., *Environ. Prot. Engin., Wroclaw Technol. Univ.,* 12, 51, 1986.
19. Gillberg, G., Gezelius, L.H., Lööf, G., Ringblom, H., and Rosenquist, K., *Proc. RILEM/ASTM/CIB Sympos*. Espoo, Finland 1977, 144.
20. Del Monte, M., Sabbioni, C., and Vittori, O., *Atmos. Environ.*, 15, 642, 1981.
21. Del Monte, M., Sabbioni, C., and Vittori, O., *Sci. Total Environ.*, 36, 369, 1984.
22. Wik, M. and Renberg, I., *Water, Air and Soil. Poll.,* 33, 125, 1987.
23. Wik, M. and Renberg, I., *Phil. Trans. Royal. Soc. London B,* 327, 319, 1990.
24. Wik, M. and Renberg, I., *Hydrobiol.,* 214, 85, 1991.
25. Standley, L.J. and Simoneit, B.R.T., *Environ. Sci. Technol.*, 24, 163, 1987.
26. Brorström-Lundén, E. and Lövblad, G., *Atmos. Environ.*, 25A, 2251, 1991.
27. Boyer, K.W. and Laitinen, H.A., *Environ. Sci. Technol.,* 9, 457, 1975.
28. Pastuszka, J. and Hlawiczka, S., *Atmos. Environ.*, 27B, 59, 1993.
29. Nord, A.G., *Geol. Fören. Stockholm Förhandl.*, 114, 423, 1992.
30. Chang, S.G., Toosi, R., and Novakov, T., *Atmos. Environ.*, 15, 1287, 1981.
31. Fassina, V., *Durabil. Build. Mater.,* 5, 317, 1988.
32. Mangio, R., Dissertation, Dept. of Inorganic Chemistry, Univ. of Göteborg, Sweden, 1991.
33. Grgić, I., Hudnik, V., Bizjak, M., and Levec, J., *Atmos. Environ.,* 26A, 571, 1992.
34. Grgić, I., Hudnik, V., Bizjak, M., and Levec, J., *Atmos. Environ.,* 27A, 1409, 1993.

15 Effects of Aerosol on Modern and Ancient Building Materials

Giuseppe Zappia

CONTENTS

INTRODUCTION

The deposition of atmospheric pollutants (gas and aerosol) on the surfaces of monuments and buildings of historical interest exposed to today's urban environment constitutes one of the main damage factors affecting cultural heritage.[1,2] A knowledge of the mechanisms causing damage to building materials due to environmental factors is of fundamental importance for both the conservation of modern buildings and to guarantee correct methods of restoration on works of historic or artistic interest.

The presence of SO_2 in today's urban atmosphere is responsible for sulfation, the most common process of surface deterioration on carbonate building stones. Calcite ($CaCO_3$), the main component of such materials, is superficially transformed into gypsum.[3] The process is triggered by the ion SO_4^{2-}, which arrives at the material surface either in the form of SO_2 by means of dry deposition (followed by oxidation *in situ*), or as a component of wet deposition, originating from atmospheric aerosols (oxidation of the SO_2 in the atmosphere). The surface damage layer (black patina) incorporates all the components of atmospheric deposition,[4] with carbonaceous particles and heavy metal oxides — characteristic components of urban aerosol — playing a prominent role in the SO_2 oxidation process.

0-87371-829-1/00/$0.00+$.50

Numerous authors have dealt with the effects of carbonaceous particulate and heavy metals on the process of atmospheric SO_2 oxidation,[5-9] but the results obtained are varied, perhaps due to differences in the experimental set-ups involved and in the extrapolation of laboratory studies, based on differing model carbons, to atmospheric conditions; thus, to date, the problem remains open.

So far, there has been a proliferation of studies on the environment-related deterioration of the stone surfaces on ancient monuments and buildings,[10] in view of their artistic and historical importance. Far less numerous are studies on the effects of atmospheric pollutants on other construction materials, both traditional (air-setting mortars, bricks) and modern (hydraulic mortars, concretes).

ANCIENT AND MODERN BUILDING MATERIALS: HISTORICAL EXCURSUS

Although the use of building materials over the centuries has often been limited to local materials, due to the high cost of transportation and subordinate to the know-how and quality controls prevalent within different geographical areas, it is nonetheless possible to trace a parallel, often interconnecting development of technologies and types of materials in correspondence with the various phases human history.

Thus, we find in prehistoric eras rudimentary shelters built by mixing a wide and varied range of natural materials (soil, mud, straw, etc.) with water. The most ancient civilizations of the Mediterranean area and Near East already used rough bricks or stones laid one above the other with no binding materials and there is no shortage of remarkable examples of this type of construction work; for example, the fortifications of Cape Soprano (Gela, Sicily), built in still well-preserved rough bricks, and the dome-shaped rooms of Mycenae, where small wedges placed between the large stone slabs ensured the stability of the joints. However, it was not until the acquisition of the process of firing common raw materials (clay and stone) that the first great revolution in construction technology came about, enabling the fabrication of bricks, tiles, gypsum, and lime. In particular, the discovery of binders opened the way to the development of mortars, which have since become a fundamental component of all forms of construction work.

The technology of mortar-making reached its peak with the Romans who, with their careful preparation techniques and introduction of pozzolan sands, obtained hydraulic mortars with excellent mechanical properties and waterproof, allowing the construction of impressive, long-lasting works that have been preserved up to the present day. The Medieval decline that followed the fall of the Roman Empire saw the abandonment of the refined methods developed by the Romans and a generalized fall in the quality of building work that lasted up to the 12th century; only in the Renaissance did it return to a level comparable to that obtained in Roman times.

The milestones marking the subsequent phases of development up to the modern era were the discovery of hydraulic lime mortars, attributed to J. Smeaton (1756), and the invention of Portland cement during the mid-19th century, commonly attributed to J. Aspdin, although the previous works of Vicat (1812) and the subsequent ones of I.C. Johnson were more crucial in this regard.

Before moving on to discuss the environmental degradation of building materials, it is worthwhile noting some brief information on the most common types. Table 15.1 presents a classification of ancient and modern building materials.

STRUCTURAL MATERIALS

BRICKS

The drying and firing of clay in order to obtain construction materials has taken place in many areas of the world since time immemorial. Clay is a sedimentary rock mainly composed of aluminium silicate hydrates; its basic property is that of forming a plastic mass when mixed with water.

TABLE 15.1
Classification of the Main Ancient and Modern Building Materials

Structural materials	Bricks	
	Stones	Marbles
		Limestones
		Sandstones
Binders	Air-setting	Gypsum
		Lime
	Hydraulic	Pozzolan
		Hydraulic lime
		Cement
	Polymers	
Conglomerates	Mortars	Jointing mortars
		Rendering mortars
	Concretes	Ancient concretes
		Cement-based

The origin of bricks is thought to lie in Asia, in the regions west of the Euphrates, where their use underwent a long period of expansion on account of the scarcity of natural stones and timber for construction. Bricks were initially utilized in their raw form (i.e., by simply sun-drying the clay after modeling). It was soon realized, however, that their mechanical resistance and durability could be much improved upon by firing in kilns. Due to the wide availability and accessibility of the raw materials, as well as the ease with which artifacts of any shape could be obtained, the clay-firing technique spread rapidly and is still used today, with little modification over the centuries.

Bricks are now produced by firing clay at 950 to 1000°C. The clay minerals first lose their combined water and decompose, giving rise to the formation of mullite ($3Al_2O_3 \cdot 2SiO_2$); the silica or alumina in excess after the constitution of mullite are found in an amorphous state within the brick, alongside other impurities of the clay.

Stones

Stones have always been the main raw material for the construction of permanent works: houses, palaces, monuments, etc.; their basic property is therefore durability, understood as the capacity of a material to maintain its properties in time. Stones have been tied in with the entire history of building, as they are employed, either directly or after processing, in all construction components: as a structural or ornamental component in masonry, in the production of binders (lime, cement), and as aggregate in the production of mortars and concretes. Because of the variety of uses to which they can be put, virtually all types of stones are utilized; however, three types prevail over others: marbles, limestones, and sandstones.

Marbles are monomineralic metamorphic rocks, composed almost entirely of calcite ($CaCO_3$); they have saccharoidal granules and may be white, veined, or polychromatic. Known and admired since antiquity, marble has always been used for works of great prestige: the facades of palaces and churches, columns, capitals, friezes, and sculptures. The most greatly admired marbles are those of Greece and Italy in which the finest works of classic Greek art and Imperial Rome were realized. The finest Greek marbles are Parian and Pentelic (used in the building of the Parthenon), while Carrara marble is the most prized among the Italian ones.

Limestones are sedimentary rocks, mainly composed of $CaCO_3$ and variable quantities of other minerals, including clay minerals. The principal formation process is based on the action of microorganisms that are able to fix the $CaCO_3$, forming shells or skeletons, the accumulation of which gives rise to calcareous deposits. One of the finest limestones is Travertine, which the Romans extracted from the Tivoli quarries for the construction of many important buildings and monuments, including the Colosseum.

The term "sandstone" refers to a very numerous group of rocks composed of silicate granules of approximately one millimeter in diameter bound by a cement made up of CaCO3. The granules are mainly formed of quartz, feldspars, plagioclases, and other minerals.

BINDER MATERIALS

AIR-SETTING BINDERS

These inorganic substances, when mixed with water, form a plastic mass that has the property of setting and hardening in the air. The most important air-settig binders are gypsum and lime, little used today although widespread in the past.

Gypsum, $CaSO_4 \cdot 0.5H_2O$, is found in nature as $CaSO_4 \cdot 2H_2O$ in various crystalline forms (selenite, sericolite, alabaster, etc.) that on firing at 120 to 150°C transform into the hemihydrate form. Today, gypsum is used almost exclusively for plasters, stuccoes, and ornamental work, while in the past it was also employed as a jointing mortar. Because it is easy to produce, it was the first binder ever used in history; the Egyptians, for example, used it as a jointing mortar in the construction of the Cheops Pyramid (2500 B.C.).

The discovery of lime came about much later on account of its far more complex production process. Although the Egyptian already prepared rudimentary forms of lime, it was not until the Greeks and Romans that lime of high quality was achieved and used on a regular basis.

Lime is obtained by firing calcareous rocks with a clay content not exceeding 5% at 900 to 1000°C. This process yields CaO (quick lime) which, with the addition of water, forms $Ca(OH)_2$ (hydrated lime), a binder that sets and hardens in the air through the action of the CO_2 transforming $Ca(OH)_2$ into $CaCO_3$.

HYDRAULIC BINDERS

Unlike air-setting binding materials, hydraulic binders do not require the presence of air in order to set, but can harden even in water. The term "hydraulic" is commonly taken to refer not only to this property but also to all the other excellent properties of these materials: low porosity, water resistance, high mechanical strength, etc. The most important hydraulic binders are lime-pozzolan, hydraulic lime, and Portland cement.

Pozzolan is a sand that is not in itself a binder. However, on mixing with $Ca(OH)_2$, it forms insoluble compounds similar to those obtained with Portland cement. This effect is principally due to the presence in pozzolan of silica (SiO_2) and alumina (Al_2O_3), which, thanks to their amorphous, vitreous state and high specific surface area, react with lime and water to form calcium silicates and aluminate hydrates. Pozzolans can be either natural or artificial. Natural pozzolans are obtained by crushing volcanic tuff or can be found already in the form of sand or fossil flour. Italy has an abundance of natural pozzolan in the regions of Campania and Latium; other countries that produce pozzolan are Greece, Germany, and the United States. Artificial pozzolans, obtained in the past using crushed bricks and tiles or finely ground ceramics (cocciopesto), are today composed of fly-ash or silica fume.

Hydraulic lime is produced by firing a marly limestone with a clay content of about 15% at 1000 to 1100°C. Alongside CaO, this process also gives rise to bicalcium silicate and monocalcium aluminate, due to the presence of silica and alumina in the clay. Firing is followed by hydration of the CaO, using only the stoichiometrically necessary amount of water to avoid hydration of the silicate and aluminate that must take place when the binder is in use. Artificial hydraulic limes can be obtained by mixing, prior to firing, more or less pure limestones with the required quantity of clay, or by mixing Portland cement with fillers.

The hydraulic capacity of limes depends on the amount of clay present in the limestone and clay can be evaluated using the hydarulicity index (I) given by the relation

$$I = \frac{P_s + P_a + P_f}{P_c + P_m}, \tag{15.1}$$

where P_s, P_a, P_f, P_c, and P_m are the percentages in weight of silicon, aluminum, iron, calcium, and magnesium oxides, respectively.

Portland cement owes its name to the resemblance of the hardened cement to Portland stone and is the most important and widely used hydraulic binder. It is made by firing at 1450°C marly limestones with a clay content of 25% or by mixing limestone and clay so as to reach the said composition. The product obtained (clinker), composed of a mixture of bi- and tricalcium silicate, tricalcium aluminate, and tetracalcium ferrite aluminate, is then cooled and ground, with the addition of approximately 3% of gypsum in order to regulate setting; in this state, the cement is ready for sale. Subsequent phases of setting and hardening are characterized by the hydration reactions of the cement components.

Polymers

Natural polymers have been used since ancient times and, over recent decades, synthetic types have been increasingly utilized for both the construction of new buildings and the restoration of old ones. However, since the study of polymeric materials does not fall within the province of this work, they receive only brief mention here.

Some natural organic substances are the oldest examples of polymers used for construction: wood as a structural and ornamental element, waxes, and animal and vegetable fats as protective substances. Today, widespread use is made of synthetic resins, particularly as consolidants and protective coatings.

Organic consolidants consist of polymers that, when dissolved in suitable solvents and after evaporation of the solvent, form a continuous film that covers the walls of the pores of a material, welding together their crystalline grains and impeding water adsorption; consolidants have a good capacity of penetration and are flexible as well as waterproof. The problems that may arise during use are: a different dilatation coefficient to that of the material, a reduction of permeability to vapor, and a diminished durability. Among the consolidants most commonly used are epoxy resins, which also constitute the most suitable materials for use as adhesives. Since epoxy resins become fragile and yellow with exposure to atmospheric agents, in particular to ultraviolet rays, use must be limited to the deepest areas of cracks, while acrylic resins or fiber-glass-reinforced polyester resins should be used on surface areas.

The protection of materials is obtained by covering their external surfaces with the finest and most uniform film possible of a polymeric material that is waterproof and cannot be altered by the substances present in the environment or in the treated material. In general, the requirements for a protective coating are similar to those prescribed for consolidants, although waterproofing, transparency, chromatic invariability, and permeability to water vapor assume even greater importance. The last of these requisites is particularly essential: a polymeric film impermeable to water vapor will prevent the treated material from drying out naturally, should water accidentally penetrate inside it. The most commonly used coatings are acrylic and silicon resins.

Polymer deterioration can be brought about both by physical agents (e.g., heat, light, high-energy radiations, and mechanical stress), and by chemical agents (e.g., oxygen, ozone, acids, bases, water, etc.). Decay reactions affecting the polymeric chain are highly complex depolymerization reactions that inevitably lead to a break in the chain. The most frequent reactions are: radicalic depolymerization, thermo-oxidative, photoxidative, and chemical-mechanical degradation and bio-degradation.

CONGLOMERATES

Conglomerates are generally marketed in the form of powders and, to assume the plastic properties necessary for use, must be mixed with water. In order to minimize shrinkage and for greater economy, they undergo the addition of materials that do not participate in the hardening of the mixture, called aggregates (sand, gravel, crushed stones, etc.), of an appropriate granulometry. If the aggregate granules are of a diameter not exceeding 5 mm, the conglomerate is called mortar; otherwise, it is referred to as concrete.

MORTARS

Mortar is a conglomerate obtained by mixing a binder and sand in water; it is mainly used for the fixing of structural components (jointing mortar) and for plastering (rendering mortar). A mortar is defined according to the type of binder adopted for its composition, whose characteristics it assumes; thus, we have air-setting mortars when the binder is lime or gypsum and hydraulic mortars when the binder is lime and pozzolan, hydraulic lime, or cement.

In the past, mortars were used only after the acquisition of the technological know-how required for the firing of natural raw materials. The most remote examples were among the Egyptians, who as early as 2500 B.C. made use of gypsum mortar in which lime impurities have been found. The deliberate utilization of air-setting lime is well-documented[11] on the island of Crete (2300 B.C.), while the Greeks and Romans also knew and extensively employed hydraulic mortars.

With regard to the latter, since ancient times, different materials have been added to lime in order to obtain hydraulic mortars. It is known that as early as the 10th century B.C., the Phoenicians and Israelites were familiar with the techniques of producing hydraulic mortars for the protection of all their hydraulic works (aqueducts, ports, water tanks, etc.), where washing used to cause the rapid decay of ordinary mortars. The drinking-water reservoirs that King Solomon commissioned in Jerusalem were protected by hydraulic mortar obtained by mixing lime and crushed ceramics. The Greeks employed pozzolanic sand obtained by adding volcanic ash from the island of Thera, today's Santorini. However, it was the Romans who were the first to fully understand the importance of pozzolan and utilized it regularly in the preparation of hydraulic mortars. They discovered that the use of sand of volcanic origin (of the type present near Pozzuoli) to substitute ordinary sand in lime mortar, caused it to become hydraulic. Thus, the term "pozzolanic" is used to refer to a type of sand able to transform lime mortar into a hydraulic mortar, although the binder used is itself air-setting. In more recent times, the Dutch were renowned for their hydraulic works for which a mixture of lime and trass were used. Trass is a volcanic tuff with properties similar to pozzolan, imported from Andernach, on the Rhine border, near Koblenz in Germany.

The next milestone in the development of mortars was the invention of hydraulic lime, a special type of lime that, independent of the presence of pozzolan, has the ability to harden under water. This did not take place until the 16th century and is attributed to the Italian architect, Andrea Palladio.[12]

Smeaton was to reach a fuller understanding of hydraulic reactions in 1756, while attempting to make a water-resistant lime. From the chemical analysis of the limestone used for the production of natural hydraulic lime, he found that the presence of clay in limestones is the decisive factor of hydraulicity. Hydraulic lime represents the link between lime and Portland cement discovered in the mid-19th century. The use of cement to prepare hydraulic mortars spread rapidly toward the end of the 19th century to assume the position of absolute predominance that it still occupies today.

Concretes

The technique of building masonry by mixing crushed stones and bricks with lime, sand, and water was known and used by the Romans. Vitruvius (*De Architectura*) describes the preparation and use of concrete (*Opus Caementitium*) adopting lime as a binder and there is no lack of extraordinary works still preserved today that were built with this technique; for example the Appian Aqueduct and the dome of the Pantheon in Rome. In the Medieval period, concrete was used almost solely as a filling between external hancings in bricks and stones, which functioned as permanent formworks. However, it was the advent of cement that gave rise to the widescale expansion of this building technique, lasting up to the present day where most cement is produced for the manufacture of concretes.

Modern concrete is a conglomerate made up of water, cement as binder, and sand and gravel as aggregates. To improve its mechanical properties, concrete is reinforced with steel bars, a combination exempt from any problems of a physical or chemical nature; in fact, steel adheres well to concrete, the thermal dilatation coefficients of the two materials are more or less the same, ensuring their adherence even with temperature variations, and, finally, the base environment set up in the concrete after the hydration reactions of the cement protects the steel from corrosion.

ENVIRONMENT-RELATED DETERIORATION OF BUILDING MATERIALS: STATE OF THE ART

The main damage product resulting from the interaction between today's atmosphere and building materials is gypsum.[13] The problems arising from gypsum formation depend largely on the situation in which it occurs: on the one hand, due to its greater solubility compared to the original compounds of the materials, once gypsum forms on a surface, it is easily washed off from artifacts that are exposed to rainfall. On the other hand, the reaction of gypsum formation leads to the growth of black surface patinas on materials with low porosity (marbles and limestones) or occurs up to a depth of approximately 1 cm in those with high porosity (sandstones and mortars).

The black patinas can be considered as the areas where the products of material deterioration and the deposition of atmospheric gas and aerosol accumulate. The color is ascribed to the presence of carbonaceous atmospheric particles, mainly soots, that are embedded within the crust during its formation.[14] Soots are carbonaceous particles produced by fossil fuel, oil, and coal combustion, including automobile exhaust fumes, and their carbonaceous matrix is composed of elemental and organic carbon.[15] Their heavy metal content (Fe, V, Ni) and morphology (high specific surface) have a catalytic effect on atmospheric SO_2 oxidation[5] and likely on the environmental sulfation of calcium carbonate.

Over recent decades, as part of the effort to ensure a more efficient protection of the architectural heritage, numerous works have appeared in the literature concerning the effects of SO_2 on carbonate stones.[16,17] However, studies dealing with the impact of SO_2 on mortars[18,19] and of aerosols on stones[4,20] reamain scarce, while works on the role of atmospheric aerosols in the deterioration of other building materials are entirely lacking.

The occurrence of gypsum formation on masonry is particularly dangerous in the case of cement mortars, concretes, and hydraulic binders in general, because two seriously damaging expansive reactions tend to take place in the presence of gypsum ($Ca\ SO_4 \cdot 2H_2O$), leading to the formation of ettringite and thaumasite[21,22]:

$$3(CaSO_4 \cdot 2H_2O) + 3CaO \cdot Al_2O_3 \cdot 6H_2O + 20H_2O \rightarrow \underset{\text{Ettringite}}{3CaO \cdot Al_2O_3 \cdot 3CaSO_4 \cdot 32H_2O} \quad (15.2)$$

$$CaSO_4 \cdot 2H_2O + CaCO_3 + CaSiO_3 \cdot H_2O + 12H_2O \rightarrow \underset{\text{Thaumasite}}{CaSiO_3 \cdot CaSO_4 \cdot CaCO_3 \cdot 15H_2O} \quad (15.3)$$

Ettringite is produced during the early hours of the hydration process and the reaction generally involves all the sulfate present in the cement; in this case, the process causes no damage as the mortar is in a plastic state during setting. Subsequently, however, if new sulfate interacts with the calcium aluminate hydrates in the binder paste, the formation of new ettringite, referred to as secondary ettringite, takes place.[23] This highly expansive reaction gives rise to severe stress within the pores of the cement structure, with spalls and cracks that can lead to the total destruction of the material.

To date, the role of sulfate-rich and sea waters in the formation of secondary ettringite in cement-based mortars is known. However, secondary ettringite formation due to environmental SO_2 attack has yet to be studied and consequently no knowledge is available on the parameters governing this process. In the case of blended cements (with natural or artificial pozzolan addition) and traditional pozzolanic binders (lime-pozzolan mortars), the relationship between the formation of ettringite from the pozzolan $A1_2O_3$ and the expansive capacity of the reaction also remains unknown.

Even less information is available on the mechanisms and kinetics of thaumasite formation. Although thaumasite was first observed as early as 1965 on damaged concretes and in repair mortars used for the conservation of the architectural heritage, so far no correct explanation has been provided for its formation process. However, it has been shown that thaumasite formation can be produced at low temperatures (2 to 5°C) when gypsum and calcium carbonate interact with CaO and SiO_2 or Ca_2SiO_4 in the presence of an excess of water.[24] Preventing the formation of ettringite and thaumasite in buildings exposed to the joint action of SO_2 and CO_2 from the polluted atmosphere requires a detailed knowledge of the thermodynamic parameters controlling the formation and stability of such compounds.

DAMAGE ON HISTORIC BUILDINGS AND MONUMENTS

With the aim of studying the environment-related damage on historic monuments and buildings, samples of black alteration patinas were collected from the most common building materials: stones (marbles, limestones, and sandstones), bricks, and mortars. Only black patina samples were selected, excluding other types of deterioration, as the crusts constitute the areas of maximum accumulation of alteration products and environmental deposition.

The samples were collected in three large cities (Rome, Milan, and Bologna) and in four maritime sites of central northern Italy (Venice, Ravenna, La Spezia, and Ancona). In the laboratory, they were dried, ground, and preserved at a temperature of 20°C in an inert environment (N_2), after which they underwent the following analytical procedures. X-ray diffractometry (XRD; Philips PW 1730) and infrared spectroscopy (FTIR; Nicolet 20 SX) were used to identify the main chemical species. The gypsum and carbonate contents of the samples were quantified by differential thermal analysis (DTA) and thermal gravimetric analysis (TGA) (Netzsch Simultane Thermoanalyze STM 429 apparatus). Carbon and sulfur were measured by combustion and IR techniques (Carbon-Sulfur Determinator LECO CS44). A specific methodology[14] was adopted for the quantification of non-carbonate carbon (C_{nc}). Finally, anions and cations were analyzed by ion chromatography (IC), using a Dionex 4500I ion chromatograph.

Figure 15.1 shows the X-ray diffractogram of a black patina removed from an ancient lime mortar, and Table 15.2 lists the mean concentrations for the main constituents of the black patinas, averaged for the single materials and site typologies. The data confirm that, as widely reported in the literature for marbles and limestones, the main damage mechanism affecting all components of masonry is the superficial transformation of $CaCO_3$ in the underlying support into gypsum. The percentage of gypsum in the brick patinas is similar to that of marbles and limestones, while the one reported for lime mortars resembles that found for sandstones.[4] It therefore appears that the degree of sulfation is more greatly influenced by the microstructure of the material than its calcium carbonate content. Finally, the degree of sulfation in marbles and limestones from the large urban centers turned out to be 18% greater than that found for the maritime sites.

The analysis of cations showed, in order of abundance after calcium, Fe, K, Al, Na, Mg, and small quantities of Sr, Mn, and Ba. The most abundant anions, after sulfates, were chlorides, nitrates, oxalates, fluorides, phosphates, and traces of bromides. In all the samples analyzed, appreciable amounts of C_{nc} were found, ranging between 0.6% and 1.8%. From these results, it would appear that the direct correlation between the contents of C_{nc} and sulfate reported[14] in the literature for black patinas on marbles and limestones can be extended also to the patinas found on other building materials.

LABORATORY EXPOSURE TESTS

A series of laboratory exposure tests in controlled atmosphere was performed on building stones (marbles and limestones), mortars, and a brick. The mortars were prepared in order to reproduce the composition adopted both in antiquity and in modern mortars, the latter being used not only in contemporary building projects but also in the conservation of historic buildings.

Three types of mortars were prepared: (1) a traditional air-setting mortar, composed of lime and sand, 1:3 ratio; (2) a traditional hydraulic mortar composed of lime, volcanic pozzolan, and sand in the ratio 1:1:6; and (3) a modern hydraulic mortar composed of cement and sand in a 1:3 ratio (all ratios are expressed in weight). The constituents used in mortar preparation were: powder of hydrated lime, natural pozzolan (from Segni, Italy), high-strength Portland cement, and siliceous sand. The fresh mortars were poured onto a glass plate and molded by hand into a 5-mm thickness. During setting, once a suitable consistency was reached, the fresh mortar was cut into sections measuring 10 × 10 × 5 mm and curing continued at environmental temperature and R.H. for 28 days. The samples then underwent chemical-physical characterization in order to determine porosity, measured by a mercury porosimeter (Carlo Erba); specific surface, measured by nitrogen adsorption (Quantachrome-Autosorb 1); and the concentration of sulfate ion by IC (Table 15.3). In order to determine their heavy metal content, elemental analyses of the lime, pozzolan, and cement were carried out by inductively coupled plasma spectroscopy (ICPS; Perkin Elmer 5500) through the digestion of samples in teflon vessels with an $HF\text{-}NO_3$ mixture at 120°C. The iron content turned out to be 0.5% for lime, 6% for pozzolan, and 2% for cement.

A part of the mortar and stone samples was used blank, while another part was utilized to prepare specimens of each material for coating with 50 μg of one of the following powders: iron oxide, activated carbon, or carbonaceous particles (soots). In order to ensure that the particles were

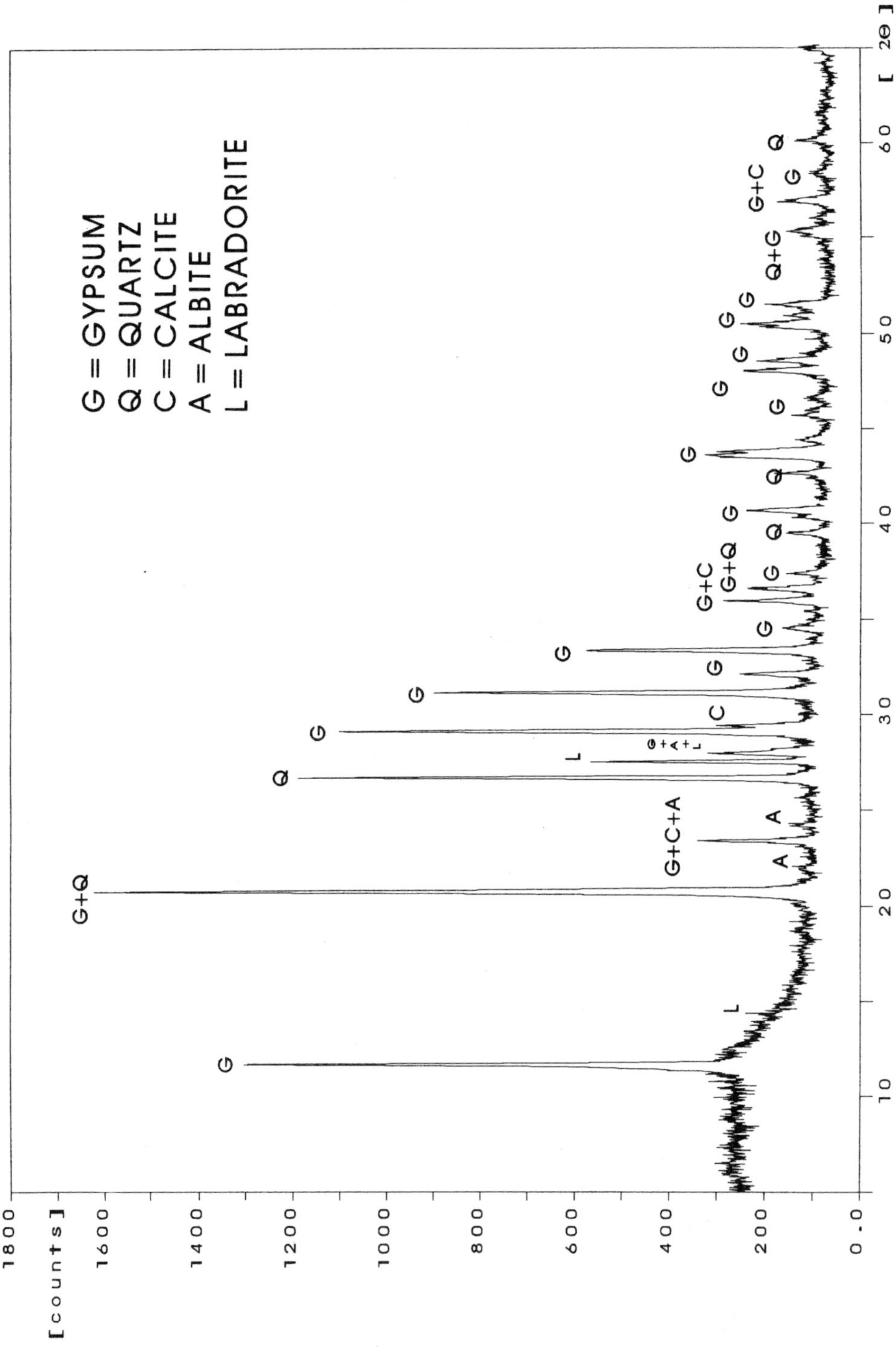

FIGURE 15.1 X-ray diffractogram of a black patina removed from an ancient lime mortar.

TABLE 15.2
Mean Concentrations of Gypsum, Carbonates, and Non-Carbonate Carbon (C_{nc}) in Black Patinas on Stones of Low Porosity (Marbles and Limestones), Stones of High Porosity (Sandstones), Bricks, and Lime Mortars

Black Patinas	Gypsum (%)	Carbonates (%)	C_{nc} (%)
Marbles and limestones			
Maritime sites	63.5	15.3	1.8
Urban sites	75.0	9.4	1.8
Mean	69.2	12.3	1.8
Sandstones (Bologna)	42.5	31.1	0.6
Bricks (Bologna)	77.2	Absent	1.0
Lime mortars (Bologna)	48.0	1.5	0.9

TABLE 15.3
Chemical-physical Parameters of the Materials Studied Prior to Exposure in the Simulation Chamber

Samples	Total Porosity (%)	Average Pore Radius (Å)	Specific Surface ($m^2\ g^{-1}$)	SO_4^{2-} ($\mu g\ g^{-1}$)
Stones (mean)	4.0	430	0.7	52
Bricks	29.1	78	22.3	56
Lime mortar	32.7	233	1.4	62
Pozzolan mortar	26.1	92	4.3	86
Cement-based mortar	13.4	94	3.0	854

similar to those actually found on Italian monuments, particles emitted by an Italian electricity generating plant were used. The carbonaceous particles were characterized to determine specific surface (2.8 $m^2\ g^{-1}$) and bulk elemental composition (CHNSO Analyzer Carlo Erba EA1108 and ICPS). The morphology and elemental composition of the single particles were analyzed with a scanning electronic microscope interfaced with an X-ray energy dispersion analyzer (SEM-EDAX); the particles revealed a characteristic spherical morphology with S, Fe, C, V, Mg, and Ni as the main elements.

For laboratory exposure tests on the various materials, a flow chamber was developed to operate at atmospheric pressure, with instrumentation for the control of the main physical and chemical parameters (temperature, relative humidity, composition, and gas flow velocity) (Figure 15.2). A series of experiments was carried out in the chamber containing filtered air at constant atmospheric pressure, 25°C temperature, 95% relative humidity, 0.5 l min^{-1} gas flow velocity, and 3 ppm SO_2 concentrations, for 150 days. Each of the materials was exposed blank and with three different powders scattered onto the specimens surfaces. The particle distribution on each sample was checked by optical microscopy in order to verify a homogeneous surface distribution and a surface concentration sufficiently low so as not to limit SO_2–material interaction. The amount of particles deposited onto the specimen surface was 50 μg. On the exposed samples, SEM-EDAX analyses were carried out (Figure 15.3). The quantitative determination of SO_4^{2-} and SO_3^{2-} ions was performed by IC; the state of hydration of the sulfite and sulfate was determined by XRD and FTIR.

The chemical species forming on the surfaces under study following interaction with SO_2 were in all cases $CaSO_3 \cdot 0.5H_2O$ and $CaSO_4 \cdot 2H_2O$ in varying quantities; thus, surface deterioration can be quantified in terms of the total sulfur that has reacted to form the two salts:

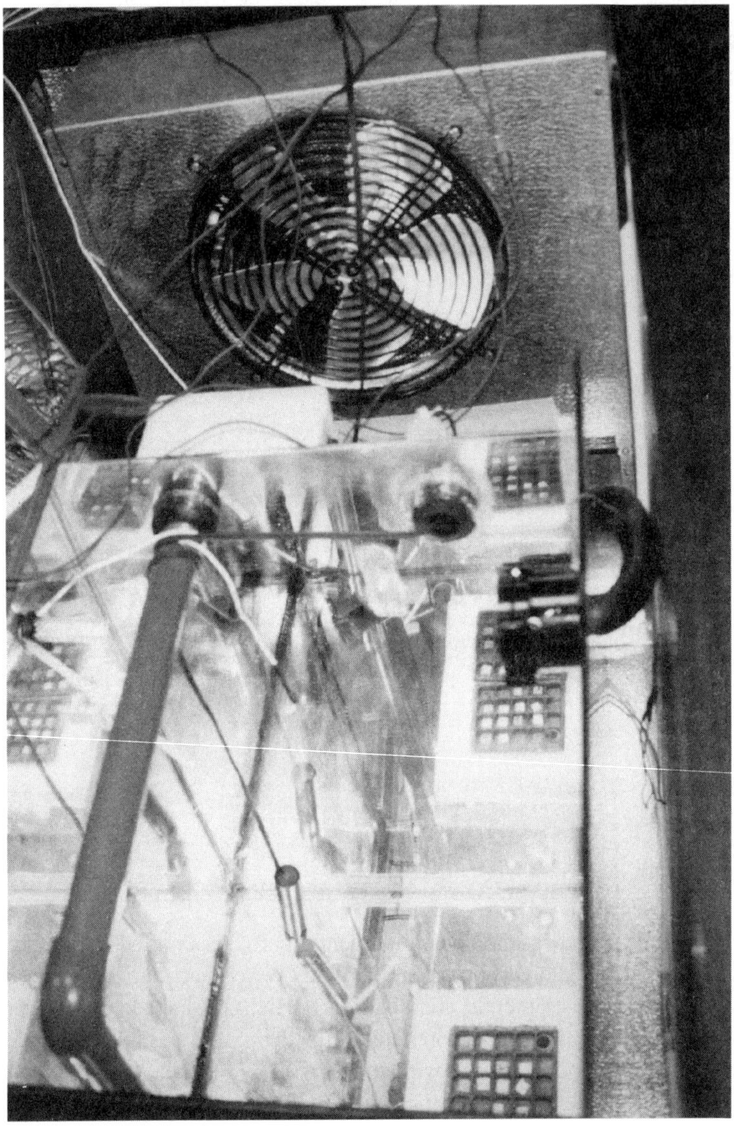

FIGURE 15.2 Interior of the simulation chamber with samples of the materials studied.

$$S_{Tot} = S_{SO_3^{2-}} + S_{SO_4^{2-}} \tag{15.4}$$

The results of XRD and FTIR agree that sulfite is present in the form of calcium sulfite hemihydrate ($CaSO_3 \cdot 0.5H_2O$), while sulfate crystallizes as the dihydrate ($CaSO_4 \cdot 2H_2O$). The values of S_{Tot}, after subtracting the blank ones, after 150 days of exposure (Figure 15.4) indicate the lower reactivity with SO_2in the bricks and stones, compared to mortars. Such reactivity is not found to be in correlation with either the $CaCO_3$ content or the physical properties (porosity, specific surface). The said parameters do influence the process which is conditioned by all these variables. Particularly surprising is the high reactivity of the pozzolan and cement mortars, which have a lower content of calcium carbonate than all of the other materials and a smaller specific surface than brick and lime mortar. This high reactivity must be attributed to the catalytic effect of the iron present in both the cement and pozzolan in considerable quantities.

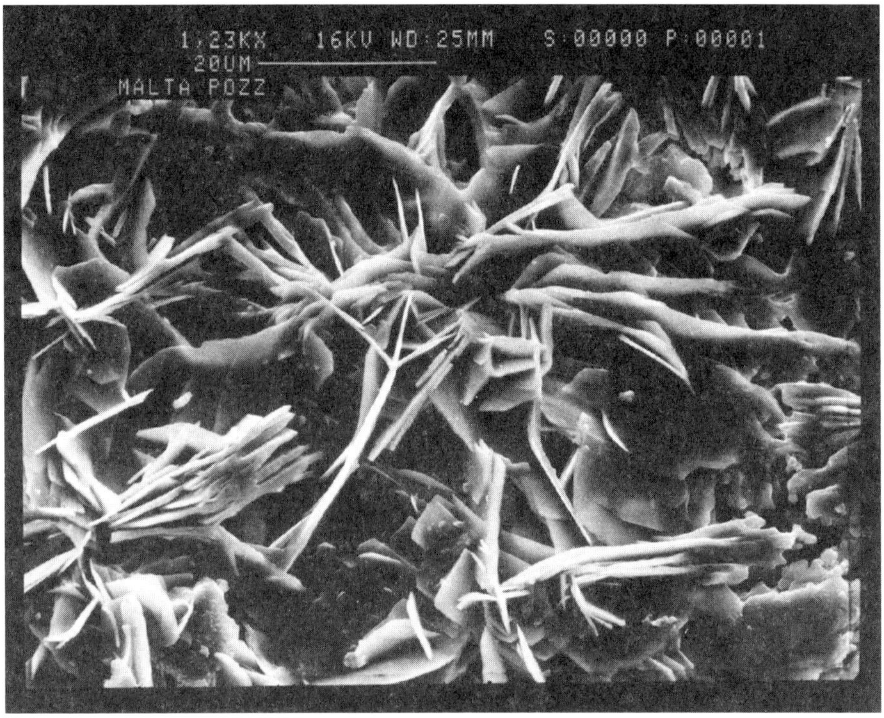

A

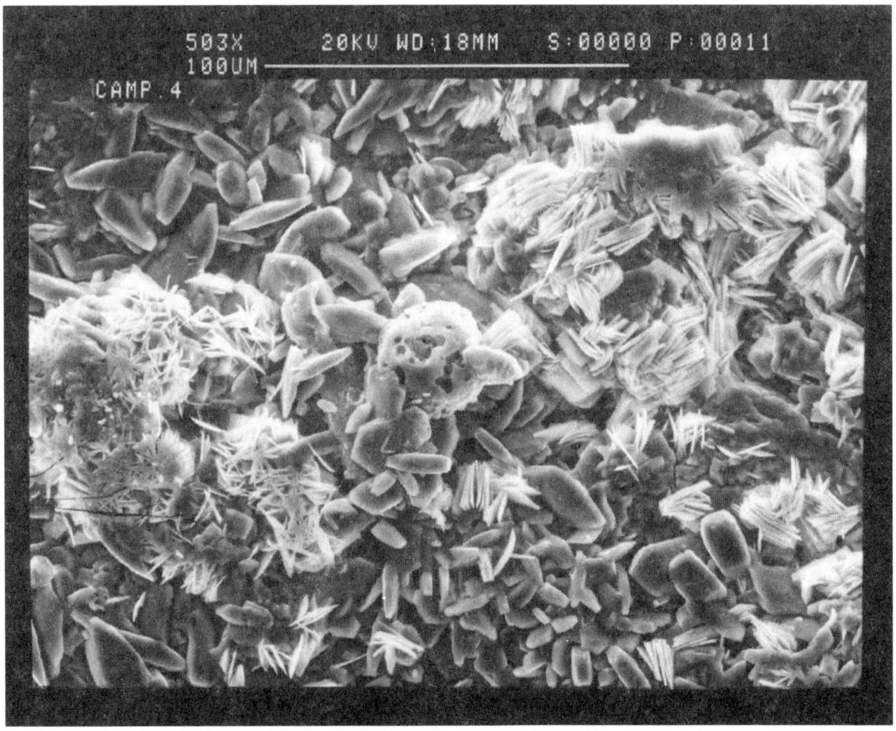

B

FIGURE 15.3 SEM micrographs of laboratory-exposed samples at 3ppm SO_2 for 60 days: (A) blank pozzolan mortar, and (B) cement mortar with carbonaceous particles (soots). Gypsum crystals are evident.

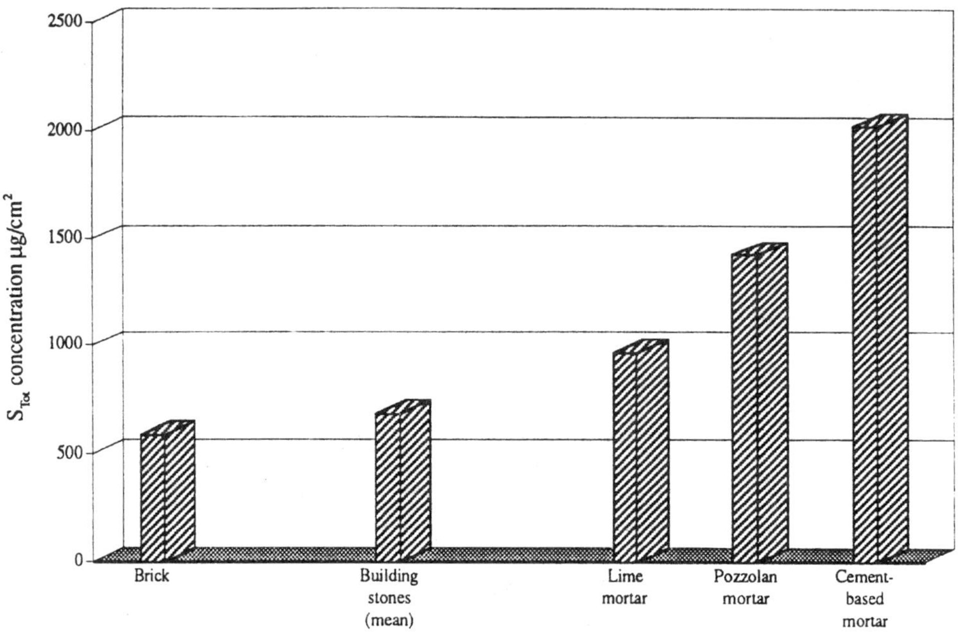

FIGURE 15.4 S_{Tot} concentrations on all materials studied after 150 days of exposure in the air at 3ppm SO_2.

Figure 15.5 shows the sulfate concentrations, blank and particle-coated, for stones and mortars studied after 150 days of exposure. The results show that the activated carbon had little influence on the formation of SO_4^{2-}, while both soots and iron oxide enhanced sulfate formation; moreover, the effect of iron oxide turns out to be far more efficient than that of soots. The results also indicate that soots play a catalytic role in the sulfation process due to their content of heavy metals, particularly iron, while the specific surface has only a very slight influence.

Concerning the possibility of the formation of ettringite and/or thaumasite in the hydraulic mortars studied, the presence of these salts on the exposed samples was not revealed by XRD analysis. Although the high R.H. and alkaline surfaces fall within the range required for the formation of the two salts, the acidity of the atmosphere within the chamber affects their stability.[25] Moreover, for the formation of the two salts, optimum temperatures of 2 to 5°C (and never higher that 20°C) are reported in the literature.[26] The verification of this possibility therefore requires different experimental conditions. Experiments for the study of such processes are presently in progress in our laboratories.

FIELD EXPOSURE TESTS

Samples of the same materials studied in the laboratory (mortars and building stones) were exposed for 12 months in the historical center of Milan, as an example of a large industrial city, and near the port of Ancona, as an example of a maritime site. The samples were mounted on special supports in such as way as to simulate (a) areas exposed to horizontal and vertical rain wash-out and (b) areas partially sheltered from rain wetting (Figure 15.6). After 12 months of exposure, the specimens were analyzed by IC in order to quantify the anions present. Analyses of some of the samples by SEM-EDAX were also performed. The micrographs reveal the presence of gypsum and carbonaceous particles on the surface of all exposed samples.

The results of chemical analyses show that the most abundant species is sulfate ion followed by nitrate, sulfite, chloride, oxalate, and fluoride, while small quantities of nitrite and phosphate were also present. Horizontally and vertically exposed areas (samples a) were more washed out

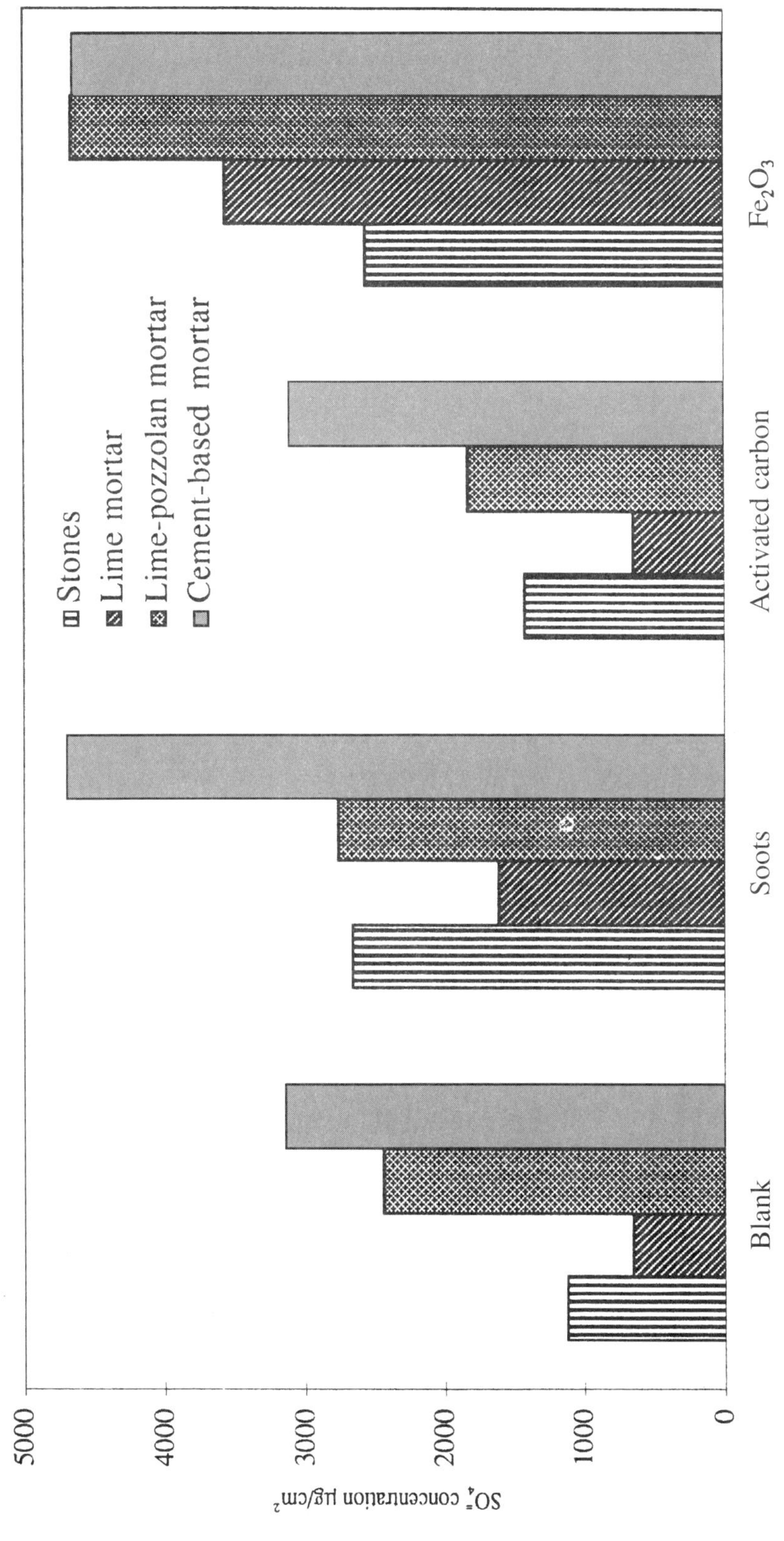

FIGURE 15.5 Sulfate concentrations on blank and particle-coated stones and mortars after 150 days of exposure in the air at 3 ppm SO_2.

FIGURE 15.6 Special supports exposed in the field with the samples of stones and mortars in Milan and Ancona.

TABLE 15.4
Anion Concentration (μg cm^{-2}) on Samples Exposed in the Field for 12 Months in Milan and Ancona

	Milan						Ancona					
Samples	SO_3^{2-}	SO_4^{2-}	NO_3^-	Cl^-	F^-	$C_2O_4^{2-}$	SO_3^{2-}	SO_4^{2-}	NO_3^-	Cl^-	F^-	$C_2O_4^{2-}$
Stones (mean)	23	606	96	63	9	15	12	259	224	157	19	10
Lime mortar	22	2697	256	154	6	24	Traces	1228	234	420	5	24
Lime-pozzolan mortar	40	3529	1	141	14	27	Absent	1667	234	481	9	33
Cement-based mortar	78	2256	113	67	13	26	9	663	122	281	21	27

than the sheltered ones (samples b). Table 15.4 shows the concentrations of the most abundant anions in the samples b. The order of reactivity found for the exposed stones and mortars confirms the chamber test results: mortars are more reactive than stones; hydraulic mortars (cement and pozzolan) are more reactive than the air-setting one (lime mortar). In comparing the two sites, the Milan samples, predictably, present higher values for all anions except chloride, which are higher at the maritime site of Ancona.

The field exposure tests confirm the findings obtained for the black patinas of ancient masonries and for the simulation chamber tests, and can be taken as a validation of these results.

CONCLUDING REMARKS

The findings of the study on the black alteration patinas of ancient masonry show how the mechanisms of surface deterioration, due to atmospheric SO_2. proposed in the literature for carbonate stones can also be extended to other masonry components. The sulfation of bricks is similar to that of stones with low porosity, while the case of lime mortars resembles that of sandstones. All the patinas contained carbonaceous particles in amounts ranging from 0.6% to 1.8%. A correlation between the C_{nc} and sulfate contents has been confirmed.

The results of the simulation chamber tests reveal how all the materials studied reacted with SO_2, forming significant quantities of calcium sulfite hemihydrate and calcium sulfate dihydrate. The reactivity of mortars was found to be clearly greater than that of both brick and stones. The pozzolan and cement mortars turned out to be the most reactive materials overall. The degree of damage, evaluated in terms of the S_{Tot} that has reacted, cannot be correlated with any single parameter, but depends rather on both chemical composition and microstructure. Alongside SO_2, the presence of soots enhances sulfate formation, an effect attributable to the presence of heavy metals, while specific surface turns out to be of negligible influence, as shown by the tests with Fe_2O_3 and activated carbon.

The field exposure tests confirmed the results obtained in the study of black patinas on ancient masonry work and in the simulation chamber.

The high reactivity of the hydraulic mortars in both the simulation and field tests underscores a problem that has thus far been wholly neglected: the deterioration of modern building materials due to environmental effects. Such effects merit attention not only in the case of modern buildings, but also in the restoration of ancient works, in view of the problems of aerosol deposition and incompatibility that can arise between the original materials and those used by restorers.[27] It is thus imperative that materials scientists and restorers be aware of these processes and include environmental sulfation among the risk factors to be considered.

REFERENCES

1. Roswall, J., *Air Pollution and Conservation*, Elsevier, Amsterdam, 1988.
2. Baer, N.S., Sabbioni, C., and Sors, A., *Science, Technology and European Cultural Heritage*, Butterworth-Heinemann, Oxford, 1991.
3. Zappia, G., Sabbioni, C., and Gobbi, G., *Mater. Engineering*, 2, 255, 1991.
4. Sabbioni, C. and Zappia, G., *Water, Air Soil Poll.*, 63, 305, 1900, 1992.
5. Novakov, T., Chang, S.G., and Harker, A.B., *Science*, 186, 259, 1974.
6. Chang, G., Toosi, R., and Novakov, T., *Atmos. Environ.*, 15, 1287, 1981.
7. Harrison, R.M. and Pio, C.A., *Atmos. Environ.*, 17, 1261, 1983.
8. Mamane, J. and Gottlieb, J., *J. Aerosol Sci.*, 20, 575, 1989.
9. Grgic, I., Hudnik, V., and Bizjak, M., *Atmos. Environ.*, 27A, 1409, 1993.
10. Winkler, E.M., *Stones, Properties, Durability in Man's Environment*, Springer, Wien, 1975.
11. Lea, F.M., *The Chemistry of Cement and Concrete*. Chemical Publishing, London, 1971.
12. Palladio, A., *Trattato di Architettura*, Venezia, 1570.
13. Zappia, G., Sabbioni, C., Pauri, M.G., and Gobbi, G., *Mater. Engineering*, 3, 445, 1992.
14. Zappia, G., Sabbioni, C., and Gobbi, G., *Atmos. Environ.*, 27A, 1117, 1993.
15. Goldberg, E., *Black Particles in the Environment,* John Wiley & Sons, New York, 1985.
16. Gauri, K.L. and Gwinn, J.A., *Durab. Building Mater.*, 1, 217, 1982.
17. Johansson, L.G., Lindqvist, O., and Mangio, R.E., *Durab. Building Mater.*, 5, 439, 1988.
18. Knöfel, D. and Bottger, K.G., *Beton. Technik*, 2, 107, 1985.
19. Zappia, G., Sabbioni, C., Pauri, M.G., and Gobbi, G., *Mater. Struct.*, 27, 469, 1994.
20. Hutchinson, A.J., Johnson, J.B., Thompson, G.E., Wood, G.C., Sage, P.W., and Cooke, M.J., *Atmos. Environ.*, 26A, 2795, 1992.
21. Metha, P.K., *Cement Concr. Res.*, 3, 1, 1973.

22. Bensted, J., *Il Cemento*, 1, 3, 1988.
23. Taylor, H.F.W., *Advances in Cement Concrete*, Ed., Grutzeck and Sarkar, 1994.
24. Varma, S.P. and Bensted, J., *Silicates Ind.*, 38, 29, 1973.
25. Van Aardt, J.H.P. and Visser, S., *Cement Concr. Res.*, 5, 225, 1975.
26. Scrivener, K.L. and Taylor, H.F.W., *Adv. in Cement Res.*, 5, 139, 1993.
27. Collepardi, M., *Mater. Struct.*, 23, 81, 1990.

16 Aerosol and Stone Monuments

Cristina Sabbioni

CONTENTS

INTRODUCTION

The deposition of atmospheric aerosols on materials of artistic interest is of paramount importance when considering problems affecting the maintenance and conservation of cultural heritage.

On the one hand, there is the very well-known effect of soiling, producing aesthetic problems on frescos and artistic objects in general. Then there are the even more damaging mechanisms of interaction between aerosols and materials, leading to: chemical processes causing the transformation of crystalline components and the formation of damage layers; physical processes with structural changes and internal mechanical stresses; biological effects with the deposition of biological particles, such as fungi, bacteria and molds; and interaction with the material itself.

Research performed over the last decade has highlighted the role played by atmospheric aerosols in the damage of stone monuments, and the main results are summarized in this chapter.

DEGRADATION OF STONE MONUMENTS

DEGRADATION OF STONES WITH LOW POROSITY

The visual features of the damage typologies that can be observed on monuments built of stone of low porosity, such as marble and limestone (i.e., carbonate rocks with less than 6% porosity), are related to the way rainwater wets the surface (Figure 16.1). *White areas* can be found where rainwater run-off removes the aerosol deposited during a dry spell, producing the dissolution of the carbonate rock so that the original white color of the stone is evidenced. *Grey areas* are typical of those surfaces completely sheltered from rain wetting, where an incoherent layer of atmospheric particles is deposited on the undamaged surface of the stone. Soiling produces visual impairment and loss of aesthetic value without producing damage, indicating that the water supplied by condensation or fog episodes is not sufficient to trigger the reactions between the deposited particulate matter and the stone materials. *Black areas* are observed on stones wetted by rainwater but sheltered from intensive run-off — they are the areas where the deposition of atmospheric gas and aerosols, as well as the products of reactions between atmosphere and materials, accumulate.

0-87371-829-1/00/$0.00+$.50

FIGURE 16.1 Monument showing the characteristic white areas (surfaces exposed to intense rain wash-out) and black crusts (sheltered to wash-out), which are the typical damage patterns of marble and limestone monuments in urban areas (detail of the Savonarola statue in Ferrara, Italy).

Protection from leaching by rain wash-out favors the formation of crusts, which are black in color due to the atmospheric particles they embed during their growth.[1]

In order to study the presence of atmospheric particles in the black areas, extensive sampling was performed on stone monuments at several urban sites in northern and central Italy: two large towns, Milan and Rome; three middle-sized towns, Bologna, Verona, and Trento; and four maritime sites, La Spezia, Venice, Ravenna, and Ancona. Care was taken to ensure that the samples collected from each town were representative of the site, rather than of a single monument alone. The samples were collected by scraping off the damage layers or removing fragments.

Once collected, the samples were dried, ground, and preserved at a temperature of 20°C in an inert environment (N_2). An analytical procedure was adopted to characterize the matrix of the damage layers and quantify the trace elements, which are of prime importance for identifying the presence of the atmospheric aerosols embedded within the damage layers.[2]

The matrix composition was detected by X-ray diffraction (XRD, Philips PW 1730), infrared spectroscopy (FTIR, Nicolet 20 SX), and thermal analysis (DTA, TGA, Netzsch Model 429). The data indicate calcium carbonate (calcite) and calcium sulfate dihydrate (gypsum) as the principal crystalline species constituting the surface layers (Figure 16.2). Their percentages vary greatly among the white and black areas, as shown in Figure 16.3, confirming the different processes giving rise to the formation of the various typologies of damage observed on stones with low porosity.

In the white areas, rain wash-out leads to the dissolution of the calcium carbonate and reprecipitation of calcite (secondary) when the rainwater evaporates. In the black areas, the transformation of the calcium carbonate into gypsum occurs due to the effect of atmospheric deposition:

$$CaCO_3 + H_2SO_4 + H_2O \Rightarrow CaSO_4 \cdot 2H_2O + CO_2. \quad (16.1)$$

Carbon and sulfur, which are the principal elements participating in the sulfation process, were measured by combustion and infrared spectroscopy (CHNSO Carlo Erba Analyzer) and a specific

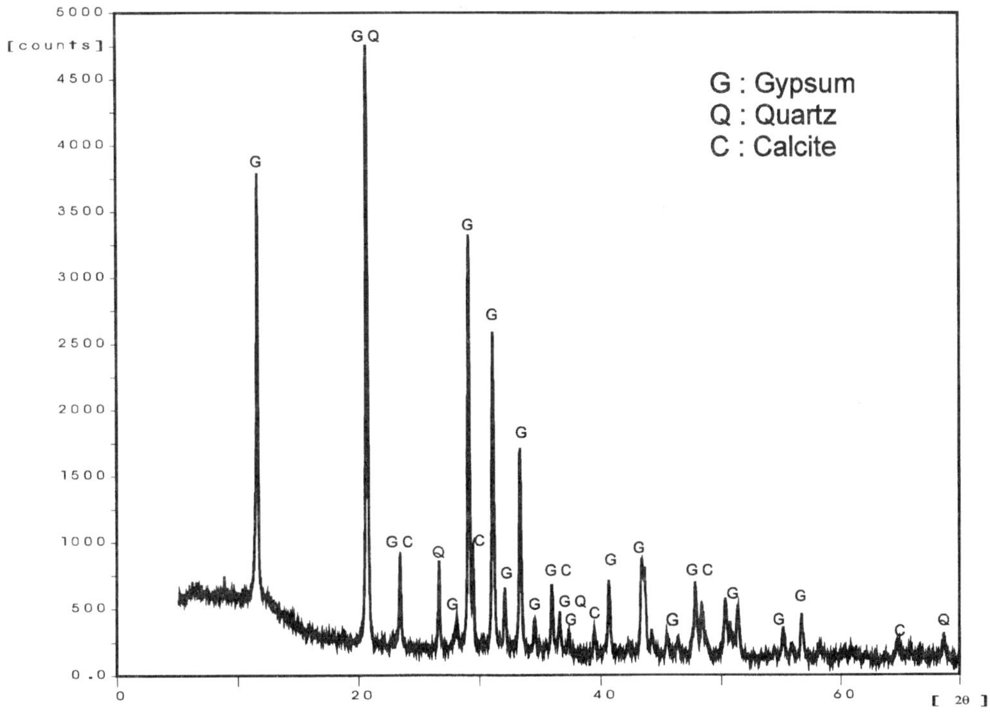

FIGURE 16.2 X-ray powder diffraction trace (CuK_α radiation) of a typical black crust showing gypsum (G), carbonate (C), and quartz (Q).

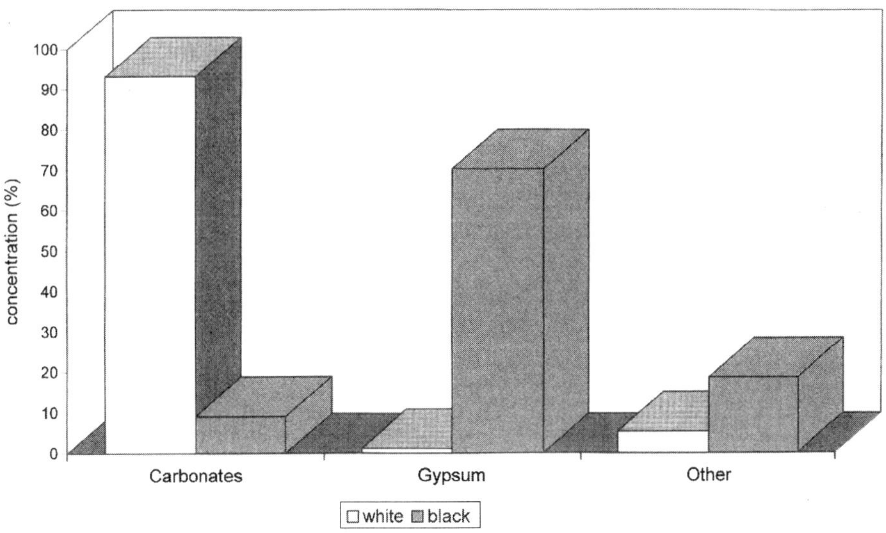

FIGURE 16.3 Mean concentrations of the main components characterizing the specimens sampled on the black and white areas, which can be observed on low-porosity carbonate stone.

TABLE 16.1
Gypsum ($CaSO_4 \cdot H_2O$), Carbonate ($CaCO_3$), and Non-carbonate Carbon (C_{nc}) Concentrations Measured in Black Crusts Sampled on Stone Monuments

Sites	Gypsum (%)	Carbonate (%)	C_{nc} (mg g^{-1})
Milan	78.4	1.4	16.0
Rome	71.0	19.6	19.5
Venice	86.6	tr	16.0
Bologna	75.6	7.1	19.7
Ravenna	56.5	6.5	28.5
Verona	66.9	8.2	23.7
Trento	57.3	18.2	14.0
La Spezia	64.0	19.3	15.1
Ancona	76.5	1.8	25.5

methodology was set up for carbon quantification. Considering the high concentrations of calcium carbonate present in the damage layers, the total carbon is not indicative for the determination of the carbon due to atmospheric deposition. In the stone damage layers, the total carbon is composed of two fractions:

$$C_t = C_c + C_{nc}, \tag{16.2}$$

where C_c is the carbonate carbon, which, considering that carbonates due to atmospheric deposition are negligible,[3] mainly originates from the carbonate rock; and C_{nc} is the non-carbonate carbon, which contains both elemental and organic carbon. In order to distinguish each component, the preliminary elimination of carbonates was performed in all the samples and, in the residual part containing only C_{nc}, the carbon was measured by the following method. The ground samples were stored in an atmosphere saturated with HCl vapors for 48 hours to obtain a slow, gradual, and complete elimination of carbonates; they were then transferred to a dryer on KOH until the complete elimination of humidity and residual HCl had taken place. The combustion of the treated samples using the CHNSO apparatus provided the C_{nc} value.[4] Table 16.1 shows the mean value of gypsum, carbonates, and C_{nc} found in the black crusts at the different sites; as can be seen, C_{nc} is the third most important component present in the black crusts analyzed, with a mean concentration of 19.8 mg g^{-1}, which is one order of magnitude higher than the C_{nc} found in the white areas (i.e., 3.97 mg g^{-1}).

Non-carbonate carbon may be of different origins: atmospheric aerosol (mainly carbonaceous particles), the underlying rock, and biological weathering (mainly oxalates). The non-carbonate carbon content of carbonate rocks is negligible[5] and a concentration of oxalic acid lower than 0.2% was found in the black crusts analyzed. Therefore, the C_{nc} data can be considered a quantitative index of the carbonaceous particles embedded in the damage layers.

A correlation of sulfate damage products with carbonaceous particles has been evidenced for a number of Italian monuments in different urban sites.[2,4]

Carbonaceous particles play an important role in the overall deterioration process. First, they are responsible for the blackening of the patina, which seriously impairs the appearance of monuments. Moreover, it has been suggested that carbonaceous particles actively participate in the sulfation process occurring on stone surfaces exposed to urban atmospheres.[6,7]

The elemental characterization of the damage layers was performed by inductively coupled plasma emission spectrometry (ICP; Perkin Elmer 5500), through the digestion of samples in Teflon vessels with an HF-HNO_3 mixture at 120°C, and by ion chromatography (Dionex 4500) after solubilization of the ground samples in bidistilled water with acid exchange resin.[8] The elemental concentrations measured in the black crusts of the studied sites show a similar order of abundance, indicating that the components contributing to the formation of the black crusts have a common origin. By way of example, the mean values found for the specimens sampled on historic buildings in Venice are reported in Figure 16.4A.

In order to highlight the different constituents due to atmospheric deposition with particular regard to aerosols, the data were processed to evaluate the enrichment factor of different elements (X) in comparison to the carbonate rock composition[5] using Ti as the normalizing element:

$$EF_{carb.}(X) = \frac{(X / Ti)\ \text{black crust}}{(X / Ti)\ \text{carb. rock}}. \tag{16.3}$$

The elements showing enrichment factors lower than or close to 1, have an origin that must be attributed to the underlying rock, as in the case of the example in Figure 16.4B for Ca, Si, Fe, Al, C, Mg, K, Mn, and Sr (alongside Ti as the normalizing element). The elements with enrichment factors greater than 1 are of non-carbonate origin and are due to atmospheric deposition, as in the case of S, Cl, Na, Pb, V, Zn, Ba, Cu, and Ni.

After identifying the elements related to atmospheric deposition, a discrimination between the natural and anthropogenic components participating in the formation of the damage layers was performed. For this purpose, the ratio between sodium and chloride was assumed as an index of marine aerosol; a value ranging between 1.3 and 1.7 was found in the maritime sites, such as Venice, Ravenna, Ancona, and La Spezia, showing a good agreement with the Cl:Na ratio of 1.8 reported in the literature for sea salt[9]; at the inland sites, such as Bologna, a Cl:Na ratio of 0.1 was obtained.[2] Thus, the sea-derived elements can be identified in the damage layers of monuments situated in the maritime sites. It is worth noting that NaCl crystallization produces mechanical stress within the stone, causing internal fractures and granular disintegration.

Furthermore, the enrichment factor with respect to the average soil composition[5] was evaluated with the aim of identifying those elements due to atmospheric pollution, using the following relation:

$$EF_{s.d.}(X) = \frac{(X / Ti)\ \text{black crust}}{(X / Ti)\ \text{crustal rock}}. \tag{16.4}$$

As can be seen in Figure 16.4C, besides Ti (the reference element), Si, Fe, Al, Na, K, and Mg have $EF_{s.d.}$ close to 1, indicating that soil dust is their major source (the origin of Cl and Na having been previously discussed). The $EF_{s.d.}$, as shown in Figure 16.4C, evidence S, C, Pb, V, Sr, Zn, Ba, Cu, and Ni as elements of anthropic origin. In fact, S and C are associated with the atmospheric emission of combustion sources (i.e., electric power plants, domestic heating systems, diesel engines); Pb is the typical tracer of gasoline combustion; V is present as a specific tracer of oil combustion (oil is the main fuel used in Italy for both electric energy production and domestic heating); Zn is related to incinerator emissions. Therefore, the presence of typical tracers of anthropogenic aerosols are evidenced within the black crusts of stone monuments.

Finally, a study on the atmospheric aerosols embedded within the damage layers was performed by scanning electron microscopy interfaced with an energy dispersive X-ray analyzer (SEM-EDX). Figure 16.5 reports a typical SEM-EDX analysis performed on a fragment of damage layer. The main particle typologies found embedded within the samples are as follows: (1) carbonaceous particles, spherical and porous, ranging from a few microns to 50 μm, emitted into the atmosphere from oil-fired industrial plants, electric generating units, and large residential heating systems; this

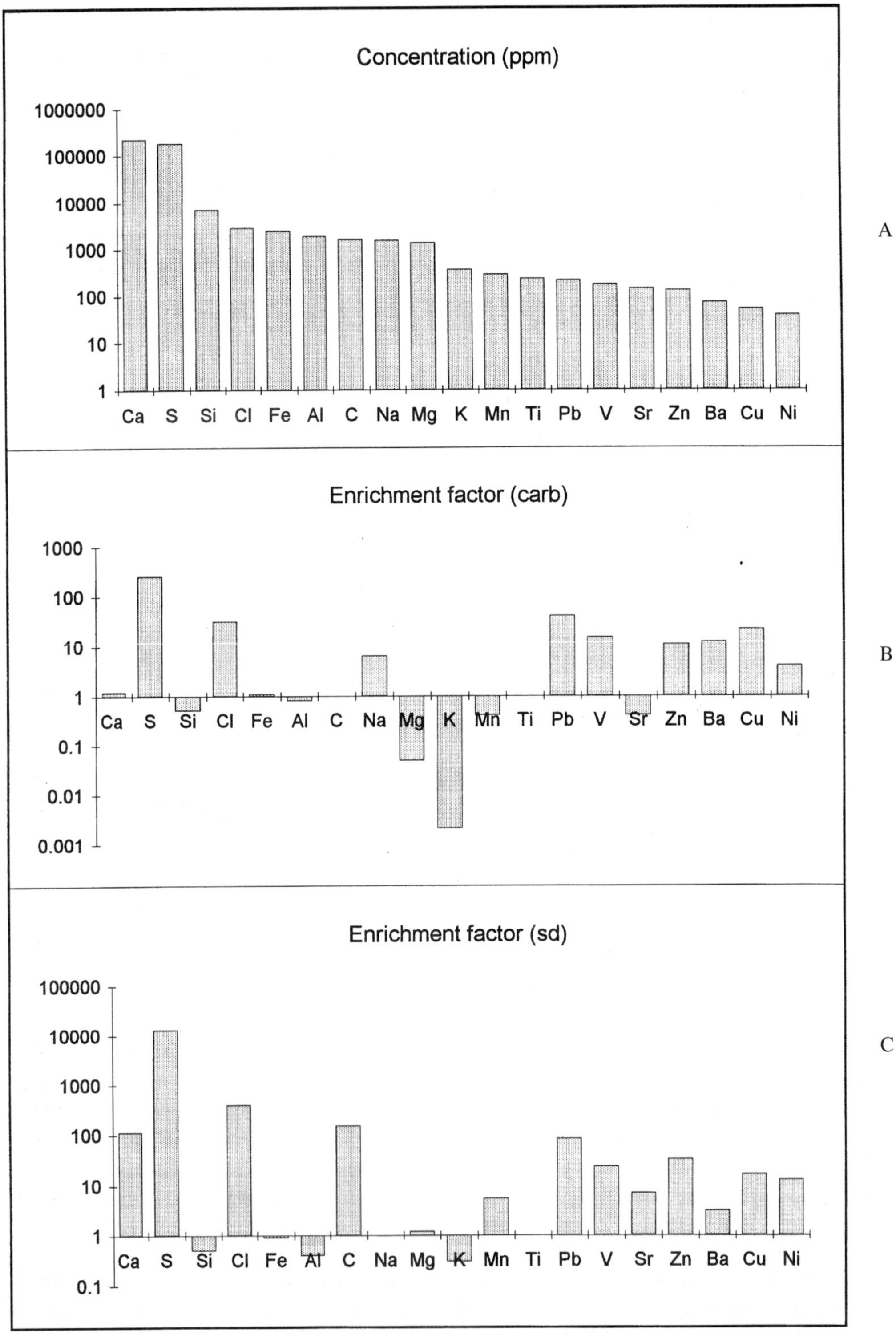

FIGURE 16.4 Mean elemental concentrations (A) measured in the black crusts sampled on historical buildings in Venice; (B) enrichment factors with respect to the carbonate rock, and (C) to the soil dust.

FIGURE 16.5 Scanning electron micrograph of a black crust on a limestone, showing the presence of gypsum crystals with laminar structure.

morphology is the one most frequently encountered in samples collected in Italy; (2) spherical particles with irregular pores or composed of an agglomeration of submicron particles, emitted by distilled oil; (3) spherical, smooth aluminosilicate particles, emitted by coal combustion; and (4) rare metallic particles, mainly composed of iron or titanium oxides. Thus, SEM-EDX analyses permit the identification of atmospheric particles emitted by specific pollutant sources embedded within the damage layers.[10,11]

Degradation of Stones with High Porosity

In the case of stones of high porosity, such as sandstone or calcarenite (total porosity, 15 to 20%) the visual features of damage patterns are completely different from those described for marble and limestone surfaces (Figure 16.6A); the surfaces are blackened homogeneously regardless of the geometry of the surface.[12]

This difference in the damage patterns must be ascribed to the different mechanisms of deposition and resuspension of atmospheric particles (particularly carbonaceous particles, responsible for the color of the crusts) occurring at the surface of the monument. Materials such as sandstone and calcarenite present high surface roughness due to their intrinsic porosity and mineralogical dishomogeneity, which prevent smoothing and polishing. Furthermore, the wetness of a surface is highly favored by the presence of pores and capillaries; in fact, condensation occurs at relative humidities below 100%. Surface roughness and high porosity facilitate the deposition of gas and particles, while at the same time reducing the removal of particles and damage products by resuspension and wash-out. Thus, on porous stones, the mechanisms of particle deposition and capture are more efficient, while those mechanisms tending to remove particles from the surface after their deposition are less efficient. These synergetic effects produce the homogeneous blackening typical of historical buildings and monuments built of porous stones such as sandstone and calcarenite.

FIGURE 16.6 (A) The blackening affects the entire surface of a sandstone monument, regardless of geometry; (B) the color of the original rock is shown only where the black crust has fallen off.

On sandstone and calcarenite, the presence of three layers can be observed (Figure 16.6B). *Layer A* is a surface layer of a few millimeter thickness with a mean composition similar to the black crusts analyzed on marble and limestone. *Layer B* is a disaggregated layer on the order of 1 cm thick, where the dissolution of the carbonate matrix occurs due to the atmospheric acid deposition, producing the decohesion of the sandy grains. *Layer C* is the original stone. When layer A is detached, the whole of layer B is also lost and the original rock (layer C) is exposed to a new damage cycle.

A large number of layers A were analyzed; the mean concentrations found for gypsum, calcite, and C_{nc} were 64.4%, 9.3%, and 1.1%, respectively, indicating that the mechanisms of their formation are similar to those occurring on the black areas of limestone, that is, sulfation of the carbonate matrix and inclusion of atmospheric particles within the damage layers during their formation on the stone surface. The mean elemental concentrations of layers A are reported in Table 16.2 and show Ca, S, and Si as the major elements for sandstones, followed by Na, Al, Fe, and K. The metals with the highest values (after Fe) are Sr, Mg, Pb, Zn, V, Cr, Ni, and Ti. Calcarenite, apart from having considerably lower concentration of Si and Al, due to the composition of this rock, presents a similar trend with regard to the order of abundance of the elements. The enrichment factors of the sandstone rock indicate S as the element with the greatest EF_{rock}, evidencing that sulfates are the predominant component in the stone damage associated to atmospheric deposition, both gaseous SO_2 and aerosol. In addition, elements with low concentration values, such as Pb, Sr, Ni, Zn, and V, have EF_{rock} values greater than 100; their origin is clearly linked to the deposition of atmospheric aerosols. Otherwise, elements presenting high concentrations, such as Al, Fe, and Mg show EF_{rock} lower than 5, indicating that their origin must be related to the underlying sandstone. Therefore, the evaluation of enrichment factors in the surface layers of porous stones highlights the presence of different components of atmospheric deposition embedded within layer A during its formation, particularly the aerosols emitted by anthropic sources.

SEM analyses confirm the presence of atmospheric particles, as in the example reported in Figure 16.7, which shows a porous carbonaceous particle, typical of oil combustion (Figure 16.7A), and a smooth particle, characteristic of coal-fired plants, deposited on a black crust of a sandstone monument (Figure 16.7B).

Biological Weathering

The effects produced by biological particles on materials in general, especially those of artistic interest, are currently of great concern.[13] Biological weathering due to the action of microorganisms, such as fungi and lichens, at the stone surfaces have been reported by Krumbein[14] and Warscheid et al.[15] Jones and Wilson,[16] in their review on the chemical activity of lichens on mineral surfaces, mention that calcium oxalates have also been identified on stones, while Alaimo et al.[16] report the presence of calcium oxalates on calcarenite without considering their origin.

Oxalate patinas have been found on a large number of monuments in Italy,[18] such as the Trajan Column (Figure 16.8A), Antonine Column, Constantine Arch, Settimio Severo Arch (Rome), Trajan Arch (Ancona), Augustus Arch (Rimini), and Basilica of St. Francesco (Assisi). XRD analyses indicate these patinas to be composed of mono- and dihydrate calcium oxalates: $Ca(COO)_2 \cdot H_2O$ (whewellite) and $Ca(COO)_2 \cdot 2H_2O$ (weddellite). The formation of these two minerals has been related to the deposition on stone surfaces of microorganisms (fungi and lichens), which are known to produce oxalic acid; the oxalic acid reacts with the calcium carbonate of the stone, leading to the precipitation of calcium oxalates:

$$CaCO_3 + (COOH)_2 \Leftrightarrow Ca(COO)_2 + CO_2 + H_2O \qquad (16.5)$$

TABLE 16.2
Mean Elemental Concentrations (ppm) Measured in the Layers A Sampled on Historical Buildings in Sandstone and Calcarenite

Elements	Sandstone (ppm)	Calcarenite (ppm)
Li	<5	<5
Be	<1	<1
Na	3552	3199
Mg	172	573
Al	1326	380
Si	89327	11679
S	119918	52506
K	316	355
Ca	199613	332558
Ti	31	51
V	34	19
Cr	14	18
Mn	153	51
Fe	653	552
Co	<5	<5
Ni	12	13
Cu	20	19
Zn	75	346
As	<10	<10
Se	<10	<10
Sr	176	280
Zr	3	4
Mo	<5	<5
Cd	4	3
Sn	<10	<10
Ba	37	30
Pb	160	37

$$\mathrm{Ca(COO)_2} \xleftrightarrow{\mathrm{H_2O}} \begin{matrix} \mathrm{Ca(COO)_2 \cdot H_2O} \text{ (whewellite)} \\ \mathrm{Ca(COO)_2 \cdot 2H_2O} \text{ (weddellite)} \end{matrix} \quad (16.6)$$

Furthermore, on carbonate stones, the filamentous hyphae of endolithic fungi penetrate into the substrate, not only along the calcite crystal planes, and produce pitting phenomena so that the stone surface appears clearly etched (Figure 16.8B).

Soil dust and carbonaceous particles have been found embedded within gypsum crusts, which were grown on the oxalate patina, indicating that the sulfation process is a phenomenon subsequent to the formation of the oxalate layers (Figure 16.9).

SIMULATION TESTS

In the literature, the laboratory tests performed on building stones have primarily concerned the effects of exposure to relative humidity and gas, such as SO_2, NO_x, and O_3.[19-21] Thus, little has been published

A

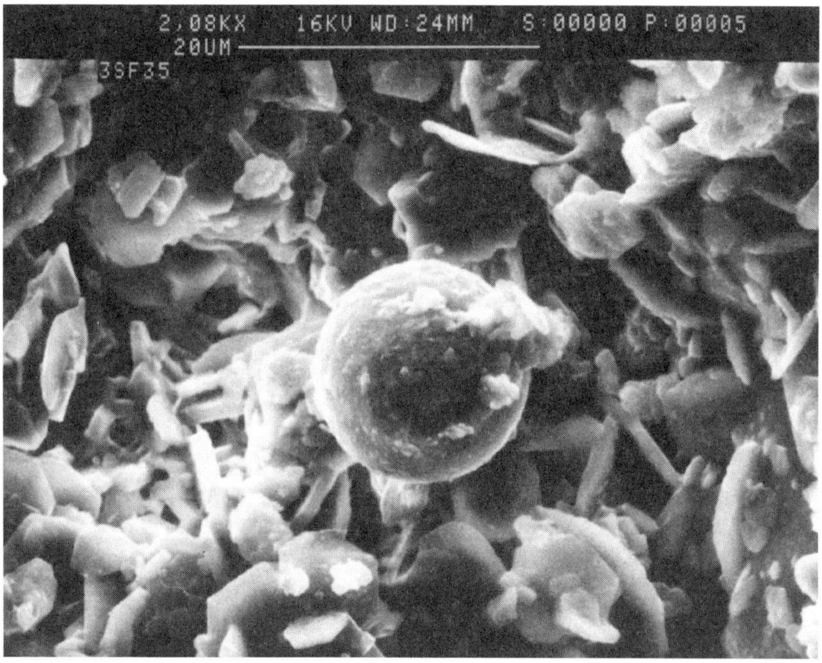

B

FIGURE 16.7 Scanning electron micrographs performed on the black crust of a sandstone monument, showing (A) a porous carbonaceous particle, typical of oil combustion, and (B) a smooth particle, characteristic of coal-fueled plants.

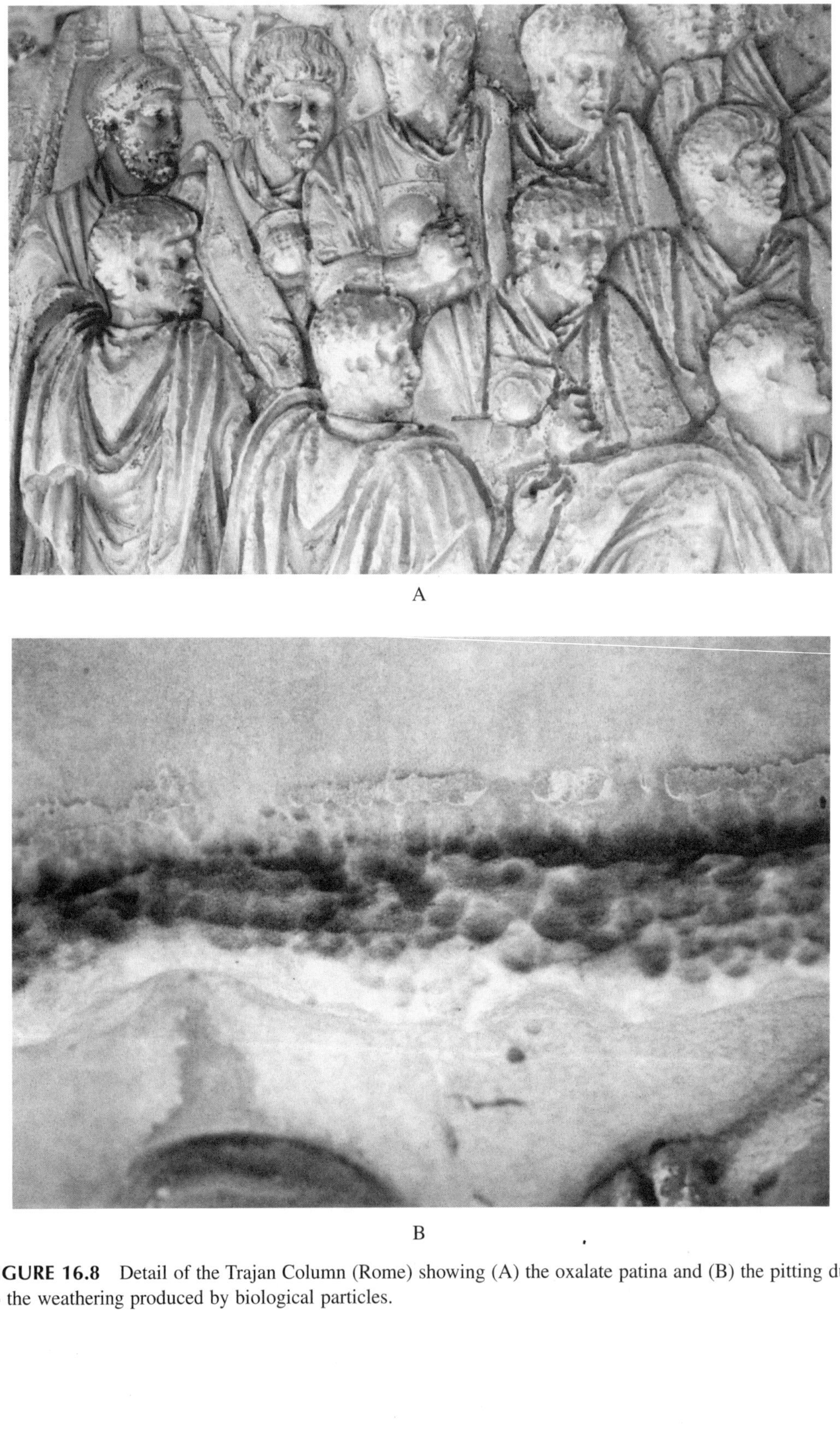

FIGURE 16.8 Detail of the Trajan Column (Rome) showing (A) the oxalate patina and (B) the pitting due to the weathering produced by biological particles.

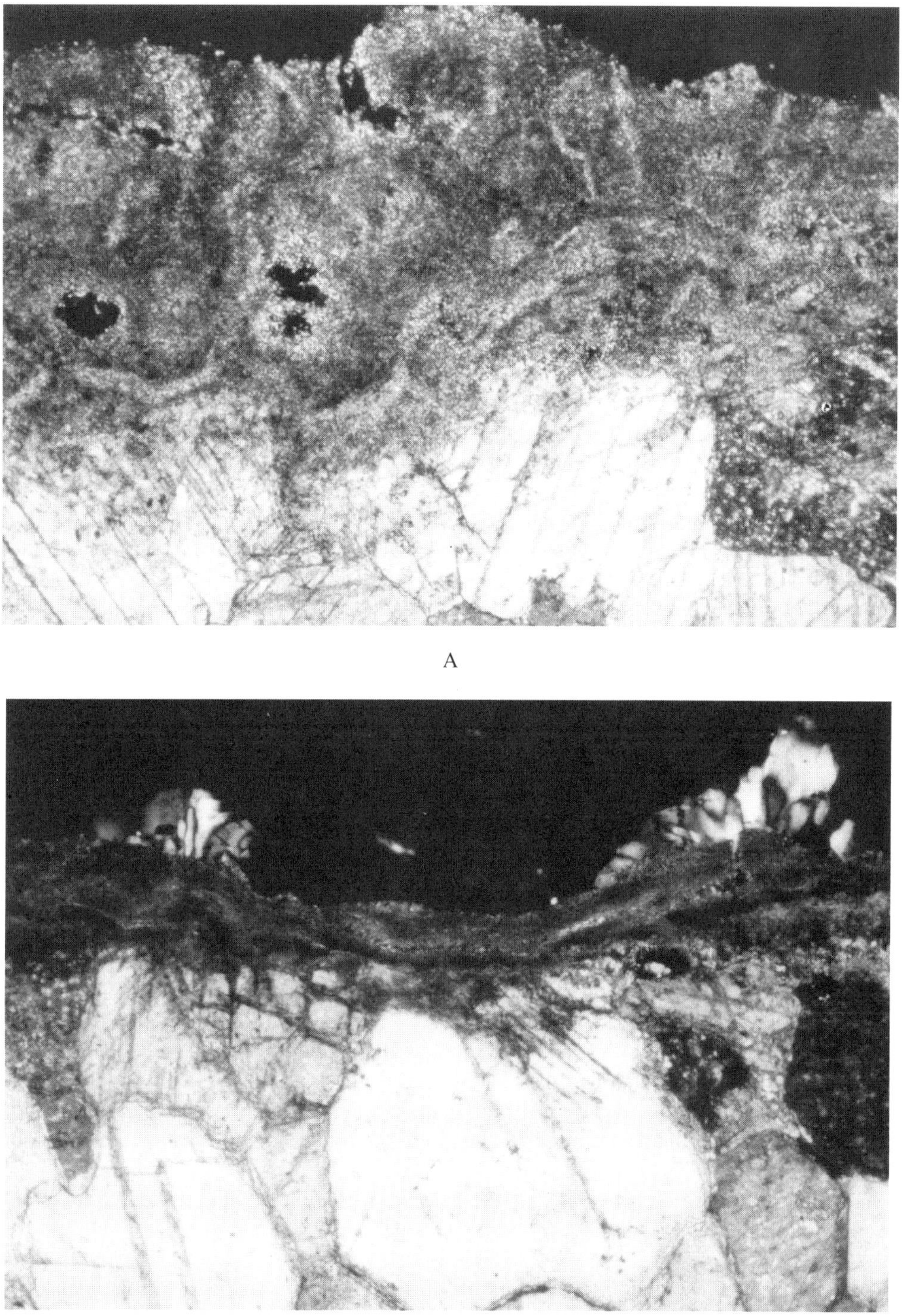

A

B

FIGURE 16.9 Optical micrographs showing (A) the oxalate patina; and (B) gypsum crystals embedding carbonaceous particles, which are found on the oxalate patina, indicating that sulfation is a transformation that occurred after the oxalate formation.

on the interaction between stones and aerosols with particular regard to carbonaceous particles. Hutchinson et al.[22] performed an experiment using fly-ash on carbonate stones and concluded that fly-ash particles play no active role in the sulfation process and actually mask the surface, providing some degree of protection.

In studies on atmospheric aerosols, numerous investigators have reported a correlation between sulfate and black carbon, underlining the importance of carbonaceous particles in atmospheric SO_2 oxidation.[23,24] However, it has not yet been established whether the catalytic effect should be attributed to the chemical composition of the particles (sulfur and heavy metal content) or to their morphology (high specific surface area).

With the aim of investigating the effects of carbonaceous particles on different stones, a number of experiments were performed in the laboratory. The tests were carried out at constant atmospheric pressure, 25°C temperature, 95% relative humidity, 3 and 0.3 ppm SO_2 concentrations, for periods ranging between 1 and 24 months. A wide range of building materials, chosen as representative examples of the stone used in northern, central, and southern Europe, was investigated: Carrara and Pentelic marble, Travertine, Portland limestone, Trani stone, and Baumberger sandstone. The stones presented a wide variety of total porosity, which ranged between 1.5 and 14.7%, and of bulk SO_4^{2-} concentration, 12 to 77 μg g^{-1}.

Each stone was exposed blank and with three types of particles (P, AC, G). Aerosol samples (P) were collected at the emission points of an oil-fueled combustion source, in order to ensure that the particles utilized in the tests were similar to those found on Italian monuments.[25] For comparison, *artificial* carbonaceous particles (i.e., particles not present in the atmospheric aerosol) were also used; active carbon (AC) was chosen for its high specific surface area, measured by means of Quantachrome-Autosorb 1 as 83 $m^2\ g^{-1}$, and graphite (G) for its high purity.

The problem of applying the particles to the stone surfaces was studied. Dispersion of the particles in a liquid phase (such as water, acetone, or organic solvents) was considered unsuitable because a dissolution of the soluble species in the particles could occur, leading to the growth of authigenic crystals. A number of tests were carried out with dispersed gas injection, but modifications in particle size and morphology were observed.

The particles were scattered onto the specimen surfaces and the samples exposed to air flow (higher than the flux operating in the chamber) so as to eliminate any excess and prevent phenomena of resuspension inside the chamber causing contamination among samples. The mean amount of particles on the specimens was 50 μg. The particle distribution on each stone specimen was checked by optical microscopy to verify a homogeneous distribution over the sample surface and a surface concentration sufficiently low so as not to limit the SO_2–material interaction. Specimens of particles deposited on passive supports were also exposed in order to check the sulfate due to the interaction between carbonaceous particles and SO_2 without any interaction with the stone.

The quantitative determination of the ions SO_4^{2-} and SO_3^{2-} was performed using a Dionex ion chromatograph (4500i series). For analysis, the samples were finely ground, dispersed in ultrapure water, and kept for 10 minutes in an ultrasound bath; finally, they were injected into the column through a nylon filter (r = 0.2 μm).

Special measures were taken to avoid the possible oxidation of sulfite into sulfate after the exposure experiments and in the column during analysis:

1. Subsequent to exposition to SO_2 and up to the time of IC analysis, the samples were preserved in an inert environment (dry N_2 UPP).
2. Prior to introducing the solution for analysis, numerous injections of a reducing solution (Na_2SO_3 solution at 200 ppm) were performed to eliminate possible internal centers of oxidation.
3. In addition, the kinetics of the $SO_3^{2-} \rightarrow SO_4^{2-}$ reaction at 20°C were studied. The results clearly show that up to 24 minutes, the concentrations of SO_3^{2-} and SO_4^{2-} are more or

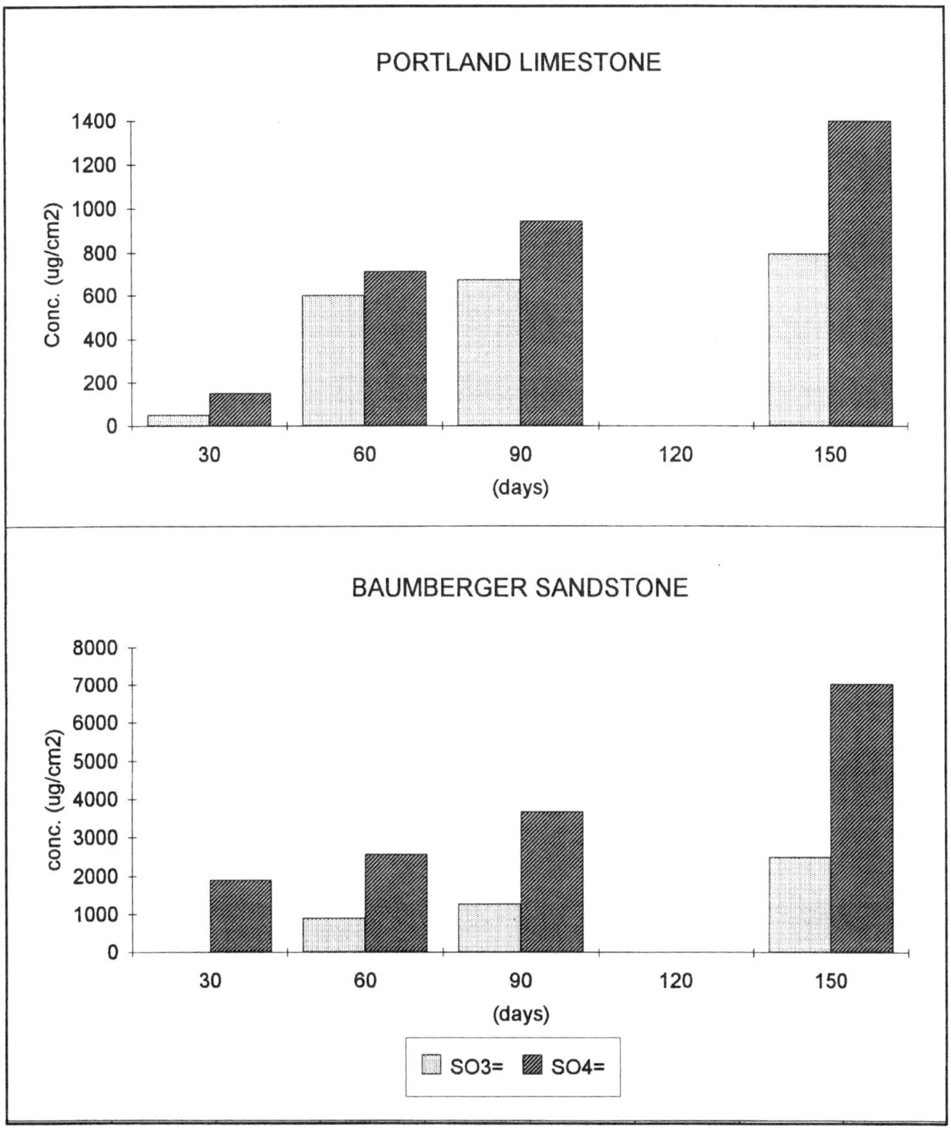

FIGURE 16.10 Sulfate and sulfite concentrations measured in Portland limestone and Baumberger sandstone specimens exposed at 3 ppm SO_2 concentration.

less constant. Thus, the total analysis time of 8 minutes for SO_3^{2-} and SO_4^{2-} turns out to fall well within the stability range of the solution.[26]

In the literature, addition of tetrachloromercurate, formaldehyde, or EDTA is mainly used to stabilize SO_3^{2-} in solution; however, such procedures are utilized for spectrophotometric analyses and are not suitable for ion chromatography measurements.[27,28] The procedure adopted permits a valid determination of sulfite and the simultaneous analysis of all the other anions present in the samples.[26]

The sulfite and sulfate concentrations of the blank Portland limestone and Baumberger sandstone exposed for 30, 60, 90, and 150 days at 3 ppm SO_2 concentration are reported in Figure 16.10. The sulfate concentrations were calculated as the difference between the SO_4^{2-} measured in the exposed samples and the SO_4^{2-} present in the bulk stones. The data underline the importance of quantifying both sulfur ions formed by interaction between SO_2 and material surfaces, in order to provide information on the chemical reactions occurring in the system. Thus, articles on simulation tests where no discrimination between sulfite and sulfate concentrations is reported[29] can only provide data on total sulfur, without any reference to the SO_2 oxidation mechanism. The data at 0.3 ppm SO_2 confirm that (1) sulfites were present in most of the samples analyzed; (2) the total sulfur concentrations showed the following sequence of SO_2-reactivity: Pentelic marble < Portland limestone < Baumberger sandstone

In order to identify the crystalline and chemical species forming on the samples during the exposure tests, analyses by X-ray diffractometry (XRD Philips PW 1730 diffractometer) and infrared spectroscopy (FTIR; Nicolet 20 SX) were performed. XRD and FTIR analyses agree in indicating sulfate as calcium sulfate dihydrate (gypsum) and sulfite as calcium sulfite hemihydrate. In almost all cases, the data show the formation of calcium sulfite as an intermediate product in the process of sulfation. Gauri et al.[19] and Amoroso and Fassina[7] report calcium sulfite dihydrate ($CaSO_3 \cdot 2H_2O$) as an intermediate product of interaction between SO_2 and $CaCO_3$; this compound was never found on our samples.

According to our results, the reactions occurring at the material surfaces are:

$$SO_2 + H_2O \Rightarrow SO_2 \cdot H_2O \tag{16.7}$$

$$CaCO_3 + SO_2 \cdot H_2O \Rightarrow CaSO_3 \cdot 0.5H_2O + CO_2 + 0.5H_2O \tag{16.8}$$

$$2CaSO_3 \cdot 0.5H_2O + O_2 + 3H_2O \Rightarrow 2CaSO_4 \cdot 2H_2O \tag{16.9}$$

$$CaCO_3 + H_2SO_4 + H_2O \Rightarrow CaSO_4 \cdot 2H_2O + CO_2 \tag{16.10}$$

In Figure 16.11 the sulfate concentrations for the Portland limestone (blank and particle-coated) are reported. The presence of particles P generally enhances SO_4^{2-} formation compared to the blank samples; the increase in sulfate concentrations can be related to the heavy metal content of carbonaceous particles.[30] Active carbon, despite its high specific surface area, and graphite do not increase and, in some cases, actually reduce the damage products.

The tests with carbonaceous particle on a passive support show that gypsum nucleated by the carbonaceous particles themselves is negligible compared to the total gypsum formed by the interaction between SO_2 and stones during the experiment (Figure 16.12A).

Finally, SEM-EDAX analyses, carried out on the exposed samples (blank and particle-coated), show that gypsum growth occurs homogeneously on stones with low porosity (Figure 16.12B); in sandstone, gypsum crystals with both laminar and globular structures are observed preferentially within pores, indicating different growth rates.

Laboratory tests indicate that carbonaceous particles produced by combustion of oil-based fuels are far from innocuous factors in the deterioration of stone but, on the contrary, contribute significantly to the sulfation of calcareous materials, producing both chemical and physical damage. In no case do they constitute protective layers. The presence of these particles in polluted atmospheres can be important in the decay of stone monuments and historical buildings and should therefore be assessed and mitigated for the preservation of cultural heritage as a whole, rather than limiting intervention to the restoration of single monuments.

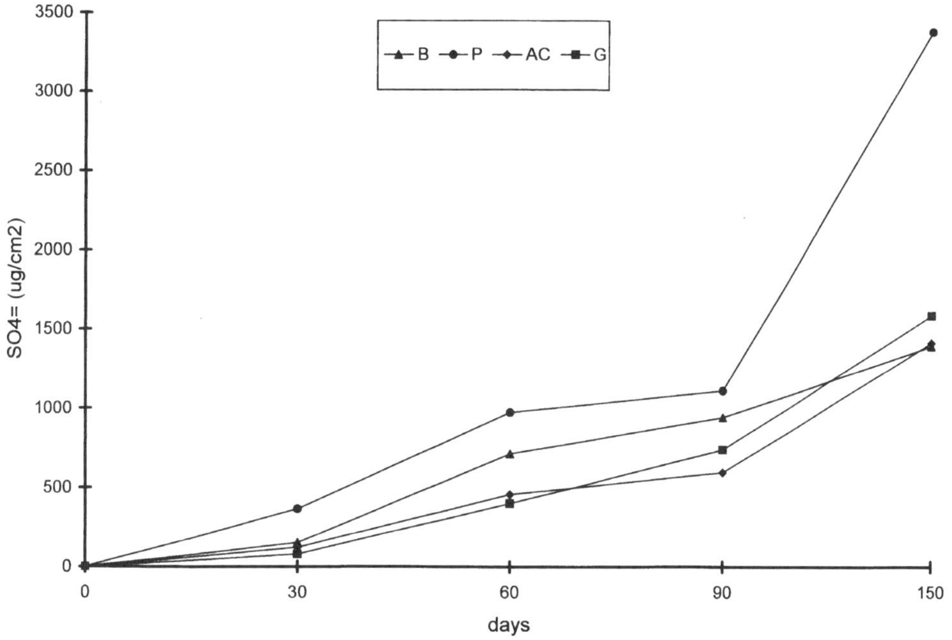

FIGURE 16.11 Sulfate concentrations for the Portland limestone, exposed to SO_2,blank [B] and coated with oil fired carbonaceous particles [P], active carbon [AC], and graphite [G].

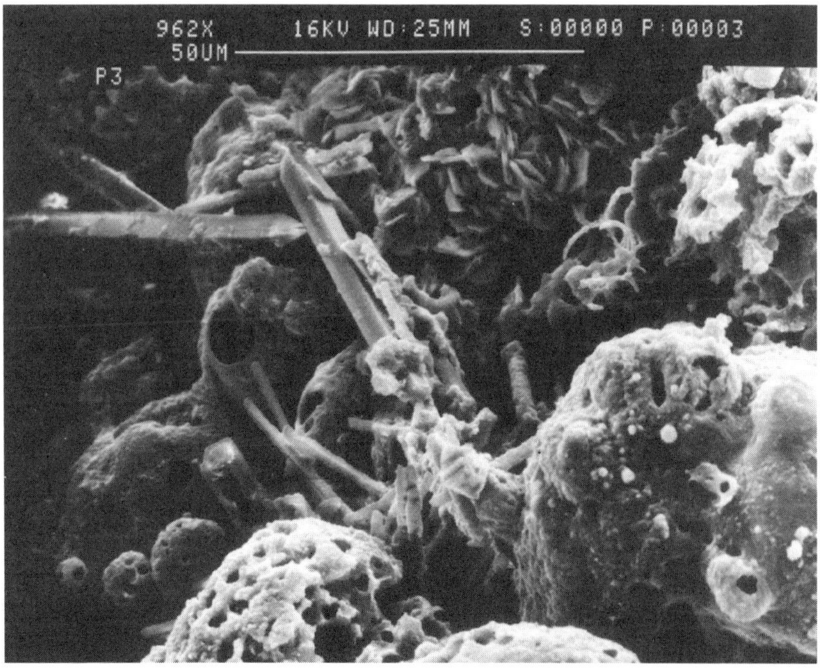

FIGURE 16.12A SEM micrograph of the carbonaceous particles, exposed at 3 ppm SO_2 for 150 days on an inert support, showing the gypsum crystals nulceated by the particles themselves.

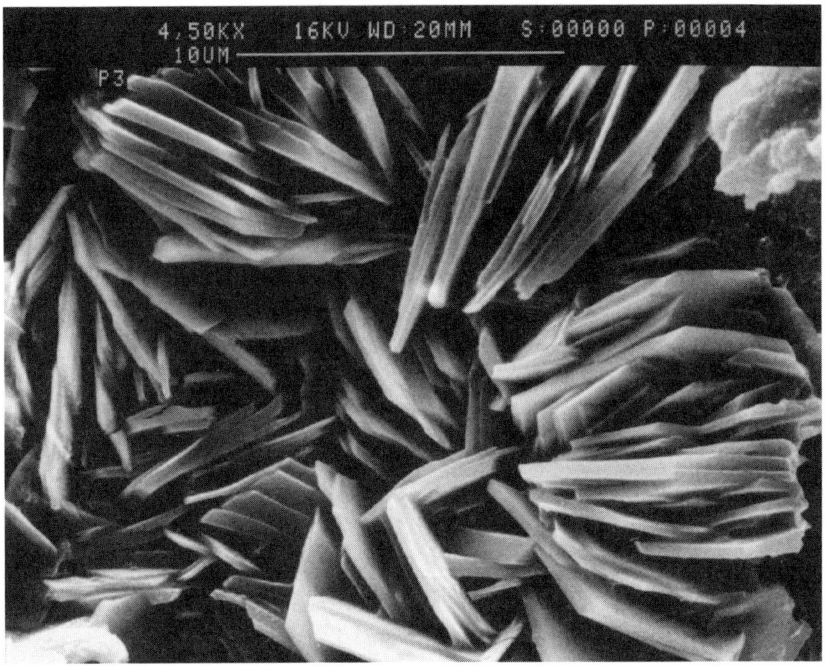

FIGURE 16.12B SEM micrograph of an exposed marble specimen, showing that gypsum growth occurs homogeneously on the stone surface with a laminar structure.

REFERENCES

1. Camuffo, C., Del Monte, M., Sabbioni, C., and Vittori, O., *Atmos. Environ.,* 30, 2253, 1982.
2. Sabbioni, C. and Zappia, G., *Sci. Total Environ.,* 126, 35, 1992.
3. Pye, K., *Aeolian Dust and Dust Deposits,* Academic Press, London, 1987.
4. Zappia, G., Sabbioni, C., and Gobbi, G., *Atmos. Environ.,* 27A, 1117, 1993.
5. Mason, B., *Principles of Geochemistry,* John Wiley & Sons, New York, 1966.
6. Del Monte, M., Sabbioni, C., and Vittori, O., *Atmos. Environ.,* 15, 645, 1981.
7. Amoroso, G. and Fassina, V., *Stone Decay and Conservation,* Elsevier, Amsterdam, 1983.
8. Gobbi, G., Zappia, G., and Sabbioni, C., *Atmos. Environ.,* 29A, 703, 1995.
9. Hidy, *Aerosol. An Industry and Environmental Science*, Academic Press, Inc., London, 1984.
10. Sabbioni, C., *Electron Micros.,* Granada, 2, 773, 1992.
11. Sabbioni,C., *Sci. Total Environ.,* 167, 49, 1995.
12. Sabbioni, C. and Zappia, G., *Water, Air Soil Pollution,* 63, 305, 1992.
13. Caneva, G., Nugari, M.P., and Salvadori, O., *Biology in the Conservation of Works of Art,* ICCROM, Roma, 1991.
14. Krumbein, W.E., *Durab. Building Mater.,* 5, 359, 1988.
15. Warscheid, T., Petersen, K., and Krumbein, W., *Studies in Conservation,* 35, 137, 1990.
16. Jones, D. and Wilson, M.J., *Int. Biodeterioration,* 21, 99, 1985.
17. Alaimo, R., Deganello, S., and Montana, G., *Miner. Petrog. Acta,* 30, 271, 1986.
18. Sabbioni, C. and Zappia, G., *Aerobiologia,* 7, 31, 1991.
19. Gauri, K.L. and Gwinn, J.A., *Durab. Building Mater.,* 1, 217, 1982.
20. Vales, J. and Martin A., *Durab. Building Mater.,* 3, 197, 1986.
21. Johansson, L.G., Lindqvist, O., and Mangio, R.E., *Durab. Building Mater.,* 5, 439, 1988.
22. Hutchinson, A.J., Johnson, J.B., Thompson, G.E., Wood, G.C., Sage, P.W., and Cooke, M.J., *Atmos. Environ.,* 26A, 2795, 1992.
23. Novakov, T., Chang, S.G., and Harker, A.B., *Science,* 186, 259, 1974.
24. Chang, S.G., Toosi, R., and Novakov, T., *Atmos. Environ.*, 15, 1287, 1981.

25. Sabbioni, C., Zappia, G., Gobbi, G., and Pauri, M.G., *Structural Repair and Maintenace of Historical Buildings*, Comput. Mechan. Publ., 235, 1993.
26. Gobbi, G., Zappia, G., and Sabbioni, C., *Atmos. Environ.,* in press, 1998
27. Dasgupta, P.K., *Atmos. Environ.*, 16, 1265, 1982.
28. Roekens, E., Bleyen, C., and Van Grieken, R., *Environ. Pollut.*, 57, 289, 1989.
29. Johnson, J.B., Hanef, S.J., Hepburn, B.J., Hutchinson, A.J., Thompson, G.E., and Wood, G.C., *Atmos. Environ.,* 24A, 2585, 1990.
30. Sabbioni, C., Zappia, G., and Gobbi, G., *J. Geophys. Res.,* 101, 19,621, 1996.

17 Calcium in the Urban Atmosphere

Marco Del Monte and P. Rossi

CONTENTS

INTRODUCTION

The atmosphere contains numerous gaseous and particulate substances of natural and anthropic origin. Such substances, lifted or emitted, are transported and, during this process, may undergo several transformations and eventually be deposited. On deposition, they interact with the environment; that is, they are affected by the soil, hydrosphere, and biosphere.

Atmospheric physics distinguishes three types of deposition: dry, wet, and occult. *Dry deposition* refers to the various processes that lead the gases and particles present in the atmosphere to deposit on the ground. *Wet deposition* involves the removal from the atmosphere of gas and particles on the part of rain, snow, and hail. *Occult deposition* is a term covering the little-known deposition processes which are therefore taken into insufficient account in geochemical balances and include the deposition on the ground of gases and particulates contained in fog[1] or cloud droplets as they intercept the ground, the latter phenomenon generally occurring in mountain areas. Fog and young cloud droplets, measuring ≈10 μm in diameter, can be transported to the ground by inertial impact, while those originating in more mature clouds have a diameter of at least 1 mm and fall to the ground by gravity.

An ecosystem characterized by such processes is the city. Here, the following conditions can generally be said to prevail:

0-87371-829-1/00/$0.00+$.50

1. There is a regional scale background of airborne pollutants, which are thus characteristic of the geographic area where the city is located.
2. The city itself constitutes a highly efficient aerial source of gases and particles.
3. At the local level, the contributions of the regional and aerial sources are combined with the effects of important point and linear sources.

Since the atmosphere is a dynamic system, the city is continually emitting and receiving gas and particles, exchanging them with the surrounding environment, a process in which thermal inversion (day-night, summer-winter) plays a crucial role.

Although the quantity of gaseous substances and particulate matter is vast, the monitoring of wet and occult deposition performed by many researchers in different geographical areas has revealed that, from a chemical point of view, the most important cations are NH_4^+, Ca^{2+}, and Na^+, while the anions that most clearly prevail are SO_4^{2-}, NO_3^-, and Cl^-.[2-9]

With regard to dry deposition — limited to suspended dusts of diameter ≈ 0.8 to 100 μm — the most important cations are Si^{4+}, Al^{3+}, and, in smaller amounts, Ca^{2+}, Fe^{3+}, Mg^{2+}, Na^+, and K^+; among the anions, alongside numerous predominating silicatic and aluminosilicatic radicals, SO_4^{2-} and CO_3^{2-} are encountered, while Cl^- and NO_3^- are virtually absent (<1 ppm). Within the city ecosystem, the deposition processes (dry, wet, occult, with or without interaction) most widely investigated are those that lead to the formation of gypsum ($CaSO_4 \cdot 2H_2O$), always mixed with carbonaceous particulate.

While the origin of SO_2 (and, then, SO_4^{2-}) is well documented, little is known on the presence, quantity, and origin of Ca^{2+} in the atmosphere, particularly in the urban environment.

In this brief chapter, the results obtained from analyses performed on wet and occult deposition, as well as suspended dust (a far from marginal fraction of dry deposition), are discussed.

The area under study is the town of Modena and its surrounding area. Modena is a town in northern Italy (180,000 inhabitants; 34 m a.s.l.) situated in the Po Valley, one of the most polluted areas in the world, due a combination of high-density population and industry with orographic and meteoclimatic conditions that are extremely unfavorable for pollutant dispersion, with a high frequency of fogs during the autumn/winter period. The town, itself polluted, is also surrounded by a number of densely industrialized areas and by other zones of intense farming activities, especially livestock breeding and crop cultivation.

While analyses of rain and fog water have already been the subject of specific works,[10,11] those relating to suspended particulate are presented here for the first time. In addition, we will also discuss the correlation between the Ca^{2+} content in rain and fog water and in particulate matter, formulating some hypotheses as to its probable origins.

SAMPLING OF SUSPENDED DUSTS, RAINS, AND FOGS AND DETERMINATION OF THEIR CALCIUM CONTENT

The sampling stations of suspended dusts (Modena, Carpi, Spezzano), rain (Modena, Carpi, Spezzano, Vignola, and Pievepelago), and fog (Modena) were chosen so as to provide a monitoring of different situations of urban and industrial pollution: Carpi (20 km north of Modena; 26 m a.s.l.) is a small town characterized by a high concentration of small industrial plants and small specialized farming activities; Spezzano (20 km southwest of Modena; 125 m a.s.l.) is a small village situated within a zone of intense industrial concentration; Vignola (25 km southeast of Modena; 120 m a.s.l.) is a small center at the foot of the Apennines with a mixed, prevalently agricultural economy; Pievepelago (100 km south of Modena; 852 m a.s.l.) is a minuscule mountain resort town in close proximity to the Apennine ridge.

The systematic sampling of suspended particulate and rainwater has been carried out in Italy for many years, with the measurement of some of the dusts and metals contained in them explicitly foreseen by Italian law (DPCM 28/03/83 and DPR 203/88) as part of a government program for air quality control in urban environments.

In addition, the sampling and chemical analysis of rainwater is carried out by the RIDEP acid deposition monitoring network, which covers the entire area of Italy under the coordination of the Ispra Research Center.

Conversely, the sampling and chemical analysis of fog water were performed at seven stations in the Po Valley, in northern Italy, an area with a very high frequency of fog episodes, for a limited time lasting from winter 1990/1991 to the same period of 1993/1994, and involving a total of four collection campaigns of 6 months each, in correspondence with the autumn-winter period.[2] The present work discusses only the data collected at Modena.

The data used in this work refer to the years 1990/1992 for suspended dust, 1989/1996 for wet deposition (those used here spanned the period from winter 1992 to winter 1993) and 1990/1994 for occult deposition. However, it is reasonable to suppose that no significant change has occurred in the considered parameters since then. The most important intervention affecting the area of study in recent years has been the extension of the methane supply network, which led to a drastic reduction in the use of oil-based fuels and a plummeting of SO_2 levels in the environment (Modena, for example, passed from an average of 275 $\mu g\ mc^{-1}$ in winter 1975/1976 to an average of 15 $\mu g\ mc^{-1}$ in winter 1993/1994). This operation, launched in the late 1970s was already complete by 1985/1990, with the eventual switch to methane of most combustion plants of public buildings and facilities.

Added to this, in the years 1975 to 1985, dust and gas filtering was enforced by law in industrial sources, leading to a decrease in suspended dusts.

Sampling of Wet Deposition

As is the case with most international monitoring networks, among the various types of samplers available, only the wet&dry one has been adopted. It consists of a container, a rain sensor, and a mechanical arm activated by the sensor, so that the container can be opened in order to obtain rain collected from a single event. This avoids the contamination of the liquid due to dry deposition. The precipitation collector, generally in ethylene or another chemically inert material, is combined with a further container for the collection of dry deposition (Figure 17.1A). Sampling was performed on a weekly basis. After pH and conductibility determinations, the collected liquid was filtered (pore ∅ 0.2 μm) and underwent chemical analysis. In particular, calcium was determined by means of ICP emission spectrometry.

Sampling of Occult Deposition

Two basic types of samplers are used — passive or active — based on the same principle.[13] The droplets of fogs (or cloud, although this is not the case here) impact upon a series of chemically inert filaments,[14] increase in diameter due to coalescence, and fall by gravity into the collector below (Figure 17.1B).

Passive samplers are more suitable for the sampling of cloud water. They can be positioned in remote places and require no electricity supply or special maintenance. The turbulent conditions characteristic of high locations facilitate the impaction of the drops onto the sampler.

The sampling of fog droplets in low-lying flat areas calls for the use of a suction pump that accelerates the flow of the droplets, thus facilitating their impaction onto the collection filaments, since the thermal inversion phenomena which give rise the fog occurrence take place at ground level and in a total absence of wind.

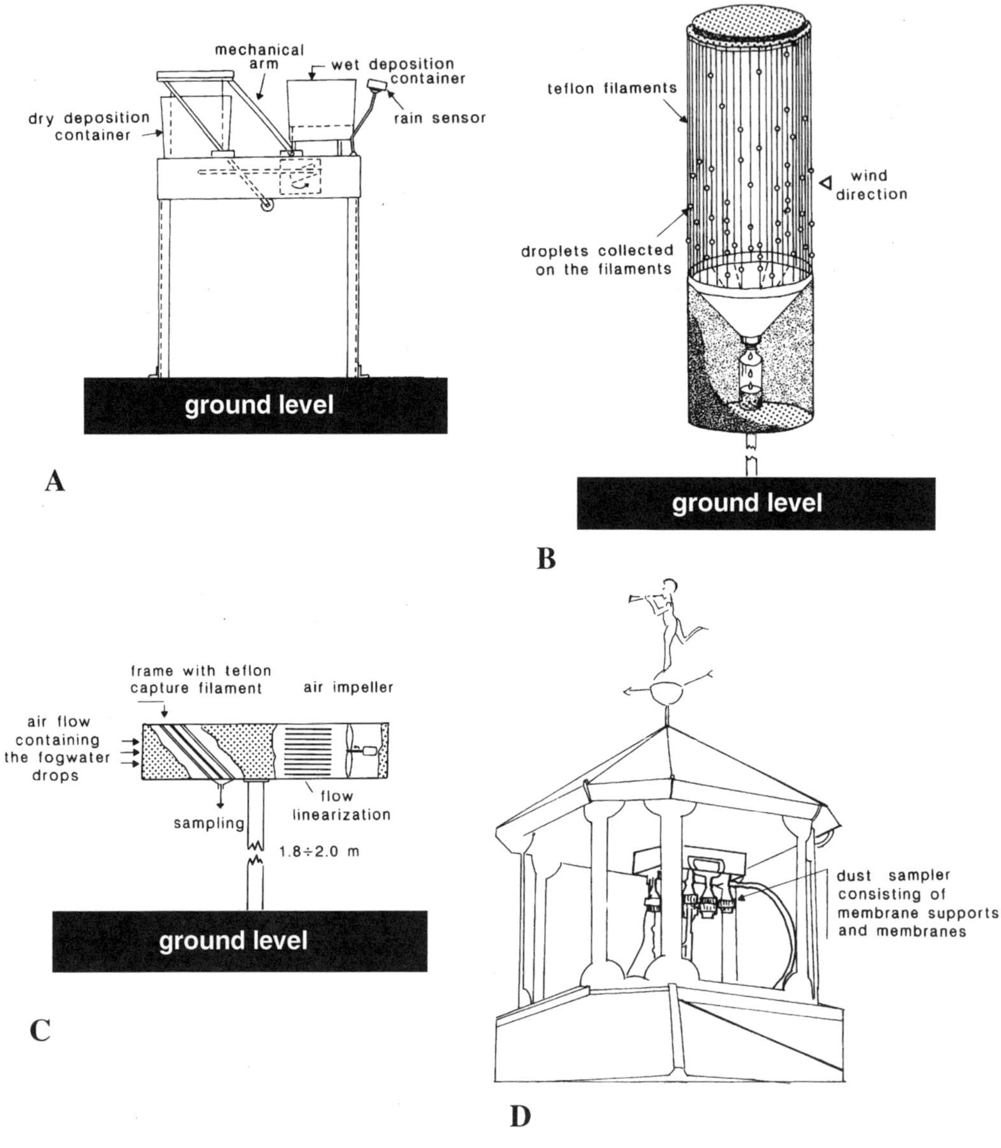

FIGURE 17.1 Samplers of wet (A) occult deposition (B and C), and suspended dusts (D). (A): wet and dry sampler composed of two containers for the collection of wet and dry deposition, a mechanical arm and rain sensor; (B): passive sampler of cloud or fog droplets impacting due to the wind on teflon filaments and collecting in the container below; (C): active sampler of fog droplets composed of an aspiration pump, three series of teflon filaments, a fog sensor, and a container below; and (D): multiple particle sampler composed of eight membrane supports and relative membranes (cellulose ester filters) linked to an aspiration pump and placed at 3 to 4 meters from the ground.

The active sampler employed for the present investigation campaign was equipped with:

- A rain sensor to prevent activation in the event of rain or snow
- An optical fog sensor to control sampler activation
- A temperature control system that switched off the sampler when the temperature went below 0°C, with the risk of freezing the teflon filaments or damaging the suction pump (Figure 17.1C).

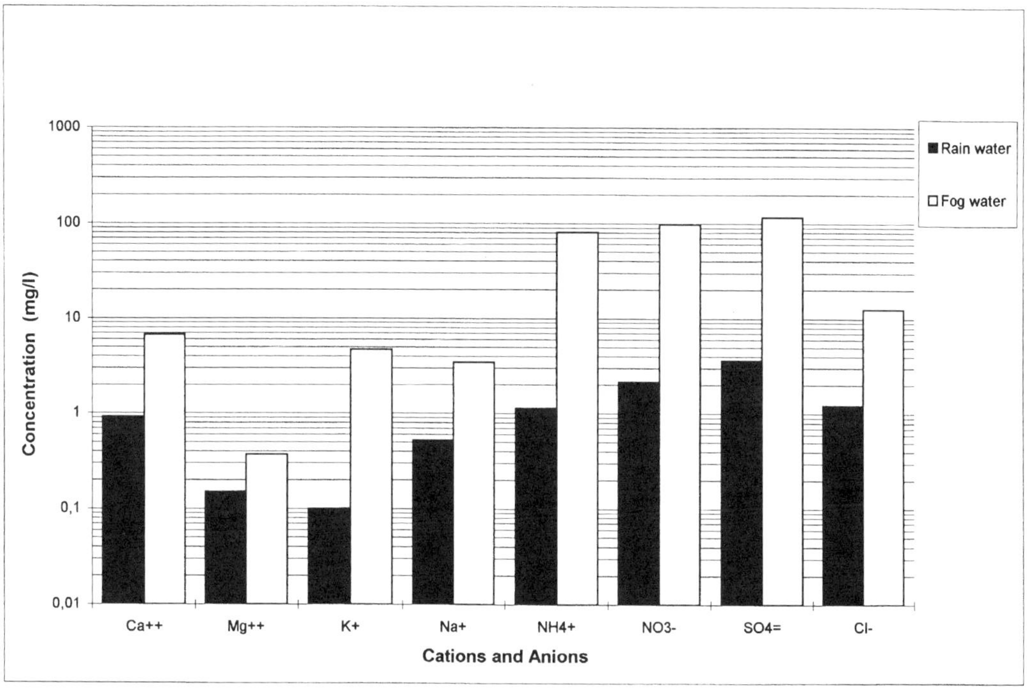

FIGURE 17.2 Fog water and rainwater.

Sampling was performed on a daily basis, from midday to the same time the following day, as this is the time of day when fog is less likely to occur and therefore no fog event was interrupted.

As was the case for wet deposition, the analysis of the collected samples began with pH and conductivity determination, filtration (pore ∅ 0.2 μm), followed by chemical analysis of cations and anions. Here too, calcium was determined by ICP emission spectrometry.

Sampling of Suspended Particulate

Sampling was performed by air aspiration (at a rate of 20 l min^{-1}) on cellulose ester filters (∅ 47 mm, pore ∅ 0.8 μm) for 24 hours. The sampling points were situated at a height of 3 to 4 meters above the ground and a distance of at least 2 meters from buildings, trees, or any other surface likely to influence significantly the representativity of the sample. During sampling, the filtration surface was kept facing downward in order to avoid direct deposition or accidental particles, as well as protecting it from atmospheric precipitation (Figure 17.1C). The total amount of dust on the filter was determined gravimetrically, after the filter had undergone conditioning in an atmosphere of controlled humidity and temperature. Since the difference in weight is of the order of a few micrograms, it is crucial that the two weightings, before and after sampling, take place in perfectly reproducible conditions. Subsequently, the filter was mineralized with nitric acid and the solution was obtained for the determination of metals by ICP emission spectrometry.

Calcium in Rainwater

Table 17.1 presents the chemical data obtained on the rainwater samples. They are discussed below and compared with those obtained from fog water (see Table 17.2 and Figure 17.2).

TABLE 17.1
Chemical Composition of Rainwater

Sample	Date	Volume (ml)	pH	Cond. (μS)	Ca^{2+} (mg l^{-1})	Mg^{2+} (mg l^{-1})	K^{+} (mg l^{-1})	Na^{+} (mg l^{-1})	NH_4^{+} (mg l^{-1})	NO_3^{-} (mg l^{-1}l)	SO_4^{2-} (mg l^{-1})	Cl^{-} (mg l^{-1})
PI 18/92	5/4/92	3750	5.09	130	0.35	0.07	0.05	0.22	0.25	3.07	6.40	1.51
VI 18/92	5/4/92	3800	4.90	22	0.42	0.04	0.10	0.17	1.10	1.93	2.59	1.45
SP 18/92	5/4/92	2600	5.89	40	0.90	0.18	0.15	0.60	0.93	1.79	4.31	1.14
MO 18/92	5/4/92	3750	6.69	45	0.88	0.13	0.31	0.43	1.20	2.19	3.41	0.96
CA 18/92	5/4/92	2520	4.57	32	0.56	0.11	0.09	0.30	1.40	1.91	5.17	1.21
PI 28/92	7/13/92	2350	4.60	25	0.69	0.13	0.10	0.28	0.42	1.14	1.92	0.90
VI 29/92	7/20/92	1580	4.65	32	0.63	0.07	0.07	0.25	1.80	3.66	3.26	0.70
SP 29/92	7/20/92	3150	5.23	35	0.97	0.17	0.09	0.44	1.79	4.45	4.48	2.04
MO 29/92	7/20/92	1400	6.15	47	1.32	0.18	0.11	0.18	3.91	7.69	5.85	1.12
CA 28/92	7/13/92	1730	6.64	28	0.74	0.10	0.09	0.25	1.62	1.90	2.72	1.01
PI 39/92	9/28/92	2700	7.15	14	0.68	0.06	0.12	0.14	0.20	1.16	1.04	0.63
VI 40/92	10/5/92	2850	6.10	11	0.88	0.11	0.08	0.67	0.54	1.05	2.07	1.08
SP 40/92	10/5/92	5000	6.31	18	1.33	0.17	0.08	0.99	0.99	1.32	3.67	1.55
MO 40/92	10/5/92	4600	4.74	12	1.44	0.14	0.04	0.59	0.78	0.90	2.15	0.85
CA 40/92	10/5/92	8000	5.37	14	0.72	0.12	0.06	0.47	0.66	0.96	2.01	1.03
PI 49/92	12/7/92	9710	4.88	13	0.34	0.09	0.03	0.58	0.06	0.63	1.17	1.18
VI 49/92	12/7/92	1450	6.31	16	0.82	0.16	0.08	0.24	0.51	1.36	1.67	0.57
SP 49/92	12/7/92	1500	4.87	16	0.64	0.05	0.07	0.17	0.81	1.37	2.01	0.56
MO 49/92	12/7/92	1020	6.47	20	1.46	0.09	0.09	0.36	0.85	1.50	2.73	0.71
CA 49/92	12/7/92	920	5.51	14	0.26	0.04	0.03	0.10	1.36	1.80	1.96	0.53
PI 10/93	3/8/93	2930	8.59	20	2.22	0.10	0.08	0.28	0.25	1.38	1.80	0.94
VI 10/93	3/8/93	3890	4.11	48	0.70	0.27	0.16	1.38	2.14	3.73	7.48	2.88
SP 9/93	3/1/93	1470	5.39	48	1.74	0.27	0.21	1.70	2.46	2.38	8.18	2.62
MO 10/93	3/8/93	2560	4.47	41	1.14	0.23	0.20	1.10	2.11	2.51	7.75	1.70
CA 10/93	3/8/93	2100	4.60	32	1.08	0.76	0.12	1.00	0.48	1.88	4.98	1.60

TABLE 17.2
Chemical Composition of Fog Water

Sample	Date	Volume (ml)	pH	Cond. (μS)	Ca^{2+} (mg l^{-1})	Mg^{2+} (mg l^{-1})	K^+ (mg l^{-1})	Na^+ (mg l^{-1})	NH_4^+ (mg l^{-1})	NO_3^- (mg l^{-1})	SO_4^{2-} (mg l^{-1})	Cl^- (mg l^{-1})
MO 4/92	4-5/2/92	68	5.94	640	12.94	0.79	4.20	16.40	92.8	73.3	187.3	23.08
MO 6/92	2/28/92	52	7.09	1340	11.38	0.70	35.11	10.57	200.8	250.7	230.9	35.32
MO 7/92	3/4/92	120	7.11	420	6.94	0.35	14.60	9.40	54.4	62.0	78.5	16.71
MO 10/92	11/23/92	160	6.30	440	11.30	0.71	2.58	2.66	68.9	69.4	111.5	8.54
MO 11/92	11/24/92	2750	6.42	245	1.20	0.16	1.09	1.04	36.6	45.4	44.0	5.94
MO 2/93	1/18/93	395	6.77	540	1.70	0.18	0.80	0.87	82.9	57.2	143.6	16.62
MO 3/93[a]	1/19/93	317	6.25	800	27.50	0.56	2.68	1.51	107.0	91.1	239.7	7.98
MO 5/93	1/22/93	930	6.54	260	3.24	0.36	1.44	1.83	41.0	44.9	65.8	9.98
MO 6/93	1/23/93	660	6.71	375	2.04	0.22	3.53	1.32	57.8	61.1	95.8	9.35
MO 7/93	1/25/93	1620	6.25	370	1.73	0.22	0.71	0.89	57.4	46.5	96.6	11.41
MO 14/93	12/21/93	1990	6.40	393	2.04	0.26	1.46	1.24	54.6	61.7	124.0	11.92
MO 15/93	12/22/93	890	6.50	625	2.48	0.25	1.49	2.69	111.0	85.4	190.5	14.99
MO 2/94	1/11/94	77	6.95	256	19.80	0.49	4.78	2.49	28.7	17.7	98.1	9.38
MO 3/94	1/15/94	132	6.13	710	9.82	0.59	5.17	3.26	99.5	127.4	171.4	12.20
MO 6/94	1/26/94	300	6.74	440	14.50	0.49	1.65	3.37	61.4	95.9	103.8	9.94
MO 7/94	2/4/94	650	6.20	770	3.14	0.31	3.90	2.67	118.0	218.6	116.3	12.18
MO 11/94	2/26/94	1140	6.17	240	1.43	0.13	0.45	0.54	35.2	60.8	39.6	3.27
MO 13/94	2/28/94	480	3.34	1310	3.66	0.43	2.72	1.81	177.0	231.0	120.0	8.73
MO 14/94	3/19/94	375	6.97	670	5.12	0.15	2.17	1.49	104.0	153.2	112.6	11.60
MO 15/94	3/25/94	405	7.44	428	12.90	0.28	1.16	1.25	56.8	97.4	79.5	6.65

[a] MO 3/93 was not utilized for data elaboration.

The saline concentration in the rainwater is rather low: 12.2 mg l^{-1} (min: 5.1; max: 27.2) when compared with those for fog water: 389.0 mg l^{-1} (min: 151.0; max: 848.0). This is in good agreement with the conclusions reported in the literature; that is,

- The "far off" origins of clouds and the very limited effect of washout of the atmosphere and capture of pollutants below clouds
- The origins of fogs (particularly irradiance fogs) near the ground and in polluted areas, not only in more concentrated form everywhere, but also reflecting the nature and concentration of pollutants near the ground

If one considers the concentration ratios of the single ions in rainwater, one sees that, unlike the case of fog water (Figure 17.2),

1. Ca^{++} is slightly higher than Na^+.
2. The correlation between Na^+ and Cl^- (in meq l^{-1}) is good.
3. Cl^- turns out to be relatively higher and only slightly lower in concentration than SO_4^{2-}, confirming what has been known for some time, (i.e., that clouds mainly form over oceans and seas and their condensation nuclei are mostly composed of NaCl (>80/90%)). It follows that the most abundant minerals deposited on the ground are, in order of abundance, calcium sulfate and sodium chloride, although the latter prevails over the former in clean areas.[10]
4. Ca^{2+} is present in concentrations similar to those of NH_4^+. On the other hand, NH_4^+, linked to crop cultivation and livestock raising, is among the principal aerodispersed cations near the the ground in all areas of human inhabitation, where it is generally more than Ca^{2+} by at least one order of magnitude. Once again, this fact is in agreement with the above-mentioned "far off" origins of clouds and small degree of washout of the underlying atmosphere on the part of raindrops.

Comparing the data obtained for rain and fog, we conclude that in rainwater, while NH_4^+ (livestock raising and farming practices), SO_4^{2-}, and NO_3^- (combustion processes) turn out to be fairly low, Ca^{++} is rather high. As there is no doubt that ammonium, sulfate, and nitrate are of anthropic origin, and there being little correlation between them and calcium, it seems probable that the calcium in rainwater is of natural origin.

Calcium in Fog Water

Table 17.2 shows the results of the analyses performed on fog water. The values of NO_3^- and SO_4^{2-} reported here include NO_2^- and SO_3^-, also present, expressed as nitrates and sulfates.

As observed above, the pollutant concentration is very high, when compared with analogous measurements in rainwater made by us and by other authors.

Considering the values for the single ions, it is possible to note that:

1. The values of NH_4^+ are very high among the cations, as are those for SO_4^{2-} and NO_3^- among the anions. Comparing our concentration averages for fog and rain, the concentration of NH_4^+ is ca. 70 times greater, NO_3^- is 45 times greater, and SO_4^{2-} 32 times greater. As mentioned above, this confirms the genesis on the ground of irradiance fogs, whose condensation nuclei are mainly composed of pollutant particles originating from crop cultivation, livestock raising, and local combustion processes (ammonium sulfate, ammonium nitrate, calcium sulfate, ammonium chloride, potassium nitrate, etc.[11])
2. Ca^{2+} is the second cation in order of abundance after NH_4^+. The mean concentration of Ca^{2+} is approximately 7 times (7.44) the value measured for rainwater. However, since

fog water is more concentrated than rainwater, Ca^{2+} is actually less abundant with respect to rainwater.
3. While Mg^{2+} and Na^{+} vary little in concentration (the concentration ratios relative to rainwater are, respectively, 2.5 and 6.7), K^{+} increases considerably (46.8).
4. Among the anions, Cl^{-} increases approximately tenfold (10.3), although its growth is far less than that registered for sulfates and nitrates.

On the basis of these observations, some hypotheses can be proposed on the origins of the elements measured, bearing in mind that, as already indicated several times, rainwater is a far less polluted system than fog water. The Mg^{2+} and Na^{+} present here in very small absolute percentage terms, can be considered mainly of natural origin.

- It is probable that the origin of Cl^{-} is partially natural and partially anthropic (solid urban waste incinerators, industrial plants, emissions during the addition of chlorine to water supplies).
- The origin of K^{+} seems to be largely anthropic (farming activities and others).
- It is possible to hypothesize a mixed origin for Ca^{2+}: part natural, part anthropic. However, we will briefly return to this subject in the conclusions. Here, we limit ourselves to pointing out that the only relatively stable mineral released by fog water evaporation in the field is calcium sulfate dihydrate ($CaSO_4 \cdot 2H_2O$[11]).

Calcium in Particulates

Turning to suspended particulates, it should be borne in mind that, because of the different sampling methods, it is not possible to make a quantitative comparison between the percentages of elements measured in dusts and those measured in rainwater and fog water. It should also be remembered that in rain and fog water, only the part that is soluble in water is measured (the remainder being filtered and removed prior to analysis). Conversely, in suspended dusts, analysis is made of everything that is etched in hot nitric (strong) acid; this includes also Na^{+}, K^{+}, Ca^{2+}, Mg^{2+}, and $Fe^{2+,3+}$ arising from the partial destruction of silicates.

For these reasons, only qualitative considerations are possible, which, as the subject of the present note, will largely regard Ca^{2+}.

Table 17.3 reports the results obtained from analyzing sample dusts.

Looking at the two diagrams in Figures 17.3 and 17.4, one sees that Figure 17.3 refers to the samples of suspended particulate collected from the various sampling sites (Modena 1, Modena 2, Modena 3, Carpi, Sassuolo) during the winter period (January 1990); and Figure 17.4 refers to suspended dusts at the same sites during the summer period (August 1991). As previously mentioned, during winter, all the main pollutant sources are very active (domestic heating systems, motor vehicles, industrial plants); while in summer, they are either absent (domestic heating) or reduced to a minimum (automobile traffic and industry).

The two diagrams, based on monthly means calculated from daily values, prove to be in agreement with the following general trend: while in January the suspended dusts exceed 150 μg mc^{-1} (x_m = 150.2; min = 136.0; max = 170.0) at all sites, in August they decrease to below 100 μg mc^{-1} (x_m = 85.6; min = 83.0; max = 94.0). This finding confirms that, as well as the certain presence of background particulate linked to soil dust, which tends not to be correlated to seasonal change, there is an abundant fraction of suspended particulate arising from anthropic pollution. This subject is briefly discussed in the conclusion of the chapter, with regard to the mineralogical composition of the particulate.

Turning now to calcium, in January 1990 it was present in an amount equal to 9.87 μg mc^{-1} (min = 7.79; max = 14.94), while it decreased to 3.29 μg mc^{-1} (min = 2.38; max = 4.12) in August 1991, a finding that shows how Ca, like all the dusts taken together, is at least partially linked to

TABLE 17.3
Calcium Concentration in Urban Atmosphere

	Giardini Mo1			Largo Garibaldi Mo2			Cavour Mo3			Carpi			Sassuolo		
	Ca ($\mu g\ m^{-3}$)	Dust ($\mu g\ m^{-3}$)	Ca/dust (%)	Ca ($\mu g\ m^{-3}$)	Dust ($\mu g\ m^{-3}$)	Ca/dust (%)	Ca ($\mu g\ m^{-3}$)	Dust ($\mu g\ m^{-3}$)	Ca/dust (%)	Ca ($\mu g\ m^{-3}$)	Dust ($\mu g\ m^{-3}$)	Ca/dust (%)	Ca ($\mu g\ m^{-3}$)	Dust ($\mu g\ m^{-3}$)	Ca/dust (%)
November 90	5.74	138.0	3.40	4.64	111.0	3.60	4.32	141.0	3.30	9.10	187.0	4.87	9.90	155.0	6.39
December 90	5.75	157.0	3.50	5.65	149.0	4.40	5.59	154.0	3.60	9.00	180.0	5.00	6.41	133.0	4.90
January 90	14.94	142.0	7.60	7.79	154.0	5.50	8.00	136.0	6.20	9.46	170.0	5.90	9.17	149.0	6.00
Mean values	8.81	146.0	4.83	6.02	138.0	4.50	5.97	144.0	4.37	9.19	179.0	5.26	8.49	146.0	5.77
July 91	4.49	108.0	4.10	4.72	106.0	4.50	2.15	87.0	2.47	4.38	100.0	5.00	5.97	112.0	5.50
August 91	3.26	87.0	3.50	2.38	83.0	2.80	3.09	94.0	3.29	3.56	79.0	4.50	4.12	85.0	4.70
September 91	4.81	113.0	4.10	4.82	117.0	4.10	4.82	91.0	5.60	4.59	113.0	4.00	7.03	127.0	5.50
Mean values	4.19	103.0	3.90	3.97	102.0	3.80	3.53	90.7	3.79	4.18	97.3	4.50	5.73	100.0	5.23
January 92	4.78	119.0	3.80	5.96	115.0	4.30	9.00	160.0	5.62	4.67	100.0	4.80	6.50	125.0	5.10
February 92	7.43	168.0	4.60	6.32	171.0	4.10	8.85	166.0	5.10	7.82	164.0	4.90	10.70	174.0	6.10
March 92	7.52	157.0	4.90	5.90	181.0	3.50	7.44	144.0	5.60	6.69	138.0	4.80	10.60	177.0	6.20
Mean values	6.58	148.0	4.43	6.27	155.7	3.97	8.63	156.7	5.43	6.60	134.0	4.83	9.27	158.3	5.80

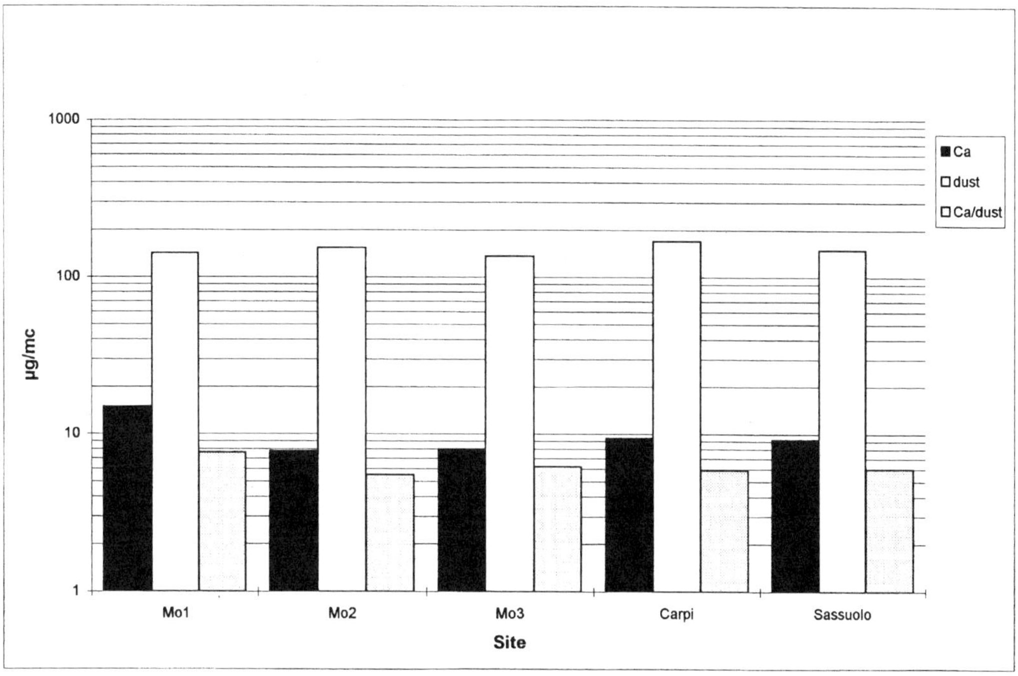

FIGURE 17.3 January 1990.

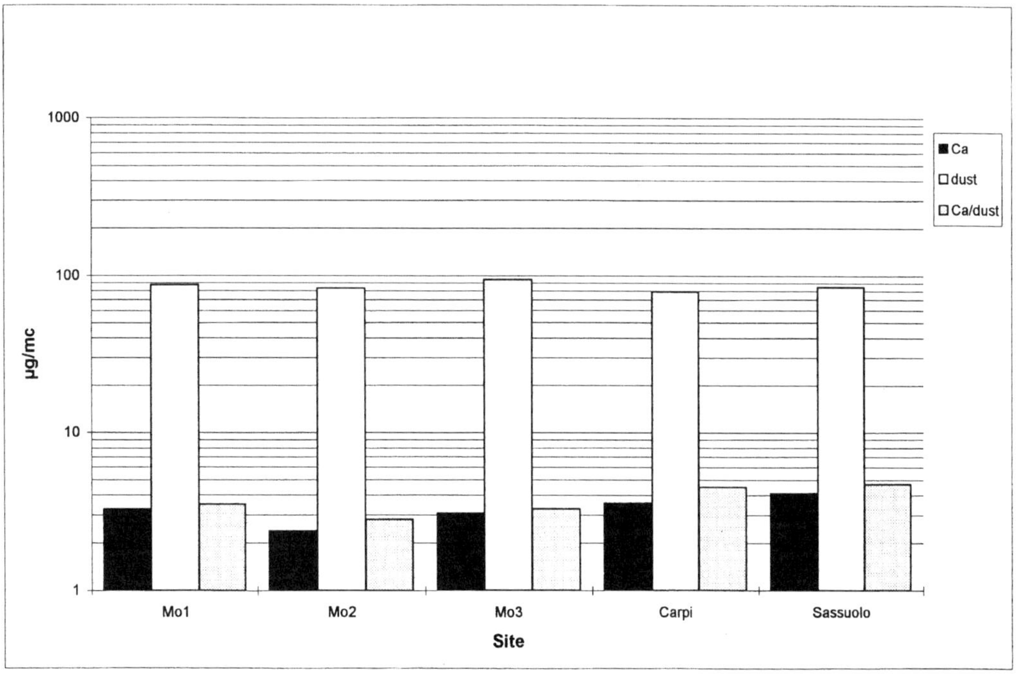

FIGURE 17.4 August 1991.

TABLE 17.4

	E 81-1 (% w/w)	E 81-2 (% w/w)	BO 1 (% w/w)	BO 2 (% w/w)	BO 3 (% w/w)
Si	28.66	18.90	23.89	26.7	21.67
Al	8.21	6.14	7.62	9.45	8.89
Fe	5.64	7.48	4.13	6.12	6.69
Ca	2.48	1.51	7.53	6.15	9.98
Mg	2.32	2.20	2.67	1.89	2.12
K	2.71	0.90	3.15	4.23	6.12
Na	0.60	1.36	0.79	0.98	0.78
P	0.08	0.09	0.17	0.09	0.12
Ti	0.44	0.60	0.58	0.12	0.56
C	n.d.	n.d.	n.d.	3.98	8.03

air pollution. Considering all of the available data, the percentage of Ca^{2+} is ~ 9% (max = 14.94; min = 4.32) in winter and ~ 4% (max = 7.03; min = 2.15) in summer, figures which also point toward a partially anthropic origin for Ca^{2+}. In fact, the values show a fair degree of crustal enrichment (see Reference 15) and are higher than those generally measured in nearby areas (studied by us) that are clean (Tyrrhenian Sea, Adriatic Sea).

CALCIUM IN RELATIVELY CLEAN AREAS

The data obtained on suspended particulate can be better understood through a comparison with those measured in nearby coastal areas that are relatively clean. The comparison presented here is based on dust samples collected during the Eolo Project of the Italian National Research Council.[15]

Table 17.4 reports the mean chemical values for the main elements encountered in dust sampled during the Eolo 81-1 and Eolo 81-2 sailings.

The percentage of Ca^{2+} in suspended dust is 2.5% less in weight, while the same campaign found that in the polluted area of Bologna, also situated in the Po Valley, 45 km southeast of Modena, it reached a value of 7.53% in weight. Similar results were obtained by us on dust samples collected in the same town during subsequent years (BO 2, Summer 1995: 6.15%; BO 3, Winter 1996: 9.98%; see Table 17.4).

These data confirm the results of the measurements shown above and unequivocally point toward a largely anthropic origin of the Ca^{2+} present in the suspended particulate of the air in polluted cities. In other words, it would seem that part of the Ca^{2+} present in the city atmosphere is generated by the city itself.

The possible sources of Ca^{2+} in urban atmospheres are numerous. In the following section, we advance only a few preliminary hypotheses concerning them, since research in this field is still in progress.

CALCIUM IN FLY-ASH AND SOOTS

Among the polluting substances that cause visible effects in the urban environment are carbonaceous particulates originating from the combustion of fossil fuels and their derivatives. They blacken the surfaces of buildings and monuments, inflicting enormous damage (not only of an aesthetic character) and incurring a collective cost of millions of dollars per year.

The best-known are the micronic fly-ash emitted by the combustion of petrol and coal in domestic heating systems (today in decline), industrial plants of various size, electricity generating

TABLE 17.5

	Fly-ash (% w/w)	Fly-ash (% w/w)	Soots (% w/w)	Soots (% w/w)
C	58.90	63.91	57.26	53.89
S	7.16	6.00	5.78	5.23
N	0.97	0.89	1.15	0.81
Ca	1.35	1.35	1.10	0.98
Na	1.60	1.57	1.58	1.43
K	0.08	0.12	0.18	0.16

plants (which are, however, generally located outside of towns and therefore contribute only to the regional background), and submicronic soot mainly emitted by diesel engines.

Table 17.5 shows the elemental composition expressed in % weight of several samples of carbonaceous fly-ash from oil combustion and soots.

As can be seen, the carbonaceous particulate emitted by the main urban combustion processes contain Ca^{2+}.

However, since the C content of bulk samples collected in an urban atmosphere is low (in Bologna, it ranges from 3.98% in weight in summer to 8.03% in weight in winter; as mentioned, the data refers to the years 1995–1996; see Table 17.4), it can be affirmed that the calcium contribution of fly-ash and soot to atmospheric particulate is very low: ~ 0.05% in summer and 0.1% in winter. This means that the various combustion processes emit Ca^{2+} into the atmosphere, but it represents an insignificant fraction of the airborne calcium of anthropic origin. The sources of Ca^{2+} must therefore be sought elsewhere.

LOCAL SOURCES OF CALCIUM

Local sources of Ca^{2+} in urban atmospheres are theoretically very numerous, although the single contribution made by each of the sources is, for now, difficult to quantify.

As seen above, the calcium derived from various combustion processes associated with carbonaceous particles, for which a quantitative estimate can be made, accounts for a negligible fraction of atmospheric calcium (0.05% in summer; 0.1% in winter). Considerable quantities of calcium are also emitted by asphalt road surfaces (bitumen plus limestone fragments), particles of which are detached, lifted, and emitted into the atmosphere by motor vehicles. Other sources of calcium are those parts of vehicles that undergo constant wear and tear: tires, brake linings, and clutch plates. Such sources are probably most efficient during the winter season when urban road traffic is at its highest. Due to weathering processes, other efficient sources of Ca^{2+} are cement mortars, plasters (composed of lime, marble powder, or sometimes chalk), and, to a lesser degree, bricks. Such sources are expected to be equally efficient in both summer and winter.

As previously, work has only just begun in this field and the few data available thus far are insufficient to allow the formulation of any conclusive hypothesis.

CALCIUM SULFATE DEPOSITION IN CITIES

Deposits of calcium sulfate in cities are referred to as blacking, black patinas, or black crusts. The black color is due to gypsum, which invariably mixes in with the carbonaceous particulate originating from various combustion processes (see "Discussion").

TABLE 17.6

	SP20 (% w/w)	SP203 (% w/w)	SP202 (% w/w)	SP201 (% w/w)	SP206 (% w/w)	SP204 (% w/w)	GA2 (% w/w)	MO1 (% w/w)	MO23 (% w/w)	MO31 (% w/w)
Ca	17.45	18.60	19.40	19.21	18.01	19.34	20.01	17.69	18.98	18.45
S tot.	16.50	20.13	19.70	19.24	20.07	19.60	18.71	19.99	20.67	20.12
S SO_4^{--}	13.04	14.81	15.30	15.24	15.01	13.84	15.70	15.89	16.09	16.89
C tot	2.53	1.16	1.41	1.53	1.23	1.58	1.58	2.67	3.02	1.98
Si	5.33	4.07	4.93	4.73	5.05	3.90	2.50	4.89	3.98	4.10
Fe	3.06	0.96	1.05	1.22	1.43	0.74	0.56	1.87	1.34	1.23
Al	1.13	0.68	0.76	0.79	0.82	0.75	0.47	1.12	1.67	0.98
Mg	0.15	0.17	0.20	0.18	0.00	0.20	0.14	0.13	0.19	0.17
Na	0.15	0.16	0.19	0.18	0.21	0.13	0.08	0.12	0.18	0.10
Ti	0.06	0.02	0.04	0.04	0.03	0.03	0.02	0.09	0.10	0.05
Pb	0.06	0.03	0.04	0.04	0.05	0.03	0.04	0.09	0.12	0.07
Ba	0.05	0.03	0.03	0.04	0.03	0.03	0.04	0.00	0.02	0.01
Cu	0.04	0.01	0.01	0.02	0.02	0.00	0.01	0.03	0.01	0.05
Zn	0.02	0.01	0.02	0.02	0.02	0.00	0.07	0.00	0.00	0.03
Mn	0.02	0.01	0.01	0.01	0.01	0.01	0.01	0.00	0.02	0.01
K	0.01	0.00	0.01	0.01	0.01	0.01	0.00	0.00	0.00	0.00
V	0.01	0.00	0.01	0.01	0.01	0.00	0.01	0.03	0.04	0.09

Table 17.6 presents the data obtained from several chemical analysis of black crusts collected from monuments in limestone, bronze, and selenite in the cities of Modena and Bologna.[16] As can be seen, the mean concentration of Ca^{2+} is ca. 19% in weight (18.7). This value is far greater than those previously observed in the various systems analyzed: rainwater, fog water, and suspended particulate.

It is clear that Ca^{2+} in these deposits is indirectly enriched on account of gypsum being a stable and relatively insoluble mineral.[17-19] All other minerals hypothesized or observed in the laboratory (chlorides, nitrates, sulfates of NH_4^+, Na^+, and K^+), being hygroscopic and deliquescent, either do not crystallize at all or are eliminated rapidly from the "black crust" system.

DISCUSSION

If we consider the mean composition of the Earth's crust, calcium is the fourth element in order of abundance ($\approx$3.6% in weight). If the calculation is limited to its surface concentration, the content decreases on average to $\approx$1.5 to 2.0% in weight. That is to say, calcium is an abundant element that originates from the "ground" and which, through numerous chemical, physical, and biological mechanisms (both natural and anthropic), enters and becomes part of other systems (e.g., hydrometeors and the atmosphere). In good agreement with the data in the literature, our measurements show that calcium is an important cation in rainwater, fogwater, and suspended particulates.

If we consider hydrometeors, Ca^{2+} turns out to be relatively more abundant in rainwater, where it is present in amounts similar to Na^+ (natural origin: sea spray) and even higher than those of NH_4^+ (anthropic origin: crop cultivation and livestock raising). This suggests that it is partially natural in origin.

In fogs, the marked increase in all ions of certain pollutant origin (NH_4^+, 70 times; K^+, 47 times; SO_4^{2-}, 32 times, and NO_3^-, 45 times) is not accompanied by a corresponding rise in calcium, which therefore appears uncorrelated with the same pollutant sources.

Conversely, the calcium content in atmospheric particulate is so high as to strongly suggest a mainly anthropic and local origin for this cation. As we have seen, this hypothesis appears to be confirmed by the seasonal variations in this element.

Also, the mineralogical composition of the sampled particulate from urban atmospheres often reveals the presence of calcite, and sometimes dolomite (and gypsum), minerals which are generally absent in the dusts collected in clean, non-urbanized areas.

One of the most interesting phenomena with regard to the Ca^{2+} in urban atmospheres is its deposition, by wet, occult (see Figure 17.5), and dry means, onto surfaces of all materials (stone, wood, bronze, glass, cement bricks, etc.), in the form of calcium sulfate dihydrate, in all cases mixed with carbonaceous particulate. It is possible that gypsum and fly-ash co-precipitate simply because they are both components in the "city system," or it could be that gypsum formation is facilitated by the presence of carbonaceous particles. A vast, rather contrasting amount of literature exists on this subject (e.g., see Reference 20 and bibliography presented therein).

The effect of such phenomena is referred to as "soiling": locally, alongside the almost ubiquitous blackening of surfaces, particular deposits form, sometimes of considerable thickness, which are known by the term "black crusts." In such deposits, Ca^{2+} is indirectly enriched compared to many other cations present in the atmosphere. In fact, it leads to the formation of a relatively insoluble mineral, gypsum. By contrast, other cations more abundant than calcium, such as NH_4^+, K^+, and Na^+, form hygroscopic and deliquescent (often merely theoretic) salts, such as sulfates, nitrates, and chlorides, which are carried out of the system by rainwater.

Through such mechanisms, significant quantities of Ca^{2+} and SO_2 (SO_4^{2-}) are removed from the atmospheric cycle to form calcium sulfate dihydrate, which is a relatively stable salt; however, this phenomenon is also difficult to quantify. One of the reasons is that part of the gypsum deposited is, in fact, dissolved and carried away by rainwater.

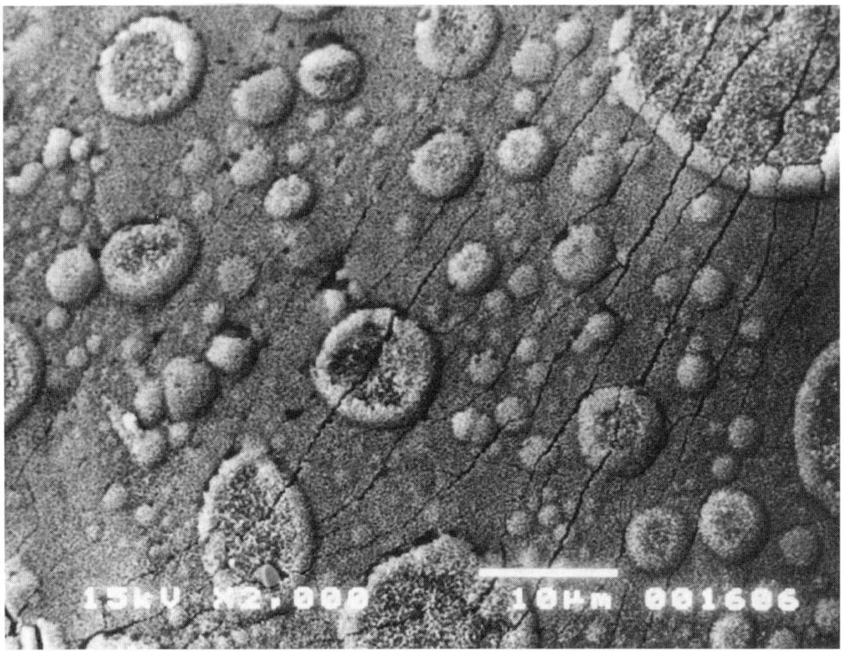

FIGURE 17.5 Fog water droplets after evaporation: the circular spots are made up of a myriad of crystals of calcium sulfate dihydrate. Despite many cations present in fog and the numerous minerals that may, at least in theory, form, the only relatively stable one remaining due to fog/surface interaction is gypsum. An analogous phenomenon takes place after the evaporation of rainwater: halite and gypsum crystallize, but only the latter remains and accumulates on the soil.

Gypsum formation, ubiquitous in polluted cities, is an important step in the geochemical cycle of calcium, but also obviously of SO_2 (SO_4^{2-} in H_2O). Whether gypsum remains on the surfaces, or is dissolved by rainwater wash-out, descending to the water table and reaching the sea, the final result, as observed above, is the subtraction of Ca^{2+} and SO_4^{2-} from the urban atmosphere. It should be remembered that, since the introduction of methane heating plants to Italy in the 1970s to 1990s, the values of SO_2 and carbonaceous particulate in the air have declined drastically. The thick black crusts due to coal combustion that formed in the past in the U.K., Belgium (see Figure 17.6), Germany, and northern France, and the thinner ones due to oil combustion also forming in the past in the southern France and Italy, no longer form today, or form much more slowly. They represent "fossils," the result of environmental conditions that have now ceased. We can therefore expect in the future a strong increase in levels of Ca^{2+} in urban atmospheres or its deposition on the ground, no longer in the form of sulfate, but presumably as carbonate.

ACKNOWLEDGMENTS

This work received the financial support of the ECC, Environmental and Energy Commission, contract ENV4-CT 95-0092 "Archeometric Study to Reconstruct the Pollution and Climate of the Past and Their Effects on Cultural Heritage." We wish to thank Prof. Adriano Ferrari and Dott. Sylvia Lincon for their invaluable assistance.

A

B

FIGURE 17.6 A black crust taken from the surface of a stone monument photographed from both sides (Church of SS. Michael and Gudula, Brussels). As known, the main mineral in these subareal sediments is gypsum ($CaSO_4 \cdot 2H_2O$), which is associated with much smaller amounts of numerous other particles, not only minerals, originating from atmospheric aerosol. The black color of the deposit comes from carbonaceous particles incorporated into the gypsum, and mainly produced in this case by coal combustion. Deposits such as these no longer form today. Urban heating, once relying on coal and petrol, has now been converted almost entirely to methane. The levels of SO_2 (followed by SO_4^{2-}) and the quantities of carbonaceous particles in the air have drastically decreased. Therefore, in the future, we expect a sharp increase in calcium in the atmosphere and/or its deposit on surfaces in the form, no longer of sulfate, but of carbonate.

REFERENCES

1. Fuzzi, S., Monitoraggio delle deposizioni secche, umide, ed occulte, *Il Controllo dell'Ambiente: Sintesi delle Tecniche di Monitoraggio Ambientale,* Pitagora Ed., Bologna, 1993.
2. Dollard, G.J, Unsworth, M.H., and Harve, M.J., Pollutant transfer in upland regions by occult precipitation, *Nature*, 302, 241, 1983.
3. Munger, J.W., Jacob, D.J., Waldman, J.M., and Hoffmann, M.R., Fogwater chemistry in an urban atmosphere, *J. Geophys. Res.*, 88, 5109, 1983.
4. Jacob, D.J., Waldman, J.M., Munger, J.W., and Hoffmann, M.R., A field investigation of physical and chemical mechanisms affecting pollutant concentrations in fog droplets, *Tellus*, 36B, 272, 1984.
5. Jacob, D.J., Waldman, J.M., Munger, J.W., and Hoffmann, M.R., Chemical composition of fogwater collected along the California coast, *Envir. Sci. Technol.*, 19, 730, 1985.
6. Johnson, C.A., Sigg, L., and Zobrist, J., Case studies on the chemical composition of fogwater: the influence of local gaseous emissions, *Atmos. Environ.*, 21, 2365, 1987.
7. Collett, Jr., J., Daube, Jr., B., Munger, J.W., and Hoffmann, M.R., Cloudwater chemistry in Sequoia National Park, *Atmos. Environ.*, 23, 999, 1989.
8. Munger, J.W., Collett, Jr., J., Daube, Jr., H., and Hoffmann, M.R., Fogwater chemistry at Riverside, California, *Atmos. Environ.*, 24B, 185, 1990.
9. Muir, P.S., Fogwater chemistry in a wood-burning community, west Oregon, *J. Air. Waste Manage.*, 41, 32, 1991.
10. Del Monte, M. and Rossi, P., Effects on materials of crystalline compounds precipitated from rainwater, *Eur. Cult. Herit. NLR*, 6, 4, 40, 1992.
11. Del Monte, M. and Rossi, P., Fog and gypsum cristals on building materials, *Atmos. Environ.*, 31, 1637, 1997.
12. Fuzzi, S., Facchini, M.C., Orsi, G. et al., The Nevalpa project: a regional network for fog chemical climatology over the Po valley basin, *Atmos. Environ.*, 30, 2, 201, 1996.
13. Hering, S.V., Blumenthal, D.L., Brewer, R.L., Gertler, A., Hoffmann, M., Kadlecek, A.J., and Pettus, K., Field intercomparison of five types of fogwater collectors, *Envir. Sci. Technol.*, 21, 654, 1987.
14. Fuzzi, S., Cesari, G., Evangelisti, F., Facchini, M.C., and Orsi, G., An automatic station for fog water collection, *Atmos. Environ.*, 24A, 2609, 1990.
15. Tomadin, L., Lenaz, R., Landuzzi, V., Mazzucotelli, A., and Vannucci, R., Wind-blown dusts over the Central Mediterranean, *Oceanologica Acta,* 7, 13, 1984.
16. Del Monte, M. and Forti, P., Interazione tra aerosol atmosferico e superfici carbonatiche: patine nere e patine bianche, Paper presented at *13th SEP-Pollution: Air pollution,* Padova 143, 1990.
17. Del Monte, M. and Vittori, O., Air pollution on stone decay: the case of Venice, *Endeavour*, 9, 117, 1985.
18. Del Monte, M., Air pollution and decay of building stone, *Analusis Magazine*, 20, 3, 20, 1992.
19. Del Monte, M. and Furlan, V., Croûtes noires, sulfatation et dégradation de la pierre, *Proc. 1995 LCP Congress,* Montreux, 349, 1996.
20. Ausset, P., Crovisier, J.L., Del Monte, M., Furlan, V., Girardet, F., Hammecker, C., Jeannette, D., and Lefévre, R.A., Experimental study of limestone and sandstone sulfation in polluted realistic conditions: the Lausanne Atmospheric Simulation Chamber (LASC), *Atmos. Environ.*, 30, 18, 3197, 1996.

18 Corrosion of Asbestos Cement Building Materials by the Action of Atmospheric Acidic Aerosols and Precipitations

Kvetoslav R. Spurny

CONTENTS

INTRODUCTION

By the action of atmospheric anthropogenic aerosols — mainly of acidic aerosols, water, and gaseous acidic pollutants, soot, as well as of their reaction products (acid rain) — several building materials are corroded and degraded. These complex processes are schematically summarized in the Figure 18.1 and are described more in detail in Chapter 17.

Asbestos cement products are subject to such corrosion as are other building materials. Nevertheless, in comparison with commonly used materials (e.g., stones, bricks, tailings, concrete, etc.), the corrosion of asbestos cement building materials is much more hazardous and harmful because this process is associated with the release and dispersion of fine, highly carcinogenic asbestos fibers into the air and rainwater.

During the corrosion of asbestos cement building materials, the cement matrix of the surface material is being destroyed and a thin layer of free and weakly bound asbestos fibers is formed. Wind can then disperse the free fibers into the ambient air. Also, rainwater can wash them from the roof and façade surface and transport them into other parts of the environment — in particular, sewage water, soil, and groundwater.

0-87371-829-1/00/$0.00+$.50

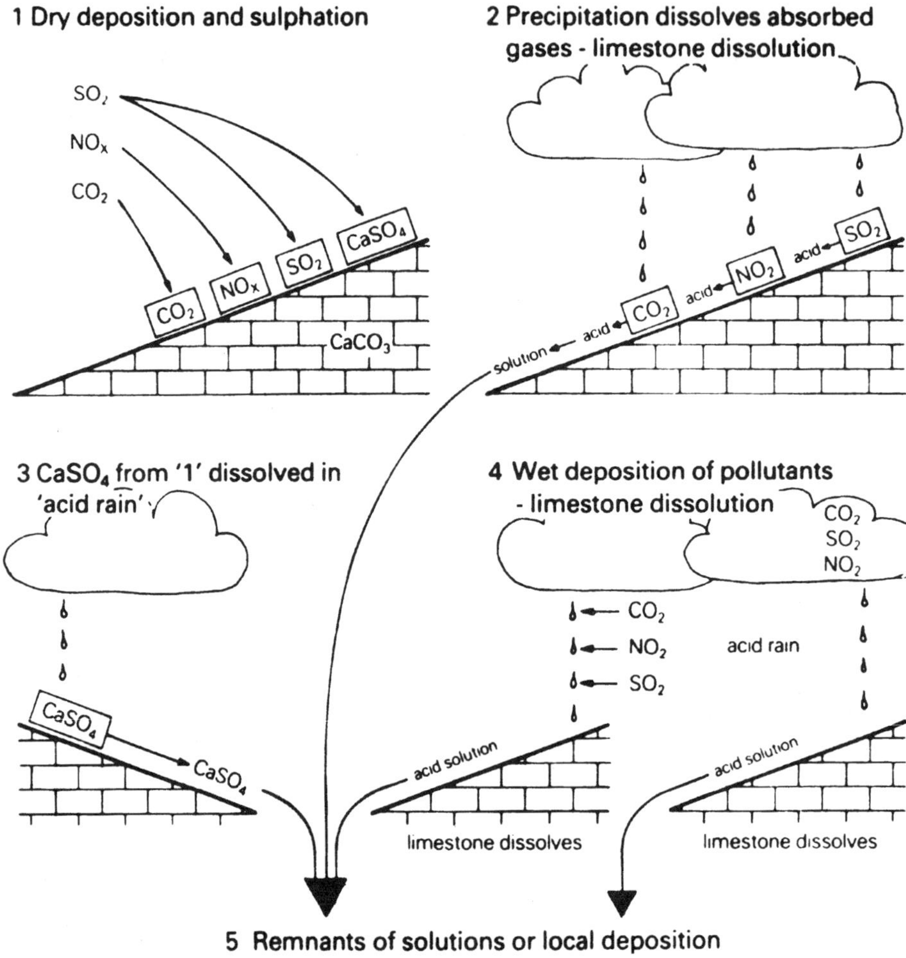

FIGURE 18.1 Scheamatic picture showing the chemical action of gaseous and aqueous air pollutants on stone and other building materials.

OBSERVATIONS AND MEASUREMENTS IN GERMANY

By the end of the 1970s, asbestos concentrations higher than normal had been measured in Germany in the ambient air in the vicinity of buildings containing asbestos cement products.[1] In more complex investigations, at the end of the 1980s, asbestos fiber emissions and ambient air asbestos fiber concentrations were measured.[2,3]

Buildings and Asbestos Cement Products

By observation and electron microscope investigation, the degradation process on building roofs and façades can be evaluated (Figures 18.2 and 18.3).

The surface of diffusion emission sources of asbestos from asbestos cement (AC) products (façade and roofing sheets) in Germany were estimated to be in the range of 10^9 m^2. For all of Western Europe, these surfaces can be as high as 10^{10} m^2.

By mechanical and electron microscope measurements, mean annual corrosion rates were estimated to lie in the range of 0.02 mm per year. This is about 0.5% of the plate thickness. The

FIGURE 18.2 Photographs of corroded asbestos cement roofing and façade shingles in Germany.

mean amount of released asbestos fibers is as high as 3 g m^{-2} per year. The corrosion of asbestos cement products is a linear process.

Sampling and Measurement

A special instrument and the procedure for sampling released fibers has been developed.[4]

The fibrous emissions were collected on Nuclepore or membrane filters, and then the number of fibers, their size distribution, and identities (i.e., composition, crystal structure) were evaluated by means of electron microscopy (SEM and TEM). The principle and application for this instrument are shown in Figure 18.4.

FIGURE 18.3 A photograph (top) and a scanning electron micrograph (bottom) of heavy corroded asbestos-cement roofing sheets.

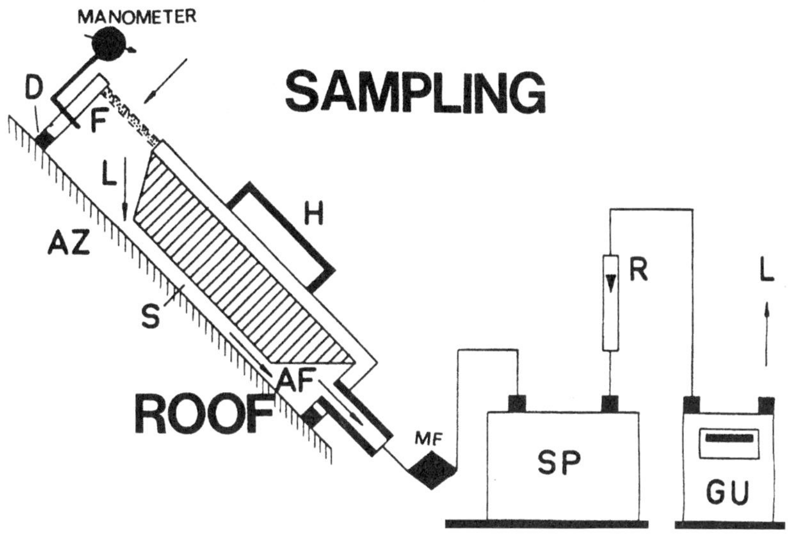

FIGURE 18.4 Schematic picture of the asbestos emission equipment. (Description given in text.)

The sampling chamber with a volume between 2 and 4 liters is placed on the surface (e.g., asbestos roofing tiles). The contact between the instrument and the surface is sealed by means of elatomer ribbons (D). Ambient air is then sucked into the chamber through filters (F). The cleaned air (L) flows over a corroded asbestos cement sheet (AZ) through the small slot (S). The flow velocities of the simulated wind in the slot (S) for the measurements lie between 1 and 5 m s^{-1}. The sampling chamber can be maintained on the measured surface by means of the handle (H), which can be as long as necessary.

Asbestos fibers released from the corroded surface are then sampled on a membrane filter (MF). The fiber concentrations and the fiber emissions are evaluated by means of an electron microscope procedure (Figure 18.5). Other instruments are used to measure the flow rate (R) and the total volume (GU). SP is a vacuum pump with a capacity of about 180 l min^{-1}.

The sampling time was 2 h. Each filter sample (Figure 18.5) was then evaluated by a scanning microscopical procedure (SEM). Individual fibers were identified by energy dispersive X-ray analysis (EDXA). Bulk analysis was provided by means of X-ray fluorescence spectroscopy. The filter samples were evaluated quantatively using a combination of the two standard procedures: ISO[5] and VDI[6].

The German (VDI) standard method that samples with gold-coated Nuclepore filters was used for the measurement of ambient air concentrations of asbestos fibers in the vicinity of buildings. The chemcial and crystallographic changes in the corroded asbestos fibers were evaluated by the application of several methods (i.e., electro microdiffraction analysis (SAED), analytical electron microscopy (ATEM), and mass spectrometry LAMMA for single particle analysis).[7-9] Asbestos fiber contamination by organic particulate pollutants (e.g., PAH) was estimated after extraction with organic solvents and analysis by GC.

Fibrous Dust Emissions

Data regarding asbestos fiber and asbestos mass emissions were obtained from 2-year measurements. The mean values of fiber emissions at wind velocities of 2 ± 1.2 m s^{-1} lay in the range of 10^5 to 10^9 fibers m^{-2} h^{-1}. The dispersion (standard deviation) of the measured values was very broad, with a mean value of about $6 \cdot 10^7$, because single measurements were done on objects with different corrosion levels.

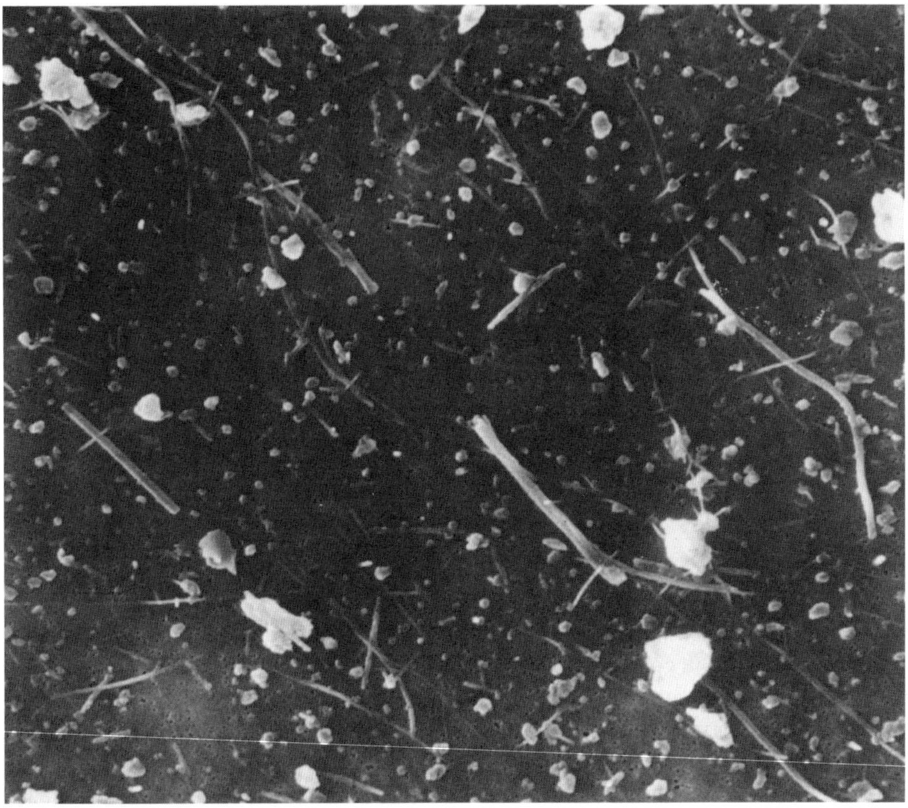

FIGURE 18.5 Scanning electron micrograph (magnification 2000X), showing dust particles and chrysotile fibers (collected on a Nuclepore filter) that were released into the atmosphere from a corroded asbestos cement roof.

The emission values in fiber numbers can be converted into mass rates (e.g., ng m^{-2} h^{-1}.[3] Considering the entire surfaces of the corroded asbestos cement building materials in Germany, the total mass of asbestos fibers emitted into the ambient air was very approximately estimated. The mean annual value was about 600 tons, with a range of about 10 to 10^4 t/a (see Figure 18.6).

Fibrous Dust Concentrations

The measurement of ambient air asbestos fiber concentrations was realized during the same 2-year period in the very vicinity of the buildings that were tested for their fibrous emissions (e.g., at distances of 0.5 to 1 m from the AC façades or roofs). The concentrations of total dust (TSP) were higher than the background dust concentrations (mean value of 150 μg m^{-3}). The mean value of ambient air concentration for asbestos fibers longer than 5 mm was 750 fibers m^{-3}. The maximum values were in the range of 10^3 fibers m^{-3}. Approximately 35% of the measured asbestos fiber (longer than 5 mm) concentrations were above 500 fibers m^{-3}, and 12% were higher than 1000 fibers m^{-3}.[10]

Leaching and Contamination

Remarkable qualitative differences in the composition of asbestos fibers sampled from the corroded AC plates were observed. The asbestos fibers were usually heavily contaminated with elements from external pollutants (e.g., S, K, Ti, Zn, Pb, V, and Sr). The corroded layer of AC plates acts as an effective "aerosol filter." Particulate air pollutants were separated from the air and deposited

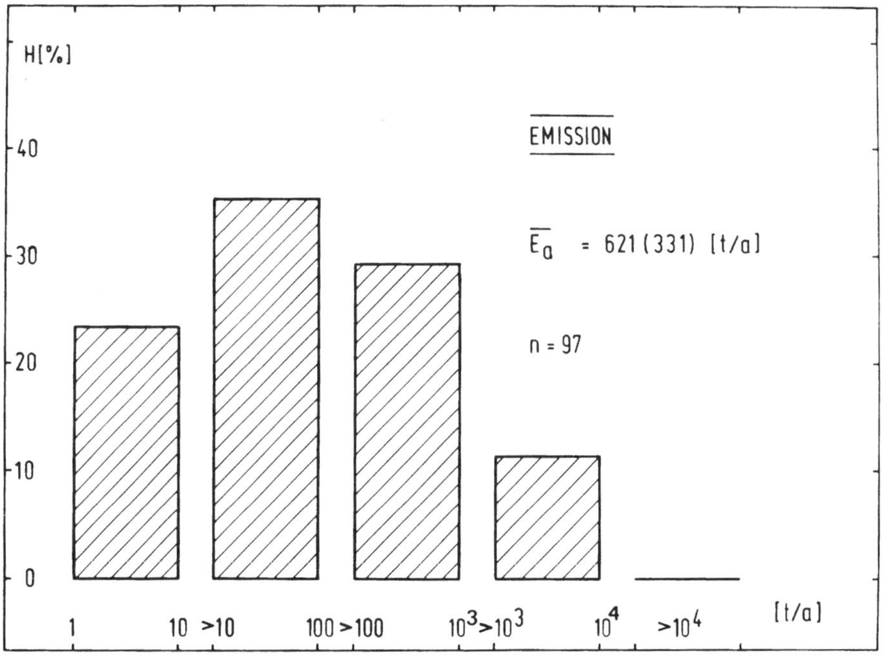

FIGURE 18.6 The distribution of asbestos fiber emissions from corroded asbestos cement roofs and facades in Germany.

on the free asbestos fibers. The fibrous material sampled from the corroded surface layer was extracted with organic solvents and the solution was analyzed by a standard GC procedure to esitmate the PAH content. An example of such an analysis is given in Table 18.1. From these results, it was concluded that the asbestos fibers released from the corroded AC products were contaminated with different inorganic and organic particulate air pollutants. The electron microscope and mass spectroscope analyses on individual fibers from the corroded sheet surface confirmed that many released crysotile fibers were leached after lengthy exposure to acid rain. The mean loss of magnesium for different analyzed chrysotile fibers was 48 ± 25%. The leaching effect was observed by analysis of individual chrysotile fibers using the LAMMA procedure. Some examples are depicted in Figure 18.7.[11]

Conclusions

Investigations in Germany have shown that asbestos cement products can corrode relatively fast by action of atmospheric pollutants and their reaction products. The corroded surface cement matrix is chemically charged into a gypsum layer, which is then dissolved and washed out by acid rain. The reamining asbestos fiber layer then becomes a source of asbestos pollution. The majority of these asbestos fibers are transported into the total environment by rainwater. A non-negligible part of these fibers is dispersed by wind into the ambient air. These contribute to the total carcinogenic potential of the polluted atmosphere and increase the health risk for the general population.[12]

OBSERVATIONS AND MEASUREMENTS IN SOUTH AFRICA

Asbestos deposits occur in three areas in South Africa: crocidolite in northwestern Cape, amosite and crocidolite in northern Transvaal, and chrysotile in eastern Transvaal (see Figure 18.8). Commercial mining of asbestos commenced in 1930 and was controlled by foreign companies that were

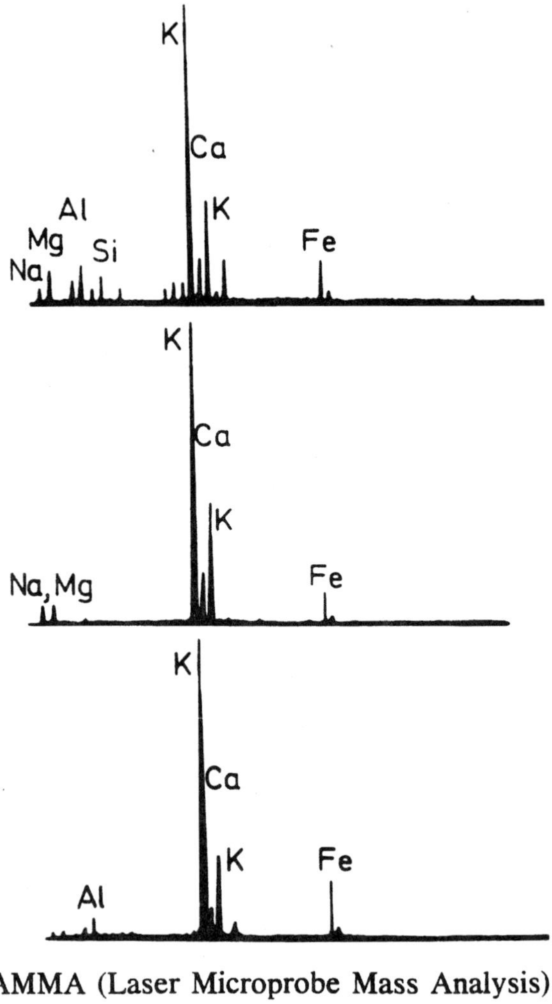

FIGURE 18.7 Mass spectroscopic analysis of three individual corroded asbestos fibers. The leaching of Mg and other elements on different fibers is evident.

not bound by law to conserve the environment or to protect the labor force against excessive inhalation of fibers. Tailings with high fiber contents were randomly dumped with no consideration to resultant fiber concentrations in the industrial and ambient atmospheres.[13-15] Since legislation concerning allowable fiber concentrations has only come into being since 1956, the older disused dumps present a problem because of the fact that the responsible companies either dissolved or withdrew from South Africa.[14]

To obtain scientific bases for future sanitation and rehabilitation procedures, laboratory investigations (e.g., in wind tunnels) and environmental measurements were realized in the 1990s.[14]

Asbestos Deposits

Tailings with high fiber content were randomly dumped with no consideration for the possible spreading of the fibers by water and wind. Water has the ability to corrode, erode, and spread asbestos tailings in the environment. When fiber-polluted water evaporates, the fiber is left behind and is susceptible to wind erosion. Polluted rainwater produces the fiber contamination of ground-

TABLE 18.1
PAH Concentration in an Asbestos Fiber Sample from the Weathered AC Roofing Plate

PAH	Concentration ($ng\ g^{-1}$ = ppb)
Fluoranthene	501.7
Pyrene	620.9
Benz[*a*]anthracene	53.0
Chrysene	219.6
Benzofluoranthene (3 isomers)	486.4
Benzo[*e*]pyrene	205.9
Benzo[*a*]pyrene	91.7
Indeno[1,2,3-*cd*]pyrene	108.8
Benzo[*ghi*]perylene	187.3
PAH (total)	2.5 ($\mu g\ g^{-1}$ = ppm)

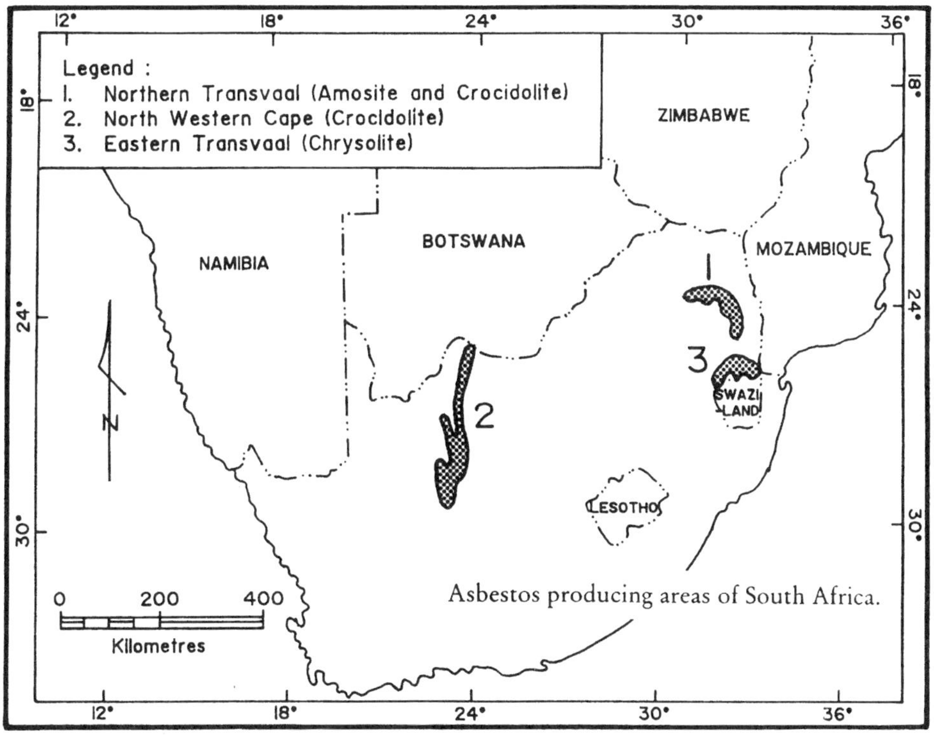

FIGURE 18.8 Schematic map showing the most important asbestos-producing areas of South Africa. (From Van de Walt, I.J. and de Villiers, A.B., *Geoökodynamik*, 14, 141-152, 1993. With permission.)

water and/or is responsible for the gradual enlargement of atmospheric asbestos pollution. Furthermore, irrigation with polluted water leads to a rise in atmospheric fiber concentration.

The tailings dumps are, of course, an important source for the emission of fine asbestos fibers into the atmosphere. The produced atmospheric asbestos fiber concentrations are, however, influenced by several factors: topography, plant growth, moisture, climatic and weather conditions, wind, etc.

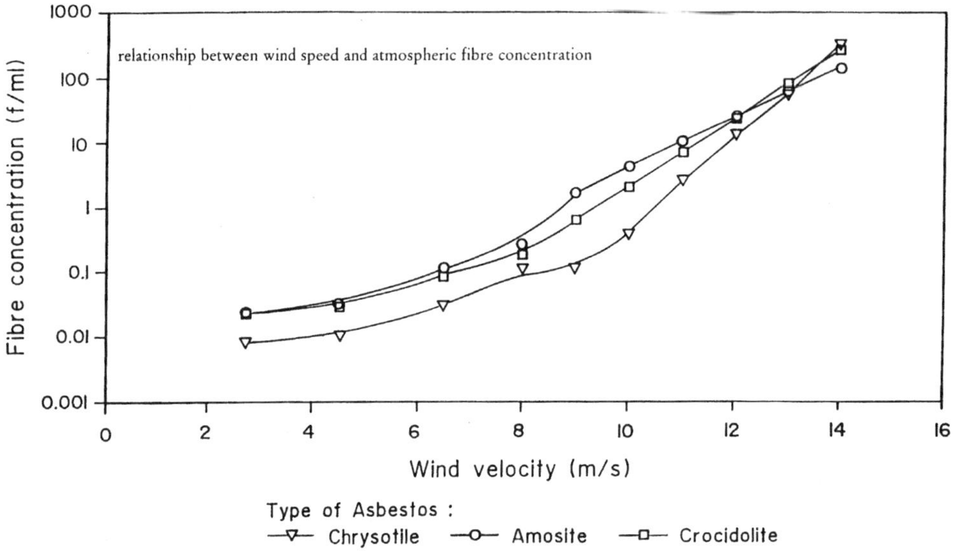

FIGURE 18.9 The relationship between wind and atmospheric fiber concentration for different types of asbestos tailings, as experimentally determined. (From Van de Walt, I.J. and de Villiers, A.B., *Geoökodynamik*, 14, 141-152, 1993. With permission.)

Water Contamination by Tailings Dumps

The total areas of different unrehabilitated tailings dumps are as large as 150 to 1500 hectares. Environmental measurements and environmental rainfall simulation experiments were performed on several tailings dumps. The measured erosion rates lay in the range of 2 g m^{-2} h^{-1} for amosite, 6 g m^{-2} h^{-1} for crocidolite, and almost 90 g m^{-2} h^{-1} for crysotile. The total amounts of estimated asbestos release into rainwater in different geographic areas range from 10 to 100 tons per year.[13]

Air Contamination by Asbestos Mine Dumps

Using laboratory measurements in a wind tunnel, and following model calculations and direct environmental measurements, it was possible to estimate atmospheric pollution by asbestos fibrous particulates.

The atmospheric concentrations produced by wind only depended on the wind velocity and lay in the range of 0.01 to 550 fibers ml^{-1}, or 10^4 to 5.10^8 fibers m^{-3} (see Figure 18.9). As much as 1000 times higher ambient-air fiber concentrations are being produced, while the tailings are being dumped by activities of people, animals, vehicles, etc.

Conclusions

There is a great concern in asbestos mining areas in South Africa regarding the amount of asbestos eroded from tailings dumps because, in time, it causes the average asbestos levels in the environment — water and air — to rise.

There are therefore areas in the asbestos-producing districts of South Africa where the average ambient-air fiber concentrations exceed the environmental standard and may cause a health hazard. The available data on fatalities due to excessive inhalation of asbestos fibers in South Africa confirms this statement.[16] Therefore, areas where asbestos tailings are regularly disturbed should be rehabilitated as soon as possible.[14]

REFERENCES

1. Spurny, K., Weiss, G., and Opiela, H., Zur Emission von Asbestfasern aus Asbestzementplatten, *Staub-Reinhalt. Luft*, 39, 422, 1979.
2. Spurny, K.R., On the release of asbestos fibers from weathered and corroded asbestos cement products, *Environ. Res.*, 48, 100, 1989.
3. Spurny, K., Marfels, H., Boose, C., Weiss, G., Opiela, H., and Wulbeck, F.J., Zu faserigen Emissionen aus abgewitterten Asbestzement-Produkten: Freisetzung in die Auβenluft, *Zbl. Hyg.*, 188, 127, 1989.
4. Spurny, K., Mönig, F.J., and Hochrainer, D., Zur Messung von Schadstoffemissionen aus Oberflächenquellen, *Staub-Reinhalt. Luft*, 45, 328, 1985.
5. Chatfield, E.J., Determination of asbestos fibers in air and water, Guideline ISO-TC 146 SC 3, WG 1, pp. 1-55. Report. Ontario Res. Foundation, Mississauga, Canada, 1984.
6. VDI-Richtlinie 3492, Messen anorganischer, faserförmiger Partkeln in der Auβenluft, REM-Verfahren, VDI-Verlag, Düsseldorf, 1991.
7. Spurny, K.R., Schörmann, J., and Kaufmann, R., Identification and microanalysis of mineral fibers by LAMMA, *Fresenius Zschr. Anal. Chem.*, 308, 274, 1981.
8. Spurny, K.R., *Physical and Chemical Characterization of Individual Airborne Particles*, Ellis Horwood, Chichester, U.K., 1986.
9. Spurny, K.R., Sampling, analysis, identification and monitoring of fibrous dusts and aerosols, *Analyst*, 119, 41, 1994.
10. Spurny, K., Marfels, H., Boose, C., Weiss, G., Opiela, H., and Wulbeck, F.J., Immissionsmessungen von faserigen Stäuben in der Bundesrepublik Deutschland, *Zbl. Bakt. Hyg. B*, 187, 136, 1988.
11. Spurny, K., Marfels, H., Boose, C., Weiss, G., Opiela, H., and Wulbeck, F.J., Zu faserigen Emissionen aus abgewitterten Asbestzement-Produkten. Physikalisch-chemische Eigenschaften der freigesetzten Asbestfasern, *Zbl. Hyg.*, 188, 262, 1989.
12. Hanβ, A., Herzig, S., and Lutz-Holzhauer, C., Abschätzung des Krebsrisikos durch Luftverunreinigungen, *Gefahrstoffe-Reinhalt. Luft*, 57, 71, 1997.
13. Van de Walt, I.J. and de Villiers, A.B., The spreading of asbestos tailings from disused tailings dumps by water, *Geoökodynamik (Bensheim, Germany)*, 14, 141, 1993.
14. Van de Walt, I.J. and de Villiers, A.B., A model to determine airborne environmental pollution from asbestos mine dumps, *Zschr. Geomorphologie, Berlin*, 40, 339, 1996.
15. Hart, H., Asbestos in South Africa, *J.S.A. Inst. Min. Metall.*, 88, 185, 1988.
16. Webster, I., Asbestos related diseases, *Cont. Med. Educ.*, 4, 71, 1986.

Part V

Aerosols in the Atmosphere

19 Characterization of Urban Aerosols in the Nagoya Area

Satoshi Kadowaki

CONTENTS

INTRODUCTION

Chemical composition and behavior of urban aerosols are very complicated because urban aerosols comprise primary aerosols derived directly from numerous anthropogenic and natural sources and secondary aerosols produced from gaseous molecules by chemical reaction in the atmosphere. The aerosol particle size is an important factor that characterizes behaviors of particles in the atmosphere. The size is also closely related to sources and formation processes of atmospheric aerosols. Therefore, knowledge concerning the size distribution and chemical composition is useful for a full understanding of urban aerosols.

A field study has continued in the Nagoya urban area since 1973 for the characterization of urban aerosols on the basis of measurements for size distribution and chemical composition. Several results were previously reported relating to sulfate and nitrate,[1–3] soil particles,[4,5] sea salt particles,[6] elemental and organic carbon,[7] and organic aerosols[8] in Nagoya urban aerosols.

This chapter summarizes what is known about major components and behavior of Nagoya urban aerosols. The origins and formation processes of the major components are also discussed. Since the experimental methods for sampling and analysis of aerosols have been published,[1–8] methodology will not de described in much detail herein.

SITE DESCRIPTION AND AEROSOL PARTICLE SIZE DISTRIBUTIONS

Nagoya is the fourth largest metropolis in Japan, with a population of approximately 2,200,000. The site's latitude is 35°10′N, its longitude is 136°58′E, and its altitude is 51.1 m. The area is about 326 km^2. Nagoya faces the Pacific Ocean on the south, and has the rolling hills and mountains to the north. All the aerosol samples were collected at the Aichi Environmental Research Center in a mixed residential/light industry area not affected by major local sources of pollution. It is located

15 km north of the coastal heavy industry area and 5 km north of the central business district of Nagoya.

Aerosol concentrations in the Nagoya urban area have a tendency to increase in the autumn, early winter, and spring and to decrease in the summer and winter. The mass concentrations range from about 20 to 250 μg m^{-3}. These differences in the concentrations are mainly due not to the change of source contributions, but to the change of meteorological conditions corresponding to the seasons in Japan.

During the period from autumn to early winter, the Japan Islands are covered with a large traveling anticyclone and the surface inversion layer frequently appears in the calm night, so the concentrations of urban aerosols in Japan generally increase during this period. The aerosol concentrations also increase in the spring because yellow sand dust called Kosa is transported by the upper winds from the arid desert regions of the Asiatic continent to Japan. In the summer, high ambient temperature and high solar radiation occur. The mean maximum August temperature and the solar radiation in August are 32.2°C and 480 ly/day in Nagoya, respectively. Although the amount of the photochemically formed aerosols is greatest in the summer, the aerosol concentrations decrease generally due to atmospheric turbulence. Also, the aerosol concentrations decrease in the winter because the northwest monsoon blows strongly from Siberia to Japan and sweeps air pollutants.

On the other hand, it is well-known that the size distribution of atmospheric aerosols is almost always multimodal in nature.[9] These size distributions are presented by number, surface, volume, or mass distribution. In this chapter, mass distributions are used because they fit in with the measurements of size distributions for the chemical compositions of urban aerosols. Atmospheric aerosols primarily have a bimodal mass distribution, with the fine particle mode of diameter between 0.1 μm and 1.0 μm, and the coarse particle mode of diameter greater than 2 μm. The profiles of size distribution for Nagoya urban aerosols are essentially bimodal, although the concentration and composition change remarkably with the meteorological conditions.

Figure 19.1 shows the representative profiles of size distribution for Nagoya urban aerosols measured using an Andersen cascade impactor (Model 21-000, 2000 INC). Profile (a) is mainly observed in the period from the autumn to early winter, and is characterized by the bimodal distribution in which both the coarse and fine particles have a high concentration. Profile (b) was observed when a large-scale Kosa was transported to Japan. It has a very large peak in the coarse-particle range and appears to be unimodal. Profile (c) is typically observed during the summer photochemical season, and is characterized by a bimodal distribution in which fine particles are predominant. Profile (d) was observed when a typhoon visited. Typhoons represent a kind of tropical cyclone, as do the hurricanes over the Caribbean Sea and the cyclones over the Indian Ocean, and visit Japan most frequently in August and September. The strong southerly winds sweep air pollutants and bring sea salt particles. This profile is characterized by the bimodal distribution in which coarse particles are predominant.

The following sections describe how the two modes differ considerably in their chemical compositions; and the behavior of Nagoya urban aerosols will be discussed on the basis of the seasonal variations of concentrations and size distributions for the chemical compositions.

SOIL PARTICLES AND THEIR ORIGINS

Wind-generated soil particles are typical primary atmospheric aerosols produced by mechanical processes from natural sources.[10] The main source for this mineral dust is the arid and semiarid regions of the Earth, which cover approximately 30% of the continental surface. Airborne mineral dust is one of the major constituents of tropospheric aerosol, and affects the change of climate on a global scale. The Sahara Desert and the arid regions of central and eastern Asia (namely, the Takla Makan, Gobi, and Ordos Deserts) are the two major sources of windblown dust in the world.

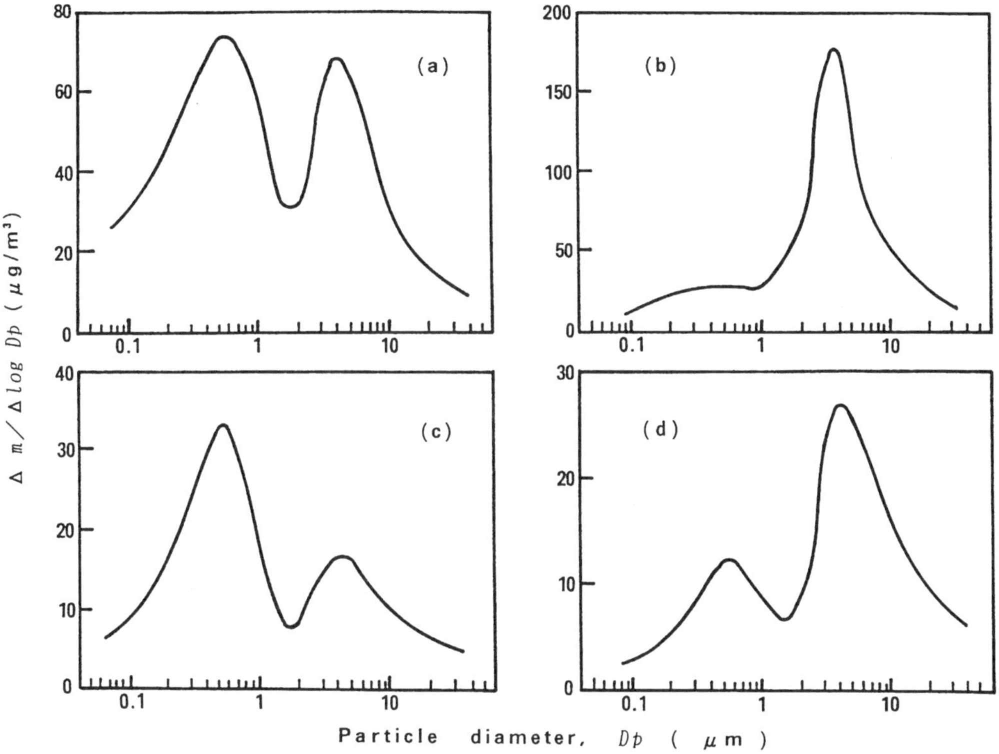

FIGURE 19.1 Representative profiles of size distribution by mass for Nagoya urban aerosols. (a) Sampled when surface inversion layers frequently appeared (1978, 12/9–12/15, TSP = 112 μg m^{-3}; (b) sampled when Kosa was clearly observed (1979, 4/11–4/17, TSP = 124 μg m^{-3}); (c) sampled when photochemical smogs frequently appeared (1974, 7/31–8/6, TSP = 82.9 μg m^{-3}); (d) sampled when Typhoon visited (1978, 7/27–8/1, TSP = 30.4 μg m^{-3}).

Sahara Desert dust is transported and deposited as "red" dust over wide areas of the Mediterranean, Europe, the mid-Atlantic, and as far as the Caribbean Sea.[11,12]

On the other hand, Asian desert dust is transported to the North Pacific Ocean[13,14] over the Japan Islands, as shown in Figure 19.2. This mineral dust generally appears to be yellow or light brown, and is called Kosa in Japan. Kosa is observed in almost all regions of Japan, most frequently in the spring when large-scale dust storms occur in the arid desert regions of North China and Mongolia.

As described in the previous section, the aerosol concentration of the Nagoya urban area increased remarkably when Kosa was observed, and the profile of the size distribution suggests that Nagoya urban aerosols are influenced by Asian desert dust. Si and Al are important constituents to consider in studying the behaviors of soil-derived aerosols in ambient air because next to oxygen, Si and Al are the two most abundant elements in the crust. This section reviews the results[4,5] obtained for the size distributions and concentrations of Si and Al in urban aerosols at Nagoya, and of the contribution of soil dust to total suspended particulate (TSP). Also, the origin of soil particles in the Nagoya urban area will be discussed.

Figure 19.3 shows typical examples of cumulative size distribution of Si and Al in Nagoya urban aerosols; the size distribution curves are shown in Figure 19.4. The solid and dashed lines in Figure 19.3 represent the measurement result sampled in the summer when Asian desert dusts were not transported to Japan and that sampled in the spring when a Kosa event was clearly observed, respectively. More than about 90% of the mass of atmospheric Si and Al were concentrated in

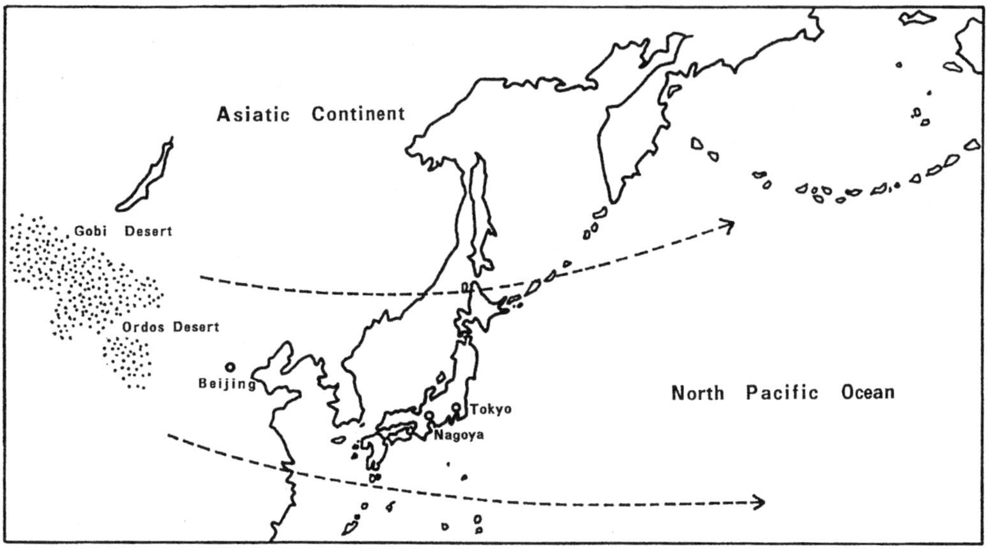

FIGURE 19.2 Approximate trajectory of Kosa (Asian desert dust) from North China to Japan and the North Pacific Ocean.

particles larger than 2 μm in diameter. The shape of the Si cumulative size distribution curve was in good agreement with that of Al, and the log probability plots gave an approximately straight line. These characteristics of Si and Al were recognized whether soil particles came flying from Asian deserts or not.

As seen in Figure 19.4, the size distributions for Si and Al were unimodal in the coarse-particle range. Although the concentrations of Si and Al sampled in a Kosa event increased by about 10 times over those sampled in the summer, the size distribution patterns did not change — except the peaks shifted slightly to a smaller size. This slight shift suggests that the large soil components of Kosa (e.g., quartz) fall out as Kosa is transported to Japan over the sea, in contrast to the small soil components (e.g., clay minerals) that remain in the atmosphere.

Table 19.1 lists the average concentrations and concentration ratios of Si and Al for the four seasons at Nagoya. The average concentrations of Si and Al in the spring were 2.3 to 3.0 times higher than those in the other seasons. However, the concentration ratio of Si:Al did not change throughout the year — in no connection with the influence of Kosa.

Rahn[15] has reviewed 26 pairs of Si:Al concentrations in various atmospheric aerosols, and reported that high a correlation was found between the concentrations of Si and Al (the overall arithmetic mean Si:Al = 2.67 with a S.D. = 0.71), and that the Si:Al ratio in atmospheric aerosols was distinctly lower than those in average rock and soil (3.41 to 4.65) and somewhat higher than those in common clays (1.04 to 2.07). Thus, he estimated that Si and Al in atmospheric aerosols have a single common source of crustal material, and that a crust–air fractionation mechanism influences the generation of crustal aerosols.

Figure 19.5 is a scatterplot for the concentration of Al against to that of Si in Nagoya urban aerosols from 1976 to 1979. Good correlation was found between the concentration of Si and that of Al, with a high correlation coefficient of 0.966 (n = 48). The arithmetic mean of the Si:Al ratio was 2.85, with a S.D. of only 0.37. This remarkable Si–Al correlation and the size distributions for Si and Al shown in Figures 19.3 and 19.4 suggest that not only the atmospheric Si and Al in the Nagoya urban area are derived almost entirely from the soil by mechanical processes, but also that the soil components of the atmospheric soil particles do not change very much. The similarity of the soil components is inconvenient for the distinction between the soil particles derived from

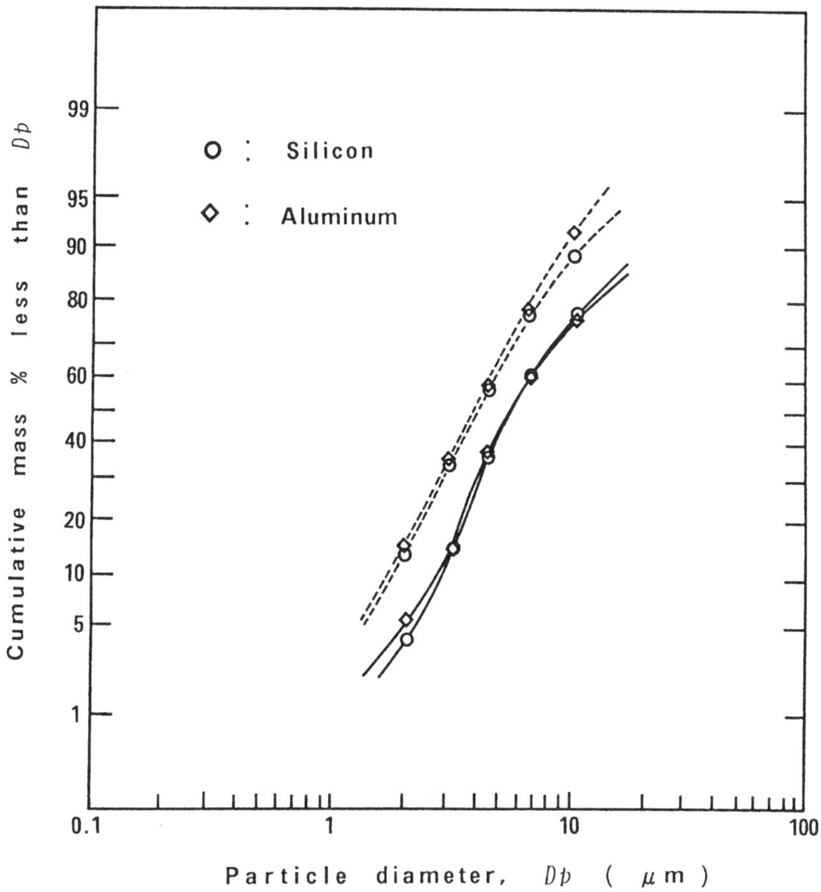

FIGURE 19.3 Log probability plot of size distributions for silicon and aluminum in Nagoya urban aerosols. ——: sampled when Asian desert dusts were not transported to Japan (1978, 8/3–8/9); ……: sampled when Kosa event was clearly observed (1979, 4/11–4/17).

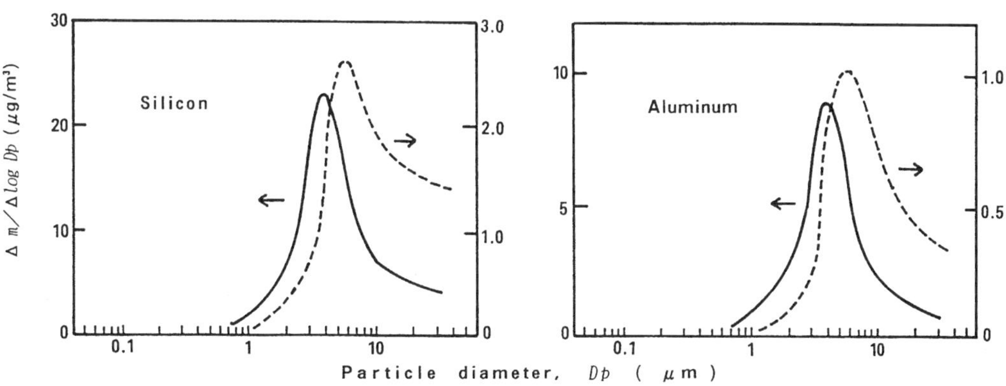

FIGURE 19.4 Size distributions for silicon and aluminum in Nagoya urban aerosols. ——: Sampled when Kosa event was clearly observed (1979, 4/11–4/17, Si = 13.1 μg m^{-3}, Al = 4.47 μg m^{-3}); ……:Sampled when Asian desert dust were not transported to Japan (1978, 8/3–8/9, Si = 2.03 μg m^{-3}, Al = 0.689 μg m^{-3}).

TABLE 19.1
Average Concentrations of Silicon and Aluminum in Nagoya Urban Aerosols and Predicted Concentration and Contribution to TSP of Soil Particles

Sampling Period[a]	No. of Samples	Silicon (μg m^{-3})	Aluminum (μg m^{-3})	TSP (μg m^{-3})	Concentration Ratio of Si:Al	Soil Particles[b] (μg m^{-3})	Contribution of Soil Particles to TSP (%)
Spring	12	5.97	2.09	89.8	2.9	34.8	38.8
Summer	12	1.97	0.721	66.5	2.7	12.0	18.0
Autumn	12	2.40	0.903	75.6	2.7	15.1	20.0
Winter	12	2.55	0.881	65.0	2.9	14.7	22.6

[a] Aerosol samples were collected by Andersen samplers for about 6-day duration from 1976 to 1979.
[b] Concentrations of soil particles are predicted using an aluminum concentration in soil of 6%.

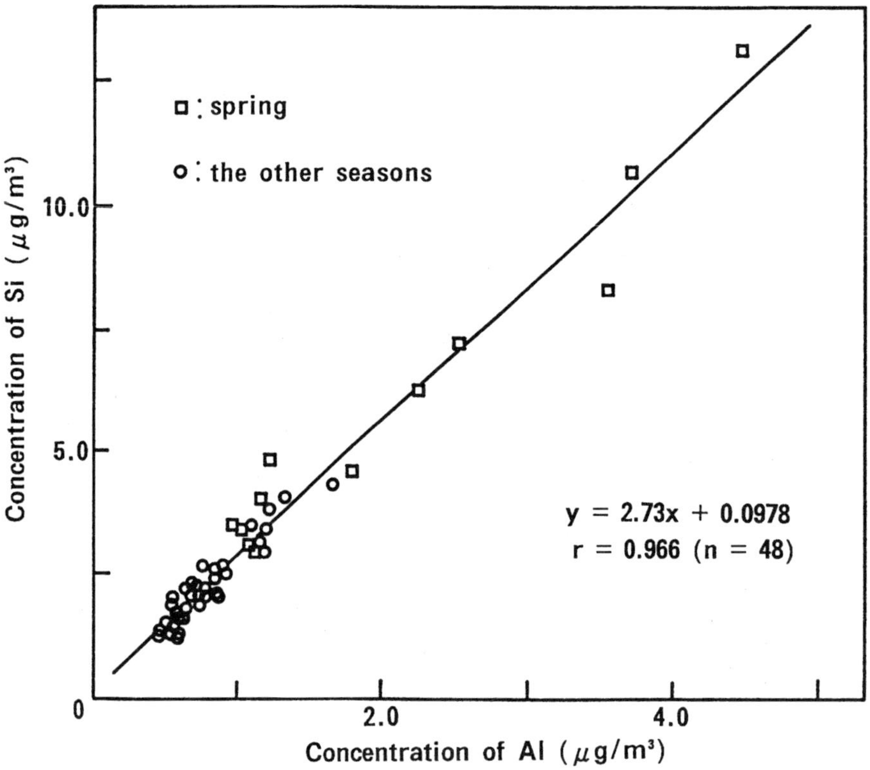

FIGURE 19.5 Scatter diagram of silicon and aluminum concentrations in Nagoya urban aerosols, 1976 to 1979.

local sources and those transported from Asian deserts. It is, however, convenient for the estimation of the concentration of soil particles and of the contribution of soil particles to TSP, using the Si or Al concentration data. Because Si is contained in primary minerals, such as quartz and mica, of large diameter as well as in clay minerals, the Al concentration is more suitable for the estimation.

Almost all Al concentrations in soils that are most abundant on the land surfaces of Japan[16,17] and in loess of China,[18,19] which is the source of Kosa, are about 6%. Also, the Al concentrations in coarse particles (>2 μm) when a Kosa event was clearly observed were about 5% on the data shown in Figure 19.5. From these Al concentrations, in the present study, the Al concentration in soil particles in Nagoya urban air is assumed to be 6%. The predicted concentration and contribution of atmospheric soil particles are summarized in Table 19.1. The average predicted concentration in the three seasons except spring was about 14 $\mu g\ m^{-3}$, and the contribution of about 20% to TSP was estimated. In the spring period influenced by Kosa, the average predicted concentration was about 35 $\mu g\ m^{-3}$, and a twofold increase of the contribution to TSP was found in comparison with the other seasons. It was therefore concluded that soil particles are one of the major components of Nagoya urban aerosols, and that long-range transport of Asian desert dust contributes to the remarkable increase of soil particle concentrations in the spring in the Nagoya urban area.

In urban areas, street dust (or road dust)[20,21] is the probable source of crustal components in ambient aerosols, as well as in soil. In the Nagoya urban area too, the atmospheric soil particles other than Kosa of Asian desert dust may be due mainly to particles caused by surface stresses from vehicular traffic on asphalt roads, because there are hardly any exposed soils in and around Nagoya. In general, the compositions of crustal components in surface soils resemble those in street dusts[21] and asphalt pavement materials.[22] However, the Ca contents in street dusts and asphalt pavement materials are higher than those in surface soils. From the result[23] summarizing the crustal compositions in surface soils, street dusts, and asphalt pavement materials in Japan, it was found that Ca:Al ratios in surface soils range from 0.13 to 0.27, which are in agreement with 0.21 for the typical estimate of Ca:Al in average soil,[24] and that Ca:Al ratios in street dusts and asphalt pavement materials range from 0.82 to 1.1 and are roughly 5 times higher than those in surface soils.

Tanaka et al.[25] reviewed the data of the National Air Surveillance Network of Japan from 1977 to 1978, and reported that Ca:Al ratios in urban aerosols distributed between 0.71 and 1.22. These ratios are similar to those in street dusts and asphalt pavement materials, but distinct from those in surface soils. On the other hand, Ca:Al ratios in both Kosa aerosols[26] and Asian desert soils,[18,25] which are the source of Kosa, are about 0.8. Also, it is of great interest from the viewpoint of the low variation in the Si:Al ratio in Nagoya urban aerosols that Kosa aerosols[26] and asphalt pavement materials[22] show a highly similar crustal composition profile.

The facts described above suggest that the soil particles in Nagoya urban air except Kosa aerosols originate mainly from the crustal material of asphalt road surfaces chipped off and resuspended by vehicular traffic. It is, however, evident that additional research is required to decide the origin of the soil particles in Nagoya urban air except Kosa aerosols.

SEA SALT PARTICLES AND CHLORINE LOSS

The ocean is the most prolific of all natural sources in the production of atmospheric particulate material.[27] Of all the components that make up the particulate material introduced into the global atmosphere each year, that of the sea salt particle is one of the — if not *the* — largest. Many investigations have been reported for atmospheric sea salt particle on the production mechanisms, the relation between wind speed and production rate, size distribution and concentration in the marine atmosphere, etc. Several review papers[28–30] on marine aerosols have also been published. However, there is not enough knowledge of sea salt particles in the polluted urban atmosphere.

Japan is surrounded by the sea on four sides, and all of the metropolitan areas in Japan face the sea. Thus, sea salt particles must have effects on the urban areas; however, the practical contribution of sea salt particles to TSP in the Nagoya urban area is not well-known. Because NO_x and HNO_3 concentrations are usually high in urban atmospheres, the change of chemical composition of sea salt particles — namely, chlorine loss from sea salt particles[31–34] — by chemical reactions in the atmosphere is an interesting problem. This section describes the results[6] of studies of the size distributions and concentrations of water-soluble sodium ion (Na^+) and chloride ion (Cl^-) in urban aerosols at Nagoya, and of the contribution of sea salt particles to TSP. Also, chlorine loss from sea salt particles will be discussed in connection with the behavior of nitrate[2] in the Nagoya urban area.

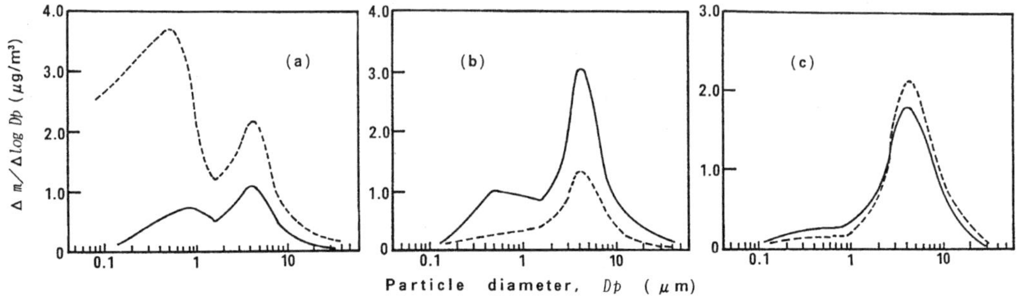

FIGURE 19.6 Representative size distributions for water-soluble sodium and chloride ions in Nagoya urban aerosols. ——: Na^+, ……: Cl^-. (a) Sampled when northwest monsoon prevailed in winter (1979, 1/19–1/26); (b) sampled when southeast monsoon prevailed in summer (1978, 8/26–8/31); (c) sampled when typhoon No. 8 visited (1978, 7/27–8/1).

It is generally accepted that sea salt particles in the marine atmosphere distribute in the coarse-particle size range.[28,30] Fujimura et al.[35] reported that the size distribution by mass of the marine aerosols over the sea near Japan was unimodal in the coarse-particle size range, and that the marine aerosols were mostly comprised of sea salt particles. It has also been reported that the Cl^-:Na^+ concentration ratio in freshly formed marine aerosols was close to that of seawater (1.8).[35,36] In polluted urban areas, Na^+ and Cl^- from anthropogenic sources are abundant, so it is necessary to make clear the question of whether Na^+ and/or Cl^- in aerosols can be used as the index for sea salt particles in the urban atmosphere as well as in clean marine atmospheres.

Figure 19.6 shows the representative profiles of size distributions for Na^+ and Cl^- in Nagoya urban aerosols. The profiles in (a) represent the results sampled in the winter when sea salt particles were transported from the Japan Sea, located about 100 km northwest of Nagoya, by the winter monsoon wind. The size distributions of both Na^+ and Cl^- were bimodal, and clearly showed that the there are considerably Na^+ and Cl^-, and not derived from sea salt, in the fine particles. It seems, however, that Na^+ and Cl^- in the coarse particles can be used as the index of sea salt particles because the Cl^-:Na^+ concentration ratios were close to 1.8 in all of the fractionated size ranges in the coarse-particle peaks of (a).

The profiles in (b) represent the results sampled in the summer when sea salt particles were transported from the Pacific Ocean by the summer monsoon wind. The Na^+ profile showed a bimodal distribution in which the coarse-particle peak was predominant as compared with (a). On the other hand, the Cl^- profile in the summer changed significantly in comparison with that in the winter; namely, the dominant fine-particle peak in (a) disappeared, and the distribution showed a unimodal in the coarse-particle size range. In the coarse-particle peak, the Cl^-:Na^+ concentration ratios were remarkably lower than 1.8 of seawater. Chlorine loss from the sea salt particles was observed. The profiles in (c) represent the results sampled in the summer when a typhoon visited, and swept air pollutants and transported sea salt particles. The size distributions of both Na^+ and Cl^- were unimodal. They agreed with those in the clean marine atmosphere,[35] although chlorine loss was observed somewhat.

Figure 19.7 shows the change of Cl^-:Na^+ concentration ratios in the coarse particles in the Nagoya urban area. The seasonal averages of the ratio are listed in Table 19.2. Figure 19.7 suggests that the ratios change periodically with the season: (1) the ratio in the winter is equal to that in seawater; (2) the ratio begins to decrease in the spring and the largest chlorine loss occurs in the summer; and (3) the ratio begins to increase in the autumn and equals to 1.8 again in the winter. The seasonal average in the winter was 1.8, and that in the summer was 0.70.

From the results of Figures 19.6 and 19.7 and Table 19.2, it is found that Cl^- cannot be used as the index of sea salt particle independently of the size range because of the chlorine loss in the

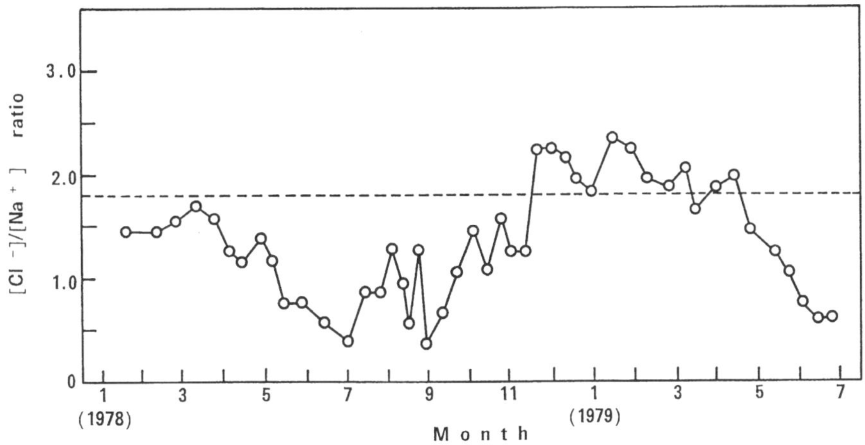

FIGURE 19.7 Variation of Cl^-/Na^+ concentration ratio in water-soluble component in Nagoya urban aerosols in coarse-particle size range (≥2.1 μm). ......:$Cl^-:Na^+$ concentration ratio in seawater (= 1.8).

TABLE 19.2
Average Concentrations of Water-Soluble Sodium Ion and Chloride Ion in Coarse Particles in Nagoya Urban Aerosols and Predicted Concentration and Contribution of Sea Salt Particles to TSP

Sampling Period[a]	No. of Samples	Sodium Ion[b] (μg m^{-3})	Chloride Ion[b] (μg m^{-3})	TSP (μg m^{-3})	Concentration Ratio $Cl^-:Na^+$	Sea Salt Particles[c] (μg m^{-3})	Contribution of Sea Salt Particles to TSP (%)
Spring	15	0.561	0.785	67.1	1.4	1.81	2.7
Summer	12	1.01	0.712	58.8	0.70	3.26	5.5
Autumn	13	0.544	0.719	68.6	1.3	1.75	2.6
Winter	12	0.334	0.612	62.2	1.8	1.08	1.7

[a] Aerosol samples were collected by Andersen samplers for about 6-day durations from 1978 to 1980.
[b] Concentrations of sodium and chloride ions in coarse-particle size range (≥2.1 μm).
[c] Concentrations of sea salt particles are predicted by use of the Na^+ concentration in coarse-particle size range and the Na^+ concentration in sea salt of 31%.

summer, but Na^+ in the coarse particles can be used as the index and to estimate the concentration of sea salt particles in the Nagoya urban area.

Seasonal concentration data for TSP and Na^+ and Cl^- in the coarse-particle size range in the Nagoya urban area from 1978 to 1980 are summarized in Table 19.2. Also, the predicted concentration of sea salt particles and the contribution to TSP are given in Table 19.2, assuming that the Na^+ concentration in sea salt particles is 31%. The predicted concentrations in Table 19.2 correspond to those of original sea salt particles before they lose their chlorine. The highest seasonal average concentration of predicted sea salt particles, 3.26 μg m^{-3}, was observed in the summer and was 3 times higher than the minimum of 1.08 μg m^{-3} predicted in the winter. The contributions of sea salt particles to TSP were estimated to be about 6% and 2% in the summer and winter, respectively.

Prospero[37] collected about 250 samples of marine aerosols in the Atlantic, Pacific, and other ocean regions, and estimated the concentrations of atmospheric sea salt particles on the basis of the Na^+ concentration in marine aerosols. He reported that sea salt particle concentrations in the marine atmosphere are relatively constant from one region to another, with arithmetic means ranging

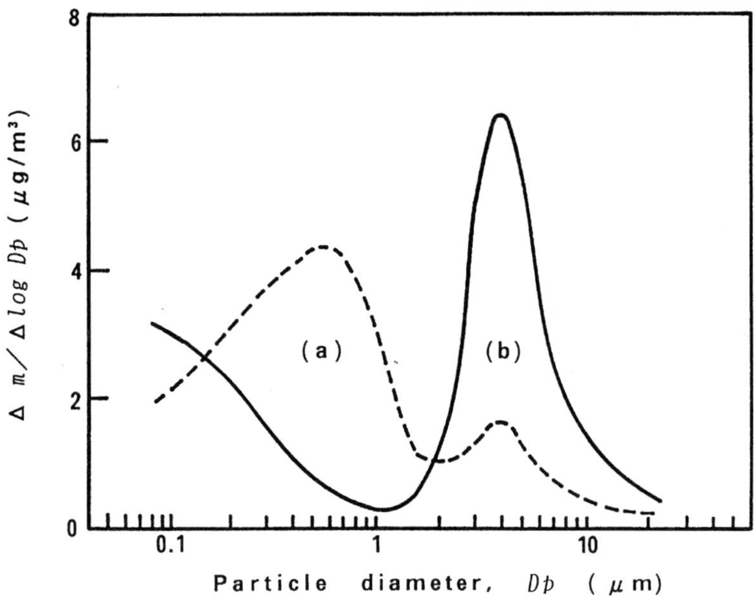

FIGURE 19.8 Representative size distributions for nitrate in Nagoya urban aerosols. (a):NH_4NO_3; (b) $NaNO_3$. ——: summer (1978, 25–31 Aug., NO_3^- = 5.11 μg m^{-3}; ……: winter (1979, 9–15 Dec., NO_3^- = 4.90 μg m^{-3}).

from about 3 μg m^{-3} to 25 μg m^{-3}. It is suggested from the comparison between the results in Table 19.2 and his own that in the summer, the concentration level of sea salt particles in the Nagoya urban air is close to that in the marine atmosphere.

The mechanism of chlorine loss from sea salt particles has been attributed as follows to a reaction of HNO_3[31,32] or NO_2[33,34] with NaCl in the atmosphere:

$$NaCl_{(aerosol)} + HNO_{3(gas)} \rightarrow NaNO_{3(aerosol)} + HCl_{(gas)} \quad (19.1)$$

$$NaCl_{(aerosol)} + 2NO_{2(gas)} \rightarrow NaNO_{3(aerosol)} + NOCl_{(gas)} \quad (19.2)$$

Both reactions were suggested by laboratory experiments, but have not been proved by field measurements as yet. In both reactions, NaCl is converted to $NaNO_3$ by chlorine loss, so that $NaNO_3$ aerosol is the key ingredient in order to investigate the mechanism of chlorine loss on the basis of field measurements.

With regard to atmospheric nitrate in the Nagoya urban area,[2] it was found that the size distribution of nitrate is bimodal and consists of a fine nitrate (NH_4NO_3) and a coarse nitrate ($NaNO_3$), and that the shapes of the size distribution curves change periodically with the season, in which the coarse nitrate of $NaNO_3$ dominates in the summer, but few in the winter. Figure 19.8 shows the representative size distributions and chemical compositions of nitrate in the Nagoya urban area. As seen Figure 19.8, the coarse nitrate of $NaNO_3$ increased in the summer when the largest chlorine loss occurred. The presence of $NaNO_3$ particles in the atmosphere has been also confirmed by mass spectrometry[38] and electron microscopy techniques.[39,40]

To quantitatively evaluate the relationship between chlorine loss and $NaNO_3$ particles, the amount of chlorine loss (μmole m^{-3}) was calculated as follows:

$$Cl_{theor} = 1.17 \times Na_{c\text{-}obs} \quad (19.3)$$

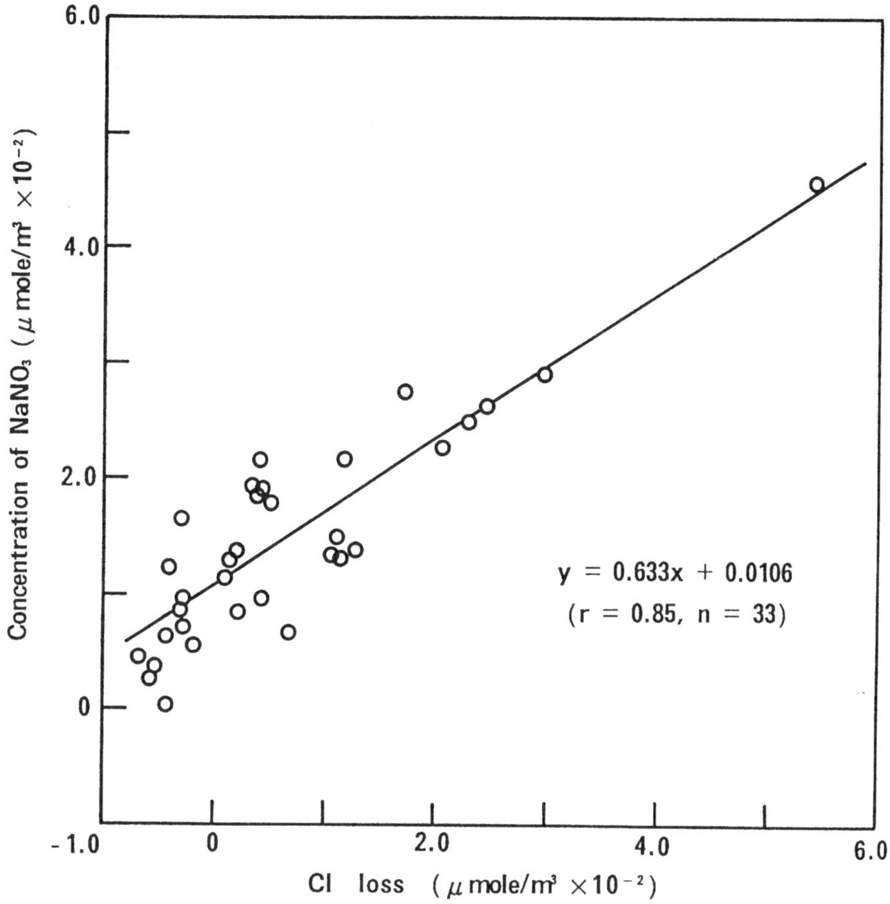

FIGURE 19.9 Relationship between calculated chlorine loss from sea salt particles and observed concentration of $NaNO_3$ particles in Nagoya urban aerosols.

$$Cl_{loss} = Cl_{theor} - Cl_{c\text{-}obs} \tag{19.4}$$

where $Na_{c\text{-}obs}$ and $Cl_{c\text{-}obs}$ are the observed concentration (μmole m^{-3}) in the coarse-particle size range of Na^+ and Cl^-, respectively, and Cl_{theor} is the theoretical concentration (μmole m^{-3}) in the coarse-particle size range of Cl^-, assuming that the $Cl^-:Na^+$ concentration ratio in atmospheric sea salt particle is equal to that in seawater (i.e., 1.8). Also, the observed nitrate concentration (μmole m^{-3}) in the coarse-particle size range is used as the concentration of $NaNO_3$ particles.

Figure 19.9 is a scatterplot of the calculated amount of chlorine loss against the observed concentration of $NaNO_3$ particles. The least-squares fit of the data (n = 33) yields a correlation coefficient of 0.85. This excellent correlation supports the conclusions that the chlorine loss is due to the reaction of Equation 19.1 and/or Equation 19.2.

On the other hand, the concentration of gaseous HNO_3 in the atmosphere usually increases during the high photochemical activity of the summer season. In the Nagoya urban area, the average concentration of gaseous HNO_3 in daytime in the summer (ca. 5 μg m^{-3}) was about 10 times higher than that in the winter.[3] In contrast to HNO_3, the average concentration of NO_2 was higher in the winter than in the summer by roughly twofold,[3] yet chlorine loss from sea salt particles hardly occurred in the winter. Therefore, it seems that, judging from these seasonal variations in HNO_3

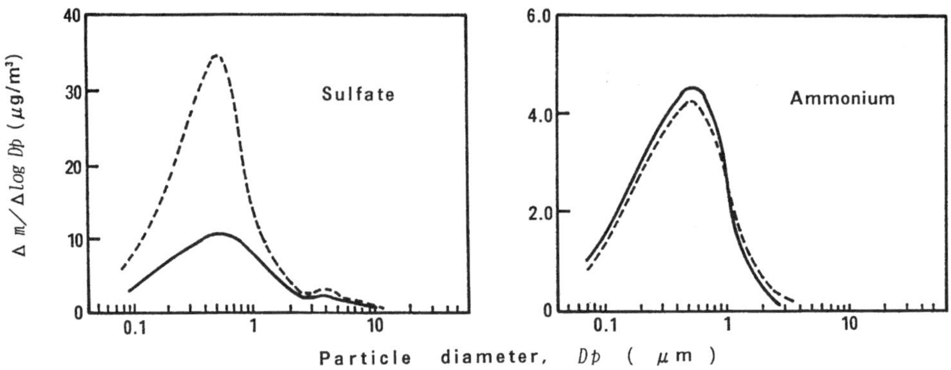

FIGURE 19.10 Size distributions for sulfate and ammonium in Nagoya urban aerosols. ——— winter (1978, 2/9–2/16, SO_4^{-2} = 11.3 μg m^{-3}; NH_4^+ = 3.78 μg m^{-3}); ……: summer (1978, 8/25–8/31, SO_4^{-2} = 24.1 μg m^{-3}; NH_4^+ = 3.75 μg m^{-3}).

and NO_2 concentrations, the Equation 19.1 is the main mechanism of chlorine loss in the Nagoya urban area. This reaction typifies a reaction between a more volatile acid salt and a less volatile acid to form the less volatile acid salt, releasing the more volatile acid; and is also of importance in discussing the formation mechanisms of secondary sulfate and nitrate.

FORMATION PROCESSES OF SECONDARY SULFATE AND NITRATE

The conversion mechanisms and rates for sulfate and nitrate formation are of considerable interest because of concern over apparent widespread environmental effects associated with acid deposition, visibility reduction, and weather and climate changes. These secondary aerosols primarily result from the oxidation of sulfur dioxide and nitrogen oxides. In order to clarify the oxidation processes and rates, numerous extensive studies have been conducted, including laboratory studies,[41–43] theoretical and kinetic computer-modeling studies,[44,45] and field studies in ambient air[46–48] and in plumes.[49,50] In this section, results[1,2] and supplementary data will be described for the size distributions and concentrations of sulfate and nitrate in Nagoya urban aerosols, and formation processes[3,51] will be proposed that can systematically explain the complicated behaviors of secondary sulfate/nitrate/chloride/ammonium aerosols in the Nagoya urban air.

It is widely accepted that sulfate compounds, which exist mainly as ammonium salts, are a major fraction of the mass of fine particles in the atmosphere, and that the oxidation of gaseous SO_2 is only one main atmospheric reaction pathway for the formation of secondary sulfate. Figure 19.10 shows the representative profiles of size distributions for sulfate and ammonium in Nagoya urban aerosols. The profiles of ammonium were unimodal in the fine-particle range, and did not change with the seasons. Also, those of sulfate were practically unimodal in the fine-particle range, although a slight peak of the coarse sulfate (derived from sea salt, soil, and etc.) was observed. Because the concentrations of sulfate increased in the summer, the peak heights in the fine-particle range differed significantly between in the summer and winter, as seen in Figure 19.10. However, the profiles of sulfate were practically unimodal in the fine-particle range throughout the year.

Table 19.3 gives the seasonal average concentrations of sulfate, ammonium, nitrate, and chloride in Nagoya urban aerosols from 1978 to 1985. As noted above in in the previous chapter section, nitrate and chloride show a bimodal size distribution, so that these concentrations were separately averaged in the coarse- and fine-particle ranges.

The sulfate concentrations clearly showed a seasonal variation in which the highest was observed in the summer. It is well-known that oxidation of SO_2 can occur not only in the gas phase (homogeneous reactions mainly with radicals such as OH and HO_2), but also in liquids or on the

TABLE 19.3
Average Concentrations of Sulfate, Ammonium, Nitrate, and Chloride in Nagoya Urban Aerosols

				Nitrate[c]		Chloride[c]		
Sampling Period[a]	No. of Samples	Sulfate[b] (μg m^{-3})	Ammonium[b] (μg m^{-3})	Coarse (μg m^{-3})	Fine (μg m^{-3})	Coarse (μg m^{-3})	Fine (μg m^{-3})	TSP (μg m^{-3})
Spring	13	7.78	2.77	1.03	2.02	0.993	1.71	64.0
Summer	13	10.1	2.50	1.35	1.34	0.760	0.874	49.3
Autumn	13	7.26	2.71	0.812	2.36	0.741	2.41	73.3
Winter	13	6.82	3.26	0.271	2.48	0.722	3.47	67.9

[a] Aerosol samples were collected by Andersen samplers for about 6-day durations from 1978 to 1985.
[b] More than 90% of the mass of sulfate and ammonium was found in the fine-particle (<2.1 μm) size range.
[c] Nitrate and chloride were found in both fine and coarse (≥2.1 μm) particle size range.

surface of existing particles (heterogeneous reactions). Therefore, fine sulfate — namely, secondary sulfate — formation from the oxidation of SO_2 in the Nagoya urban area must be enhanced under the summertime condition of high photochemical activity and high relative humidity. The field measurement results at Nagoya on the sulfur conversion ratio[3] suggest that the oxidation rate of SO_2 in the summer increases by about 2 times than in the winter. The seasonal variation of sulfate concentration can be explained by the increase of secondary sulfate formation under summertime conditions. In contrast to sulfate, the ammonium concentrations were almost constant, independent of season, although the seasonal average in the winter was somewhat higher than that in other seasons.

It is also well-known that oxidation of NO_2 can occur in the gas phase (homogeneous reactions mainly with radicals such as OH and HO_2) and in liquids or on the surface of existing particles (heterogeneous reactions). The field measurement results at Nagoya on the nitrogen conversion ratio[3] also suggest that the oxidation rate of NO_2 in the summer increases by about 4 times than in the winter. Therefore, the seasonal variation of fine nitrate concentrations was expected to be similar to that of sulfate, but it actually showed an inverse seasonal variation, as given in Table 19.3, in which the minimum was observed in the summer.

The difference in the seasonal variation between secondary nitrate and sulfate seems to be due to the difference in vapor pressure between H_2SO_4 and HNO_3, which are produced in the oxidation processes of SO_2 and NO_2. H_2SO_4 has a very low pressure (e.g., 10^{-4}–10^{-6} torr at 20 °C), so it can either condense on existing aerosol particles or nucleate to form new particles. On the other hand, the vapor pressure of HNO_3 at 20°C is approximately 10^5 to 10^7 times higher than that of H_2SO_4. Therefore, gaseous HNO_3 can react to form secondary aerosols on the surfaces of or within existing aerosol particles, but it can hardly nucleate to form new particles under normal circumstances.

As stated in the previous chapter section, secondary nitrate exists as ammonium nitrate. Particulate NH_4NO_3 can dissociate reversibly according to

$$NH_4NO_3(s) \Leftrightarrow HNO_3(g) + NH_3(g). \tag{19.5}$$

Stelson et al.[52] studied the atmospheric equilibrium for Equation 19.5 using thermodynamic data. They found that the $NH_3/HNO_3/NH_4NO_3$ equilibrium is very sensitive to temperature, and that at a high temperature (over the range 20 to 30°C), the dissociation of particulate NH_4NO_3 proceeds.

Another chemical process, which decreases the fine nitrate concentrations in the summer, is probably that of photochemically formed H_2SO_4 reacting with nitrate to form volatile HNO_3,[53] as shown by

$$NO_3^-{}_{(particulate)} + H_2SO_4 \rightarrow HNO_{3(gas)} + HSO_4^-{}_{(particulate)}. \quad (19.6)$$

By the chemical mechanisms described above, the seasonal variation of fine nitrate concentrations can be explained, in which the minimum occurs in the summer, although the oxidation rate of NO_2 and the gaseous HNO_3 concentration are high in the summer.

As shown by Equation 19.1, part of the gaseous HNO_3 reacts with atmospheric sea salt (NaCl) particles and forms $NaNO_3$ particles, releasing gaseous HCl. This reaction is probably the main mechanism of the chlorine loss clearly observed in the summer. The chemical mechanism of Equation 19.1 is the same as Equation 19.6, because HCl is more volatile than HNO_3. It seems that the particle size of $NaNO_3$ formed in Equation 19.1 is the same size of sea salt particle reacted with gaseous HNO_3, and the $NaNO_3$ particles cause the coarse nitrate that increases in the summer, as shown in Figure 19.8 and Table 19.3.

Less focus has been placed on secondary chloride than sulfate and nitrate. However, Harrison and Pio[54,55] reported that chloride in the fine-particle range is present as NH_4Cl and is a major component of secondary aerosols as well as sulfate and nitrate. The major source of HCl (precursor gas of secondary chloride) in Japan's atmosphere is probably municipal incineration.[56] In the Nagoya urban area, the seasonal average concentrations (see Table 19.3) show that fine chloride is more abundant than fine nitrate in winter.

As shown in Figure 19.6 and Table 19.3, fine chloride particles were predominant in the winter, but were few in the summer. This seasonal variation is basically the same as fine nitrate shown in Figure 19.8 and Table 19.3. Like NH_4NO_3, particulate NH_4Cl can also dissociate reversibly according to

$$NH_4Cl(s) \Leftrightarrow HCl(g) + NH_3(g). \quad (19.7)$$

Pio and Harrison[57,58] studied the atmospheric equilibrium for Equation 19.7 using thermodynamic data. They found that the behavior of particulate NH_4Cl was remarkably similar to that of particulate NH_4NO_3, and that equilibrium concentrations of HCl were 1.5 to 2 times higher than HNO_3 levels for the same NH_3 concentration. The findings can explain the resemblance and a little difference in the seasonal variation of concentrations between fine chloride and fine nitrate.

Secondary sulfate, nitrate, and chloride exist mainly as ammonium salts such as $(NH_4)_2SO_4$, NH_4NO_3, and NH_4Cl. However, in the atmospheric gaseous NH_3-deficient case or in the formed H_2SO_4-rich case, H_2SO_4 is partially neutralized to various acidic sulfates, such as NH_4HSO_4 (ammonium bisulfate) and $(NH_4)_3H(SO_4)_2$ (letovicite).[59] The results of concentrations in Table 19.3 indicate that the neutralization of H_2SO_4 by atmospheric gaseous NH_3 is consistently incomplete. Figure 19.11 shows a plot of the $(NH_4^+ - NO_3^-, Cl^-):SO_4^{2-}$ mole ratio in the fine particles as a function of the sum of the SO_4^{2-}, NO_3^-, and Cl^- concentrations (in units of $\mu mol\ m^{-3}$) in the fine particles on 52 aerosol samples collected at Nagoya from 1978 to 1985. The $(NH_4^+ - NO_3^-, Cl^-):SO_4^{2-}$ mole ratios indicate the $NH_4^+:SO_4^{2-}$ mole ratio of the secondary sulfates because the only possible chemical compositions of secondary nitrate and chloride are NH_4NO_3 and NH_4Cl, respectively. As seen from Figure 19.11, the $(NH_4^+ - NO_3^-, Cl^-):SO_4^{2-}$ mole ratio decreases with an increase in the sum of the SO_4^{2-}, NO_3^-, and Cl^- concentrations. The average $(NH_4^+ - NO_3^-, Cl^-):SO_4^{2-}$ mole ratio was 0.91. These data suggest that secondary sulfates in Nagoya urban aerosols are present as a mixture of $(NH_4)_2SO_4$, NH_4HSO_4, $(NH_4)_3H(SO_4)_2$, H_2SO_4, etc., and that the average chemical composition is nearly NH_4HSO_4. Assuming that the chemical compositions of secondary sulfate, nitrate, and chloride are NH_4HSO_4, NH_4NO_3, and NH_4Cl, respectively, the predicted concentrations of secondary aerosols and total ammonium are given in Table 19.4, as calculated from observed concentrations of sulfate, nitrate, and chloride in the fine particles. The predicted total ammonium concentrations agreed with the observed concentrations, except in the autumn.

The contributions of these secondary aerosols to TSP are also summarized in Table 19.4. The contributions increased in the order of sulfate, chloride, and nitrate. The seasonal total contributions

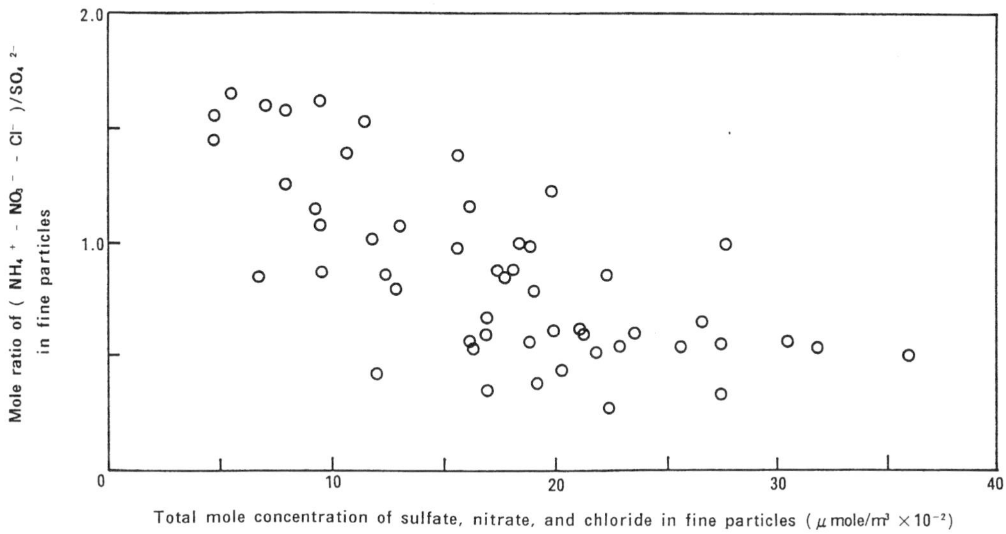

FIGURE 19.11 Plot of $(NH_4^+ - NO_3^-, Cl^-):SO_4^{2-}$ mole ratio in fine particles as a function of the sum of SO_4^{2-}, NO_3^-, and Cl^- concentrations in the units of μ mol m^{-3} in fine particles at Nagoya, 1978–1985.

TABLE 19.4
Predicted Concentrations of Particulate NH_4HSO_4, NH_4NO_3, and NH_4Cl in Nagoya Urban Aerosols and Contribution of Secondary Sulfate, Nitrate, and Chloride Aerosols to TSP

	Predicted Concentration (μg m^{-3})				Observed NH_4^+	Contribution to TSP (%)			
Season	NH_4HSO_4	NH_4NO_3	NH_4Cl	Total NH_4^-	(μg m^{-3})	Sulfate	Nitrate	Chloride	Total
Spring	9.32	2.61	2.58	2.91	2.77	14.6	4.1	4.0	22.7
Summer	12.1	1.73	1.32	2.72	2.50	24.5	3.5	2.7	30.7
Autumn	8.70	3.05	3.63	3.27	2.71	11.9	4.2	5.0	21.1
Winter	8.17	3.20	5.23	3.76	3.26	12.0	4.7	7.7	24.4

ranged about from 21% to 31%. The highest occurred in the summer and arose from the increase of secondary sulfate formation under the summertime condition. In the winter, chloride was predominant, as well as sulfate.

Based on the considerations described above, a schematic diagram of chemical mechanisms for sulfate, nitrate, and chloride formation in urban air is shown in Figure 19.12. This diagram can systematically explain the complicated behaviors of the secondary sulfate/nitrate/chloride/ammonium aerosols observed in the Nagoya urban air.

ELEMENTAL CARBON AND ORGANIC CARBON

It has become apparent in recent years that carbonaceous aerosols are a significant contributor to the mass of atmospheric aerosols. Recent studies on the chemical composition revealed that carbonaceous aerosols comprise about 20 to 30% of the total aerosol mass,[60–62] and account for approximately 30 to 50% of the fine-particle mass fraction.[63–65] The carbonaceous material in atmospheric aerosols is mainly comprised of organic compounds and elemental carbon (EC). The latter is sometimes referred to as "black carbon." The mass of the other carbonaceous materials

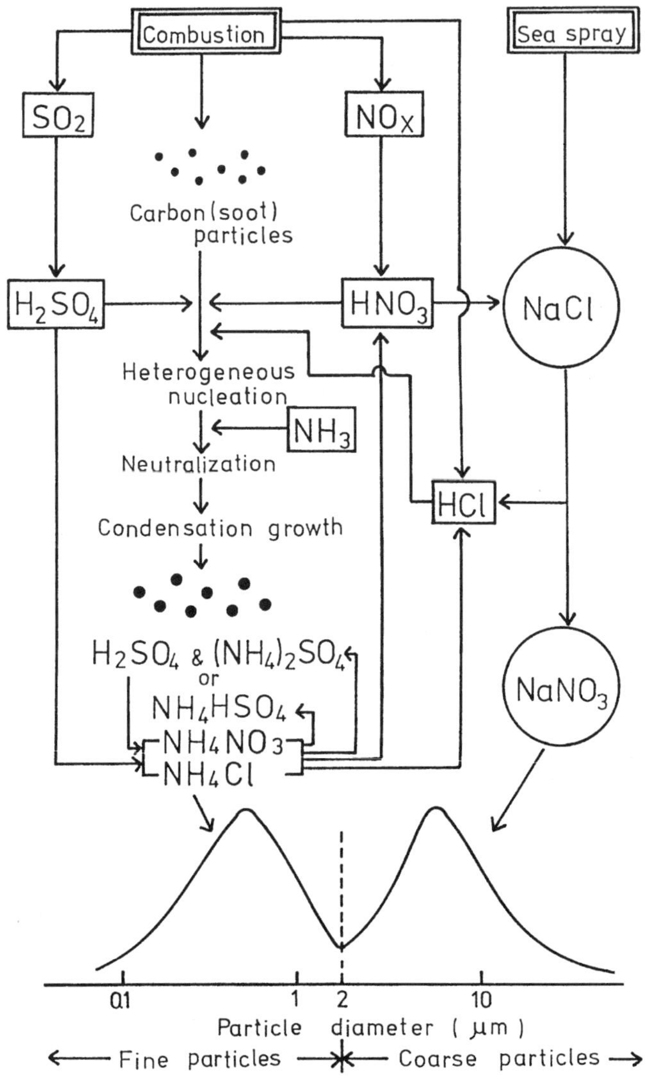

FIGURE 19.12 Schematic diagram of suggested mechanisms for sulfate, nitrate, and chloride formation in urban air at Nagoya.

(e.g., carbonate carbon) is negligibly lower than those of organic compounds and EC.[60] Despite the major component of atmospheric aerosols, information on carbonaceous aerosols is sparse in comparison to other species such as sulfate and nitrate. However, recent advances in analytical measurement techniques have permitted routine determination of EC and organic carbon (OC) concentrations in ambient aerosol samples.[66] Application of new techniques for carbon analysis is making up for the lack of data on carbonaceous aerosols. This section discusses the results[7] obtained for the concentrations of EC and OC and of their contributions to TSP in Nagoya urban air. From these results, the origin of organic compounds in urban aerosols can be identified and further discussed.

In our technique for carbon analysis,[67] it is necessary for the determination with a high accuracy that aerosol samples be collected uniformly on quartz fiber filters; thus, a standard high-volume sampler was used for sampling instead of an Andersen cascade impactor. In general, TSP concentrations measured by a standard high-volume sampler are higher than those by an

TABLE 19.5
Average Concentrations of Carbonaceous Aerosols and the Ratios to TSP in the in Nagoya Urban Area

Sampling Period[a]	No. of Samples	Elemental Carbon (EC) (μg m^{-3})	Organic Carbon (OC) (μg m^{-3})	TSP (μg m^{-3})	EC/TSP (%)	OC/TSP (%)	Concentration Ratio of TC:EC[b]
Spring	13	10.7	15.2	114	9.4	13.3	2.4
Summer	15	9.77	12.6	87.2	11.2	14.4	2.3
Autumn	15	17.9	22.0	119	15.0	18.5	2.2
Winter	12	13.7	15.1	102	13.4	14.8	2.1
Annual av.		13.0	16.2	106	12.3	15.3	2.3

[a] Aerosol samples were collected by hi-vol samplers for 24-h durations from 1984 to 1986.
[b] TC (Total Carbon) is the sum of EC and OC.

Andersen cascade impactor. For the size distributions of carbonaceous aerosols, it was reported that >80 to 90% of the EC mass[68,69] and >70% of the OC mass[68] were concentrated in the fine-particle size range. Similar results were obtained for Nagoya urban aerosols collected by an Andersen cascade impactor at the sacrifice of the determination accuracy.[70]

Seasonal concentration data for EC, OC, and TSP in Nagoya urban air from 1984 to 1986 are summarized in Table 19.5. The seasonal average concentrations ranged from ~10 to 18 μg m^{-3} for EC and from ~13 to 22 μg m^{-3}, for OC. Both the concentrations of EC and OC had a tendency to increase in autumn months. The highest seasonal average concentration of EC was observed in the autumn and was roughly 2 times higher than the minimum ocurred in summer. Identical behavior was observed for OC. The concentration ratios of EC:TSP and OC:TSP are also given in Table 19.5. The annual average EC:TSP and OC:TSP ratios were 12.3% and 15.3%, respectively; these ratios showed that carbonaceous aerosols account for approximately 30% of TSP mass loading and they are one of the most abundant components of Nagoya urban aerosols, like soil particles and secondary inorganic aerosols such as sulfate.

With regard to the origin of atmospheric EC and OC (i.e., organic aerosol), almost all of them, except OC produced by mechanical processes and by atmospheric reactions from gaseous precursors, are probably emitted from combustion sources like fossil fuel combustion and biomass burning. In the Nagoya urban area, a large quantity of fossil fuel is consumed for the generation of electric power, and fuel for automobiles, industrial production, and home use. The consumption of fossil fuel is almost constant throughout the year because the percentage for home heating in the winter is slight (ca. 5%) as compared with total annual consumption. For domestic heating purposes in the winter, kerosene, natural gas, and electricity are mainly used; wood stoves are not used. The largest combustion source of biomass burning is probably municipal incineration, in which a great deal of biomass (e.g., vegetable garbage, paper, wood) is burned independently of season. From these considerations on the sources of EC and OC, it is inferred that a major cause of the seasonal changes in the concentrations of EC and OC is not a change in emissions from sources, but a change in meteorological conditions such as atmospheric stability and photochemical activity.

EC is not formed in the atmosphere by reactions involving gaseous precursors. It is produced only in combustion processes as a primary pollutant. In contrast to EC, OC is not only directly emitted from sources, but can also be produced by atmospheric reactions from gaseous precursors as a secondary pollutant. The ratio of primary to secondary organic aerosols is important in the estimation of the origin and in the development of strategies for control of aerosol carbon air quality. Several investigators[71–75] reported that photochemical gas-to-particle conversion plays an important role in secondary organic aerosol production, and that the greater part of organic aerosols is secondary during photochemical smog season. On the other hand, measurement results[68,76,77] con-

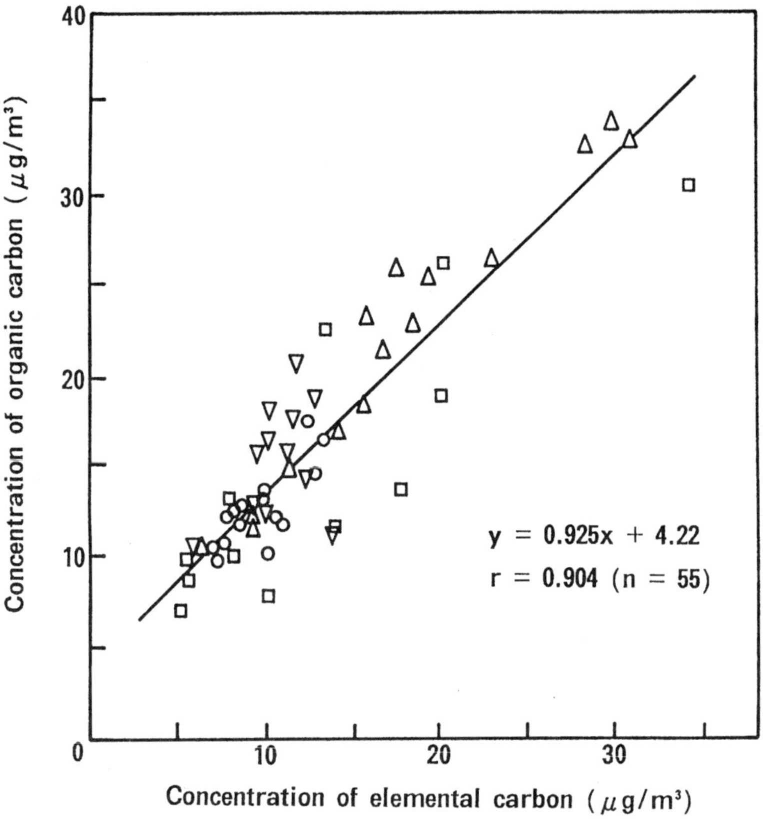

FIGURE 19.13 Scatter diagram of elemental carbon and organic carbon concentrations for 24-h samples collected in the Nagoya urban area, 1984–1986: ▽: spring, ○: summer, △: autumn, and □: winter.

trary to the above conclusion were also reported; that is, no relationship between photochemical activity indicated by ozone level and secondary organic aerosol production.

Primary fine organic aerosols should be emitted from combustion sources at a nearly constant rate for EC. For concentration ratios of carbonaceous aerosols in primary emissions, TC:EC or EC:TC is usually used, rather than OC:EC. TC (total carbon) is defined as the sum of EC and OC. For example, Cass et al.[78] estimated the TC/EC ratio in primary aerosol emissions to be ~3.2 averaged over all sources in Los Angeles; and the overall EC:TC ratio was estimated by Wolff et al.[79] at 0.37 (2.7 for the TC:EC ratio) for primary emissions in Denver. Therefore, concentration ratios of carbonaceous aerosols (e.g., TC:EC) in ambient air are useful as an indicator of secondary organic aerosol production because if a large fraction of organic aerosol is contributed by photochemically produced secondary organic aerosols, a seasonal peak in the TC:EC ratio should be observed during the summer photochemical season, reflecting enhanced secondary organic aerosol production.

In Table 19.5, the seasonal averages of TC:EC in Nagoya urban aerosols are listed. The seasonal averages ranged from 2.1 to 2.4, and the annual average was 2.3. There was no significant seasonal variation in the ratio of TC:EC ratio. This result suggests that secondary organic aerosols are not overwhelming contributors to organic aerosols in the Nagoya urban area. Gray et al.[64,80] reported similar field data, in which the concentration ratio of TC:EC in Los Angeles showed little seasonal dependence and averaged 2.6 over an annual cycle. Figure 19.13 is a scatterplot for the concentration of EC against that of OC. The least-squares fit of the data (n = 55) yields a correlation coefficient, r, of 0.904. This excellent EC–OC correlation also supports the view that the composition of

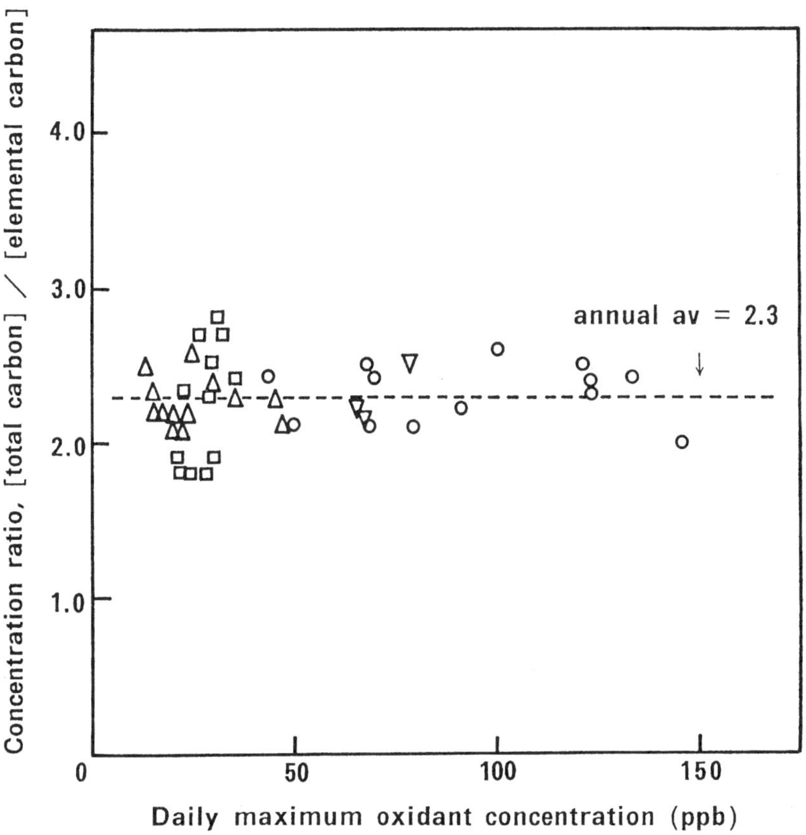

FIGURE 19.14 Relationship between daily maximum oxidant concentration and concentration ratio of total carbon to elemental carbon (TC:EC) in the Nagoya urban area, 1984–1986. Symbols are described in Figure 19.13. The dashed line corresponds to the annual average of TC:EC.

carbonaceous aerosols in the Nagoya urban area is almost constant throughout the year, even during the summer photochemical season.

In order to further clarify the contribution of photochemical processes to secondary organic aerosol production, the relationship between the TC:EC concentration ratio and the daily maximum oxidant concentration, which is viewed as an indicator of the photochemical activity, is shown in Figure 19.14. As seen in the figure, the TC:EC ratio is approximately constant under the wide photochemical activity conditions, ranging in daily maximum oxidant concentration from about 10 to 150 ppb. This result indicates that gas-to-particle conversion by photochemical reactions hardly contributes to the total organic aerosols in the Nagoya urban area, although secondary organic aerosols may be present.

From the field data shown in Table 19.5 and Figures 19.13 and 19.14, it was found that organic aerosols in the Nagoya urban area are primary in origin, and that secondary organic aerosols hardly contribute to the total organic aerosol mass. However, in order to conclude the exact origin of organic aerosols in the Nagoya urban area, another approach is required in addition to measurements of particulate carbon concentration.

CHEMICAL COMPOSITION OF ORGANIC AEROSOLS

In the previous section, it was shown that organic carbon comprises about 15% of the total aerosol mass in the Nagoya urban area; thus, organic aerosols are a major component of Nagoya urban aerosols. It is very interesting and important to determine what compounds make up the organic aerosols because the determination can provide useful information about sources, fate, and behavior of atmospheric organic aerosols. However, the composition of organic aerosols has been poorly understood up until now, except for a few compound groups, such as polycyclic aromatic hydrocarbons and *n*-alkanes. This is partly because of the complexity of the composition, consisting of roughly thousands of individual compounds, and partly because of the many difficulties in identifying and quantifying the great portion of individual compounds which are high molecular weight and highly polar.

There are two approaches to clarify the sources, fate, and behavior of organic aerosols. The first approach is the rough classification and quantification of organic aerosols according to physical or chemical properties. The second approach is the use of tracer compounds that can be quantitated relatively easily in comparison with high molecular weight and highly polar organics. In this section, the results[8,81,82] and supplementary data will be described for the two approaches for organic aerosols in Nagoya urban area.

As an example of the first approach, solvent extraction procedures have been used by several investigators. Grosjean[83] compared the ability of various solvents and solvent mixtures to extract organic compounds from aerosol samples collected at Pasadena, CA, and found that successive and binary mixture extractions gave identical organic carbon extraction efficiency results (95 to 100%). He also reported that the extraction efficiencies of the polar solvents, ethanol and acetone, increased with an increase in the ozone concentration, (i.e., photochemical activity), while those of nonpolar and moderately polar solvents, isooctane and methylene chloride, did not change.

Based on Grosjean's paper,[83] Appel et al.[71,84] applied similar techniques to Los Angeles aerosols to estimate the relative contributions of primary and secondary organic aerosols. In their procedure, primary organics were estimated from the carbon extracted with cyclohexane, and secondary organics by successive extraction with benzene and methanol/chloroform minus the primary organics. Daisey et al.[72] also estimated the contribution of secondary organics in New Jersey aerosols using similar techniques in which aerosol samples were extracted sequentially with increasingly polar solvents (cyclohexane, methylene chloride, and acetone) to extract nonpolar (NP), moderately polar (MP), and polar (P) organic fractions, respectively. Both their results[71,72,84] found that much of the organic aerosol material consisted of polar organic compounds, and the estimated contributions of secondary organics to total organic aerosols in the summer was about 70% in Los Angeles[84] and 15 to 36% in New Jersey.[72]

The above estimations of secondary organics depend on the assumption that the concentrations of primary and secondary organics are equated with concentrations of nonpolar and polar compounds, respectively. This assumption, however, has been denied by the results that there was no significant difference in the polar-to-nonpolar ratio between summer (photochemical episode) and winter (low photochemical activity),[77] and that primary organics from combustion sources also contained polar organic compounds.[77,85] Therefore, it is now realized that the polar fraction of organic aerosols is not entirely secondary in origin, and that the solvent extraction techniques cannot distinguish secondary organics from primary organics.

In our result[81] using the solvent extraction procedure of Daisey et al.[72] applied to Nagoya urban aerosols, the annual average NP:MP:P ratio was 20:14:66, and so the organic aerosol was mainly composed of polar organics. However, no significant seasonal variations in the ratio were observed as Gundel and Novakov[77] had reported. Furthermore, a new problem in the solvent extraction procedure was found: the artifact formation of organics occurred during the Soxhlet extraction with acetone due to a reaction of the acetone and nitrate (i.e., NH_4NO_3) in aerosol samples.

Pre-ashed (600°C for 3 h in air) quartz fiber filters (8 × 10 in. Palfflex 2500 QAST) were spiked with known amounts of NH_4NO_3 dissolved in aqueous solution. After air-drying, the filters were extracted with acetone for 8 h in a Soxhlet apparatus. The extracts were dehydrated with anhydrous sodium sulfate, filtered, and then reduced to about 2 ml on a rotary evaporator at low temperature (<40°C). The extractable masses were determined by weighing the residues of the concentrated

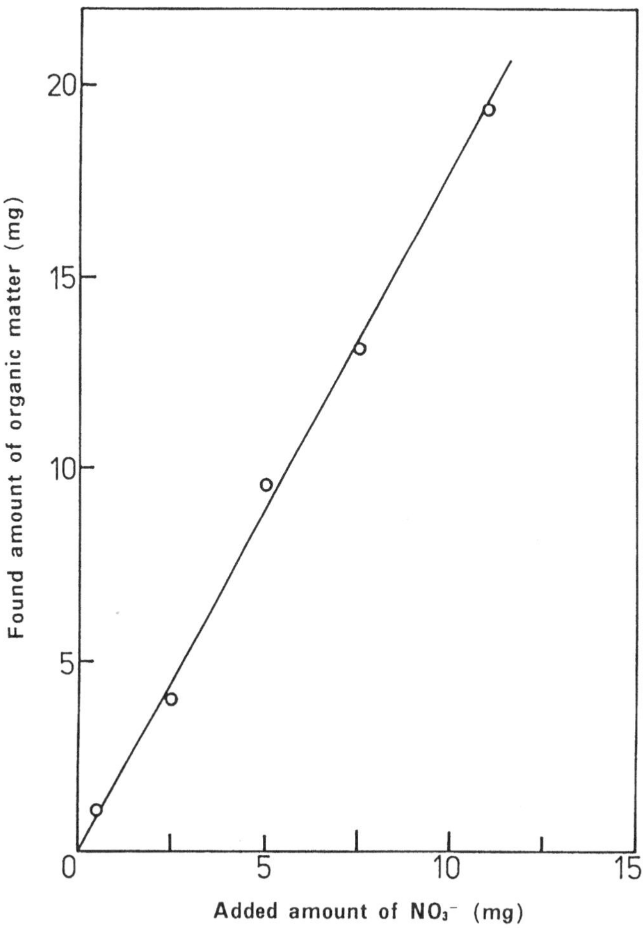

FIGURE 19.15 Relationship between amounts of added NO_3^- and those of found organic matter when a quartz fiber filter spiked with NH_4NO_3 is extracted with acetone for 8 h in a Soxhlet apparatus.

extracts after removal of the solvent. The acetone was tinged with brown after about 1 h from the start of Soxhlet extraction. The color of the acetone deepened with an increase in the spiked amounts of NH_4NO_3. Figure 19.15 represents the relationship between the amounts of added NO_3^- and those of the residues of extracts, which are evaluated as organic matter in the solvent extraction procedure. As seen in Figure 19.15, the found amounts of organic matter were approximately two times that of the added amounts of NO_3^-; thus, it was concluded that artifact organics were formed in the extraction procedure.

Figure 19.16 shows two total ion chromatograms (TICs) of (a) and (b), where (a) is the TIC of organics in the acetone extract of Nagoya urban aerosols after the extractions with cyclohexane and methylene chloride, and (b) is the TIC of the artifact organics formed in the acetone extraction procedure. These TICs are similar in shape. The ☆ symbol shows a peak of which the mass spectrum of (a) agreed with that of (b), although the peaks could not be identified. The peaks marked with ★ were identified with 2,2,6,6-tetramethyl-1-nitroso-4-piperidinone. The mass spectrum and structural formula are shown in Figure 19.17.

As stated above, a number of problems exist with the first approach using solvent extraction procedures. So the second approach is now applied to the characterization of organic aerosols in the Nagoya urban area. Gas chromatography/mass spectrometry (GC/MS) is indispensable to the second approach, so that an appropriate preseparation procedure is also necessary for analyses of

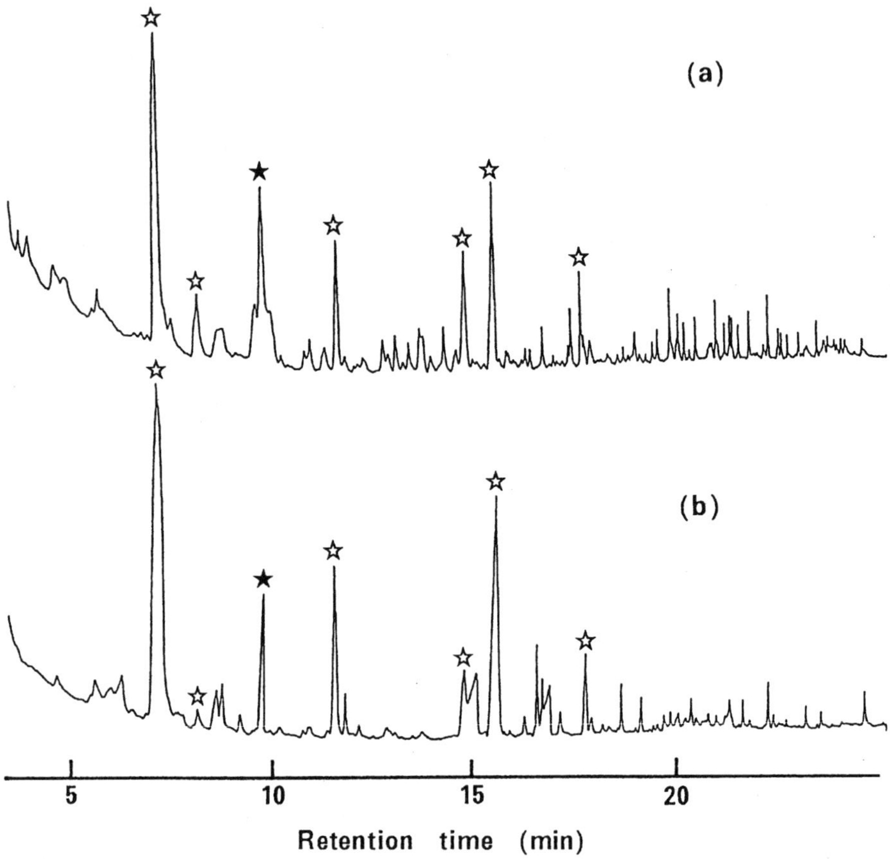

FIGURE 19.16 Comparison of total ion chromatogram (TIC) of acetone extract of Nagoya urban aerosols with that of reaction products of acetone and NH_4NO_3. (a) TIC of acetone extract of Nagoya urban aerosols after the extraction with cyclohexane and methylene chloride; (b) TIC of reaction products of acetone and NH_4NO_3. Symbols ☆ and ★ show peaks for which the mass spectrum of (a) agrees with that of (b).

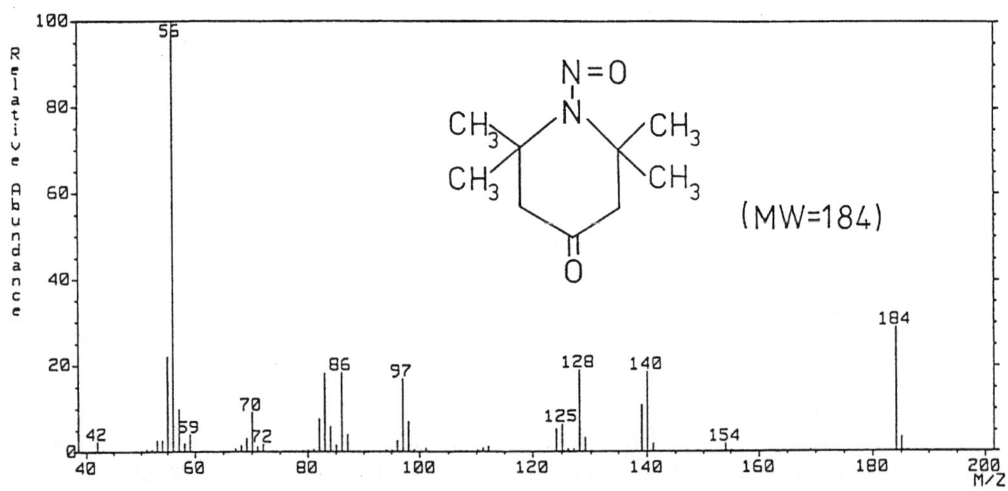

FIGURE 19.17 Mass spectrum of peak of ★ shown in Figure 19.16. The mass spectrum was identical to that of 2.2.6.6-tetramethyl-1-nitroso-4-piperidinone.

TABLE 19.6
Organic Compounds in Nagoya Urban Aerosols Identified by Mass Spectra

Fraction of Figure 19.18	Compound Class	Peak No. in Figure 19.18 and Organic Compounds Identified
(a)	*n*-Alkanes	20:Eicosane, 21:Heneicosane, 22:Docosane, 23:Tricosane, 24:Tetracosane, 25:Pentacosane, 26:Hexacosane, 27:Heptacosane, 28:Octacosane, 29:Nonacosane, 30:Triacontane, 31:Hentriacontane, 32:Dotriacontane, 33:Tritriacontane, 34:Tetretriacontane, 35:Pentatriacontane, 36:Hexatriacontane, 37:Heptatriacontane
(b)	Polycyclic aromatic hydrocarbons	1:Fluoranthene, 2:Pyrene, 3:Benzo[ghi]fluoranthene, 4:Benz[a]anthracene, 5:Chrysene, 6:Benzo[b]fluoranthene and Benzo[k]fluoranthene, 7:Benzo[e]pyrene, 8:Benzol[a]pyrene, 9:1,3,5-Triphenylbenzene, [a] 10:Indeno[1,2,3-cd]pyrene, 11:Benzo[ghi]perylene
(c)	Alcohols	1:1-Hexadecanol, 2:1-Octadecanol, 3:1-Eiconsanol, 4:1-Docosanol, 5:Bis(2-ethylhexyl)phthalate, [b] 6:1-Tetracosanol, 7:1-Hexacosanol
(d)	Phthalic acid esters	1:Unknown, 2 and 3:Diheptyl phthalate, 4:Bis(2-ethylhexyl)phthalate
(e)	Carboxylic acids	9:Pelargonic acid, 10:Capric acid, 11:undecylic acid, 12:Lauric acid, 13:Tridecylic acid, 14:Myristic acid, 15:Pentadecylic acid, 16:Palmitic acid, 17:Margaric acid, 18:Stearic acid, 19:Palmitic acid, 20:Arachidic acid, 21:Heneicosanoic acid, 22:Behenic acid, 23:Tricosanoic acid, 24:Lignoceric acid, A:Adipic acid, B:Suberic acid, C:Azelaic acid, D:Sebacic acid

[a] This compound is not a polycyclic aromatic hydrocarbon (PAH), but fractionated in the PAH fraction.
[b] A part of bis(2-ethylhexyl)phthalate is fractionated in the alcohol fraction.

tracer organic compounds in aerosols by GC/MS. For Nagoya urban aerosols, a pre-separation procedure was developed using column chromatography with Florisil.[82] This procedure fractionated organic compounds in aerosols into seven discrete compound classes: alkylcyclohexanes; *n*-alkanes; polycyclic aromatic hydrocarbons (PAHs); ketones and esters; alcohols; phthalic acid esters; and carboxylic acids. Carboxylic acids were esterified with a mixture of 1*M* HCl/methanol and methyl acetate in the Florisil column, and then eluted as the methyl esters.

Figure 19.18 shows mass chromatograms of organic compounds in Nagoya urban aerosols fractionated by Florisil column chromatography. For the PAH fraction, TIC was used because PAHs do not have the common fragment ion. In this example, no compounds were detected in the alkylcyclohexanes and ketones and esters fractions. As seen Figure 19.18, about 60 organic compounds listed in Table 19.6 were systematically identified in the fractions by GC/MS.

n-Alkanes are injected into the atmosphere by both natural and anthropogenic sources. It is recognized that the distribution profiles of *n*-alkanes derived from contemporary biological sources (land plant wax, soil, etc.) show a saw-tooth pattern with a strong odd-carbon-number predominance, while those derived from fossil fuel components (petroleum, coal, etc.) show a pattern with no carbon number predominance.[86,87] Therefore, *n*-alkane profiles have been used to explain the origin of atmospheric organic aerosols.

From the field measurement results[8] of concentrations, size distributions, and vapor-to-particle partitioning for *n*-alkanes in Nagoya urban aerosols, it was found that particulate *n*-alkanes in Nagoya mainly originate from anthropogenic combustion sources of fossil fuels and contemporary biological materials, such as municipal incineration, and that the summer combustion of fossil fuels hardly contributes as a source of particulate *n*-alkanes due to changes in the partitioning behavior.

For the present, the second approach gives more useful information for the characterization of organic aerosols than the first approach. Studies on the PAH and carboxylic acid fractions are now in progress. In the near future, advances in the characterization of organic aerosols will be made by these studies.

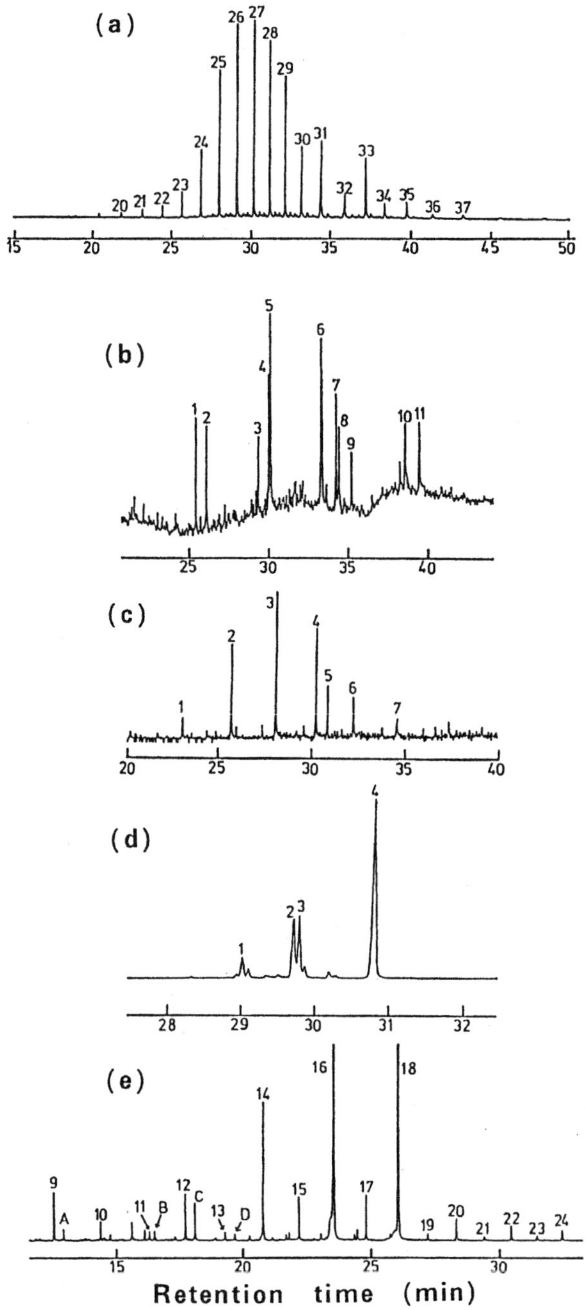

FIGURE 19.18 Mass chromatograms of organic compounds in Nagoya urban aerosols fractionated by Florisil column chromatography. (a) *n*-Alkane fraction, $m/z = 85$; (b) polycyclic aromatic hydrocarbon fraction (chromatogram of (b) is total ion chromatogram); (c) alcohol fraction, $m/z = 97$, (d) phthalic acid ester fraction, $m/z = 149$; (e) carboxylic acid fraction derivatized to methyl esters, $m/z = 74$. For peak identification, see Table 19.6.

REFERENCES

1. Kadowaki, S., *Atmos. Environ.*, 10, 39, 1976.
2. Kadowaki, S., *Atmos. Environ.*, 11, 671, 1977.
3. Kadowaki, S., *Environ. Sci. Technol.*, 20, 1249, 1986.
4. Kadowaki, S., *J. Chem. Soc. Jpn.*, 1977, 1911, 1977.
5. Kadowaki, S., *Environ. Sci. Technol.*, 13, 1130, 1979.
6. Kadowaki, S., *J. Chem. Soc. Jpn.*, 1980, 141, 1980.
7. Kadowaki, S., *Environ. Sci. Technol.*, 24, 741, 1990.
8. Kadowaki, S., *Environ. Sci. Technol.*, 28, 129, 1994.
9. Willeke, K. and Whitby, K.T., *J. Air Pollut. Control Assoc.*, 25, 529, 1975.
10. Gillette, D., *Ann. N.Y. Acad. Sci.*, 338, 348, 1980.
11. Schütz, L. and Jaenicke, *J. Appl. Meteorol.*, 13, 863, 1974.
12. Schütz, L., *Ann. N.Y. Acad. Sci.*, 338, 515, 1980.
13. Duce, R.A., Unni, C.K., Prospero, J.M., and Merrill, J.T., *Science,* 209, 1522, 1980.
14. Uematsu, M., Duce, R.A., Nakaya, S., and Tsunogai, S., *J. Geophys. Res.*, 90, 1167, 1985.
15. Rahn, K.A., *Atmos. Environ.*, 10, 597, 1976.
16. Kawaguchi, K., *Dojogaku,* Asakura Syoten, Tokyo, 1965, 12.
17. Takai, Y. and Miyoshi, H., *Dojo Tsuuron*, Asakura Syoten, Tokyo, 1977, 13.
18. Inoue, K. and Yoshida, M., *Nippon Dojo Hiryo Gaku Zasshi*, 49, 226, 1978.
19. Tanaka, S., Tajima, M., and Hashimoto, Y., *J. Chem. Soc. Jpn.*, 1986, 713, 1986.
20. Sehmel, G.A., *Atmos. Environ.*, 7, 291, 1973.
21. Batterman, S.A., Duzbay, T.G., and Baumgardner, R.E., *Atmos. Environ.*, 22, 1821, 1988.
22. Yanaka, T., Urushiyama, Y., Kitajima, E., Fukuzaki, N., Tamura, R., and Maruyama, T., *J. Environ. Lab. Assoc.*, 10, 18, 1985.
23. Kadowaki, S., Influence of Kosa on the air pollution in Japan, *Kosa: Air and Water Science* (edited by Water Research Institute, Nagoya University), Kokin Syoin, Tokyo, 1991, 256.
24. Bowen, H.J.M., *Environmenral Chemistry of the Elements*, Academic Press, London, 1979, 333.
25. Tanaka, S., Onoue, T., Hashimoto, Y., and Otoshi, T., *J. Jpn. Soc. Air Pollut.*, 24, 119, 1989.
26. Mizohata, A. and Mamuro, T., *J. Jpn. Soc. Air Pollut.*, 7, 289, 1978.
27. Blanchard, D.C., *Prog. Oceanogr.*, 1, 71, 1963.
28. Blanchard, D.C. and Woodcock, A.H., *Ann. N.Y. Acad. Sci.*, 338, 330, 1980.
29. Warnek, P., *Chemistry of the Natural Atmosphere*, International Geophysics Series, Vol. 41, Academic Press, New York, 1988.
30. Fitzgerald, J.W., *Atmos. Environ.*, 25A, 533, 1991.
31. Robbins, R.C., Cadle, R.D., and Eckhardt, D.L., *J. Meteorol.*, 16, 53, 1959.
32. Martens, C.S., Wesolowski, J.J., Harriss, R.C., and Kaifer, R., *J. Geophys. Res.*, 78, 8778, 1973.
33. Schroeder, W.H. and Urone, P., *Environ. Sci. Technol.*, 8, 756, 1974.
34. Finlayson-Pitts, B.J., *Nature*, 306, 676, 1983.
35. Fujimura, M., Yano, N., and Hashimoto, Y., *J. Chem. Soc. Jpn.*, 1978, 456, 1978.
36. Gordon, C.M., Jones, E.C., and Larson, R.E., *J. Geophys. Res.*, 82, 988, 1977.
37. Prospero, J.M., *J. Geophys. Res.*, 84, 725, 1979.
38. Cronn, D.R., Charlson, R.J., Knights, R.L., Crittenden, A.L., and Appel, B.R., *Atmos. Environ.*, 11, 929, 1977.
39. Mamane, Y. and Pueschel, R.F., *Atmos. Environ.*, 14, 629, 1980.
40. Mamane, Y. and Mehler, M., *Atmos. Environ.*, 21, 1989, 1987.
41. Spicer, C.W., *Environ. Sci. Technol.*, 17, 112, 1983.
42. Damschen, D.E. and Martin, L.R., *Atmos. Environ.*, 17, 2005, 1983.
43. Meagher, J.F., Olszyna, K.J., and Luria, M., *Atmos. Environ.*, 18, 2095, 1984.
44. Altshuller, A.P., *Atmos. Environ.*, 13, 1653, 1979.
45. Calvert, J.G. and Stockwell, W.R., *Environ. Sci. Technol.*, 17, 428A, 1983.
46. Weiss, R.E., Larson, T.V., and Waggoner, A.P., *Environ. Sci. Technol.,* 16, 525, 1982.
47. Grosjean, D., *Environ. Sci. Technol.*, 17, 13, 1983.
48. Huntzicker, J.J., Hoffman, R.S., and Cary, R.A., *Environ. Sci. Technol.*, 18, 962, 1984.
49. Daivis, D.D., Heaps, W., Philen, D., and McGee, T., *Atmos. Environ.*, 13, 1197, 1979.

50. Forrest, J., Garber, R.W., and Newman, L., *Atmos. Environ.*, 15, 2273, 1981.
51. Kadowaki, S., Behavior of Urban Aerosols on the Basis of Size Distribution and Chemical Composition Measurements, Ph.D. thesis, Nagoya University, Nagoya, Japan, 1981.
52. Stelson, A.W., Friedlander, S.K., and Seinfeld, J.H., *Atmos. Environ.*, 13, 369, 1979.
53. Harker, A.B., Richards, L.W., and Clark, W.E., *Atmos. Environ.*, 11, 87, 1977.
54. Harrison, R.M. and Pio, C.A., *Environ. Sci. Technol.*, 17, 169, 1983.
55. Harrison, R.M. and Pio, C.A., *Atmos. Environ.*, 17, 1733, 1983.
56. Hiraoka, M., Takeda, N., and Fujita, K., *J. Environ. Pollut. Control*, 15, 1102, 1979.
57. Pio, C.A. and Harrison, R.M., *Atmos. Environ.*, 21, 1243, 1987.
58. Pio, C.A. and Harrison, R.M., *Atmos. Environ.*, 21, 2711, 1987.
59. Tani, B., Siegel, S., Johnson, S.A., and Kumar, R., *Atmos. Environ.*, 17, 2277, 1983.
60. Mueller, P.K., Mosley, R.W., and Pierce, L.B., *J. Colloid Interface Sci.*, 39, 235, 1972.
61. Pierson, W.R. and Russell, P.A., *Atmos. Environ.*, 13, 1623, 1979.
62. Shah, J.J., Johnson, R.L., Heyerdahl, E.K., and Huntzicker, J.J., *J. Air Pollut. Control Assoc.*, 36, 254, 1986.
63. Groblicki, P.J., Wolff, G.T., and Countess, R.J., *Atmos. Environ.*, 15, 2473, 1981.
64. Gray, H.A., Cass, G.R., Huntzicker, J.J., Heyerdahl, E.K., and Rau, J.A., *Sci. Total Environ.*, 36, 17, 1984.
65. Sexton, K., Liu, K.S., Hayward, S.B., and Spengler, J.D., *Atmos. Environ.*, 19, 1225, 1985.
66. Wolff, G.T. and Klimisch, R.L., Eds., *Particulate Carbon: Atmos. Life Cycle*, Plenum Press, New York, 1982.
67. Sakai, Y. and Kadowaki, S., *J. Jpn. Soc. Air Pollut.*, 21, 396, 1986.
68. Wolff, G.T., Groblicki, P.J., Cadle, S.H., and Countess, R.J., *Particulate Carbon: Atmos. Life Cycle*, Wolff, G.T. and Klimisch, R.L., Eds., Plenum Press, New York, 1982, 297.
69. Venkataraman, C., Lyons, J.M., and Friedlander, S.K., *Environ. Sci. Technol.*, 28, 552, 1994.
70. Sakai, Y. and Kadowaki, S., Size distribution of carbonaceous materials in atmospheric particulates, *Annual Meeting of Japan Society of Air Pollution,* Tokyo, Japan, 1985, 611.
71. Appel, B.R., Colodny, P., and Wesolowski, J.J., *Environ. Sci. Technol.*, 10, 359, 1976.
72. Daisey, J.M., Morandi, M., Lioy, P.J., and Wolff, G.T., *Atmos. Environ.*, 18, 1411, 1984.
73. Pratsinis, S., Ellis, E.C., Novakov, T., and Friedlander, S.K., *J. Air Pollut. Control Assoc.*, 34, 643, 1984.
74. Grosjean, D., *Sci. Total Environ.*, 32, 133, 1984.
75. Turpin, B.J. and Huntzicker, J.J., *Atmos. Environ.*, 25A, 207, 1991.
76. Rosen, H., Hansen, A.D.A., Dod, R.L., and Novakov, T., *Science*, 208, 741, 1980.
77. Gundel, L.A. and Novakov, T., *Atmos. Environ.*, 18, 273, 1984.
78. Cass, G.R., Boone, P.M., and Macials, E.S., *Particulate Carbon*: *Atmos. Life Cycle*. Wolff, G.T. and Klimisch, R.L., Eds., Plenum Press, New York, 1982, 207.
79. Wolff, G.T., Countess, R.J., Groblicki, P.J., Ferman, M.A., Cadle, S.H., and Muhlbaier, J.L., *Atmos. Environ.*, 15, 2485, 1981.
80. Gray, H.A., Cass, G.R., Huntzicker, J.J., Heyerdahl, E.K., and Rau, J.A., *Environ. Sci. Technol.*, 20, 580, 1986.
81. Kadowaki, S. and Yamamoto, H., *Bull. Aichi Environ. Res. Ctr.*, 14, 1, 1986.
82. Kadowaki, S., *J. Chem. Soc. Jpn.*, 1991, 392, 1991.
83. Grosjean, D., *Anal. Chem.*, 47, 797, 1975.
84. Appel, B.R., Hoffer, E.M., Kothny, E.L., Wall, S.M., Haik, M., and Knights, R.L., *Environ. Sci. Technol.*, 13, 98, 1979.
85. Kawamura, K. and Kaplan, I.R., *Environ. Sci. Technol.*, 21, 105, 1987.
86. Bary, E.E. and Evans, E.D., *Geochim. Cosmochim. Acta*, 22, 2, 1961.
87. Simoneit, B.R.T. and Mazurek, M.A., *Atmos. Environ.*, 16, 2139, 1982.

20 Analysis of Atmospheric Aerosols in Large Urban Areas with Particle Induced X-ray Emission

Javier Miranda

CONTENTS

INTRODUCTION

Air pollution in urban areas is a growing problem, due to the increase in their dimensions, and the effects it may have on public health, other living species, buildings, and visibility. Moreover, there is a worldwide tendency toward having larger cities. Thus, according to the World Almanac[1] (1994), there were 45 cities with more than 4 million inhabitants in 1985, going up to 52 in 1991. This process gives origin to a reinforcement in health problems as the number of pollutant sources becomes greater in concentrated areas, and more people are exposed to the emissions of those sources. The pollutants produced in large urban areas are also transported to surrounding zones, or even remote sites, thus worsening the air quality in those places also.

Among air pollutants, atmospheric aerosols are a special kind that is receiving increased attention because, besides the aforementioned effects, they may be responsible for a global temperature decrease produced by radiation reflection.[2] In particular, fine aerosols (those having dimensions below 2.5 μm) are capable of scattering solar radiation efficiently, while additionally they can enter the respiratory tract and be deposited deeply in the lungs.

Atmospheric aerosols are normally studied with one or more analytical techniques,[3] which include atomic absorption spectrometry (AAS), ion chromatography (IC), neutron activation analysis (NAA), inductively-coupled plasma spectrometries (ICP), X-ray fluorescence (XRF), and particle-induced X-ray emission (PIXE). None of them is individually capable of giving full characterization of the aerosols. However, for analysis of elements having atomic numbers larger or equal to 13 (i.e., heavier than Mg), PIXE is the most suitable technique.[4,5] This is due to the possibility of analyzing all those elements at the same time, with very good sensitivity (a few nanograms of the elements per cubic meter of air), and non-destructiveness of the sample. It can

0-87371-829-19/00/$0.00+$.50

TABLE 20.1
Typical Sources and Elements Found in Urban Aerosols

Source	Elements
Soil	Al, Si, K, Ca, Ti, Mn, Fe, Sr
Smoke	K
Automobile (gasoline)	C, Br, Pb
Fuel oil	V, Ni
Coal burning	S
Industry, smelters	Cr, Mn, Fe, Cu, Zn, As, Se, Pb
Sea salt	Na, Cl
Refuse, domestic incineration	As, Br, Zn
Sulfate	H, N, O, S

also be used with other techniques that provide complementary information, before, simultaneously, or after the application of PIXE.

In this chapter, a review of studies carried out to determine elemental concentrations in aerosols from large urban areas (arbitrarily taken as those having more than 4 million inhabitants) using PIXE is presented. It supersedes the data given in a previous paper.[6]

URBAN AEROSOLS

The properties of aerosols in an urban area are not the same, in general, as those of airborne particles in rural or remote zones. This is because of the different types of emitting sources. In what can be called a "typical" urban area, there is a mixture of discharges from automobiles, industry, houses, local dust, organic emissions, and others with external origins, like sea salt or volcanic eruptions. These aerosols may have been produced directly as particles or as the result of chemical reactions among polluting gases,[7] with different chemical and physical properties according to their origin. This helps to identify their source through elemental analysis of aerosol samples. Table 20.1 provides examples of elements found in aerosols from urban areas, and some of their common emitting sources. Some of the elements cited in Table 20.1 cannot be measured with PIXE, but their analyses are possible with other techniques, before, simultaneously, or after the application of PIXE.[4]

BASIC PRINCIPLES OF PIXE ANALYSIS

PIXE is based on a process of characteristic X-ray emission after bombardment of a sample by an ion beam, in a similar fashion to that of X-ray fluorescence or electron probe microanalysis.[4] Commonly, a proton beam with energies between 2 MeV and 5 MeV is used to excite the atoms present in a sample such that characteristic X-rays of the elements are emitted. Afterward, these X-rays reach a detector for an energy-dispersive spectrometry. Other ions (such as helium or oxygen) can also be used. Figure 20.1 displays a typical experimental set-up for PIXE studies of aerosol samples. The ion beam, produced by a particle accelerator impinges on the sample, producing the characteristic X-rays. When these X-rays hit the detector (usually of the lithium-drifted silicon, or Si(Li), type), an electronic signal is produced and then processed sequentially by a preamplifier, an amplifier, and a multichannel analyzer (MCA). Ion beam currents are measured by means of a Faraday cup behind the sample. The spectra collected by the MCA can be stored for on-line or later deconvolution to determine the elemental contents in the sample.

The probability (or cross-section) for producing the characteristic X-rays using ion beams is higher than that occurring when X-rays or electrons are used as the primary excitation source, thus

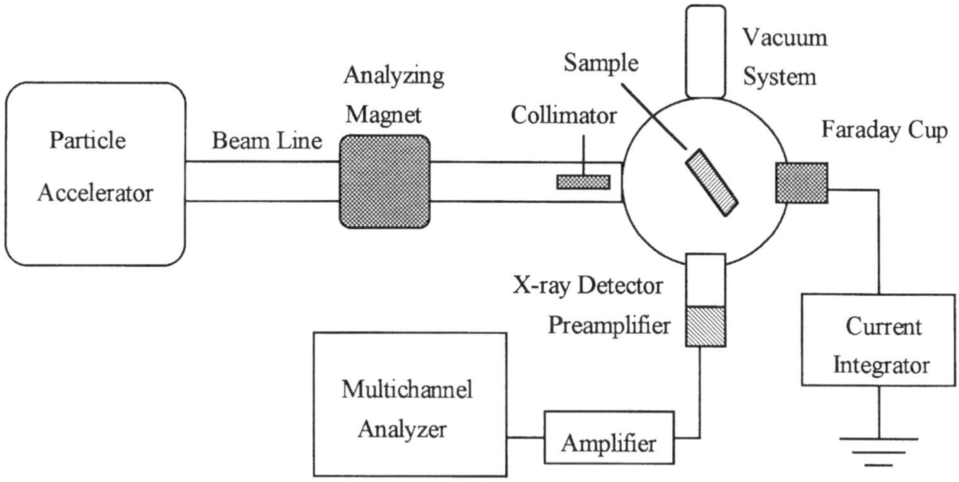

FIGURE 20.1 Diagram of an experimental set-up for PIXE analysis of atmospheric aerosol samples. The ion beam goes through the collimator and impinges on the filter containing the deposited particles, to reach the Faraday cup behind the sample, where the beam charge is integrated. The analyzing magnet selects mass and energy of the beam particles.

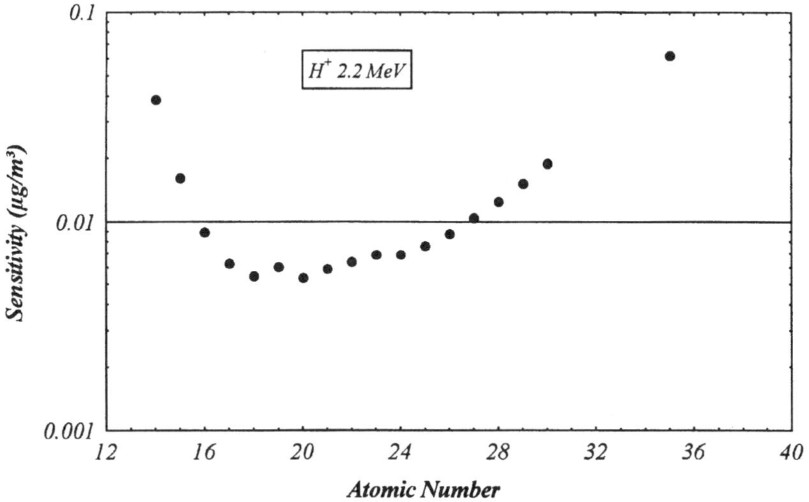

FIGURE 20.2 Sensitivity curve for aerosol analyses with PIXE, corresponding to a 2.2-MeV proton beam, with a 3-mm diameter, and an Si(Li) detector with a resolution of 180 eV at 5.9 keV located at a 90° angle with the beam direction.

making PIXE more sensitive than the latter two methods for most of the elements of interest in aerosol analyses. Ordinarily, elements with atomic numbers higher than 12 (i.e., heavier than Mg) can be measured. Sensitivity of the method is dependent on many factors, including type of ion, beam energy, beam diameter, detector, and sample matrix. An example of a sensitivity curve for aerosol analyses is presented in Figure 20.2, corresponding to a 2.2-MeV proton beam with a 3-mm diameter, and an Si(Li) detector with a resolution of 180 eV at 5.9 keV located at a 90° angle with the beam direction.[8] In this particular case, the best sensitivity is achieved for the element Ca. This can be changed by placing filters in front of the detector window or increasing the beam energy.

SAMPLING PROCEDURES

When sampling aerosols for PIXE analysis, it is necessary to have the samples in the form of a thin target, in order to reduce the effect of particle energy loss and X-ray absorption in the sample itself. Thus, it is desirable to obtain uniform deposits on thin filters. There are a number of commercially available filters that may be suitable for PIXE analysis, although Teflon and polycarbonate filters are the most popular, by far. The reason is that Teflon filters are originally hydrogen-free, and it is possible to carry out simultaneous proton elastic scattering analysis (PESA) for hydrogen content measurements.[9] On the other hand, polycarbonate membrane filters (such as Nuclepore, by Costar Corp.), also have certain advantages. Teflon filters consist of a mesh of thin fibers, and aerosol particles are deposited in the middle of this mesh. Because of this, it may be necessary to compute some X-ray absorption corrections in the filter; moreover, fluorine in the filter also emits gamma-radiation detected by the Si(Li) detector, producing a background in the spectra that deteriorates the sensitivity of the method. Finally, Teflon filters are more brittle, expensive, and radiation-sensitive than polycarbonate filters. Disadvantages of the latter kind of filters are the presence of hydrogen (thus, PESA analysis is not convenient), and the adherence of particles onto their surface is sometimes limited, making necessary to apply a layer of grease to fix the particles. Additionally, several authors have reported bromine contamination of these filters.[10,11]

There have been important developments of sampling equipment that provide for the possibility of obtaining samples appropriate for PIXE analysis. A deeper review of this subject is presented by Akselsson.[12] There is a wide variety of sampling devices used for the application of PIXE in urban areas, although the most common is the Stacked Filter Unit (SFU) in its two designs (Davis and Gent), which allow particle-size separation in two sequential filters: one for particles between 15 μm (Davis) or 10 μm (Gent) and 2.5 μm, and the other with sizes below 2.5 μm. Also, cascade impactors (Batelle type, for example) or similar mechanisms have been used to obtain a finer size resolution. Although the information resulting from the latter sampling methods is very complete, their use is limited to short periods, due to lengthy elemental analysis. Of course, other samplers with no particle-size selection can be used (e.g., like the Hi-Vol), but PIXE analysis is inconvenient in this case because normally thick filters (such as fiberglass) are utilized. Other procedures are then necessary to extract the particles from the filter so that they can be deposited onto another one appropriate for PIXE.

PIXE STUDIES AND RESULTS

The total number of large urban areas (with populations above 4 million inhabitants) where aerosol studies with PIXE have been carried out and published is 14. Table 20.2 shows these megacities, the date when the studies were done, and the type of sampling device and filter used for collecting the aerosols to be analyzed by PIXE. As can be seen, in most of the works, Nuclepore filters were employed, while there is no clear tendency in the collection equipment. This presents a serious problem for direct comparisons because of the different particle-size selection. This fact cannot be overlooked, as the comparisons are required to evaluate the aerosol behavior not only in large urban areas, but in any other study in this field. Furthermore, the use of samplers with no particle-size information is inconvenient for making such comparisons. Although not given in Table 20.2, sampling protocols (i.e., days and hours for sampling) were seldom the same, and published results sometimes refer to 24-hour, 8-hour, or 6-hour averages; automatic samplers (like the DRUM or the streaker) can even give information for shorter periods. Also, samples were taken daily, once or twice a week. No standard has been imposed in this regard.

Table 20.3 gives the experimental parameters used for the analyses of the aerosol samples, including accelerator type, particle used, and beam energy. Unfortunately, two works did not report

TABLE 20.2
Large Urban Areas, Sampling Devices, and Filters for Aerosol Studies Using PIXE

City	Year	Sampling Device	Type of Filter	Ref.
Beijing	1980	Cascade impactor	Nuclepore, Mylar	26, 32
	1983–1984	8-Stage cascade	Mylar, Gelman film	29
Buenos Aires	1988	Open impactor	Nuclepore	18
Dhaka	1993–1994	Gent SFU	Nuclepore	14
Kyoto-Osaka	1986–1992	SFU; rotating stage cascade	Nuclepore	33, 34
Los Angeles	1973	Multiday impactor	Mylar, nuclepore	35
	1983, 1987–1988	Dichotomous	Teflon	13
	1986	Davis SFU; DRUM	Nuclepore, Mylar, Teflon	35
Madrid	1992	Cyclonic	Millipore cellulose	22
Mexico City	1988–1990	Open impactor	Nuclepore	15, 19, 36
	1990–1991	Davis SFU	Nuclepore, Teflon	28, 37
	1993	Davis SFU	Nuclepore, Teflon	8
	1993	Davis SFU	Nuclepore	38
	1995	Davis SFU	Nuclepore	11
Milan	1977–	Cascade impactors	Nuclepore	39–41
	1990	Integral sampler	Nuclepore	42
New York City	1977	Streaker; cascade	Nuclepore; Mylar	20
Rio de Janeiro	1990	Hi-Vol, SFU	Fiberglass, Nuclepore	25
St. Petersburg	1989	Total; 2-stage streaker	Kimfoil, Nuclepore	43
Santiago	1979–1981	Hi-Vol	Fiber glass-Kapton	44
	1987	Davis SFU	Nuclepore	10
	1989	Davis SFU	Nuclepore	10
	1993	Gent SFU	Nuclepore	16
São Paulo	1976–1977	Cascade impactors	Mylar	21
	1983	Gent SFU	Nuclepore	23, 24
	1989	Gent SFU	Nuclepore	45, 46
	1990	Hi-Vol, Gent SFU	Nuclepore	25
Tokyo	1989	Mini-SFU	Nuclepore	47

on the experimental conditions for PIXE.[13,14] Again, no standard analytical procedures were followed in most of the cases, except in those where the analysis was done by the same laboratory. This restriction is not as important as the sampling procedure because elemental concentrations can be accurately measured within a large range of beam energies and particles. The only obstacle in this regard is sensitivity, which may change with these parameters. In a few cases, laboratory intercomparisons have been performed,[15,16] supporting the results. Additionally, the lack of trustworthy reference materials for PIXE analysis of aerosols is still a problem.[17]

Most of the research groups used other complementary techniques, and some tried to compare the results obtained with PIXE and the other methods. However, in several cases, full multi-elemental capabilities of PIXE analysis were not used, focusing on only one element: for example, in Buenos Aires[18] and Mexico City[19] for Pb, and New York City[20] for S. Moreover, in one of the latter cases,[18] the analysis was performed with 50 MeV ^{12}C and ^{16}O ions, instead of protons — which is the most common beam.

The results of the studies of aerosols in large urban areas based on PIXE are as dissimilar as the cities themselves. Geographical and climatic characteristics, together with the traffic density and industrialization of the region, modify the composition of aerosols. Although direct comparisons are impossible to do in a general manner, Table 20.4 shows the concentrations measured in atmospheric aerosols for a few representative elements and urban areas. Where relevant, fine and

TABLE 20.3
Experimental Parameters for PIXE Analysis of Aerosols in Large Urban Areas

City	Year	Ref.	Beam	Accelerator	Other Techniques
Beijing	1980	26, 29, 32	3 MeV H^+	Tandetron EAC, Florida	
	1983–1984	Yin, 1992	NA	NA	INAA
Buenos Aires	1988	18	50 MeV ^{12}C, ^{16}O	Tandem CNEA	^{12}C, ^{16}O ions
Dhaka	1993–1994	14	NA	NA	
Kyoto-Osaka	1986–1992	33, 34	2.0 MeV H^+	Tandem Kyoto Univ.	
Los Angeles	1973	35	4.5 MeV H^+	Cyclotron, UCD	FAST, PESA
	1983, 87-88	13	NA	NA	XRF
	1986	35	4.5 MeV H^+	Cyclotron, UCD	PESA
Madrid	1992	22	2.55 MeV H^+	Tandem Lund Univ.	Gases
Mexico City	1988–1990	15, 19, 36	4.5 MeV H^+	Cyclotron, UCD	LIPM
	1990–1991	28, 37	4.5 MeV H^+	Cyclotron, UCD	PESA, LIPM
	1993	38	3.2 MeV H^+	Van de Graaff, Florida St. Univ.	
	1993-95	8, 11	2.2 MeV H^+	Van de Graaff, U.N.A.M.	PESA, XRF
Milan	1977–	39–41	2.8 MeV H^+	Tandem Van de Graaff/CISE	XRF, AAS
	1990	42	3.0 MeV H^+	Van de Graaff, Florence Univ.	
New York City	1977	20	NA	Tandetron EAC, Florida	
Rio de Janeiro	1990	25	2.4 MeV H^+	Cyclotron, Univ. Gent	AAS
St. Petersburg	1989	43	3 MeV H^+	Tandetron EAC, Florida	XRF
Santiago	1979–1989	10	2.2, 4, 6 MeV H^+	Cyclotron, Univ. of Chile	
	1993	16	2.2 MeV H^+	Cyclotron, Univ. of Chile	AAS, INAA, IC
São Paulo	1976–1977	21	3 MeV H^+	Tandetron EAC, Florida	
	1983	23, 24	2.4 MeV H^+	Cyclotron, Univ. Gent	
	1989	45, 46	2.4 MeV H^+	Cyclotron, Univ. Gent	AAS
	1990	25	2.4 MeV H^+	Cyclotron, Univ. Gent	
Tokyo	1989	47	2.4 MeV H^+	Van de Graaff, Nagoya Univ.	

coarse fractions have been added to balance with those studies that did not make any particle-size separation. Whenever data from more than one site was published, one of them having concentrations close to the average was chosen; in a few works, the concentration values in tabular form were not published.[20-25] The elements named usually have different origins (see Table 20.1). Thus, it is useful to note the contributions of the various sources to the aerosols. The S concentrations (sulfate or coal burning) are in most of the cases between 3 $\mu g\ m^{-1}$ and 5 $\mu g\ m^{-1}$, although extreme cases are Milan (of the order of 10 $\mu g\ m^{-1}$), and St. Petersburg and Mexico City in 1995 (below 1 $\mu g\ m^{-1}$); however, care must be taken in the latter case, as this figure refers to a 24-h average, in contrast with 6-h averages in other studies. As for other elements, there seems to be no clear pattern in their contents. V, as a tracer of fuel oil, in many cases lies below the minimum detectable limits. On the other hand, Fe, normally associated with soil dust, may be highly influenced by the geographical characteristics of the zone. Thus, Mexico City, situated in a normally dry region, has relatively high Fe concentrations, enhanced by traffic-removed dust. Beijing data was possibly affected by an episodic industrial emission,[26,27] explaining the very high contents of Fe. The presence of Zn, clearly anthropogenic, is definitely dependent on the urban zone. Therefore, in Mexico City, with the highest concentrations,[28] it must have an industrial origin (the sampling site was in the middle of an important industrial zone); while in Beijing, it was attributed to refuse incineration.[29] Finally, Pb is the tracer of traffic-related aerosols. It is interesting to see that Pb contents are not connected to the population of the city (and possibly the number of motor vehicles), but, rather, is influenced by the quality of the gasoline consumed there.

TABLE 20.4
Concentrations Measured in Some Urban Areas for Selected Elements (μg m^{-3})

City	Year	S	V	Fe	Zn	Pb
Beijing	1980	3.99	<0.093	10.3	0.304	0.182
Buenos Aires	1988	NA	NA	NA	NA	1.5
Dhaka	1993–1994	2.08	NA	1.05	0.40	0.49
Kyoto	1986–1992	1.09	0.009	0.56	0.088	0.044
Los Angeles	1973	4.01	0.025	1.01	0.183	1.50
	1983	~3.8	NA	~0.4	~0.1	~0.6
	1986	2.5	NA	NA	0.244	0.266
Madrid	1992	~0.9	NA	~0.6	~0.04	~0.15
Mexico City	1991	4.86	0.150	3.72	2.37	1.23
	1993	4.91	0.048	2.25	0.177	0.10
	1995	0.605	0.008	0.114	0.057	0.046
Milan	1981	16.5	0.30	8.8	1.1	2.8
	1983–1984	10.7	0.079	4.3	0.785	1.6
New York	1977	>2.46	NA	NA	NA	NA
St. Petersburg	1989	0.618	<0.008	0.211	0.029	<0.012
Santiago	1979–1989	1.41	0.03	0.69	NA	NA
	1984	4.89	NA	0.26	0.31	NA
	1987	1.02	NA	1.56	2.58	NA
	1993	3.94	0.03	1.41	0.24	NA
São Paulo	1989	2.90	0.031	1.96	0.238	0.10
Tokyo	1989	5.5	NA	2.05	0.16	~0

The use of size-resolving sampling equipment allows the study of the behavior of different elements according to particle dimensions. Thus, some anthropogenic elements have been associated with the fine fractions, while other "natural" ones have been found mostly in the coarse fractions. PIXE is particularly useful for analyzing the small amounts of matter in the finest fractions deposited on the filters obtained in the size-resolving samplers.

The multi-elemental information provided by PIXE, together with its high sensitivity (or, in other words, high analytical speed), can be used to apply multivariate statistical techniques, such as principal component analysis (PCA), or the extension of PCA known as absolute principal component analysis (APCA).[30] Another very popular method is the chemical mass balance (CMB) technique. All of them have been employed in several urban areas to identify possible emitting sources and their contribution to the mass measured with PIXE, developing *receptor models*.[31] Table 20.5 shows the urban areas where these statistical techniques were used, after PIXE analysis. Also, some of the sources identified in the studies are given.

CONCLUSIONS

A description of how PIXE can be used for aerosol characterization in large urban areas was presented. Despite its powerful capabilities, the number of megacities in the world where PIXE was applied is still very small. Thus, it can be stated that there is an open field to study airborne particles in large cities.

According to the data presented in this chapter, there is no evident relationship between city proportions and air quality. Several urban areas present low elemental contents, while others with comparable geographical characteristics and less population exhibit high concentrations.

TABLE 20.5
Pollutant Sources Identified in Large Urban Areas Using Multivariate Statistics

Urban area	Ref.	Method	Fine	Coarse
Beijing[a]	29	CMB	Soil, coal, limestone, traffic, refuse, oil, marine	
Los Angeles[b]	13	APCA	Traffic, soil, sulfate, marine, fuel oil	
Madrid[a]	22	APCA	Traffic dust, traffic gas, refuse	
Mexico City	8	PCA	Soil, fuel oil, traffic	Soil, fuel oil, traffic
St. Petersburg	43	PCA	Sulfur	Dust
São Paulo	46	APCA	Industry, combustion, soil, Cu, Mg	Soil, industry, combustion, marine

[a] No size separation was considered in the multivariate analysis.
[b] No size separation was used in the sampling.

Moreover, there is an urgent need for standard sampling and analytical procedures. Comparisons among different studies are strongly limited due to the wide variety of samplers, sampling protocols, filters, elemental standards, reference materials, etc. used in the various studies.

The application of receptor models must also become a regular procedure to characterize atmospheric aerosols in these urban areas, as this would provide a very powerful tool to reach a more complete knowledge of the problem.

Finally, information from smaller urban areas should be used to understand, in several instances, the problem of aerosols in megacities, although there is no obvious extension from those smaller cities to the larger ones.

REFERENCES

1. *The World Almanac 1994,* Pharos Books, New York, 1994, 774.
2. Cahill, T.A., *Nucl Instr. Meth.,* B109/110, 402, 1996.
3. Maenhut, W., *Int. J. PIXE,* 2, 609, 1992.
4. Johansson, S.A.E. and Campbell, J.L., *PIXE: A Novel Technique for Elemental Analysis*, John Wiley & Sons, Chichester, 1988.
5. IAEA, *Final Report of the IAEA Advisory Group Meeting on Accelerator-based Analytical Techniques for Characterization and Source Identification of Aerosol Particles,* International Atomic Energy Agency, Vienna, 1995.
6. Miranda, J., *Nucl Instr. Meth.,* B109/110, 439, 1996.
7. Manahan, S.E., *Environ. al Chemistry*, 6th. ed., Lewis Publishers, Boca Raton, 1994.
8. Miranda, J., Andrade, E., López-Suárez, A., Ledesma, R., and Cahill, T.A., *Atmos. Environ.,* 30, 3471, 1996.
9. Kusko, B.H., Cahill, T.A., and Eldred, R.A., *Proc. of the APCA 81st Annu. Meet., Dallas, 19-24 June*, APCA, PA, 1988, 88 53.4.
10. Romo-Kröger, C.M., *Environ. Poll.*, 68, 161, 1990.
11. Miranda, J., Crespo, I., and Morales, M.A., *Proc. V Latin-American Seminar on X-ray Analysis*, Cosquín, Argentina, 19-23 November, 46, 1996.
12. Akselsson, K.R., *Nucl. Instr. and Meth.,* B3, 425, 1984.
13. Ehrman, S.H., Pratsinis, S.E., and Young, J.R., *Atmos. Environ.*, 26B, 473, 1992.
14. Khaliquzzaman, M., Biswas, S.K., Tarafdar, S.A., Islam, A., and Khan, A.H., Bangladesh Atomic Energy Commission Report AECD/AFD-CH/3-44 (Bangladesh Atomic Energy Commission, Dhaka), 1995.
15. Aldape, F., Flores M., J., Díaz, R.V., Morales, J.R., Cahill, T.A., and Saravia, L., *Int. J. PIXE,* 1, 355, 1991.

16. Dinator, M.I., Morales, J.R., Romo-Kröger, C.M., Toro, P., Aguila, C., Cassorla, V., Cortes, E., Gras, N., Marín, S., Olave, S., Rojas, X., and Klockow, D., *Proc. of the 10th World Clean Air Congress,* Helsinki, in press, 1995.
17. Wätjen, U. and Cavé, H., *Nucl Instr. Meth.,* B109/110, 395, 1996.
18. Caridi, A., Kreiner, A.J., Davidson, J., Davidson, M., Debray, M., Hojman, D., and Santos, D., *Atmos. Environ.*, 23, 2855, 1989.
19. Aldape, F., Flores M., J., Díaz, R.V., Miranda, J., Cahill, T.A., and Morales, J.R., *Int. J. PIXE,* 1, 373, 1991.
20. Bauman, S.E., Ferek, R., Williams, E.T., and Finston, H.L., *Nucl. Instr. Meth.,* 181, 411, 1981.
21. Bouéres, L.C.S. and Orsini, C.M.Q., *Nucl. Instr. Meth.,* 181, 417, 1981.
22. Climent-Font, A., Swietlicki, E., and Revuelta, A., *Nucl. Instr. Meth.,* B85, 830, 1994.
23. Orsini, C.Q., Netto, P.A., and Tabacniks, M.H., *Nucl. Instr. Meth.,* B3, 462, 1984.
24. Tabacniks, M.H., Orsini, C., and Artaxo, P., *Nucl. Instr. Meth.,* B22, 315, 1987.
25. Tabacniks, M.H., Orsini, C.Q., and Maenhut, W., *Nucl. Instr. Meth.,* B75, 262, 1993.
26. Winchester, J.W. and Bi, M.T., *Atmos. Environ.,* 18, 1399, 1984.
27. Winchester, J.W., Wang, M.X., Lu, W.X., Ren, L.X., and Hong, Z.S., *Nucl. Instr. Meth.,* B3, 503, 1984.
28. Miranda, J., Cahill, T.A., Morales, J.R., Aldape, F., Flores M., J., and Díaz, R.V., *Atmos. Environ.,* 28, 2299, 1994.
29. Yin, S., Pingsheng, L., Zhaohui, H., Ming, Z., Shaojin, Y., Yinan, Y., Qinfang, Q., and Bingru, C., *Int. J. PIXE,* 2, 593, 1992.
30. Maenhut, W. and Cafmeyer, J., *J. Trace Microprobe Tech.*, 5, 135, 1987.
31. Henry, R.C., Lewis, C.W., Hopke, P.K., and Williamson, H.J., *Atmos. Environ.*, 18, 1507, 1984.
32. Winchester, J.W., Wang, M.X., Ren, L.X., Lu, W.X., Hansson, H.C., Lannesfors, H., Darzi, M., and Leslie, A.C., *Nucl. Instr. Meth.*, 181, 391, 1981.
33. Kasahara, M., Choi, K-.C., and Takahashi, K., *Int. J. PIXE,* 2, 665, 1992.
34. Kasahara, M., Yoshida, K., and Takahashi, K., *Nucl. Instr. Meth.,* B75, 240, 1993.
35. Cahill, T.A., Surovik, M., and Wittmeyer, I., *Aerosol Sci. Technol.,* 12, 149, 1990.
36. Flocchini, R.G., Cahill, T.A., Shadoan, D.J., Lange, S.J., Eldred, R.A., Feeney, P.J., Wolfe, G.W., Simmeroth, D.C., and Suder, J.K., *Environ. Sci. Technol.*, 10, 76, 1976.
37. Aldape, F., Flores M., J., Díaz, R.V., and Crumpton, D., *Nucl. Instr. Meth.,* B75, 304, 1993.
38. Miranda, J., Cahill, T.A., Morales, J.R., Aldape, F., and Flores M., J., *Atmósfera*, 5, 95, 1992.
39. Aldape, F., Flores M., J., García G., R., and Nelson, J.W., *Nucl Instr. Meth.,* B109/110, 502, 1996.
40. Caruso, E., Braga Marcazzan, G.M., and Redaelli, P., *Nucl. Instr. Meth.*, 181, 425, 1981.
41. Caruso, E., Braga Marcazzan, G.M., and Redaelli, P., *Nucl. Instr. Meth.,* B3, 498, 1984.
42. Braga Marcazzan, G.M., Caruso, E., Cereda, E., Redaelli, P., Bacci, P., Ventura, A., and Lombardo, G., *Nucl. Instr. Meth.,* B22, 305, 1987.
43. Braga Marcazzan, G.M., Cavicchioli, C., Lucarelli, F., Redaelli, P., Ventura, A., and Marchioni, M., *Nucl. Instr. Meth.,* B75, 230, 1993.
44. Winchester, J.W., Ivanov, V.A., Prokofyev, M.A., Zhukovski, D.A., Shlikhta, A., Zhvalev, V.F., Zhukovski, A.N., Stroganov, D.M., Nelson, J.W., Bauman, S.E., and Nakhgaltsev, L.N., *Nucl. Instr. Meth.,* B49, 351, 1990.
45. Morales, J.R. and Romo-Kröger, C.M., Nucleotécnica 3, 45, 1983.
46. Andrade, F., Orsini, C., and Maenhut, W., *Nucl. Instr. Meth.,* B75, 308, 1993.
47. Andrade, F., Orsini, C, and Maenhut, W., *Atmos. Environ.,* 28, 2307, 1994.
48. Katoh, T., Amemiya, S., Tsurita, Y., Masuda, T., Koltay, E., and Borbély-Kiss, I., *Nucl. Instr. Meth.,* B75, 296, 1993.

21 Chemical Characteristics and Temporal Variation of Size-Fractionated Urban Aerosols and Trace Gases in Budapest

Imre Salma, Willy Maenhaut, Éva Zemplén-Papp, and János Bobvos

CONTENTS

INTRODUCTION

Considerable air pollution occurs in large cities throughout the world.[1] Sources of the air pollution include different industrial emissions, transportation, domestic and household activities and natural processes in the urban area, and some other external sources.[2-4] The pollution affects many people inasmuch as these urban areas usually accommodate a large number of inhabitants. It also has a great impact on the local environment (e.g., on the biosphere, buildings, and materials) and visibility, and can even influence the surrounding rural areas.[5] The coincidence of the pollution sources, population, and/or geographical and climatic conditions in large cities can multiply the adverse effect on the public. Although the air pollution and smog problems are very complex, a small set of compounds has been identified as major contributors to the phenomenon. They are called *criteria pollutants,* and serve as indicators of the air quality in pollution control policy. Criteria pollutants usually include NO_x, SO_2, CO, tropospheric O_3, total or fractionated suspended airborne particulate matter, and Pb. (Rn has also been recognized in indoor air pollution.) Most of the monitoring of emissions, concentrations, transport, and effects of the air pollution have been directed toward the criteria pollutants. For each criteria pollutant, a maximum concentration — including also an adequate margin of safety — was established at different levels by national authorities and international bodies (e.g., National Ambient Air Quality Standards by the U.S. Environmental Protection Agency, guidelines of the UN World Health Organization, or the air quality directives of the EU).

0-87371-829-1/00/$0.00+$.50

Above threshold concentrations, adverse effects on human health can occur. The criteria pollutants are reviewed periodically and the standards are adjusted according to the latest scientific information. Research into the criteria pollutants and their synergism has revealed that it is the elevated levels of airborne particles that are mainly responsible for the increased health risk to inhabitants of large cities.

URBAN ATMOSPHERIC AEROSOLS

Concerns over atmospheric aerosols in general, and in particular in urban air, have received special attention in recent years. It has been asserted that the increased risk of respiratory-, cardiovascular-, and cancer-related death, alterations in the body's defense system against foreign materials, and other respiratory problems can be linked, in particular, to higher levels of particulates.[6] Some studies show statistical associations between Total Suspended Particulates (TSP) and increased sickness or aggravation of existing diseases, even at levels within current national air quality standards. Observation of the aerosols in urban air is already included in most regulatory work. Unfortunately, aerosol particles are usually treated regardless of their size and/or chemical composition, and are generally characterized by measuring their mass concentration (total mass per unit volume). However, it is increasingly recognized that ecological impact, health effects, as well as biogeochemical or other behavior of the aerosols are determined mainly by *both* their chemical composition and their physical properties (mainly size distribution and morphology). For example, aerosol particles with an equivalent aerodynamic diameter (EAD) larger than about 10 μm may dominate the particulate mass, but they are quite effectively removed in the upper respiratory system by its defense mechanisms, and, therefore, they are not important in terms of human health hazard.

As to the size of the atmospheric particles, their mass size distribution covers several orders of magnitudes and is generally multimodal.[7] However, two main, separate, and essentially independent size classes — coarse and fine — can very clearly be distinguished far from gaseous sources. These two fractions differ strongly from each other in terms of sources and, consequently, in their chemical composition and physical properties. The division between the two size fractions is typically around 2 μm EAD. The coarse size fraction is mostly associated with natural processes (mechanical disintegration of materials), while the fine size fraction is mainly produced by antropogenic sources and/or high-temperature processes. The fine particles have a more pronounced effect on public health than the coarse ones because they are deposited deeply into the lungs. Atmospheric residence times are also strongly dependent on the particle size.

The chemical composition of atmospheric aerosols is exceedingly complex. The wide variety of formation, dynamic transformation, and removal processes has the effect that both the concentration and the composition of the aerosols display a pronounced variability in time and space. The situation is particularly complicated in highly polluted (urban) airsheds. The atmospheric concentrations of many toxic species and trace elements are more elevated in urban areas than in rural ones. Determination of the aerosol composition, and in particular of the size-differentiated composition, can give invaluable information on the sources and source processes, and thus provide clues on how to reduce the atmospheric levels of the particulate species. Hence, in contemporary atmospheric aerosol research, collection and characterization in (at least) two different and well-defined size fractions are highly desirable.

AIR QUALITY MONITORING IN BUDAPEST

Budapest, with almost 2 million inhabitants, is the largest city in Hungary.[8] An automated monitoring network has operated in Budapest since 1974 by the Municipal Institute of State Public Health Officer Service[9] in order to follow up and oversee urban air quality. At present, NO, NO_2, SO_2, CO, TSP, and meteorological parameters are measured at eight monitoring stations, and

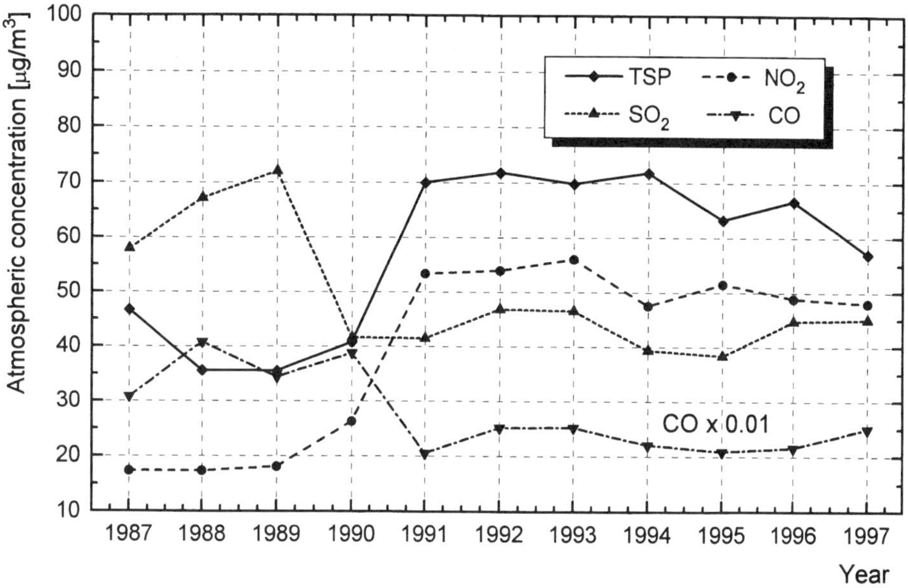

FIGURE 21.1 Annual mean atmospheric concentrations for NO_2, SO_2, CO, and TSP, averaged over eight automated monitoring stations of ÁNTSz FI throughout Budapest, for the period 1987 through 1997.

tropospheric O_3 and total hydrocarbons are recorded at two of these stations.[10] Four monitoring stations are located in the downtown areas; the others in the outer parts of town. In addition, deposited dust samples are collected monthly at 51 locations, while samples for NO_2 and SO_2 are collected on a daily basis at 27 sites.[10] The air quality of several European cities in Austria, Germany, Sweden, Norway, Poland, the Czech Republic, Switzerland, Hungary, Italy, Belgium, Spain, and Croatia has been compared.[11] From the results, it was concluded that the annual mean concentrations for the criteria pollutants in Budapest for 1996 were higher than the average pollution of the cities, and that the difference was significant, in particular for SO_2, CO, and TSP. Long-time trends in annual mean atmospheric concentration for some criteria pollutants at 298K and 101.3 kPa through the last decade averaged over the eight monitoring stations[9] are displayed in Figure 21.1. Concentrations of SO_2 and CO generally show a tendency to decrease, similar to international trends; and between 1989 and 1991, they decreased by factors of 2 and 1.7, respectively. The change can be explained by improvements and technological changes in the domestic heating and transportation sectors, partly by shifting from coal-fired power plants to gas-fired ones, and by a significant decrease in industrial production. Atmospheric concentrations of NO_2 and TSP increased by factors of about 3 and 2, respectively, during the same period. This is most likely related to the rapid increase in the number of automobiles. (Interestingly, the time period coincides with the basic changes in the political and economic system of Hungary.) Furthermore, it should be noted that the annual mean concentration of TSP was regularly and considerably above the annual national limit of 50 μg m^{-3} in the last 7 years.[12] It is thus evident that Budapest faces a problem of air pollution, and that it is the aerosols that are one of the most important and acute contributors to this pollution.

To further improve our knowledge of the air pollution, and of all variables involved, the pollution monitoring work in Budapest should be complemented by research on urban aerosols. Total atmospheric aerosols (together with the concentrations of S and N compounds) at different receptor sites in Budapest and at regional background stations in Hungary have been investigated.[13,14] Because of the fundamental role of particle size, as indicated briefly in the previous section, the present study was initiated. A campaign of size-fractionated aerosol sampling was performed at two urban residential sites, and complemented with simultaneous measurements of some criteria pollutants. The goals of the study were to characterize the atmospheric levels, multi-elemental composition, and time trends of the urban aerosols in separate fine and coarse size fractions; to investigate the atmospheric levels, diurnal variation, and time trends of criteria pollutants; and to compare these characteristics. Identification of the major source types of the aerosol fractions and trace gases and assessment of the relative contribution from these sources will eventually be accomplished by multivariate receptor modeling.

SAMPLE COLLECTION FOR AEROSOL RESEARCH

The size-fractionated aerosol samples were collected by Gent-type stacked filter units (SFUs).[15] Separation of the aerosol particles into two size fractions was achieved by sequential filtration through two Nuclepore filters of different pore size.[16] The principle behind such a separation has been explained.[17-19] Two 47-mm-diameter Nuclepore polycarbonate filters with pore sizes of 8 μm (Apiezon-coated) and 0.4 μm are placed in an NILU-type, open-face, stacked filter cassette in series. In the Gent variant of the SFU, the cassette itself is inserted into a cylindrical container, which is provided with a greased pre-impaction plate at the inlet. The sampled air is drawn through the SFU by means of a dry-running (oilless) pump, and the sampling line is further equipped with a flow control valve, a vacuum gage, an air flow rate meter, and a volume meter. A programmable 24-hour time switch allows for interrupted sampling. The sampler is designed to operate at a flow rate of 15 to 16 l min^{-1}. At this value, the pre-impaction stage intercepts particles larger than about 10 μm EAD (for 293K and 101.3 kPa), and the first filter with 8-μm pore size has a 50% collection efficiency at about 2 μm EAD. Consequently, the aerosol particles are separated into a coarse (about 10–2 μm EAD) and a fine (< 2 μm EAD) size fraction. The atmospheric concentrations of some criteria pollutants were measured by commercial equipment; that is, Monitor Labs (ML) Models 8830 (IR absorption spectrometry for CO), 8841 (chemiluminescent analyzer for NO and for NO_2 with Molycon converter), 8850S (UV fluorescence analyzer for SO_2), 8810 (UV photometry for O_3), and FH62IN (β-absorption on glass fiber paper for the mass of the TSP). The sampling flow rate for the ML instruments was 0.5 l min^{-1}, except for the TSP measurement where it was set to 1 m^3 h^{-1}. Relative uncertainty in the atmospheric concentrations obtained by the commercial equipment in routine analysis is estimated to be less than 5%.

The collection and measuring campaign was conducted from April 9 to May 17, 1996, during the non-heating season.[20] One of the two sampling locations was chosen downtown in a small park at Széna Square (latitude 47°30.6′ N, longitude 19°1.8′ E, altitude 115 m above sea level, a.s.l.) at the automated monitoring station no. 2 of the ÁNTSz FI. This site is affected by heavy nearby traffic. The SFU sampler was set up on the roof of the monitoring station with its intake facing down at 4.5 m above the ground. Daily aerosol samples were taken with planned and regular interruptions. The starting time was typically about 7:30 in the morning, and the mean sampled volume was 18.6 m^3. Besides the size-fractionated aerosol samples, atmospheric concentrations of NO, NO_2, SO_2, CO, and TSP, and temperature, pressure, humidity, and wind speed and direction were measured and recorded every half hour. The concentration of O_3, which is used in the present work for illustrative purposes only (see later and Figure 21.4), was recorded not on the Széna Square, but at another location with generally higher NO and NO_2 concentrations and similar diurnal variation and temporal characteristics. Median values for the daily temperature, pressure, and humidity during the sampling campaign were 292 K, 100.3 kPa, and 68%, respectively. The second sampling site was located on the western border of Budapest within the wooded campus of the Central Research Institute for Physics of the Hungarian Academy of Sciences (KFKI, latitude 47°29.3′ N, longitude 18°57.3′ E, altitude 424 m a.s.l.). Here, the SFU sampler was set up in a tree about 1.8 m above the ground. The starting time of collections at KFKI was typically about 6:30 in the morning, and the mean sampled volume was 21.9 m^3. Local temperature and pressure data were also obtained as daily averages. Median values for the daily temperature and pressure during

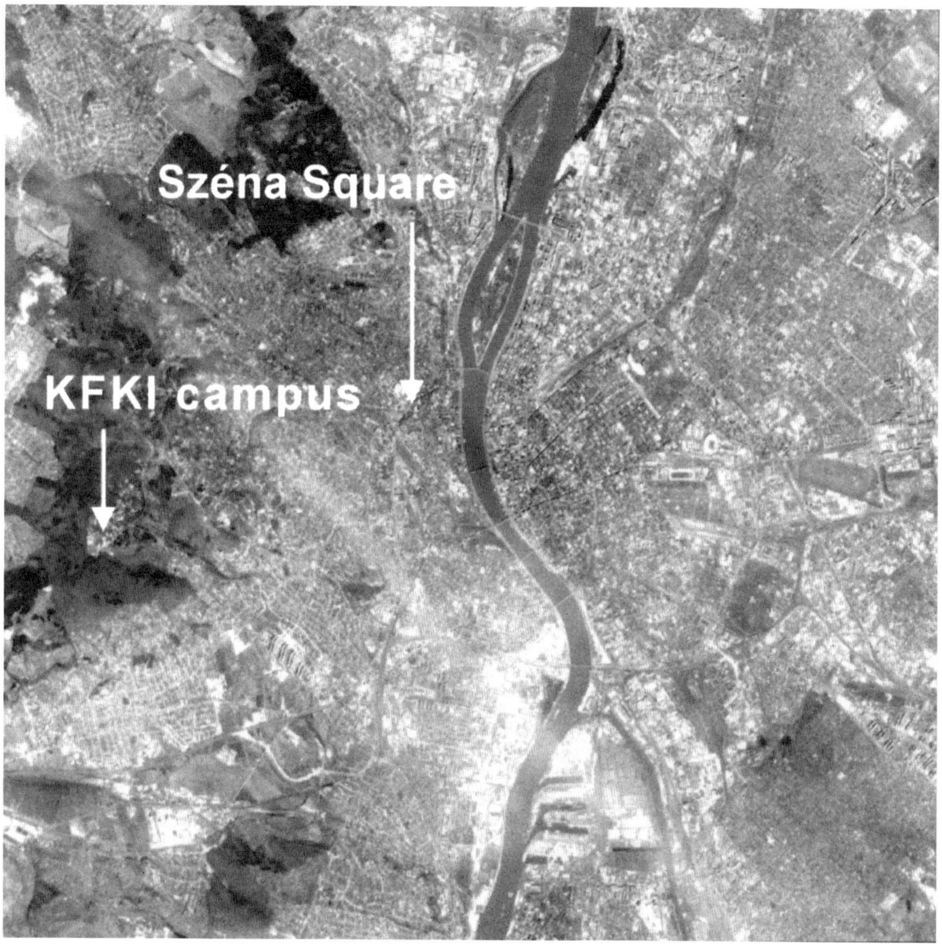

FIGURE 21.2 Satellite spot pan of Budapest (no. 75–254) on April 18, 1996, with indicated location of the sampling sites at Széna Square and the KFKI campus. (Copyright CNES, 1996, distributed by Spotimage, processed by FÖMI RSC. With permission.)

the sampling campaign were about 288 K and 96.1 kPa, respectively. A total of 33 sample pairs and four field blanks were collected at each sampling site and then stored in plastic Petrislide dishes. The field blanks are samples that were collected by doing all the same manipulations and operations as for real samples, including even drawing air through them for about 1 min. The weather was slowly warming with occasional rain showers during the sampling period. The prevailing wind direction in Budapest is northwest. A satellite spot pan of Budapest (no. 75–254) on April 18, 1996, with indicated locations of the sampling sites at Széna Square and KFKI campus is presented in Figure 21.2

ANALYTICAL METHODS

Each SFU filter was weighed before and after sampling using a microbalance to obtain the particulate mass (PM). During actual gravimetry, static electricity was eliminated. The filters were also analyzed for black carbon (BC) by a commercial smoke stain light reflectometer (Diffusion System, model 43) calibrated with other filters for which the BC loading had been determined.[21]

A quarter section of each SFU filter was analyzed by particle-induced X-ray emission (PIXE) analysis with the experimental set-up of the Institute for Nuclear Sciences at the University of

Gent, Belgium.[22,23] The samples mounted onto target rings were bombarded in a vacuum chamber with a beam of protons of 2.4-MeV energy supplied by a compact isochronous cyclotron. The beam area and beam current were 0.54 cm^2 and about 150 nA, respectively. After passage through the sample (which is considered to be infinitesimally thin), the proton beam was dumped in a Faraday cup for charge integration. The induced X-ray spectra were measured using a composite filter by an Si(Li) detector, and were typically accumulated for a preset charge of 60 μC. The spectra were fitted for 29 elements (i.e., Na, Mg, Al, Si, P, S, Cl, K, Ca, Ti, V, Cr, Mn, Fe, Ni, Cu, Zn, Ga, Ge, As, Se (in the fine fraction only), Br, Rb, Sr, Y (in the fine fraction only), Zr, Mo, Ba, and Pb) with the computer program AXIL-84.[24] Elemental amounts were derived using experimental calibration factors and field blank values; and finally, atmospheric concentrations (in ng m^{-3}) were calculated. So far, no corrections for volatile (fine) particle loss or for particle size effect have been applied.

One half of each SFU filter was analyzed by instrumental neutron activation analysis (INAA) at the Budapest Research Reactor. The procedure consists of a short- and a long-time activation and two γ-ray spectrometric measurements after each irradiation. Analytical calculations are performed by the k_0 standardization method, and are in progress. Publication of the experimental details, methodology, and results of the INAA measurements is in preparation. A comparison of PIXE and INAA applied in atmospheric aerosol studies has been performed.[25] Since the elemental amounts determined by the INAA can only be combined with the PIXE data later, this chapter focuses on the analytical results obtained by PIXE as far as the aerosol is concerned.

TEMPORAL VARIATION OF CRITERIA POLLUTANTS

Diurnal variations of the atmospheric concentration of CO and SO_2, of NO, NO_2, and tropospheric O_3, and of TSP are presented in Figures 21.3 to 21.5, respectively. The data points indicate half-hour concentration values averaged over the entire sampling period (i.e., from April 10 to May 17 (over 38 days)). The diurnal variation of CO reflects typical city driving patterns. The concentration exhibits one maximum during the morning rush hours at about 7:30, and two maxima (a smaller maximum at about 17:00 and another broad one at about 20:00) in the evening. SO_2 data are rather constant, showing a slight elevation only at about 11:00. The data in Figure 21.4 can be explained in terms of the simplified atmospheric nitrogen dioxide photolytic cycle.[26] The concentration of NO rises as early-morning traffic emits its load of NO; it is soon oxidized to NO_2, leading to decreased concentrations of NO (a pronounced peak appears at 7:00) and to increased concentrations of NO_2. As the sun's intensity increases in the morning, atmospheric photochemical reactions take place, including photolysis of NO_2 and formation of O_3; thus, the concentration of NO_2 begins to drop (resulting in a peak at 8:00) while the concentration of O_3 rises. Ozone then can convert NO to NO_2, and it is so effective in this reaction that as long as O_3 is present, the NO concentration does not rise throughout the rest of the daylight although there may be new NO emissions.[26] In the evening, the cycle reverses. Figure 21.5 shows that the concentration of TSP increases rapidly in the morning from about 5:30, reaches a value of about 100 μg m^{-3} at 7:30, and remains at that level with small variations until late evening when it starts to decrease. Half-hour concentration data of TSP and CO averaged separately over the workdays (28 days) and holidays (10 days) within the sampling campaign are shown in Figures 21.6 and 21.7, respectively. It can be seen that TSP values for workdays and holidays are quite similar during night from 23:30 until about 4:00. For workdays, the concentration starts to increase 1.5 h earlier (at about 5:00) than for holidays, and reaches significantly higher values. In contrast, from 20:00 until 23:00, TSP data are much higher on holidays than on workdays. The trace gases (with the exception of SO_2) exhibit differences between workdays and holidays similar to those for TSP, as illustrated by CO in Figure 21.7. In addition to these differences, in case of CO, the first peak in the afternoon that is present on workdays is completely absent on holidays. Ratios of daily average of the half-hour concentration data averaged over the workdays to that over the holidays for NO, NO_2, SO_2, CO, and TSP are

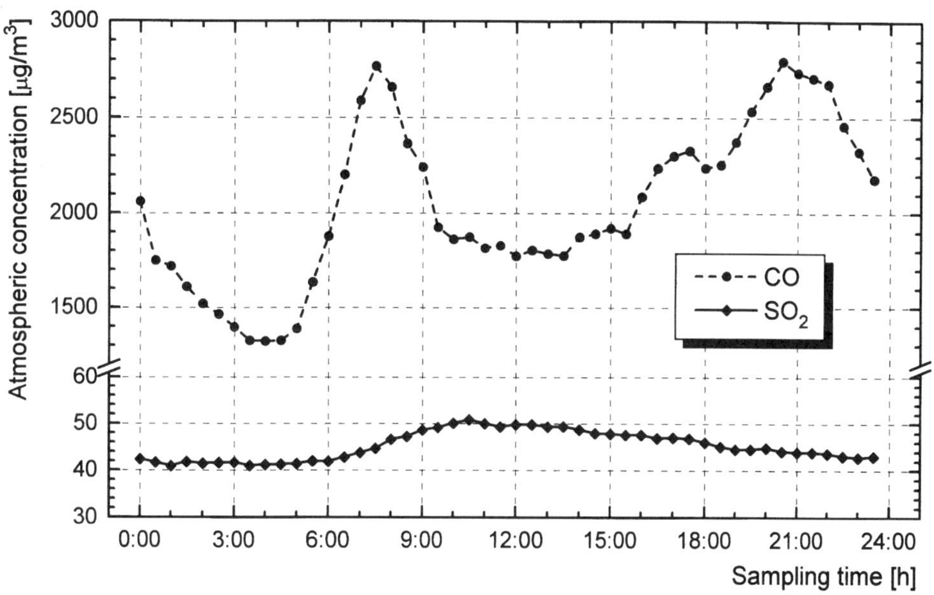

FIGURE 21.3 Diurnal variation of the atmospheric concentration of CO and SO_2 at Széna Square averaged over the sampling period April 10 to May 17, 1996 (38 days). (From Reference 20. With permission.)

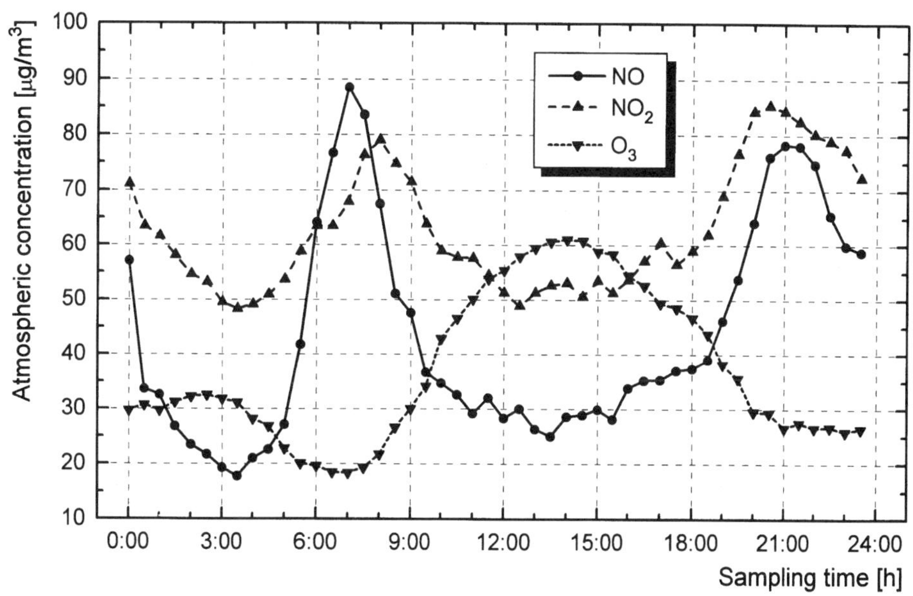

FIGURE 21.4 Diurnal variation of the atmospheric concentration of NO and NO_2 at Széna Square and O_3 averaged over the sampling period April 10 to May 17, 1996 (38 days). (From Reference 20. With permission.)

1.4, 1.2, 1.0, 1.1, and 1.3, respectively. The ratio for CO is surprisingly close to 1, which can be explained by the huge peak at late evening on holidays (see Figure 21.7). Summarizing, in the late evening from about 20:00 until 23:30, the air pollution at Széna Square was more serious during the weekends than during the workdays. Most likely, it is related to vehicle circulation. Many people return in their cars from weekend houses, and vacation to Budapest on holidays in the evening hours, passing through the city center.

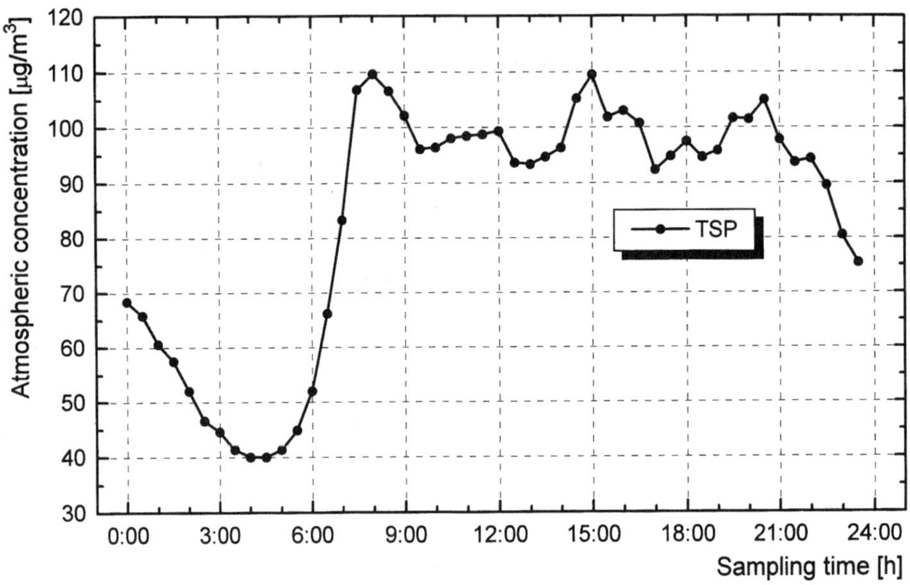

FIGURE 21.5 Diurnal variation of the atmospheric concentration of TSP at Széna Square averaged over the sampling period April 10 to May 17, 1996 (38 days). (From Reference 20. With permission.)

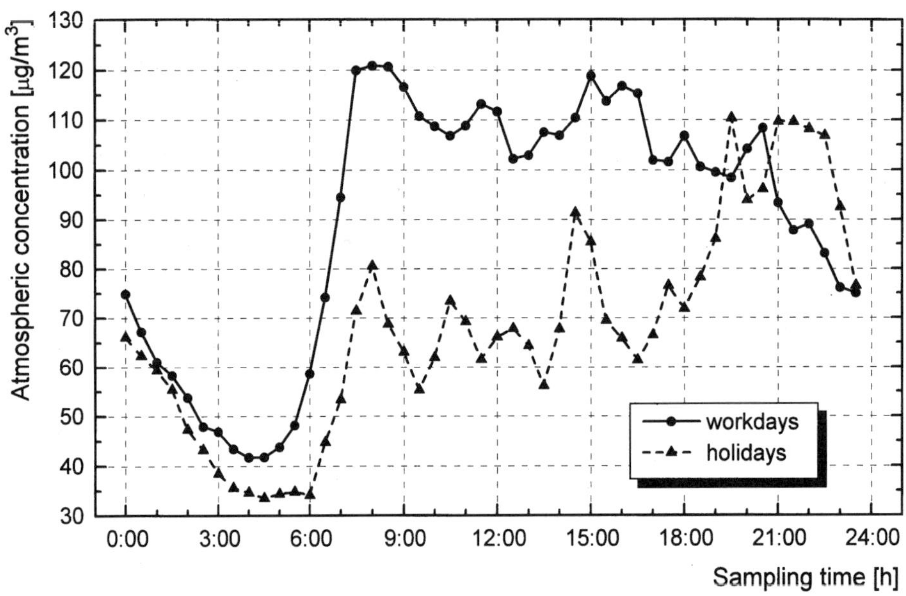

FIGURE 21.6 Diurnal variation of the atmospheric concentration of TSP at Széna Square averaged separately over the workdays (28 days) and holidays (10 days). (From Reference 20. With permission.)

The daily average atmospheric concentrations for the measured criteria pollutants as a function of the sampling date (time trends over the sampling period) are presented in Figure 21.8. Again, the concentration data for SO_2 are quite constant, while the other pollutants exhibit more variation. For these other pollutants, a good correlation between the time trends can be found, suggesting that they have common sources. The median atmospheric concentrations for NO, NO_2, SO_2, CO, and TSP over the sampling period are 43, 62, 45 μg m^{-3}, 1.98 mg m^{-3}, and 81 μg m^{-3}, respectively.

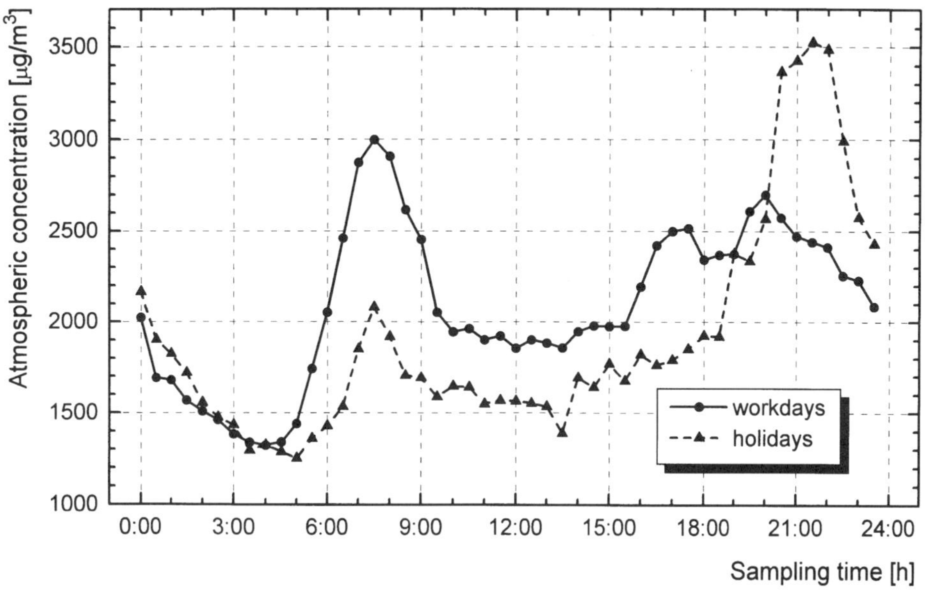

FIGURE 21.7 Diurnal variation of the atmospheric concentration of CO at Széna Square averaged separately over the workdays (28 days) and holidays (10 days). (From Reference 20. With permission.)

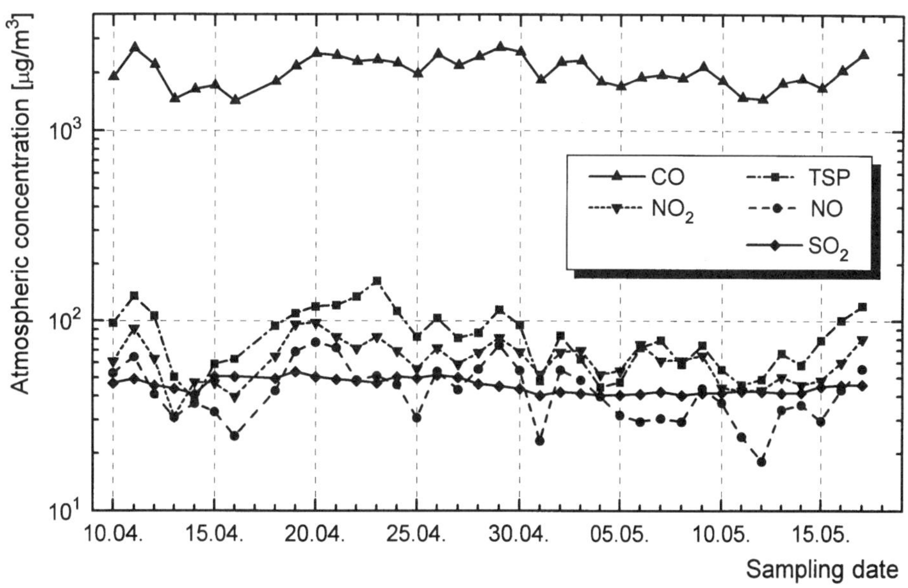

FIGURE 21.8 Atmospheric concentrations as a function of sampling date (time trends) for NO, NO_2, SO_2, CO, and TSP at Széna Square from April 10 to May 17, 1996. (From Reference 20. With permission.)

CHEMICAL AND TEMPORAL CHARACTERISTICS OF URBAN AEROSOL

The median atmospheric concentrations for 31 aerosol species in the coarse and fine aerosol fractions at Széna Square and the KFKI campus are shown in Table 21.1. In the coarse fraction, all median concentration data for the KFKI campus are lower than at Széna Square, and the

TABLE 21.1
Median Atmospheric Composition in the Coarse and fine Aerosol Fractions (in ng m^{-3}) at Széna Square and the KFKI Campus

Aerosol	Coarse-Size Fraction		Fine-Size Fraction	
Species	Széna Square	KFKI Campus	Széna Square	KFKI Campus
PM	41×10^3	20×10^3	24×10^3	17×10^3
BC	2100	300	7400	1780
Na	220	86	184	118
Mg	260	77	40	21
Al	850	250	97	68
Si	3300	790	340	210
P	53	22	5.6	0.72
S	760	210	1560	1600
Cl	120	32	13.3	9.2
K	380	120	134	126
Ca	2400	490	190	90
Ti	83	21	8.4	5.2
V	2.1	0.44	1.03	0.92
Cr	4.5	.99	1.76	0.46
Mn	28	5.6	6.3	3.5
Fe	1700	240	300	87
Ni	2.1	0.48	0.79	0.67
Cu	29	1.54	9.8	1.81
Zn	50	8.2	29	17.4
Ga	1.45	0.26	0.20	0.11
Ge	0.74	0.04	0.09	0.04
As	1.42	0.54	0.43	0.96
Se	—	—	0.40	0.36
Br	9.3	0.84	8.3	2.3
Rb	1.23	0.48	0.38	0.26
Sr	6.4	1.60	0.90	0.36
Y	—	—	0.45	0.115
Zr	2.9	0.70	0.58	0.25
Mo	0.52	0.134	0.49	0.25
Ba	20	0.46	4.2	UL[a]
Pb	35	3.7	46	14.8

[a] UL = upper limit (26 upper limits in the set of 33 concentration data).

difference is typically a factor of about 3 to 4. For a few species (i.e., Cu, Ge, Br, Ba, and Pb), the difference is about one order of magnitude, suggesting that in or near the city center, there may be a substantial source for these elements. In the fine fraction, the concentration data for the KFKI campus are also generally lower than at Széna Square, but the difference is much smaller; the downtown-to-suburban ratio is typically about 2. In some cases (i.e., for PM, S, K, V, Ni, As, and Se) the concentrations are almost equal to or higher at the KFKI campus than at Széna Square. The similarities and differences in aerosol composition of the two sites are illustrated in Figure 21.9, which shows median crustal rock enrichment factors (EFs), calculated relative to Mason's average crustal rock[27] with Al as the reference element. Most elements in the coarse size fraction have EFs close to 1, suggesting that they are attributable mainly to soil (and road) dust dispersal and resuspension. (The small median EF for coarse Ba at the KFKI campus can be explained by

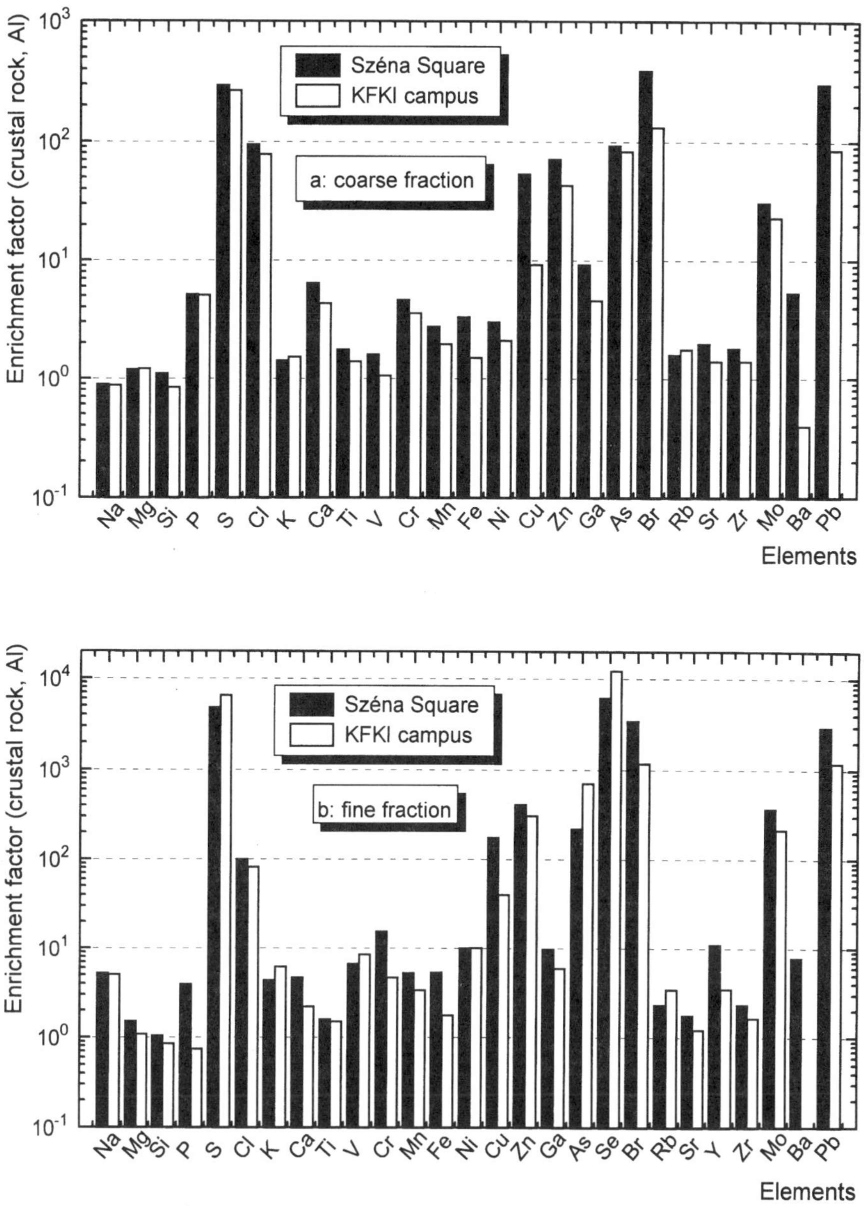

FIGURE 21.9 Median crustal enrichment factors, calculated relative to Mason's crustal rock with Al as reference element, in the coarse (a) and fine (b) aerosol size fractions at Széna Square and the KFKI campus. (From Reference 20. With permission.)

having 13 upper limits [detection limits[28]]) in the concentration data. The mean EF for Ba is 1.27. Note that upper limit values were set equal to zero when calculating medians, but were left out when calculating means.) Nevertheless, some elements (i.e., S, Cl, Cu, Zn, As, Br, Mo, and Pb) are significantly enriched. The same elements and Se exhibit large (10^2–10^4) EFs in the fine size fraction, indicating their possible anthropogenic origin. For several elements (i.e., S, K, V, Ni, As, Se, and Rb), the fine fraction EFs are higher at the KFKI campus than at Széna Square, but this is due to the fact that the concentration differences between the two sites are larger for Al (and the other crustal elements) than they are for the typical pollutant elements. The mean fine-to-coarse

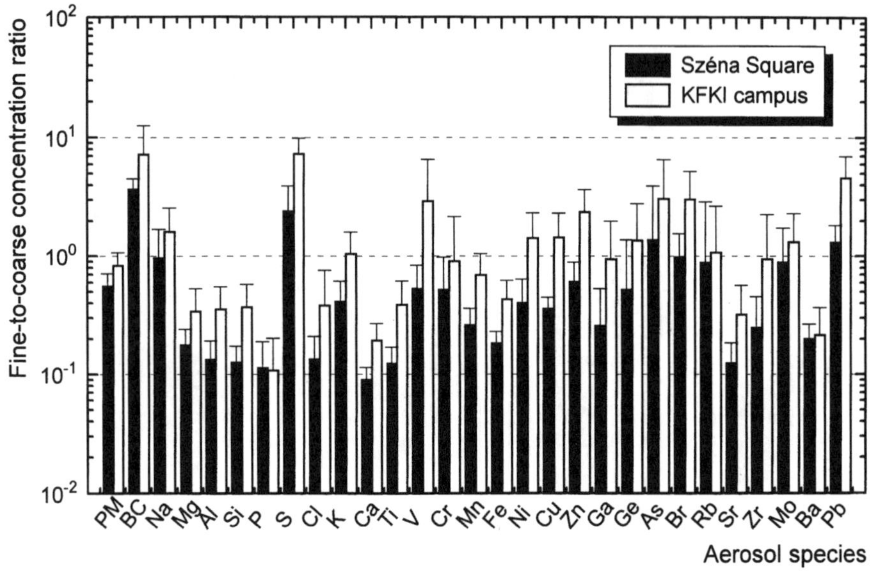

FIGURE 21.10 Mean fine-to-coarse concentration ratios for 29 aerosol species at Széna Square and the KFKI campus. Error bars indicate one standard deviation. (From Reference 20. With permission.)

concentration ratios (F:C ratios) for both sampling sites are displayed in Figure 21.10. The F:C ratios for Széna Square are typically below 1 (except for BC, S, As, and Pb), showing that the elements are present predominantly in the coarse aerosol fraction. For typical crustal elements such as Al, Si, Ca, Ti, Mn, Fe, and Sr, the F:C ratio is about 0.1. The F:C ratios for the KFKI campus are larger than for Széna Square (except for P, for which the two ratios are almost equal); and for many species, including BC, Na, S, K, V, Ni, Cu, Zn, Ge, As, Br, Rb, Mo, and Pb, the F:C ratio is even above 1. These elements are present mainly in smaller aerosol particles. Most likely, the differences in F:C ratios are due to the 309-m difference in altitude between the two sampling sites, to the wooded suburban character of the KFKI campus, and to the shorter atmospheric residence times of the coarse particles.

Figure 21.11 displays time trends for some typical crustal elements (i.e., Al, Si, Ca, Ti, Mn, and Fe) at Széna Square (a) and at the KFKI campus (b). It is evident that the individual time trends show significant fluctuation with time. At each of the two sampling sites, the elements depicted appear highly correlated with each other, suggesting that a single dominant crustal component can be expected in multivariate receptor modeling. In contrast, when comparing the time trends from the two sites with each other, there seems to be little relationship between the crustal element data from both locations. Nevertheless, during certain periods (e.g., from April 10 to 20), some correlation can be found. (Interestingly, unusually cold weather was observed after April 10, which became milder with stormy wind only after April 15, and a real, warm spring arrived after April 20.) Time trends for Br, Pb, BC, and CO in the fine fraction (markers for internal combustion engine emissions) for Széna Square are shown in Figure 21.12. As can be seen, there is quite a good correlation between Br, Pb, and BC, but CO exhibits a somewhat different pattern from the three aerosol species. This might mean that there is another meaningful source for the CO emissions besides transportation. The correlation coefficients between Br and Pb in the fine-size fraction are 0.77 (sample number $N = 33$) and 0.73 ($N = 33$) at Széna Square and the KFKI campus, respectively, while the mean fine Br:Pb concentration ratios (calculated on a sample by sample basis) are 0.19 ± 0.04 and 0.18 ± 0.07. These ratios are significantly smaller than the Br:Pb mass ratio in fresh automotive exhaust (the ethyl ratio is 0.386),[29] but this is due mainly to volatilization of fine Br during the PIXE bombardment.[25] Further interpretation can be performed after the INAA results

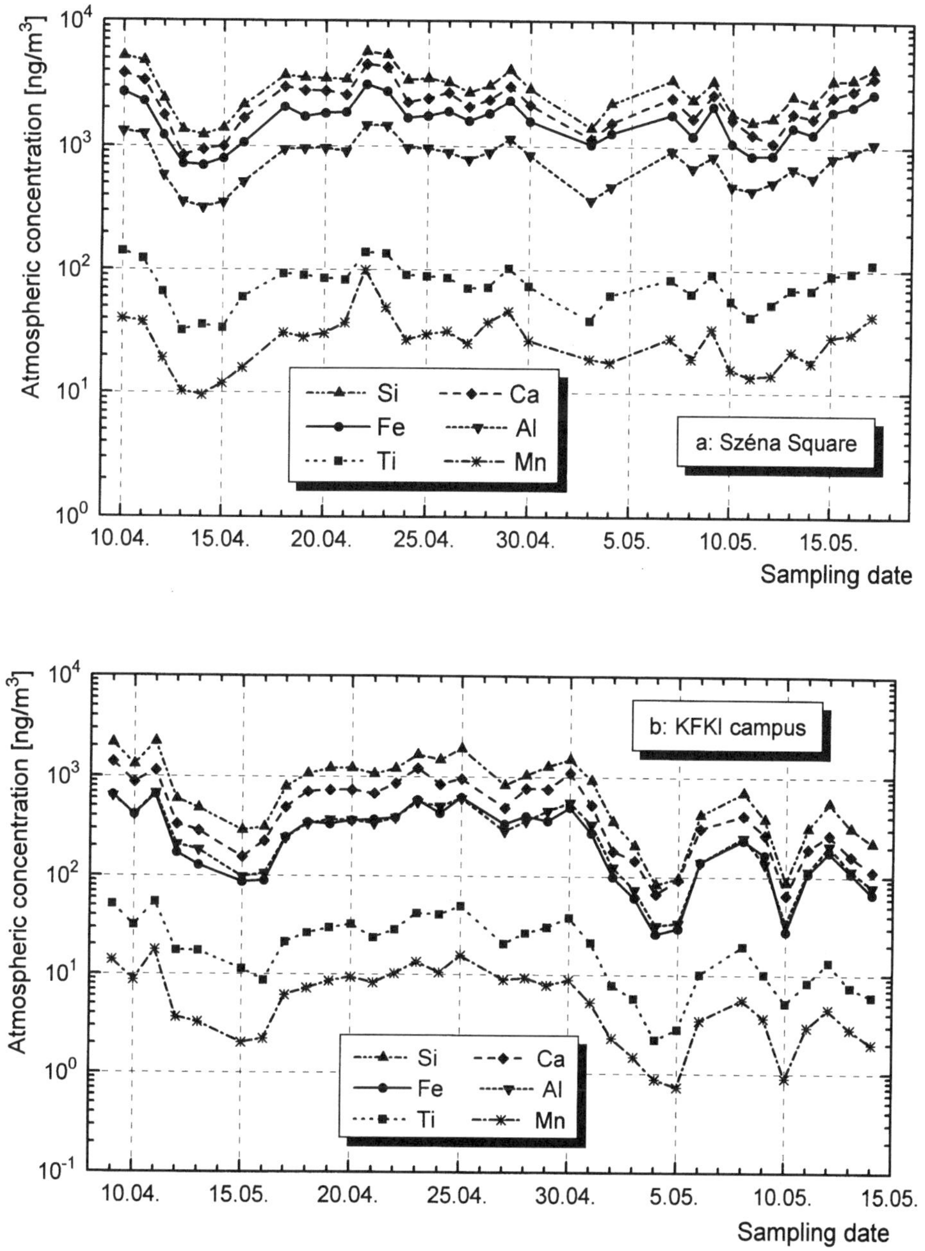

FIGURE 21.11 Time trends for Al, Si, Ca, Ti, Mn, and Fe in the coarse size fraction at Széna Square (a) and the KFKI campus (b). (From Reference 20. With permission.)

become available. Figure 21.13 presents time trends at Széna Square for species that are related to fossil fuel combustion. Sulfur, As, and Se exhibit significant fluctuation with time, and tend to be correlated with each other, but SO_2 shows completely discrepant behavior. Moreover, both the diurnal variability (cf. Figure 21.3) and time-trend fluctuation for SO_2 during the sampling campaign are quite small, which seems to indicate that its dominant sources are more widespread and/or at a greater distance. The average annual variation of SO_2 in Budapest (and also at the regional stations) is, however, significant, and shows a minimum in summer. In 1980, the mean downtown atmospheric

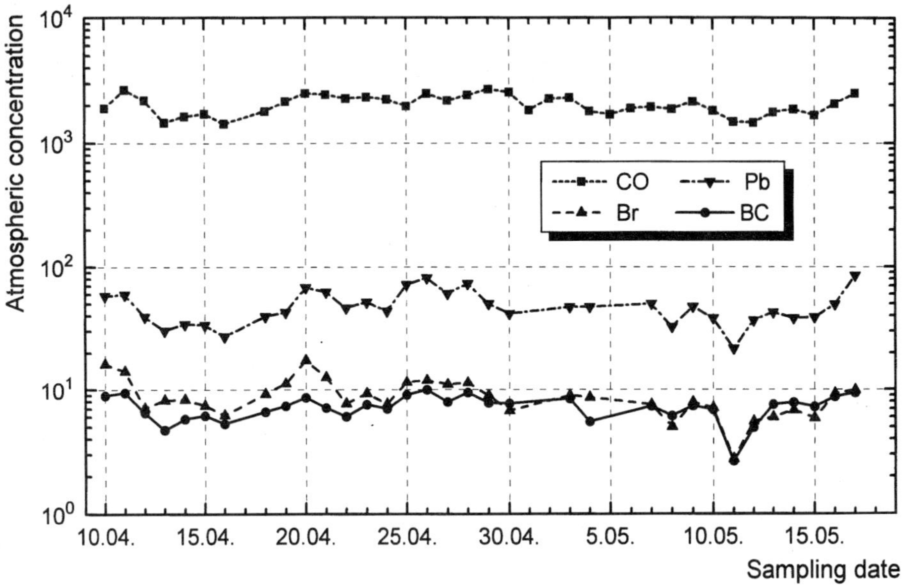

FIGURE 21.12 Time trends for Br, Pb, and black carbon (BC) in the fine size fraction and for CO at Széna Square. (The concentration units are μg m^{-3} for BC and CO and ng m^{-3} for Br and Pb.) (From Reference 20. With permission.)

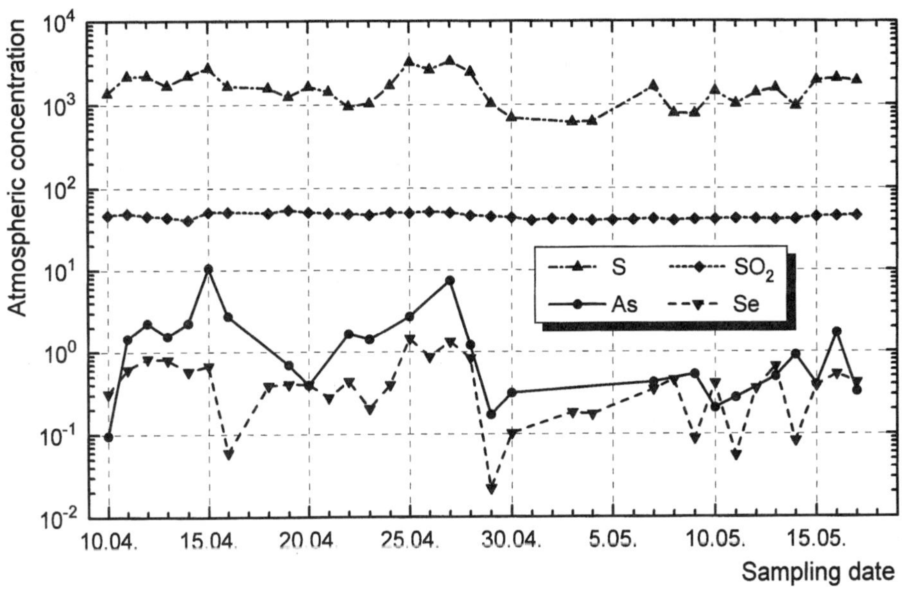

FIGURE 21.13 Time trends for S, As, and Se in the fine size fraction and for SO_2 at Széna Square. (The concentration units are μg m^{-3} for SO_2 and ng m^{-3} for the other species.) (From Reference 20. With permission.)

concentration for SO_2 in Budapest (although at different sites than in the present chapter study) over a long period was 147 μg m^{-3}, and the downtown-to-suburban concentration ratio was 2.3.[13] It was also concluded that there were correlations between the average SO_2 concentrations at different regional stations in Hungary. Similarly, in the present work, the fine fraction S data at Széna Square and those at the KFKI campus were highly correlated (with a correlation coefficient of 0.88, N = 27). In contrast, for the other aerosol species, the correlation between the two sampling sites was weak or not significant. The high correlation for fine aerosol S is attributed to the fact that this species is a secondary aerosol constituent that is formed from a fairly constant pool of precursor SO_2 gas; its concentration in the Budapest area is essentially controlled by the gas-to-particle conversion rate, the fine sulfate deposition velocity, and meteorological conditions (fumigation, surface inversions). The other aerosol species measured are virtually all primary aerosol constituents. The lack of correlation for these species between the two sites indicates that their sources are rather local and that these sources exhibit different and variable intensity near each of the two sites. This is definitely the case for the coarse crustal aerosol component, but probably also for most of the typical anthropogenic elements. As a result, the spatial variability of the urban (aerosol) air pollution is considerable.

The crustal EFs and the differences in time trends for the various aerosol species suggest that several natural and anthropogenic components (dominant source types) can be anticipated in the coarse and fine fraction data sets, and that urban aerosols in Budapest are a mixture of locally generated and transported aerosols. The identification of the various source types and the assessment of their contribution to urban air pollution will be performed by multivariate receptor modeling after all INAA measurements are completed and the PIXE and INAA results have been combined into a final analytical data set. The relative contribution of local and distant sources, the relationships between the aerosol source types and the atmospheric trace gases, and various meteorological parameters will then also be examined.

ACKNOWLEDGMENTS

This work was funded by the Hungarian Scientific Research Fund (OTKA) under Contract F014962. The assistance of Mrs. J. Márton and A. Lágler in the sample collection and preparation is gratefully acknowledged. Willy Maenhaut is indebted to the Fonds voor Wetenschappelijk Onderzoek-Vlaanderen for research support.

REFERENCES

1. De Koning, H.W., Kretzschmar, J.G., Akland, G.G., and Bennett, B.G., *Atmos. Environ.*, 20, 101, 1986.
2. Hopke, Ph.K., Lamb, R.E., and Natusch, D.F.S., *Environ. Sci. Technol.*, 14, 164, 1980.
3. Aldape, F., Flores, J.M., Diaz, R.V., Miranda, J., Cahill, T.A., and Morales, J.R., *Int. J. PIXE,* 1, 373, 1991.
4. Miranda, J., *Nucl. Instrum. Meth.*, B109/110, 439, 1996.
5. Winchester, J.W., Wang, M., Ren, L., Lu, W., Hansson, H., Lannefors, H., Darzi, M., and Leslie, A., *Nucl. Instrum. Meth.*, 181, 391, 1981.
6. Reichhardt, T., *Environ. Sci. Technol.*, 360A, 29, 1995.
7. Whitby, K.T., *Atmos. Environ.*, 12, 135, 1978.
8. Hungarian Central Statistical Office, *Statistical Yearbook of Budapest*, Budapest, in Hungarian, 1998.
9. ÁNTSz FI, *Annual Report of the Municipal Institute of State Public Health Officer Service,* Budapest, in Hungarian, 1987–1997.
10. Ministry for Environment and Regional Policy, *Air quality in Budapest*, Budapest, in Hungarian, 1997.
11. Sameh, F. and Hager, W., Air Quality Data 1996, Austrian and International Comparison of Cities and Regions, Magistrat der Landeshauptstadt Linz, No. 2/97, Linz, Austria, 1997.
12. National Standard, Hungarian Bureau of Standards, Budapest, *MSZ-21854, Requirements of Cleanness of Ambient Air,* in Hungarian, 1990.
13. Mészáros, E. and Horváth, L., *Atmos. Environ.*, 18, 1725, 1984.
14. Molnár, A., Mészáros, E., Bozó, L., Borbély-Kiss, I., Koltay, E., and Szabó, Gy., *Atmos. Environ.*, 27A, 2457, 1993.

15. Maenhaut, W., François, F., and Cafmeyer, J., The "Gent" stacked filter unit sampler for the collection of Atmos. aerosols in two size fractions: description and instructions for installation and use, in *Applied Research on Air Pollution Using Nuclear-Related Analytical Techniques,* IAEA Report NAHRES-19, Vienna, 1994, 249.
16. Spurný, K.R., Lodge, J.P., Frank, E.R., and Sheesley, D.C., *Environ. Sci. Technol.,* 3, 453, 1969.
17. Cahill, T.A., Asbaugh, L.L., Barone, J.B., Eldred, R., Feeney, P.J., Flocchini, R.G., Godart, C., Shadoan, D.J., and Wolfe, G., *J. Air Pollut. Control Assoc.,* 27, 675, 1977.
18. Heidam, N.Z., *Atmos. Environ.,* 15, 891, 1981.
19. John, W., Hering, S., Reischl, G., Sasaki, G., and Goren, S., *Atmos Environ.,* 17, 373, 1983.
20. Salma, I., Maenhaut, W., Zemplén-Papp, É., and Bobvos, J., *Microchem. J.,* 58, 291, 1998.
21. Andreae, M.O., Andreae, T.W., Ferek, R.J. and Raemdonck, H., *Sci. Total Environ.,* 36, 73, 1984.
22. Maenhaut, W., Selen, A., Van Espen, P., Van Grieken, R., and Winchester, J.W., *Nucl. Instrum. Meth.,* 181, 399, 1981.
23. Maenhaut, W. and Raemdonck, H., *Nucl. Instrum. Meth.,* B1, 123, 1984.
24. Maenhaut, W. and Vandenhaute, J., *Bull. Soc. Chim. Belg.,* 95, 407, 1986.
25. Salma, I., Maenhaut, W., Annegarn, H.J., Andreae, M.O., Meixner, F.X., and Garstang, M., *J. Radioanal. Nucl. Chem.,* 216, 143, 1997.
26. Masters, G.M., *Introduction to Environ.al Engineering and Science*, Prentice-Hall, London, 1991.
27. Mason, B., *Principles of Geochemistry,* 3rd ed., Wiley, New York, 1966.
28. Currie, L.A., *Anal. Chem.,* 40, 587, 1968.
29. Harrison, R. and Sturges, W., *Atmos. Environ.,* 17, 311, 1983.

22 Trace Elements in Atmospheric Pollution Processes: The Contribution of Neutron Activation Analysis

Mario Gallorini

CONTENTS

INTRODUCTION

Trace elements (TE) released in the atmosphere from anthropogenic sources represent part of the pollutant agents that may be responsible for serious risk to public health.[1] Depending on their

0-87371-829-1/00/$0.00+$.50

concentrations and chemical and physico-chemical forms, adverse and/or toxic effects are well-established for elements such as As, Cd, Cr, Hg, Pb, and Ni that can be released in the air from different pollution processes. The concentrations in the atmosphere of urban and industrialized areas of these elements and many others are increasing with the increase in anthropogenic activity.[2–5] Thus, the control of their concentration in the air is essential for those studies devoted to: risk threshold assessment, identification of pollution sources, long-distance transportation, environmental impact, long-term exposure to low levels, etc.[6–11] The analysis of TE for atmospheric pollution studies is an exacting task. The monitoring of air samples and/or related materials such as fly-ash, suspended particles, air particulates, dry depositions, fumes and others is comprehensive of hundred of determinations, often carried out with very small samples (i.e., a few micrograms, as in the filters). For such purposes, the optimal analytical technique required should meet the main following criteria: high sensitivity; multi-element capability for as many elements as possible; high accuracy and precision; instrumental performance; the possibility of analyzing microsamples; and reproducibility at different concentration ranges.

Neutron activation analysis (NAA) possesses these requirements and is universally accepted as one of the most reliable analytical tools for trace and ultratrace element determinations. Its use in TE and atmospheric pollution-related studies has been and remains extensive, as demonstrated by several specific works and detailed reviews.[12–18]

Here, the application of this nuclear technique in solving a series of different analytical problems related to air pollution processes is reported. Examples and results are given for the following topics: control of the emissions from municipal waste incinerators; analysis of fly-ashes, suspended particles, fumes, and vapors; granulometric distribution profiles; characterization of dry depositions; and urban air monitoring.

THE ANALYTICAL TECHNIQUE

Neutron activation analysis consists of thermal neutron bombardment of a given material, followed by measurement of the induced radioactivity. In general, the analysis is performed by gamma-spectroscopy of the resulting radionuclides produced in the unknown material and in the corresponding standards, irradiated under the same conditions.[12] The analysis can be performed instrumentally (INAA) or, depending on the matrix to be analyzed and the elements to be determined, may necessitate, to achieve maximum sensitivity, selective radiochemical procedures subsequent to neutron irradiation (RNAA; radiochemical neutron activation analysis).

INAA

The general procedure is very simple and consists of (1) sealing known amounts of samples and standards in irradiation vials or containers; (2) irradiation at the same neutron flux and for the same time (possible neutron flux variation along the irradiation vials should be monitored); (3) gamma-spectroscopy by semiconductor detectors and computerized multichannel analyzer of the irradiated samples using the same counting geometry; and (4) data evaluation.

Typical sensitivity ranges obtainable for more than 40 elements are given in Table 22.1.[19] The calculations are made on the basis of 1 hour of irradiation at a neutron flux of 10^{12} n·cm^{-2}·sec^{-1} and 1 hour of delay before counting with 100 disintegrations per minute as a typical lower activity limit.

Depending of the irradiation time, the availability of higher neutron fluxes (10^{3} n·cm^{-2}·sec^{-1} is common in many nuclear research reactors), and the possibility of performing epithermal neutron irradiation,[20–21] the sensitivity for many elements can be increased. More detailed information on the limits of detection in INAA can be found in the specific literature, especially for air dust analyses.[22–24]

Table 22.2 lists the elements and their corresponding radionuclides normally used in the analysis of environmental samples by NAA. According to nuclear characteristics, short-, medium-, and long-

TABLE 22.1
Typical Sensitivity Ranges Obtainable by NAA in Trace Elements Analysis

Elements	Sensitivity (g)
In, Eu, Dy	10^{-11}–10^{-12}
Mn, Lu	10^{-10}–10^{-11}
Co, Br, I, Sm, Ho, Hf, Re, Ir, Au, Th, U	10^{-9}–10^{-10}
Na, Cl, Cu, Ga, Ge, Se, As, Pd, Sb, Te, Ba, La, Pr, Nd, Er, Yb, Ta, W, Pt,	10^{-8}–10^{-9}
K, Sc, Ni, Rb, Sr, Y, Nb, Ru, Cd, Sn, Gd, Tb, Tm, Os, Hg	10^{-7}–10^{-8}

Source: From Reference 19. With permission.

term irradiation times are selectively chosen for different groups of elements. Short half-life radionuclides are usually obtained with irradiation times varying from a few seconds to a few minutes; while for the medium/long halflife radionuclides, several hours of irradiation are necessary. In the case of short-lived radionuclides, cyclic activation can be performed to increase the sensitivity of selected elements such as Se, Cu, O, and S.[25–26]

As can be seen, INAA covers almost all the elements of analytical interest in atmospheric pollution-related studies, but elements such as Be, B, Pb, Bi, and Tl cannot be conveniently determined because these elements do not offer the optimal nuclear characteristics required for their determination by this technique.

RNAA

Because the relationship between the composition of material subjected to neutron irradiation, it may occur that the determination of selective trace elements is prevented by high matrix background activity and/or by interfering radionuclides. Such is the case, for example, in trace elements determination in biological materials where the gamma-ray spectrum is dominated by radionuclides such as ^{24}Na, ^{32}P, or ^{32}Br which “cover” the gamma-lines of many trace radioelemets.

RNAA represents, for these cases, the only way to obtain maximum sensitivity for trace analysis and is based on the selective separation, after irradiation, of the interfering radionuclides or the isolation of the radionuclides of interest from the irradiated matrix. The typical radiochemical separation procedures involve a series of chemical treatments carried out in radiochemical laboratory, and generally consist of the dissolution of the irradiated sample followed by the chemical separation (distillation, ion exchange chromatography, solvent extraction, precipitation, and others).[27–31] The main advantages offered by this technique include performing the chemical treatment without contamination (no pure reagents are necessary) and the possibility of adding inactive carriers of the elements under investigation to better perform the overall procedure (separation yield). In the case of samples related to atmospheric pollution studies, RNAA can be applied only in select cases where specific ultratrace elements have to be determined in complex matrices such as bottom and fly-ash or soils, as shown later.

APPLICATIONS

NAA has been applied at the CNR Center of Radiochemistry and Activation Analysis of Pavia (Italy) to carry out research in TE evaluation in: (1) emissions from different incineration plants of solid urban waste, (2) the corresponding suspended particles granulometric distribution and vapor phase, (3) the characterization of dry depositions collected in urban and industrialized areas, and (4) the monitoring of airborne particulate matter. The results obtained refer not only to the total

TABLE 22.2
Nuclear Parameters of the Most Commonly Used Radionuclides in Trace Element Analysis by NAA

Elements	Isotope	Half-life[a]	Main Gamma-rays (keV)
	Short-Lived Radionuclides		
Al	^{28}Al	2.31 min	1778.9
V	^{52}V	3.75 min	1434.4
Cu	^{66}Cu	5.1 min	1039.0
Ti	^{51}Ti	5.79 min	320.0
Mg	^{27}Mg	9.46 min	844, 1014.1
Br	^{80}Br	17.6 min	617.3
I	^{128}I	25.4 min	442.7
Cl	^{38}Cl	37.3 min	1642.0; 2166.8
In	^{116m}In	54.0 min	129.34; 1097.1
Mn	^{56}Mn	2.58 h	846.9; 1810.7
	Medium/Long-Lived Radionuclides		
Eu	^{152m}Eu	9.35 h	121.8; 841.6; 963.5
Cu	^{64}Cu	12.5 h	511.0
Zn	^{69m}Zn	13.8 h	438.7
Ga	^{72}Ga	14.3 h	630; 834.1; 1860.4
Na	^{24}Na	15.0 h	1368.4; 2753.6
W	^{187}W	24.0 h	479.3; 685.7
As	^{76}As	26.3 h	559.2; 657.0
Br	^{82}Br	35.8 h	554.3; 776.6
La	^{140}La	40.3 h	486.8; 1595.4
Sm	^{153}Sm	47.1 h	103.2
Hg	^{197}Hg	65.0 h	77.6
Mo	^{99}Mo	66.0 h	140 (from ^{99m} Tc)
Cd	^{115}Cd	2.30 d	527.7 (336.6 from ^{115m}In)
U	^{239}Np	2.35 d	106.1; 288.2
Au	^{198}Au	2.70 d	411.8
Sb	^{122}Sb	2.75 d	564.0; 692.5
Th	^{233}Pa	27.0 d	311.8
Cr	^{51}Cr	27.8 d	320.0
Ce	^{141}Ce	32.5 d	145.4
Fe	^{59}Fe	45.1 d	1098.6; 1291.5
Hf	^{181}Hf	44.6 d	133.1; 482.2
Hg	^{203}Hg	46.9 d	279.1
Sb	^{124}Sb	60.9 d	602.6; 1690.7
Ni	^{58}Co	71.3 d	810.3
Sc	^{46}Sc	83.9 d	889.4; 1120.3
Ta	^{182}Ta	115.1 d	1121.2; 1221.6
Se	^{75}Se	121.0 d	136.0; 264.6
Zn	^{65}Zn	245.0 d	1115.4
Ag	^{110m}Ag	253.0 d	657.8; 884.5; 1384.0
Cs	^{134}Cs	2.07 y	795.8
Co	^{60}Co	5.2 y	1173.1; 1332.4

[a] min (minutes); d (days); y (years).

TE concentrations, but also to their particles size distribution, seasonal and daily variations, sampling procedures, and possible element-to-element correlation.

Equipments and Facilities

Nuclear Reactor

The TRIGA MarkII (General Atomic, U.S.) research nuclear reactor of the University of Pavia was utilized for all neutron irradiations. According to the type of analysis, three different irradiation procedures have been used. The pneumatic irradiation facility "Rabbit" (samples transfer time 3.5 s) was used for short irradiations (between 30 s and 5 min) at a neutron flux of 5 10^{12} $n \cdot cm^{-2} \cdot sec^{-1}$. Long-term irradiations (4 to 40 h) were performed in the "Lazy Susan" rotating irradiation position or in the reactor central thimble facility at neutron fluxes of 10^{12} and 10^{13} $n \cdot cm^{-2} \cdot sec^{-1}$, respectively.

Gamma Spectrometry

For the gamma spectra evaluation, HPGe (gamma-x) detectors (ORTEC, U.S.) coupled to computerized multichannel analyzers (ORTEC ADCAM, U.S.) were used. The relative efficiencies of the detectors ranged between 25 and 40%, with resolutions of 1.9 to 2.0 on the 1332.4 keV gamma-line of ^{60}Co. For medium and long countings, automatic sample changers were coupled to the systems and typical counting times ranged between 3000 and 25,000 s.

Atomic Absorption Spectrometry

Both flame (FAAS) and electrothermal (ETAAS) atomic absorption spectroscopy have been used in the analyses. The instrumentation consisted of two different equipment pieces: Shimadzu AA660 — GFA4B and Perkin-Elmer 1100 B.

Dry Depositions

Atmospheric dry depositions were collected by a wet and dry bulk type sampler (MTX S.r.l., Bologna, Italy). The system consisted of two polyethylene vessels to sample the wet and dry depositions separately.

Suspended Particles

Isokinetic sampling for total suspended particles from incinerator emissions was performed in the stack fumes using a stainless steel ASI 316 probe with a "anticorodal" filter holder, coupled to a condenser, pump, and volumetric counter in series. The gas speed was meanwhile measured with a Pitot probe joined to a water micromanometer and a thermometer. Fiberglass filters were 47 mm in diameter (Spectrograde Gelman, U.S.). The overall system was made by the Stazione Sperimentale per i Combustibili (SSC), S. Donato Milanese, Italy.

Multistage Impactors

For the granulometric sampling of suspended particles in stacks, a stainless steel eight-stage cascade impactor (Andersen Mark III, U.S.) with spectrograde fiberglass filters was isokinetically employed (stage cut from <0.56 μm to >10.5 μm). In the case of atmospheric particulate collection, a low-pressure multistage impactor (Berner-Germany) with cellulose filters (MilliPore®) was used at the rate of 30 l m^{-1} with final sampled volumes ranging between 170 to 200 m^3 (eight stages with effective cutoff diameters (ECD) from 0.08 to 11.3 μm).

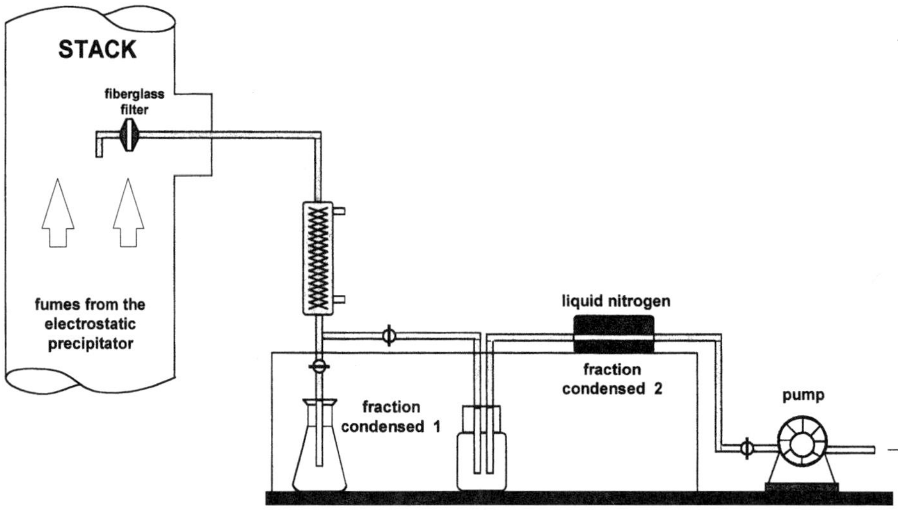

FIGURE 22.1 Collection of trace metals present in the vapor phase released from urban waste incinerators.

Vapor Phase

For vapor phase sampling in the stack, a specific apparatus was developed and made in collaboration with SSC laboratories. The system, as depicted in Figure 22.1, consisted of three different parts: (1) a Pyrex glass probe (positioned in the stack's pipe) was used as a 0.45-µm filter (quartz fiber) holder to retain the particles, (2) the passing gases and vapors were first condensed in a quartz water-cooled condenser, and (3) the residual vapors were frozen in a quartz tube cooled at liquid nitrogen temperature. Except for the Pyrex filter holder, all system components were made of quartz and connected to a suction pump in series.

INCINERATION PLANTS

Two types of urban solid waste incinerators, both located in northern Italy and serving metropolitan areas with more than 500,000 inhabitants each, were investigated. The two plants have different fume treatment systems (dry and wet), as shown in Table 22.3, which also reports their main characteristics.

Materials and Methods

Samples

Emissions from Urban Solid Waste Incineration Plants

Four different emission materials were collected for TE analysis. Three were related to the solid phases (bottom-ash, fly-ash, and suspended particles), while the fourth consisted of the vapor phase released from the stack. In particular,

1. Bottom-ash (BA): collected from incinerator 1, the BA consisted of material obtained by joining and cooling, by water treatment, the combustion residues from the combustion chambers and the fly-ash from the electrostatic precipitators. Three portions of about 10 kg each of this slug wree collected, dried at 105°C, ground, and sieved in sequence to a final powder of 170–180 mesh and then mixed in one homogeneous material.[32] The

TABLE 22.3
Solid Urban Waste Incinerators under Investigation

Plant Characteristics	Plant 1	Plant 2
Treated waste (tons/day)	660	240
Number of furnaces	3	2
Postcombustion chamber	No	Yes
Fumes and ash treatment	Dry electrostatic precipitation (electrofilters)	Water scrubber system and electrostatic precipitation
Total fume flow ($m^3 h^{-1}$)	112,000	110,000
Fume temperature (°C)	250–300	80–100
Total suspended particles in fumes released at the stack ($mg\ m^{-3}$ on dry fumes basis)[a]	75–130	7–15
O_2 in Fumes (%)	9.5–13.5	16.0–17.5
H_2O vap. in fumes (%)	15–17	9–14
Stack height (m)	80	54

[a] Collected isokinetically.

same sampling procedure was carried out at three different times over 1 year, obtaining three different final batches. The analyses were performed on a series of different subsampled amounts of 80 to 100 mg each.

2. Fly-ash (FA): collected from the settling tanks of three different electrostatic precipitators of the same incinerator and in the same time periods. Samples of about 5 kg each were collected and treated in the same manner as used for BA. Identical subsampling procedures (80 to 100 mg each sample) were adopted. For both BA and FA, homogeneity tests were performed by INAA to determine the distribution of more than 15 elements in 12 different samples of about 100 mg of the two materials. The overall homogeneity ranged between 10 and 15%.
3. Suspended particles: were isokinetically collected (for the incineration plants 1 and 2) from the pipe connecting the electrostatic precipitators and the stack. Both whole filter equipment and multistage cascade impactor were used. Between four and ten different sampling runs for every collection period were utilized for the analysis of the total suspended particles matter, while at least three different cascade impactor collections were carried out for TE granulometric distribution measurements.
4. Vapor phase: Samples of the vapor phase released from the incinerators stacks (plant 1 and 2) were collected by a homemade apparatus (Figure 22.1) where the gases flowing through the stack were trapped by condensation at low temperature (see details in the equipment and procedure sections). The vapor samples in the resulting solutions were obtained by dissolving the trapped condensation in 1*N* high-purity HNO_3.

Dry Depositions

Atmospheric dry deposition samples were collected monthly over a long period of time (about 2 years) in a large urban and industrialized area located in the suburbs of a northern city in Italy.[33] The samples consisted of a few milligrams of solid dust, ranging from 10 to 15 mg total weight.

Air Samples

Air samples for both total atmospheric particulate and for granulometric distribution were collected for TE determination. The following samples were analyzed:

1. Total air particulate: Air dust samples were collected using cellulose filters (MilliPore, 0.45 μm, 4.7 cm diameter) by a computerized sampling device with a collection rate of about $1m^3\ h^{-1}$. Series of samplings, depending on the different studies, were carried out in downtown Milan (the largest city in northern Italy). The air samples were collected downtown at 2 m height from ground in the courtyard of the Department of Physics of the University of Milan during the winters of 1993 and 1995 (February).
2. Particle sizes fractionation: An eight-stage, low-pressure impactor for granulometric distribution was used to collect air samples onto cellulose filters at a rate of about $30\ l\ m^{-1}$. The sampling was performed in nonindustrialized areas (Ispra, Northern Italy)

Soils

Samples of soils surrounding the dry particulate sampling points were superficially collected (8 samples of 100 g for each point), dried at 105°C, mixed, and sieved at 50 mesh; samples of about 100 mg were used for analysis.

Blanks

In TE analysis, the problem of blank contribution, derived from materials, reagents, and/or pre-analytical procedures, is of fundamental importance. In the case of samples collected onto filtering materials, the main contaminations arise from the filtering membranes.[34–37] In particular, for the analysis of airborne particulate matter, the trace element contents in the filters may be, for many elements, of the same order of magnitude of the concentrations of the elements present in the particulate sample itself. A careful analysis of the field blanks must always be carried out to correct the final measurements. In this work, all filtering membranes have been analyzed in the same way as the real samples, and particular attention has been given to the blanks arising from the filters used in both total and granulometric atmospheric particulate samplings. As an example, Table 22.4 reports the results obtained in the analysis of two series of four filters (MilliPore) sampled from two different batches.

The quite large scattering in the standard deviation found for many elements is probably due to the dishomogeneity among the filters. Some authors have found significant variations between the same type of filter coming from different batches[13] in accordance with the fact that impurities may be not homogeneously distributed. As seen from Table 22.4, the corrections for blank contribution were necessary for almost all the elements considered, and particulary for Br, Cl, Cr, Fe, Mn, Mg, Ti, Ni, and Zn. Especially in the analysis of air particulates collected in rural, low-polluted areas, the correction for blanks from filters may give, for some elements, negative values.[38] In this case, to reduce the field blank contribution, a higher amount of air dust should be collected, thereby encreasing the sampling time.

The blank contribution from the irradiation vials was also tested. This was necessary only for "short-lived" determinations because for the other elements the samples, after irradiation, were transferred into the non-irradiated counting vials.

All reagents and chemicals used in the analytical procedures and for the dissolution of the samples for the AAS measurements have been also tested for blank content and taken into account.

Standards, Reagents, and Quality Control

All primary standards were obtained by dissolving high-purity metals, salts, or compounds in high-grade acids. The resulting solutions were used for both NAA and AAS measurements. In particular, the NAA series of multi-element standards were prepared by taking into consideration the nuclear characteristics (see "Analyses") of the elements to be analyzed. As comparator standards, the following standard reference materials of the National Institute of Standard and Technology (U.S.) were used: SRM Coal 1632a and 1632, SRM Urban Dust 1648, SRM Coal Fly Ash 1633, and SRM Apple Leaves 1515. Except for reagents used in radiochemical separations, all the reagents were of high-grade purity (HNO_3 and HCl Aristar® from BDH, U.K.; $HClO_4$, H_2SO_4 Ultrapure from Carlo Erba, Italy).

TABLE 22.4
Blank Values of 35 Trace Elements Determined by INAA in the Filtering Membranes Used to Collect the Air Particulate

Element	ng per filter[a]	Element	ng per filter[a]
Ag	<0.5	I	12 ± 8
Al	1100 ± 400	La	0.6 ± 0.4
As	1.8 ± 1.0	Mg	2180 ± 850
Au	0.2 ± 0.03	Mn	28 ± 6
Ba	<150	Mo	<3
Br	58 ± 42	Ni[b]	26 ± 15
Cd[b]	<0.5	Pb[c]	12 ± 8
Ce	2.1 ± 2.0	Rb	0.9 ± 0.5
Cl	7730 ± 446	Sb	2.4 ± 1.5
Co	2.2 ± 0.7	Sc	<0.05
Cr	40 ± 25	Se	0.9 ± 0.7
Cs	0.6 ± 0.3	Ta	<0.1
Cu[b]	82 ± 51	Th	<0.1
Eu	<0.05	Ti	140 ± 95
Fe	230 ± 180	U	<0.5
Hf	<0.5	V	<0.8 ± 0.4
Hg	<0.5	W	<0.5
Zn	647 ± 400		

Note: MilliPore HAWP04700 — 0.45 μm (47 mm ∅).

[a] Data obtained from the analysis of four filters of each type.

[b] Determined also by ET-AAS and comprehensive of reagent blanks.

[c] Determined exclusively by ET-ASS and comprehensive of reagents blanks.

Analyses

INAA

INAA has been used for the determination of As, Au, Ba, Br, Ce, Cd, Cl, Cr, Cu, Co, Cs, Hf, Hg, K, I, La, Mn, Na, Ni, Rb, Sc, Se, Sm, Sn, Sb, Ta, Th, V, and Zn. Two different series of irradiations were usually performed to carry out the determination of the corresponding short- and medium/long-lived radionuclides. In some cases, RNAA was applied to determine As, Cd, Cr, Hg, and Se. ETAAS was employed for all Pb determinations and, in some cases, for Cu, Cd, Sn, and Ni analyses; FAAS was mainly used for the analysis of elements such as Na, K, Ba, and Mg. In many cases, intercomparison determinations between INAA and AAS were carried out for the analysis of Cd, Cu, Ni, Sn, Zn, Ba, Mg, and K.

Determination of "Short-Lived"

Cl, Cu, I, Mn, Mg, Ti, and V were determined using the corresponding short-lived radionuclides. The samples and standards were sealed in polyethylene vials, irradiated by the pneumatic facility for times varying between 30 and 300 s, allowed to decay (from 1 to 60 min), and then gamma-counted for 200 to 500 s. Blanks of the polyethylene vials were determined in the same way.

Determination of "Medium/Long Lived"

All other elements were determined by irradiating the samples and standards sealed in polyethylene or quartz vials (for liquid samples and multistandard solutions) in the rotating facility or in the central thimble of the reactor. Irradiation times varied between 4 and 40 h. After irradiation, the samples were transferred to plastic counting containers expressly made for the automatic samples changers. Accurate amounts of the irradiated multistandard solutions were spiked onto cellulose powder or onto non-irradiated samples to simulate the same "counting geometry" of the irradiated samples. Depending on the half-life of the radionuclides of interest and the activity of the samples, different decay and counting times were used. The medium-lived of the corresponding As, Au, Br, Cd, Na, La, Mo, Sb, U, W radionuclides were usually counted after 1 or 2 days from the end of irradiation for 3000 s. The long-lived radionuclides were counted several days after the irradiation and for counting times ranging between 10,000 and 25,000 s.

RNAA

RNAA was utilized for the determination of Hg, Cd, As, Se, Cr in matrices of bottom- and fly-ash where the high radioactivity background present in the irradiated samples, in many cases, prevented their instrumental determination. The radiochemical separation consisted of fusion-dissolution of the irradiated matrices with a simultaneous separation of ^{197}Hg and a subsequent separation of the ^{76}As, ^{75}Se, ^{51}Cr by inorganic-ion exchange chromatography and of ^{115}Cd by solvent extraction.[39] These kinds of samples, originating from high-temperature combustion procedures, contain large amounts of silica, oxides (and their mixture), incomplete oxidized carbon, and fused aluminosilicate, and are characterized by high insolubility. The key point of the separation was to carry out a rapid alkaline fusion of the sample in a closed system that allowed the simultaneous destruction of the matrix (with an easy subsequent dissolution) and the quantitative separation of ^{197}Hg by distillation.

RESULTS AND DISCUSSION

EMISSION FROM MUNICIPAL INCINERATORS

Urban waste incineration produces three different solid emissions: bottom-ash, fly-ash, and suspended particles. The characterization of these materials for TE content has been evaluated by following their distribution and the relative concentration variation. Table 22.5 shows the results obtained in the determination of 21 elements, along with the relative standard deviations and the concentration ranges. The results refer to the years 1989 to 1990 and were obtained from the analysis of series of samples collected, as previously described, from Plant 1 over 1 year. The quite good relative standard deviation shows that the homogeneity and the representativeness of the samples obtained in collecting and treating the BA and the FA materials were acceptable. The concentration ranges reflect the typical variable nature of the refuse burned and their values are comparable with those obtained in other investigations.[40–42] An interesting distribution between the three different emissions can be observed for several elements. The concentrations of Ta, Th, V, and Cu show a quite constant trend with no appreciable variation. Elements such as Ba, Co, Cr, Fe, Mn, and Sc seem to decrease in concentration in suspended particles, while As, Cd, Cs, Hg, Ni, Pb, Sb, Se, Sn, and Zn show, in this fraction, a remarkable enrichment which, for many of them, reaches more than 50% of the total solid emission amount. In fact, As, Cd, Cs, Hg, Pb, Sb, Se, and Zn may be considered more volatile or related to compounds with more mobility and/or associated to the finest particles in the fumes. This underlines that, even if associated with solid emissions, the suspended particles represent the real impact on the atmosphere for both amount released and physical form.

The TE concentrations in the suspended particles are also related to the type of pollution control used in the incinerator. Table 22.6 shows the results obtained for 13 elements in the intercomparison

TABLE 22.5
Determination of 21 Elements in the Solid Emissions from Refuse Incinerators (Plant 1)

Element	Bottom-Ash (BA) Av±SD	Bottom-Ash (BA) Range	Fly Ash (FA) Av±SD	Fly Ash (FA) Range	Susp. Particles (SP) Av±SD	Susp. Particles (SP) Range	% Related to Susp. Particles
As	40 ± 7	20–65	26 ± 4	18–51	110 ± 25	60–230	62
Ba (mg)	2.7 ± 0.18	2.3–3.2	2.1 ± 0.28	1.7–2.5	0.98 ± 0.2	0.52–1.1	17
Cd	100 ± 15	65–170	270 ± 40	90–420	440 ± 60	210–880	54
Cs	3.7 ± 0.2	3.0–4.8	8.8 ± 0.4	6–13	15.5 ± 3.0	10–220	55
Co	38 ± 2	35–45	53 ± 3	35–68	22 ± 5	10–35	20
Cr (mg)	1.55 ± 0.25	1.2–1.8	2.65 ± 0.2	2.2–2.9	1.1 ± 0.2	0.6–1.5	20
Cu (mg)	1.8 ± 0.15	1.0–2.3	2.25 ± 0.17	0.9–3.18	1.7 ± 0.2	0.9–3.0	29
Fe (mg)	38 ± 2	30–48	25.6 ± 1.5	18.5–32	11.3 ± 1	9–12.1	15
Hg	3.5 ± 0.1	3.0–3.9	5.2 ± 0.5	3.2–6.0	35 ± 8	19–75	80
Mn (mg)	1.7 ± 0.1	1.5–2.0	2.14 ± 0.23	1.6–2.8	1.12 ± 0.45	0.5–1.7	22
Ni	380 ± 30	270–430	345 ± 50	300–460	450 ± 100	220–600	38
Pb (mg)	3.8 ± 0.3	2.9–4.4	9.5 ± 0.7	6–14	10 ± 1.5	5–14.2	43
Rb	71 ± 10	42–100	146 ± 16	70–205	145 ± 45	80–280	40
Sb	180 ± 25	75–290	340 ± 35	210–560	665 ± 180	380–1100	56
Sc	4.2 ± 0.5	3.5–6	3.7 ± 0.3	3–5	2.2 ± 0.5	1.5–2.5	21
Se	5.1 ± 1.0	3.5–9.0	13.2 ± 1.5	7–19	12 ± 1.5	9–20	39
Sn (mg)	0.85 ± 0.09	0.6–1.05	1.7 ± 0.15	1.3–2.4	2.8 ± 0.3	2.3–5.1	52
Ta	1.8 ± 0.2	1.0–2.7	1.9 ± 0.2	1.1–2.8	1.0 ± 0.3	0.3–1.3	21
Th	5.8 ± 0.3	5.2–6.3	6.2 ± 0.5	4.0–7.1	5.5 ± 0.7	2.3–6.6	13
V	90 ± 12	60–110	80 ± 8	60–130	77 ± 10	50–100	45
Zn (mg)	11.9 ± 1.5	9.5–12.8	26.7 ± 4.0	23.5–28	43.5 ± 11	26–71	53

Note: Concentration given in $\mu g\ g^{-1}$ unless otherwise (mg) noted. Data referred to the incinerator Plant 1. Mean concentration values and relative standard deviations obtained from 10 independent samplings and analyses related to one collection period. Ranges obtained from series of data collected over three different periods during the same year.

between the two different plants (1 and 2) with dry and wet fumes treatments, respectively. Different concentration trends can be observed for many elements in the suspended materials; while As, Cd, Cs, Hg, Pb, and Se have higher concentration in the particulate of the Plant 2, other elements such as Br, Co, Cr, Sb, and Th show lower values. It appears that the suspended particles in the wet fumes treatment contain higher concentration of the more volatile elements or their compounds characterized by a major mobility. Nevertheless, the total release to the atmosphere, if expressed in $\mu g\ m^{-3}$, is much larger for the Plant 1, the total particulate per cubic meter being more than one order of magnitude in comparison to that of the Plant 2.

Since the major contribution in the emission of TE into the air is related to the fumes and the suspended matter, it may be very important to know to which particle sizes the TE are related. A series of results on the granulometric distribution, as well as the total concentration in the suspended particles, were obtained for some elements with high concentration in suspended matter, and are given in Table 22.7. Here, five elements (i.e., Cd, Cr, Pb, Sb, and Zn) were used to compare the emissions from the two different incinerators. The corresponding percentage distribution with the stage cut is also given. As has already been observed by other authors,[43–45] the values of Cd, Zn, and Pb from both plants show a considerably higher percentage in the finest fraction (<0.56 μm), and their distribution trend can be considered quite similar, as shown in Figure 22.2. A different behavior can be observed for Cr and Sb, as depicted in Figure 22.3. Both elements show different granulometric distributions that depend on the type of plant, with more evidence for Cr in the

TABLE 22.6
Analytical Results of Suspended Particles Released from Different Urban Waste Incinerators

	Plant 1[a]			Plant 2[a]		
Element[a]	(μg g^{-1}) Mean	Range	(μg m^{-3}) Mean	(μg g^{-1}) Mean	Range	(μg m^{-3}) Mean
As	110	60–230	15.4	160	50–380	1.12
Br	750	420–1300	105	194	87–270	1.35
Cd	440	210–880	61.6	1570	950–2350	10.98
Co	22	10–35	3.1	3.4	0.5–5	0.024
Cr	1100	600–1500	154	660	450–830	4.62
Cs	15	10–22	2.1	33	15–400	0.23
Fe (%)	1.13	0.9–1.2	0.16	1.2	1.0–1.4	0.008
Hg	35	19–75	4.9	104	18–190	0.73
Pb (%)	1.0	0.5–1.4	0.14	4.1	2.5–5.6	0.03
Sb	665	380–1100	93	481	220–800	3.36
Se	12	9–20	1.68	99	50–160	0.69
Th	5.5	2.3–6.6	0.77	1.4	0.8–1.6	0.01
Zn (%)	4.3	2.6–7.1	0.60	4.8	2.4–8.7	0.03

Note: Mean concentration values obtained from ten indipendent samplings and analyses related to one collection period. Ranges obtained from series of data collected over three different periods during the same year.

[a] Concentrations in in μg g^{-1} and in μg m^{-3} unless indicated as (%).

Plant 1, where a higher percentage is associated to the largest particles. Figure 22.4 shows the granulometric distributions of Sn and As, together with other minor elements such as Co, Hf, Ce, and Sc in the suspended particles released from the Plant 2. In this case, to compare the relative concentrations, the fractionation is given in terms of μg m^{-3}.

As concerns the emission related to the vapor phase, a specific study was carried out to evaluate the presence of those volatile elements (or their compounds) such as As, Cd, Hg, Pb, Sb, Se, and Zn. The samplings were performed using the expressly developed apparatus as previously described in Figure 22.1. The samples analyzed consisted of the two different liquid fractions obtained by the condensation of the vapors; that is (1) the condensate from the refrigerator and the nitric acid of the impinger, and (2) the nitric acid washing solution from the liquid nitrogen trap. This additional, latter trapping system was found very effective for the quantitative collection of all the elements considered, as shown in Table 22.8 which reports the percentage of the individual recovery.

Table 22.9 reports the results obtained in the analysis of both suspended matter and the gas phase of the two different incinerators. To compare the different behaviors, all the concentrations were expressed in μg m^{-3} and the percentage associated to the vapor phase, for each element considered, is also reported. The concentrations of As, Sb, Cd, Pb, and Se in the vapor phase can be considered comparable for both plants, while the concentrations in the suspended particles of Plant 1 are, except for Se and Hg, much higher. Interesting variations between the concentration in the suspended particles and in the vapor phase can be observed. While the percentage associated with the vapor phase of the volatile forms of As, Se, and Hg is very similar in both cases, significant differences can be observed for the other elements. In the case of Cd, Zn, and Pb, despite their high concentration in the suspended matter of Plant 1, the corresponding percentages in the vapor phase appear lower in comparison with those of the Plant 2. Considerable differences can also be observed for Sb and Zn, which seem to enrich the vapor phase of the plant with wet fume treatment.

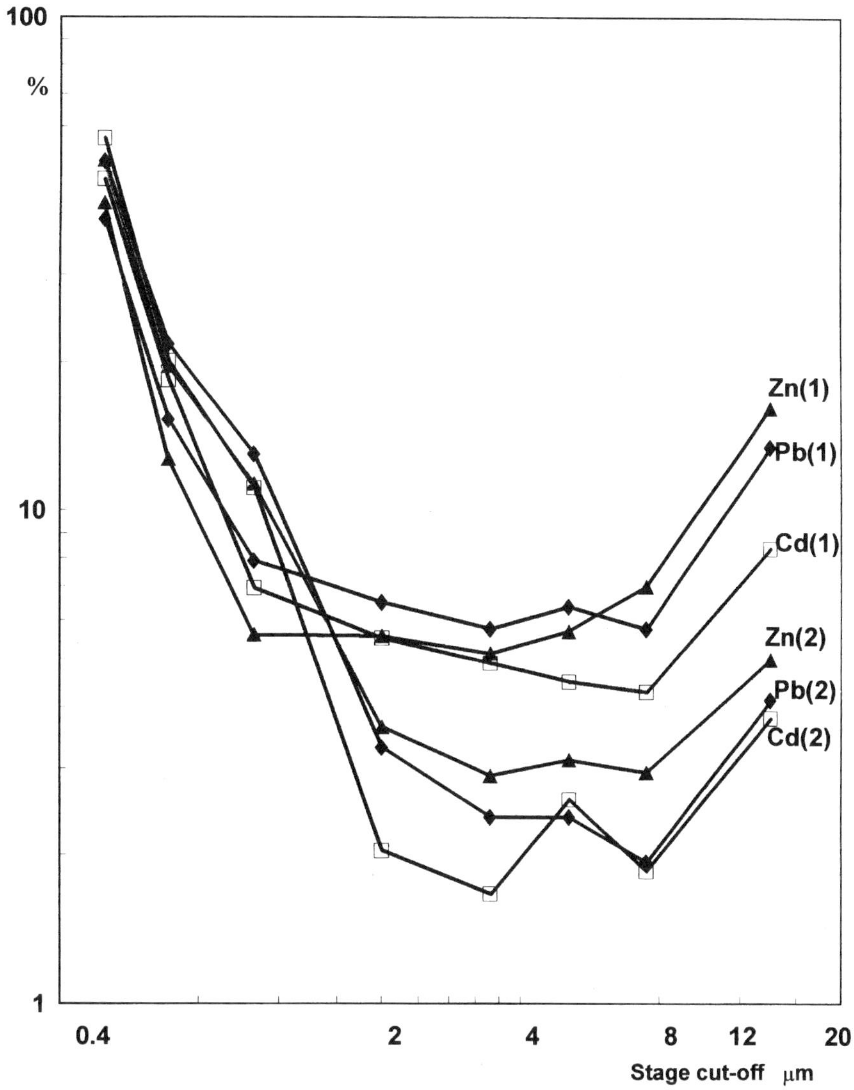

FIGURE 22.2 Granulometric distribution Pb, Cd, and Zn in the suspended particles released from the incinerators 1 (dry) and 2 (wet).

In fact, the concentrations of Zn and Sb in the suspended particles of Plant 1 drop to a lower level in the vapor phase, respectively 10 times and 5 times more than those of Plant 2.

On the contrary, Hg concentrations rise quite differently from the suspended particles to the vapor phase, being much higher in the case of Plant 1. In both cases, however, mercury is present almost quantitatively in vapor form, with a concentration more than 3 times higher when released from the incinerator with no water scrubber system.

Atmospheric Dry Depositions

Samples of dry depositions collected, as previously reported, in suburban areas of Bologna, a large industrialized city in northern Italy, were analyzed for the contents of 15 elements. This was done to follow their concentrations for a period of 2 years (1990–1991) and to obtain informations on their seasonal variations.

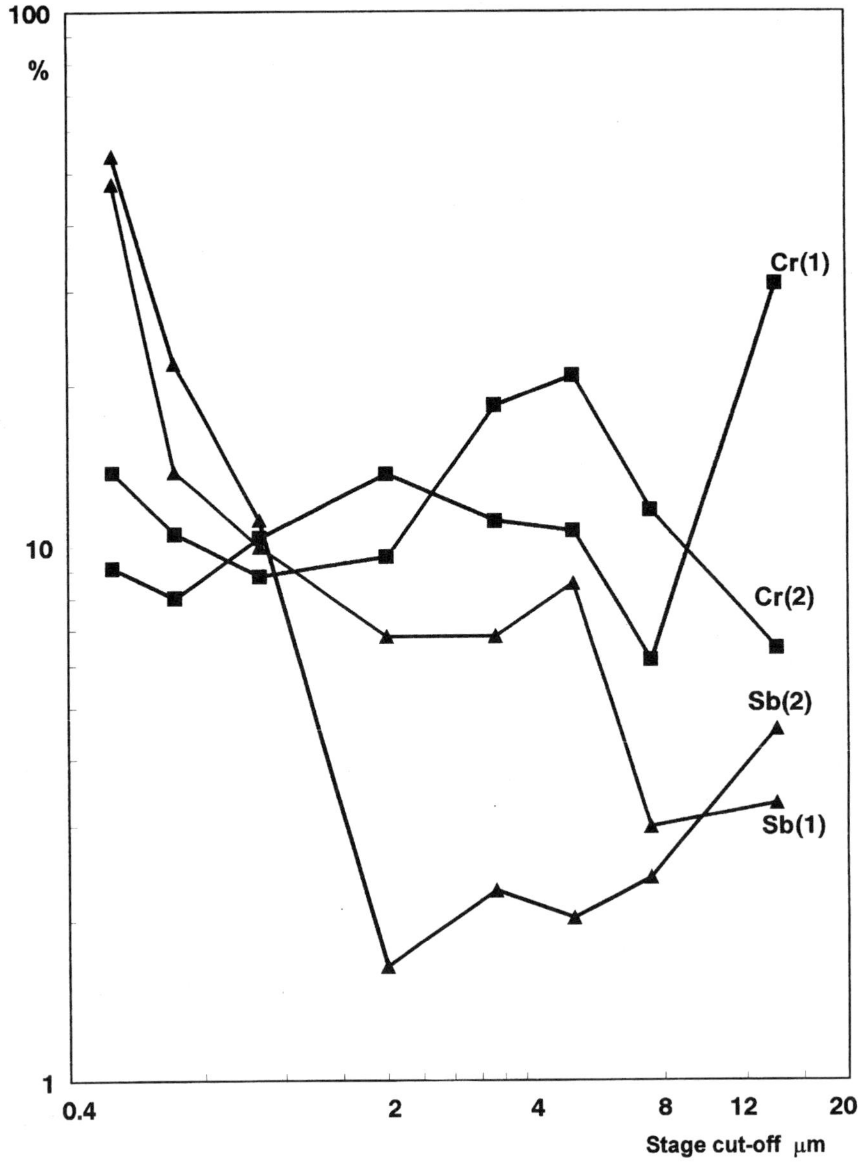

FIGURE 22.3 Granulomteric distrihbution of Cr and Sb in suspended particles released from the incinerators 1 (dry) and 2 (wet).

In addition, series of analyses of the soils surrounding the sampling points were also performed to calculate the enrichment factor (EF) for some elements of toxicological interest. As other authors have done,[46] scandium was chosen as the crustal reference element and used in the EF equation: EF = CX/CSc (ambient): CX/CSc (background), where CX is the concentration of the X element whose enrichment was determined.[1]

The results of the concentrations determined in winter and summer seasons, together with the relative EF calculated, are given in Table 22.10. Both potentially toxic elements and elements related to natural sources were determined to compare their concentration ranges and possible seasonal trend. The elements of anthropogenic origin with corresponding EF values significantly greater than 1 were: As, Br, Ce, Cr, Co, Mn, Sb, V, and Zn.

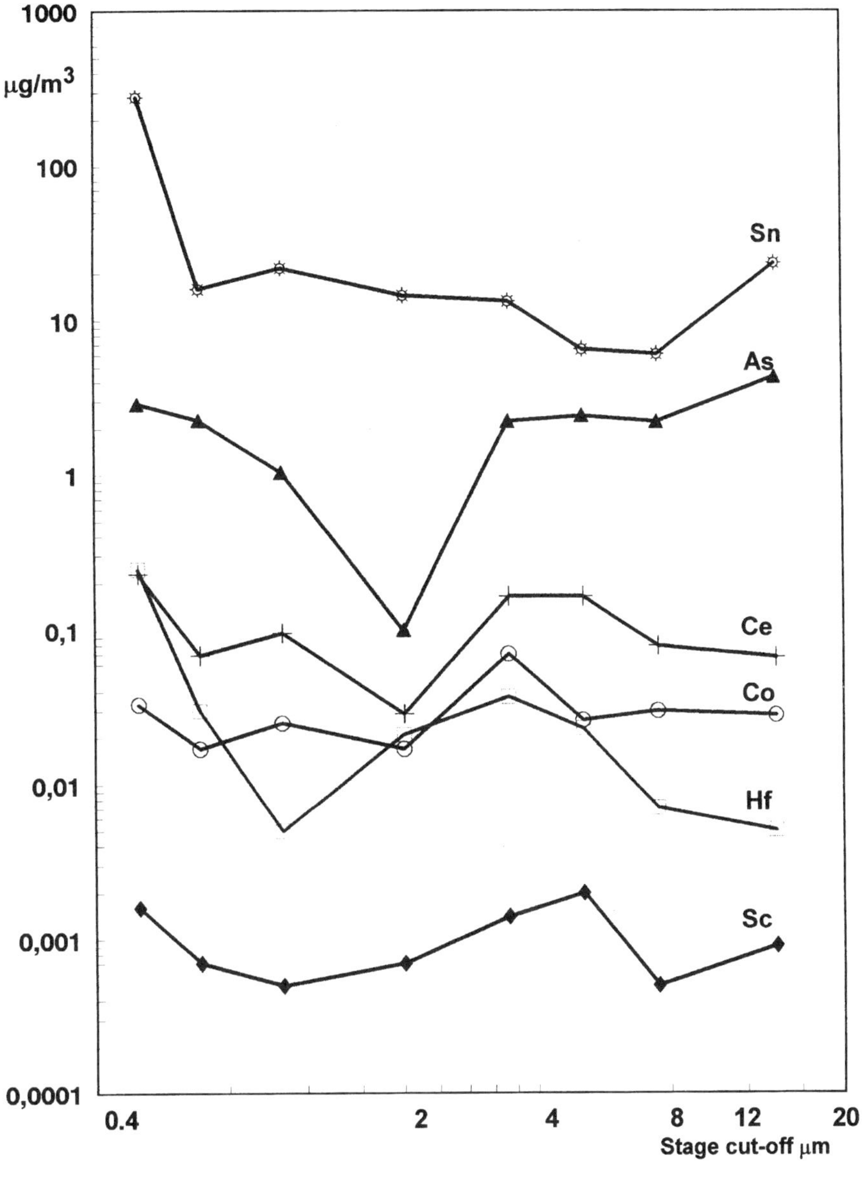

FIGURE 22.4 Granulometric distribution of some trace elements in suspended particles released from the incinerator 2 (wet fume treatment).

Many elements (such as As, Br, Zn, and V) show evident differences in the concentrations between the winter and the summer months. In particular, the vanadium concentration, which increases from a minimum of 55 to a maximum of 670 µg g^{-1} during the winter, may identify its main source as domestic oil heating. This can be also observed by comparing the corresponding EFs. All the other elements may be associated with natural sources, with concentrations lying in the ranges of the soil composition and with no significative seasonal variations.

In Figure 22.5, to compare the different monthly distributions, the concentrations (in µg g^{-1}) of some of the most polluting elements have been plotted on a semi-logarithmic scale against the time in months. The concentrations of zinc and vanadium increase during the winter months with a quite similar trend; a seasonal dependence might be considered. In the case of arsenic, chromium,

TABLE 22.7
Granulometric Distribution of Some Elements in Suspended Particles Released from Two Different Solid Waste Incinerators

Stage	Cd				Zn				Pb				Sb				Cr			
	1		2		1		2		1		2		1		2		1		2	
(μm)	μg m^{-3}	%	μg m^{-3}	%	μg m^{-3}	%	μg m^{-3}	%	μg m^{-3}	%	μg m^{-3}	%	μg m^{-3}	%	μg m^{-3}	%	μg m^{-3}	%	μg m^{-3}	%
>10.5	4.1	8.4	0.41	3.8	372	16.2	31.2	4.9	136	13.5	25.5	4.1	0.31	3.3	0.36	6.1	19.7	31	0.58	6.4
6.5–10.5	2.1	4.3	0.20	1.8	161	7.0	18.4	2.9	58.1	5.8	11.9	1.9	0.28	3.0	0.19	3.2	3.9	6.1	1.05	11.6
4.4–6.5	2.2	4.5	0.28	2.6	131	5.7	19.5	3.1	64.4	6.4	14.7	2.4	0.80	8.5	0.16	2.7	6.8	10.7	1.90	21
3.0–4.4	2.4	4.9	0.18	1.7	118	5.1	18.1	2.9	58.1	5.8	14.0	2.2	0.64	6.8	0.18	3.0	7.1	11.2	1.65	18.3
1.9–3.0	2.7	5.5	0.22	2.0	128	5.5	22.8	3.6	65.6	6.5	20.4	3.3	0.67	7.1	0.13	2.2	8.7	13.7	0.85	9.4
0.9–1.9	3.4	6.9	1.20	11.1	129	5.6	10.9	11.3	79.6	7.9	80.4	13.0	0.94	10	0.89	15	6.6	10.3	0.80	8.8
0.6–0.9	9.0	18.4	2.17	20.1	292	12.7	123	19.7	154	15.3	134	21.7	1.30	14	1.75	29.5	5.1	8.0	0.95	10.5
<0.56	23.0	47.0	6.13	56.8	967	42.1	321	51.3	392	38.9	315	51.1	4.50	48	2.26	38.2	5.8	9.1	1.24	13.7
Total	63 ± 17		14 ± 4		2710 ± 407		695 ± 190		1295 ± 250		690 ± 115		12 ± 4		6.7 ± 3.6		73 ± 15		9.8 ± 5	

Note: 1, plant with dry fume treatment; 2, plant with wet fume treatment. Granulometric concentration values are the mean of two independent samplings. Total concentration values refer to four independent samplings carried out the same day of cascade impactor collection. The total efficiency of the cascade impactor ranged from 77 to 90%, compared to the total suspended particles sampling.

TABLE 22.8
Recovery of As, Cd, Hg, Pb, Sb, Se, and Zn in Vapor-Phase Sampling Using the Apparatus of Figure 22.1

Element	Fraction (a) (Condenser + Impinger) % Recovery	Fraction (b) Liquid Nitrogen Trap % Recovery
As	85 ± 12	15 ± 9
Cd	60 ± 15	38 ± 12
Hg	65 ± 18	35 ± 10
Pb	80 ± 25	19 ± 8
Sb	88 ± 20	15 ± 7
Se	85 ± 15	16 ± 6
Zn	68 ± 20	32 ± 11

Note: Results obtained from five independent samplings.

TABLE 22.9
Trace Elements in Vapor and in Suspended Particles Released from Municipal Incinerators

	Plant 1 (Dry Fume Treatment)			Plant 2 (Wet Fume Treatment)		
Element	Susp. Particles ($\mu g\ m^{-3}$)	Vapor Phase ($\mu g\ m^{-3}$)	% Related to Vapor Phase	Susp. Particles ($\mu g\ m^{-3}$)	Vapor Phase ($\mu g\ m^{-3}$)	% Related to Vapor Phase
As	8.8 ± 3.5	1.0 ± 0.7	10	4.7 ± 3.2	0.5 ± 0.4	9.6
Sb	16.4 ± 4.6	0.08 ± 0.02	0.5	3.2 ± 1.5	0.09 ± 0.4	2.7
Cd	63 ± 14	1.2 ± 0.5	2	16.1 ± 6.1	1.3 ± 0.9	7.4
Pb	1533 ± 130	57 ± 19	3.6	431 ± 86	36 ± 28	7.7
Se	0.97 ± 0.50	2.1 ± 0.7	68	1.2 ± 1.9	1.6 ± 0.6	57
Zn	2830 ± 332	55 ± 20	2	629 ± 250	111 ± 64	15
Hg	1.3 ± 0.3	151 ± 37	>99	0.9 ± 0.5	41 ± 15	98

Note: Suspended particles collected isokinetically into the stack. Vapor phase collected by vapor condensation and subsequent liquid nitrogen trap. Results obtained from five independent collection runs. Both samplings (suspended particles and vapor phase) were carried out on the same days.

and antimony, this behavior is less evident and the large scatter of their concentrations with monthly variation seems not strictly related to seasonal variation. However, the higher concentrations in the winter periods should be related to the weather conditions, which are in that region, characterized by heavy fog and humidity. These conditions may increase the fallout by condensation of the elements associated with volatile forms or to the finest particles.[47]

Air Monitoring

Increasing air pollution in industrial and urban areas makes the control of air quality mandatory to evaluate the risk of possible toxic effects on public health. It has been found that several toxic micro-compounds are released into the city air, mainly from vehicle exhausts and domestic heating

TABLE 22.10
Determination of Some Trace Elements in Dry Depositions of Suburban Bologna (Italy) and Their Enrichment Factors

	Winter (October–March)			Summer (April–September)		
Element	Mean	Range	EF	Mean	Range	EF
As	15.2 ± 9.0	5–42	5	8.1 ± 4.0	2.6–22.3	2.6
Br	170 ± 66	23–278	357	77 ± 31	19–98	166
Ce	18 ± 11	15–65	2.0	33 ± 18	19–70	3.7
Cr	93 ± 26	38–145	3.7	101 ± 39	43–140	3.8
Co	15.4 ± 8.5	5–38	4.0	17 ± 10	6–33	4.5
Fe (%)	1.5 ± 0.3	0.7–3.0	1.7	1.9 ± 0.4	0.5–2.9	2.2
Hf	2.0 ± 1.3	0.4–5.3	1.6	2.2 ± 1.0	0.5–4.7	1.7
La	11.6 ± 2.8	8–26	1.4	15 ± 3	5–25	1.7
Mn	580 ± 165	114–831	3.2	567 ± 177	382–868	3
Rb	38 ± 12	18–95	1.0	48 ± 7	40–91	1.2
Sb	16.7 ± 7.1	10–41	45.0	15.6 ± 7.0	7–51	41
Sc	2.8 ± 0.8	1–5	1	2.9 ± 0.8	1.7–6	1
Th	3.9 ± 0.7	1–8	1.1	3.6 ± 0.5	2.6–6.7	1.1
V	448 ± 153	90–670	17	132 ± 49	55–273	5
Zn	1050 ± 470	487–1970	104	643 ± 128	345–850	62

Note: All concentrations in in $\mu g\ g^{-1}$, unless otherwise noted (%). Concentration data obtained from three independent samplings carried out monthly at each collection point over a period of 18 months (1990–1991). EF calculated using Sc as the reference crustal element. EF data obtained on the basis of the medians of the results from the analysis of dry depositions and soils samples.

processes. Among these pollutants, many metals and elements can be included as potentially toxic agents, depending on their concentration into the air, their chemical form, and the fact that are normally associated with the inhalable finest particles (<10 μm).

Monitoring campaigns to follow the TE concentrations in city as well rural air have been carried out for many years and are still in progress in research laboratories all over the world. The literature reports hundred of specific works, and some of them are listed in the references as examples.[48–55] Here, a series of determinations of TE in the air particulate of a large city are reported. In Italy, Milan is the most industrialized town, with almost 2 million inhabitants living in the metropolitan area; it is located in Lombardy, which is a flat region at foot of the Alps. The local weather is characterized, during the winter season, by extended periods of heavy fog and thermal inversion. Under these conditions, the air pollution reaches warning levels and, in many cases, automotive traffic restrictions have to be ordered by city authorities. Several monitoring campaigns have been carried out to analyze the city air particulate using the sampling procedures described above. In particular, two series of air filter samples were collected in the month of February in 1993 and in 1995 in downtown Milan. In the first collecting period, the samplings were performed every 4 hours for 14 days (February 8 to 21) to obtain information on both total concentrations and daily variations. In the second period (February 15 to 24), the air was sampled every 24 h for 10 days. In both periods, the weather conditions were quite similar, except four days (February 12–16, 1993), in which the weather was characterized by heavy fog with stagnant air and thermal inversion. All filters were analyzed by INAA for all TE except Pb, which was determined by AAS. As mentioned above, the series of results obtained from the first collecting period were also used to follow the daily concentration profiles of many elements.[56] Figure 22.6 depicts the concentration profiles of

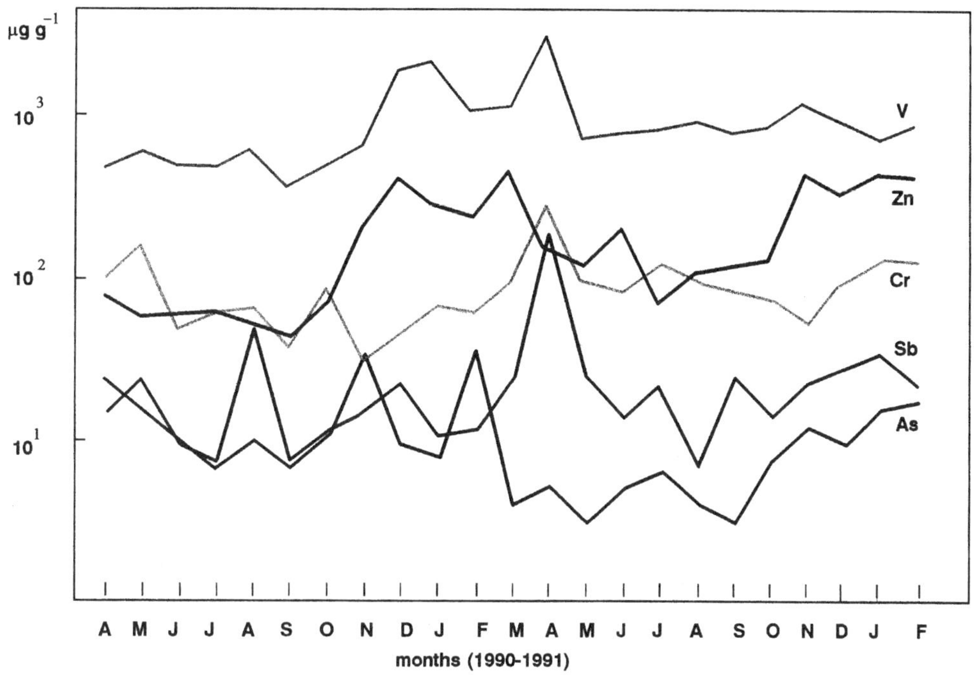

FIGURE 22.5 As, Sb, Cr, V, and Zn concentrations in dry particles vs. monthly variation.

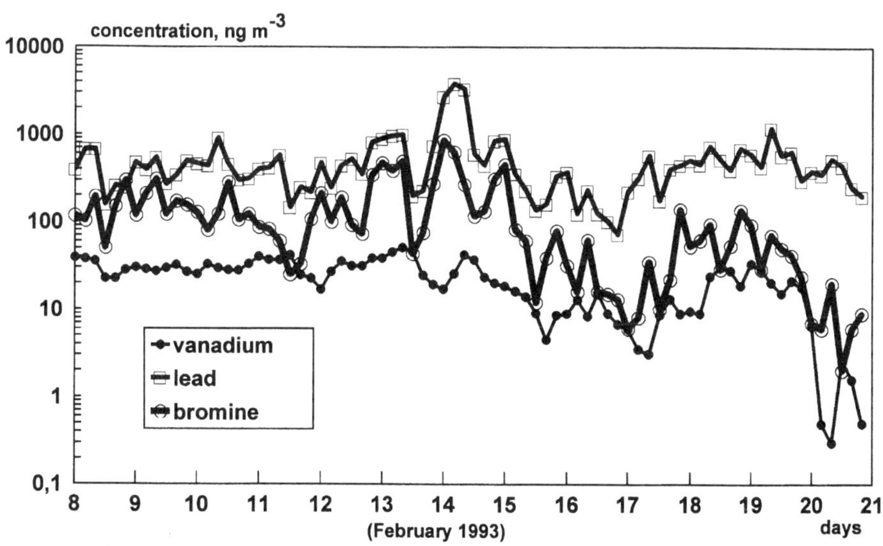

FIGURE 22.6 Daily concentration profiles of vanadium, lead, and bromine in atmospheric particulates of Milan.

V, Pb, and Br, which are related to the main emission sources (Pb and Br from gasoline, and V from fossil fuels, domestic heating, and diesel engines). The data show a significant correspondence between Pb and Br, while the V concentration trend appears quite constant within the period considered. For all three elements, the highest values were found during the 4-day period (February 12–16) with the particular weather conditions mentioned above. To better evidence the Pb and Br relationship, their concentration profiles have been plotted, together the Pb:Br ratio, as shown in

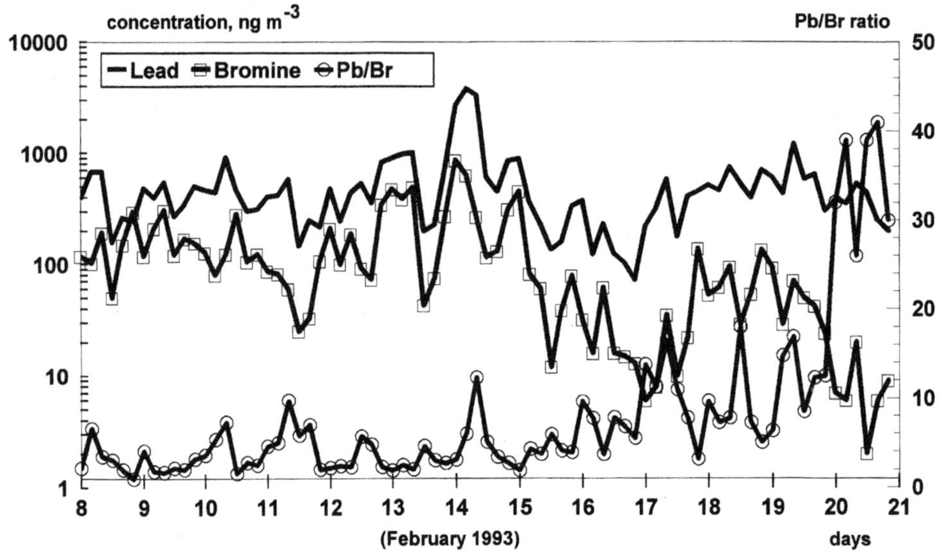

FIGURE 22.7 Daily concentration profiles of lead and bromine and the Pb:Br ratio in atmospheric particulates of Milan.

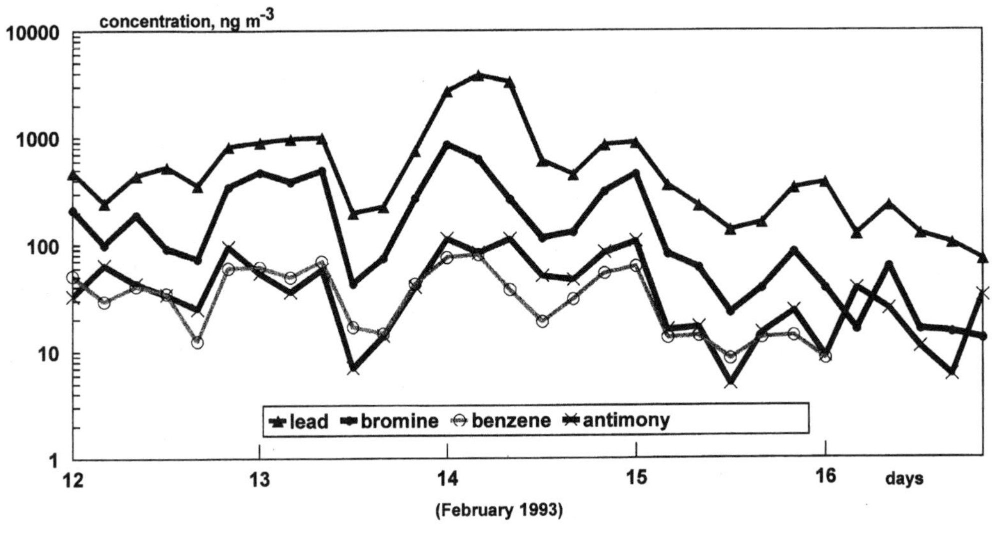

FIGURE 22.8 Daily concentration profiles of lead, bromine, antimony, and benzene in atmospheric particulates of Milan (concentraion of benzene given in μg m^{-3}).

the Figure 22.7. In this figure, the scattering of the results observed during the last 3 days may be due to the rapid weather change to clear and cooler conditions caused by the winds from the Alps. From data collected during the same period, it was also possible to compare the Pb and Br concentration profiles with those of benzene[57] and Sb. Figure 22.8 reveals these profiles in the higher pollution period (February 12–16), with the concentrations of benzene expressed in μg m^{-3}. The correlation between their variations is so evident that it can be extrapolated hour to hour. It may be thought that Sb is strictly correlated with the emission from automotive traffic.

Comparing the results obtained in the first monitoring campaign (February 1993) with those of the second (February 1995), a similar behavior was found for both the total concentrations and

TABLE 22.11
Concentration Ranges of Trace Elements in the City Air of Milan Determined by INAA: Sampling Periods, February 1993 and February 1995

	Milan Air			Literature Values in Other Cities				
Element	Ave.	Min.	Max.	Europe[ab]	Europe[c]	U.S.[a]	U.S.[c]	South Pole[d]
Ag	1.0	0.4	2.1	0.85–4.2	—	1.5–2.4	—	<0.41
As	4.9	0.5	11	1–700	5–330	1.5–30	2–2320	0.007
Au	0.12	0.005	0.6	(0.015)	—	—	—	—
Ba	105	22	146	—	—	91	—	—
Br	145	12	872	24–433	—	32–1720	—	1.4
Cd	6.0	0.3	85	1.3–27	0.4–260	—	0.2–7000	—
Ce	1.8	0.3	2.8	0.4–14	—	0.8–13	—	<0.004
Cl	1350	550	2700	153–7063	—	366–1500	—	2.6
Co	2.9	0.2	16	0.25–6.7	0.4–18.3	0.14–2.6	0.2–83	0.0005
Cr	70	2	264	4.2–16	3.7–227	2.1–290	2.2–124	0.04
Cs	1.2	0.2	2.7	0.2–0.6	—	(0.6)	—	0.0001
Cu	47	6	320	17–64	13–2760	—	3–5140	0.029
Eu	0.05	0.02	0.08	0.014–0.1	—	0.03–0.09	—	$2 \cdot 10^{-5}$
Fe	1870	300	3300	520–3500	294–13000	247–13800	130–13800	0.62
Hf	0.1	0.05	0.4	0.02–0.056	—	(0.3)	—	—
I	8.5	4.5	13.7	0.24	—	0.5–5.4	—	
La	1.5	0.4	1.8	0.2–3.4	—	0.5–9.1	—	0.0004
Mg	980	720	2710	—	1100	—	—	
Mn	43	6	275	13–390	23–850	27–390	4–488	0.013
Ni	38	3.8	140	3.8–32.5	0.3–1400	4–30	1–328	—
Pb[d]	575	75	4000	235–365	10–9000	—	30–96270	—
Sb	33	4	114	0.5–51	2–470	3.5–25	0.5–171	$8 \cdot 10^{-4}$
Se	0.6	0.3	1.6	0.3–3.5	0.01–127	1.2–2.2	0.2–30	—
Sc	1.6	0.5	5.7	0.1–0.8	—	0.1–3.1	—	0.00016
Sm	0.1	0.03	0.3	0.02–0.4	—	0.02–0.8	—	$8 \cdot 810^{-5}$
Ti	60	40	85	—	—	36–180	—	0.10
Th	0.03	0.01	0.08	0.05–0.095	—	0.02–0.42	—	0.00014
V	27	0.3	61	(18)	11–73	7–725	0.4–1460	0.0013
Zn	196	30	357	80–200	160–8340	58–741	15–8328	0.033
W	1	0.08	4	(<0.4)	—	(0.65)	—	
	—							

Note: Values in parenthesis are simple data. Concentrations are in ng m^{-3}.

[a] Values obtained from Reference 14. With permission.
[b] Values obtained from Reference 58. With permission.
[c] Values obtained from Reference 1. With permission.
[d] Values obtained from Reference 59. With permission.
[e] Pb determined in this work by AAS. Values in parenthesis are single data.

the daily variations. For all other elements analyzed, the concentrations found in the two different periods were very similar and always inside the ranges obtained during the first series of analyses. Table 22.11 gives the average of the concentrations (in ng m^{-3}) of 30 elements and the corresponding minimum and maximum values found. The same table also reports some of the literature range values of different cities in Europe[1,14,58] and in the U.S. and the data of the South Pole obtained by Zoller et al.[59] The ranges found were typical of a large industrialized city. The highest concentration

TABLE 22.12
Total Concentrations and Granulometric Distribution of Some TE in Rural Areas in Northern Italy (Ispra)

Element	Total Mean Conc.	Range	Granulometric Distribution (stage cut-off in μm) (0.08)	(0.17)	(0.35)	(0.7)	(1.4)	(2.8)	(5.7)	(11)
As	1.5	0.5–3	0.022	0.14	0.61	0.38	0.25	0.07	0.04	<0.01
Br	24.7	11–55	0.48	2.64	8.80	6.77	5.50	2.10	0.82	0.23
Co	0.73	0.1–1.2	0.05	<0.05	0.16	0.11	0.12	0.12	0.13	<0.05
Cr	8.3	0.9–12	<0.1	<0.1	1.1	1.54	1.35	1.80	1.96	0.57
Fe	574	150–950	<10	<10	<10	54.7	117	180	172	50
Sb	4.7	0.1–8	<0.1	0.32	1.30	1.15	0.78	0.64	0.47	0.10
Sc	0.038	0.02–0.1	<0.005	0.005	<0.005	<0.005	0.009	0.01	0.013	0.006
Se	1.10	0.5–1.8	<0.05	<0.05	0.51	0.55	<0.05	<0.05	<0.05	<0.05
Zn	140	20–202	0.28	0.40	15.4	30.0	66.8	27.0	17.0	2.76

Note: Preliminary results obtained from three independent samplings carried out in December 1995. Concentrations given in ng m^{-3}.

values were obtained, in both periods, during similar weather conditions characterized by fog and thermal inversion; these concentrations dropped to much lower values (minimum of the range values) as soon the weather conditions changed. Appreciable daily variations were found only in dependence on the weather parameters, while small differences were registered during working hours, nights, and weekend days. In conclusion, the results showed that, as in any large city, the main pollution sources could be identified as automotive traffic and domestic heating. Nevertheless, especially in Milan, the pollution peaks were strictly dependent on weather conditions, as noted in the Figure 22.8 where the higher values of pollutants were found on the Saturday, February 14, and Sunday, February 15, when commuting traffic was absent. For some trace elements, the results found in atmospheric particulate of downtown Milan can be compared with those of the surrounding rural non-industrialized areas. For this purpose, air particulate measurements were carried out during the same winter period (December 6–10, 1995) in a region 80 km north of Milan near Lake Maggiore (Ispra). During that sampling, a preliminary granulometric distribution of TE was also carried out. The results are shown in Table 22.12, which gives the values of the total concentrations found, as well as the granulometric distributions (expressed in ng m^{-3}). Except for Zn, all values are in the ranges of the minimum values found in Milan air, in agreement with the much lower pollution present in that area. It may be interesting to note, that these preliminary results showed a significant enrichment in the finest respirable fraction particulate (0.25–0.35 μm) for those "volatile" elements such as As, Br, Sb, and Se. Further studies have confirmed this trend.[60,61]

CONCLUSIONS

The studies described above are only a small demonstration of the potential offered by NAA in research fields related to trace elements and atmospheric pollution. Wherever it may be necessary to obtain wide spectral information on trace elements and their concentrations in processes related to air monitoring, the technique plays an irreplaceable role. As previously shown, more than 30 elements can routinely be followed with, for many of them, detection limits superior to other trace analytical techniques. Furthermore, the multi-elemental analyses can be performed on very small samples, typical of air particulate collection. At present, no other analytical technique can offer the same suitable performance in carrying out, for example, multi-element analysis at ppb levels on a few micrograms of dust present in one stage of a multicascade impactor. The examples reported herein have been chosen to underline a typical approach to atmospheric pollution problems, starting with the study of potential emission sources such as the waste incineration and continuing with the monitoring of the air fallout and atmospheric particulate. Unfortunately, NAA cannot give information on the chemical form nor the oxidation state of the trace metal object of study, but is fundamental in furnishing all the qualitative and quantitative information — which always represents the initial most important point in this kind of research.

ACKNOWLEDGMENTS

I would like to thank Dr. Ivo Allegrini, Director of the Istituto sull'Inqinamento Atmosferico of the CNR (Rome), and Dr. Antonio Rolla of the Stazione Sperimentale per i Combustibili (Milan) for their collaboration. Part of this work was financially supported, with the help of Dr. Alfredo Liberatori, by the Stratetic Project PS Aree Metropolitane ed Ambiente of the Italian National Research Council (CNR). A special thanks to Pierangelo Borroni, who helped in drawing the tables and figures.

REFERENCES

1. Schroeder, W., Dobson, M., Kane, D.M., and Johnson, N.D., Toxic trace elements associated with airborne particulate matter: a review, *J. Air Poll. Control Assoc. (JAPCA),* 37, 1267, 1987.
2. WHO/UNEP, 1992 Urban Air Pollution in Megacities of the World, World Health Organization, United Nations Environmental Programme, Blackwell, Oxford, 1992, 1900.
3. Fox, D.L., Air pollution, *Anal. Chem.*, 59, 280R, 1987.
4. Melissa, H.E., Ed., *Analysis of Airborne Particles by Physical Methods,* CRC Press, Boca Raton, FL, 1978.
5. Oikawa, K., *Trace Analysis of Atmos. Samples*, Halsted Press Book, John Wiley & Sons, New York, 1977.
6. Stern, C., Ed., *Air Pollution*, 3d ed., Monitoring and Surveillance of air Pollution, Academic Press, New York, 1976.
7. deBruin, M., vanWijk, P.M., vanAssema, R., and deRoos, C., The Use of Multielement Concentration Datasets Obtained by INAA in the Identification of Sources of Environmental Pollutants. *J. Radioanal. Nucl. Chem., Articles,* 112/1, 199, 1987.
8. Landsberger, S., Zhang, P., and Chatt, A., Analysis of the Arctic aerosol a ten year period using various neutron activation analysis methods, *J. Radioanal. Nucl. Chem.,* 217, 11, 1997.
9. Wang, C.F., Chang, E.E., Chiang, P.C., and Aras, N.K., Analytical Procedures on Multielement Determinations of Airborne Particles for receptors Model Use, *Analyst*, 120, 2521, 1995.
10. Linton, R.W., Loh, A., Natusch, D.F., Evans, C.A., and Williams, P., Surface predominance of trace elements in airborne particulate, *Science*, 191, 852, 1976.
11. Clement, R.E., Yang, P.W., and Koester, C.J., Environmental analysis – (review), *Anal. Chem.*, 69, 251R, 1997, 1900.
12. De Soete, D., Gijbels, D., and Hoste, R., *Neutron Activation Analysis,* Wiley Interscience London, P.J. Elving-I.M. Kolthoff, Ed., 1972.
13. Dams, R., Application of Mulielemental Neutron Activation Analysis in Environmental Research, *Instrumentelle Multyelementanalyse,* Sansoni, B., Ed. VCH-Verlagsgesellschaft Weiheim, West Germany, 1985, 55.
14. Alian, A. and Sansoni, B., A review on activation analysis of air particulate matter, *J. Radioanal. Nucl. Chem., Articles*, 89/1, 191, 1985.
15. Alian, A. and Sansoni, B., Activation analysis in air particulate matter, *Activation Analysis,* Zeev B. Alfassi, Ed., 1990, 503.
16. Kolesov, G.M., Neutron activation analysis of environmental materials, *Analyst*, 120, 1457, 1995.

17. Kronoborg, D. and Steiness, E., A routine procedure for multielement analysis of atmospheric particulates by INAA, *J. Radiochem. Radioanal. Lett*., 21, 379, 1975.
18. Cercasov, V., Pantelica, A., Salgean, M., and Schreiber, H., Application of INAA and XRFA in a comparative environmental study, *J. Radioanal. Nucl. Chem.,* 204, 173, 1996.
19. Bowen, H.J.M., Activation analysis, *Radiochemical Methods in Analysis*, Coomber, D.I., Ed., Plenum Press, New York, London, 1975.
20. Alfassi, Z.B., Activation with nuclear reactors, *Activation Analysis*, Vol. I, Alfassi, Z.B., Ed., CRC Press, Vol Boca Raton, FL, 1990.
21. Alian, A., Born, H.J., and Kim, J.I., Thermal and epithermal neutron activation analysis using the monostandard method, *J. Radioanal. Chem.,* 15, 535, 1973.
22. Aras, N.K., Zoller, W.H., Gordon, G.E., and Lutz, G.J., IPAA of atmospheric particulate material, *Anal. Chem*., 45, 1461, 1973.
23. Biegalski, S.R. and Landsberger, S., Improved detection limits for trace elements on aerosol filters using compton suppression counting and epithermal irradiation techniques, *J. Radioanal. Nucl. Chem, Articles*, 192/2, 195, 1995.
24. Heindryckx, R. and Dams, R., Evaluation of three different procedures for neutron activation analysis of elements in atmospheric aerosols using short lived isotopes, *Radiochem. Radioanal. Lett*., 16/4, 209, 1974.
25. Spyrou, N.M., Cyclic activation analysis — a review, *J. Radioanal. Nucl. Chem*., 112, 277, 1981.
26. Guinn, V.P., Cyclic nuclear activation analysis, *Radiochem. Radioanal. Lett*., 44, 133, 1980.
27. Girardi, F. and Pietra, R., Multielement and automated radiochemical separation procedures for activation analysis, *At. Energy Rev.,* 14, 521, 1976.
28. Gallorini, M., Greenberg, R.R., and Gills, T.E., Simultaneous determination of As, Sb, Cd, Cr, Cu and Se in environmental materials by radiochemical neutron activation analysis, *Anal. Chem*., 50, 1479, 1978.
29. Greenberg, R.R., Gallorini, M., and Gills, T.E., Cadmium analysis by radiochemical neutron activation analysis, *Environm. Health Perspect.,* 28, 1, 1979.
30. Gallorini, M. and Gills, T.E., Modified procedure for the determination of Hg and noble metals in geological environmental related samples, *Trans. Am. Nucl. Soc.,* (TANSAO 32-832), 32, 177, 1979.
31. Pietra, R., Sabbioni, E., Gallorini, M., and Orvini, E., Environmental, toxicological and biomedical research on trace metals: radiochemical separations for neutron activation analysis, *J. Radioanal. Nucl. Chem.,* 102, 69, 1986.
32. Schmitt, B.F., Segebade, C.H.R., and Fusban, H.U., Waste incineration ash a versatile environmental reference material, *J. Radioanal. Chem*, 60/1, 99, 1980.
33. Morselli, L., Zappoli, S., Gallorini, M., and Rizzio, E., Characterization of trace elements in dry deposition by neutron activation analysis, *Analyst*, 113, 1575, 1988.
34. Spurny, K. and Fiser, J., Staub Renheit, *J. Staub,* 30, 249, 1970.
35. Bogen, J., Trace elements in atmospheric aerosol in the Heidelberg area, measured by INAA, *Atmos. Environ.,* 7, 1117, 1973.
36. Ali, A.E. and Bacso, J., Investigation of different types of filters for atmospheric trace elements analysis by three analytical techniques, *J. Radioanal. Nucl. Chem.,* 209, 147, 1996.
37. Dams, R., Nuclear activation techniques for the determination of trace elements in atmosphericaereosols, particulates and sludge samples, *Pure & Appl. Chem.,* 64, 991, 1992.
38. Zeisler, R., Haselberger, N., Makarewicz, M., Ogris, R., Parr, R.M., Stone, S.F., Walkovic, O., Valkovic, V., and Wehrstein, E., *J. Radioanal. Nucl. Chem.,* 217, 5, 1997.
39. Gallorini, M., Orvini, E., Rolla, A., and Burdisso, M., Destructive neutron activation analysis of toxic elements in suspended materials released from refuse incinerators, *Analyst*, 106, 328, 1981.
40. Greenberg, R.R., Zoller, W.H., and Gordon, G.E., The contribution of refuse incineration to urban aerosols, *Proc. 4th Joint Conf. on Sensing of Environ. Poll. - 1977,* Am. Nucl. Soc. Washington D.C., 1978.
41. Law, S.L. and Greenberg, R.R., Characterization of municipal incinerator effluents, *Progress in Anal. Chem*., I.L. Simmons and G.W. Ewing Eds., Vol. 8, Plenum, New York, 1976, 55.
42. Morselli, L., Zappoli, S., and Militerno, S., The presence and distribution of heavy metals in municipal solid waste incinerators, *Toxicol. Environ. Chem.,* 37, 139, 1993.

43. Greenberg, R.R., Gordon, C.E., Zoller, W.H., Jacko, R.B., Neuendorf, D.W., and Yost, K.J., Composition of particles emitted from the Nicosia municipal incinerator, *Environ. Sci. Technol.,* 12, 1329, 1978.
44. Wadge, A., Hutton, M., and Peterson, P.J., The concentration and the particles size relationships of selected trace elements in fly ashes from U.K. coal-fired power plants, *Sci. of the Total Environm.*, 54, 13, 1986.
45. Ragaini, R.C. and Ondov, J.M., Trace elements emissions from western U.S. coal-fired power plants, *J. Radioanal. Chem.,* 37, 679, 1977.
46. Kist, An. A. and Bertman, E.B., Atmospheric trace elements collected with the use of passive sorption sampling, *J. Radioanal. Nucl. Chem., Articles,* 192/2, 249, 1995.
47. Ondov, J.M., Divita, F., Jr., and Suarez, A., Size-spectra and growth of particles bearing As, Se, Sb and Zn in Washington D.C. Aerosol by INAA, *J. Radioanal. Nucl. Chem., Articles,* 192, 215, 1995.
48. Schramel, P.S., Samsahl, K., and Pavlu, J., Determination of 12 selected microelements in air particles by neutron activation analysis, *J. Radioanal. Chem.,* 19, 329, 1974.
49. Salmon, L., Atkins, D.H.F., Fisher, E.M.R., and Law, D.V., Retrospective analysis of air samples in the U.K. 1957–1974, *J. Radioanal. Chem.,* 37, 867, 1977.
50. Olmez, I. and Aras, N.K., Trace elements in the atmosphere determined by nuclear activation analysis and their interpretation, *J. Radioanal. Chem.,* 37, 671, 1977.
51. Chuang, L.S., Kwong, L.S., and Yeh, S.J., Nondestructive multielement determination of air particulates in Hong Kong, *J. Radioanal. Chem.,* 49, 103, 1979.
52. Bahal, B.M. and Pepelnik, R., Multielemental analysis of a Milanese air-dust samples by 14 MeV neutron activation, *J. Radioanal. Nucl. Chem., Articles,* 97, 359, 1986.
53. Chutke, N.L., Ambulkar, M.N., Garg, A.N., and Aggarwal, A.L., Instrumental neutron activation analysis of ambient air dust particulate from metropolitan cities in India, *Environ. Poll.*, 85, 67, 1994.
54. Chung, Y.S., Chung, Y.J., Jeong, E.S., and Cho, S.Y., Study on air pollution monitoring in Korea using instrumental neutron activation analysis, *J. Radioanal. and Nucl. Chem.,* 217, 83, 1997.
55. Querol, X., Alastuey, A., Lopez-Soler, A., Boix, A., Sanfeliu, T., Martynov, V.V., Piven, P.I., Kabina, L.P., and Soushov, L.P., Trace elements contents in atmospheric suspended particles: inferences from instrumental neutron activation analysis, *Fresenius J. Anal. Chem.,* 357, 934, 1997.
56. Gallorini, M., Trace elements monitoring in atmospheric pollution processes, *Microchemical J.,* 51, 127, 1995
57. Personal communication, Data obtained from the Istituto sull'Inquinamento Atmosferico of the CNR (Natl. Research Council), Roma (Italy).
58. Krivan, V. and Egger, V., Multielementalnalyse von Schwebstauben der Stadt Ulm und Vergleich der Luftbelastung mit anderen Regionen, *Fresnius Z. Anal. Chem.,* 325, 41, 1986.
59. Zoller, W.H., Gladney, W.H., and Duce, R.A., Atmospheric concentration and sources of trace metals at the south pole, *Science,* 183, 198, 1974.
60. Gallorini, M., Borroni, P.A., Bondardi, M., and Rolla, A., Trace elements in the atmospheric particulate of Milan and suburban areas: a study carried out by INNA, *J. Radioanal. Nucl. Chem.*, 235, 241, 1998.
61. Rizzio, E., Giaveri, G., Arginelli, D., Gini, L., Profumo, A., and Gallorini, M., Trace elements in total content and particle sizes distribution in the air particulate matter of a rural-residential area in north Italy investigated by INAA, *Sci. Total Environ.*, 226, 47, 1999.

23 Urban and Rural Organic Fine Aerosols: Components Source Reconciliation Using an Organic Geochemical Approach

Alexandra Gogou and Euripides G. Stephanou

CONTENTS

INTRODUCTION

The chemical composition of fine particulate matter in urban and rural atmospheres, is controlled to a significant extent by emissions from terrestrial, marine, and various anthropogenic sources. An important portion of aeolian particulates consists of organic anthropogenic and biogenic compounds related to several emissions. Long-chain *n*-alkanes, *n*-alkanols, *n*-alkanals, 2-alkanones, *n*-alkanoic acids, *n*-alkanoic acid salts, α,ω-dicarboxylic acids (of higher plant origin and/or photo-oxidation products of anthropogenic cyclic olefins and biogenic unsaturated fatty acids), polycyclic aromatic hydrocarbons (PAHs), and their oxygenated and nitrated derivatives have been detected in urban and rural, as well as, in remote marine aerosols.[1-16]

The study of the chemical composition of air particulate matter is of major importance because anthropogenic compounds, and some of their atmospheric transformation products, often represent a high potential hazard to human health,[14] but also because the atmosphere constitutes a conveyor belt for naturally emitted compounds and anthropogenic chemicals to the ocean.[2,17-23]

As petrogenic hydrocarbons, products of pyrolytic processes (such as PAHs), and various polar organic compounds constitute a very important fraction of organic aerosols, significant effort has been made to reconcile their presence with specific emission sources. On the other hand, particular

0-87371-829-1/00/$0.00+$.50

effort has recently been given[24-30] to the characterization of organic matter in major urban sources of emission of fine carbonaceous aerosols.

In order to study the environmental presence and occurrence of anthropogenically and naturally emitted compounds, organic geochemical parameters have been quite successfully used. Among these parameters, one can consider:

1. Specific organic compounds found in the geological environment, which are called *molecular markers:* Molecular markers such as hopanes and steranes[31] have been extensively used, not only for oil-spill correlations,[32-34] but also in evaluating petrogenic inputs in tropospheric aerosols.[7,8,35-37]
2. The *Carbon Preference Index* (CPI),[38] which is a single value to express the ratio of odd-carbon-numbered to even-carbon-numbered *n*-alkanes in a given sample: CPI values greater than unity indicate the relative contributions of *n*-alkanes from natural sources, while CPI values equal to or lower than unity indicate significant petrogenic hydocarbon input. The CPI has been used to differentiate biogenic from petrogenic or anthropogenic *n*-alkanes in organic aerosols.[4-7,10,11,13,36,39]
3. The "signature" of *natural wax n-alkanes* (from higher plants), introduced by Simoneit.[7] Since it is known that petrogenic *n*-alkanes usually have a CPI = 1, the concentrations of wax *n*-alkanes were calculated by subtraction of the average of the next higher and lower even-carbon-numbered homolog concentrations.
4. Although CPI has probably received the greatest attention among other formulations, its use is been often criticized because of limitations due to computantional artifacts rather than to peculiarities of natural abundances in homologous series. An equation has been derived, by Scalan and Smith,[40] to compute the ratio of the relative concentrations of homologs that contain an odd number of carbon atoms to those that contain an even number, called *odd-even predominance* (OEP). Plots of OEP vs. *n*-alkane chain length varied among unrelated crude oils or rock extracts, but were similar for related samples. To our knowledge, this parameter has not been used for components source reconciliation of organic aerosols.
5. The *Unresolved Complex Mixture* (UCM) of branched and cyclic hydrocarbons: This diagnostic parameter is considered an indication of petrogenic hydrocarbon inputs related to unburned petroleum emissions from vehicular traffic.[4-7,10,11,13,39]
6. *Diagnostic concentration ratios of PAHs*[16,41-45] have been used to reconcile their presence in air particles, with potential emission sources.[13,14,19,22,44,45]
   a. The *sum of concentrations of nine major non-alkylated compounds* (fluoranthene, pyrene, benz[a]anthracene, chrysene, benzofluoranthenes, benzo[a]pyrene, benzo[e]pyrene, indeno[cd]pyrene, and benzo[ghi]perylene) expressed as CPAHs, has often been used as a characteristic value for PAHs produced by combustion.[19,44] The ratio of CPAHs to the total concentration of PAHs (CPAHs:TPAHs) is denotative of the extent of pyrogenic to petrogenic PAHs.
   b. The *ratio of methyl-phenanthrenes to phenanthrene (MP:P),* has been used for source identification of PAHs.[14,19,44] MP:P ratios between 1 and 8 represent evidence for enhanced mobile sources or unburned fossil organic material contribution, while ratios below 1 are typical for emissions from stationary combustion sources where fuel is burning at higher temperatures.[44]
   c. The *Methylphenthrene Index* (MPI1) has been used as a hydrocarbon internal maturity parameter.[43] The influence of temperature on the yields of phenanthrene and its methylated derivatives was evaluated, and the corresponding MPI1 values were calculated (using the data provided by Adams et al.[55]). MPI1 decreases as combustion temperature, during cellulose pyrolysis, increases (MPI1(600°C) 0.64; MPI1(700°C) 0.67; MPI1(800°C) 0.60; MPI1(900°C) 0.08; MPI1(1000°C) 0.04). MPI1 has also been used, as a mobile or stationary source diagnostic parameter.

d. Benzo[a]anthracene to [benzo[a]anthracene+chrysene,tryphenylene] (BA:BA+CT), benzo[*e*]pyrene to [benzo[*e*]pyrene+benzo[*a*]pyrene] (BeP:BeP+BaP), fluoranthene to [fluoranthene + pyrene] (Fl:Fl+Py) and indeno[1,2,3-*cd*]pyrene to (benzo[*ghi*]perylene+indeno[*1,2,3-cd*]pyrene) (IP:IP+BgP) ratios are usually used for source reconciliation.[41,42] These ratios should be cautiously evaluated if one wants to assess the different sources of PAHs present in aeolian particulates, by comparing them with the corresponding ratios calculated from well-characterized emission sources.

CPI and OEP formulations can also be used to reconcile polar organic compounds with their emission sources[7,24-30,36] or origins.[40]

Analytical chemical techniques, with subsequent use of the above-mentioned organic geochemical parameters, have been reported for the determination of molecular markers in airborne particles.[4,15,16] Using these analytical methods, the aeolian particles total organic extract is derivatized (methylated or/and sililated) to produce the GC-amenable, less polar derivatives of fatty acids and alkanols, and then the compounds of the derivatized extract, are separated by thin layer chromatography,[4] by liquid column chromatography,[15] or directly by high-resolution gas chromatography.[16] Taking into consideration that the concentrations of various polar compounds often differ greatly from one other, problems are expected if, after fractionation, they are found in the same fraction. Therefore, we used a method[11] in which fatty acids are separated from neutral compounds using a liquid chromatographic column containing specially prepared silica gel, and then neutral compounds are fractionationated by silica gel flash chromatography into very distinct nonpolar, semi-polar and polar compound fractions. The fractionated organic compounds were analyzed by high-resolution gas chromatography (HRGC) and gas chromatography/mass spectrometry (GC/MS).

In this chapter, the application of an organic geochemical approach to study the molecular compositions of the nonpolar and polar fractions of organic aerosols is reported. Samples were taken from a urban site (city of Heraclion) and a rural arid ("remote") site (Finokalia), both situated on the northern coast of the Island of Crete (Greece) in the eastern Mediterranean. Urban aerosol samples were taken outdoors, as well as indoors, for comparison purposes. Organic geochemical parameters, such as CPI, wax *n*-alkanes signature, and OEP were used for compound source reconciliation. PAHs present in the examined aerosol samples were assigned to specific sources using diagnostic ratios and parameters. CPI and OEP were also utilized to determine the origin of polar compounds, such as alkanols, alkanals, alkanones, and alkanoic acids.

EXPERIMENTAL METHODS AND CALCULATIONS

SAMPLING OF AIR PARTICLES

Particulate material was collected on a pre-extracted 20 × 25 cm quartz filter, having a collection efficiency higher than 99% for particles with radius larger than 0.3 μm at the 90 m^3 h^{-1} flow rate used. Filters were mounted in a high-volume air sampling system (Model GMWL-2000, General Metals Works, Ohio 45002, U.S.). Urban samples were collected for 24 to 36 h (2500 to 3000 m^3 air sampled) on a 15-m high building situated in the center of the city of Heraclion, Island of Crete (Greece). Rural samples were collected on a 20-m high tower, specifically constructed for this project in an arid coastal location in the northern part of the same island. The samples were stored frozen (–30°C) in pre-cleaned glass flasks sealed with teflon tape and covered with aluminium foil.

MATERIALS

All solvents, "Pestanal" grade, were purchased from Riedel-de-Haen (Seelze, Germany). Standard compounds were purchased from Ehrenstorfer (Augsbourg, Germany). Silica gel (70–230 mesh and 230–400 mesh) was from Merck (Darmstadt, Germany). Soxhlet cartridges were from Schleicher and Schuell (Dassel, Germany). Glass fiber filters and quartz fiber filters were delivered by Whatman (Maidstone, U.K.).

All materials used (silica gel, glass and cotton wool, paper filters, anhydrous sodium sulfate, etc.) were extracted in a Soxhlet overnight, and kept dry until use. Quartz fiber filters were backed overnight at 550°C. The quartz filters were then kept in a dedicated clean glass container, with silica gel, to avoid contamination and humidity.

Fractionation, Derivatization, and Identification

The detailed description of analytical procedure used for extraction, separation, and analysis of the main lipid fractions is given in Reference 11. A general scheme of the analytical procedure is given in Figure 23.1.

Calculations

1. The wax *n*-alkanes concentration (WNA) was calculated according to the following[7]:

$$\text{wax } C_n = C_n - 0.5\left[\left(C_{n+1}\right) + \left(C_{n-1}\right)\right] \tag{23.1}$$

   Negative values of C_n were taken as zero.
2. The odd carbon preference indices (CPI) for *n*-alkanes, *n*-alkanols, and *n*-alkanoic acids were calulated according to Reference 36 as follows:
   a. Whole range for *n*-alkanes:

$$CPI_1 = \Sigma C_{13}\text{–}C_{35} / \Sigma C_{12}\text{–}C_{34} \tag{23.2}$$

   b. Split range for bacterial, algal *n*-alkanes:

$$CPI_2 = \Sigma C_{11}\text{–}C_{25} / \Sigma C_{10}\text{–}C_{24} \tag{23.3}$$

   c. Split range for higher plant wax *n*-alkanes:

$$CPI_3 = \Sigma C_{27}\text{–}C_{35} / \Sigma C_{26}\text{–}C_{34} \tag{23.4}$$

   d. CPI for *n*-alkanols and *n*-alkanoic acids:

$$CPI = \Sigma C_{12}\text{–}C_{34} / \Sigma C_{13}\text{–}C_{35} \tag{23.5}$$

3. Running OEP ratios were computed from the following equation[40]:

$$\text{OEP for } C_n = \left[(\text{Cn-2} + 6\text{Cn} + \text{Cn+2})/(4\text{Cn-1} + 4\text{Cn+1})\right]^{(-1)^{(n-1)}} \tag{23.6}$$

   OEP values are plotted vs. the carbon chain length to construct the OEP curves.
4. The methylphenanthrene index (MPI) is calculated as follows[43]:

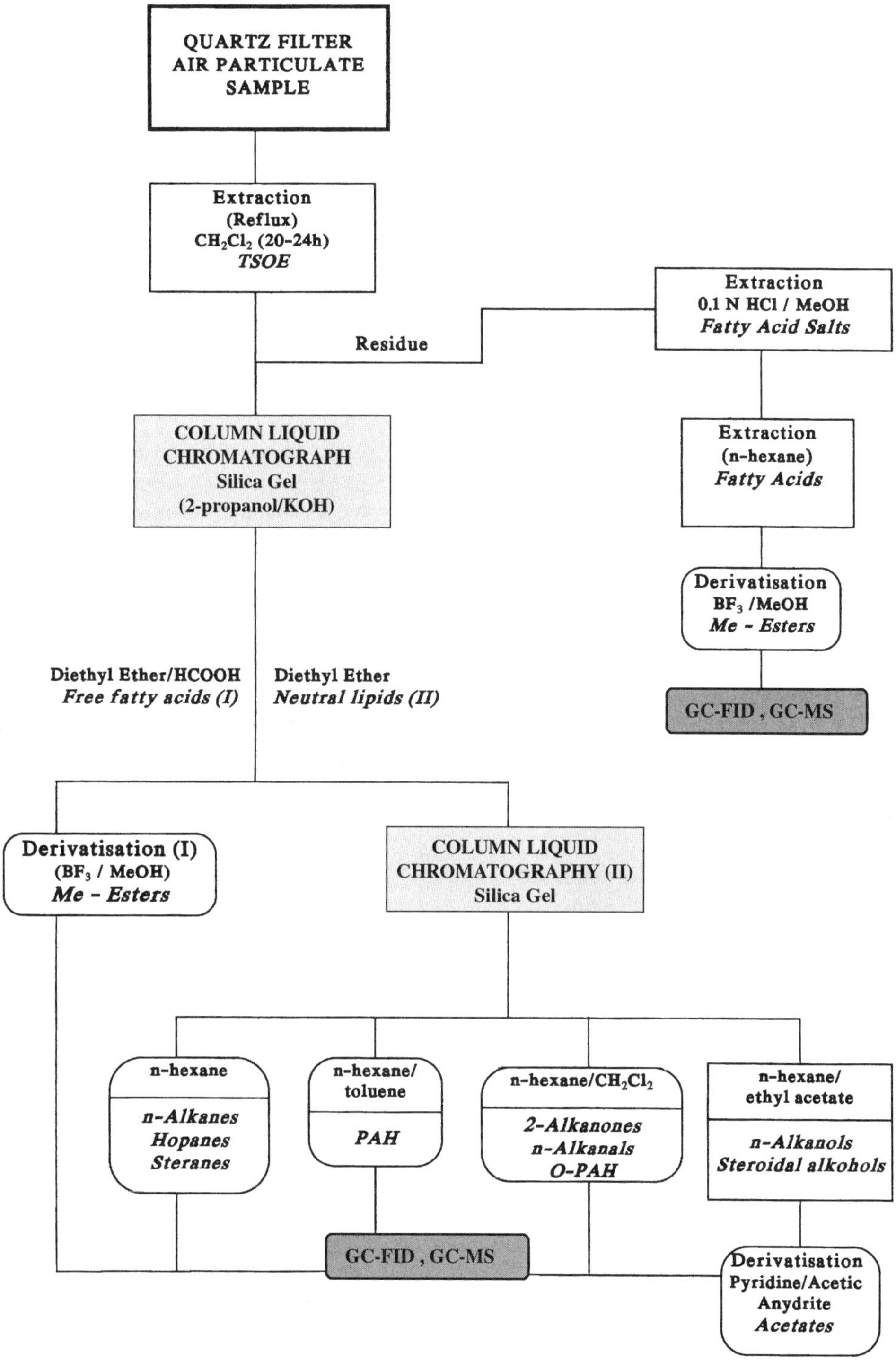

FIGURE 23.1 General scheme of chemical analysis of organic aerosols.

$$\text{MPI} = \left[1.5(2\text{-MP} + 3\text{-MP})/(\text{P} + 1\text{-MP} + 9\text{-MP})\right], \tag{23.7}$$

where, MP = corresponding methylphenathrene concentration, and P = phenanthrene concentration.

GENERAL AEROSOL CHARACTERISTICS

General samples characteristics, such as sampling time, wind regime, total suspended particulates (TSP), and total solvent organic extract (methylene chloride, TSOE) are given in Table 23.1. Higher TSP concentration values (average 89.4 μg m^{-3}, with standard deviation (S.D.) 3.6 μg m^{-3}) occurred in the urban samples, compared to that in rural samples (average 23.5 μg m^{-3}, S.D. 8.4 μg m^{-3}). The highest TSP values occurred in urban indoor aerosol samples. The TSP values in urban outdoor samples remained stable under different sampling wind regimes and seasons. This was not the case for rural samples, where the sampling season and the wind regime (to a lesser degree) appear to have an influence on this parameter. The total solvent organic extract (TSOE) represented a low amount of TSP (6.8 to 15.9%) in urban samples, and was also relatively stable with respect to different wind regimes. The lowest TSOE concentration value was obtained from a sample collected during strong rain (ua11). Indoor aerosol samples provided the highest proportion of TSOE with respect to TSP (23.92–66.67%). The TSOE concentration levels in rural samples ranged from 0.73 to 8.9 μg m^{-3} (Table 23.1). Levels in the same range were measured in rural and remote areas in the Western United States[4] and over the Amazon.[7] There is a seasonal trend in the aerosol proportion of solvent extractable organic material in rural samples. Samples collected in autumn and early spring contain more extractable organic material (6.40 to 23.30%, Table 23.1) than those collected during late spring (3.80 to 4.20%, Table 23.1). Calculation of backward trajectories of air masses provided a better interpretation of the above results.[46]

TABLE 23.1
General Characteristics of Urban and Rural Aerosols in the Eastern Mediterranean

Sample	Date	Wind Direction	TSP [μm m^{-3}]	TSOE [μm m^{-3}]	TSOE:TSP [%]
Urban Outdoor					
ua4	4/25/1992	N	88.33	12.60	14.28
ua5	5/2/1992	SW	92.82	14.80	15.94
ua8	8/28/1992	NW, W	84.82	13.20	15.61
ua11	11/30/1992	W, SW	91.73	6.20	6.80
Urban indoor					
ia1	11/20/1993		155.1	37.1	23.92
ia2	11/23/1993		774.4	516.3	66.67
Rural outdoor					
ra611	11/6/1993	S	27.77	2.55	9.20
ra911	11/9/1993	S	35.57	2.27	6.40
ra293	3/29/1994	N	37.03	8.62	23.30
ra303	3/30/1994	N	22.50	2.55	11.30
ra313	3/31/1994	NW	22.15	2.40	11.00
ra65	5/6/1994	N, NW	19.24	0.80	4.20
ra75	5/7/1994	W, NW	19.32	0.50	4.00
ra85	5/8/1994	W, NW	19.29	0.77	4.00
ra115	5/11/1994	NW	15.68	0.59	3.80

Presented in Table 23.2 are the values of the concentration ranges, concentration diagnostic ratios, and other parameters (CPI, homolog with the highest concentration (C_n*max*), wax *n*-alkanes (WNA), etc.) of all compound classes studied in the urban and rural aerosol samples.

NONPOLAR LIPIDS: ALIPHATIC HYDROCARBON FRACTION

Characteristic gas chromatograms of the aliphatic fraction of urban and rural samples are shown in Figure 23.2. All chromatograms were dominated by *n*-alkanes (NA). The determination of the homologous *n*-alkanes in the aliphatic fraction of the aerosol samples allowed the determination of their relative distribution, C_{max}, UCM, CPI, WNA content, and OEP curves.

In outdoor urban samples, the concentrations varied from 75.87 to 316.54 ng m^{-3} (Table 23.2). These values are in the same range as those measured in Barcelona,[13,48] but are higher than those measured in different locations in Los Angeles.[49] Late spring and summer samples contained higher *n*-alkane concentrations, than those collected in late autumn and early spring. The highest *n*-alkanes concentrations were determined in indoor aerosols (371.6 to 2789.4 ng m^{-3}). The concentrations measured in rural aerosols varied from 7.06 to 24.33 ng m^{-3}. These concentrations are of the same order as those measured over the western Mediterranean by Saliot and Sicre,[3,5,6,47] higher than those measured by Mazurek in the arid southeastern United States,[8] and much higher than those measured by Gagosian in the North Pacific.[2,20]

Characteristic relative homolog distribution diagrams for the collected aerosol samples are given in Figure 23.3. Homolog distributions in urban outdoor samples range from C_{11} to C_{34} (Figure 23.3A). C_{29}, C_{31}, and C_{33} are diagnostic components of land-plant waxes, while C_{max} at lower carbon numbers may indicate major input from petrogenic or microbial sources.[34]

The sampling season had an apparent influence on the distribution of *n*-alkanes in urban aerosol samples. *n*-Alkanes in early and late spring and late autumn samples (ua4, ua5, ua11) maximized at C_{25}, while the summer sample (ua8) maximized at C_{29} (Figure 23.3A). Homolog with more than 27 carbons in the sample suggest predominant vascular plant wax alkane inputs. Urban aerosol samples exhibited low CPI_1 values (1.29 to 1.64, Table 23.2). These CPI values indicate some incorporation of biogenic material, but they reflect a clear petrogenic input.

Urban samples chromatograms (Figure 23.2A) exhibited an intense envelope of unresolved complex mixture (UCM) of branched and cyclic hydrocarbons, in contrast to those of rural samples, which exhibited a very weak or even absent UCM envelope (Figure 23.2B). The ratio of UCM concentration to the corresponding resolvable *n*-alkanes concentration (UCM/NA, Table 23.2) is a criterion of petrogenic input assessment.[7] High UCM:NA values (8.63–14.02) were obtained in all samples, indicating a significant petroliferous input. The ratios measured in these samples are lower than those measured in Barcelona,[13,48] but they also express an important contamination by petroleum products.

Most oils contain isoprenoid hydrocarbons like pristane (Pr) and phytane (Ph) and some molecular markers like hopanes and steranes.[29,50] Pristane and phytane were present in all analyzed urban samples. In addition, hopanes and steranes were also present in all urban samples (Figure 23.4). Pentacyclic triterpenes, characterized structurally by the hopane skeleton,[51] are ubiquitous biomarkers.[51] Naturally occurring hopanes generally have a 17β(H), 21β(H) stereochemistry and occur only with R configuration at the C_{22} position,[49] while hopanes related to petroleum have the thermodynamically more stable 17α(H), 21β(H) configuration and also give rise to the R and S epimers at the C_{22} position.[31,50] Figure 23.4A shows the *m/z* 191 mass chromatogram of the aliphatic fraction of an urban aerosol extract investigated in this study. The chemical structures of the compounds (indicated with arabic numbers in Figures 23.4A and B) were determined using their mass spectra and as well as their elution order. The presence of these hopanes has also been confirmed in exhaust extracts from non-catalyst, catalyst-equipped automobiles and heavy-duty diesel trucks.[25,35] Similar *m/z* 191 mass chromatograms have been determined in aerosol extracts reported in other studies.[7,8,35,36] In all urban aerosol extracts (Figure 23.4A), analyzed in this study,

TABLE 23.2
Homolog Ranges (C_m–C_n) Concentrations, Homolog with the Maximum Concentration ($C_{n\,max}$), Carbon Preference Indices (CPI), and Other Diagnostic Parameters of Organic Compounds Determined in Fine Organic Aerosols

Compound Classes		Urban Aerosols Outdoor	Urban Aerosols Indoor	Rural Aerosols Outdoor
I.	Aiphatics			
	(Cm-Cn); Cn max	(11–34); 25,27,29,31	(15–34); 31	(11–40); 29,31
	Concentration [ng m^{-3}]	75.9–316.5	371.6–2789.6	7.1–24.3
	CPI1; CPI2	1.3–1.6; 1.2–1.6	2.9–9.3; 1.3–1.4	1.6–3.3; 1.2–23.0
	UCM:NA	8.6–14.0	1.9–2.1	0.0–5.8
	WNA[%]	13.9–25.0	54.2–68 9	29.9–50.32
II.	PAHs			
	Concentration [ng m^{-3}]	21.4–59.0	21.0–1545.9	0.2–2.0
	CPAHs/TPAHs	0.43 (S.D. 0.04)	0.70 (S.D. 0.07)	0.70 (S.D. 0.10)
	MP/P	3.19 (S.D. 0.65)	2.94 (S.D. 1.02)	0.64 (S.D. 0.22)
	MPI1	0.78 (S.D. 0.07)	N.D.	0.42 (S.D. 0.11)
	BA:BA+CT	0.33 (S.D. 0.02)	0.19 (S.D. 0.01)	0.16 (S.D. 0.05)
	BeP:BeP+BaP	0.72 (S.D. 0.02)	0.64 (S.D. 0.09)	0.84 (S.D. 0.09)
	FI:FI+Py	0.42 (S.D. 0.06)	0.34	0.73 (S.D. 0.11)
	IP:IP+BgP	0.29 (S.D. 0.02)	0.34	0.52 (S.D. 0.03)
III.	*n*-Alkanals			
	(Cm-Cn); Cn max	(9–32); 26,28	(13–31); 23	(15–30); 26
	Concentration [ng m^{-3}]; CPI	5.4–6.7; 0.8–1.4	47.8; 0.9	0.9–3.7; 2.6–14.8
IV.	*n*-Alkanones			
	(Cm-Cn); Cn m~	(15–32); 23, 29, 31	(21–31); 31	(10–31); 29, 27
	Concentration [ng m^{-3}]; CPI	1.9–2.6; 1.3–1.8	13.8; 1.6	0.4–2.1; 2.2–6.2
V.	*n*-Alkanols			
	(Cm-Cn); Cnm~x	(12–30); 26, 26	(12–30); 14, 28	(12–30); 26
	Concentration [ng m^{-3}]; CPI	16.7–30.5; 5.0–10.4	36.8–2027.3; 1.9–3.6	2.7–16.7; 8.7–50.6
VI.	PAHs	5.0–8.1	28.8	0.0–0.05
VII.	Fatty acids			
	a) Saturated			
	(Cm-Cn); Cn max	(9–30); 16	(8–26); 16	(9–32); 16,18
	Concentration [ng m^{-3}]; CPI	128.1–205.2; 6.9–9.5	661.3–12300.0; 11.1–37.9	1.0–20.1; 5.0–10.0
	b) Unsaturated			
	Cn:1; Cn max	16:1, 18:1; 18		16:1, 18:1; 18
	Concentration [ng m^{-3}]	0.0–6.6	N.D.	0.0–0.5
	c) oxo- and di-acids			
	(Cm-Cn); Cn max	(6–12); 9	(6–12); 9	(6–24); 9
	Concentration [ng m^{-3}]	20.2–48.3	125.1–350.0	3.0–6.2
VIII.	Fatty acid salts			
	a) Saturated			
	(Cm-Cn); Cmax	(12–30); 16	(9–24); 16	(12–28); 16, 18
	Concentration [ng m^{-3}]; CPI	39.3–50.4; 3.9–21.1	111.0–3012.0; 8.4–8.9	1.1–7.2; 5.0–7.0
	b) Unsaturated			
	Cn:1; Cmax	16:1, 18:1; 18	16:1, 18:1; 18	16:1, 18:1; 18
	Concentration [ng m^{-3}]	0.1–0.5	0.0–340.1	0.0–0.9
	c) oxo- and di-acids			
	(Cm-Cn); Cn max	(6–26); 9	(6–12); 9	(8–22); 9
	Concentration [ng m^{-3}]	17.8–35.0	0.0–110.9	0.1–0.6

Note: The unsaturated homolog are indicated with their carbon number and number of double bonds ($C_{n:1}$). S.D.: Standard Deviation.

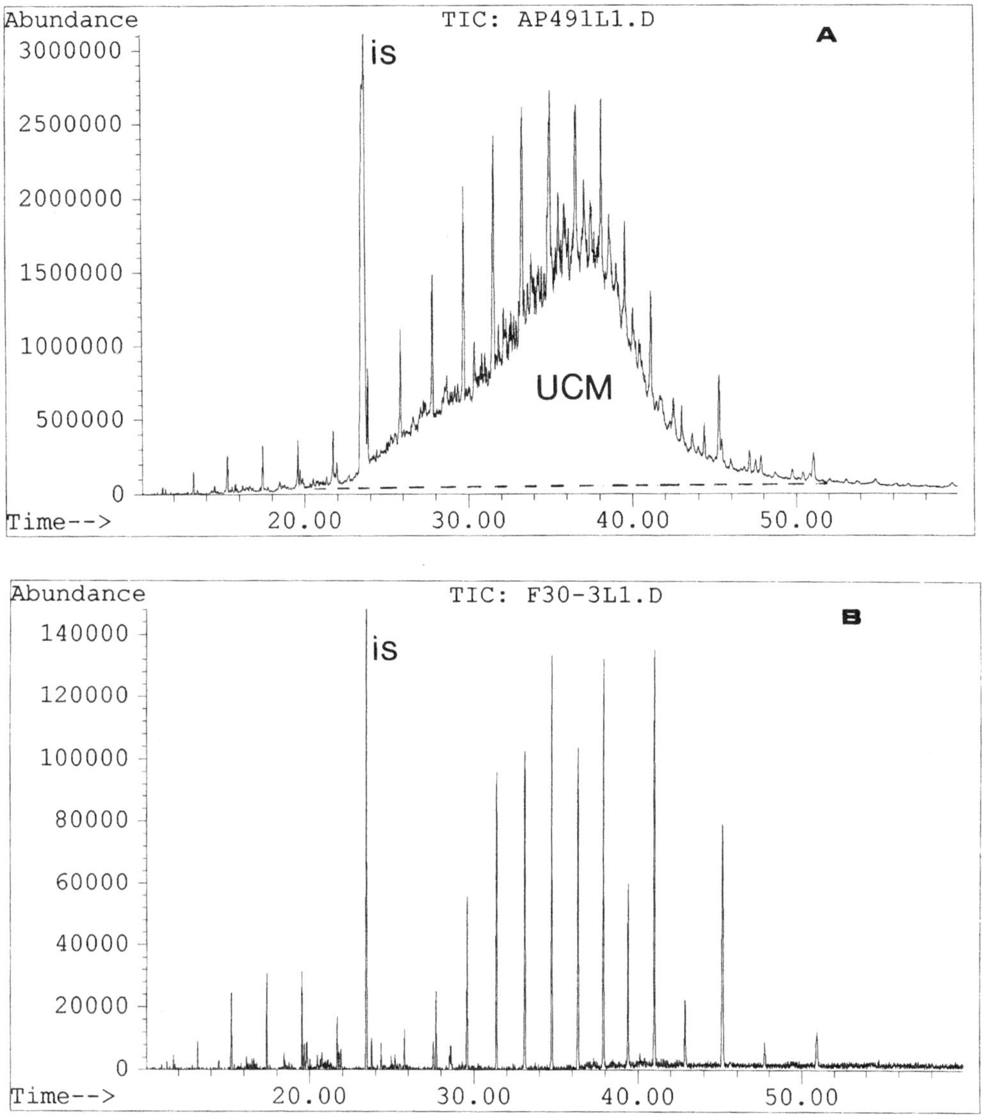

FIGURE 23.2 Total ion chromatograms (TIC) of the aliphatic fraction of a urban (A) and a rural (B) aerosol. UCM: unresolved complex mixture; is: internal standard; time in minutes.

17α(H),21β(H)-29-norhopane ($C_{29}H_{50}$) was measured in higher relative proportion than 17α(H),21β(H)-29-hopane ($C_{30}H_{52}$). This pattern was the same in the *m/z* 191 mass chromatogram from marine sediment extracts of the same area,[32] but is different ($C_{30}H_{52}$ is of higher proportion than $C_{29}H_{50}$) in aerosols and automobile exhaust extracts studied in the United States.[7,8,25,49] This is probably due to the different source of petroleum products used. Steranes, which are derived from sterols and are not found in living organisms,[51] produce in their mass spectra a number of very characteristic ions that can be used to detect their presence in complex mixtures and environmental samples.[25,31,51] Figure 23.4A, shows a typical sterane distribution (illustrated by the *m/z* 217 mass chromatogram) for the aliphatic fraction of an urban aerosol extract obtained in this study. Although it is difficult to identify all the sterane isomers by examining only their mass spectra (some co-elute), according to published mass spectra and ion chromatograms for crude oils[31,51] and

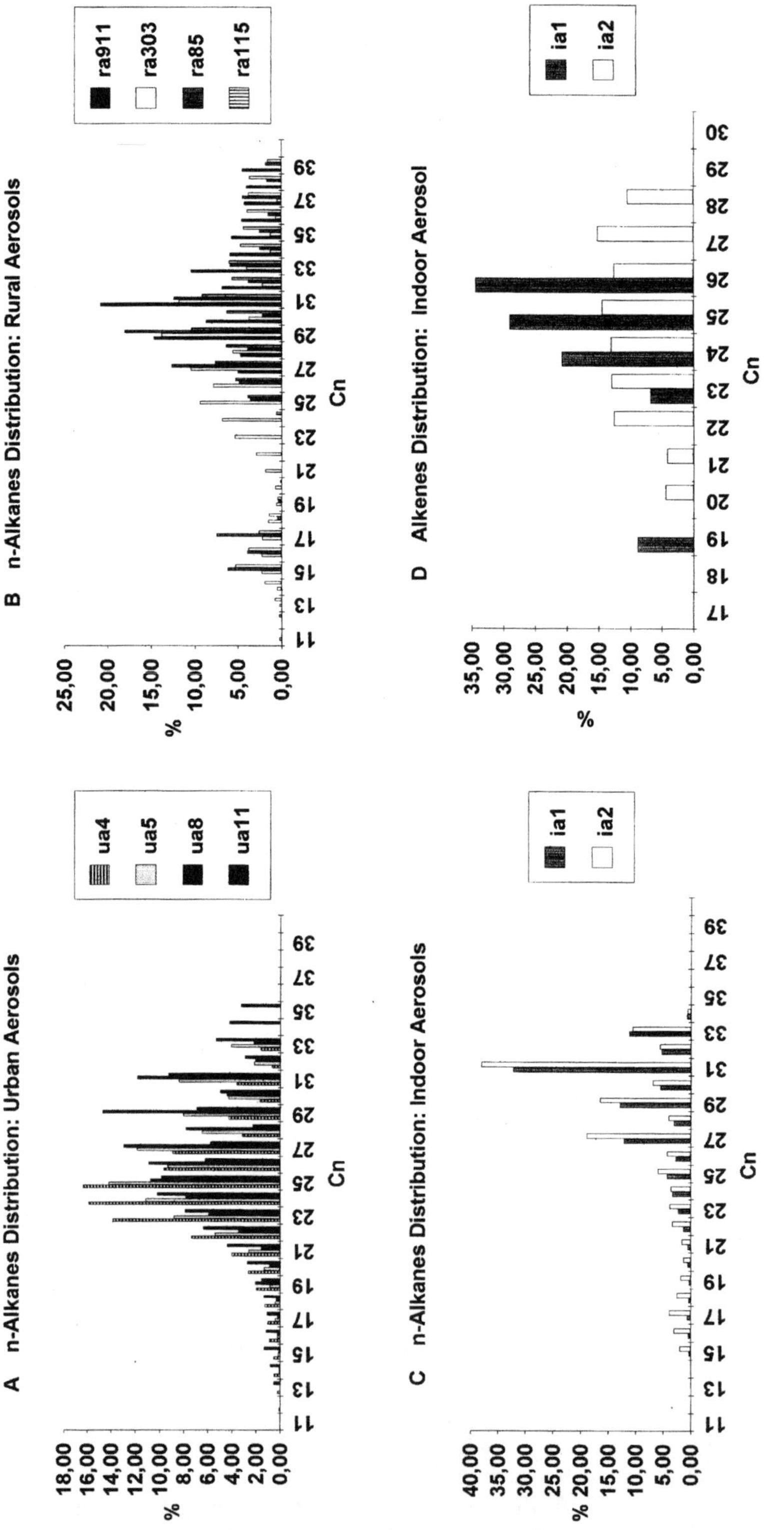

FIGURE 23.3 *n*-Alkanes distribution in urban outdoor (A), rural outdoor (B), and urban indoor (C) aerosols. Distribution of alkenes in urban indoor aerosols (D).

particulate exhaust emissions,[7,23] the compounds determined by Rogge et al.,[25] namely the 5α(H),14β(H),17β(H)-configured steranes, could also be identified. These features confirm a clear anthropogenic input related to automobile exhausts and unburned petroleum emissions from vehicular traffic. In order to better evaluate the relative input from various sources, the concentrations of wax terrestrial *n*-alkanes (WNA) were calculated according to Simoneit.[7] The ranges of their proportion (in % of total *n*-alkanes; WNA%) are given in Table 23.2, for all examined samples. Their relative distributions vs. carbon number are presented in Figure 23.5. In urban outdoor aerosol samples, there was a higher input in late spring (ua5, 19.37%) and summer (ua8 24.98%), with wax *n*-alkanes distribution maximizing at C_{29} and C_{31} (Figure 23.5A). In early spring and late autumn samples, wax *n*-alkanes represented a lower proportion (13.91–15.15%) of the total *n*-alkanes, with distribution maximizing at C_{25} (ua4, early spring) and C_{31} (ua11, late autumn). As wax lipids help plants prevent the loss of water from leaves due to respiration, plants need from their environment higher amounts of energy to biosynthesize compounds of higher carbon numbers. Therefore, samples collected during warmer seasons contain biogenic *n*-alkanes with higher carbon numbers.

The relative *n*-alkanes homolog distribution, for rural samples, is given in Figure 23.3B. The sampling season and wind regime seem to have influenced the distribution of *n*-alkanes in rural samples. Samples collected in November under very weak (almost calm) south winds (ra611 and ra911, Table 23.1), ranged from C_{25} to C_{40}, had a unimodal distribution, and maximize at C_{31} (Figure 23.5B, ra911). Samples collected in March (ra293, ra303, and ra313) under intensive north and northwest winds (Table 23.1), and May (ra65, ra75, ra85, and ra115) under north, west, and northwest winds (Table 23.1), ranging from C_{11} to C_{40}, have a bimodal distribution and maximized at C_{15}, C_{16}, and C_{17} in the first mode, and at C_{29} in the second mode (samples ra303, ra85, ra115; Figure 23.5B). The homolog distributions (Figure 23.3B), C_{max}, WNA, and CPI values (Table 23.2), confirm a stronger input of biogenic *n*-alkanes in rural, than in urban samples. The mean value of WNA in rural samples is 35.9 (S.D. of 10.1%), while in the urban samples is 18.4 (with S.D. of 5.0%). The CPI_1 values were definitely higher for rural samples, having a mean value of 2.16 (S.D. of 0.59) for CPI_1, while the corresponding value for urban aerosols is 1..43 (S.D. of 0.15). The lower standard deviation values of the samples show that the input of wax *n*-alkanes in an urban environment (local input) is more stable than in the rural environment (remote input depending on the origin of air masses). *n*-Alkanes in rural samples with $C_n < C_{25}$ had a mean CPI_2 value of 1.48 (S.D. of 0.22), with two exceptions, where a relatively high CPI_2 value (1.90, ra75) and a very high CPI_2 value were observed (<20, ra85). In these samples, the distribution in the C_{11} to C_{23} range maximized at C_{17}, with second higher homolog C_{15} (ra85, Figure 23.3B), and at C_{17} (ra75). These data indicate important input from marine bacteria and algae, as the above compounds are their well-known constituents.[51] The other samples are characterized by a mixed petroleum and marine algae *n*-alkanes input. In addition, lower UCM:NA ratios were observed in rural samples (0.0–5.8) than in urban samples (Table 23.2), indicating an even lower contribution of petroleum hydrocarbons to these rural samples. Pristane and phytane were determined in the rural aerosol samples, but hopanes (Figure 23.4B, *m/z* 191) and steranes (Figure 23.4B, *m/z* 217) were detected only in some samples. Although the use of backward air mass trajectories is neccessary for better interpratation of the results,[46] it is clearly perceptible that the sampling wind regime clearly influences the composition of the aliphatic fraction of rural more so than of the urban coastal aerosols.

A stronger odd/even predominance existed in indoor urban samples than in the corresponding outdoor ones (see CPI_1 values, Table 23.2). The of wax *n*-alkanes ranged from 54.22 to 68.93% (Table 23.2), and their distribution maximized at C_{31} (Figure 23.5C). CPI_1 values and the *n*-alkane distribution suggest a strong input of biogenic *n*-alkanes.[27] As the indoor samples were taken in two different indoor spaces, where smoking (ia1) or heavy smoking (ia2) was occurring, other molecular markers such as *iso*- and *anteiso*-alkanes[29] and alkenes[53] were considered. These compounds are enriched in cigarette smoke particles and show a concentration pattern characteristic of tobacco leaf surface waxes.[29] The mass fragmentogram (*m/z* 85+99, Figure 23.6A) and the relative

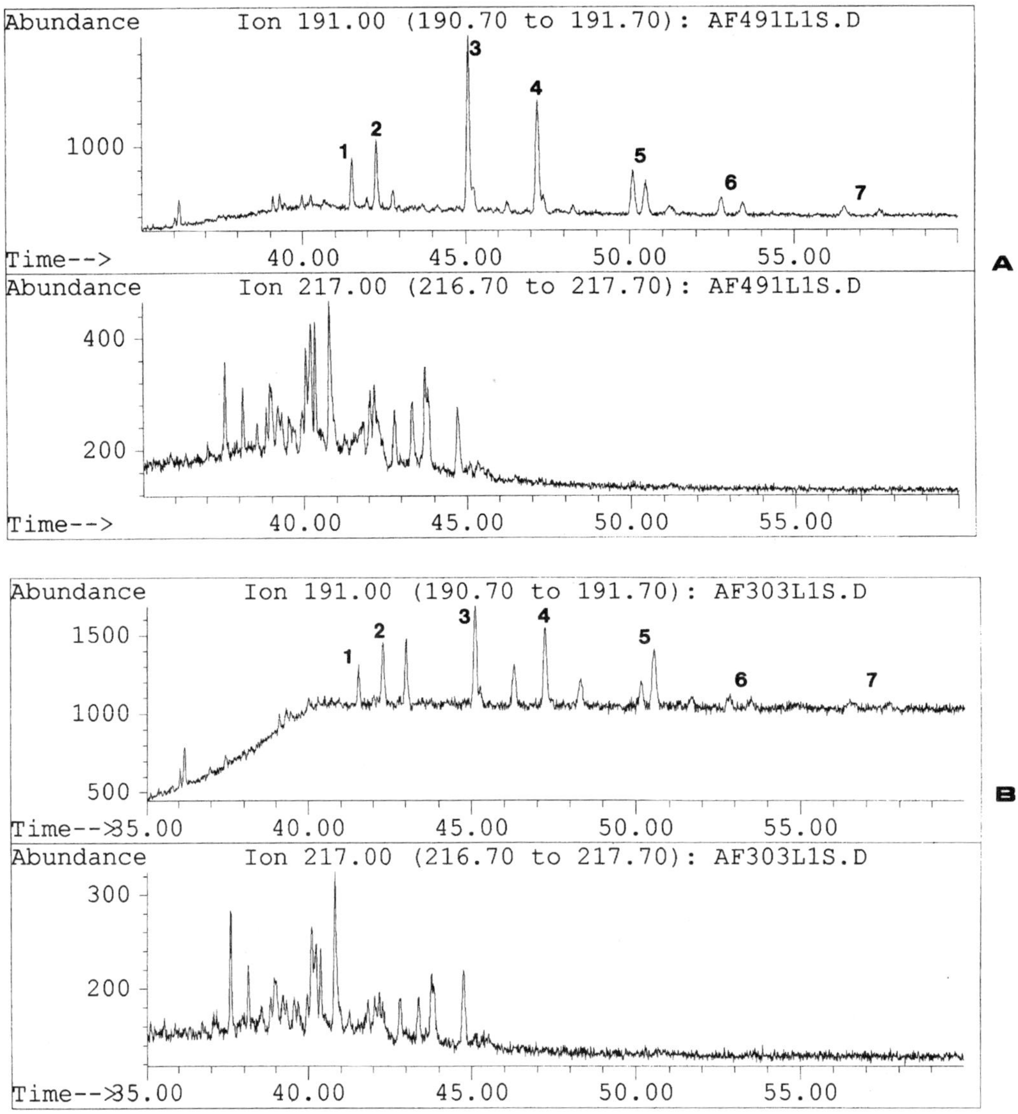

FIGURE 23.4 Ion chromatograms of hopanes (*m/z* 191) and steranes (*m/z* 217) in urban outdoor (A) and rural outdoor (B) aerosols. 1: 18a(H)-22, 29, 30-trisnorbopane ($C_{27}H_{46}$, Ts); 2: 17α(H)22, 29, 30-trisnorhopane ($C_{27}H_{46}$, Tm); 3: 17α(H), 21β(H)-30-norhopane ($C_{29}H_{50}$); 4: 17α(H), 21β(H)-30-hopane ($C_{30}H_{52}$; 5: 22S & 22R-17α(H), 21β(H)-30-homobopane ($C_{31}H_{54}$); 6: 22S & 22R-17α(H), 21β(H)-30,31-bishomohopane ($C_{32}H_{56}$); 7: 22S & 22R-17α(H), 21β(H)-30,31,32–trishomohopane ($C_{33}H_{58}$).

concentration distributions of the *iso*- and *anteiso*-alkanes determined in this study (Figure 23.7A), and in cigarette smoke particles[29] (Figure 23.7B), are supportive of the predominant input of *n*-alkanes from cigarette smoke in indoor samples. The OEP curves calculated from indoor *n*-alkanes concentrations obtained in this study and those calculated for *n*-alkanes in cigarette smoke (from *n*-alkane concentrations given in Reference 29), have almost identical shapes (Figure 23.7C) and they further support the origin of indoor *n*-alkanes. The distribution of linear alkenes determined in both indoor samples is given in Figure 23.3D. The sample ia2 was collected in the cafeteria of the University of Crete building only during the opening hours (continuing input of cigarette smoke),

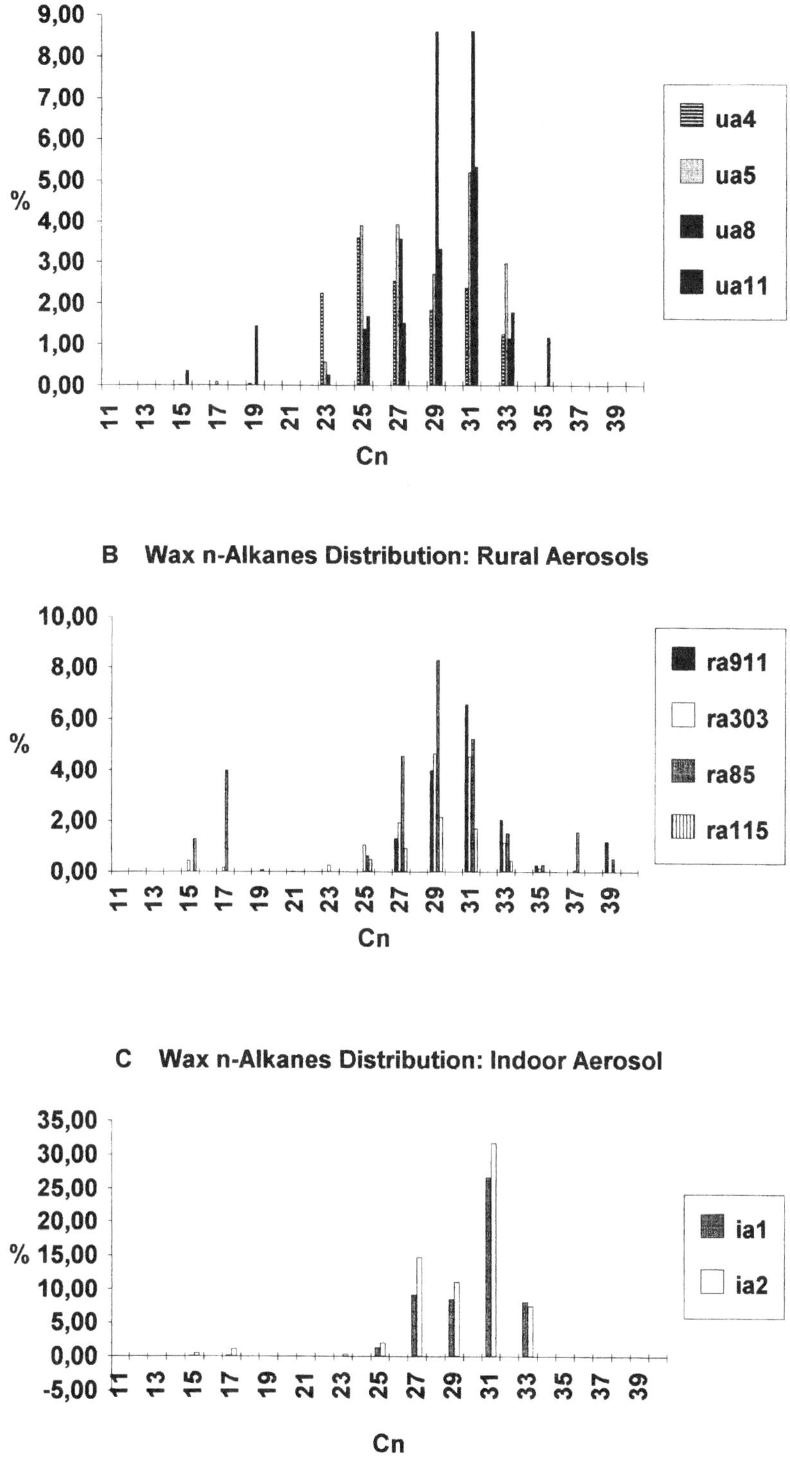

FIGURE 23.5 Wax *n*-alkane distribution urban outdoor (A), rural outdoor (B), and urban indoor (C) aerosols.

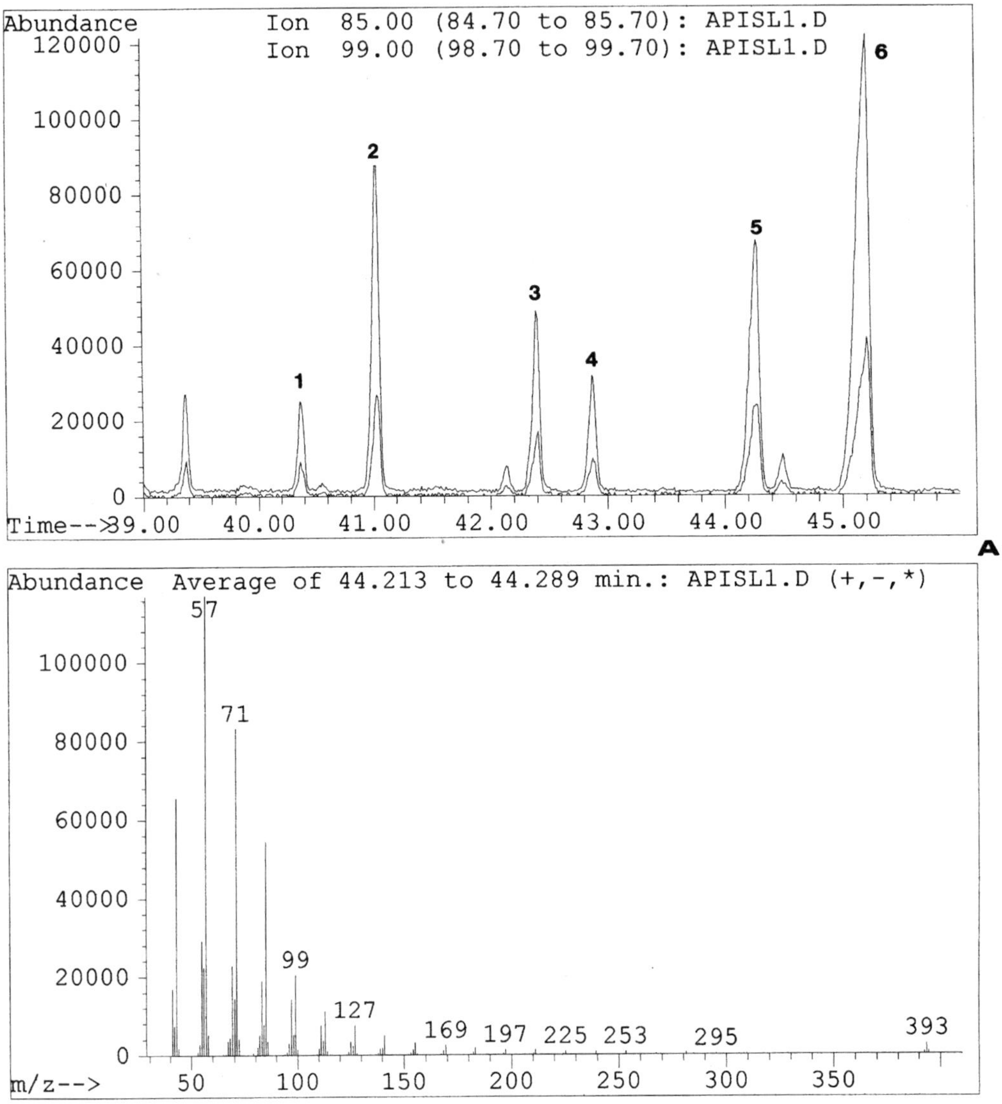

FIGURE 23.6 Ion chromatograms of *iso-* and *antelso*-alkanes (*m/z* 85+99) in urban indoor (A) and urban outdoor (B) aerosols: 1: isononacosane; 2: nonacosane; 3: anteisotriacontane; 4: triacontane; 5: isohentriacontane; 6: hentriacontane. The mass spectra of isohentriacontane determined in indoor (A) and outdoor (B) aerosols.

while the sample ia1 was collected in the central lobby after the building was closed. The weathering of alkenes of lower carbon numbers is probably due to their reactivity to oxidation over time, but a sampling artifact cannot be excluded.

The same characteristic *m/z* 85+99 mass chromatogram for outdoor aerosol samples, and compound mass spectrum are shown in Figure 23.6B. Table 23.3, are also given diagnostic ratios of concentrations of *iso-* and *anteiso*-alkanes measured in indoor and outdoor samples examined in this study. By comparing the results obtained in this study, and those published by Cass and coworkers,[27,29] we can confirm that cigarette smoking is a source of aliphatic hydrocarbons in the urban outdoor atmosphere. The ratios of concentrations of the most abundant *n-*, *iso-*, and *anteiso*-alkanes indicate that the *iso-* and *anteiso*-alkanes found in the urban atmosphere originate likewise

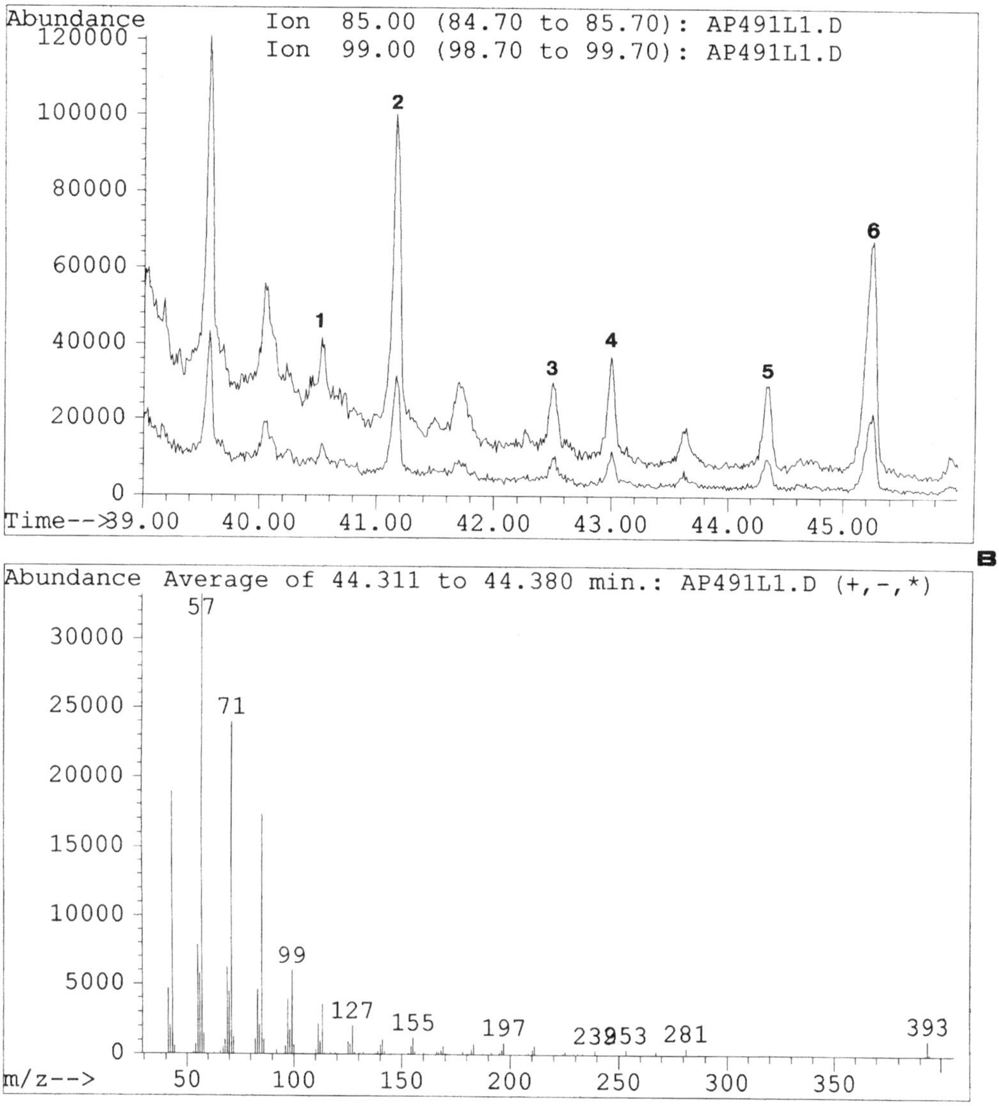

FIGURE 23.6 (*Continued.*)

from cigarette smoke. Although the ratios obtained in this study should be compared with caution to those obtained by Cass and co-workers in Los Angeles,[27,29] because of the differences that exist between types of vegetation and tobacco smoked in two different areas, some conclusions can be drawn. The concentrations ratio of *anteiso*-C_{32} to *iso*-C_{33} is 0.35 for leaf abrasion products,[27] while it is is 2.00 for cigarette smoke.[29] This pronounced difference is reflected by the same ratio measured in all urban and indoor aerosol samples collected in this study (1.40–1.75 for ua4, ua5, ua8; 2.43 and 2.66 for ia1 and ia2; Table 23.3). A pronounced difference is also noticed when the concentrations ratio of *iso*-C_{33} to *iso*-C_{31} is considered (1.87 for leaf abrasion products, 0.50 for cigarette smoke, and 0.24–0.47 for indoor and outdoor samples; Table 23.3). In addition, if we assume that the most abundant *iso*- and *anteiso*-alkanes (*anteiso*-C_{30}, *iso*-C_{31}, *anteiso*-C_{32}, and *iso*-C_{33}, Table 23.3) found in our samples, were due mostly to the leaf surface abrasion products, then the sum of their concentration could be used to calculate the expected concentrations of the corresponding *n*-alkanes (*n*-C_{30}, *n*-C_{31}, *n*-C_{33}, *n*-C_{33}) by using the ratio of the summed concentrations of *iso*- and

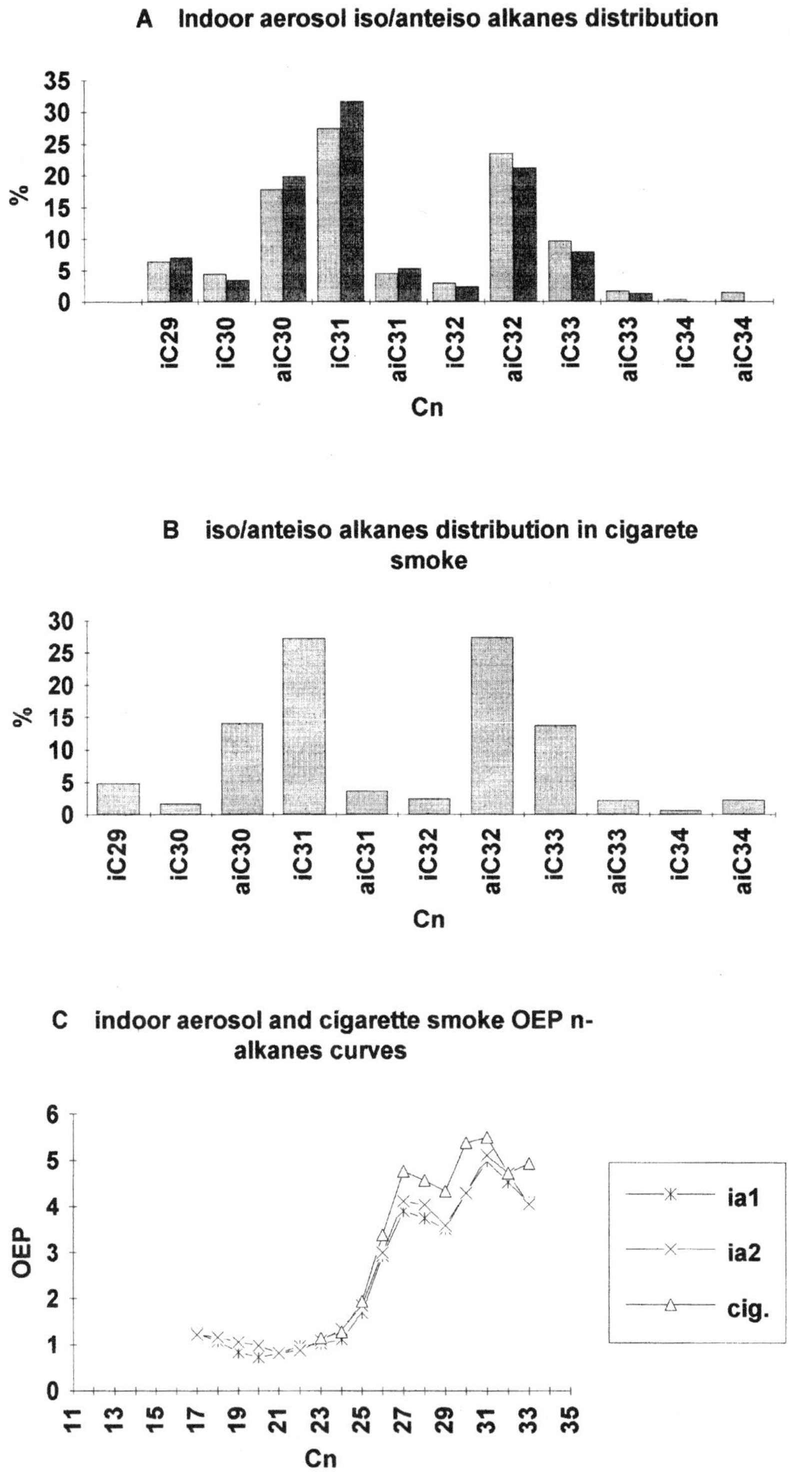

FIGURE 23.7 *Iso-* and *anteiso*-alkane distribution in urban indoor aerosol (A), in cigarette smoke.[29] (B): *n*-Alkane OEP curves from indoor aerosols and cigarette smoke.

TABLE 23.3
Concentrations and Diagnostic Concentration Ratios of *iso-* (*i* C_m), anteiso-(*a* C_m), and normal (*n* C_m) Alkanes Determined in Plant Leaves, Cigarette Smoke, and Air Particles (ia: indoor aerosols, ua: urban outdoor aerosols)

Compound	Leaves (μm g^{-1})	Cig. Smoke (μm per cig.)	Indoor ia1 (ng m^{-3})	Indoor ia2 (ng m^{-3})	Outdoor ua4 (ng m^{-3})	Outdoor ua5 (ng m^{-3})	Outdoor ua8 (ng m^{-3})
AC_{30}	19.7	40.7	38.7	387.8	0.60	4.71	4.98
I C_{31}	65.7	78.7	59.7	617.2	1.58	3.31	5.37
a C_{32}	42.7	79.1	51.1	413.3	1.05	1.72	1.85
I C_{33}	123.0	39.5	21.1	155.0	0.37	0.98	1.28
a C_{32}/I C_{33}[a]	0.35	2.00	2.43	2.66	1.40	1.75	1.43
I C_{33}/I C_{31}[a]	1.87	0.50	0.35	0.25	0.47	0.30	0.24
$\Sigma_I + aC_m/\Sigma_n C_m$[a]	0.02	0.65	0.59	0.70	0.62	0.20	0.48

[a] Ratios of concentrations: no units.

anteiso-alkanes ($\Sigma i{+}aC_m$) to the sum of concentrations of *n*-alkanes (ΣnC_m, calculated from data concerning leaf abrasion products[27]; Table 23.3). The estimated *n*-alkane concentrations should be 10- to 30-fold higher than those actually determined in this study (Table 23.3). Forest fire and burning of vegetative waste may also constitute sources of *iso-* and *anteiso*-alkanes to the atmosphere. In the larger area of Heraclion, during the period of sampling, forest fires were not reported. The examination of the above reported parameters show that in April (ua4), the *iso-* and *anteiso*-alkanes have distributions corresponding more closely with those expected for cigarette smoke, than with those of May (ua5) and August (ua8). As the outdoor urban sampling location was in an area where many administrative buildings exist and numerous persons gather outdoors (the cigarette consumption per capita in Greece is the third highest in the world), with a relatively lower wind speed (ua4), this similarity of *iso-* and *anteiso-alkanes* aerosol distribution with the corresponding indoor aerosol (Table 23.3) is not unexpected. *Iso-* and *anteiso*-alkanes were not detected in rural samples.

SEMIPOLAR COMPOUNDS: PAH COMPOSITION AND CHARACTERISTICS

Figure 23.8 shows the PAH (for compounds numbers, see Table 23.4) distribution in different urban and rural aerosol samples. Collective parameters and PAH concentrations diagnostic ratios are given in Table 23.2.

Urban samples had total PAHs concentrations (TPAHs), that were two orders of magnitude higher than those of the rural samples (Table 23.2), and contained perylene, while rural samples did not (Figure 23.8). On the other hand, retene was present only in one urban aerosol sample (collected in August), but present in all rural aerosol samples (excepting those taken during calm, ra611 and ra911; Figure 23.8B). Retene is an incomplete combustion product of compounds with the abietane skeleton, and it can be used as a tracer for forest fires during the dry season, as well as an indicator of residential coniferous wood combustion.[54] Although the qualitative composition of PAHs in urban aerosols is very similar to those of the rural ones, the relative abundances of specific components within mixtures display discernable differences. For example, urban samples contained mostly 3- and 4-ring compounds and their methylated derivatives (Table 23.4). Benzo[ghi]perylene and coronene are the most abundant components in all urban aerosol samples. Rural samples collected during north and northwestern wind events (ra303, ra585, ra115) contained

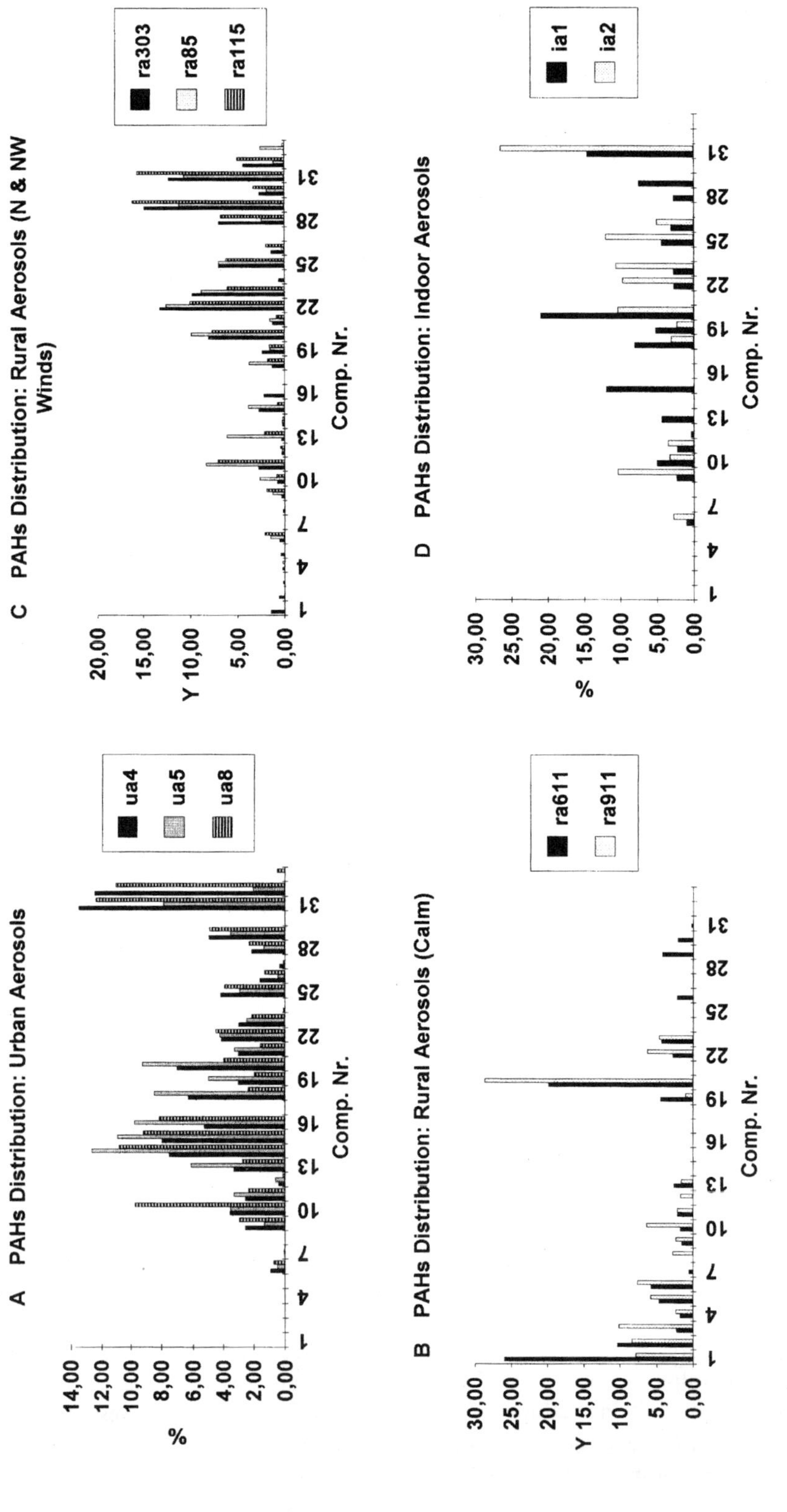

FIGURE 23.8 PAH distributions in urban outdoor (A), rural outdoor under calm wind regime (B), rural outdoor under north and northwestern winds **(C)**, and urban indoor (D) aerosols.

TABLE 23.4
Polycyclic Aromatic Hydrocarbons (PAHs) Compound Number (Comp. Nr.), Number of Rings (No. of Rings), and Molecular Weight (M.W.), Determined in Air Particles

PAH Compound	Comp. No.	No. of Rings	M.W.
Naphthalene	1	2	128
Methynaphthalenes	2	2	142
WAcenaphthene	3	3	154
Fluorene	4	3	166
Methyfluorenes	5	3	180
Phenanthrene	6	3	178
Anthracene	7	3	178
Dimethylfluorenes	8	3	194
Me-penathrene/anthracene	9	3	192
Di-Me-phenanthrene/anthracene	10	3	206
Fluoranthene	11	4	202
Acephenanthrylene	12	4	202
Pyrene	13	4	202
Tri-Me-phenanthrene	14	3	220
Methyipyrenes+benzofluorenes	15	4	216
Dimethyl-202	16	4	230
Benzo[ghi]fluoranthene	17	4	226
4(H)-cyclopenta[cd]pyrene	18	4	226
Benzo[a]anthracene	19	4	228
Chrysen+tphenylene	20	4	228
Me-Chrysene	21	4	242
Benzo[bj]fluoranthene	22	5	252
Benzo[k]fluoranthene	23	5	252
Benzo[a]fluoranthene	24	5	252
Benzo[e]pyrene	25	5	252
Benzo[a]pyrene	26	5	252
Perylene	27	5	252
Indeno[7,1,2,3-cdefjchrysene	28	6	276
Indeno[1 52,3-cd]pyrene	29	6	276
Dibenzo[a,h]anthracene	30	6	276
Benzo[ghi]perylene	31	6	276
Coronene	32	6	300
Retene	33	3	234

chiefly 4-, 5-, and 6-ring compounds of pyrolytic origin Table 23.4), with chrysene (+triphenylene), benzo[bj]fluoranthenes, indeno[c,d]pyrene, and benzo[ghi]perylene being the major individual compounds. Rural samples collected during calm (ra611, and ra911) contained mostly 3-ring compounds, whereas 4-, 5-, and 6-ring compounds (Table 23.4) were rather absent (Figure 23.8B). Indoor aerosol samples contained (in number) less PAHs and had different compound distributions from urban and rural samples, with chrysene (+triphenylene), benzo[e]pyrene, and benzo[ghi]perylene being the most abundant homologs.

The compositional differences between different samples are better expressed by PAH concentration diagnostic ratios. Table 23.2 shows that these ratios differentiate very well the PAH compositional differences existing between urban and rural samples. Urban samples were characterized by a relative value stability of these ratios (Table 23.2). The mean value of CPAHs:TPAHs was

0.43 (S.D. of 0.04) in urban samples and 0.70 (S.D. of 0.10) in rural samples. These values point out that combustion PAHs are the major components in rural aerosols in comparison to urban ones, where petrogenic PAHs are very abundant. CPAHs:TPAHs values in urban samples were within the same range of CPAHs:TPAHs values calculated (from data provided by Rogge et al.[25]) for non-catalyst (0.41) and catalyst-equipped (0.51) automobiles, and heavy-duty diesel trucks (0.30). Significant differences were found in the MP:P and MPI1 values of urban and rural samples (Table 23.2). The mean MP:P value of urban samples was 3.19 (S.D. of 0.65), and the corresponding of MPI1 is 0.78 (S.D. of 0.07). These values characterized unburned fossil PAH mixtures[19,44] in the urban samples. The MP:P and MPI1 values in urban samples analyzed in this study (Table 23.2), are similar to those calculated (from data provided by Rogge et al.[25]) for non-catalyst-2.44 and catalyst-equipped (2.16) automobiles, and heavy-duty diesel trucks (2.57), and those calculated from data reported by Bayona et al.[14] for the city of Barcelona. The values for Barcelona were 1.10 to 3.48 for MP/P and 0.59 to 0.98 for MPI1. In rural samples, the corresponding MP:P and MPI1 values (Table 23.2; mean values for MP:P 0.64 (S.D. of 0.22) and MPI1 0.42 (S.D. of 0.11)) are well within the typical range for combustion-derived products.[14,55] These average MP:P and MPI1 values suggest that either unburned fossil PAHs are not atmospherically well dispersed or, more likely, they are quantitatively dominated in rural samples by PAHs of combustion origin during long-range transport. A correlation has been found between MP:P and MPI1 of these samples and literature data.[14] The MP:P and MPI1 values confirmed the indications provided by the CPAHs:TPAHs values.

In order to further assess the different sources of PAHs present in the examined aerosol samples, a comparison can be made between the diagnostic ratios (BA:BA+CT, BeP:BeP+BaP, Fl:Fl+Py and IP:IP+BgP; see Introduction for their definition; Table 23.2) calculated for the samples, and the same ratios computed from well-characterized emission sources (ratios calculated from References 25, 41, 42). Consideration of these ratios should be taken with caution because their use assumes only minor modifications occurred from emission of the PAHs and no sampling artifacts (especially in the case of the reactive BaP). Thus, the BA:BA+CT ratio (0.33; S.D. 0.02) indicates automobile exhaust and more probably from catalytic automobiles (0.33; calculated from data provided by Rogge et al.[25]) as the most important source of PAHs in urban samples. The BeP:BeP+BaP ratio has a mean value of 0.72 (S.D. 0.02), which corresponds to a faster decay of benzo[*a*]pyrene, and indicates an origin from a more distant source. The Fl:Fl+Py mean value was 0.42 (S.D. 0.06) for urban samples. This value is very similar to vehicular emissions and especially to those of catalytic automobiles (0.44; calculated from data provided by Rogge et al.[25]). The IP:IP+BgP values reported are for: cars 0.18, diesel 0.37, and coal 0.56.[41,42] This ratio measured in urban samples had a mean value of 0.29 (S.D. 0.02), indicating mixed combustion sources. In the rural samples, the corresponding IP:IP+BgP values are comparable to the values calculated for coal soot (0.56) and wood burning (0.62).[41,42]

POLAR LIPIDS: CARBONYL AND HYDROXY COMPOUNDS

The concentration ranges of *n*-alkanals and 2-alkanones were at least one order of magnitude lower than the other polar compounds, such as *n*-alkanols and alkanoic acids (Table 23.2).

The *n*-alkanals, especially those with carbon atoms number higher than 20, are of biogenic origin.[56] The chain length distribution of alkanals in the range C_{20}–C_{32} (Figure 23.9A) was usually very similar or identical to that of alkanols (Figure 23.10A), for the homolog with $C_n > 20$. The CPI (eve*n-to-odd*) values for the alkanals of the urban aerosols (0.8–1.4; Table 23.2), were not as high as expected for biogenic compounds. If we consider the range C_{20}–C_{32}, these values were higher, indicating more biogenic input. The homologs with lower carbon atoms number (C_9–C_{19}) may originate by oxidation of alkanes and/or microbial sources or mixed anthropogenic processes.[56,57] For the rural samples, the CPI values ranged from 2.6 to 14.8 for homolog with more than 20 carbon atoms — close to the corresponding values for the long-chain *n*-alkanols. This was also reflected in the homolog distributions (Figures 23.9B and Figure 23.10B) and, as well, in the OEP plots, indicating a close relationship (common origin) of these two compound classes (Figure 23.11). The most abundant compound class determined among the neutral oxygenated lipids is, by far, that of *n*-alkanols. The homologs ranged from C_{11} to C_{33}, with a strong even-carbon-number preference (CPI 5.0–10.4 for urban samples and 8.7–50.6 for rural samples; Table 23.2, and Figures 23.10A and B). This very strong *even-to-odd* predominance and the presence of major amounts of C_{26} and C_{28}, suggests higher plant waxes as the source of these aerosol lipids. The distributions obtained here are similar to those obtained in other areas such as the North Pacific,[2] in many locations in the U.S.,[4] rural sites of Australia,[64] and the open western Mediterranean region.[65]

The homologs up to C_{20} are believed to be characteristic of vascular plant waxes,[62] while those lower than C_{20} may originate from microbial or marine sources.[4,63] In the samples collected in November 1993 (ra611, ra911), where low concentrations of only two long-chain homologs (C_{28} and C_{26}) were detected, there is a dominant pattern of homologs from C_{12} to C_{20}, with the most abundant being C_{16} and C_{18}. This pattern suggests microbial or/and marine contribution.

Generally in rural samples taken under the calm wind regime (for example, ra911, Figures 23.9B and 23.10B), the biogenic *n*-alkanals and *n*-alkanols (with $C_n > 20$) were absent or in very low concentrations; while for the samples collected during N and NW winds events (ra85, ra115), these compounds were the most abundant. This indicates that biogenic lipids from higher plants are transported to this arid area, where the local input probably contains lipids from microbial and/or marine origin.

Indoor aerosols contained *n*-alkanals and *n*-alkanols with very low CPI values (Table 23.2, Figures 23.9C and 23.10C). Rogge and co-workers[29] reported similar distribution patterns for *n*-alkanols, but they did not report the presence of *n*-alkanals, in cigarette smoke. The presence, the CPI (CPI < 1, Table 23.2) and the distribution pattern of n-alkanals (Figure 23.9B) could indicate their probable source, the oxidation of the alkenes, emitted from cigarette smoke.

2-Alkanones ranged from C_{15} to C_{33} (Figure 23.12), with odd-to-even predominance (CPIs, Table 23.2). They are considered to originate from *in situ* microbial formation from *n*-alkanes in the β-position.[58] Their homolog distribution and C_{max} (Figures 23.12A, B, and C) within the range C_{25}–C_{33}, similar to the corresponding *n*-alkane ones especially for the rural samples (Figure 23.3B) and supports the above hypothesis. Often, the two distributions do not show the same odd-to-even predominance and the major homolog is not the same.[59] Another possible explanation for the presence of these lipids in aerosols is direct emission from higher plant waxes, where they occur as a minor constituent[60] and/or from wind-blown soil dust, where they occur in significant concentrations.[61] The latter supports the hypothesis of that microbial activation is of great importance in this kind of environment and may cause some alterations to their distribution, compared to the precursors.

Alkan-2-ones were not detected in rural samples collected in November 1993 in Finokalia (ra611, ra911) under calm wind regime. This fact supports again the hypothesis of long-range transport of these lipids.

ORGANIC ACIDS: CARBOXYLIC ACIDS AND THEIR SALTS

The chromatographic profile of this fraction (metyl esters ion chromatogram m/z 74; Figure 23.13A) shows that the compounds ranged from C_{10} to C_{32}. The compound distribution is characterized by a strong *even-to-odd* CPI (Table 23.2), indicating a definite biogenic origin. The homologs (C_{20} are attributed to microbial sources, while those (C_{20} show a pronounced plant origin.[4]

In all samples analyzed in this study, a series of α,ω-dicarboxylic acids and ω-oxocarboxylic acids were determined (Table 23.2 and ion chromatogram Figure 23.13B). It has been proposed

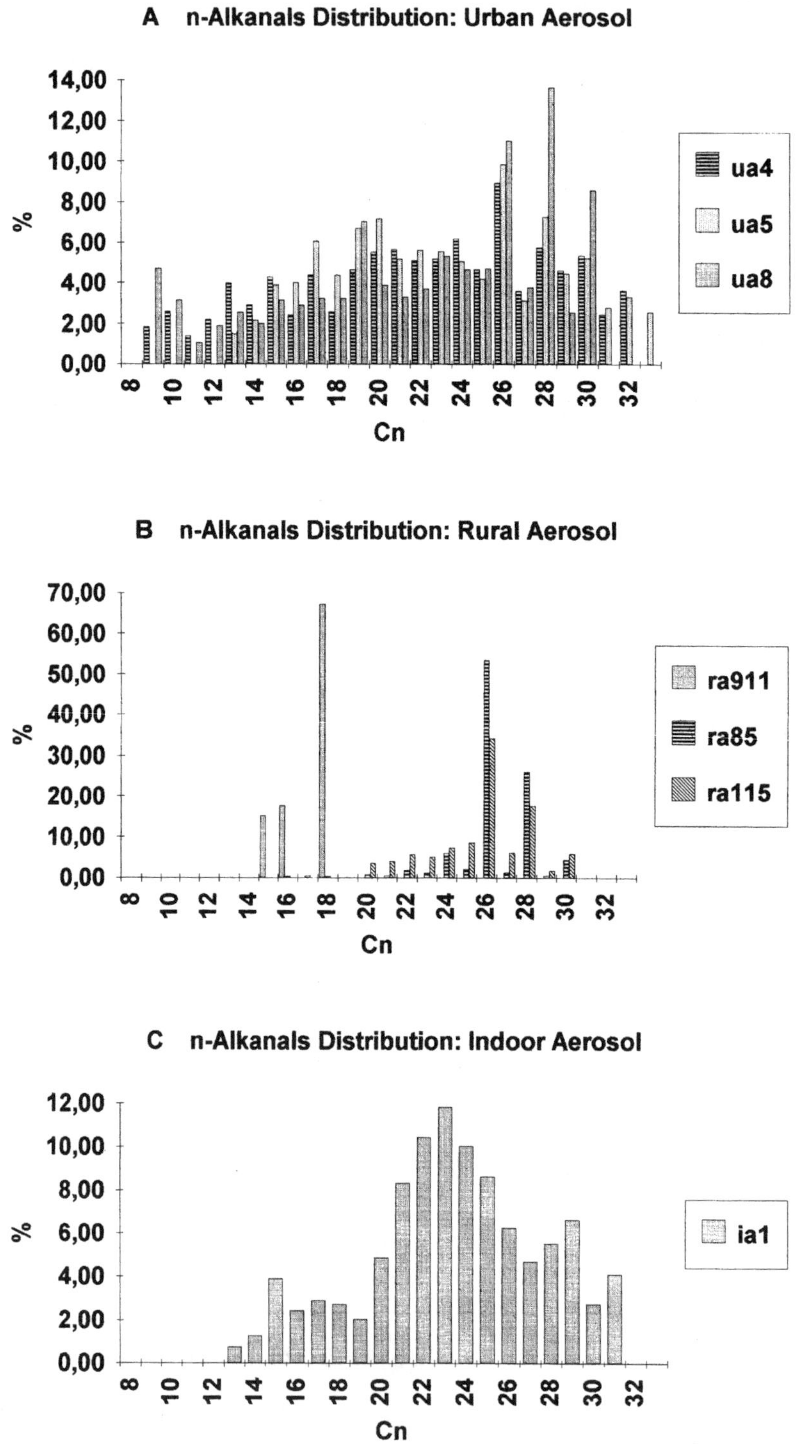

FIGURE 23.9 *n*-Alkanal distribution urban outdoor (A), rural outdoor (B), and urban indoor (C) aerosols.

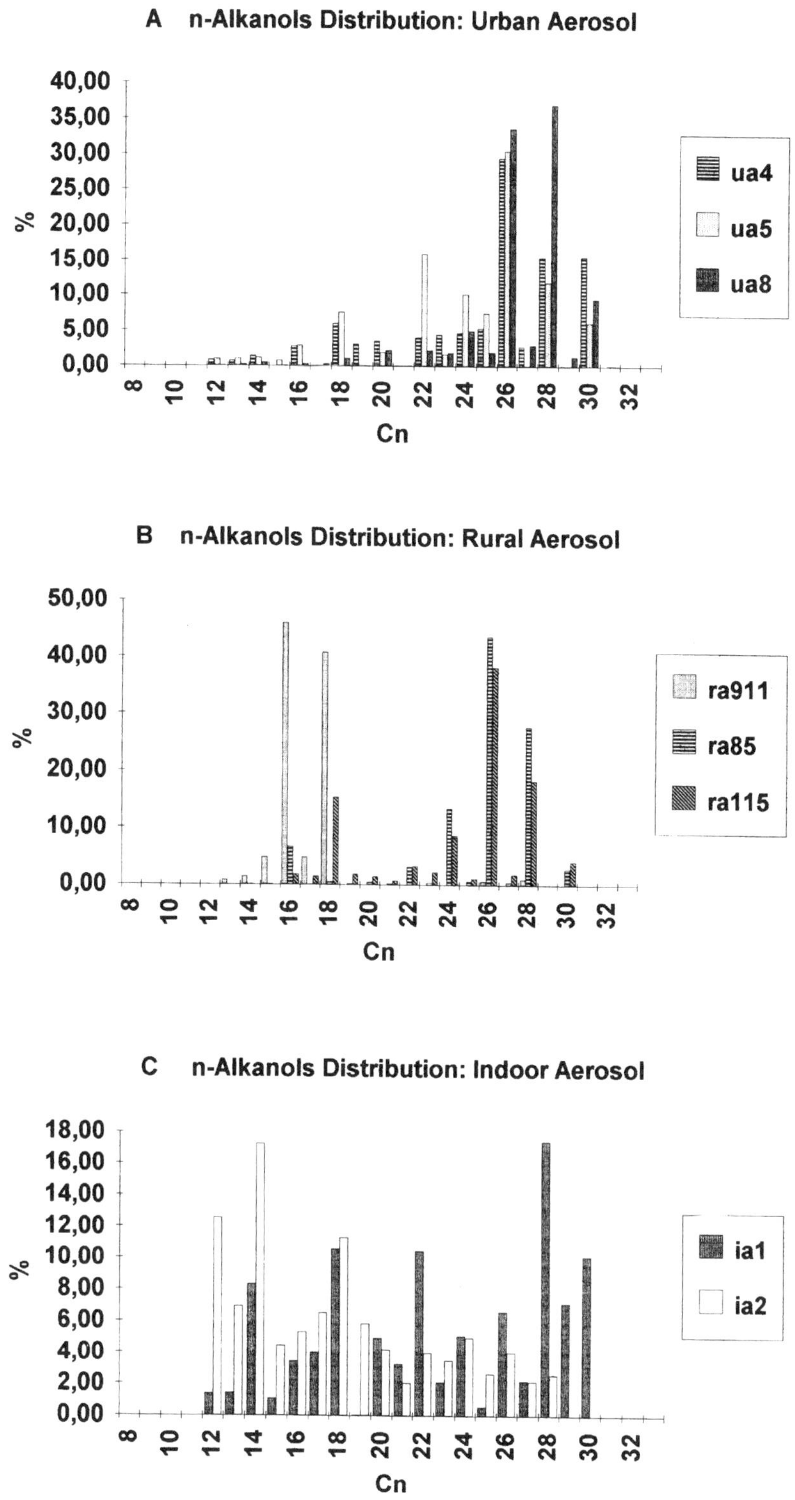

FIGURE 23.10 *n*-Alkanol distribution in urban outdoor (A), rural outdoor (B), and urban indoor aerosols.

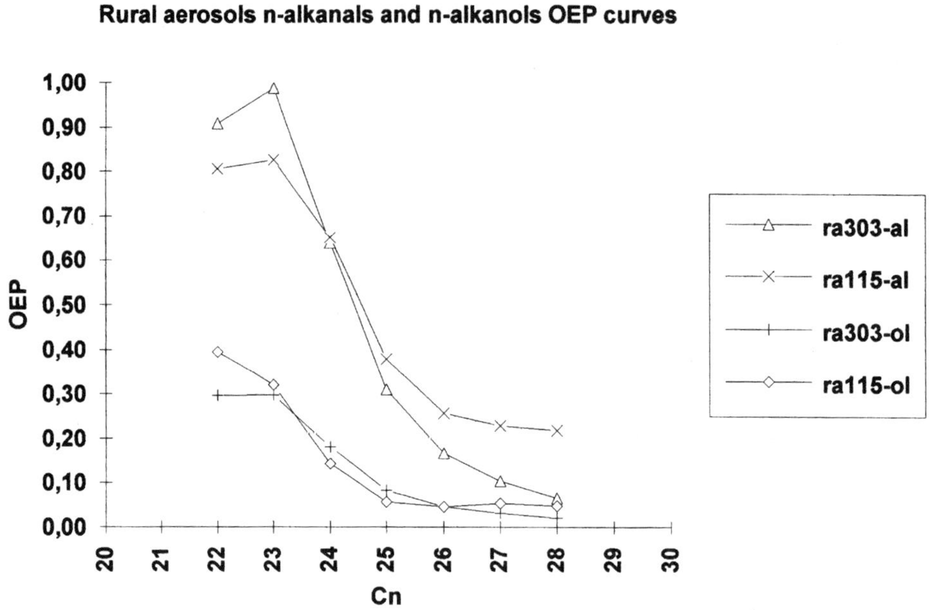

FIGURE 23.11 *n*-Alkanals (ra.-al) and *n*-alkanols (ra.ol) OEP curves from rural aerosols.

that these compounds are photo-oxidation products of cyclic olefins[66] and of unsaturated fatty acids.[12] The mass spectra of these compounds have been thoroughly presented.[67] The C_5 and C_6 homologs are formed by the oxidation of cyclic olefins, while the C_8 and C_9 ones, which were the most abundant, are formed by the photo-oxidation of unsaturated carboxylic acids such as oleic (C_{181}) and linoleic (C_{182}) acids. In the urban and the remote samples, the latter compounds were present in very low concentrations or even absent. It is interesting to note that the highest α,ω-dicarboxylic acids concentrations occurred simultaneously with the lowest concentration of their precursors (oleic and linoleic acids), and relate to highest ozone concentrations (160 μg m^{-3}) determined in the urban sampling area.[12]

The alkanoic acid salts showed very similar gas chromatographic profiles and compound distribution to the free alkanoic acids, ranging between C_{10} and C_{30}. Their CPI values (from 3.9 to 21.1 for the urban, and from 5.0 to 7.0 for rural samples) indicate a definite biogenic origin. The alkanoic acid salts fraction is dominated by the lower carbon number homologs, which most likely originated from marine sources, but also shows a terrestrial plant wax source in lesser amounts.

In the urban aerosols, unsaturated fatty acids in the salt form were also detected, but at lower concentrations than in the free carboxylic acid fraction. In these samples, the ω-oxocarboxylic and α,ω-dicarboxylic acids were determined in the salt form in higher proportion than in the free form. (Table 23.2). Conversely, in the remote aerosols, when the proportion of unsaturated fatty acids is compared to the photo-oxidation products, it was found to be higher in the salt than in the free form, suggesting that the unsaturated fatty acid salts are more protected from photo-oxidation processes.

As carboxylic acids are emitted from an important variety of sources in the environment, we believe that their presence cannot be reliably reconciled with a specific origin.

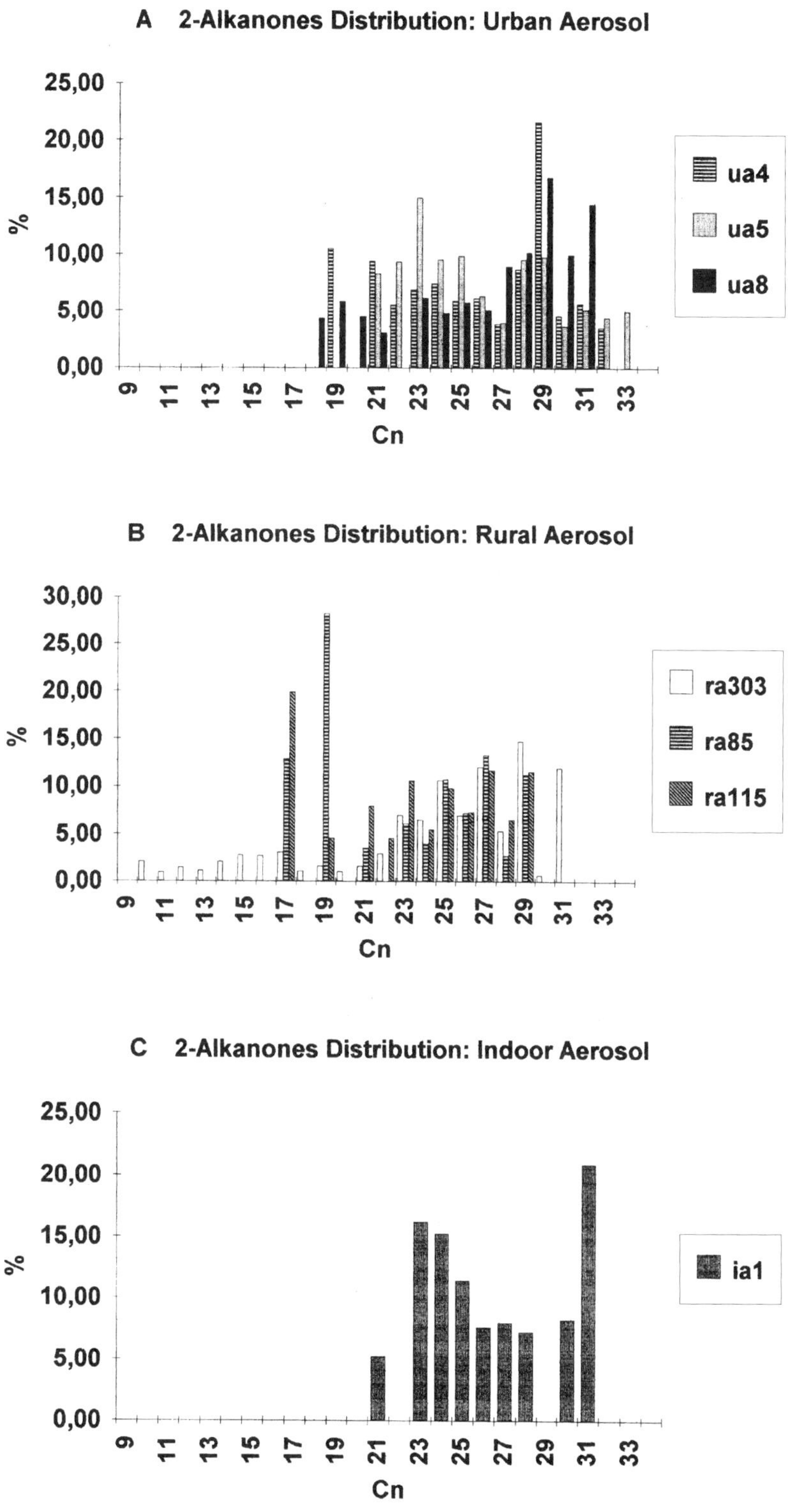

FIGURE 23.12 2-Alkanone distribution in urban outdoor (A), rural outdoor (B), and urban indoor (C) aerosols.

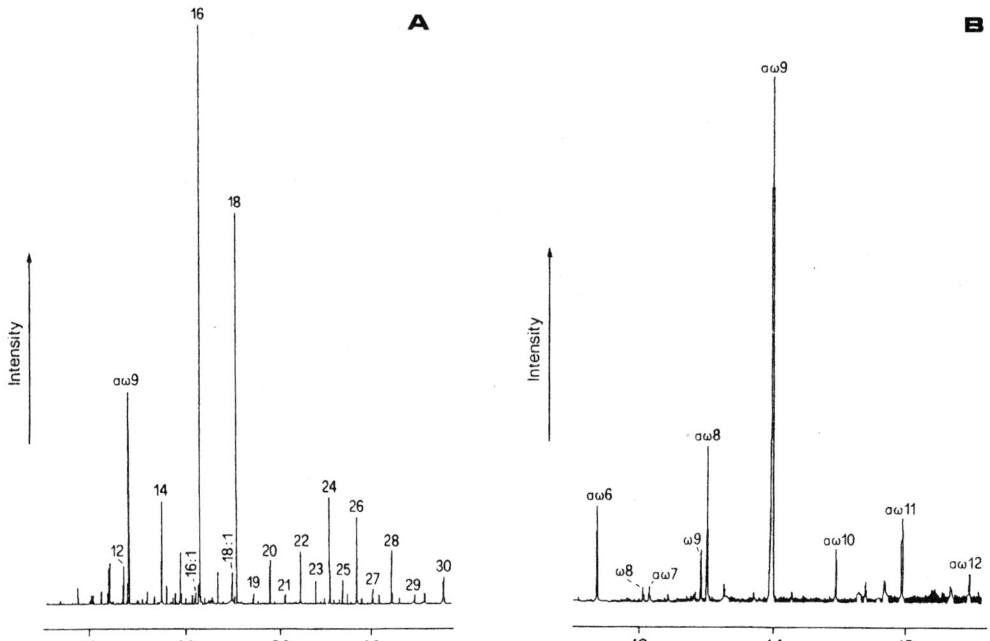

FIGURE 23.13 A: *m/z* 74 ion chromatogram of the methylated acidic extract of an urban outdoor aerosol sample. n: carbon numbers of fatty acid methyl esters; n:1 Carbon numbers of unsaturated acid methyl esters, and α,ω9: 1,9-nonanedioic dimethyl ester. B: Combined ion chromatograms for *m/z* 97, 111, 112 (specfic ions for ω-oxo and α,ω-dicarboxylic fatty acids) of the same (A) methylated acidic fraction (ωn: carbon numbers of ω-oxo and α,ωn: carbon numbers of α,ω-dicarboxyhc acids).

OXYGENATED COMPOUNDS: O-PAHs

In Figure 23.14 are presents the selected ion chromatograms for specific oxygenated PAHs (O-PAHs), extracted from the total ion chromatogram of the fraction containing the less-polar oxygenated lipids. These compounds are generally formed by photoinduced reactions of parent PAH. Higher concentrations of O-PAHs were reported during summer.[14] These higher O-PAH concentrations were attributed to higher concentrations of oxidant species in urban atmospheres during this season. The total concentrations of O-PAHs (Table 23.2) were lower than those of PAHs (Table 23.2), although they are relatively higher than those reported in other areas.[14] In the urban samples, polycyclic aromatic ketones, carboxyaldehydes, and quinones were the most representative compounds. The compounds determined in the urban aerosols were (Figure 23.14): 1:9H-fluoren-9-one (*m/z* 180), 2: anthracene-9,10-dione (*m/z* 180), **3**: 4H-cyclopenta[def]phenanthrene-4-one (*m/z* 204), 4: anthracene-9-carboxyaldehyde (*m/z* 206), 5: phenanthrene-9-carboxyaldehyde (*m/z* 206), 6: anthracene-9,10-dione (*m/z* 208), 7: C_1-phenthrenecarboxyaldehydes (m/z 220), 8: 7H-Benz[de]anthracene-7-one (*m/z* 230), 9: 1-pyrene carboxyaldehyde (*m/z* 230), 10: C_2-phenthrenecarboxyaldehydes (*m/z* 234), and 11: benz[a]anthracene-7,12-dione (*m/z* 258). The above compounds were determined by means of their mass spectra and by comparison of their retention times, with available standards.

The polycyclic aromatic ketone (PAK) 7H-benz[de]anthracene-7-one, and the quinone benza[a]anthracene-7,12-dione were the most abundant compounds in the late spring urban outdoor aerosol sample (ua5) and, and the C_1 and C_2-phenanthrene carboxyaldehydes were the most abundant compounds in the summer aerosol sample (ua8). Pitts and co-workers[68] determined that benz[a]anthracene-7,12-dione and 7H-benz[de]antracene-7-one were the major reaction products

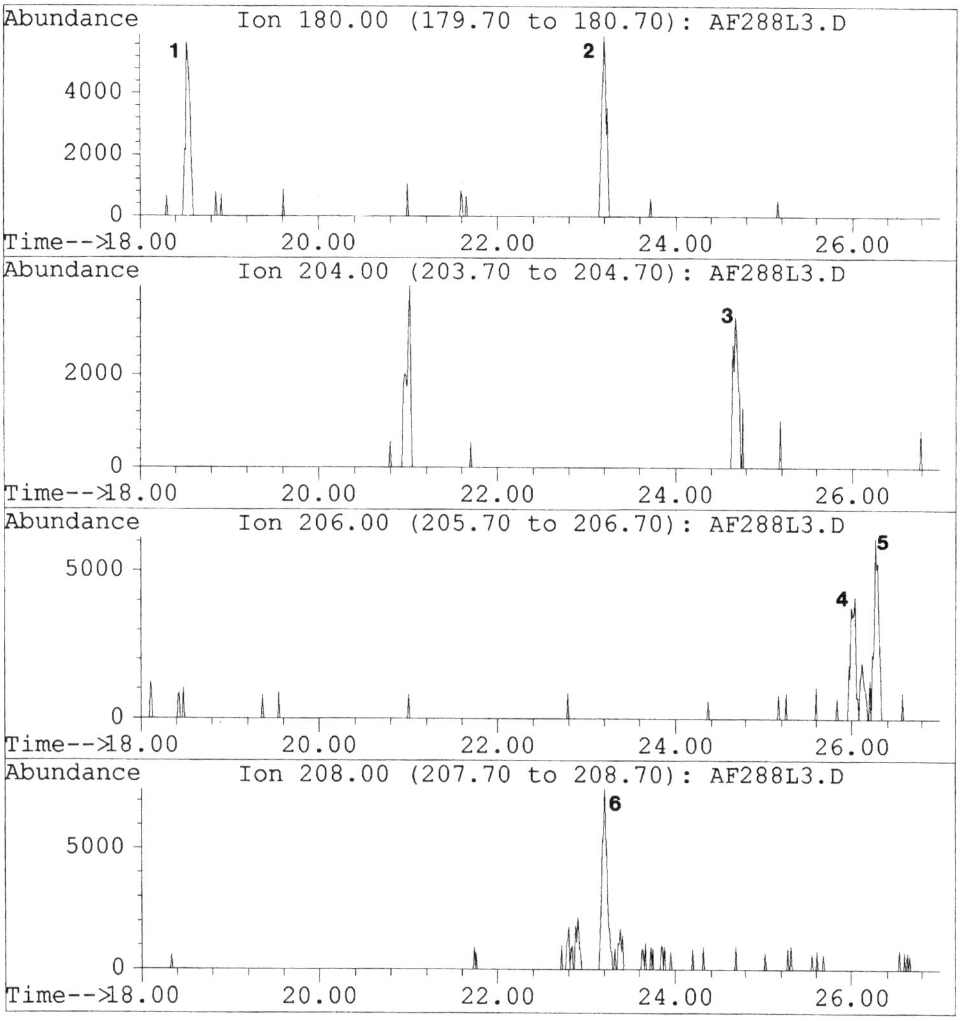

A

FIGURE 23.14 Ion chromatograms for O-PAHs determined in urban outdoor aerosols. **1**: 9H-fluoren-9-one (*m/z* 180), **2**: anthracene-9,10-dione (*m/z* 180), **3**: 4H-cyclopenta[def]phenanthrene-4-one (*m/z* 204), **4**: anthracene-9-carboxyaldehyde (*m/z* 206), **5**: Phenanthrene-9-carboxyaldehyde (*m/z* 206), **6**: anthracene-9,10-dione (*m/z* 208), **7**: C_1-phenthrenecarboxyaldehydes (*m/z* 220), **8**: 7H-benz[de]anthracene-7-one (*m/z* 230), **9**: 1-pyrene carboxyaldehyde (*m/z* 230), **10**: C_2-phenthrenecarboxyaldehydes (*m/z* 234), and **11**: benz[a] anthracene-7,12-dione (*m/z* 258).

of benzo[a]anthracene and benzo[a]pyrene, respectively, with ozone. The presence of phenanthrene carboxyaldehydes has been reported in diesel particulates[69] and is in good agreement with the reported values of of alkylated phenanthrenes in the PAH fraction of urban aerosols.

In only one rural sample (ra85) was benz[a]anthracene-7,12-dione determined. Polyaromatic quinones (PAQ) are suposed to be stable to atmospheric conditions.[70] One should not exclude that the presence of this PAQ is related to transport in the rural aerosol samples, while the presence of the other O-PAHs in the urban environment is rather connected to local input (automobiles). As O-PAHs can be formed during sampling,[68] one should be very careful in reconciling their presence with specific sources.

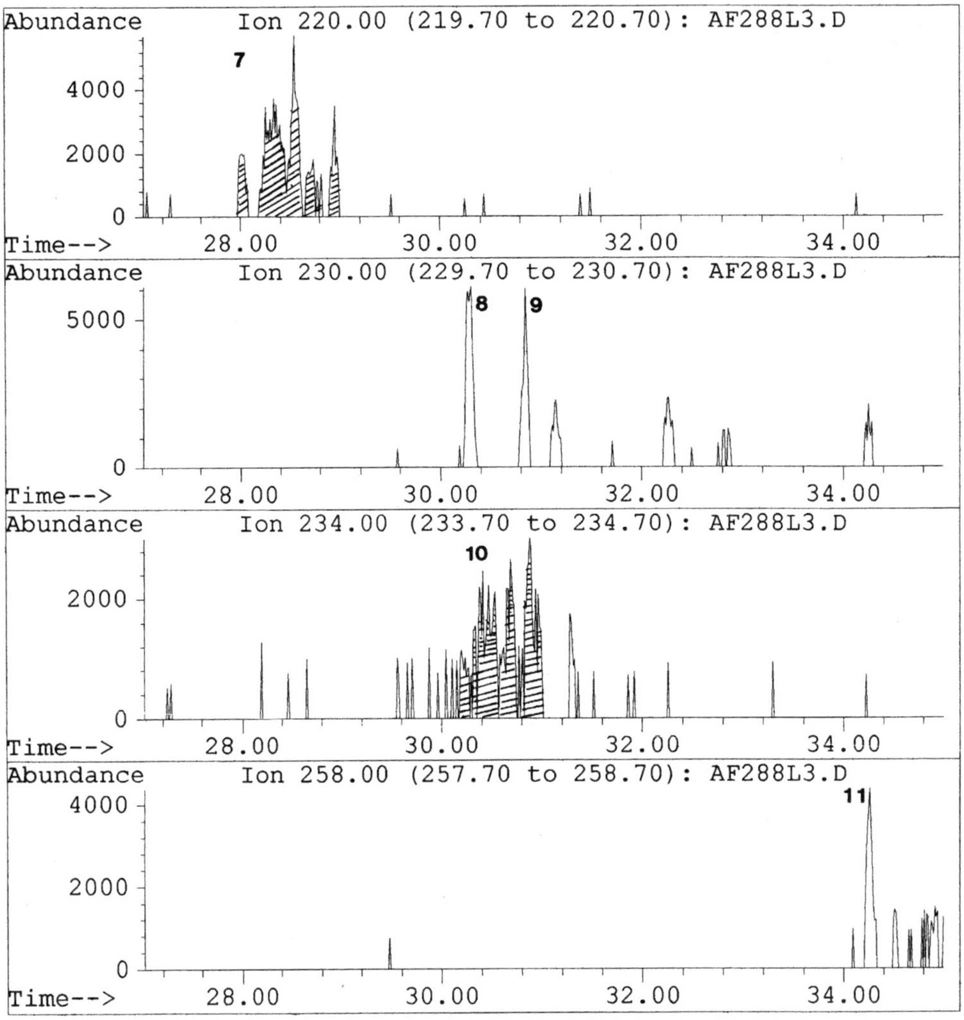

B

FIGURE 23.14 (*Continued.*)

ACKNOWLEDGMENTS

This research was supported by the European Commission CT92-0824 SCIENCE program. We thank Nikolaos Stratigakis for technical support. This contribution is dedicated to Professor Manfred Hesse, Institute of Organic Chemistry, University of Zurich, on the occasion of his 65th birthday.

REFERENCES

1. Simoneit, R.T., *Advances in Organic Geochemistry, 1979,* Douglas, A.G. and Maxwell, J.R., Eds., Pergamon Press, Oxford, 1980, 343.
2. Gagosian, R.B., Peltzer, E.T., and Zafiriou, O., *Nature*, 291, 312, 1981.
3. Marty, J.C. and Saliot, A., *Nature*, 298, 312, 1982.
4. Simoneit, R.B.T. and Mazurek, M., *Atmos. Environ.*, 16, 2139, 1982.

5. Sicre, M.A., Marty, J.C., Saliot, A., Aparicio, X., Grimalt, J., and Albaiges, J., *Atmos. Environ.*, 21, 2247, 1987.
6. Sicre, M.A., Marty J.C., and Saliot, A., *J. Geophys. Res.*, 95, 3649, 1990.
7. Simoneit, R.T., Cardoso, J.N., and Robonson, N., *Chemosphere*, 21, 1285, 1990.
8. Mazurek, M.A., Cass, G.R., and Simoneit, B.R., *Environ. Sci. Technol.*, 25, 684, 1991.
9. Simo, R., Colom-Altes, M., Grimalt, J.O., and Albaiges, J., *Atmos. Environ.*, 25A, 1463, 1991.
10. Stephanou, E.G., *Atmos. Environ.*, 26A, 2821, 1992.
11. Stephanou, E.G. and Stratigakis, N., *J. Chromatogr.*, 644, 141, 1993.
12. Stephanou, E.G. and Stratigakis, N., *Environ. Sci. Technol.*, 27, 1403, 1993.
13. Gogou, A., Stephanou, E.G., Stratigakis, N., Grimalt, J.O., Simo, R., Aceves, M., and Albaiges, J., *Atmos. Environ.*, 28, 1301, 1994.
14. Bayona, J.M., Casellas, M., Fernandez, P., Solanas, A.M., and Albaiges, J., *Chemosphere*, 29, 441, 1994.
15. Canton, L. and Grimalt, J.O., *J. Chromatogr.*, 607, 279, 1992.
16. Hildemann, L.M., Mazurek, M.A., Cass, G.R., and Simoneit, B.R.T., *Environ. Sci. Technol.*, 25, 1311, 1991.
17. Atlas, E. and Giam, C.S., *Science*, 211, 163, 1981.
18. Wade, T.L., *Atmos. Environ.*, 17, 23111, 1983.
19. Prahl, F.G., Crecellus, E., and Carpenter, R., *Environ. Sci. Technol.*, 18, 687, 1984.
20. Gagosian, R.B. and Peltzer, E.T., *Org. Geochem.*, 10, 661, 1986.
21. McVeety, B.D. and Hites, R.A., *Atmos. Environ.*, 22, 511, 1988.
22. Grimalt, J.O., Albaiges, J., Sicre, M.A., Marty, J.C., and Saliot, A., *Naturwissenschaften*, 75, 39, 1988.
23. Fernandez, P., Grifoll, M., Solanas, A.M., Bayona, J.M., and Albaiges, J., *Environ. Sci. Technol.*, 26, 817, 1992.
24. Rogge, W.F., Hildemann, L.M., Mazurek, M.A., Cass, G.R., and Simoneit, B.R.T., *Environ. Sci. Technol.*, 25, 1112, 1991.
25. Rogge, W.F., Hildemann, L.M., Mazurek, M.A., Cass, G.R., and Simoneit, B.R.T., *Environ. Sci. Technol.*, 27, 636, 1993.
26. Rogge, W.F., Hildemann, L.M., Mazurek, M.A., Cass, G.R., and Simoneit, B.R.T., *Environ. Sci. Technol.*, 27, 1892, 1993.
27. Rogge, W.F., Hildemann, L.M., Mazurek, M.A., Cass, G.R., and Simoneit, B.R.T., *Environ. Sci. Technol.*, 27, 2700, 1993.
28. Rogge, W.F., Hildemann, L.M., Mazurek, M.A., Cass, G.R., and Simoneit, B.R.T., *Environ. Sci. Technol.*, 27, 2736, 1993.
29. Rogge, W.F., Hildemann, L.M., Mazurek, M.A., Cass, G.R., and Simoneit, B.R.T., *Environ. Sci. Technol.*, 28, 1375, 1994.
30. Hildemann, L., Klinedinst, D.B., Klouda, G.A., Currie, L.A., and Cass, G.R., *Environ. Sci. Technol.*, 28, 1565, 1994.
31. Albaiges, J. and Albrecht, P., *Int. J. Envir. Analyt. Chem.*, 6, 171, 1979.
32. Psathaki, M., Zourari, M., and Stephanou, E.G., *Organic Micropollutants in the Aquatic Environ.: Proceedings of the Fifth European Symposium*; Angeletti, G. and Bjorseth, A., Eds., Kluwer Academic, Dordecht, The Netherlands, 1988, 121.
33. Killops, S.D. and Howell, V.J., *Chem. Geol.*, 91, 65, 1991.
34. Kvenvolden, K.A., Hostettler, F.D., Rapp, J.B., and Carlson, P.R., *Mar. Pollut. Bull.*, 26, 1993.
35. Simoneit, R.B.T., *Sci. Total Envir.*, 36, 61, 1984.
36. Simoneit, B.R., *J. Atmos. Chem.*, 8, 251, 1989.
37. Stephanou, E., *Fresenius Z. Anal. Chem.*, 339, 780, 1991.
38. Bray, E.E. and Evans, E.D., *Geochim. Cosmochim. Acta*, 22, 2, 1961.
39. Leuenberger, C., Czuczwa, J., Heyderdahl, E., and Giger, W., *Atmos. Environ.*, 22, 695, 1988.
40. Scalan, R.S. and Smith, J.E., *Geochim. Cosmochim. Acta*, 34, 611, 1970.
41. Grimmer, G. and Hildebrandt, A., *Zbl. Bakt. Hyg. (I. Abt. Orig.)*, B161, 104, 1975.
42. Grimmer, G., Jacob, J., and Naujack, K.W., *Fresenius Z. Analyt. Chem.*, 314, 13, 1983.
43. Radke, M., Welte, D.H., and Willisch, H., *Geochim. Cosmochim. Acta*, 46, 1, 1982.
44. Takada, H., Onda, T., and Ogura, N., *Environ. Sci. Technol.*, 24, 1179, 1990.
45. Nielsen, T., *Atmos. Environ.*, 22, 2249, 1988.

46. Gogou, A., Stratigakis, N., Kanakidou, M., and Stephanou, E.G., *Org. Geochem.*, 25, 79, 1996.
47. Sicre, M.A., Marty J.C., Lorre A., and Saliot, A., *Geophys. Res. Letters*, 17, 2161, 1990.
48. Aceves, M. and Grimalt, J.O., *Atmos. Environ.*, 27B, 251, 1993.
49. Rogge, W.F., Ph.D. Thesis, California Institute of Technology, Pasadena, 1993.
50. Didyk, B.M., Simoneit, B.R.T., Brassel, S.C., and Eglinton, G., *Nature*, 272, 216, 1978.
51. Philp, R.P., *Fossil Fuel Biomarkers: Applications and Spectra*; Elsevier: Amsterdam, 1985, 23.
52. Simoneit, B.R.T., *Atmos Environ.*, 18, 51, 1984.
53. Stedman, R.L., *Chem. Rev.*, 68, 153, 1968.
54. Standley, L.J. and Simoneit, B.R.T., *Environ. Sci. Technol.*, 21, 163, 1987.
55. Adams, J.D., La Voie, E.J., and Hoffmann, D., *J. Chromatogr. Sci.*, 20, 274, 1982.
56. Stephanou, E.G., *Naturwissenschaften*, 76, 464, 1989.
57. Killinger, A., *Arch. Mikrobiol.*, 73, 160, 1970.
58. Hollerbach, A., *Grundlagen der Organischen Geochemie,* Springer Verlag, 1985, 50.
59. Volkman, J.K., Farrington, J.W., Gagosian, R.B., and Wakeham, S.G., *Adv. Org. Geochem.*, 228, 1981.
60. Scora, R.W., Muller, E., and Gultz, P.G., *J. Agric. Food Chem.*, 34, 1024, 1986.
61. Morrison, R.I. and Bick, W.J., *Sci. Food Agric.*, 18, 351, 1967.
62. Eglinton, G. and Hamilton, R.J., *Chemical Plant Taxonomy,* T. Swain, Ed., Academic Press, 1963, 187.
63. Weete, J.D., *Chemistry and Biochemistry of Natural Waxes,* P.E. Kolattukudy, Ed., Elsevier, 1976, 349.
64. Simoneit, B.R.T., Crisp, P.T., Mazurek, M.A., and Standley, L.J., *Environ. Intern.*, 17, 405, 1991.
65. Simo, R., Grimalt, J.O., Colom-Altes, M., and Albaiges, J., *Fr. J. Anal. Chem.*, 339, 757, 1990.
66. Hatakeyama, S., Tanokaka, T., Weng, J.H., Bandow, H., Takagi, H., and Akimoto, H., *Environ. Sci. Technol.*, 19, 935, 1985
67. Stephanou, E.G., *Naturwissenschaften*, 79, 128, 1992.
68. Pitts, J.N., Lokensgard, D.M., Ripley, P.S., van Cauwenberghe, K.A., van Vaeck, L., Schaffer, S.D., Thill, A.J., and Belser, W.L., *Science*, 210, 1347, 1980.
69. Yu, M.L. and Hites, R.A., *Anal. Chem.*, 53, 951, 1981.

24 Elimination of Diesel Soots Using Oxidation Catalysts

Vincent Perrichon and P. Mériaudeau

CONTENTS

0-87371-829-1/00/$0.00+$.50

INTRODUCTION TO DIESEL POLLUTION CONTROL

The first diesel engine appeared in 1925 in trucks and in 1935 in small cars. Today, due to fuel economy and long life, diesel engines have become the standard for buses and heavy trucks. They are also widely used in taxis for the same reasons. In Europe, more and more customers are giving preference to diesel engines.[1,2] In some countries, like France, a fiscal policy of lower taxes on diesel fuel compared to gasoline and a general improvement in diesel technology have increased during recent years the share of the diesel engines in the passenger car market to more than 40%, as shown in Figure 24.1. The total production of multicylinder diesel engines throughout the world, which was 8 million units in 1989,[1] is constantly increasing.

In parallel with the growth of the diesel engine vehicle population, concern regarding the environmental impact of diesel emissions has begun to rise. A typical average exhaust gas composition of a diesel engine is given in Table 24.1.

In addition to these gaseous emissions, there is production of soot particles and some liquid hydrocarbons, generally condensed on the particulate matter.

From an environmental point of view, gaseous diesel emissions are nearly the same as those obtained from a spark ignition gasoline engine. The three controlled major air pollutants are carbon monoxide (CO), unburnt hydrocarbons (HC), and nitrogen oxides NO_x. However, in the case of diesel emissions, the levels of CO and HC are much lower.[4] The unburnt hydrocarbons, which are responsible for the odor, have varying molecular weights and contain oxygenated and polynuclear aromatic hydrocarbons (PAHs). Concerning the other gaseous emissions and in relation to the greenhouse effect, a potential advantage of diesel compared to gasoline engines is the lower carbon dioxide formation: 25% less CO_2 emitted, taking into account all the processes from the refinery to the engine efficiency.[5] However, under actual conditions, the use of diesel engines has three main drawbacks.

1. The presence of sulfur in diesel fuel leads to the formation of sulfur oxides in the exhaust, with sulfuric acid condensed on the solid particles. Together with the nitrogen oxides, they contribute to the formation of acid rains, which are highly damaging to the global environment.
2. The exhaust contains particulate matter, the composition of which varies with driving conditions. It is essentially composed of soot (carbon), heavy hydrocarbons, and sulfates.[1] Indeed, these particulates are responsible for the formation of smoke, which has a detrimental effect on visibility, contamination of the soils, and attack the building stones. They are also considered to create serious health problems due to their mutagenic character and ability to cause cancer diseases.[6,7] The amount of atmospheric haze from smoke and particulates coming from diesel exhaust is debatable. Particles less than 1 micron average diameter are called "aerosols" because they float in the air until they are washed out by rainfall. However, their importance in pollution can be underlined by considering that for each liter of fuel, 2 to 4 g of particulates are emitted into the environment.[8] As a consequence, in Europe, the particulate emissions from diesel engines can be estimated to be almost 300,000 tons per year.[7]
3. A third problem concerns NO_x removal. Unlike gasoline emissions, diesel engine exhausts are oxygen rich, and the conventional three-way catalysts do not reduce the nitrogen oxides under a high excess of oxygen.

Diesel pollutant emissions control is an important challenge for the near future. Specific regulations have been elaborated and notable progress has been achieved in reducing diesel pollution. However, the great increase in the number of diesel vehicles has made the situation even worse and the need for technological solutions urgent. To improve the situation, several countries have introduced into legislation more and more severe standards on the level of diesel pollutants. In 2000 and probably

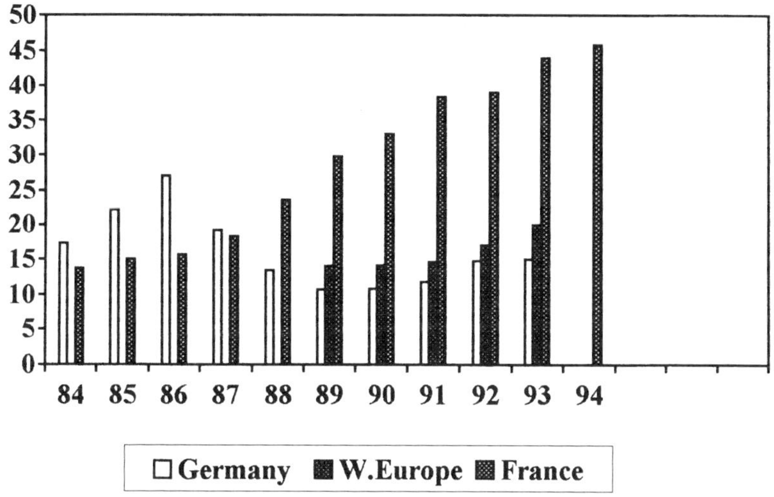

FIGURE 24.1 Percentage of diesel cars in Europe.

TABLE 24.1
Main Gaseous Components of Diesel Emissions

Compound	N_2	CO_2	H_2O	CO	O_2	NO_x	H_2	HC
(% by volume)	Balance	10.7	10	0.035	6	0.03	0.01	0.1

Source: Adapted from Reference 3.

2005, stringent new regulations will appear in the EU. However, the ways to meet these standards remain questionable, particularly for the removal of the diesel soot particulates.

This chapter focuses on particulate control in diesel exhaust gas. Although not exhaustive, it summarizes the main features of the properties, the control, and the possible solutions to reduce particulate pollution. The accent is put on the use of catalytic materials in the prevention and after-treatment of the diesel exhaust gas. The reader can find several reviews in this domain.[1,9,10] The recent book by Degobert[11] on car pollution is also a major source of information. The present chapter discusses recent developments that are significant in this area. It must be noted that the problem of NO_x reduction will be considered only incidentally, even if it may be accentuated in the future, due to possible changes in the combustion conditions necessary to decrease soot formation.

CHARACTERISTICS OF DIESEL SOOTS

Formation

Incomplete combustion of fuel and lube oil results in partially oxidized products such as carbon monoxide and oxygenates. However, in the flame region where oxygen is highly deficient, and at high temperature and pressure, the hydrocarbons can undergo a series of transformations leading progressively to highly hydrogen-deficient hydrocarbons.[12,13] They are at the origin of soot particles. In case of aromatization, the condensation of several cycles leads also to the formation of polyaromatic hydrocarbons, which are usually trapped on the soot, together with unburnt oxygenated hydrocarbons, and constitute the major constituents of the particulate matter that can be filtered in diesel exhaust gas. In general, with various engine loads, the particulates emissions decrease with increasing excess air ratio.[14] In addition, a large fraction of the matter collected is mineral and corresponds to the sulfate formed through the total oxidation of the organic sulfur and reaction with the water formed during the combustion.

Morphology and Texture

As outlined above, the composition of particulates is complex, and their physical texture will depend greatly on the conditions of collection. In general, the diesel soot particulates consist of primary particulates ranging from 10 to 60 nm in size.[9,15,16] These primary particulates aggregate in agglomerates of around 100 to 1000 nm.[17]

The reported bulk density of soot is very low, 0.075 g cm^{-3}, and can be lower (0.02–0.06 g cm^{-3}), depending on pretreatment (washing) of the soot.[15]

The surface area of soot particles ranges from 19 $m^2\,g^{-1}$ to 300 $m^2\,g^{-1}$.[18,19] It can be changed by desorption of water and volatile compounds. For example, it increased from 35 $m^2\,g^{-1}$ to 270 $m^2\,g^{-1}$ just by increasing the outgassing temperature from 100 to 600°C.[16,20] Most of the surface areas are between 100 and 200 $m^2\,g^{-1}$.[15,21]

Chemical Composition

Several investigations have shown that diesel exhaust particulates are a complex mixture of chemical compounds.[22-24] More than 10,000 chemical compounds could be identified. In a modelized approach, each particle can be described as a solid soot nucleus on which is condensed a quasi-liquid film of polynuclear aromatic (PAH) and unburnt hydrocarbons. A representative diesel soot contains 70 wt % C, 20 wt % O, 3 wt % S, 1.5 wt % H, and <1 wt% N.[20] The mineral matter can appear as another solid phase mixed with the soot and can be covered by sulfates insofar as metallic particles can catalyze the oxidation to sulfates.[1] Moreover, the sulfate is usually associated with water.[25]

The soot fraction can be considered as a graphitic "hard" carbon, containing only a very small proportion of hydrogen. The organic fraction itself consists of components from the fuel and the engine lubricating oil that are adsorbed on the large surface area of the soot during the cooling of the exhaust gas.[21] It is quantified as either volatile or soluble organic fraction, usually called SOF. The amount of volatile is determined by thermal gravimetric analysis in an inert atmosphere with a maximum temperature around 750°C.[21,26,27] The SOF is determined by extraction with a solvent such as methylene chloride, toluene, acetone, or water.[21]

The amount of volatiles is a function of the operating conditions of the engine. At idle running, the soot contains 45 to 55% volatiles; whereas at 65 km h^{-1}, it contains 15%.[26-28] When diesel soot is heated under nitrogen, the hydrocarbon emission is maximal at 210 to 220°C, followed by a tailing emission up to 450°C.[16] The amount of SOF decreases with increasing exhaust temperature.[15] In general, it varies between 10 and 50%,[9] 1.8 and 10.5%,[29] 3 and 20%,[15] or 10 and 20%.[21] A detailed analysis of the volatile matter was given by Hunter et al.[29]

The amount of inorganic material in soot is in the few percent range: 2 to 6 wt%[21] or 0.6 to 2 wt%.[9,30] The major components determined were Fe, Ba, Ca, Zn, and Ni. The minor components were S, P, Cr, and Cu. These elements originate from the diesel fuel and the crankcase oil. For example, the presence of calcium sulfate dihydrate ($CaSO_4 \cdot 2H_2O$) was identified by X-ray powder diffraction analysis on the residue obtained after regeneration of a particulate trap.[31] The calcium presence originates from the lubricating oil, which can combine with the sulfur from the diesel fuel.

Physiological Properties

The colored smoke due to the oxygenates and hydrocarbon emissions is well-known for having a nasty smell. However, a more important health problem comes from the submicron particulates that contain on their porous surface different kinds of heavy hydrocarbons. Among them, the polycyclic aromatic hydrocarbons (PAH and nitroPAH) are known for their mutagenicinity[23] and carcinogenic properties.[1,32] The small size of the particles facilitates their passage into the lungs alveoli. Lung and bladder cancers were found more numerous in people exposed to diesel-powered vehicles like taxi drivers (cited in Reference 33).

POLLUTION CONTROL AND REGULATIONS

Diesel Pollution and Its Evolution

Some 50 million vehicles circulated throughout the world in 1950; that is, ten times less than today. In 40 years, there will probably be 1 billion vehicles. To avoid an asphyxia due to exhaust gas pollution, this evolution has given birth and justified increasingly stringent regulations in those countries concerned with this pollution problem. The first air pollution laws for gasoline motor vehicles were passed in California in 1959 and the first exhaust control devices were installed on California cars in 1966.[34] The United States and Japan took measures in this domain in 1966. They were followed later by the countries in western Europe. Primary attention was first given to CO and HC emissions. Then, the standards were extended to NO_x and particulates, the latter concerning more specifically the diesel engines. A detailed presentation of the different regulations throughout the world was given by Degobert.[11] The more recent points of interest are summarized below.

The Control Methods

In the U.S., Europe, and Japan, precise standardized emissions tests have been established that try to mimic the specific driving conditions in each country and include both the circulation in cities and high speeds on freeways. The definition of these tests have changed through the years, and it is rather difficult for non-specialists to compare the data, often given with different units, of one country with that of another. The difference in the transient tests also increases the difficulty and makes the comparison often impossible. It may happen that models of cars meeting the U.S. 83 requirements have emission levels higher than the limits given by the current European standards, although they are supposed to be equivalent.[35] The European Transient Test (Cycle A), representing a distance of 11.18 km, contains 13 cycles divided in two segments: the ECE low-speed urban segment (12 cycles during 800 s, average speed = 19 km h^{-1}) and the EUDC high-speed segment (1 cycle of 400 s, average speed = 62.6 km h^{-1}), including a top speed of 120 km h^{-1}. This last feature differentiates the European driving cycle from that of the U.S. because it introduces vehicles speeds much higher than the U.S. highway cycle.

During the test, the emissions are collected through filters in special bags. The analysis is performed at the end on the total recovered products. The diesel particulate standard corresponds to the weight increase of the filter maintained at a temperature lower than 52°C. It includes the soot, the SOF, water with sulfuric acid, and some inorganic materials. The results are usually expressed in grams per kilometer, rather than the previous used "g/test."

Present and Future Regulations

The first EEC directive related to emission control was issued in 1970 (70/220/EEC). At that time, the standards recommended for cars were very high compared with what is ruled now: about 30 g km^{-1} CO and 2.5 g km^{-1} HC for the first ECE test. During recent years, after discussion and successive proposals, more stringent emissions standards were decided by the European Parliament in order to have regulations equivalent to the emission standards of the U.S.[36] Table 24.2 gives the values decided by the U.S. Clean Air Act Amendments of 1990. The particulate matter had to decrease to 0.05 g km^{-1} in 1994.

TABLE 24.2
Limit Values of Exhaust Emissions for Diesel Cars (in g. km^{-1}) [specific U.S. (FTP), Japan (10.15 mode), or European (Cycle A) transient tests]

Standards	Year	HC + NO_x	CO	Particulates
U.S. 83	1987	0.26 + 0.62	2.2	0.12
U.S. 93	1995	0.16 + 0.62	2.2	0.05
Japan	1994	0.5	—	0.2
EURO I	1992	0.97	2.72	0.14
EURO II	1996	IDI 0.7[a]	1.0	IDI 0.08[a]
EURO III	2000	0.56	0.4	0.05

[a] IDI: Indirect Injection Diesel.

The European standards are also given in Table 24.2. Following the propositions of EURO II and EURO III, strong limitations have appeared for 1996 and 2000, both on gaseous and particulates emissions.

Similar standards regulations, expressed by power unit, were decided for medium- and heavy-duty vehicles and also for buses.[11]

PREVENTION OF SOOT FORMATION

The reduction of particulates emissions is a complicated problem that may require multiple answers and probably new technologies. Two main approaches will be discussed. The first one is to avoid the formation of the soot, the second one being to treat the soot after its formation. Prevention can be realized in different ways, which are summarized below. Most of them depend on the improvement of catalysts and catalytic engineering.

Improvement in Gas Oil Composition

The influence of diesel fuel composition on the different exhaust emissions has already been established.[11] Egebäck et al.[23] have shown a clear relationship between fuel properties and particulate emissions, the heavier fuels emitting more particles than lighter fuels. High levels of polycyclic aromatic hydrocarbons (PAHs) are obtained from fuels containing aromatic hydrocarbons. In addition, the sulfur content has a direct influence on particulate composition and SO_x emissions.

As a consequence, one way to lower pollutant emissions is to improve the refining process to produce cleaner fuels. In several countries, like California and Sweden, a limitation on aromatics has been introduced, with a maximum content inferior to 20 or even 5%, which requires very severe dearomatization processes. It is also essential to desulfurize more deeply the fuels in order to lower the sulfur content present in diesel and gas oil. For example, in Japan, reducing the sulfur content to a level less than 0.2 wt% was fixed, and equipment for desulfurization to the level of 0.05 wt% is now under construction.[37] Similarly, the 1989 EPA standards of 0.11 g/mile particulates for buses in 1991 and for trucks in 1994 required severe rules for all U.S. refineries, which now have to produce a 0.05 wt % sulfur content diesel fuel, with no more than 10 vol% in aromatics. In Europe (EEC), the maximum sulfur content in diesel fuel was established at 0.05 wt% in January 1996, but could be lower to meet the exhaust emission standards imposed in certain countries. For example, in Sweden, a lower limit of 0.001 wt% was imposed, together with severe limitations on aromatics content (less than 5 vol%). In order to meet these new specifications, the fabrication of fuels by deeper hydrodesulfurization and dearomatization processes is already effective,[38] but new processes are required as well.[39] It involves extra cost, but is worth it for the environment and indispensable if the vehicle is equipped with a catalytic after-treatment. For example, the use of oxidation catalysts as a means of reducing the HC levels and a part of the particulate matter, requires a diesel fuel with a sulfur level as low as possible.

Another possibility was successfully tried. It consists of producing diesel fuels using the Fischer-Tropsch synthesis. The main advantages are that sulfur must be eliminated before the synthesis and that the hydrocarbons formed are essentially linear paraffin without aromatics.[40] Only limited facilities in the Fischer-Tropsch domain are available in the world (South Africa) and, for the moment, the cost of this production limits its application for cars.[41] A related process with the same objective is the Shell Middle Distillate Synthesis (SMDS), in which natural gas is converted to middle distillates through a combination of an improved Fischer-Tropsch chemistry and a special hydroconversion process.[42] The SMDS gas oil has a high cetane number and would lead to substantially reduced particulate emissions.[43]

There is also a potential interest to use vegetable products as chemical bases to produce diesel fuels or fuel additives. For example, diesters from vegetable oil are an excellent substitute, with pollutant emissions comparable to that of diesel fuel, or even lower for particulates.[11,44] However, due to the higher cost price compared to that of gas oil, its general use as a substitute is unlikely in the near future without lower tax incentives.

The use of low-molecular-weight alcohols, such as methanol or ethanol, could also be interesting because they produce only small amounts of particulates. However, they have too low a cetane number, which requires engine-deep modifications and, compared to diesel oil, they result in higher emissions of unburnt hydrocarbons and aldehydes or other oxygenated by-products.[45] This last problem could be overcome by the use of an oxidation catalytic converter, as studied by Pettersson et al.[46]

Improvement in Lubricant Oil

Although lubricating oil volatilizes at temperatures slightly higher than diesel fuel, it largely contributes to the formation of the SOF. The fraction associated with the lube oil can represent one fourth of the total particulate weight[25] or even half of the SOF.[47] Hence, more thermostable lubricant oils that work at less consumption are needed.

In addition, we noticed that transition metals are present in lube oil, particularly calcium. The deposit of oxide or sulfate compounds on the walls of the catalytic device could eventually mask or even poison the active phase. Consequently, there is a need to develop new types of oils with lower calcium content.

Improving the Combustion of the Fuel

Improving the Combustion of the Engine

Particulate emissions decrease with increasing excess air. However, improving the combustion of the fuel results in a higher temperature, which leads to increased NO_x emissions. Nevertheless, some improvements in diesel combustion may appear most desirable. They can be accomplished by redesigning the combustion chamber, eventually with optimization of the prechamber for indirect injection. Direct-injection engines using electronic control and intercooler turbo-charging, high pressure fuel injection up to 2200 bar will also be used in the future for passenger cars to lower the amount of released particulates.[2,10,48-50] Water or organic additives can also modify the combustion and decrease the particulate emission levels.[11] Nevertheless, the complete particulate control within the engine does not seem to be a likely possibility.

Catalytic Additives to the Diesel Fuel

In order to promote fuel combustion, fuel additives can be employed. The catalyst is then produced directly during the combustion, in close contact with the fuel droplets. More than inorganic metallic salts, organometallic compounds were preferred. For example, metallic naphtenates of barium and calcium resulted in little soot production, as did ferrocene and ferric acetylacetonate.[51] In the same manner, cerium in the fuel improves combustion efficiency (less smoke) and reduces particulate buildup (carbon). However, the use of metal additives without a filter in the exhaust line is insufficient in itself, because a substantial fraction of the additive will end up in the exhaust, which may be unsafe for human health. In fact, as presented below, the major effect of these additives could be to help the regeneration of the filter trap.

TREATMENT OF SOOT AFTER ITS FORMATION

The second way to decrease soot in diesel exhaust gas is to trap or (and) burn it, continuously or discontinuously. This type of solution seems remote in the near future, because better quality of the fuel and improvement of the combustion will probably be insufficient to realize the abatement of the pollution required by future adopted standards.

FILTERS

In theory, the first efficient mode to remove a maximum of soot particles would be to filter the smoke. Ceramic filters appeared in the early 1980s.[52] Made of cordierite (magnesium aluminosilicate), a monolith is a cellular ceramic honeycomb with square-shaped cells typically having 300 to 400 channels per square inch of frontal area.[53] A system of plugs force the gas to flow through the thin porous walls. When it is put in the exhaust pipe, it can trap more than 90% of the emitted particles. Metallic filters, consisting of knitted stainless steel wire, with a graduated structure in order to trap finer and finer particles as the exhaust flows from the outside toward the center, were also tested.[12,13] Other different kinds of texture and shape were evaluated: packed bed, sintered ceramic, fibers, and foam ceramic.[8,54] The geometric configuration and the container assembly are also important parameters in particulate collection efficiency and thermomechanical characteristics.[54]

The essential problem in filtering is that the filter becomes very rapidly saturated and the increase in the back-pressure progressively stops the normal running of the engine. Plugging is more rapid on monoliths than on ceramic foams.[9] The disposal of collected soots in a non-polluting manner cannot be solved easily. Discontinuous regenerations are thus needed to overcome this difficulty. The principle is to periodically burn the accumulated soot by the excess oxygen present in the exhaust gas. It was shown that the amount of pollutants emitted in such a regeneration mode is negligible compared to emissions during the accumulation.[31] However, the temperature needed to oxidize the carbon present in the particulates is almost 600 to 650°C, whereas the usual temperature of the exhaust gas is near 300°C on roads, and even much lower (<200°C) for passengers cars in the cities, where the pollution problem is most critical. Two possible series of solutions were tried to regenerate the filters. They are based either on an increase of the exhaust temperature or on a lowering of the ignition temperature of the soot. The most promising techniques will probably combine these two solutions. Let us first consider the various techniques involving an increase in the temperature of the exhaust gas.

Electrical Heating

Electrical heating aims to increase the temperature of exhaust gas or the filter itself. It has been tested by Niura et al.[55] They developed a system with heating wires attached on the front surface of the filter. They found that the propagation of soot combustion in the radial direction was difficult and could give rise to serious thermal damage. The problem could be solved using high-density packaging of the heating wires, which in terms of energy consumption is very unfavorable. Microwave irradiation of the trapped particulates, which offers the advantage of selective energy absorption by the soot, can be also used.[56]

Hydrocarbon Combustion

A burner can be used to burn the particulates by injecting fuel or propane gas into the trap, thus igniting the soot. This approach is based on the fact that a high content of volatile fraction in the soot decreases the ignition temperature.[26,27] Thus, the deliberate injection of hydrocarbons into the tap was tried with some success to promote low-temperature regeneration.[57] However, according to Niura et al.,[55] a diesel fuel burner situated before the filter and able to heat the filter at more than 600°C gave unreliable results. This is due in particular to frequent failure of the spark igniter because of unavoidable soot deposits. After some time, the diesel fuel retained by the filter could burn and, on occasion, severe damage by hot spots or even melting of the filter could occur.

Air Throttling

The group of Prof. Pattas has realized a regeneration system that consists of limiting the stream flux by a throttle valve as soon as the back-pressure becomes too high.[58,59] The resulting overpressure raises the exhaust gas temperature up to the required level for filter regeneration. At 2 to 3 bar exhaust overpressure, temperatures of at least 550°C can be obtained. This system of forced regeneration was tested on buses in Athens. A by-pass device between two filters allowed permanent driving of the bus.[60] However, exhaust overthrottling leads to a substantial reduction in engine net power output at high speeds and, consequently, higher fuel consumption.

In fact, the system of air throttling is highly improved by adding to the gas oil a cerium organo-soluble compound that is dissolved in the fuel, in order to decrease the ignition temperature from 600 to less than 500°C, a temperature much easier to obtain with the normal driving conditions of a bus.[61] The system of air throttling, combined with the addition of cerium-based catalysts, was developed by Rhône-Poulenc (now Rhodia) and has been tested on buses. It has been successfully run for 5 years in Athens and will equip experimental buses in Paris. However, its price is certainly too high to equip small cars. Another drawback of the system is that it is necessary to periodically "air wash" the filter from the cerium dioxide that accumulates in the filter, a procedure which cannot be easily envisaged for private diesel cars.

The main problem with this discontinuous regeneration is that the combustion of a large amount of carbon is highly exothermic, and the resulting high temperatures may cause severe damage to the filter: melting of the ceramic in some cases, cracks and destruction of the tightness. Security devices are then necessary to avoid these overheatings, which requires an additional cost.

Before surveying the principles of the catalytic filters, it should be noted that the design of a filter is critical; it can influence the oxidizing trapping efficiency more than its composition. Thus, Hoffmann and Rieckmann[8] studied several kinds of diesel soot filters (packed bed, sintered ceramic, fibers, foam ceramic) and several materials with high melting points (Al_2O_3, SiC, ZrO_2, industrial ceramics). A packed bed of alumina particulates was found to be the most active filter in soot combustion. Catalytic coating with oxides (V, Cu, Ce) accelerated the reaction. However, the manner of filtration and the nature of the soot filter was found more important for the overall result than the nature of the catalyst coating.[8]

Oxidation Catalysts

A more elegant and ideally cheaper solution would consist of continuous combustion of the soots using oxidation catalysts. The role of a catalyst is to initiate the combustion of the trapped particulates below the thermal ignition temperature. The reactivity pattern of the different reactive species for oxidation in the exhaust gas is the following: CO > gaseous hydrocarbons > unburnt and adsorbed hydrocarbons (SOF) > solid carbon. In an ideal scheme, the oxidation catalyst would be active enough to continuously oxidize the solid carbon for all driving conditions (i.e., even at low temperatures (150–250°C). Moreover, it must realize the total oxidation of the hydrocarbons with 100% selectivity to CO_2 (i.e., without CO formation). The main drawbacks in using an oxidation catalyst are that (1) it is also able to oxidize the sulfur to SO_3, which then remains trapped on the particulates as sulfates; and (2) it can form additional nitrogen oxides if the reaction temperature becomes high enough. In the urban part of the European test, the temperature of the gas remains quite low, (e.g. 100–250°C). However, under high-speed conditions, the temperature can increase to 550–600°C, temperature for which the activity for the formation of SO_3 and NO_x can be quite high. As a result, the catalyst must be tailored to work over a broad range of temperatures, but also with conditions that are often quite different.

Four types of catalytic approaches were developed. The two more extensive approaches concern the deposition of a catalyst on a "flow-through" monolithic reactor or on a filter surface. In the two others, which also require the use of a filter, a catalytic component is added either directly in the fuel or sprayed on the soot before trapping in the filter. We present below the main features of these catalytic devices, and the next section will summarize the numerous studies performed to select the best active catalytic formulas.

Flow-Through Type Oxidation Catalysts

The use of catalyzed flow-through monoliths constitutes a recent advance to meet the air quality standards. It mimics the technology of the catalystic monolith used for gasoline engines, but keeping only the oxidizing function. This type of device was introduced for passenger cars in 1989.[50] It aims to burn the SOF part of the particulates, as well as CO and gaseous hydrocarbons.[1] However, due to the low temperatures of diesel exhaust and the low contact time, the conversion of soot is not possible. Because the know-how of "three-way catalysts," which are based on the catalytic role of the noble metals, numerous catalytic systems were tried based on precious-metal catalysts, particularly on platinum (see below). They were found active enough at low temperature to oxidize the gaseous effluents (CO and HC) and most of the SOF part of the particulates. However, platinum is typically a catalyst for the oxidation of SO_2 in SO_3.[62] Thus, the formation of sulfuric acid on the particulates may become an important drawback for high temperatures which occur at high speed.[53,63] A partial replacement of platinum by palladium and modifications of the washcoat can minimize such a formation.[25,48-50] Fouling of the catalyst due to carbonaceous soot must also be minimized. Diesel oxidation catalyst technology has been demonstrated in steady-state and transient engine tests and is now commercialized in most diesel vehicles.[64]

Catalytic Filter

In a catalytic filter, the exhaust gases are forced through the channel wall pores on which a catalyst has been coated in order to burn the soot deposit in a continuous manner. A major problem is related to the design of the catalytic filter. In addition to the problems of avoiding back-pressures as for filters, another problem concerns mass transfer between the catalytic surface and the particulates, which have submicron size. Sharp angle changes in the gas flow and the creation of turbulence will increase the probability of impingement of the particulates onto the catalytic sites, provided that it is not covered by an unburned protective carbon layer. The design of the catalytic filters was also extensively studied. They include alumina-coated wire mesh, ceramic fibers,[26,27] ceramic honeycombs, ceramic or metallic foams.[8,65] Finally, all the questions of chemical engineering regarding the heat transfer and reactants diffusion processes also exist in these systems.

We will discuss the possible developments that must be found to resolve this challenge. Let us say here that many attempts were realized to oxidize the particulates through catalytic trap oxidizers.[8,12,13,26,27,54,65] The main difficulty that is not really overcome is to have a global regeneration system active enough to avoid the build-up of particulate layers.

Addition of Oxidation Catalyst Precursors in the Fuel

Laboratory studies have shown that ignition temperatures of the soots trapped in the filter can be decreased by up to 250° by incorporating metal additives in the fuel. As noted before, the first function of the additive is to prevent soot formation by improving the combustion itself in the chamber; it is also to lower the ignition temperature in the filter by a supposedly better intimate contact between the carbon and the catalyst. Copper, lead and manganese with a concentration of 0.1 to 0.2 g metal l^{-1} were found to improve soot burning.[14,66] Copper additives appear to be more efficient than manganese, iron, and lead additives.[26,27,67] Calcium and barium additives were also tested.[68] Moreover, as cited above, the system of air throttling to remove the soot trapped in a filter is highly improved by adding to the gas oil a cerium organo-soluble compound that is dissolved in the fuel. However, filter life could be limited by ash clogging due to metallic compounds. In addition, for widespread use, it must be demonstrated that the fuel additive has no long-term toxicological effect on human health.

Injection of Oxidizing Agents into the Exhaust Gas

The method consists of injecting an oxidant onto the particulates after their formation in order to induce the soot combustion at low temperature. Thus, as in the case of the additives in the fuels, the sprayed soot particles are in very close contact with the oxidation catalyst.[55] Only metal-containing compounds were found sufficiently active. Hardenberg et al.[69] observed that iron acetylacetonate was very effective because it lowered the regeneration temperature to 210°C, compared to 230 and 310°C, respectively, for cuprous chloride and copper perchlorate.

It must be noted that periodic injection of the additive eliminates the problem of catalyst deterioration. Moreover, after a period of utilization, the filter walls themselves become catalytically active due to the deposition of layers of the active components. However, the durability of such systems is dubious and thus far unproved, particularly due to plugging problems.

CATALYTIC COMBUSTION

Literature Survey on Catalytic Oxidation of Graphite

The main challenge to be solved is to find a catalyst with enough activity to oxidize the soot and hydrocarbons at 200°C or lower. As pointed out in the previous section, catalytic soot combustion is a rather complicated problem because it is a solid (catalyst)–solid (reactant)–gas (reactant) reaction. Consequently, the physical contact between the soot and the catalyst is a crucial parameter. Similar systems have been well-investigated in the recent past (e.g., those relative to the catalytic oxidation of graphite). We will briefly review the most salient conclusions obtained from catalytic oxidation of graphite or coke before considering diesel soot combustion.

Catalyzed oxidations of carbon by oxygen have been studied using different types of catalysts, such as noble metals,[69,70] oxides,[71] and alkalis.[72]

1. First of all, it has been clearly established that the catalyst–graphite or catalyst–carbon interface is a key parameter and it is probably why most of the studies have been performed on catalysts directly deposited onto the reactant support in order to ensure the best possible contact between the two solids.

The use of *in situ* electron microscopy and of finely divided metals dispersed on the basal plane surface of natural graphite crystals provide evidence that the oxidation of the graphite occurred either via the formation of etch pits for Pt or via channeling of the basal plane for MoO_3.[73] The mobility of the catalyst is an important factor and, due to the high temperatures required to oxidize graphite, some catalysts are acting as liquids.

2. Experimental and thermodynamic considerations have shown that the elements able to adsorb oxygen dissociatively and form stable oxides are generally active for the oxidation, but they should not be converted into carbides (which are inactive) through their interaction with the carbonaceous support at high temperature.[74]
3. The reactivity of the material depends on the H:C ratio — the higher H:C is, the higher the reactivity.[75]
4. The major part of metal oxide catalysts exhibiting catalytic properties for carbon oxidation are able to undergo cyclic redox processes on the carbon surface during the oxidation reaction and this is why the oxygen transfer mechanism has been proposed for such solids; it is assumed that the catalyst exists in two states (reduced and oxidized); the higher oxidation state is reduced by the carbon at the carbon–catalyst interface, and the gas-phase oxygen reoxidizes it. For copper-based catalysts[76]:

$$Cu + 1/2\ O_2 \rightarrow CuO$$

$$CuO + 1/2\ C \rightarrow 1/2\ CO_2 + Cu$$

Most of the elements of the periodic table have been tested in the catalytic oxidation of carbon. It is rather difficult to classify these elements as to order of reactivity because it is the specific activity that must be considered, and the dispersion (or surface area) of the catalyst is difficult to measure in experimental reaction conditions. Nevertheless, among the oxides, Cu is the most active: $Cu >> Cr > Fe > Mn > Co \approx Ni \approx Zn$.[77,78]

Among the group IIIB and IVB elements, Tl and Pb are highly active, the addition of IA or IIA elements (Ca, Na, K) having a beneficial effect.[79,80]

Among the noble metals, Pd is the most active catalyst.[81]

It appears from this very brief review that the results of the literature relative to the catalytic combustion of carbon or graphite can help elucidate a catalyst formulation able to combust the diesel soot, keeping in mind that some elements, even if they are active, have to be discarded because they are noxious (e.g., Tl).

Catalytic Combustion of Diesel Soot

Among the published results, some were obtained with model systems (the most abundant part of the literature). Others were obtained in real experimental conditions, using either a burner or a diesel engine. We will first report on the findings obtained for model systems, and then on those related to more realistic conditions.

Model Systems

In order to simplify the experimental procedure, mechanical mixtures of diesel soots and of catalysts were made. It is obvious that by using this procedure, a much better contact is obtained between the soot particulate and the catalyst grain. The soot–catalyst interface is a crucial parameter[28] and reproducible results were obtained by mixing the two components with a ball milling system, the time of milling being an important parameter. For a constant milling time, reproducible results were obtained,[65] allowing the comparison of different catalyst formulas.[19]

The rate of soot combustion was measured either gravimetrically via ATG/DTG experiments or via analysis of CO/CO_2 formed, these gases being detected either by gas chromatography or mass spectrometry. This simple method allows the classification of the different catalysts by order of activity. Thus, among the different solids studied the order of activity is: $V_2O_5 > La_2O_3 > CeO_2 > CuO > TiO_2 > Al_2O_3$.[19,20,65]

This classification was obtained in the absence of SO_2 in the combustion gases ($N_2 + O_2$); but in the presence of 100 ppm SO_2, the CeO_2 activity was strongly depressed.

It is generally reported that mixed oxides are more active than pure oxides, showing a synergistic effect. Among the most active mixed oxides are CuO-V_2O_5[65,82] and CuO-Nb_2O_5.[65] The addition of an element from group IA improves the catalytic properties of such mixed oxides.[18] As an example, CuO-V_2O_5-K_2O/Al_2O_3 or CuO-Nb_2O_5-K_2O/La_2O_3 can be considered as potentially interesting catalysts for the soot combustion: combustion temperatures as low as 300°C have been obtained.[83-87]

The combustion of soot on such catalysts generally generates CO_2, H_2O, and some amount of CO. In order to have complete combustion, traces (0.1–0.5 wt%) of noble metals like Pt have been added.[88,89] It is also worth noting that they are primarily active in oxidizing the hydrocarbons that desorb from the soot.[20,82]

Having summarized the model catalysts that might be suitable for exhaust gas purification, we will now consider how they can be utilized in practice.

Catalytic Combustion on Ceramic Diesel Particulate Filter Containing Catalyst

Since ceramic filters[12,13] in the form of monoliths are recognized as being the most effective systems for trapping soot particles, they have been utilized for supporting different catalysts, either noble metals on oxides,[53] mixed oxides,[90] or a combination of the two.[89,91] In any case, it is possible to combust the soot particles at all driving modes because of the low temperature of the exhaust gases. Under these conditions, soot could accumulate and cause an increase in the pressure by plugging the monolith pores.

So, as mentioned above, an additional energy source must be used to increase the temperature of the filter containing the catalyst up to 300 to 350°C, a temperature for which the activity of the catalyst becomes important for soot combustion. At lower temperatures, it is only part (or all) of the SOF that is removed by the catalytic trap, thus resulting in decreased pollution.

It is also noteworthy that despite the fact that the sulfur content of diesel fuel has significantly decreased in the past few years, this still is an important parameter. The presence of an oxidation catalyst causes the formation of SO_3, which is generally trapped on soot particles and can represent, in some cases, nearly 30 wt% of the solid particulates.[53]

Moreover, if it is possible to obtain an exhaust gas containing a minor amount of dry soot particulates because, for example, of high-pressure injection system (see previous section), another noxious contaminant is produced at higher levels: NO_x. Before surveying the question of its elimination, it must be noted that attempts were made to use soot particulates as reducing agent of NO in the presence of perovskite-type oxides as catalysts.[92,93]

It was also proposed to use nitrogen dioxide rather than activated oxygen to oxidize soot.[94] In this respect, a new technology was recently developed in which a platinum/alumina catalyst was mounted before the ceramic filter to fully oxidize NO_x in NO_2, and a continuous regeneration of the filter could be obtained at a temperature of 275°C.[95]

The NO_x Problem

It has been shown that the decrease of soot content in the exhaust gas of diesel engines is done at the expense of a higher production of SOF (which are easier to remove at low temperatures than the soot particulates) and high NO_x production. The removal of these NO_x compounds, in the presence of a high oxygen content, is not straightforward. The diesel vehicle engine covers A:F ratios from around 19:1 at full load to 120:1 at idle. At no time does there exist a reducing environment to convert NO_x, and the possible selective catalytic reduction of NO_x (SCR) by ammonia cannot be taken into consideration from an applied viewpoint. Consequently, other solutions must be explored.

In this respect, direct decomposition and selective reduction by hydrocarbons are potential ways to remove NO_x, and they have been widely studied in the recent years.[96] For example, it has been reported that mixed oxides such as Ba_xCu_yO could be efficient for decomposition and, recently, Iwamoto et al.[97] have shown that Cu/ZSM-5 are quite effective in removing NOx.[97] Nevertheless, these catalysts quickly lose their activity due to strong adsorption of oxygen, and it is not presently known how such catalysts would behave for NO_x removal in the presence of soot particles, SOF, and SO_2. It can be envisaged that if Cu-based catalysts are not able to work properly in the presence of SOF and soot, a two-stage system could help to solve this question: the first catalytic converter removing SOF and soot, the second one having to decompose NO_x into N_2 and O_2 in the presence of H_2O, CO_2, O_2 coming from the first particulate filter containing catalyst.

In a different approach, it was reported that Cu/ZSM-5 catalysts are active in the selective reduction of NO_x by hydrocarbons under oxidizing conditions.[98] Since this finding, a large number of investigations have been carried out to test other catalysts. Several types of metal/ZSM-5 catalysts, acid supports, and mixed oxides were found to be fairly active SCR catalysts.[96] Typical hydrocarbons are ethene, propene, and propane. The presence of oxygen is necessary to obtain high conversion levels. It was also shown that platinum-based systems can catalyze the reduction of NO_x by hydrocarbons in well-defined operating conditions and also on vehicles.[3] However, efforts must continue to find more selective and stable catalysts.

CONCLUSIONS

Although the development of new engines or new fuels will find a growing place in the next 50 years, gasoline and diesel engines will for many years dominate the worldwide transport market. Concerning diesel engines, modifications of the engine, the evolution toward direct injection for top- and middle-range vehicles, and the introduction of low sulfur diesel fuel will contribute to reaching new standards. However, at present, the use of after-treatment technologies has proved the necessity to obtain emissions levels below the allowed limits. This is illustrated by flow-through catalytic converters that allow the combustion of SOF, together with CO and gaseous hydrocarbons. In the future, this technology will have to move toward the oxidation catalytic filter, which today appears the only way to get rid of particulates. The actual results are promising but must be highly improved to meet the challenge of the very stringent environment control in the years to come.

However, the complexity is so high that the catalytic filter alone cannot solve the problem. Various measures must be taken in parallel. For example, sulfate formation must be decreased or minimized through the use of a diesel fuel containing a very limited amount of sulfur (0.005%, or even 0.001%). The working catalyst temperature must be increased: it is actually impossible to realize combustion at around 150°C, which is the temperature of the exhaust emissions in the traffic jams of the urban cities. A convenient and economic external heating device must also be developed. Concerning oxidant catalytic additives to diesel fuel, together with a catalytic filter trap, their use could be reliable only if the concentration is very low (to avoid a plugging of the filter with the deposits of the oxidized salt). Finally, efforts to improve diesel engine combustion and the possible modifications of exhaust gas devices will continue to be essential in developing environmentally safe diesel engine vehicles.

ACKNOWLEDGMENTS

The authors are grateful to J. van Doorn, A. Bellaloui, and Y. Shibin for numerous discussions throughout recent years. They are also indebted to M. Chevrier and C. Gauthier for providing valuable information on the subject.

REFERENCES

1. Lox, E.S, Engler, B.H and Koberstein, E., *Catalysis and Automotive Pollution Control II*, Crucq, A., Ed., Studies in Surface Science and Catalysis, Vol. 71, Elsevier, Amsterdam, 1991, 291.
2. Eyzat, P., *Catalysis and Automotive Pollution Control III*, Frennet, A. and Bastin, J.M., Eds., Studies in Surface Science and Catalysis, Vol. 96, Elsevier, Amsterdam, 33, 1995.
3. Engler, B.H., Leyrer, J., Lox, E.S., and Ostgathe, K., *Catalysis and Automotive Pollution Control III*, Frennet, A. and Bastin, J.M., Eds., Studies in Surface Science and Catalysis, Vol. 96, Elsevier, Amsterdam, 1995, 523.
4. Cayot, J.F., *SAE Paper 930941,* 1993, 199.
5. Delmon, B., *Appl. Catal. B1,* 139, 1992.
6. Lewtas, J., *Toxicological Effects of Emissions from Diesel Engines,* Elsevier, Amsterdam, 1982.
7. Walsh, M.P., *Catalysis and Automotive Pollution Control,* Crucq, A. and Frennet, A., Eds., Studies in Surface Science and Catalysis, Vol. 30, Elsevier, Amsterdam, 1987, 51.
8. Hoffmann, U. and Rieckmann, T., *Chem. Eng. Technol.,* 17, 149, 1994.
9. Goldenberg, E. and Degobert, P., *Rev. Inst. Fr. Pétrole,* 41, 797, 1986.
10. Saito, K. and Ichihara, S., *Catalysis Today,* 10, 45, 1991.
11. Degobert, P., *Automobile et Pollution,* Technip Ed., Paris, 1992.
12. Enga, B.E., *Platinum Metals Rev.,* 26, 50, 1982.
13. Enga, B.E., Buchman, M.F., and Lichstenstein, I.E., *SAE Paper 820184,* 1982, 35.
14. Ise, H., Saitoh, K., Kawagoe, M., and Nakayama, O., *SAE Paper 860292,* 1986, 185.
15. Murphy, M.J., Hillenbrand, L.J., Traysor, D.A., and Wasser, J.H., *SAE Paper 810112,* 1981, 1.
16. Ahlstrom, A.F. and Odenbrand, C.U.I., *Carbon,* 27, 475, 1989.
17. Springer, K.J. and Stahman, R.C., *SAE Paper 770716,* 1977.
18. Yuan, S., Mériaudeau, P., and Perrichon, V., *Appl. Catal.B.,* 3, 319, 1994.
19. van Doorn, J., Varloud, J., Mériaudeau, P., Perrichon, V., Chevrier, M., and Gauthier, C., *Appl. Catal. B1,* 117, 1992.
20. Ahlstrom, A.F. and Odenbrand, C.U.I., *Appl. Catal.,* 60, 143, 1990.
21. Otto, K., Sieg, M.H., Zinbo, M., and Bartosiewicz, L., *SAE Paper 800336*, 1980.
22. Yamane, K., Chikahisa, T., Murayama, T., and Miyamoto, N., *SAE Paper 880343,* 1988.
23. Egebäck, K.E., Mason, G., Rannug, U., and Westerholm, R., *Catalysis and Automotive Pollution Control II,* Crucq, A., Ed., Studies in Surface Science and Catalysis, Vol. 71, Elsevier, Amsterdam, 1991, 75.
24. Westerholm, R., Almen, J., Hang, L., Rannug, U., Egeback, K.E., and Gragg, K., *Environ. Sci. Technol.,* 25, 332, 1991.
25. Stein, H.J., Hüthwohl, G., and Lepperhoff, G., *Catalysis and Automotive Pollution Control III,* Frennet, A. and Bastin, J.M., Eds., Studies in Surface Science and Catalysis, Vol. 96, Elsevier, Amsterdam, 517, 1995.
26. McCabe, R.W. and Sinkevitch, R.M., *SAE Paper 870009,* 1987, 1.
27. McCabe, R.W. and Sinkevitch, R.M., *SAE Paper 860011,* 1986, 41.
28. Neeft, J.P.A., van Pruissen, O.P., Makkee, M., and Moulijn, J.A., *Catalysis and Automotive Pollution Control III*, Frennet, A. and Bastin, J.M., Eds., Studies in Surface Science and Catalysis, Vol. 96, Elsevier, Amsterdam, 549, 1995.
29. Hunter, G, Scholl, J., Hibbler, J., Bagley, S., and Leddy, D., *SAE Paper 811192,* 1981, 15.
30. Hillenbrand, L.J. and Traysor, D.A., *SAE Paper 811236,* 1981, 1.
31. Li, H., Westerholm, R., Almén, J. and Grägg, K., *Fuel,* 73, 11, 1994.
32. Courtois, Y.A., Festy, B., and Vanrell, B., *Proc. Int. S.I.A. Seminar, Diesel engines: Prospects 1990–2000, Soc. Ingénieurs de l'Automobile, Lyon,* 1990.

33. Taschner, K., *Catalysis and Automotive Pollution Control II,* Crucq, A., Ed., Studies in Surface Science and Catalysis, Vol. 71, Elsevier, Amsterdam, 1991, 17.
34. Starkman, E., Sawyer, R., and Caretto, L., *California Engineer,* March, 35, 1970.
35. Cucchi, C. and Hublin, M., *Catalysis and Automotive Pollution Control II,* Crucq, A., Ed., Studies in Surface Science and Catalysis, Vol. 71, Elsevier, Amsterdam, 1991, 41.
36. Henssler, H., *Catalysis and Automotive Pollution Control II,* Crucq, A., Ed., Studies in Surface Science and Catalysis, Vol. 71, Elsevier, Amsterdam, 1991, 35.
37. Tanaka, T., *Seminar in Villeurbanne,* 1994.
38. Sogaard-Andersen, P., Cooper, B.H., and Hannerup, P.N., *NPRA Annual Meeting AM-* 92-50, 1992, 1.
39. de Jong, K.P., *Catal. Today,* 29, 71, 1996.
40. Bartholomew, C.H., *Catal. Lett.,* 7, 303, 1990.
41. Mills, G.A., *Fuel,* 73, 1243, 1994.
42. Eilers, J., Posthuma, S.A., and Sie, S.T., *Catal. Lett.,* 7, 253, 1990.
43. Ansorge, J., Cooke, J. and Eilers, J.L., *Natural Gas Conversion Symposium, Oslo, Abstracts,* 1990, 134.
44. Guibet, J.C. and Martin, B., *Carburants et Moteurs, Technip, Paris,* 903, 1987.
45. Goodrich, R.S.., *Chem. Eng. Prog.,* 78, 29, 1982.
46. Pettersson, L.J, Järas, S.G, Andersson, S. and Marsh, P., *Catalysis and Automotive Pollution Control III,* Frennet, A. and Bastin, J.M., Eds., Studies in Surface Science and Catalysis, Vol. 96, Elsevier, Amsterdam, 855, 1995.
47. Williams, P.T., Abbass, M.K., and Andews, G.E., *SAE Paper 890825,* 1989.
48. Zelenka, P., Kriegler, W., Herzog, P.L., and Cartellieri, W., *SAE Paper 900602,* 1990.
49. Zelenka, P., Ostgathe, K., and Lox E., *SAE Paper 902111,* 1990.
50. Engler, B.H., Lox, E.S., Ostgathe, K, Cartellieri, W. and Zelenka, P., *SAE Paper 910607,* 1991.
51. Wong, C., *Carbon,* 26, 723, 1988.
52. Howitt, J.S. and Mortierth, *SAE Paper 810114,* 1981.
53. Cooper, B.J. and Roth, S.A., *Platinum Metals Rev.,* 35, 178, 1991.
54. Kiyota, Y., Tsuji, K., Kume, S., and Nakayama, O., *SAE Paper 860294,* 1986, 203.
55. Niura, Y., Ohkubo, K., and Yagi, K., *SAE Paper 860290,* 1986, 163.
56. Garner, C.P. and Dent, J.C., *SAE Paper 890174,* 1989.
57. Ludecke, O.A. and Dimick, D.L., *SAE Paper 830085,* 1983.
58. Pattas, K.N., Kikidis, P.S., Aidarinis, J.K., Patsatzis, N.A., and Stamatellos, A.M., *SAE Paper 860136,* 1986, 127.
59. Pattas, K.N., Stamatellos, A.M., Patsatzis, N.A., Kikidis, P.S., Aidarinis, J.K., and Samaras, Z.C., *SAE Paper 860293,* 1986, 195.
60. Pattas, K.N., Samaras, Z.C., and Kikidis, P.S., *SAE Paper 870252,* 1987, 113.
61. Pattas, K.N. and Michalopoulou, C.C., *SAE Paper 920632,* 1992.
62. Xue, E., Seshan, K., van Ommen, J.G., and Ross, J.R.H., *Appl. Catal. B.,* 2, 183, 1993.
63. Ball, D.J. and Stack, R.G., *Catalysis and Automotive Pollution Control II,* Crucq, A., Ed., Studies in Surface Science and Catalysis, Vol. 71, Elsevier, Amsterdam, 1991, 337.
64. Voss, K.E., Lampert, J.K., Farrauto, R.J., and Rice, G.W., *Catalysis and Automotive Pollution Control III,* Frennet, A. and Bastin, J.M., Eds., Studies in Surface Science and Catalysis, Vol. 96, Elsevier, Amsterdam, 499, 1995.
65. Watabe, Y., Irako, K., Miyajima, T., Yoshimoto, T., and Murakami, Y., *SAE Paper 830082,* 1983, 45.
66. Dainty, E.D., Lawson, A., Vergeer, H.C., Manicom, B., Kreuzer, T.P., and Engler, B.H., *SAE Paper 870014,* 1987, 57.
67. Wiedemann, B. and Neumann, K.H., *SAE Paper 850017,* 1985.
68. Miyamoto, N., Hou, Z., and Ogawa, H., *SAE Paper 881224,* 1988, 1.
69. Baker, R.T.K. and Sherwood, R.D., *J. Catal.,* 61, 378, 1980, and references herein.
69. Hardenberg, H.O., Daudel, H.L., and Erdmannsodörfer, H.J., *SAE Paper 870016,* 79, 1987.
70. Inui, T., Otawa, T., and Takagami, Y., *J. Catal.,* 76, 84, 1982.
71. Mc Kee, D.W., *J. Catal.,* 108, 480, 1987; and references of the same author herein.
72. Takarada, T., Nabatama, T., Ohtsuka, Y., and Tomita, A., *Energy and Fuels,* 1, 308, 1987.
73. Baker, R.T., *Carbon and coal gaseification,* J.L. Figueiredo and J.A. Moulijn, Eds., Nato A.S.I. Series E105, Martinus Nijhpff Publishers, Dordrecht, The Netherlands, 1986, 231.
74. Baker, R.T.K., Chludzinski, Jr., Dispenzire, N.C., and Murrell, L.L., *Carbon,* 21, 579, 1983.

75. Furimksy, E., *Fuel Procd. Technol.,* 19, 203, 1988.
76. Mc Kee, D.W., *Carbon* 8, 131, 1970.
77. Moreno-Castilla, C., Utrilla, J.R., Peinado, A.L., Morales, J.F., and Garzon, J.L., *Fuel,* 64, 1220, 1984.
78. Mc Kee, D.W., *Carbon* 8, 623, 1970.
79. Wagner, R. and Mühler, H.J., *Fuel,* 68, 251, 1989.
80. Marsh, H. and Adair, R.R., *Carbon,* 13, 327, 1975.
81. Pennemann, B. and Auton, R., *J. Catal.,* 118, 417, 1989.
82. Ahlstrom, A.F. and Odenbrand, C.U.I., *Appl. Catal.,* 60, 157, 1990.
83. Ciambelli, P., Corbo, P., Parrella, P., Scialo, M., and Vaccaro, S., *Thermo Chimica Acta,* 162, 83, 1990.
84. Ciambelli, P., Parrella, P., and Vaccaro, S., *Catalysis and Automotive Pollution Control II,* Crucq, A., Ed., Studies in Surface Science and Catalysis, Vol. 71, Elsevier, Amsterdam, 1991, 323.
85. Gu, Q., Li, B.Q., Chen, H.D., and Kuang, R.Z., *Catalysis and Automotive Pollution Control III,* Frennet, A. and Bastin, J.M., Eds., Preprints, Vol. 2, ULB, Bruxelles, 1994, 375.
86. Mériaudeau, P., Perrichon, V., and Bellaloui, A., *Eur. Pat. Appl.: 543,716,* 1993.
87. Bellaloui, A., Varloud, J., Mériaudeau, M., Perrichon, V., Lox, E., Chevrier, M., Gauthier, C., and Mathis, F., *Catal. Today,* 23, 421, 1996.
88. Inui, T. and Otawa, T., *Appl. Catal.,* 14, 83, 1985.
89. Homeier, E.H. and Joy, III, G.C., U.S. patent: 4,759,918, 1988.
90. Ciambelli, P., Corbo, P., Scialo, R., and Vaccaro, S., Ital. Patent A 40421/1988, 1988.
91. Marinangeli, R.E., Homeier, E.H., and Molinaro, F.S., *Catalysis and Automotive Pollution Control,* Crucq, A. and Frennet, A., Eds., Studies in Surface Science and Catalysis, Vol. 30, Elsevier, Amsterdam, 1987, 457.
92. Duriez, V., Monceaux, L., and Courtine, P., *Catalysis and Automotive Pollution Control III,* Frennet, A. and Bastin, J.M., Eds, Studies in Surface Science and Catalysis, Vol. 96, Elsevier, Amsterdam, 137, 1995.
93. Teraoka, Y., Nakano, K., Kagawa, S., and Shangguan, W.F., *Appl. Catal.,* B, 5, L181, 1995.
94. Cooper, B.J. and Thoss, J.E., *SAE Paper 890404,* 1989.
95. Hawker, P.N., *Platinum Metals Rev.,* 39, 1, 1995.
96. Iwamoto, M., *Catal. Today,* 29, 29, 1996.
97. Iwamoto, M. and Yahiro, H., *Catal. Today,* 22, 5, 1994.
98. Shelef, M., *Chem. Rev.,* 95, 209, 1995.

25 The Influence of Morphological Restructuring of Carbonaceous Aerosol on Microphysical Atmospheric Processes

S. Nyeki and I. Colbeck

CONTENTS

INTRODUCTION

Recent studies have suggested that increased anthropogenic emissions of aerosols may induce perceptible climate changes.[1,2] In certain regions, the global average forcing due to sulfate[2,3] and biomass[4] aerosols has been estimated as opposite and potentially comparable to the forcing by anthropogenic greenhouse gases. As a consequence, it has been postulated that this phenomenon may have delayed the detection of greenhouse warming and partially offset the effect over large regions of the northern hemisphere.[5,6] However, the effect of atmospheric aerosols on climate change is difficult to quantify due to the sporadic nature of emissions and the relatively low atmospheric lifetimes, from hours to days, in comparison to most greenhouse gases whose lifetimes are reckoned in years. While sulfate aerosols are recognized as the dominant contributor of tropospheric aerosols over and near industrialized regions,[2,7] smoke aerosols containing elemental carbon (EC) are regarded with increasing importance on a global basis.[8] Such concern has stemmed partly from the possible effects of Nuclear Winter scenarios in which large amounts of smoke are injected into the atmosphere (see review by Turco et al.[9]). Although EC aerosols without organic/inorganic components are uncommon in the atmosphere, it is customary to refer to this component of combustion aerosols in considering their microphysical/chemical properties. Such an approach is validated by the inertness of EC. In the discussion below, the term "soot" is used to represent an aerosol with an almost pure EC content, typical of many combustion smokes.

In considering the influence of combustion aerosols on climate, it is generally recognized that the direct effects (the absorption and scattering of radiation) are of secondary importance.[7,10] to the

0-87371-829-1/00/$0.00+$.50

climatic effects of greenhouse gases and sulfate aerosols. Of greater concern are the potential indirect effects (modification of cloud radiative properties, albedo, and lifetimes) of smoke aerosols on the number of cloud condensation nuclei (CCN).[11,12]

Smoke aerosols may be nominally divided into those from: (1) biomass burning, generally with a higher organic carbon (OC) content than EC and of compact morphology, and (2) the combustion of fossil fuels: (a) liquid petroleum derived smoke, with a higher EC content than OC and of tenuous or fractal morphology and (b) coal-derived smoke, with a high EC content, but of angular/compact morphology.

There are several further reasons, apart from those already mentioned, for the increased concern about smoke aerosol emissions into the atmosphere.

1. EC is the only major light-absorbing species commonly found in atmospheric aerosols apart from hematite (Fe_2O_3), whose geographical extent and absorption in the visible spectrum is low and can thus be overlooked. While combustion smoke both scatters and absorbs radiation, the high specific scattering coefficient for soot smokes (~10 m^2 g^{-1}) is of concern because it may result in modification of the vertical atmospheric temperature profile and a reduction in the radiation reaching the surface. Global estimates of the climatic effect have not yet been conducted, but it is thought that soot aerosols may contribute to a positive forcing of the Earth–atmosphere system.[5] Such analysis warrants further investigation, although it has been noted above that direct effects are of secondary importance to indirect effects.
2. Smoke from both biomass and fossil fuel sources possesses long atmospheric residence times due to its relative inertness and a mass mean diameter (0.2 to 0.4 μm) within the accumulation size range (0.1 to 1 μm). As a consequence, transportation over distances greater than 1000 km is not uncommon.
3. The surface area of the particles allows condensation and heterogeneous chemistry to occur, which in turn can influence the balance of gaseous species.

The fate of smoke aerosols, once released into the atmosphere, is poorly understood but involves various atmospheric processes, including humidity and temperature variations, aging, and photochemical reactions. The following discussion examines the significance of the modification of smoke aggregate morphology by atmospheric processes, such as evaporation/condensation (or humidity) cycling. The role of aggregate morphology on indirect aerosol effects remains uncertain, while the influence on direct effects has been increasingly investigated.[13-15] The resultant compaction or restructuring of aggregates has been highlighted in Nuclear Winter studies[9,16] as altering the radiative and microphysical properties, although the relevance of such a process remains to be further considered and quantified. In this discussion, a literature survey was conducted to establish such basic parameters as the EC global burden, aggregate morphology (described as a fractal dimension), and the atmospheric lifetime. Global smoke emissions to the atmosphere are first considered and quantified. The microphysical interaction of aggregates in the atmosphere is influenced by their morphology and can be described in terms of fractal theory. The use of a fractal dimension (D) enables more accurate modeling of aggregate sedimentation and optical characteristics. Representative values of D, in a review of various combustion aerosols, are discussed next. Once emitted to the atmosphere, aggregates are subject to removal processes, of which wet deposition is dominant over dry deposition. Through the influence of humidity cycling in the atmosphere, aggregate morphological restructuring occurs. Finally, evidence from the Kuwaiti oil fires is examined as a case study to assess the relevance of aggregate morphology on a microphysical atmospheric processes. At each stage, values of D are suggested for typical carbonaceous aggregates. Such an analysis may help to reduce parameter uncertainties in the use of general circulation models (GCM).

GLOBAL COMPARISON OF SMOKE AND TOTAL AEROSOL EMISSIONS TO THE ATMOSPHERE

Although soot emissions are mainly restricted to industrial and forest clearing regions, their inert properties and accumulation mode size allow long-range transport. Soot, in nanogram quantities, has been measured in the Arctic[17] and Antarctic,[18] while other typical concentrations are 0.2 to 0.8 $\mu g\ m^{-3}$ in remote areas of the northern hemisphere, 0.1 to 1 $\mu g\ m^{-3}$ in U.S. rural areas, and >10 $\mu g\ m^{-3}$ in some urban areas.[20]

At present, simulations using GCMs either neglect aerosols or only consider average values. Modeling of the global EC distribution for climate change studies is beginning to be investigated, due to the secondary importance of EC in comparison to sulfate aerosol, as noted previously. The importance of dynamical processes involving size distribution, chemical composition, and morphology thus remain to be considered in GCMs incorporating EC aerosols, although the geographic soot inventory by Penner et al.[21] was a step in this direction.

The relevance of smoke emissions to the atmosphere, in relation to global aerosol emissions, is illustrated in Table 25.1. Values found in the literature are somewhat variable, although more recent estimates are comparable. In order to suitably assess the figures, the results of Crutzen and Andreae,[22] d'Almeida et al.,[23] and Ghan and Penner[24] are used in the discussion below, mainly due to their internal consistency and their representation of median values. These values, indicated in bold in Table 25.1, give biomass and fossil fuel smoke emissions of 36 to 154 and 22.5 to 24.0 Tg y^{-1} (1 Tg = 1×10^{12} g), of which the EC contents are 6.4 to 28 and 5.1 to 6.0 Tg y^{-1}, respectively.

As tenuous carbonaceous aggregates only result from the combustion of petroleum fuels, their contribution to the global burden is of further interest in this discussion. Further analysis of the burdens is aided by the detailed fossil fuel emission estimates in Table 25.2.[24] Fossil fuels have been separated into two categories — petroleum and solid fuels — which not only reflect their different compositions and sources, but also the different morphologies of the combustion particles. Total smoke aggregate emissions amount to 2.52 to 3.88 Tg y^{-1}, of which the EC content is 0.91 to 1.81 Tg y^{-1}. These burdens may be taken as representative in further discussion of the effect of aggregates emitted to the atmosphere. The high EC content of aggregates from petroleum fuels (in particular, diesel) is emphasized in these figures. To put the EC aggregate burden into context, it represents an average 0.45% of anthropogenic emissions (excluding biomass) and 0.09% of total particulate emissions. In the latter calculation, a total atmospheric burden of 1500 Tg y^{-1} has been used[22] as representative of nuclei and accumulation mode sizes, as it omits the contribution from the short-lived coarse mode.

Such fractions are relatively small, and were it not for the high specific absorption coefficient, surface catalytic properties, and chemical inertness of EC, then they would be deemed insignificant. These figures should, however, be interpreted cautiously as the EC values in Table 25.1 represent the EC mass component of the total smoke mass. Treating EC as a separate or external component of the same smoke aerosol may, for instance, simplify radiative calculations but may not satisfactorily describe the effect of the non-EC components on other microphysical/chemical properties (such as CCN characteristics). Considerations of the internal/external aggregate composition have been studied in the context of optical properties[25] and photoelectric properties,[26] but remain to be further characterized. In order to simplify the problem, it is general practice to consider the EC as a chemically and physically separate component.

The above EC aggregate burdens are, of course, subject to many uncertainties, as well as representing globally averaged values rather than regional values. The majority of aggregate emissions occur in the industrialized regions of the northern hemisphere. In a European study, diesel particles were reported as contributing an average 80% of total EC emissions in 1978,[27] increasing to an average 88% in 1987.[28] This illustrates that, over localized regions, anthropogenic emissions may exceed natural emissions, despite the latter being an order of magnitude greater on a global basis.

TABLE 25.1
Global Emission Source Strengths for Atmospheric Smoke Aerosols, in Tg year^{-1}

Aerosol Component		Source Strength	Ref.
Biomass:	Smoke content	**36–154**	22
		104	10
		25–79[a]	24
	Smoke EC content	**6.4–28**	22
		19	10
		5.7	11
		1.3	80
Fossil fuel:	Smoke Content	**22.5–23.9**[a]	24
	Smoke EC Content	7.8[b]	22
		5.1–6.0[a]	24
		3.5	80
EC, all sources (biomass and fossil fuel)		**20–30**	22
		3–22	44
		<22	10
Anthropogenic, all sources (excluding biomass)		**185–415**	23
Total particulate mass (all sources)		**1500**	22
		3454–14013[c]	23

Note: 1 Tg = 1 × 10^{12} g. Figures in bold are used in the discussion as representative values.

[a] Fine particulate range.
[b] Inferred value from the mid-range biomass and total EC values.
[c] Includes contribution from the coarse aerosol mode range (>2 μm).

TABLE 25.2
EC and Smoke Emissions from Fossil Fuel Combustion (Tg y^{-1})

Fuel Type	EC Emissions	Total Smoke Fine Particulate Emissions
Petroleum Fuels		
Petrol	0.01–0.07	0.07–0.40
Residual fuel oil	0.01	0.25
Natural gas	0.02–0.025	0.50–0.63
Diesel fuels	0.87–1.70	1.70–2.60
Subtotal	0.91–1.81	2.52–3.88
Solid Fuels		
Coal etc.	4.20	20.0
Total	5.1–6.0	22.5–23.9

Note: After Reference 24 for 1986 world consumption figures.

COMBUSTION AEROSOL FORMATION

Smoke aerosols formed from the combustion of petroleum fuel produce tenuous complex morphologies, composed of branches of primary spherules (generally <50 nm in diameter). Such clusters then agglomerate to submicron sizes and may further grow into aggregates of agglomerates with mode geometric diameters in the 0.2- to 0.4-μm range. In contrast, biomass aerosols have a globular or "bulk" morphology, possibly due to the higher organic/inorganic content, and grow to mode geometric diameters of 0.3 to 0.5 μm.

Aerosol morphology can be described by a number of characteristic diameters (e.g., aerodynamic, mobility, and volume equivalent diameters) or by a dynamic shape factor. However, as spherical geometry is rare in both natural and anthropogenic aerosols, the concepts of fractality[29] have found widespread use in aerosol science, as well as in a number of other fields.[30-32]

Application of fractal theory allows an aggregate to be described by the power-law equation:[33]

$$N = \varepsilon\left(d_x/d_o\right)^D \tag{25.1}$$

where N is the number of primary spherules in the cluster, d_X is the characteristic diameter and represents either the mobility equivalent diameter or twice the radius of gyration, d_o is the primary spherule diameter, ε is a constant depending on the aggregation mechanism and the definition of the characteristic diameter, and D is the fractal dimension measured in three-dimensional space.

For an object to be classified as fractal, D should be less than its spatial dimension (e.g., less than 3 in three-dimensional space). Typical fractal aerosols will have a non-integer value of D between 1 and 3, indicating a non-idealized shape with a large surface area/volume ratio. For fractal clusters than have formed under similar conditions, the value of D also remains similar over a range of sizes and may hence be used to characterize formation mechanisms and burning conditions.

Most determinations of the fractal dimension are based on electron or optical microscopy, in which several methods (see Reference 34) use digitized images to determine the fractal dimension in two-dimensional space (D_{2D}). Such methods are, however, only suitable for aggregates with $D < 2$. For those with $D > 2$, the projection onto a plane always results in $D_{2D} = 2$ and hence *in situ*, three-dimensional methods are generally favored.

The simulation of aggregate formation has been described previously (see References 35 and 36), and is only summarized briefly here. The introduction of the cluster-cluster model allowed the simulation of realistic aggregate morphologies in which D values were comparable to those obtained by experiment. In three-dimensional models, aggregate growth by diffusion-limited cluster-cluster aggregation (DLCC) results in a typical mean value $D \sim 1.80$, while growth by ballistic-limited cluster-cluster aggregation (BLCC) gives $D \sim 1.95$. The DLCC mechanism is more common, and experimental values for various combustion fuels are illustrated in Table 25.3. The methods give a range of values: transmission electron microscopy (TEM) gives $D_{2D} \sim 1.5$–1.8, light scattering (LS) $D \sim 1.6$ to 1.8, the modified Millikan cell (MMC) $D \sim 1.9$ to 2.2, and the in situ ensemble (IE) method $D \sim 1.4$ to 2.0.

The wide range of fractal dimensions reported in Table 25.3, typically around a central range 1.7 to 1.8, illustrates the influence of measurement methods and experimental conditions on the final morphology and aggregate size. If the influence of measurement methods is considered first, then it should be noted that the TEM method gives the two-dimensional value D_{2D}, while the other methods give the three-dimensional value D. The relation between D_{2D} and D is unclear, although simulations have generally shown that $D_{2D} < D$ by ~0.09 to 0.1.[34] While this is not reflected in the results of Zhang et al.,[37] who compared TEM and LS methods, the disparity might be explained by the larger error margin encountered in the former method. Other *in situ* techniques, such as the MMC and IE methods, report values of 1.9 to 2.2 for supermicron, aged butane smoke. The higher values suggest a possible dependency of D on aggregate size and/or pre-measurement restructuring

TABLE 25.3
Fractal Dimensions of Laboratory Smoke Aerosols, Determined by Various Methods

Type of Fuel	Aggregate Diameter (μm)	Primary Sphere Size (nm)	Aging Time	Method	Fractal Dimension D	Ref.
Acetylene/air	<1.0	30	ISC	TEM	1.5–1.6[a]	81
Diffusion flame	5.5–12	30			1.82[a]	
Methane/O_2	1–5	20	IFM	LS	1.62 ± 0.06	37
Pre-mixed			ISC	TEM	1.72 ± 0.1[a]	
Ethene/air	<0.9	10–40	ISC	TEM	1.62 ± 0.04[a]	33
Diffusion					1.74 ± 0.06[a]	
Diesel pre-mixed	~0.6	50	ISC	TEM	1.83 ± 0.06[a]	40
					1.90 ± 0.07[b]	
Methane/O_2 pre-mixed	<0.12	30	IFM	LS	1.6–1.8 ±0.15	82
Ethene/air Diffusion	<0.3	4–10	IFM	LS	1.49	83
Methane/O_2 pre-mixed	<0.14	~8–10	IFM	LS	1.79	84
Butane/air	≤10	50	0.25–5 h	IE	1.4 ± 0.11[c]	39
Diffusion					1.96 ± 0.1	
Butane/air Diffusion	≤10	50	0.5 h	MMC	1.87–2.19 ±0.03	85

Note: TEM = transmission electron microscope, LS = light scattering, MMC = modified Millikan cell, IE = *in situ* ensemble method, ISC = *in situ* collected – but measurement of D at some later undetermined time, IFM = in-flame measurement. Fractal dimensions are in three-dimensional space unless stated otherwise.

[a] The fractal dimension (D_{2D}) in two-dimensional space.
[b] The value D_{2D} = 1.83, corrected for three-dimensional space.
[c] Higher air/fuel ratio – unspecified value.

to be responsible, rather than a BLCC aggregation mechanism. Despite the wide range of conditions, an analysis of Table 25.3 indicates no significant dependence of D on aggregate size and, hence, pre-measurement restructuring is likely to have occurred.

The influence of experimental conditions on the fractal dimension, especially the relatively low aging times indicated in Table 25.3, is more difficult to assess. Few studies have examined the effect of the air/fuel (A/F) ratio on aggregate morphology using fractal theory. It is well known that the fraction of organic to elemental carbon may vary from exclusively EC to mainly OC emissions,[38] which suggests a lower D for higher A/F ratios. This influence is reflected in the findings of Colbeck and Wu,[39] where butane diffusion flames gave D = 1.96 and a value D = 1.40 for a higher unspecified A/F ratio, and would suggest that D decreases with increasing A/F ratio. This premise is seemingly not supported in Table 25.3 by comparing the results for pre-mixed and diffusion flames. For instance, Nelson et al.[40] report D = 1.90 (a D_{2D} result corrected for three-dimensional space) for stoichometrically produced diesel smoke. The effect of the A/F ratio on morphology remains to be comprehensively investigated.

Among other variables, primary spherule size, aggregate size, and fuel type have been found to influence D to a lesser extent. For example, an investigation by Köylü and Faeth[41] on eight different fuel types found diffusion flames to give D_{2D} in the range 1.70 to 1.79, suggesting a comparative independence on fuel type, although differing A/F ratios were not investigated.

In conclusion, the wide range of D values in Table 25.3 may be largely attributed to different experimental methods and conditions, while the influence of the A/F ratio is unclear. Combustion of a range of fuels under pre-mixed and air diffusion conditions results in aggregates with a value

D ~ 1.8 to 1.9 when initially released to the atmosphere. A typical value of D for diesel aggregates is more difficult to assess due to the shortage of published results, but a value D = 1.90 ± 0.07 (Reference 40) may be nominally used as representative for diesel aggregates. In contrast, EC particles from biomass and solid fuel combustion have a "bulk" or compact angular morphology, which predominantly have a fractal dimension close to D ~ 3. While coal combustion particles have a porous fractal surface,[87] this does not influence the aerodynamic drag as for tenuous aggregates.

Once released to the atmosphere, aggregates are subject to sink processes, divided into wet and dry deposition, and are further discussed below.

THE REMOVAL OF AGGREGATES BY ATMOSPHERIC DEPOSITION PROCESSES

The residence lifetime of smoke aggregates in the atmosphere depends on four principle factors: the frequency of precipitation episodes, the ambient concentration of aerosols, the aggregate size distribution, and the efficiency of removal mechanisms.[42] The first two aspects are difficult to quantify and are dependent on location. For example, typical ambient aerosol concentrations in urban areas can exceed 1000 cm^{-3}, while concentrations in remote marine areas may be less than 10 cm^{-3}.

The last two aspects can be considered together. The atmospheric lifetime of aerosols with typical diameters of 0.1 μm is less than a few days due to Brownian coagulation, while aerosols larger than ~1 μm are removed efficiently by sedimentation and precipitation scavenging. Aerosols in the intervening accumulation mode have the longest residence times of 1 to 5 days. A similar residence time has been estimated for EC[42] and biomass smoke aerosols[11] (3 to 7 days). The above estimates are for temperate conditions, while the effect of rainy climates was estimated by Ogren and Charlson[42] to reduce the EC lifetime to under 40 h. In the latter investigation, accumulation mode EC was conjectured to remain as an externally mixed aerosol for up to 1/2 day in urban areas and for over 1 month in clean areas.

Removal of the accumulation mode through wet deposition is initiated by in-cloud nucleation scavenging. Several humidity cycles are possible, during which the droplet scavenges further aerosols. The number of cycles has been estimated at between 10 and 20 by Charlson and Ogren.[43] Even if smoke aerosols are activated as CCN, their removal from the atmosphere is not necessarily implied. In this manner, it is possible that the aerosol mass increases, soluble components are absorbed into an internal mixture, and the morphology becomes more compact. The details of the complex processes whereby hydrophobic aerosols become hydrophilic will not be examined in detail here. Instead, the fact that wet deposition dominates dry deposition as an atmospheric removal process is assumed.[5,9,42,44] Although the ultimate removal of virtually all particles by wet deposition, complex scavenging pathways comprising coagulation, phoretic, impaction, and nucleation scavenging processes may be considered as intermediary events. Hence, it can be argued that all atmospheric aggregates removed within a 1- to 7-day timespan would be subject to varying degrees of humidity processing. In the consideration below, the activation of smoke aerosols as CCN is considered in the context of whether there is a subsequent change in aggregate morphology, as opposed to the effect on overall cloud properties. Such alteration of the morphology is shown to have a minor effect on dry deposition characteristics.

THE REMOVAL OF AGGREGATES BY ATMOSPHERIC WET DEPOSITION

The effect of humidity on atmospheric aerosols and combustion smoke from Nuclear Winter scenarios has been widely reported in the literature, whereas the effect on vehicle smoke emissions has been characterized to a lesser extent. Fossil fuel, and especially petroleum smoke emissions,

TABLE 25.4
The Variation of Cloud Condensation Nuclei/Condensation Nuclei (CCN/CN) Ratio with Type of Smoke

Type of Fuel and/or Aggregate	CCN/CN Ratio (after aging times ~10–80 h)	Tendency of the CCN/CN Ratio after Generation	Ref.
Flaming vegetation and forest fires	0.8–1.0	Slowly increasing	12
Coagulated crude oil (0.55% S v/w) and wood smokes	<0.80	Increasing	49
Diesel/petrol exhaust fumes (0.6/0.4 ratio)	0.1	n/a	86
High-sulfur (2–3% S v/w) crude oil	<0.6	Increasing	49
Low-sulfur (0.55% S v/w) crude oil	0.1–0.2	Relatively stable	49
Acetylene	0.2–0.75	Increasing	54
JP-4 aviation fuel	~0.01–0.03	Relatively stable	54

dominate the northern hemisphere EC burden, as opposed to biomass emissions for the southern hemisphere. Their characterization and fate in the atmosphere is therefore of importance.

The activation of condensation nuclei (CN) to form CCN depends on a number of variables, which include the maximum supersaturation attained in the cloud, aerosol size, and composition. At relative humidity values up to ~98%, pure graphitic aerosols (50–100 nm) have been shown to exhibit no humidity growth, in contrast to similarly sized petrol smoke, which just begins to exhibit growth at ~95%.[45] With increasing relative humidity, the CCN activation of combustion aerosols is more widely investigated and can be characterized as a CCN/CN ratio normally around 101% relative humidity (1% supersaturation), typical of cloud conditions. Table 25.4 summarizes CCN/CN ratios for various fuels. Ratios range from ~0.8 to 1.0 for flaming vegetation and forest fires, down to ~0.01 to 0.1 for a pool fire of JP-4 aviation fuel. Of particular interest here are the low ratios found for high EC content aggregate aerosols (JP-4 fuel, low-sulfur crude oil, and a diesel/petrol mixture), which demonstrate their hydrophobic activity. Increased ratios, in the range >0.6 to 0.8, are found for high-sulfur crude oil and a coagulated wood smoke/low-sulfur crude oil mixture. Such increased CCN activity has been found to increase with aggregate size[46] and organic (e.g., CH_3COO^- and $HCOO^-$) and inorganic (e.g., Cl^-, NH_4^+, and K^+) composition. Higher organic/inorganic concentrations in biomass samples would explain their higher CCN activity, although heterogeneous reactions on soot aggregates with their higher surface areas may be important in their subsequent activity.

These results emphasize the influence of initial CCN activity on composition and imply that aggregates are not at first efficiently removed by wet deposition. The removal of atmospheric pollution has been studied by Pruppacher and Klett,[47] in which it was found that an hour or less of steady rain can remove most pollutants. Similar analyses have been conducted on Nuclear Winter scenarios, having investigated the emission and removal of smoke from the atmosphere as a consequence of numerous, intense fire sources. The same physical/chemical processes can also be applied with prudence to atmospheric smoke from day-to-day activities (vehicular traffic, biomass burning, etc.). In their review, Turco et al.[9] assessed the amount of smoke that would be promptly scavenged from urban fires. About 10 to 25% was estimated to be promptly removed, a range supported in a literature survey by Penner and Molenkamp,[48] although consideration of several other scavenging pathways in the latter investigation supported a wider 10–90% range. If the former range is used as being nominally representative, then this implies the prompt removal of 0.09 to 0.45 Tg y^{-1} of the 0.91 to 1.81 EC aggregate burden estimated previously. Although such a 10–25% range is more appropriate for prompt removal of smoke from intense sources, as opposed to smoke emission and removal on synoptic scales, it is similar to the 10% CCN activity found for a diesel/petrol mixture (Table 25.4). Such prompt removal will of course depend on location and prevailing weather patterns, and will further determine whether long-range transport occurs.

As aggregates age, the CCN activity has been found to increase in most cases (Table 25.4). Notable exceptions are JP-4 and low-sulfur fuel, which remain relatively stable, while the trend for the diesel/petrol mixture is not available. The increased activity can be attributed to a number of processes: (1) heterogeneous reactions occurring on the aggregate surfaces, (2) increasing coagulation/aggregate size, or (3) the loss of inactive aggregates, hence increasing the CCN/CN ratio. In their experiments, Rogers et al.[49] held the loss of inactive aggregates and gas-to-particle formation responsible.

As a consequence of humidity cycling, the morphology of the tenuous aggregates becomes more compact and "spheroidal," resulting in an increase of D. Jullien and Meakin[50] concluded that such a phenomenon would be important in the consideration of subsequent aggregate atmospheric lifetimes and optical properties. Various laboratory experiments have demonstrated a number of different methods by which aerosols can be restructured. Among the mechanisms investigated have been: (1) temperature gradient and mechanical shear in colloids,[51] (2) the tempering of silver aerosols (D ~ 2.18) at 80 to 270°C into coalesced spheres for which D ~ 3,[52] and (3) electrostatic induced restructuring of single in situ butane smoke aggregates in discrete steps to values as high as D ~ 2.2.[53] Another mechanism that may be of importance for larger aggregates in the supermicron range is the effect of wind shear. Such investigations on aerosols under atmospheric conditions are unknown to the authors.

Humidity-induced restructuring under high humidities or low supersaturations, however, remains more thoroughly investigated than other mechanisms to date. An investigation by Hallett et al.[54] noted a moderate restructuring effect on acetylene smoke, while JP-4 fuel exhibited virtually no effect. Compaction was attributed to capillary forces acting upon branched chains, facilitated by the condensation of humidity on hydrophilic components of the aggregate. Later studies used fractal analysis to quantify the degree of restructuring as a function of relative humidity. Table 25.5 illustrates the available literature data on the restructuring of combustion aggregates, as a result of various mechanisms. Colbeck et al.[55] observed an increase in the fractal dimension from D ~ 1.8 to 2.0–2.5 for butane smoke aggregates. Figure 25.1 illustrates the effects of various relative humidities and supersaturations on the resulting morphology of aged smoke particles in comparison with samples aged for only 4 hours. A curve fit analysis of their data (Figure 25.2) suggests an exponential increase of D with relative humidity (point at relative humidity = 98% omitted from analysis). For typical cloud supersaturations of 1%, a value D ~ 2.0 is implied for up to five humidity cycles. A further humidity cycling study by Huang et al.[56] investigated the restructuring of diesel aggregates in an environmental scanning electron microscope. Increases in D_{2D} for three different sulfur content fuels were observed, where the high-sulfur case (0.084% wt. S) surprisingly exhibited no significant increase and was attributed to possible pre-sampling restructuring.

These studies illustrate the ease with which aggregates can be restructured by humidity cycling. The tenuous nature of supermicron aggregates has been further demonstrated in the restructuring of individual particles suspended in a modified Millikan cell. Various degrees of compaction could be induced with limiting values of D as high as ~2.2. As each aggregate exhibited different limiting values, it may suggest that restructuring mechanisms are not only dependent on external influences (i.e., humidity cycling, etc.), but also on individual aggregate morphology. To what extend humidity-induced restructuring is relevant to atmospheric processes would require further investigation to measure D under realistic conditions.

Direct *in situ* atmospheric observations of humidity-induced restructuring are difficult to obtain and remain sparse. One study observed the narrowing of a smoke aerosol size distribution after having passed through a capping cloud.[57] This suggested coagulation scavenging of the smallest size fraction (<0.1 μm) and possible restructuring of the largest (supermicron range). Most evidence for humidity processing has come from laboratory and ground-based measurements.

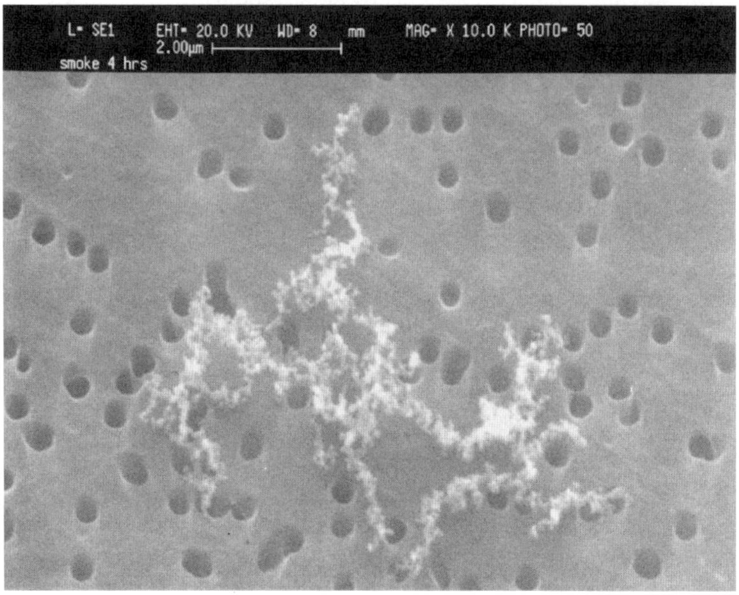

A

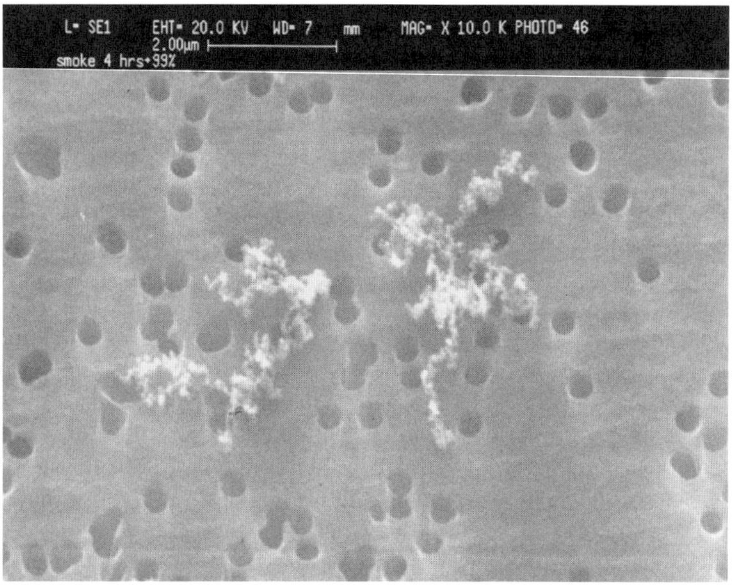

B

FIGURE 25.1 Electron micrographs of (A) smoke aged for 4 hours, (B) smoke aged for 4 hours and then subjected to 99% relative humidity. .

Trials on the humidity processing of atmospheric samples attributed[58] a decrease in size to the restructuring of freshly emitted smoke. In the same trials, laboratory experiments indicated that aggregate size may be important in the degree of final restructuring obtained. While size was not directly measured, the degree of restructuring increased for larger smoke aggregates (all submicron), a result supported in later restructuring experiments on supermicron aggregates.[53] A more recent humidity processing study[45] on petrol and diesel-derived smoke demonstrated a small degree

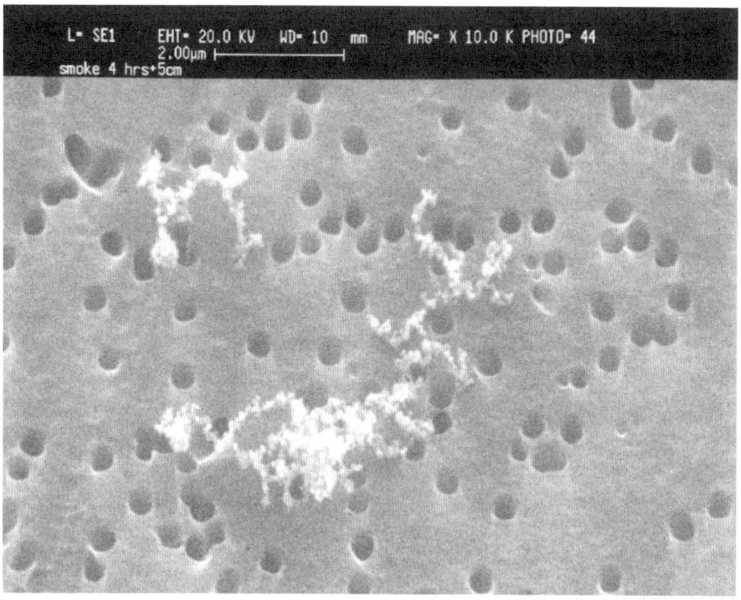

C

FIGURE 25.1 (*Continued*) (C) smoke aged for 4 hours and then subjected to 2% supersaturation.

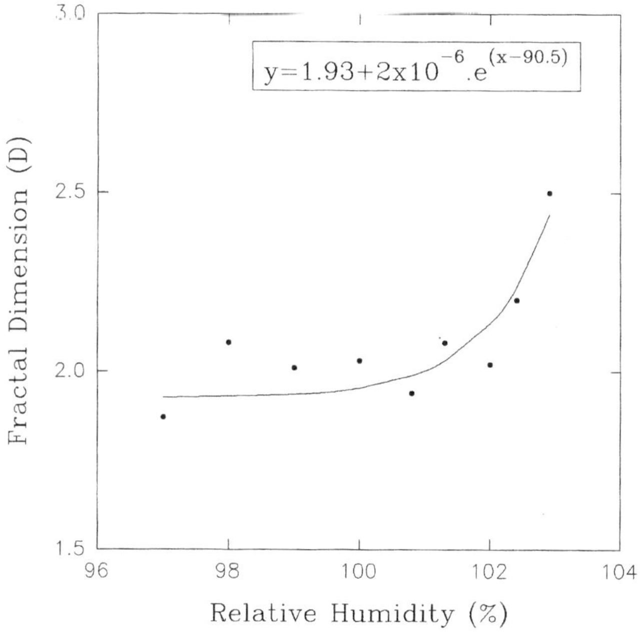

FIGURE 25.2 Mean values of D for butane smoke aged for 4 hours under high humidities and low supersaturations. The value D = 2.08 at 98% humidity has been omitted from the curve fit. (Data from Reference 55.)

TABLE 25.5
Morphological Restructuring of Combustion Aggregates, as Measured by Fractal Analysis, for Various Fuel and Restructuring Mechanisms

Type of Fuel and/or Aggregate	Aggregate Feret Diameter (μm)	D before Restructuring	D after Restructuring	Restructuring Mechanism	Ref.
Laboratory					
Butane	≤9	~1.78	~2.0–2.5	Humidity processing	55
Diesel					
(0.034% wt. S)	1.7–1.9	1.56[a]	1.75–1.82[a]	Humidity processing	56
(0.32% wt. S)	1.5–1.7	1.40[a]	1.48–1.54[a]		
(0.84% wt. S)	1.0	1.46[a]	1.46–1.48[a]		
Butane	6–32	~1.81–1.96	~1.99–2.19	Electrostatic	53
Atmosphere					
Combustion	0.2–2.6	1.7–1.8 and 1.35–1.6[b]	1.38–1.89[a]	Various humidity, aging, initial combustion conditions, etc.	61
Aggregates	0.4–2.1		D>2 (uncoated)[c]		
	~1.1		D>2 (coated)[c]		

Note: DLCC = diffusion-limited cluster-cluster aggregation and TTA = tip-to-tip aggregation. Fractal dimensions are in three-dimensional space unless stated otherwise.

[a] The fractal dimension (D_{2D}) in two-dimensional space.
[b] Conjectured initial range; DLCC mechanism dominant but lower values explained by mixture of DLCC (D = 1.7–1.8) and TTA (D = 1.35–1.6) mechanisms.
[c] As only D_{2D} values were measured, larger values than D = 2 were reported as D > 2.

of restructuring for aggregates with a mobility diameter of ~100 nm at humidities as low as 50 to 60%. This may give an indication of the initial restructuring possible under typical humidity conditions encountered in the boundary layer. Restructuring at <95% relative humidity is also implied for the results of Colbeck et al.[55] in Figure 25.2, where it is suggested that D increases from an initial value of 1.8–1.9 to ~1.9–1.95. Such conclusions, however, need more verification due to the overall sparseness of data and diverse experimental conditions.

The possible degree of cloud-processing induced restructuring has been hinted at by the absence of tenuous combustion aggregates in atmospheric samples.[59,60] This, however, was not the case in a field study by Katrinak et al.,[61] who divided carbonaceous aggregate morphologies analyzed by TEM into three categories: (1) D_{2D} in the range 1.35 to 1.89 (mean length 1.1 μm), where the lowest value was explained in terms of the tip-to-tip aggregation model (a DLCC variant with electrostatic forces) and the upper range by DLCC aggregation; (2) $D_{2D} = 2$ (i.e., D > 2 for uncoated aggregates); and (3) the same as (2) but coated with organic/inorganic substances. The latter two categories indicate restructuring, but to what degree cannot be determined by the two-dimensional technique used. Various restructuring mechanisms were postulated, including humidity processing, aging, and combustion conditions. Some aggregates exhibited coatings, determined by electron energy-loss spectroscopy to contain sulfate and nitrate compounds, which may have been present during formation or formed subsequently through coagulation/photochemical activity.

The most intermediate effect of restructuring would be to alter the aggregate radiative properties. Whether this is significant on a global basis remains to be assessed. Traditionally, Mie theory has been used to predict aerosol optical properties, which is directly applicable to spherical particles and may be extended to other regular shapes (e.g., cylinders, ellipsoids). It is, however, not directly

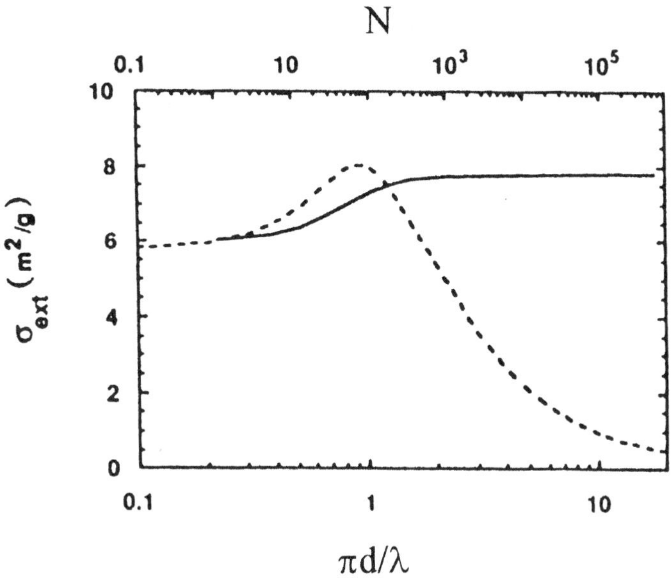

FIGURE 25.3 Specific extinction vs. number of primary particles N for λ= 630 nm, d_o = 45 nm, refractive index m = 1.55 + 0.78i and D = 1.75. The dotted line is the Mie theory prediction for the volume-equivalent diameter of the aggregate of N primary particles with the same refractive index. (From Dobbins, R.A., *Atmos. Environ.*, 28, 889, 1994. With permission.)

applicable to fractal clusters or other complicated particle shapes. The erroneous application of Mie theory has led to predictions of optical properties that are strongly dependent on the extent of particle aggregation. However, because the coagulating carbonaceous particles are not spherical, they maintain a larger effective surface area than would spheres, and thus the optical properties remain largely unaffected by the increasing particle size. Experimental observations[15,62] indicate that the specific extinction coefficient remains essentially constant with increasing size parameter, $\pi d/\lambda$ (λ = wavelength of light, d = particle diameter), indicated in Figure 25.3 by the solid line, as opposed to the eventual decrease exhibited by Mie theory (dotted line).

There are several techniques available for calculating the optical properties of aggregates.[13-15] A rigorous discussion of the techniques is beyond the scope of this chapter, and the reader is referred to Reference 63 for further details. The numerical techniques chiefly used are the method of moments[14,64] and the coupled dipole moment.[16,65] The application of such methods indicates a decrease in the specific absorption coefficient with increasing fractal dimension, as observed experimentally.[62] As $D \rightarrow 3$, the radiative characteristics of aggregates are expected to approach Mie theory.

In summary, the typical final restructured value D ~ 2.0–2.2 for laboratory results may be compared to a value D ~ 2.2 for simulations.[50,66] In their model, DLCC aggregates restructured in discrete steps to D ~ 2.2, after which further increase was not possible. The situation in the atmosphere is more complex and there probably exists a spectrum of D values in the range 1.8 to 2.0–2.2, dependent mainly on aggregate age and weather systems encountered. An average value D ~ 2.0 for cloud-processed aggregates is suggested from Figure 25.2 for radiative calculations involving fractal theory. As most atmospheric particles are ultimately removed by wet deposition, most aggregates can be assumed to have been restructured to this degree.

The Removal of Aggregates by Atmospheric Dry Deposition

Although wet deposition is the dominant removal process, it is still informative to consider aggregate dry deposition. The tenuous morphology of aggregates results in a lower sedimentation velocity than an aerosol of equivalent volume. This would nominally imply an enhanced atmospheric lifetime that with increasing restructuring, would tend toward the lifetime of the same volume-equivalent aerosol. A model describing the sedimentation velocity of aggregates as a function of D, among other parameters, was formulated by Berry.[67] If a mass mean diameter of 0.3 μm is used as representative of an EC aggregate, this then corresponds to about 220 primary spherules of 50 nm diameter. Using a spherule density of 2 g cm^{-3} for the values of D = 1.8, 2.0, and 3.0, the resulting sedimentation velocities are 51, 69, and 170 m y^{-1}, respectively. The enhanced lifetime, in comparison to the volume-equivalent aerosol for which D = 3, is clearly demonstrated and would be of consequence in a stable atmosphere. However, the effect of atmospheric turbulence and circulation prescribes the use of the dry deposition velocity as a more appropriate parameter to quantify deposition through the viscous air layers adjacent to a surface. Aerosols with a geometric diameter larger than ~1 μm sediment according to their size and density, whereas those in the accumulation size range are influenced primarily by the "roughness" of the deposition surface, the wind speed, and the atmospheric stability. Aerosols in the accumulation size range exhibit a minimum in the deposition velocity in the approximate range 0.001 to 0.1 cm s^{-1}, and equivalent to 315 to 31,536 m y^{-1}. The higher deposition velocity therefore illustrates a relative independence of composition and morphology. In this manner, the morphology of accumulation mode aggregates, whether restructured or unrestructured, plays a minor role in determining dry deposition velocities.

Various models have investigated the relative importance of wet and dry deposition. In a model of the EC atmospheric life cycle, Ogren and Charlson[42] concluded that only in remote regions would dry deposition account for more than one third of the removal flux. More recently, a GCM on the distribution of biomass smoke aerosols[21] was found to be sensitive to the wet removal efficiency and the dry deposition velocity adopted. A value of 0.1 cm s^{-1} for a 0.4 μm diameter was used for a variety of atmospheric conditions and represented an average for all smoke morphologies and densities. In general, however, dry deposition is considered to be of minor importance compared to wet deposition in the atmospheric removal of smoke aerosols,[9,21,44] while the restructuring of aggregates is not expected to influence the dry deposition efficiency.

KUWAITI OIL FIRES

The Kuwaiti oil fires of 1991 provided an opportunity to study the long-range transport of smoke aggregates in the ensuing supercomposite plume. The environmental impacts have been considered in many studies (see References 68 and 69) and, thus, the discussion below will mainly consider the smoke microphysical findings.

The individual plumes in the oil fires were of three distinct types: black plumes with a high smoke content (20–25% wt. EC), pool fires with an even higher smoke content (~50% wt. EC), and white plumes with a high organic salt content due to aspiration of brine from gas wells (~4% wt. EC).[70,71] A recent assessment of the event[72] estimated an overall EC production of ~18,700 tons per day at the peak of the fires.

A number of morphological studies were also conducted during the oil fires. Parungo et al.[73] conducted impactor and X-ray fluorescence spectroscopy. Smoke collected from the plumes with an aerodynamic diameter, d_{AE} > 2.5 μm, was dominated by soil-derived particles of which a small fraction was found to have coagulated with combustion aggregates in the size range d ~ 10–30 μm.[74] Their low concentration and high sedimentation velocity restricted their influence to the downwind vicinity. Aerosols with d_{AE} < 2.5 μm were characterized by salt, sulfates, organic material, and soot, where the sulfate coatings were found to increase farther from the fires.

Fractal measurements of morphology have not appeared in the literature, although evidence for the non-sphericity of aggregates was presented by Weiss et al.[75] The use of a modified integrating nephelometer fitted with a bipolar electrostatic field caused the alignment of aggregates within the sampling volume and hence led to a variation in the extinction coefficient. Smoke from black plumes was found to be highly non-spherical in contrast to that from white plumes, which consisted of a larger fraction of aspirated brine.

The modification of aggregate morphology due to humidity cycling, etc., was not investigated in these studies, so an assessment of the role that restructuring plays, if any, is hampered. The high CCN activity noted 200 km downwind would suggest restructuring to have eventually occurred. At this distance, Hudson and Clarke[76] found that 70% of the aerosols were active at a supersaturation of ~1%, similar to the findings of Daum et al.[77] The high active fraction can be compared to the lower values in Table 25.4 and can be explained by the presence of hydrophilic coatings, as reported by Parungo et al.[73] Although no precipitation episodes occurred in the region at that time, the increased probability of subsequent removal by precipitation and the reduction of optical absorption due to humidity processing was therefore postulated to have reduced the potential for a Nuclear Winter effect.[70]

The effect of combustion aerosols was concluded to have been limited to the regional scale,[70] while the overall global impact was regarded as insignificant[71,77] due to lower smoke emissions than expected and confinement of the plume to the lower/mid-troposphere (<5000 m altitude). Some evidence was found, however, for long-range smoke transport to as far as Hawaii,[78,79] among other locations, and correlation with unusual rainfall patterns in India and China.[73]

CONCLUSIONS

In this chapter, the relevance of aggregate emissions on atmospheric microphysical/chemical processes has been assessed. Elemental carbon aggregate emissions have been quantified and their CCN activity and morphology investigated in a literature survey. The lifetime of EC in the atmosphere is on average between 1 and 7 days and is predominantly influenced by wet deposition. An EC burden of 0.91 to 1.81 Tg y^{-1}, from petroleum fuels, is emitted to the atmosphere, where the tenuous aggregate morphology is characterized by a fractal dimension D ~ 1.8. The remainder of the EC burden is due to biomass and solid fuels at 10.6 to 32.2 Tg y^{-1} and has a "bulk" or compact angular morphology with a fractal dimension closer to D ~ 3. An assessment of the microphysical/chemical interaction with the atmosphere has concluded that:

1. Due to the ultimate removal of the majority of aggregates by wet deposition, a spectrum of fractal dimension values probably exists due to morphological restructuring. Few comprehensive humidity processing experiments have been conducted, but evidence exists for restructuring to already occur at <95% relative humidity, followed by further restructuring to D ~ 2.0–2.2 at supersaturations of around 1%.
2. EC aggregate aerosols are not expected to exhibit enhanced atmospheric lifetimes due to their tenuous morphology. Deposition to a surface, described more aptly by the deposition velocity as opposed to the sedimentation velocity, is relatively independent of morphology for sizes in the accumulation mode. The lower sedimentation velocity for aggregates, in comparison to their volume-equivalent counterparts, becomes insignificant when considering the vertical transport of aggregates by atmospheric turbulence and circulation.
3. The release of aggretates with D ~ 1.8 to the atmosphere is probably followed by restructuring to a range D ~ 1.8 to 2.0–2.2 due to humidity cycling. The main effect of morphological restructuring is to reduce the optical absorption by aggregates, and is of interest as radiative calculations of the climatic effect of smoke particles have not yet been made.

The significance of restructuring on the optical properties of aggregates remains to be further investigated. In order to further assess the microphysical/chemical interaction of aggregate emissions to the atmosphere, investigations of smoke aging under different atmospheric conditions will be necessary. Laboratory experiments have, for example, demonstrated humidity-induced restructuring, but to what extent this occurs in the atmosphere remains to be further assessed.

REFERENCES

1. Grassl, H., What are the radiative and climatic consequences of the changing concentration of atmospheric aerosol particles?, *The Changing Atmosphere*, Rowland, F.S. and Isaksen, I.S.A., Eds., John Wiley & Sons, Chichester, 1988.
2. Charlson, R.J., Schwartz, S.E., Hales, J.M., Cess, R.D., Coakley, J.A. Jr., Hansen, J.E., and Hofmann, D.J., Climate forcing by anthropogenic aerosols, *Science*, 255, 423, 1992.
3. IPCC, *Climate Change 1992, Supplementary Report*, Houghton, J.T., Callander, B.A., and Varney, S.K., Eds., University Press, Cambridge, 1992.
4. Penner, J.E., Dickinson, R.E., and O'Neill, C.A., Effects of aerosol from biomass burining on the global radiation budget, *Science*, 256, 1432, 1992.
5. IPCC, *Climate Change 1994*, Houghton, J.T., Meira Filho, L.G., Bruce, J., Lee, H., Callander, B.A., Haites, E., Harris, N., and Maskell, K., Eds., University Press, Cambridge, 1995.
6. Engardt, M. and Rodhe, H., A comparison between patterns of temperature trends and sulfate aerosol pollution, *Geophys. Res. Lett.*, 20, 117, 1993.
7. Penner, J., Charlson, R.J., Hales, J.M., Laulainen, N., Leifer, R., Novakov, T., Orgen, J., Radke, L.F., Schwartz, S.E., and Travis, L., Quantifying and Minimizing Uncertainty of Climate Forcing by Anthropogenic Aerosols, DOE/NBB–0092T, Environ. Sci. Div., Washington, D.C., 20585, 1993.
8. Kiehl, J.T. and Briegleb, B.P., The relative role of sulfate aerosols and greenhouse gases in climate forcing, *Science*, 260, 311, 1993.
9. Turco, R.F., Toon, O.B., Ackerman, T.P., Pollack, J.P., and Sagan, C., Nuclear winter: physics and physical mechanisms, *Annu. Rev. Earth Planet Sci.*, 19, 383, 1991.
10. Andreae, M.O., Biomass burning, *Global Biomass Burning*, (Edited by Levine J.S.), MIT Press, Cambridge, MA, 1991.
11. Penner, J.E., Ghan, S.J., and Walton, J.J., The role of biomass burning in the budget and cycle of carbonaceous soot aerosols and their climate impact, *Global Biomass Burning*, Levine, J.S., Ed., MIT Press, Cambridge, MA, 1991.
12. Rogers, C.F., Hudson, J.G., Zielinska, B., Tanner, R.L., Hallett, J., and Watson, J.G., Cloud condensation nuclei from biomass burning, *Global Biomass Burning,* Levine, J.S., Ed., MIT Press, Cambridge, MA, 1991.
13. Berry, M.V. and Percival, I.C., Optics of fractal clusters such as smoke, *Optica Acta*, 33, 577-591, 1986.
14. Iskander, M.F., Chen, H.Y., and Penner, J.E., Resonance optical absorption by fractal agglomerates of smoke aerosols, *Atmos. Environ.*, 25A, 2563, 1991.
15. Dobbins, R.A., Mulholland, G.W., and Bryner, N.P., Comparison of a fractal smoke optics model with light extinction measurements, *Atmos. Environ.*, 28, 889, 1994.
16. Nelson, J., Test of a mean field theory for the optics of fractal clusters, *J. Mod. Opt.*, 36, 1031, 1989.
17. Clarke, A.D. and Noone, K.J., Soot in the arctic snowpack: a cause for perturbation in radiative transfer, *Atmos. Environ.*, 19, 2045, 1985.
18. Hansen, A.D.A., Bodhaine, B.A., Dutton, E.G., and Schnell, R.C., Aerosol black carbon measurements at the South Pole, *Geophys. Res. Lett.*, 15, 1193, 1988.
19. Cadle, S.H. and Dasch, J.M., Wintertime concentrations and sinks of atmospheric particulate carbon at a rural location in northern Michigan, *Atmos. Environ.*, 22, 1373, 1988.
20. Grosjean, D., Particulate carbon in the Los Angeles air, *Sci. Total Environ.*, 32, 133, 1984.
21. Penner, J.E., Eddleman, H., and Novakov, T., Towards the development of a global inventory for black carbon emissions, *Atmos. Environ.*, 27A, 1277, 1993.
22. Crutzen, P.J. and Andreae, M.O., Biomass burning in the tropics: impact on atmospheric chemistry and biogeochemical cycles, *Science*, 250, 1669, 1990.
23. d'Almeida, G.A., Koepke, P., and Shettle, E.P., Eds., *Atmospheric Aerosols: Global Climatology and Radiative Characteristics*, Deepak, Hampton, 1991.
24. Ghan, S.J. and Penner, J.E., Smoke, effect on climate, *Encycl. Earth System Science*, Academic, San Diego, Vol. 4, 191, 1992.

25. Ackerman, T.P. and Toon, O.B., Absorption of visible radiation in atmosphere containing mixtures of absorbing and non-absorbing particles, *Appl. Optics*, 20, 3661, 1981.
26. Steiner, D., Burtscher, H., and Gross, H., Structure and disposition of particles from a spark ignition engine, *Atmos. Environ.*, 26A, 997, 1992.
27. Ogren, J.A. and Charlson, R.S., Wet deposition of elemental carbon and sulphate in Sweden, *Tellus,* 36B, 262, 1984.
28. OECD, Energy Statistics 1986/1987, OECD, Paris, 1989.
29. Mandelbrot, B.B., *The Fractal Geometry of Nature*, Freeman, San Francisco, 1983.
30. Kaye, B.H., *A Random Walk through Fractal Dimensions*, VCH, Weinheim, 1989.
31. Avnir, D., *The Fractal Approach to Heterogeneous Chemistry: Surfaces, Colloids, Polymers*, John Wiley, Chichester, 1990.
32. Colbeck, I. and Nyeki, S., Optical and dynamical investigations of fractal clusters, *Sci. Prog.*, 76, 149, 1994.
33. Megaridis, C.M. and Dobbins, R.A., Morphological description of flame-generated materials, *Combust. Sci. Technol.*, 71, 95, 1990.
34. Cleary, T.G., Samson, R., and Gentry, J.W., Methodology for fractal analysis of combustion aerosols and particle clusters, *Aerosol Sci. Technol.,* 12, 518, 1990.
35. Jullien, R., Botet, R., and Mors, P.M., Computer simulations of cluster-cluster aggregation, *Faraday Discuss. Chem. Soc.*, 83, 125, 1987.
36. Smirnov, B.M., The properties of fractal clusters, *Phys. Rep.*, 188, 1, 1990.
37. Zhang, H.X., Sorensen, C.M., Ramer, E.R., Olivier, R.J. and Merklin, J.F., *In situ* optical structure factor measurements of an aggregating soot aerosol, *Langmuir*, 4, 867, 1988.
38. Roessler, D.M., Faxvog, F.R., Stevenson, R., and Smith, G.W., Optical properties and morphology of particulate carbon: variation with air/fuel ratio, *Particulate Carbon Formation During Combustion*, (Edited by Siegla, D.C. and Smith, G.W.), Plenum Press, New York, 1981.
39. Colbeck, I. and Wu, Z., Measurement of the fractal dimensions of smoke aggregates, *J. Phys. D*, 27, 670, 1994.
40. Nelson, J.A., Crookes, R.J., and Simons, S., On obtaining the fractal dimension of a 3D cluster from its projection on a plane — application to smoke agglomerates, *J. Phys. D*, 23, 465, 1990.
41. Köylü, O.U. and Faeth, G.M., Structure of overfire soot in burning turbulent diffusion flames at long residence times, *Combust. Flame*, 89, 140, 1992.
42. Ogren, J.A. and Charlson, R.S., Elemental carbon in the atmosphere: cycle and lifetime, *Tellus*, 35B, 241, 1983.
43. Charlson, R.J. and Ogren, J.A., The atmospheric cycle of elemental carbon, *Particulate Carbon*, (Edited by Wolff, G.T. and Klimisch, R.L.), Plenum, New York, 1982.
44. Turco, R.F., Toon, O.B., Ackerman, T.P., Pollack, J.P., and Sagan, C., Nuclear winter global consequences of multiple nuclear explosions, *Science*, 222, 1283, 1983.
45. Weingartner, E., Baltensperger, U., and Burtscher, H., Growth and structural change of combustion aerosols at high relative humidity, *Environ. Sci. Technol.*, 29, 2982, 1995.
46. Hagen, D.E., Trueblood, M.B., and White, D.R., Hydration properties of combustion aerosols, *Aerosol Sci. Technol.*, 10, 63, 1989.
47. Pruppacher, H.R. and Klett, J.D., *Microphysics of Clouds and Precipitation*, Reidel, Boston, 1980.
48. Penner, J.E. and Molenkamp, C.R., Predicting the consequences of nuclear war: precipitation scavenging of smoke, *Aerosol Sci. Technol.*, 10, 51, 1989.
49. Rogers, C.F., Hudson, J.G., Hallett, J., and Penner, J.E., Cloud droplet nucleation by crude oil smoke and coagulated crude oil/wood smoke particles, *Atmos. Environ.*, 25A, 2571, 1991.
50. Jullien, R. and Meakin, P. Simple models for the restructuring of three-dimensional ballistic aggregates, *J. Colloid Int. Sci.*, 127, 265, 1989.
51. Kantor, Y. and Witten, T.A., Mechanical stability of tenuous objects, *J. Phys. (France)*, 45, 675, 1984.
52. Schmidt-Ott, A., *In situ* measurement of fractal dimensionality of ultrafine particles, *Appl. Phys. Lett.*, 52, 954, 1988.
53. Nyeki, S. and Colbeck, I., Fractal dimension analysis of single, *in situ*, restructured carbonaceous aerosols, *Aerosol Sci. Technol.*, 23, 109, 1995.
54. Hallett, J., Hudson, J.G., and Rogers, C.F., Characterization of combustion aerosols for haze and cloud formation, *Aerosol Sci. Technol.*, 10, 70, 1989.

55. Colbeck, I., Appleby, L., Hardman, E.J., and Harrison, R.M., The optical properties and morphology of cloud-processed carbonaceous smoke, *J. Aerosol Sci.*, 21, 527, 1990.
56. Huang, P.-F., Turpin, B.J., Pipho, M.J., Kittelson, D.B., and McMurry, P.H., Effects of water condensation and evaporation on diesel chain-agglomerate morphology, *J. Aerosol Sci.*, 25, 447, 1994.
57. Scope 28, *Environmental Consequences of Nuclear War. Volume I. Physical and Atmospheric Effects*, 2nd ed., John Wiley, Chichester, 1989, 63.
58. Kütz, S. and Schmidt-Ott, A., Characterization of agglomerates by condensation-induced restructuring, *J. Aerosol Sci.*, 23, S357, 1992.
59. Liousse, C., Cachier, H., and Jennings, S.G., Optical and thermal measurements of black carbon aerosol content in different envrionments: variation of the specific attenuation cross-section, sigma (σ), *Atmos. Environ.*, 27A, 1203, 1993.
60. Zhang, H.Q., Turpin, B.J., Hering, S.V., and Stolzenburg, M.R., Mie theory evaluation of species contributions to wintertime visibility reduction in Grand Canyon, *J. Air Waste Man. Assoc.*, 44, 153, 1994.
61. Katrinak, K.A., Rez, P., Perkes, P.R., and Buseck, P.R., Fractal geometry of carbonaceous aggregates from an urban aerosol, *Environ. Sci. Technol.*, 27, 539, 1993.
62. Colbeck, I., Hardman, E.J., and Harrison, R.M., Optical and dynamical properties of fractal clusters of carbonaceous smoke, *J. Aerosol Sci.*, 21, 765, 1989.
63. Lakhtakia, A. and Mulholland, G.W., On two numerical techniques for light scattering by dielectric agglomerated structures, *J. Res. National Institute of Standards and Technology*, 98, 699, 1993.
64. Ku, J.C. and Shim, K.H., A comparison of solutions for light scattering and absorption by agglomerated or arbitrarily-shaped particles, *J. Quant. Spectrosc. Radiat. Transfer*, 47, 201, 1992.
65. Singham, S.B. and Bohren, C.F., Scattering of unpolarized and polarized light by particle aggregates of different size and fractal dimension, *Langmuir*, 9, 1431, 1993.
66. Meakin, P. and Jullien, R., The effects of restructuring in the geometry of clusters formed by diffusion limited ballistic, and reaction-limited cluster-cluster aggregation, *J. Chem. Phys.*, 89, 246, 1988.
67. Berry, M.V., Falling fractal flakes, *Physica D*, 38, 29, 1989.
68. JGR Special Section, The Kuwaiti oil fires, *J. Geophys. Res.*, 97, (D13), 1992.
69. Brimblecombe, P., Introduction: atmosphere surrounding the Kuwait oil fires, *Atmos. Environ.*, 28, 2137, 1994.
70. Cofer, W.R., Stevens, R.K., Winstead, E.L., Pinto, J.P., Sebacher, D.I., Abdulraheem, M.Y., Al-Sahafl, M., Mazurek, M.A., Rasmussen, R.A., Cahoon, D.R., and Levine, J.S., Kuwaiti oil fires: compositions of source smoke, *J. Geophys. Res.*, 97, 14521, 1992.
71. Hobbs, P.V. and Radke, L.F., Airborne studies of the smoke from the Kuwait oil fires, *Science*, 256, 987, 1992.
72. Husain, T., Kuwait oil fires: modeling revisited, *Atmos. Environ.*, 28A, 2211, 1994.
73. Parungo, F., Kopcewicz, B., Nagamoto, C., Schnell, R., Sheridan, P., Zhu, C., and Harris, J., Aerosol particles in the Kuwait oil fire plumes: their morphology, size distribution, chemical composition, transport and potential effect on climate, *J. Geophys. Res.*, 97, 15867, 1992.
74. Cahill, T.A., Wilkinson, K., and Schnell, R., Composition analyses of size-resolved aerosol samples taken from aircraft downwind of Kuwait, Spring 1991, *J. Geophys. Res.*, 97, 14513, 1992.
75. Weiss, R.E., Kapustin, V.N., and Hobbs, P.V., Chain-aggregate aerosols in smoke from the Kuwait oil fires, *J. Geophys. Res.*, 97, 14527, 1992.
76. Hudson, J.G. and Clarke, A.D., Aerosol and cloud condensation nuclei measurements in the Kuwait plume, *J. Geophys. Res.*, 97, 14533, 1992.
77. Daum, P.H., Al-Sunaid, A., Busness, K.M., Hales, J.M., and Mazurek, M., Studies of the Kuwait oil fire plume during midsummer 1991, *J. Geophys. Res.*, 98, 16809, 1992.
78. Bodhaine, B.A., Hams, J.M., Ogren, J.A., and Hofmann, D.J., Aerosol optical properties at Mauna Loa observatory — long range transport from Kuwait, *Geophys. Res. Lett.*, 19, 581, 1992.
80. Horvath, H., Atmospheric light absorption — A review, *Atmos. Environ.*, 27A, 293, 1993.
81. Samson, R.J., Mulholland, G.W., and Gentry J.W., Structural analysis of soot agglomerates, *Langmuir*, 3, 272, 1987.
82. Gangopadhyay, S., Elminyawi, I., and Sorensen, C.M., Optical structure factor measurements of soot particles in a premixed flame, *Appl. Optics*, 30, 4859, 1991.

83. Bonczyk, P.A. and Hall, R.J., Measurement of the fractal dimension of soot using UV laster radiation, *Langmuir*, 8, 1666, 1992.
84. Sorensen, C.M., Cai, J., and Lu, N., Light scattering measurements of monomer size, monomers per aggregate and fractal dimension for soot aggregates in flames, *Appl. Optics*, 31, 6547, 1992.
85. Nyeki, S. and Colbeck, I., The measurement of the fractal dimension of individual *in situ* soot agglomerates using a modified Millikan cell technique, *J. Aerosol Sci.*, 25, 75, 1994.
86. Hallett, J., Gardiner, B., Hudson, J.G., and Rogers, C.F., *AAAR Meeting*, Albuquerque, NM, 1985.
87. Friesen, W.I. and Mikula, R.J., Fractal dimensions of coal particles, *J. Colloid Int. Sci.*, 120, 263, 1987.

26 Atmospheric Contamination by Fibrous Aerosols

Kvetoslav R. Spurny

CONTENTS

INTRODUCTION

The problem of atmospheric contamination with finely dispersed fibrous mineral aerosols has, from a global point of view, two important aspects. While in highly developed industrial areas, like the U.S. and Europe, the air pollution by fibrous particulates has decreased since the 1980s and the potential health risk of fibrous air pollutants is relatively low for the general population, the situation is different in less-developed countries and regions.

Asbestos, its mining, processing, and use, is the most important source of all environmental pollution by fine mineral fibers with harmful health effects on human beings.

Fortunately, asbestos import, and industrial and other applications in highly developed countries, has already been prohibited. Nevertheless, in the majority of less-developed countries (e.g., in Russia, China, several countries in Eastern Europe, Asia, Africa, and South America), the production and industrial applications of asbestos, especially in the building industries, are large and often increasing.

Atmospheric contamination by asbestos and other mineral fibers, as well as their potential health risk for the general population, are well documented in the U.S., Europe, Canada, and Japan.

0-87371-829-1/00/$0.00+$.50

On the other hand, very little is known about atmospheric concentrations and health risks in the above-mentioned underdeveloped countries. It has to be presumed that both asbestos fiber concentrations in the atmosphere and the health risk for the population are relatively high.

What is an Atmospheric Fibrous Aerosol?

Fibrous aerosols are aerodispersed systems consisting of fine fibrous particles. They belong to the group of aerosols with non-spherical particles. This aerosol group differs in its dynamic behavior from the classical aerosols consisting of isometric or spherical particles. A spherical aerosol particle is sufficiently defined by only one geometric parameter — its diameter or radius. A fibrous particle and its dynamic behavior depend on the particle thickness (D_F) and the particle length (L_F), as well as on the orientation angle in space and time. The shape of a single particle and its chemical composition — as well as its mineralogical characterization — can be identified by measurement and analytical procedures. For this reason, the atmospheric concentrations of fibrous aerosols are rarely measured as the mass. Measurements of fiber number concentrations (e.g., as fibers m^{-3} of air) are preferred. Electron microscope methods and single-particle analysis procedures must be used for the measurement and identification of fibrous aerosols.

Industrial and environmental hygienists and toxicologists proposed and later standardized a special definition for biologically active fibrous aerosols. This definition is based on two characteristic parameters of a single fiber: the fiber dimension and its durability. The harmful health effects of inhaled fibrous aerosols (fibrosis and cancer) are correlated with both the geometrical parameters — D_F and L_F — and fiber durability (bioavailability in the tissue).

Fibrous aerosols consisting of mineral fibers with $D_F < 3$ μm, $L_F \geq 5$ μm, $L_F/D_F \geq 3$, which are not "leachable and soluble" in biological liquids, are respirable (thoracic) and carcinogenic. Such fibers are also designated "WHO fibers," and the World Health Organization (WHO) has recommended such a definition. Two philosophies thus exist for the measurement of airborne mineral fibers. The basic environmental philosophy recommends that one measure and identify airborne fibers of all dimensions. The "health risk related philosophy" proposes to measure the concentration of the "WHO-fibers" only.

By considering the chemical composition and mineralogical structure of fibers, fibrous aerosols can be divided in two large groups: aerosols of asbestos and zeolite fibers, and aerosols of man-made fibers. These two groups differ substantially in their physico-chemical properties and biological effects. While aerosols of the first group are generally carcinogenic for humans, the aerosols of man-made fibers have lower — much lower or zero — carcinogenic potential.

Inhalation of fibrous aerosols of asbestos and erionite by man and animals initiates neoplastic diseases, such as mesothelioma and carcinoma of the bronchus. This fact has been well-proved and documented by epidemiological as well as animal inhalation studies. Heavy exposure to asbestos fibers in asbestos mining and application industries has resulted in increasing mortality of asbestos workers, due to mesothelioma and lung cancer, since about the 1960s.[1,2]

Several hundred to several thousand people die each year of mesothelioma or lung cancer in countries in which asbestos was or is still used in large amounts. An example of such a situation in the U.K. is illustrated in Figure 26.1. Asbestos was imported and used in the U.K. between 1900 and 1995. Mesothelioma disease and deaths will continue until about 2050, having a maximum in about 2020.[1]

Similar tendencies exist in other countries that previously or presently import or use asbestos. As many as several hundred thousand people worldwide die each year from asbestos-related diseases.

Similar to asbestos, erionite fibers have been proved to be carcinogenic in humans and animals after inhalation. Nevertheless, in this case, the exposed cohorts of humans are relatively small.[3]

Asbestos is being replaced by man-made mineral fibers (MMMF) — for example, glass, ceramic fibers, etc. — which are produced worldwide at the rate of several million tons per year. In animal

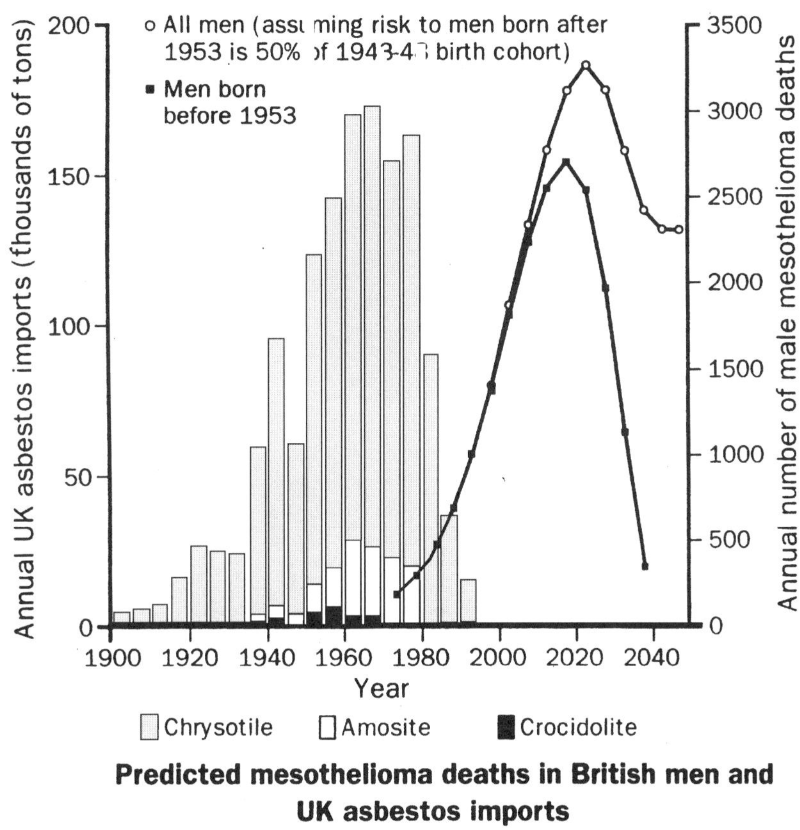

FIGURE 26.1 Results of current trends in U.K. mesothelioma incidents. (From Peto, J., Hodgson, J., Matthews, F. et al., *Lancet*, 345, 535, 1995. With permission.)

experiments, some of them also proved carcinogenic. However, no cases of cancer have been observed as yet in workers in the mineral fiber production and application industries.[4]

EMISSION SOURCES OF FIBROUS AEROSOLS

Fine mineral fibers are emitted into the atmosphere from natural as well as anthropogenic sources. For this reason, fibrous aerosols are ubiquitous air contaminants and their concentrations in the atmosphere are quite variable.

Natural Sources

Asbestos

All varieties of asbestos fibrous minerals (e.g., chrysotile, amosite, actinolite, anthophyllite, crocidolite, and tremolite) are present as veins in several rocks used for mining.[5] Asbestos and other natural mineral fibers are also dispersed in soils and may be released into the atmosphere, for example, by agricultural activities, erosion, weathering, wind, etc.

Asbestos fibers have been reported in rocks and soils in many countries; for example, in the U.S. (California, Maryland), Bulgaria (Avren), Czech Republic (Pelhrimov), Austria (Rechnitz), West Germany (Sauerland), Australia, Finland, Greece, Corsica, Cyprus, Russia, India, Italy, and Turkey.[6,7]

FIGURE 26.2 Rock pieces from a quarry in Germany containing actinolite fibers. (From Spurny, K.R., Pott, F., Huth, F. et al., *Staub-Reinhalt. Luft*, 39, 386, 1979. With permission.)

Endemic pleural calcification (cartilage-like plaques on costal pleura) could be found in the general population in some of these regions. These pleural plaques are related to continuous asbestos exposure. Several rocks (e.g., diabase rocks of several quarries may contain veins of asbestos, mainly of amphiboles). The gravel of such rocks is often used for road paving (Figure 26.2). Fiber measurements in ambient air in the vicinity of such quarries and roads showed that these activities were emission sources of fine asbestos fibers into the atmosphere.[8-10]

Zeolites

Since 1975, Baris et al. have reported asbestos and zeolite deposits in certain parts of Anatolia (Turkey).[3] The volcanic tuff containing these deposits was used for building houses (Figure 26.3). Samples of tuff material collected in the Karain area, for example, contained fibrous zeolite–erionite (Figure 26.4). Such fibers are released into the out- and indoor air. The inhabitants of these regions are exposed to the fibrous aerosols thus produced. Epidemiological studies have confirmed an increased incidence of malignant pleural mesothelioma of the population in these regions. This incidence was almost 1000 times higher than expected.[3,11]

Anthropogenic Emission Sources

The two most important fibrous mineral materials used in industrial applications are asbestos and MMMF (man-made mineral fibers). Asbestos is of natural origin and obtained by mining. MMMF is produced by industrial processes.

Asbestos

Asbestos (from the Greek *amiantus*) is the common name given to a number of naturally occurring, hydrated inorganic mineral silicates that possess a crystalline structure, are incombustible in air, and

FIGURE 26.3 Karain village in Anatolia (Turkey): the picturesque rock dwellings containing fibrous erionite. (From Baris, Y.I., *Asbestos and Erionite Related Chest Diseases*, Publ. House of the Hacettepe University, Ankara, Turkey, 1987. With permission.)

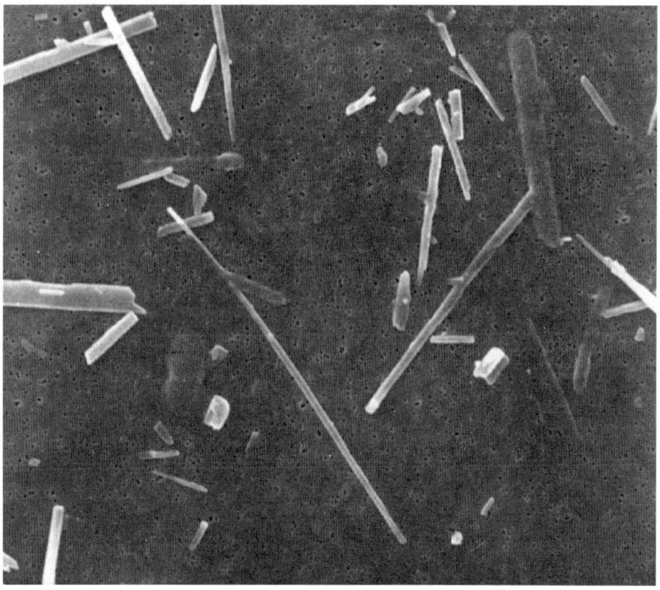

FIGURE 26.4 Scanning electron micrograph (Mag. 4000X) of erionite fibers detected in tuff from Karain.

are separable into filaments.[5,12] Asbestos has been known and used since antiquity — for as long as 4500 years. Many centuries before Christ, Finnish peasants mixed it in pottery and used it to seal cracks in their log huts. The ancient Greeks used it to make wicks for their lamps. The ancient Romans wove asbestos fibers into fabrics to make towels, nets, and even head coverings for women.

In the 18th century, asbestos remained little more than a curiosity. This changed, however, with the advent of the Industrial Age in the 1800s when industry realized its potential use. In the early

FIGURE 26.5 Workers stuffing fibrous material of crocidolite asbestos into large can for transporation. (From Vojakovic, R., *Asbestos*, Asbestos Disease Society of Australia Publication, Perth, Australia, 1997. With permission.)

1900s, doctors in Europe knew that asbestos workers were dying from respiratory ailments.[12] Cases of asbestosis (lung fibrosis) in the asbestos industry have been reported since 1900, and suggestion of a causal relationship between exposure to asbestos and cancer of the lung was reported in 1935. Early in the 1960s, due to man's increased use of asbestos products, the asbestos disease problem spread from the occupational area into the general environment.

If Greeks and Romans worked with asbestos — and they did — then asbestos fiber emissions were produced, and there probably was asbestos-related disease in Greece and Rome. Large deposits of asbestos exist in the Ural Mountains in Russia, the Alps in northern Italy, in Canada, the U.S., South Africa, India, Rhodesia, and Australia.

The mining, milling, and all other industrial procedures used in asbestos production, as well as asbestos transport, applications, etc., are accompanied by emissions of fine fibrous dust into the breathing zone of the workers and also into the atmosphere. There is evidence and documentation that these emissions were extremely high in the past, and they are probably still quite high in less-developed countries. Figures 26.5 and 26.6 document such situations in the past. Figure 26.5 shows asbestos workers in Australia processing the dust of the "blue" asbestos (crocidolite) in the vicinity of the Wittenoom mine (western Australia). Two children are shown playing with the "marvelous" blue dust as if it was harmless sand (Figure 26.6).

The asbestos exposures of workers, as well as their families living nearby the mine, should be high, and many died of mesothelioma (cancer of the pleura or the peritoneum). The concentrations of asbestos fibers in the ambient air were not measured at that time but in observing the pictures, it must be assumed that they were very high — maybe in the range of milligrams per cubic meter.

Epidemiological evidence accumulated over past decades indicates the presence of asbestos fibers in the lungs of the deceased asbestos workers, as well as in the lungs of most city dwellers.[13]

FIGURE 26.6 Two three-year-old children playing in the neighborhood of the Australian Wittenoom asbestos mine in 1953. (From Vojakovic, R., *Asbestos*, Asbestos Disease Society of Australia Publication, Perth, Australia, 1997. With permission.)

Although heavy exposure to asbestos is known to result in cancer, the health risk of smaller doses of asbestos is not clear.

Worldwide consumption of asbestos in the 1970s lay in the range of about 5 megatons per year; the estimated mean emission factors were approximately 10 kg ton^{-1}. This means that each year, approximately 50,000 tons of asbestos fibers escaped into the lower atmosphere. Major man-made sources of airborne asbestos fibers include the mining and milling of asbestos materials as well as their transportation, the manufacture of over 3000 different commercial products containing asbestos (such as vehicle brake linings and roof tiles), the use of asbestos in the building industry, the demolition and wasting of asbestos-containing structures, and the disposal of asbestos waste materials. Figure 26.7 shows schematically the most important emission sources, transportation cycle, and population exposure in the total environment.

Anthropogenic contamination of soils by asbestos also exists. Soils became contaminated with asbestos during and after industrial asbestos processing. Later use of such contaminated areas (e.g., for house-building or agricultural purposes) can release fine asbestos fibers into the atmosphere.[14] A large surfacial source of fine asbestos fiber emissions into troposphere is the weathering and corrosion of asbestos cement products. They contain as much as 12% asbestos. As a result of constant exposure to meteorologic influences (e.g., acid rain, sunshine, wind, frost) and to atmospheric pollutants, the surface of asbestos cement products (e.g., roof and façade tiles) corrode and weather (Figure 26.8). Cement particles, asbestos fibers, and agglomerates of both particles and fibers (Figure 26.9) are released from the plate surface and become dispersed in rainwater and in the air. These surfacial emission sources of fine asbestos fibers into the atmosphere are large. For example, in Europe, the entire surface area was estimated to be as large as 10^{10} m^2 or more.[15] The residence time of fine asbestos fibers in the atmosphere could range from weeks to months.

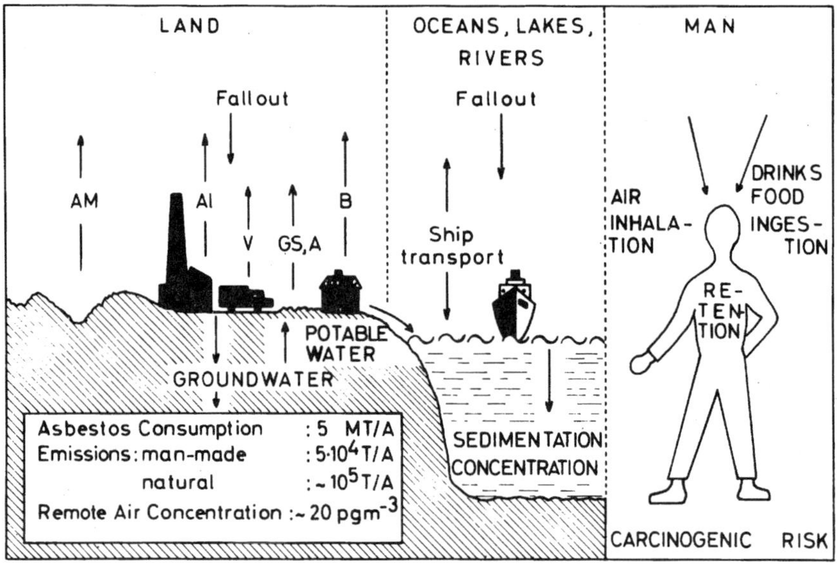

FIGURE 26.7 Schematic diagram of asbestos sources and the cycle of asbestos in the environment. Abbreviations are: AM = asbestos mining; AI = asbestos industry; V = vehicles; GS = geological sources such as soils and rocks; A = agriculture; B = building industry. (T/A = tons per annum.)

Man-made mineral fibers (MMMF), the second group of fibrous material, are commercially produced and used on a broad basis. Production levels have already reached over 5 megatons per year worldwide.

Man-Made Mineral Fibers (MMMF)

As previously noted, MMMF are considered inorganic fibrous material, capable of replacing asbestos in industrial and other applications, and in most cases are relatively harmless.[4,13]

The production of MMMF has, of course, a much shorter history than asbestos. Mineral wools (slag- and rockwool) and glass and ceramic fibers are the most important representatives of synthetic mineral fibers. Slagwool was first manufactured commercially in 1880 and was mostly coarse fibers. Today, MMMFs are produced by modern technology, in a wide range of specifications, for countless industrial purposes. The industry has expanded rapidly to meet the need for more thermal insulation and to provide fibers for many types of composites. Fibers with diameters over, but also less than, 3 μm, are being produced. As already mentioned, carcinogenic effects of MMMFs in humans have not been reported as yet. Studies on animals have shown, however, that mesothelial tumors of the pleura and peritoneum could be induced by intrapleural and intraperitoneal implantation of some MMMFs, especially some types of ceramic fibers.[16]

The fabrication of MMMFs, as well as their application and installation in buildings, are the most important emissions sources of fibrous aerosols generated by dispersion of fine MMMFs.

PRINCIPLES OF SAMPLING, MEASUREMENT, AND IDENTIFICATION

Fibrous dusts and aerosols must be distinguished from isometric ones. Fibrous particle shape is therefore the most important characteristic to be identified. Microscopic methods are thus the most suitable tools to do that. As the fibrous particle dispersed in the air can be very thin (less and much

FIGURE 26.8 Photographs of weathered and corroded asbestos-cement roofing shingles.

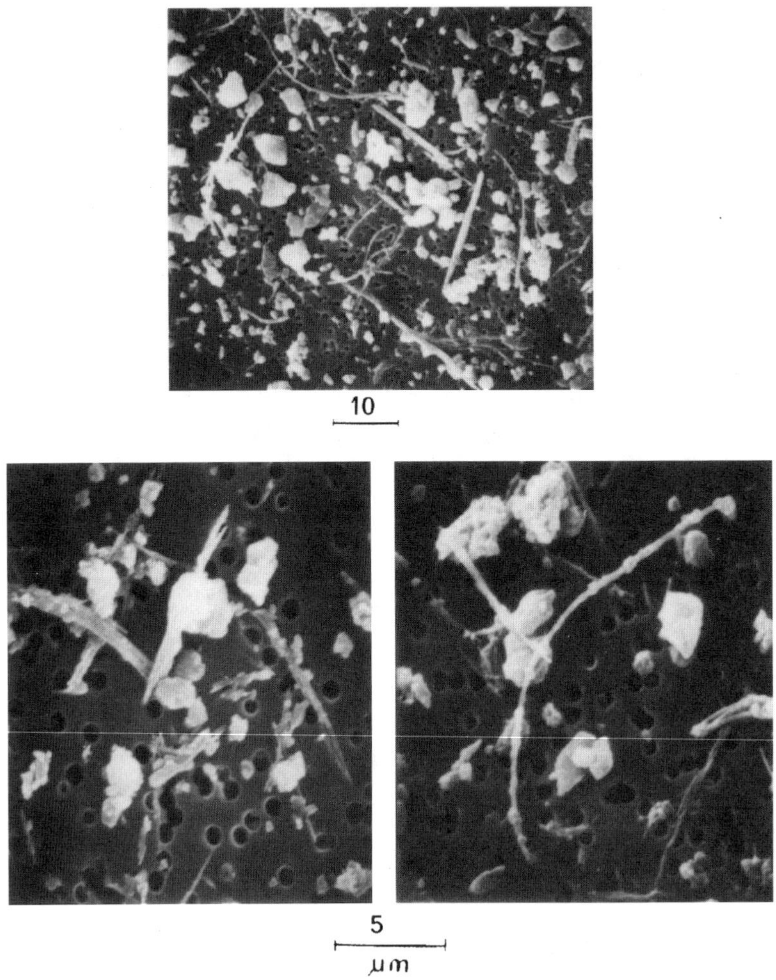

FIGURE 26.9 Scanning electron micrographs of cement particles and asbestos fibers released after corrosion into the rainwater.

less than 1 µm; see Figure 26.10), they cannot be seen by optical microscopy, and electron microscope methods must be applied.[17]

A convenient and widely used method for collecting fibers for their counting, sizing, and identification is filtration. Membrane filters (MF) and Nuclepore filters (NPF) are the filters of choice for sampling fibrous aerosols.[18]

Scanning electron microscopy (SEM), and transmission electron microscopy (TEM), and identification techniques such as energy dispersive X-ray analysis (EDXA) and selected area electron diffraction (SAED) are currently well proved, tested, and standardized procedures that can be recommended (Figures 26.11 and 26.12). Other modern methods for single-particle analysis (e.g., laser mass spectroscopy) are also useful methods (Figure 26.13). The identified elements, their combinations, and the electron diffraction patterns are very useful finger-printing for the chemical and mineralogical characterization of single fibrous particles.

TEM+EDXA+SAED is the best and, in the majority of highly developed countries, is the standardized reference methodology for the counting, sizing, and identification of fibrous particles.[19] Fibrous aerosols generated from the fabrication and application of MMMFs are, in comparison with asbestos aerosols, coarser, having thicker fibers (Figure 26.14) and they are

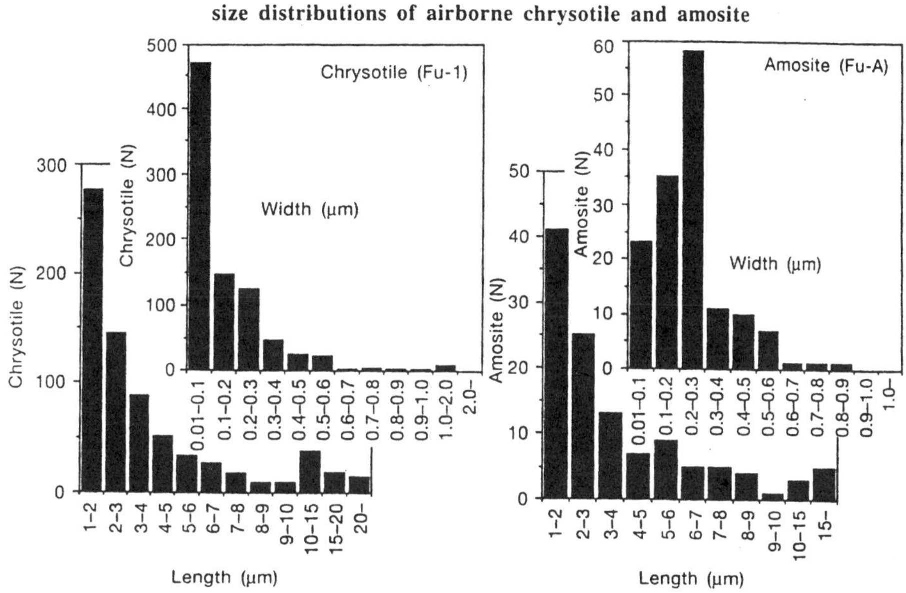

FIGURE 26.10 Size distribution of asbestos fibrous aerosols obtained by measurements in Japanese asbestos industries. (From Kohyama, N. and Kurimori, S., *Industrial Health (Japan)*, 34, 185, 1996. With permission.)

predominantly amorphous. The SEM+EDXA methodology is therefore sufficient in the majority of cases. Distinction and finger-printing of the MMMFs can be realized by the elemental analysis of single fibers. The most characteristic elements of these fiber types include Ca, Mg, Na, K, B, Al, and Zr.

AMBIENT AIR CONCENTRATIONS

Atmospheric fibrous aerosols consist in the first approximation of asbestos and other mineral fibers. Asbestos fibers were and still are the most important group of mineral fibers dispersed in the atmosphere.

Asbestos Fiber Aerosols

Because asbestos is exceptionally resistant to thermal and chemical degradation, it persists in the environment and can be widely redistributed into the atmosphere by natural forces (wind, etc.) and human means. By different processing technologies, especially milling, asbestos particles and agglomerates are discharged in the form of free fine fibers that may remain in the atmosphere for long periods of time, travel great distances, and result in exposure of many people. Studies of atmospheric pollution in the vicinity of asbestos mines and mills showed small amounts of asbestos aerosols as far away as 30 km from the emission source.[20]

Early Measurements

Prior to the 1980s, the ambient air concentrations of asbestos fibers were indicated — similar to the concentrations of all atmospheric particulates — as the mass/air volume (e.g., in ng m^{-3}). The number of microscopically counted fibers was converted approximately into the mass. There existed several conversion factors, as, for example, 1000 fine atmospheric fibers = 1 ng. Average reported concentrations of asbestos in several U.S. cities varied between 0.1 and 70 ng m^{-3} in the 1970s.[21]

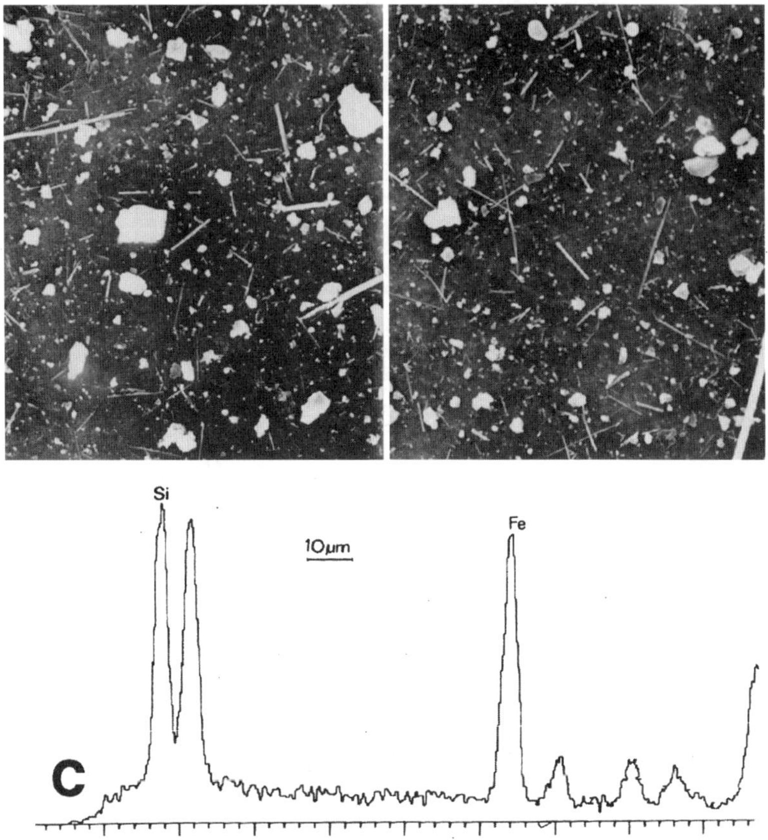

FIGURE 26.11 Crocidolite asbestos aerosol samples and their characterization by the SEM + EDXA method (Si and Fe are the characteristic elements).

The potential exposure of the U.S. urban population at that time is illustrated in Figure 26.15. The asbestos concentrations in the air near industrial facilities were higher and lay in the range of about 10 to 400 ng m^{-3}.[5]

Much lower asbestos concentrations in the general ambient air were reported in several European cities in the 1970s. Arithmetic mean air concentrations were of the order of 1 to 3 ng m^{-3}.[5]

Measurements Since the 1980s

In measurements since the 1980s, when the SEM and TEM methods were standardized and routinely used, the ambient air concentrations of asbestos were indicated as the number of fibers per air volume (e.g., as fibers ml^{-1}, fibers l^{-1}, and fibers m^{-3}). Measurements of asbestos fibers in ambient air nearby natural and anthropogenic emission sources, in ambient air of urban and rural (remote) areas, as well as in the higher troposphere, were reported.[22-26]

Asbestos Fiber Concentrations Near Industrial Emission Sources

The asbestos fiber concentrations around industrial emission sources depend on several parameters, especially on the distance and wind direction.

Asbestos fiber concentrations in ambient air were measured in the vicinity of a large asbestos-cement factory located in western Germany between 1981 and 1983. The concentrations were detected monthly in three sampling sites in the downwind direction. The total mineral fiber concentrations decreased exponentially with decreasing distance from the source. Fiber concentrations

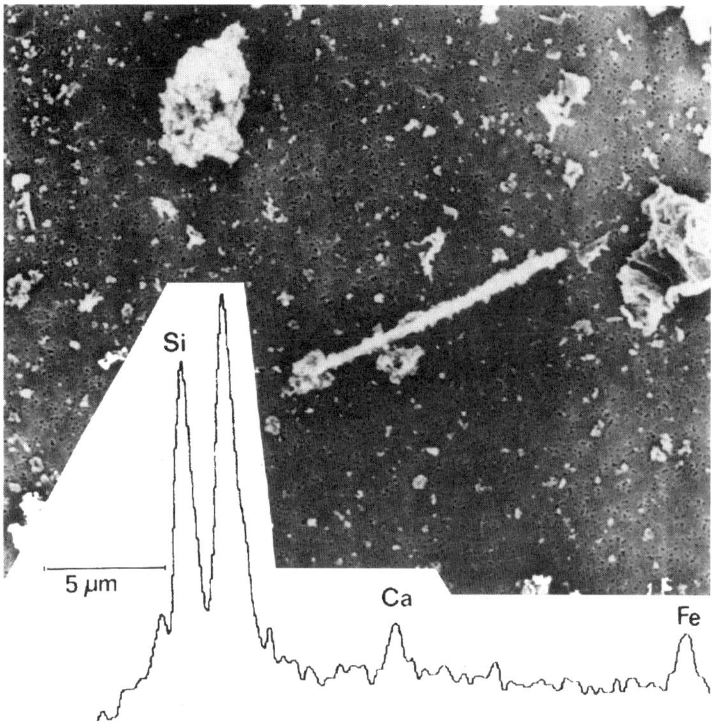

FIGURE 26.12 SEM + EDXA characterization of fibrous particles sampled in the air over a corroded asbestos cement roof. The chrysotile fiber is contaminated by cement microparticles, which are characterized by the Ca-peak.

as high as 90,000 fibers m^{-3} were measured at a distance of 300 m from the source, while concentrations of 60,000 fibers m^{-3} were found at a distance of 1000 m. A similar trend was observed by measuring the concentrations of all asbestos fibers. Within a distance of 700 m, asbestos fiber concentrations of about 11,000 fibers m^{-3} decreased to a concentration of 4000 fibers m^{-3} (annual mean). Even higher differences were obtained by evaluation of concentrations of asbestos fibers with fiber lengths more than 5 μm. The original asbestos concentrations of 2100 fibers m^{-3} were diminished to 600 fibers m^{-3} within a distance of 700 m. The "peak concentrations) of asbestos fibers longer than 5 μm sometimes reached values of over 15,000 fibers m^{-3}. The dimensions of measured asbestos fibers also decreased substantially with the increasing distance from the emission source (see also Figure 26.30).[22]

Similar investigations were done in Italy in 1985.[23] Asbestos fiber concentrations were measured at several locations near a large asbestos-cement factory in northern Italy. Mineral and asbestos fiber concentrations ranged from about 1 to 230 fibers per liter air (f l^{-1}), that is, 1000 to 230,000 fibers m^{-3}. About 15% of all measured fibers was asbestos. Wind direction and distance from the source were of major importance. Figure 26.16 shows mineral fiber concentrations measured at different sites and different distances around the factory. In southern Germany, asbestos fiber concentrations (fibers longer than 5 μm) in the range between 15,000 and 40,000 fibers m^{-3} were measured in the vicinity of an industrial emission source.[24]

Even higher asbestos fiber concentrations were obtained by measurements in areas close to industrial sources in Japan.[25] Total (all fiber sizes) asbestos fiber concentrations between 11,000 and 850,000 fibers m^{-3} were detected in the ambient air near factories making asbestos slate-board. In a town adjacent to a chrysotile mine, the measured asbestos fiber concentrations were as high as 350,000 to 800,000 fibers m^{-3}. Surrounding a serpentine quarry, asbestos fiber concentrations

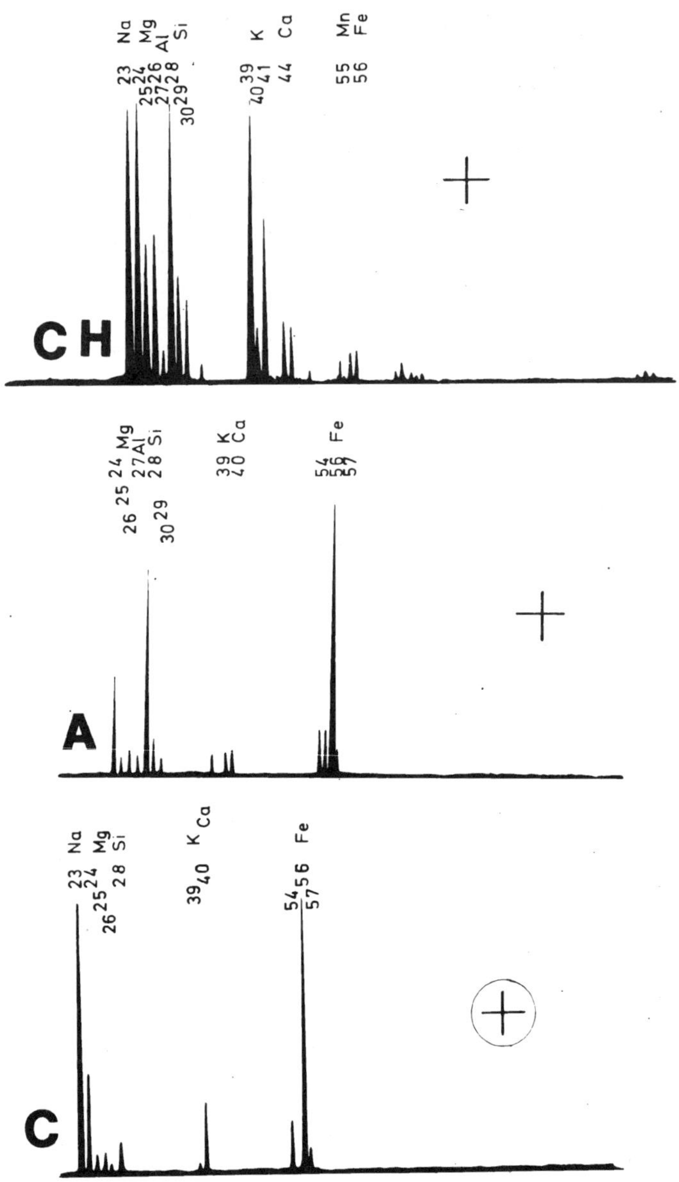

FIGURE 26.13 Positive mass spectra of single fibers of chrysotile (CH), amosite (A), and crocidolite (C) asbestos. The differential finger-printing done by the combination of element peaks is evident.

ranged from 10^5 to 10^7 fibers m^{-3}. Figure 26.17 shows chrysotile fiber concentrations (fibers per liter air) measured at distances between 10 and 1000 m of this serpentine quarry.

Urban and Rural Regions

Several non-industrial sources of emitted asbestos fibers do exist in cities and sometimes also in rural and remote areas. Fine asbestos fibers are emitted from corroded asbestos-cement roofing and façade tiles and plates, brake and clutch linings of automobiles, building demolitions, waste dumps, etc.

Vicinity of Weathered and Corroded Asbestos-Cement Materials

As already mentioned, asbestos-cement products used in the building of houses and other objects weather and corrode by the influence of weather conditions and air pollutants (see Figure 26.8 and Reference 15).

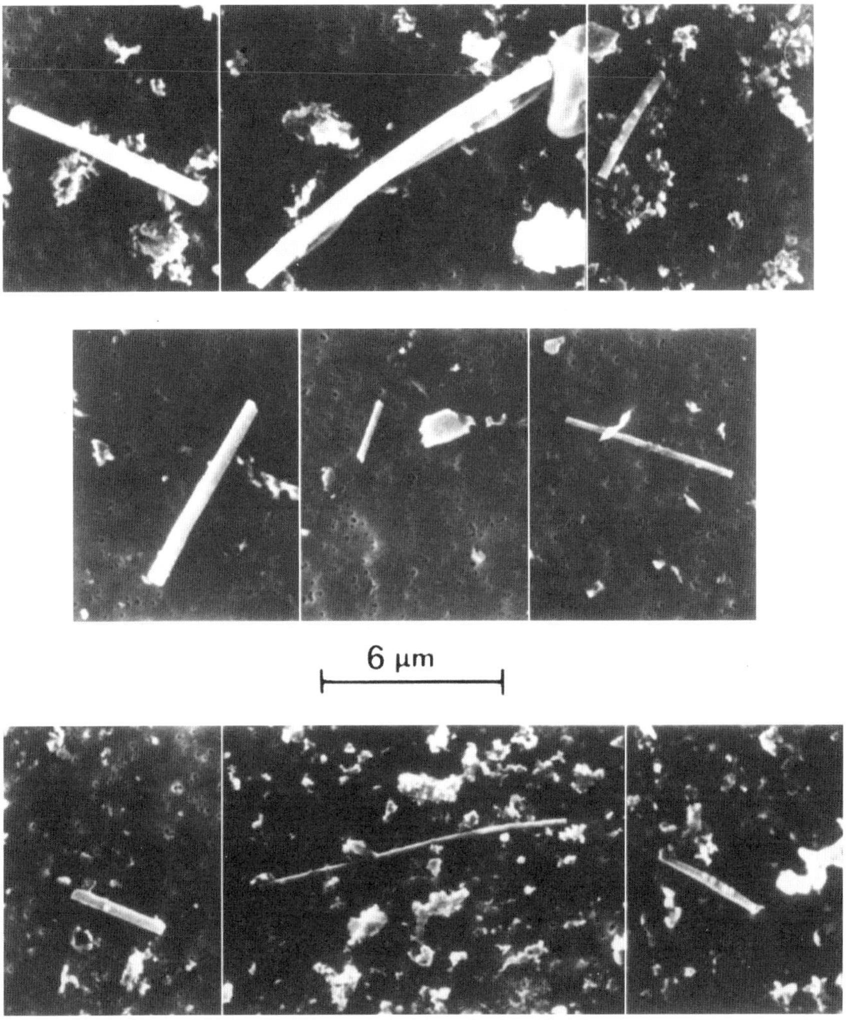

FIGURE 26.14 Scanning electron micrographs of MMMF particles sampled in ambient air.

Ambient air concentrations of fibrous aerosols emitted from corroded asbestos-cement products were measured in the vicinity of buildings in Germany from 1983 to 1986.[15] The sampling was carried out at a distance of 0.5 to 1 m from the asbestos-cement façades and roofs. The distribution of the measured fiber concentrations is shown in Figure 26.18. Single particle sizes and element compositions are summarized in Tables 26.1 and 26.2. Asbestos fiber concentrations (for fibers longer than 5 μm) in the range 200 to 1200 fibers m^{-3} have been measured. The concentrations of all asbestos fibers ranged from 10^2 to 10^4 fibers m^{-3}. When one considers that the total surface of diffusion emission sources of asbestos-cement façade and roofing sheets in Germany may be as large as 10^9 m^2 and that children like to play nearby such buildings, one can assume that these measured asbestos fiber concentrations pose an additional risk — not only for the occupants of such houses, but, since the background level of ambient air asbestos fibers is increasing, also for the general population.

Additionally, the asbestos fibers released from the corroded and weathered asbestos-cement sheets have been found to be heavily contaminated (see also Figure 26.12) by elements from external air pollutants as, for example, Ca, S, K, Ti, Zn, Pb, Sr, and by several PAH (polycyclic aromatic hydrocarbons). As much as 2.5 μg PAH was determined in 1 g of the released material (Figure 26.9).

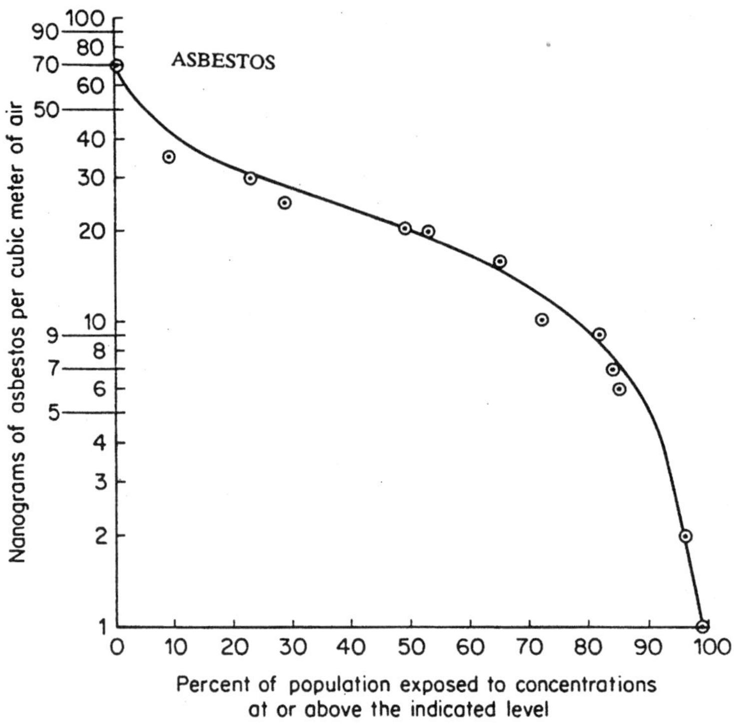

FIGURE 26.15 Percentage of urban population in the U.S. who breathe various levels of asbestos mass in ambient air. (From Michaels, L. and Chissick, S.S., Eds., *Asbestos*, John Wiley & Sons, Chichester, 1979. With permission.)

Another source of asbestos fiber emissions has been discovered in several recreation parks in Germany.[26] Giant slides were constructed in these parks to render it possible to sled in the summertime. The slides are used mainly by children. The asbestos-cement panels used for the construction of the slides are not mechanically resistant. On the surfaces, asbestos-cement brush-up is produced and fine asbestos fibers are released by wind dispersion into the breathing zone of the "sled-drivers" (Figure 26.19). The measured ambient air concentrations of all asbestos fibers was determined to be in the range between 1000 and 15000 fibers m^{-3}. Values between 900 and 8000 fibers m^{-3} were measured in the case of asbestos fibers longer than 5 μm, at a distance of 0.5 m over the slide surface.[26]

Brake and clutch linings of automobiles contained 10 to 70% asbestos, mainly chrysotile, prior to the 1980s. They were therefore considered to be a non-negligible source of asbestos fibers emitted into the ambient air. This could be confirmed by measurements performed in the urban ambient air in some German cities.[24,27,28] Concentrations of asbestos fibers longer than 5 μm were as high as 1000 fibers m^{-3} on important street crossings in downtown areas. Asbestos fibers of all sizes were measured in the ambient air of 19 cities and regions in Japan between 1982 and 1983.[25] The sampling sites were situated in residential (A), shopping (B), inland industrial (C), harbor (E) and agricultural (F) areas, in areas of small factories processing asbestos (G), on freeways (H), main roads (I), around disposal sites for waste materials (J), around buildings being demolished (K), in dockyards (L), around automobile repair shops (M), inside school rooms (N), on isolated islands in the Pacific Ocean (O,P). (Q is the contamination level in sample preparation.) The measured concentrations in fibers per liter air (f l^{-1}) are summarized in Figure 26.20.

Asbestos fiber concentrations in ambient air during the demolition of buildings containing asbestos insulation and around waste deposal sites in Germany were also measured in the 1980s.[29,30]

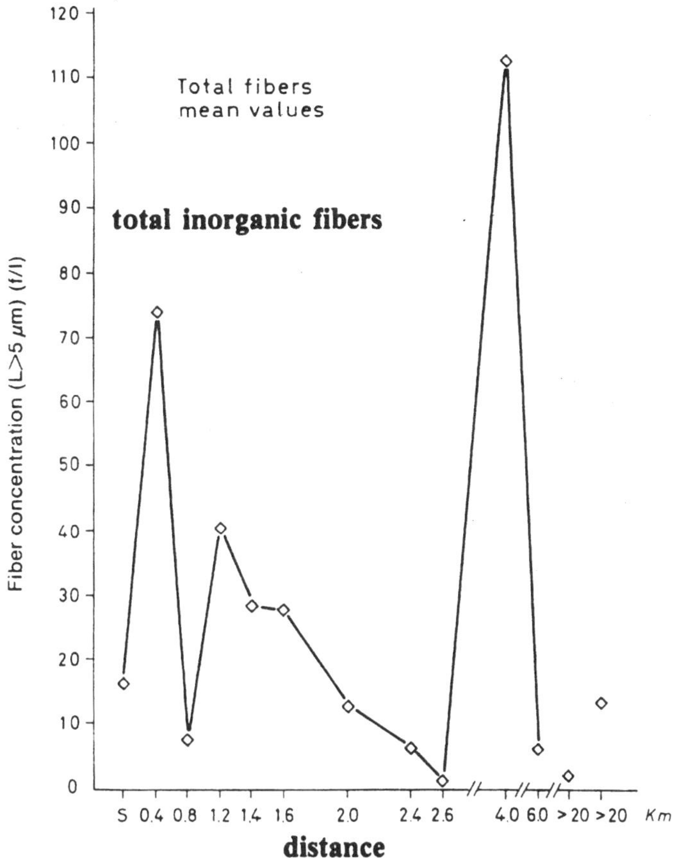

FIGURE 26.16 Mean concentrations of total inorganic fibers in relation to the distance from source. (From From Marconi, A., Cecchetti, G., and Barbieri, M., *IARC Sci. Publ. (Lyon)*, 90, 336, 1989. With permission.)

Measurements have been performed by application of different techniques for building demolition and roof renovation. During blasting operations, significant concentrations of asbestos fibers in the range 1000 to 50,000 fibers m^{-3} (for all fibers), and in the range 500 to 25,000 fibers m^{-3} (for fibers longer than 5 μm) were measured in ambient air at distances up to 200 m.

During building demolitions by means of dredgers and caterpillars, these concentrations lay in the range up to 770 fibers m^{-3} (fibers longer than 5 μm) at a distance of 70 m from the emission source. During manual removal of asbestos-cement roofs and façades, the ambient air concentration of asbestos fibers (fibers longer than 5 μm) were up to 6 times higher than the background concentrations at distances of 50 m from the buildings.

Building refuse dumps at a garbage dump cause high emissions of asbestos fibers into the atmosphere. Short time concentrations of all asbestos fibers as high as 8000 fibers m^{-3} were indicated. Waste and garbage dumps with free deposited asbestos and asbestos-containing materials (after building and sanitation and demolition) are also important sources of asbestos fibers.[29] Measurements in the vicinity of two large waste dumps were performed in Germany in 1986. Significant asbestos fiber concentrations were measured on both sites. The concentrations of asbestos fibers (fibers longer than 5 μm) were often higher than 1000 fibers m^{-3} (with maximal values in the range of 3000 fibers m^{-3}).[30]

Indoor and outdoor asbestos fiber concentrations were also measured in California (U.S.) after an earthquake struck (Richter scale 7.1) in 1989.[31] Concentrations of asbestos fibers (fibers longer

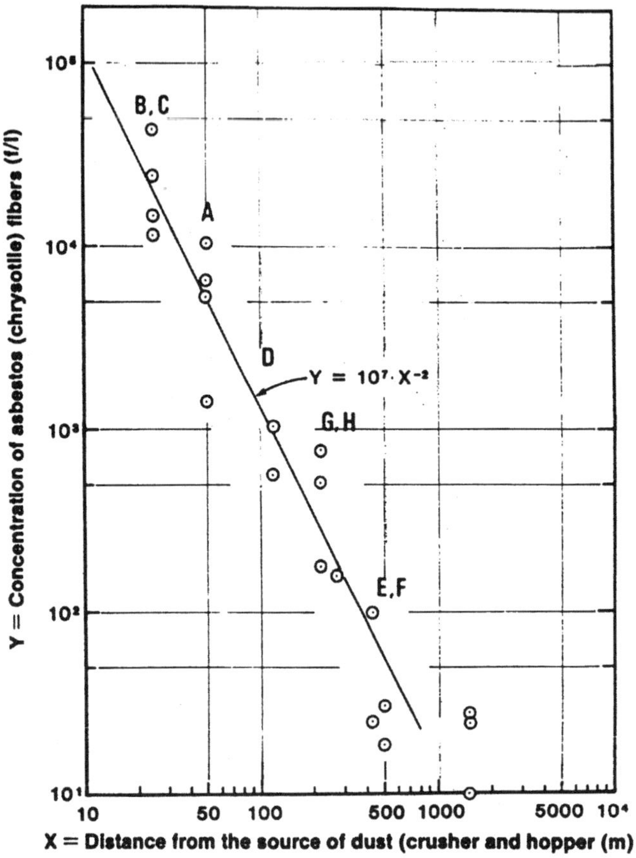

FIGURE 26.17 Concentrations of chrysotile asbestos fibers around the serpentine quarry. (From Kohyama, N., *IARC Sci. Publ. (Lyon)*, 90, 262, 1989. With permission.)

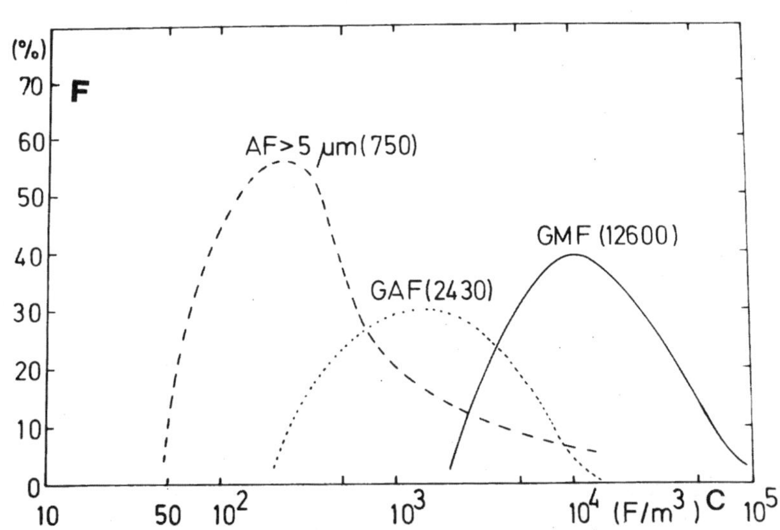

FIGURE 26.18 Ambient air fiber concentrations (C) of asbestos fibers in the vicinity of buildings with corroded asbestos-cement products: frequency (F) distribution and mean values; for asbestos fibers longer than 5 μm (AF), for all asbestos fibers (GAF), and for all mineral fibers (GMF).

TABLE 26.1
Mean Size Values of Emitted Fibers (GMF)

Fiber Size	$\bar{x}$	SDσ	$\bar{x}_{min}$	$\bar{x}_{max}$
Fiber length L_F (μm)	4.0	2.0	1.7	9.5
Fiber diameter D_F (μm)	0.22	0.09	0.1	0.5
Ratio L_F/D_F	25.1	11.7	10.2	66.0
L_F (max) (μm)	23.7	23.1	3.5	140
D_F (max) (μm)	1.07	0.6	0.5	3.0
L_F/D_F (max)	250	193	27	910

TABLE 26.2
Elements Distribution in Single Asbestos Fibers

Elements (%)	$\bar{x}$	SDσ	$\bar{x}_{min}$	$\bar{x}_{max}$
Si	28.4	8.9	13.8	48.8
Si, Fe	14.9	20.3	0.7	50.0
Mg, Si	28.2	26.3	4.3	50.0

than 5 μm) lay in the range of 10^3 to 10^4 fibers m^{-3} at indoor sites, and in the range of 150 fibers m^{-3} at outdoor sites.

As mentioned, soils can be contaminated by anthropogenic asbestos. Measurements of ambient air concentrations of asbestos fibers have been performed in the vicinity of such areas in Germany after the agricultivations of such soils.[14] The ambient air concentrations of all asbestos fibers were high; they lay between 7×10^4 and 8×10^5 fibers m^{-3}. Concentrations of asbestos fibers longer than 5 μm ranged from 300 to 15,000 fibers m^{-3}.

Asbestos fiber concentrations in ambient air of remote areas were measured in Germany at the end of the 1970s.[13] The obtained values (for all asbestos fibers) were in the majority less or much less than 100 fibers m^{-3}. Aerosol probes were sampled at a mountain observatory over 2900 m a.s.l. in a clean-air region (remote station Zugspitze, Bavaria, Germany) in June 1984.[32] The filter probes were evaluated for the content of fibrous particulates by the TEM (transmission electron microscopy) procedure (Figure 26.21). The obtained results showed that asbestos as well as other mineral fibers are present in the air at higher tropospheric elevations. The established concentrations lay between 0.2 and $2 \cdot 10^3$ fibers m^{-3} for all asbestos fibers, and between 0.7 and $1.7 \cdot 10^4$ fibers m^{-3} for all mineral fibers. In 98% of cases, the sampled mineral fibers were shorter than 1 μm and thinner than 0.1 μm.

Urban Areas Far from Significant Asbestos Emission Sources

Investigations of fibrous pollutants in ambient air have been done in several German towns and regions.[33-36] The concentrations of fibrous mineral aerosols were measured mainly in residential areas, far or very far from important asbestos emission sources. The only sources for potential asbestos emissions were asbestos-cement façades and automobile brakes.

More than 50% of asbestos fiber concentrations (fibers longer than 5 μm) were less than 100 fibers m^{-3}. Concentrations over 600 fibers m^{-3} were very rare (Figure 26.22) and were measured on street crossings and very close to asbestos-cement façades.[34]

The total particulate pollution (GS, total suspended particles) of all measured regions was also low (Figure 26.23A). The annual mean value was 58 + 40 μg m^{-3}. The concentrations of asbestos

FIGURE 26.19 Photographs of a giant slide, showing dust sampling and brush-up (OB).

fibrous aerosols (all fiber lengths; see Figure 26.23B), as well as of short non-asbestiform mineral fibers (Figure 26.23C) were in the majority of cases less than 200 fibers m^{-3}.[35]

Furthermore, the concentrations of fibrous aerosols showed very inhomogeneous random distributions in the troposphere. In the majority of cases, the annual mean concentrations reached low or very low values (Figure 26.22). The monthly measured concentrations were irregularly distributed. This was true for asbestos as well as for all types of mineral fibers. Examples are shown in Figures 26.24 and 26.25. The annual distribution of asbestos fiber concentrations (all fibers) obtained during measurements in large cities (Mannheim [M] and Karlsruhe [K]) are shown in Figure 26.24. Similar distributions for the measurements of all mineral fibers (in the regions of the towns of Mannheim [M], Karlsruhe [K], and Reutlingen [R] and on a remote station in the Black Forest [B], elevation of 900 m) are illustrated in Figure 26.25.[36] These measurements have confirmed that fibrous mineral aerosols, including asbestos, are ubiquitous in the atmosphere. Nevertheless, their ambient air concentrations undergo important variations in space and time.

Erionite Fibers

Fine fibrous erionite, dispersed in ambient air, has been observed and measured in some villages in central Turkey. This region is affected by a high incidence of mesothelial and lung tumors. This type of natural mineral fibers has also been confirmed as the etiological agent for these diseases.

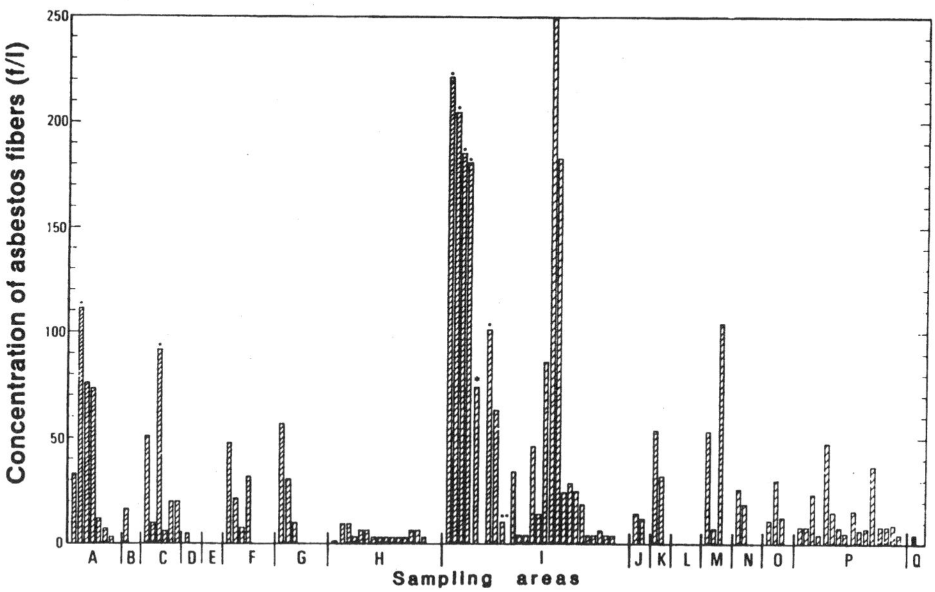

FIGURE 26.20 Concentrations of asbestos fibers (all sizes) measured at various sites in Japan during 1982 and 1983. (From Kohyama, N., *IARC Sci. Publ. (Lyon)*, 90, 262, 1989. With permission.)

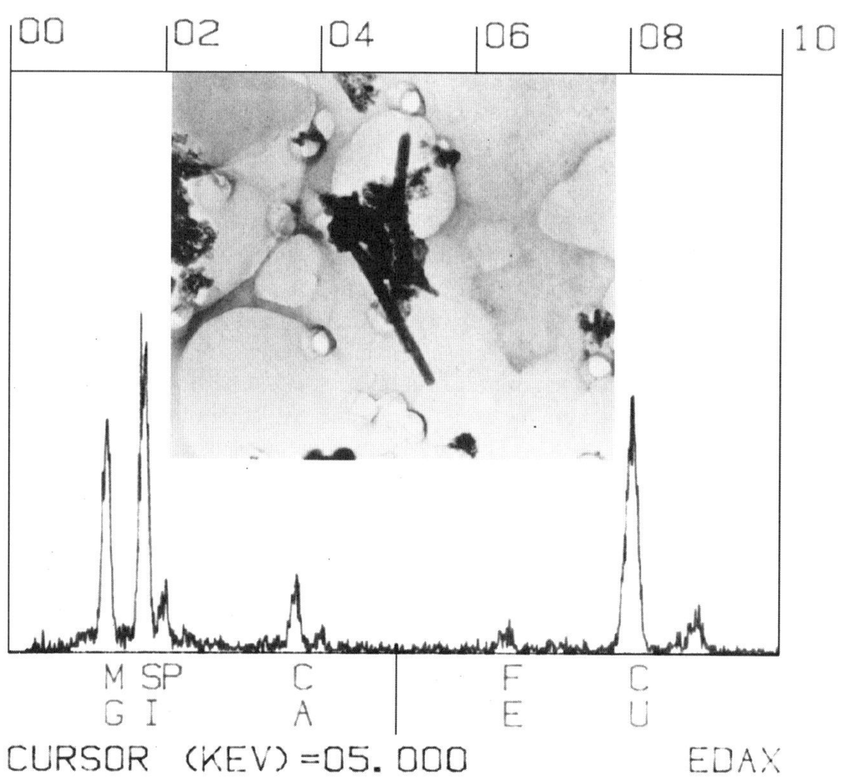

FIGURE 26.21 Transmission electron micrograph and element spectrum (EDXA) of very fine asbestos fibers, sampled on the remote station of Zugspitze (21,000X magnification).

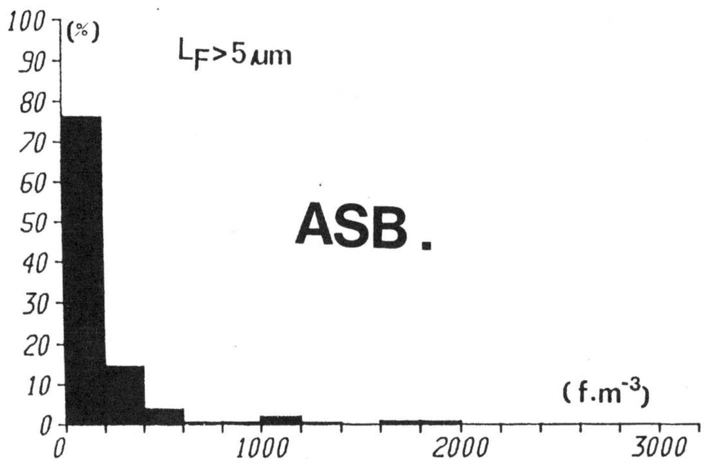

FIGURE 26.22 Distribution of measured asbestos fiber concentrations (for fibers longer than 5 μm) in the urban atmosphere.

From animal experiments, fibrous erionite appears to be the most powerful carcinogenic mineral fiber thus far known.[3,6,11] Epidemiological investigations were performed in this region during the period between 1979 and 1983. Ambient-air erionite fiber concentrations were also measured during this time period. The obtained concentration values ranged from 1000 to 30,000 fibers m^{-3}, with mean values in the range between 6000 and 9000 fibers m^{-3}. The mortality rates per 100,000 person-years for pleural mesotheliomas and lung cancer correlating to the measured ambient-air concentrations of fibrous erionite aerosols lay in the range between 350 and 1770, depending on the exposure dose and time.[3,11]

Man-Made Mineral Fibers (MMMF)

There are relatively few publications dealing with the measurement of synthetic mineral fibers in ambient air.[37-42] In a study performed in the mid-1980s, concentrations of glass fibers were determined in some rural and urban areas in Germany, using the analytical TEM procedure.[38] Glass fiber concentrations in the range of 50 fibers m^{-3} in rural air, and between 1000 and 2000 fibers m^{-3} in urban ambient air were detected. For similar measurements in ambient air of Paris, glass fiber concentrations in the same range were found.[39]

Levels of MMMF in the air were measured during the insulation of lots and after disturbance of the insulation wools in the U.K.[40] The TEM procedure was used in these measurements. Relatively very high concentrations were found; they lay between 5×10^4 and 7×10^5 fibers m^{-3} (for all fiber lengths).

Glass fiber concentrations were measured in ambient air around a large fiberglass manufacturing plant and in a rural area in the U.S.[41] In the majority of cases, the measured concentrations of glass fibers longer than 5 μm were as small as 10 to 100 fibers m^{-3}.

MMMF, including glass fiber, concentrations in ambient air were measured in the early 1990s in Germany.[42] Fiber (longer than 5 μm) concentrations ranged between 400 and 2000 fibers m^{-3}. Concentrations of all other non-synthetic mineral fibers were much higher (see also Figure 26.31).

DISPERSITIES OF MEASURED FIBROUS AEROSOLS

Fiber dimensions — fiber length (L_F), fiber diameter (D_F), and the aspect ratio (L_F/D_F) — are very important parameters that are involved in carcinogenic mechanisms and characterize the emission

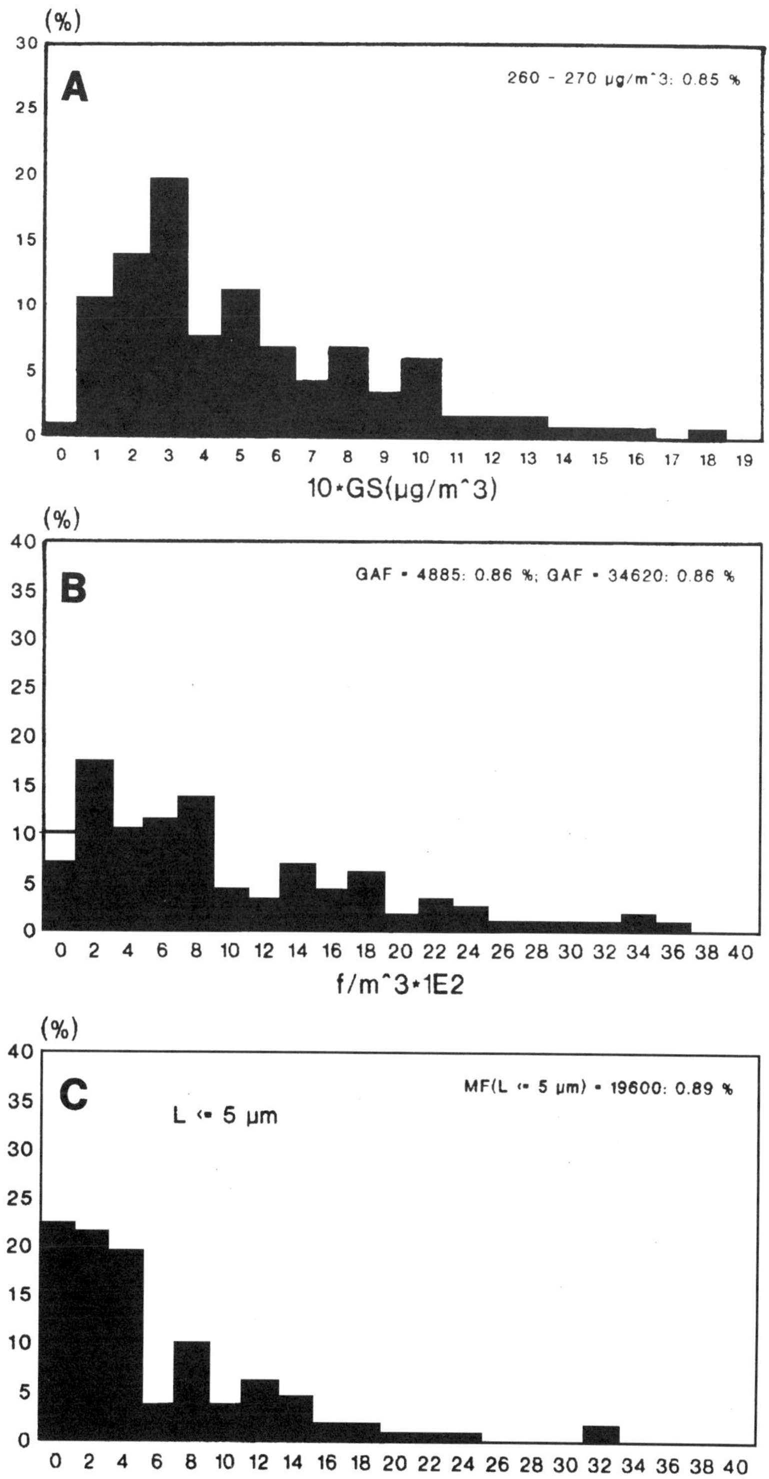

FIGURE 26.23 Measured concentrations of all suspended particulates (A), all asbestos (B), and short mineral fibers (C) in the ambient air.

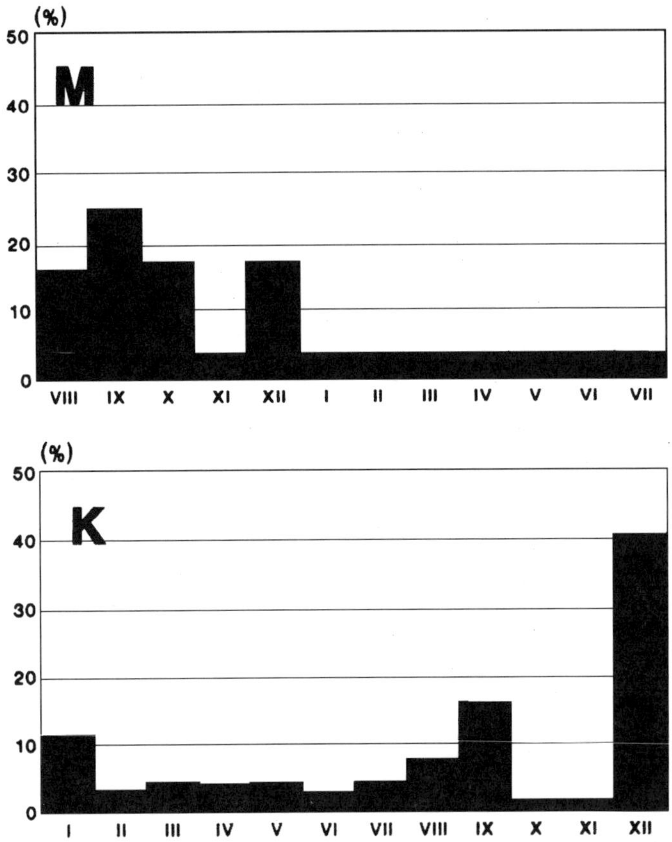

FIGURE 26.24 Annual distributions of all asbestos fibers in the atmosphere for the towns of Mannheim (M) and Karlsruhe (K).

source, fiber type, etc. Furthermore, they are important characteristics in the evaluation of the health risks for the population. Each fibrous aerosol is therefore described by its concentration and fiber size distribution. The fibrous particle can be described by means of two parameters: D_F and L_F, and therefore the obtained fiber size distribution is bivariate.

The carcinogenic potency of a durable inorganic fiber is a function of its size (D_F, L_F); it increases with increasing L_F and decreasing D_F. There is a continuous transition from the non-carcinogenicity of fibers that are too short, too thick, or too soluble to have any carcinogenic potency, to fibers that have the optimum L_F and D_F and that additionally are very persistent. The fiber size dependence is well-defined by, for example, the Pott's hypothesis shown in Figure 26.26.[43,44] Because of this, the fibers measured by SEM and TEM can be also converted to "carcinogenic fiber sizes," which respect the fiber carcinogenic potency, shown in Figure 26.26. Then, longer and thinner fibers become a greater percentage. The bivariate, microscopically measured fiber size distributions (Figure 26.27A) can be transformed into size distributions respecting the carcinogenicity potency of each fiber size interval (Figure 26.27B). The fibers of each size interval are weighted by the carcinogenic factor C_f.[17]

The bivariate fiber size distributions partly characterize, or are finger-prints, of the emission sources. In Figures 26.28 to 26.30, such examples are illustrated. Fibers emitted, for example, from a waste disposal site (Figure 26.28) have relatively high carcinogenic potency, as these asbestos fibers are "long and thin."[30] The differences in fiber size distributions of difference sources are shown in Figure 26.29. On a rural site (R), only fibers shorter than 5 μm were sampled. On a busy

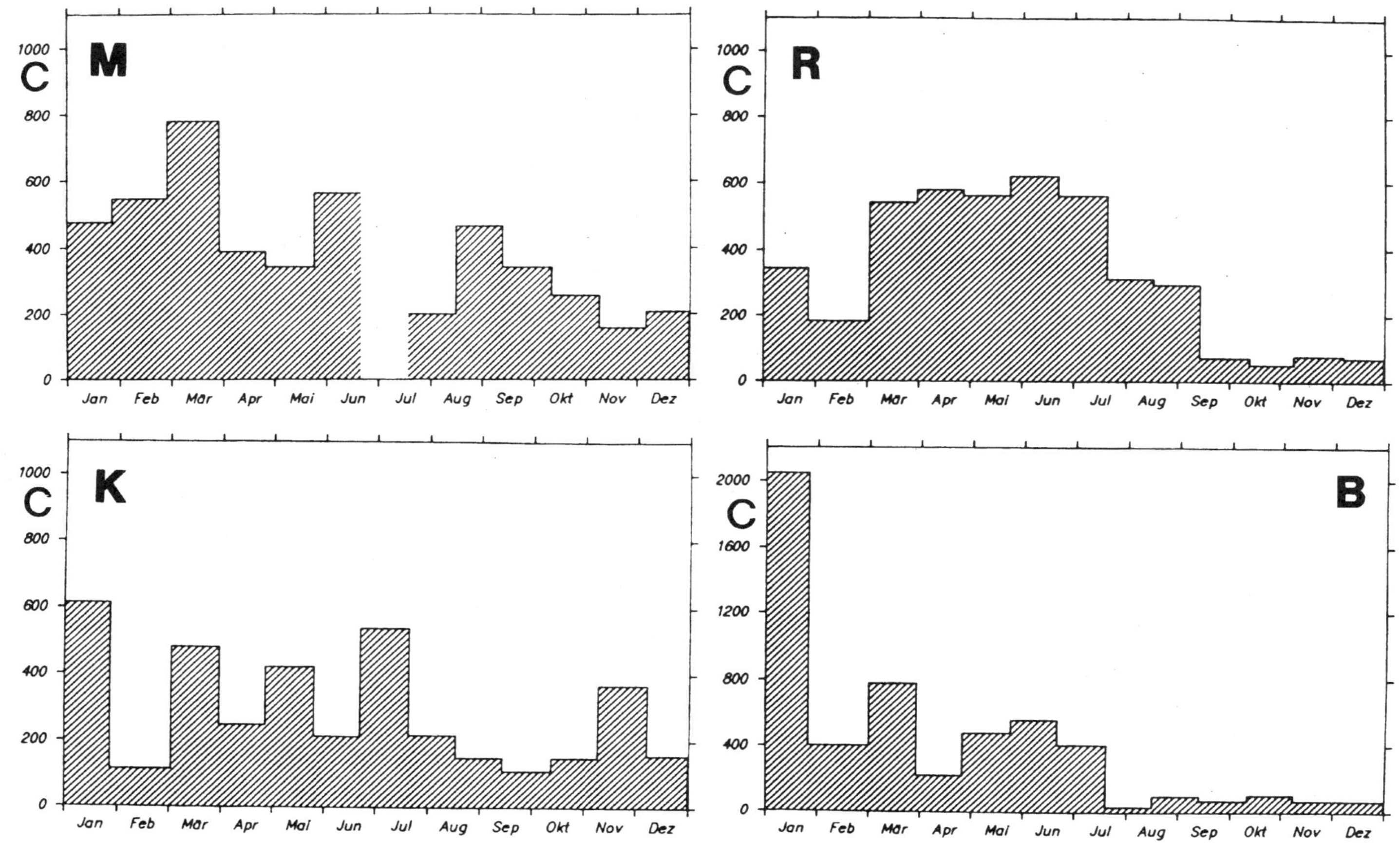

FIGURE 26.25 Annual distributions of all mineral fibers in the atmosphere, measured in Mannheim (M), Karlsruhe (K), Reutlingen (R) and the Black Forest (B). Fiber concentrations C in fibers m^{-3}.

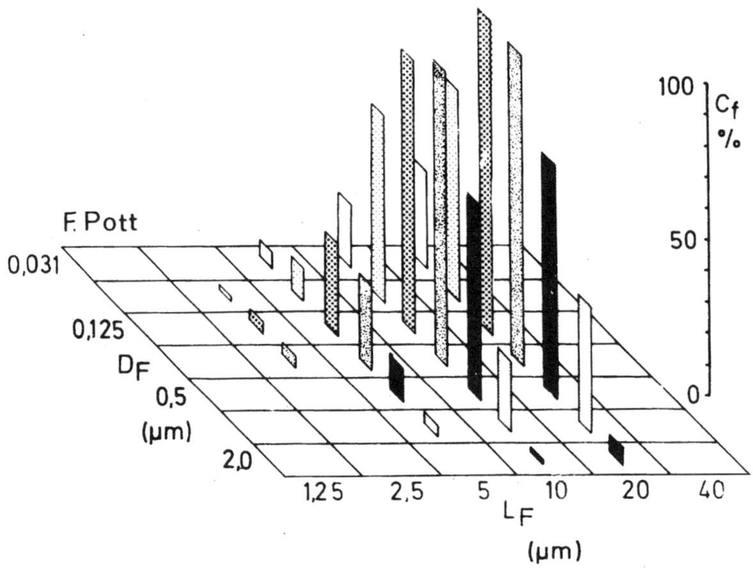

FIGURE 26.26 Chart depicting a hypothesis of carcinogenic potency (carcinogenic factor C_f) of cylindrical fibers as a function of their lengths (L_F) and diameters (D_F).

street crossing, longer asbestos fibers were found, and on a site close to an industrial emission source (AI), fibers as long as 100 μm were counted.[24] From fiber size measurements at different distances from an industrial asbestos emission source, a rapid decrease in fiber length was confirmed (Figure 26.30).[22]

While asbestos fibers sampled in ambient air are generally thinner than 1 μm, the diameters of the other mineral fibers are thicker — until 3 μm or greater (Figure 26.31).[42]

HEALTH RISK EVALUATION

The mechanisms of fiber carcinogenesis have been intensely investigated.[45] Unfortunately, they are still not fully understood. Nevertheless, the role played by the physical and chemical properties of inhaled mineral fibers is well-known. As mentioned, fiber dimensions are of basic importance (Figure 26.26). The next decisive parameter is fiber durability (insolubility) and, of course, the cell tissue incorporated fiber dose. The mechanism of fiber carcinogenesis is sometimes designated as a "3-D mechanism" that is, a mechanism that correlates to the three most important parameters — fiber dimension, fiber durability, and fiber dose.

Besides the solubility of a fiber incorporated in the cell, the chemical fiber composition is also involved in the chemical and biochemical processes at the interface between the fiber surface and the cell. Fine and chemically durable mineral fibers containing transition metals, such as Fe, and incorporated in cells, react with O_2 by generating oxygen radicals (e.g., $OH^\cdot$), which may be causally involved in the carcinogenic effect.[46]

The health risk for the general population due to inhalation of fibrous mineral aerosols present in the ambient air is, therefore, a function of the fiber size distribution, fiber chemical composition, fiber concentration, and exposure time. By estimating the health risk of inhaled fibers, the entire lifetime exposure is supposed. Then, the health risk is correlated with the concentration of the thoracic (respirable) size fraction of durable fibers in the ambient air. There is no question that such a correlation does exist. It has been derived by epidemiological studies and animal inhalation experiments.[47]

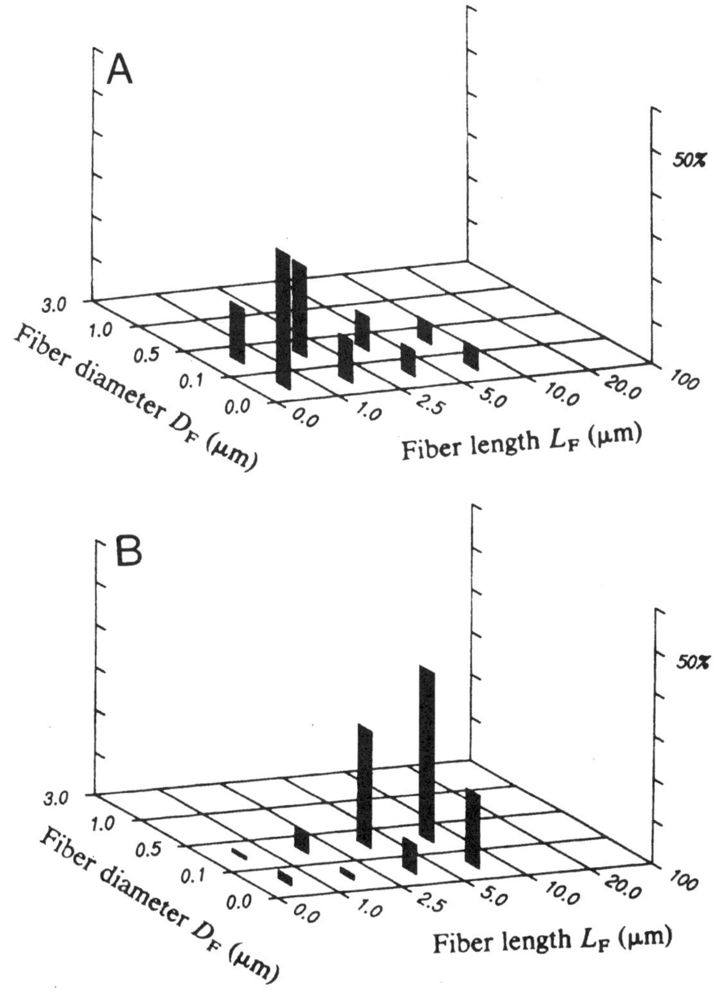

FIGURE 26.27 Diagrams illustrating the measured (A) and the "carcinogenic" (B) bivariate size distributions of asbestos fibers sampled in ambient air.

The incidence and levels of durable fibers and ferruginous bodies (FB) in the lungs of the general population is another reliable indicator for the whole life human exposure. The FBs are coated fibers; they are composed of iron-containing protein derived from hemoglobin that is precipitated on and around a fiber (Figure 26.32). The existence of a correlation between the level of FBs in lung tissue and the exposure to mineral fibers has been well-documented by long-time investigations in Japan.[48] The incidence of FBs and mineral fibers were obtained by microscopical analysis in autopsied or resected lungs of the general population during a period of almost 45 years (1937–1981). The major cores of FBs were found to be asbestos (amosite, crocidolite, tremolite, and chrysotile). The found correlation is shown in Figure 26.33. The curves well demonstrate a correspondence between the amounts of imported asbestos in Japan and the levels of FBs found in the lung tissues of the general population. There is no doubt that the increased asbestos import and its industrial use was followed by increasing concentrations of respirable asbestos fibers in ambient air and therefore also by increasing dose of fibers deposited in the lungs of the inhabitants.

The measured concentrations of asbestos and other mineral fibers in ambient air of highly developed countries produced an additional carcinogenic health risk for the general population.[49]

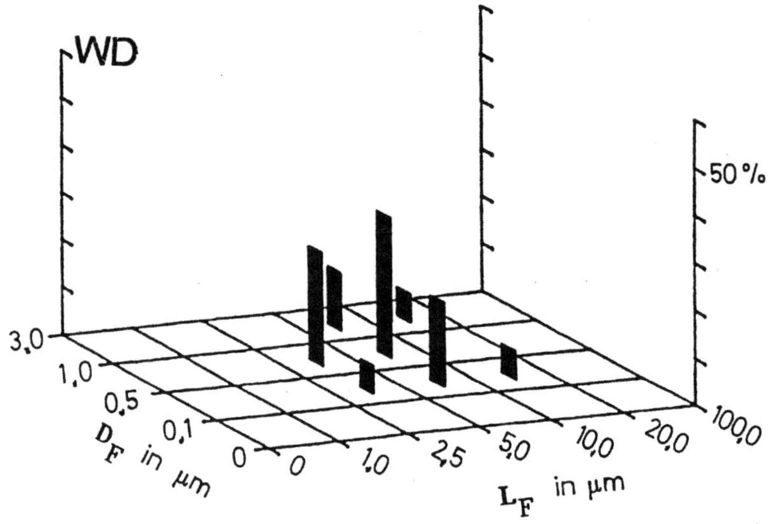

FIGURE 26.28 Diagram illustrating the bivariate size distribution of asbestos fibers sampled in ambient air in the vicinity of a large waste disposal site in Germany.

Pollution measurements made in terms of the mass of asbestos per cubic meter of air are useless, as they include thick non-respirable and minute non-carcinogenic fibers. For a realistic determination of the ambient air concentrations of asbestos and other mineral fibers, electron microscopic methods have to be used.[50]

It is assumed that there is no threshold for the induction of cancer after inhalation of fine mineral fibers, and that the risk is proportional to the inhaled dose. The risk of cancer after inhalation of asbestos fibers by workers was firstly evaluated by means of epidemiological studies on several workplaces in the asbestos industry.[51,52] Dependencies between exposure and mortality rates were plotted. The cancer risk caused by exposures in ambient air (i.e., by exposures to low and very low asbestos fiber concentrations) were approximately estimated by extrapolating the risk obtained for the workplaces.

The risk of death from lung cancer and mesothelioma due to exposure to chrysotile asbestos in asbestos textile industries (lifetime exposure to 10^6 fibers longer than 5 µm) was found to be as high as 25 deaths per 1000 men.[53,54] Using the results from the epidemiological studies on asbestos workplaces, a model for the relationship between the incidence of asbestos-related cancer and the exposure was developed.[50] The tumor incidence I(t) and age t due to exposure L (fibers ml^{-1}) from the age t_1 to the age t_2 is given by the equation

$$I(t) = 0.62 \times 10^{-10} \times L\left[\left(t - t_1\right)^4 - \left(t - t_2\right)^4\right]. \tag{26.1}$$

For ambient air concentrations of respirable asbestos fibers, the estimated health risk for the general population may lay in the range 10^{-6} to 10^{-4}. In the case of synthetic mineral fibers, this risk is much smaller, probably being in the range of 10^{-6}.[55]

Generally speaking, there do not exist any outdoor Ambient Air Quality Standards for asbestos and mineral fibers — nor in the U.S., Canada, Europe, Japan, or in any other country. Only some proposals were in the discussions in some countries (e.g., in Germany).[54] For asbestos emissions, the emitted concentrations should be less than 0.1 mg m^{-3}. The ambient air concentrations of asbestos fibers (longer than 5 µm) should be essentially lower than 1000 fibers m^{-3}. Therefore,

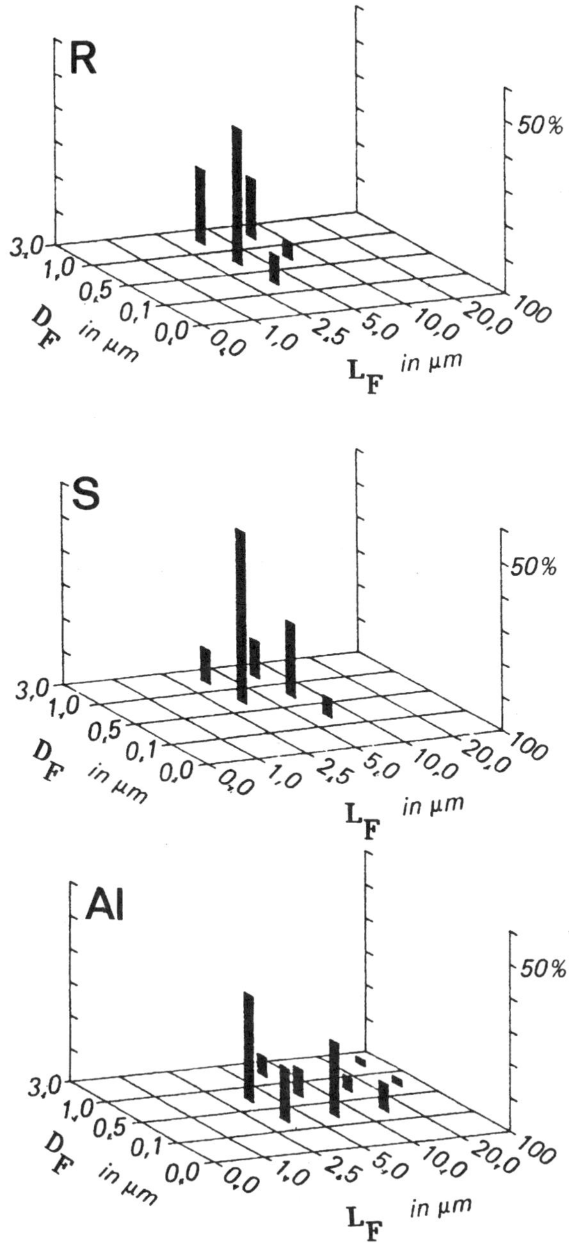

FIGURE 26.29 Diagrams illustrating the bivariate size distributions of asbestos fibers sampled in ambient air in rural (R), street crossing (S) sites, and on a site near an industrial asbestos emission source (AI).

asbestos fiber concentrations no higher than 500 fibers m^{-3} should be tolerated. At such concentrations, the expected mean lifetime cancer risk lies at approximately 10^{-5}.

By considering the possible risk for the general population in highly developed countries, only a low percentage of inhabitants living in the vicinity of industrial asbestos emission sources were or are exposed to a higher cancer risk. In the countries of the Third World, of course, much higher levels of risk are to be expected.[56]

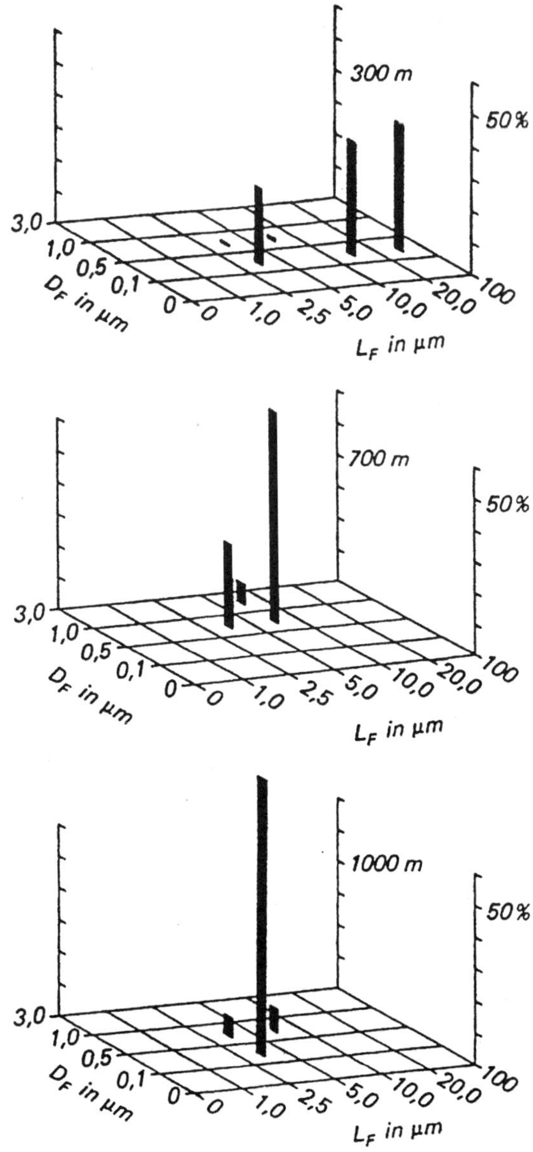

FIGURE 26.30 Diagrams illustrating the bivariate size distributions of asbestos fibers sampled in ambient air around an industrial asbestos emission source at distances of 300, 700, and 1000 m.

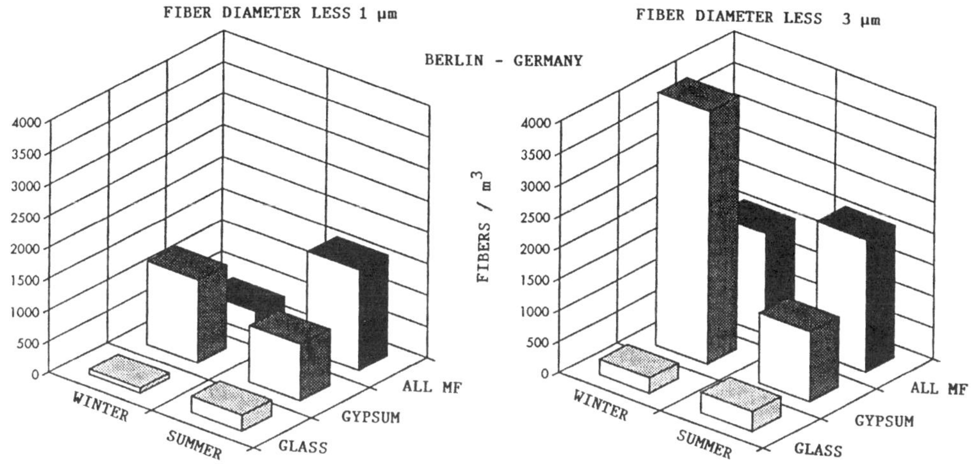

FIGURE 26.31 Ambient air concentrations of all, gypsum, and glass fibers measured in Berlin.

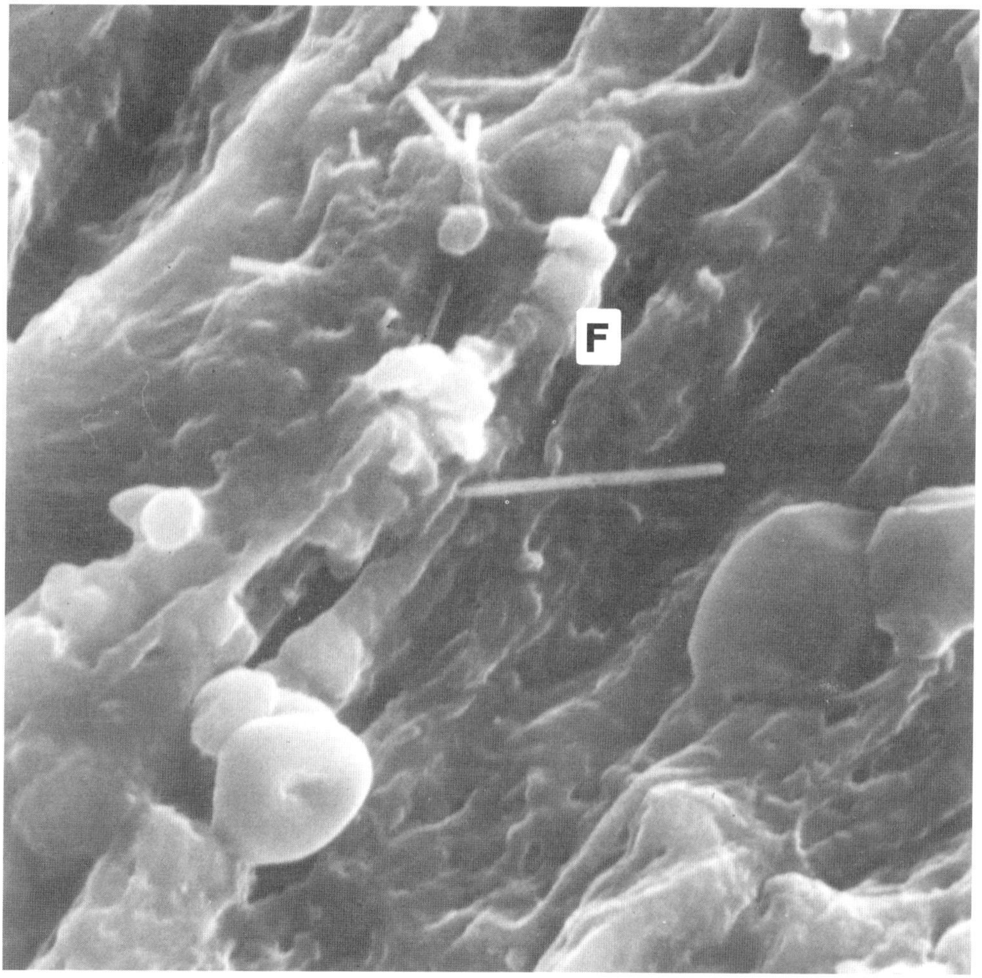

FIGURE 26.32 Scanning electron micrograph showing the inner structure of the human lung tissue with incorporated asbestos fibers, including a ferruginous body (F).

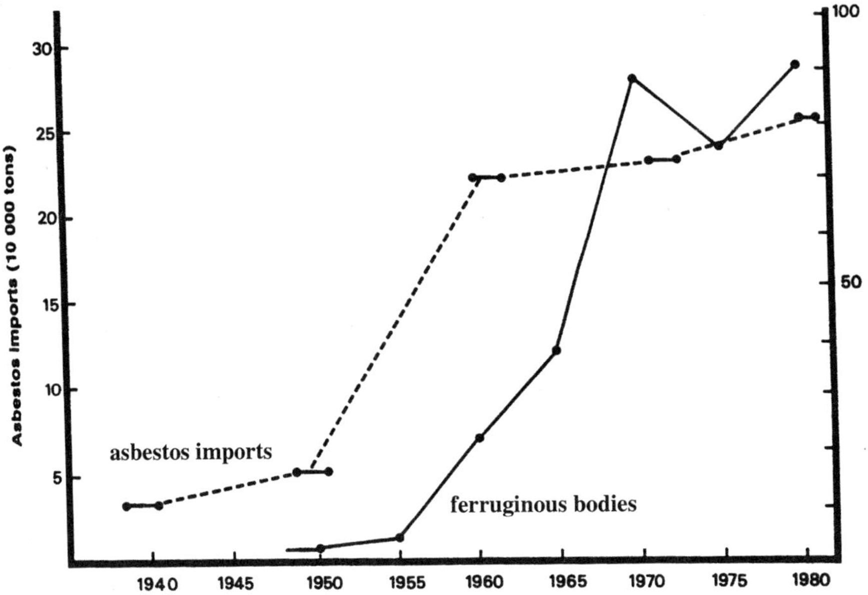

FIGURE 26.33 Incidence of FBs and annual amount of asbestos imports in Japan. (From Shishido, S., Iwai, K., and Tukagoshi, K., *WHO, IARC Sci. Publ.*, 90, 229, 1989. With permission.)

REFERENCES

1. Peto, J., Hodgson, J., Matthews, F. et al., Continuous increase in mesothelioma mortality in Britain, *Lancet*, 345, 535, 1995.
2. Price, B., Analysis of current trends in U.S. mesothelioma incidence, *Am. J. Epidemiol.*, 145, 211, 1997.
3. Baris, Y.I., *Asbestos and Erionite Related Chest Diseases*, Publ. House of the Hacettepe University, Ankara, Turkey, 1987.
4. Lidell, D. and Miller, K., Eds., *Mineral Fibers and Health*, CRC Press, Boca Raton, FL, 1991.
5. Michaels, L. and Chissick, S.S., Eds., *Asbestos*, John Wiley & Sons, Chichester, 1979.
6. Spurny, K.R., Natural fibrous zeolites and their carcinogenicity, *Sci. Total. Environ.*, 30, 147, 1983.
7. Bignon, J., Peto, J., and Saracci, R., Eds., Non-occupational exposure to mineral fibres, *IARC Sci. Publ.*, 90, Lyon, France, 1989.
8. Rohl, A.N., Langer, A.M., and Selikoff, I.J., Environmental asbestos pollution related to use of quarried serpentine rock, *Science*, 196, 1319, 1977.
9. Carter, L.J., Asbestos. Trouble in air from Maryland rock quarry, *Science*, 197, 237, 1977.
10. Spurny, K.R., Pott, F., Huth, F. et al., Identifizierung und krebserzeugende Wirkung von faserförmigen Aktinolith aus einem Steinbruch, *Staub-Reinhalt. Luft*, 39, 386, 1979.
11. Simonato, L., Baris, R., Saracci, R. et al., Relation of environmental exposure to erionite fibers to risk of respiratory cancer, Bignon, J., Peto, J., and Saracci, R., Eds., Non-occupational exposure to mineral fibres, *IARC Sci. Publ.*, 90, 398, 1989.
12. Vojakovic, R., *Asbestos*, Asbestos Disease Society of Australia Publication, Perth, Australia, 1997.
13. Spurny, K.R., Stöber, W., Opiela, H., and Weiss, G., On the evaluation of fibrous particles in remote ambient air, *Sci. Total Environ.*, 11, 1, 1979.
14. Spurny, K.R., Anthropogenic asbestos in soils. *Zschr. Pflanzenernährung und Bodenkunde*, 156, 177, 1993.
15. Spurny, K.R., On the release of asbestos fibers from weathered and corroded asbestos cement products, *Environ. Res.*, 48, 100, 1989.
16. Liddell, D., Exposure to mineral fibers and human health, D. Liddell and K. Miller, Eds., *Mineral Fibers and Health*, CRC Press, Boca Raton, FL, 1991, 1-9.

17. Spurny, K.R., Sampling, analysis, identification and monitoring of fibrous dusts and aerosols, *Analyst*, 119, 41, 1994.
18. Spurny, K.R., Ed., *Physical and Chemical Characterization of Individual Airborne Particles*, Ellis Horwood, Chichester, U.K., 1986.
19. Kohyama, N. and Kurimori, S., A total sample preparation for the measurement of airborne asbestos and other fibers by optical and electron microscopy, *Industrial Health (Japan)*, 34, 185, 1996.
20. Laamanen, A., Noro, L., and Raunio, V., Observations of atmospheric air pollution caused by asbestos, *Ann. N.Y. Acad. Sci.*, 132, 240, 1965.
21. Murchio, J.C., Cooper, W.C., and DeLeon, A., Asbestos Fibers in Ambient Air of California, CA-Air Resources Board, Berkeley, CA, 1973.
22. Marfels, H., Spurny, K., Boose, Ch. et al., Measurements of fibrous dusts in ambient air of Germany: Measurements in the vicinity of an industrial source, *Staub-Reinhalt. Luft*, 44, 259, 1984.
23. Marconi, A., Cecchetti, G., and Barbieri, M., Airborne mineral fibre concentrations in an urban area near an asbestos-cement plant, J. Bignon, J. Peto, and R. Saracci, Eds., Non-occupational Exposure to Mineral Fibres, *IARC Sci. Publ. (Lyon)*, 90, 336, 1989.
24. Iburg, J., Marfels, H., and Spurny, K., Measurements of fibrous dusts in ambient air of Germany: Measurements on three different locations in the region of Bayreuth, *Staub-Reinhalt. Luft*, 47, 271, 1987.
25. Kohyama, N., Airborne asbestos levels in non-occupational environment in Japan, J. Bignon, J. Peto, and R. Saracci, Eds., Non-occupational Exposure to Mineral Fibres, *IARC Sci. Publ. (Lyon)*, 90, 262, 1989.
26. Spurny, K.R., Marfels, H., Boose, Ch. et al., Measurements of fibrous aerosols in ambient air in Germany: Fibrous ambient air concentrations and brush up from a giant slide, *Wissenschaft und Umwelt (Germany)*, 3, 131, 1988.
27. Spurny, K.R., Measurements of fibrous dusts in ambient air in Germany: Ambient air concentrations of asbestos fibers produced by brake linings of automobiles, *Zbl. Hyg.*, 193, 193, 1992.
28. Marfels, H., Spurny, K.R., Boose, Ch. et al., Measurements of fibrous dusts in ambient air of Germany. Measurements on a busy crossing of a large town, *Staub-Reinhalt. Luft*, 44, 410, 1984.
29. Marfels, H., Spurny, K.R., Boose, Ch. et al., Measurements of fibrous dusts in ambient air of Germany. Measurements during building demolition and waste disposals. *Staub-Reinhalt. Luft*, 47, 219, 1987.
30. Marfels, H., Spurny, K.R., Boose, Ch. et al., Measurements of fibrous dusts in ambient air of Germany. Asbestos in ambient air of waste dumps, *Staub-Reinhalt. Luft*, 48, 463, 1988.
31. Van Orden, D.R., Lee, R.J., Bishop, K.M. et al., Evaluation of ambient asbestos concentrations in buildings following the Loma Prieta earthquake, *Regul. Toxicol. Pharmacol.*, 21, 117, 1995.
32. Spurny, K.R., Marfels, H. and Schörmann, J., Measurements of fibrous dusts in ambient air of Germany. Fibrous aerosols in the higher stratospheric altitudes, *Staub-Reinhalt. Luft*, 49, 165, 1989.
33. Marfels, H., Spurny, K.R., Jaekel, F. et al., Asbestos fiber measurements in the vicinity of emittents, *J. Aerosol Sci.*, 18, 627, 1987.
34. Marfels, H., Spurny, K.R., Boose, Ch. et al., Measurements of fibrous dusts in ambient air of Germany. Asbestos fiber measurements in ambient air of Lower Saxonia, *Wissenschaft und Umwelt (Germany)*, 2, 75, 1988.
35. Spurny, K.R., Measurements of fibrous dusts in ambient air of Germany. Asbestos fiber measurements in ambient air in Rheinland-Pfalz, *Wissenschaft und Umwelt (Germany)*, 3/4, 93, 1991.
36. Spurny, K.R., Measurements of fibrous dusts in ambient air of Germany. Asbestos fiber measurements in ambient air of Baden-Württenberg, *Wissenschaft und Umwelt (Germany)*, 3/4, 99, 1991.
37. Meek, M.E., Environmental measurements of man-made mineral fibers, Liddell, D. and Miller, K., Eds., *Mineral Fibers and Health*, CRC Press, Boca Raton, FL, 1991, 89.
38. Höhr, D., Investigations by means of TEM. Fibrous particles in ambient air, *Staub-Reinhalt. Luft*, 45, 171, 1985.
39. Gaudichet, A., Petit, G., Billon-Galland, M.A., and Dufour, G., Levels in atmospheric pollution by man-made mineral fibers in buildings, J. Bignon, J. Peto, and R. Saracci, Eds., Non-occupational Exposure to Mineral Fibres, *IARC Sci. Publ. (Lyon)*, 90, 291, 1989.
40. Jaffrey, S.A.M.T., Rood, A.P., Llewellyn, J.W., and Wilson, A.J., Levels of airborne man-made mineral fibers in dwellings in the U.K., J. Bignon, J. Peto, and R. Saracci, Eds., Non-occupational Exposure to Mineral Fibres, *IARC Sci. Publ. (Lyon)*, 90, 319, 1989.

41. Switala, E.D., Harlan, R.C., Schlaudecker, D.G., and Bender, J.R., Measurement of respirable glass and total fiber concentrations in the ambient air around a fiberglass wool manufacturing facility and rural area, *Regul. Toxicol. Pharmacol.*, 20, S76, 1994.
42. Schnittger, J., Measurement and identification of inorganic fibers in ambient air, *VDI-Berichte (Germany) Düsseldorf*, 1075, 283, 1993.
43. Pott, F., Some aspects of the dosimetry of the carcinogenic potency of asbestos and other fibrous dusts, *Staub-Reinhalt. Luft*, 38, 486, 1978.
44. Pott, F., The fiber as carcinogenic agent, *ZBL. Bakt. Hyg. B.*, 184, 1, 1987.
45. Brown, R.C., Hoskins, J.A., and Johnson, N.F., Eds., Mechanisms in Fibre Carcinogenesis, *NATO ASI Ser.*, Vol. 223, Plenum Press, London, 1991.
46. Mossman, B.T., and Marsh, J.P., Evidence supporting a role for active oxygen species in asbestos-induced toxicity and lung disease, *Environ. Health Perspect.*, 81, 91, 1989.
47. Spurny, K.R., Carcinogenic effect related to the fiber physics and chemistry, Brown, R.C., Hoskins, J.A., and Johnson, N.F., Eds., *Mechanisms in Fibre Carcinogenesis*, NATO ASI Ser., 223, 103, 1991.
48. Shishido, S., Iwai, K., and Tukagoshi, K., Incidence of ferruginous bodies in the lungs during a 45-year period and mineralogical analysis of the core fibers and uncoated fibers, J. Bignon, J. Peto, and R. Saracci, Eds., Non-occupational Exposure to Mineral Fibres, *IARC Sci. Publ. (Lyon)*, 90, 229, 1989.
49. Peto, J., Fibre carcinogenesis and environmental hazards, J. Bignon, J. Peto, and R. Saracci, Eds., Non-occupational Exposure to Mineral Fibres, *IARC Sci. Publ. (Lyon)*, 90, 457, 1989.
50. Doll, R., The quantativie significance of asbestos fibers in the ambient air, *Experimentia*, 51, 213, 1987.
51. Enterline, P.E., Cancer produced by non-occupational asbestos exposure in the U.S., *J. Air Pollut. Control Assoc.*, 33, 318, 1983.
52. Weil, H. and Hughes, J., Asbestos as public health risk, *Ann. Rev. Publ. Health*, 7, 171, 1986.
53. Fischer, M. and Meyer, E., The assessment of the health risk from asbestos fibers in Germany, *VDI-Berichte, Düsseldorf*, 475, 325, 1983.
54. Schenk, H., Abschätzung des asbestbedingten Erkrankungsrisikos, *VDI-Berichte, Düsseldorf*, 1075, 181, 1993.
55. Grimm, H.G., Kanzerogene Risiken durch künstliche Mineralfasern. *Staub-Reinhalt. Luft*, 46, 105, 1986.
56. Red, Battling over asbestos in the Third World, *Environ. Health Perspect.*, 105, 1178, 1997.

27 Atmospheric Contamination by Agroaerosols

Kvetoslav R. Spurny

CONTENTS

INTRODUCTION

The modern history of crop protection is little more than a hundred years old,[1] but over this period of time, enormous advances and some notable mistakes have been made. It was expected that pests and diseases of plants would be virtually eliminated using modern synthetic agrochemicals. Unfortunately, in several cases, pest and disease resistance have occurred. In addition, the problems of environmental pollution by agrochemicals (water, soil, air, agricultural products, etc.) are steadily growing.[2] Possible effects on humans, animals, and other species are non-negligible. Atmospheric pollution by agrochemicals appears to be important, but remains neither very well-established nor well-understood. Nevertheless, there seem to be two environmental areas where agrochemical air pollutants may play an important, hazardous role: (1) farmers and farm-workers who apply the toxic sprays may inhale the dangerous gaseous and particulate agrochemical pollutants, and (2) several of these pollutants, particularly the halogenated compounds that are transported into the polar atmospheric regions, may be possibly involved in chemical reactions of ozone depletion.

TOXIC SPRAYS

Toxic agrochemicals (TA) — in particular, agrochemical sprays — have been used for different applications as pesticides, herbicides, fungicides, etc., prior to but predominantly after the beginning of the 20th century. Major expansion of their application started in the 1960s (Figure 27.1).[1] During

0-87371-829-1/00/$0.00+$.50

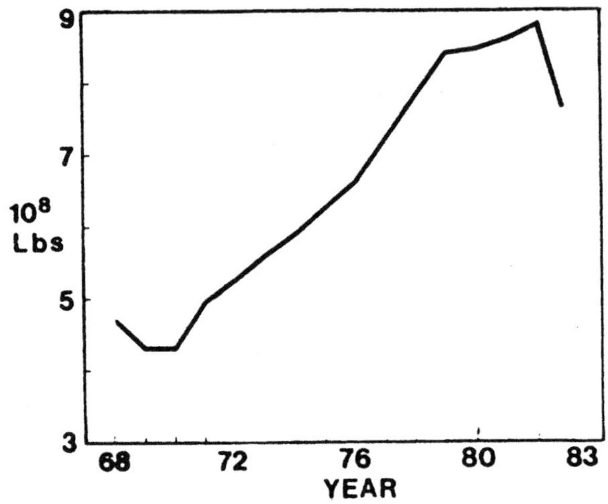

FIGURE 27.1 Agricultural use of pesticides in the U.S. (From Marco, J.G., Hollingsworth, R.M., and Plimmer, J.R., *Regulation of Agrochemicals*, American Chemical Society, Washington, D.C., 1991, 1. With permission.)

the first application period (i.e., up until the middle of the 20th century), only small, basic investigations were performed.[1-3]

Aerosols produced by different types of generators and used in agriculture are sprays with coarse particle dispersion (particle sizes in the range of 100 μm). The practical task is to efficiently deposit these sprays on crops, plants, trees, etc. and to protect them by means of their toxic ingredients against different harmful insects, illness, etc.

Aerosol physicists and chemists have engaged in the development of generation techniques, sampling and measurement, deposition studies, analytics, etc.[4-7]

Numerous groups and types of synthetic agrochemicals are presently available and used in worldwide crops protection (see Table 27.1).[8-11] Up until the 1970s, the only interest was to produce sprays with high deposition and high efficiencies in plant protection.

Agrochemical sprays can be introduced directly into the environment in the liquid phase, as a disposition of solution, or in the solid phase (e.g., as a powder, dust, etc.).

Spray physics evaluates the production, transport, and separation in the atmospheric environment. The primary particle size is coarse, and the particle size distributions are broad (Figure 27.2). The agricultural sprays are polydisperse. Spray clouds are difficult to measure and portions of the spray are invisible and lie in the size range of general aerosols. Spray droplets travel as a spray cloud, which moves away from the generation source, and expands, and the droplets are subjected to turbulence and gravitational forces.

SECONDARY ENVIRONMENTAL EFFECTS

It has been recognized that the application of TA as agricultural sprays can have a very negative impact on the total environment and produce additional health risks for the general population.[12,13] During spray application, secondary aerosols and gaseous toxic air pollutants are produced. The fine particle fraction of the sprayed liquids or solids escape into the atmosphere and are transported toward residential and clean-air areas. Besides, volatile particles and droplets evaporate and/or produce fine secondary aerosols and vapors that are toxic or carcinogenic and can be inhaled by operators and by the general population, children, etc. The loss of agrochemical aerosols into the air depends on the vapor pressure of the active ingredient, the application technique, the climate,

TABLE 27.1
The Most Important Groups of Synthetic Agrochemicals: Insecticides, Herbicides, and Fungicides

Hydrocarbons
Halogen derivatives of aliphatic hydrocarbons
Halogen derivatives of alicyclic hydrocarbons
Halogen derivatives of aromatic hydrocarbons
Nitro compounds
Amines and salts of quaternary ammonium bases
Alcohols, phenols, and ethers
Aldehydes, ketones, and quinones
Aliphatic carboxylic acids and their derivatives
Alicyclic carboxylic acids and their derivatives
Aryloxalkyl carboxylic acids and their derivatives
Derivatives of carbonic acid
Derivatives of carbamic acid
Derivatives of thio- and dithiocarbamic acids
Derivatives of urea and thiourea
Mercaptans, sulfides, and their derivatives
Thiocyanates and isothiocyanates
Derivatives of sulfuric and sulfurous acids
Sulfonic acids and their derivatives
Derivatives of hydrazine and azo compounds
Organic mercury compounds
Organotin compounds
Organophosphorus compounds
Arsenic compounds
Heterocyclic compounds with one heteroatom in the ring
Heterocyclic compounds with two heteroatoms in the ring
Heterocyclic compounds with three heteroatoms in the ring
Inorganic pesticides

etc. Their degradation and transformation in the atmosphere are driven primarily by solar irradiation. The transformation products of TA may be more toxic than their parent compounds.[14]

The major part of the spray cloud is ineffective for agricultural subjects. Delivery losses are 60 to 80%, and volatization losses are about 3 to 10%. The transport and delivery patterns of spray droplets in spray clouds are determined by turbulent diffusion and sedimentation. A calculated transport of a pesticide spray is shown in Figure 27.3.

EFFECTS ON THE ATMOSPHERE

The atmosphere can also be contaminated by means of spray applications. During the spray application, secondary aerosols and gaseous toxic air pollutants are produced.

Generation Sources of Secondary Agrochemical Aerosols

The application techniques of agrochemical sprays are well-developed, and several types of sprayers, which can be used for terrestrial or aerial plant and tree protection treatment, are commercially available.[2,7] These techniques need further improvement so as to be more effective for the deposition of sprays on plants, trees, etc. and for producing less environmental contamination.

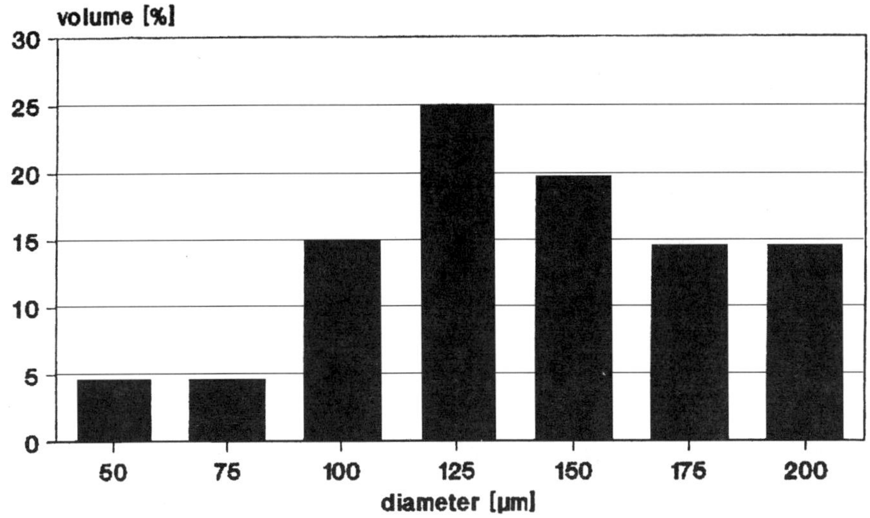

FIGURE 27.2 An example of droplet size distribution by a conventional spray generator. (From Marco, J.G., Hollingsworth, R.M., and Plimmer, J.R., *Regulation of Agrochemicals*, American Chemical Society, Washington, D.C., 1991, 1. With permission.)

Existing spray techniques are, unfortunately, relatively inefficient. The majority of sprayed solutions or dispersed powders escape into the surrounding environment. The first component of the biosphere where sprayed agrochemicals are released during the treatment of plants, trees, etc. is atmospheric air. Through atmospheric air, the chemicals pass to the target trees, plants, soil, etc. but some of them are carried away by wind, horizontally or vertically, over different distances (see also Figure 27.3). The portion of the spray-generated droplets that is carried away is designated as the spray drift.

Drift

Spray drift can be defined as the portion of the output from an agricultural crop sprayer that is deflected away from the target area by the action of wind.

Agricultural sprays can be applied directly from nozzles on vehicle-mounted spray booms, in the airstream that transports the spray from the nozzle to target (e.g., in orchards), or from an aerial spraying system.

The detailed mechanism by which droplets become detrained from spray clouds generated on a moving vehicle are often complex and still not fully understood.[15-18] The interaction between a spray cloud and the surrounding air has been long recognized as one of the important events influencing spray dispersal from aerial applications. Several useful simulation models have been developed and proved.[19-21] These models and validating measurements have shown that the turbulence in the propeller slip stream is an important factor in determining spray droplet trajectories from aerial applications. The droplet sizes in the drift cloud lie in the range of less than 100 µm, with the mean at about 20 µm. The aerial spray cloud development, including droplet size fractionation and drift, are illustrated in Figures 27.4 and 27.5. The drift, as well as several other mechanisms (e.g., a droplet rebounding from leaf surfaces, droplet evaporation, etc.), contribute to air pollution by agrochemicals (Figure 27.6). The evaporation of droplets plays an important role in determining the droplet size in spray clouds, and in the formation of secondary aerosol clouds. Since almost all TA are volatile, the spray drift consists of aerosols and vapors. Evaporation causes entry of pesticide vapor in the air and can result in pesticide contamination in the atmosphere. Physico-chemical models can describe the evaporation of the pesticide out of both an individual

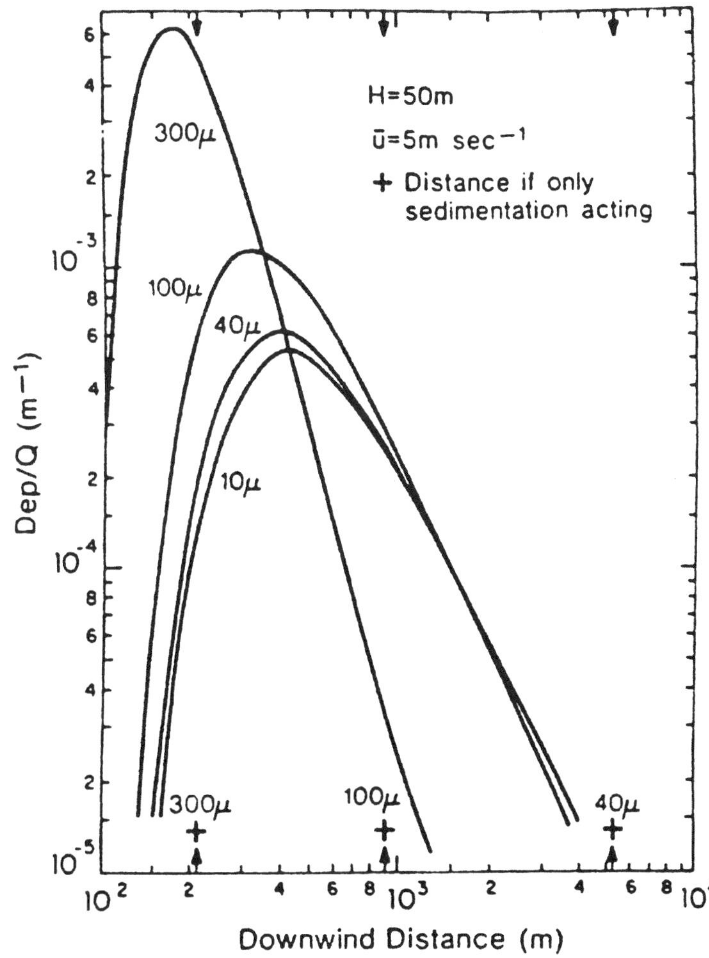

FIGURE 27.3 Calculated transport of droplets generated by a spray generator. (From Marco, J.G., Hollingsworth, R.M., and Plimmer, J.R., *Regulation of Agrochemicals*, American Chemical Society, Washington, D.C., 1991, 1. With permission.)

droplet and a droplet cloud. Polydisperse aerosol pollutants then occur in the atmospheric environment. Examples of experimental evaporation kinetics measurements are shown in Figure 27.7.[22] The characteristic evaporation times of single droplets with initial diameters of 2 μm are plotted as a function of the air temperature (1/T). The produced secondary agrochemical aerosols (SAA) undergo further changes in the air (e.g., chemical transformation and degradation) that are driven primarily by solar interaction (Figure 27.8).[14,23,24]

Air/Sea Interface

The recycling of organic substances — including agrochemicals — by the oceans was found to make an important contribution to the transport of anthropogenic and biogenic components into the global atmosphere.[25,26] The ocean/air exchange is most significant when these weather contributions (e.g., storms) that produce the greatest quantity of aerosol are present.

This process is another important generating source of SAA, which can then be transported into the global atmosphere. Agrochemicals produced by sea surfacial actions are also transported into the atmosphere in the gas phase or as liquid aerosols, depending on their chemical structure

FIGURE 27.4 Aerial spraying of pesticides produced by (A) the U.S. Air Force C-130 H Hercules, and (B) the Microair airplane in Scotland. (From Matthews, G.A. and Hilsop, E.C., Eds., *Application Technology for Crop Protection*, CAB International, Wallingford, U.K., 1993, 1. With permission.)

and physical characteristics (water surface activities, volatility, water solubility, hydrophobicity, etc.) These dependencies have been studied[25] for several organochlorinated pesticides in the laboratory. Tyrrhenian seawater samples were used to produce artificial aerosols without and with the addition of pesticides. Examples of some results are shown in Figure 27.9. The aerosols were generated from untreated (SW) and filtered seawater before (SW BAS) and after (SW

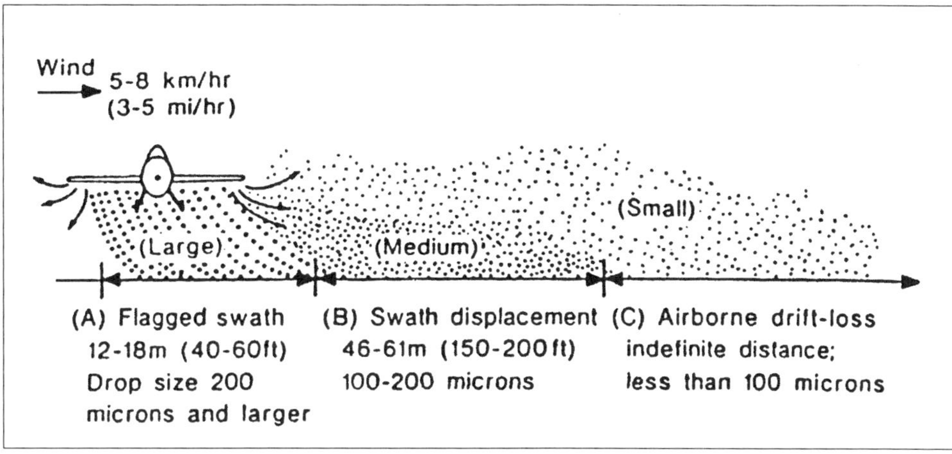

FIGURE 27.5 Schematic picture showing the separation process in agrochemical spray clouds, including the formation of the drift. (From Goodhas, L.D., *J. Econ. Enthomol.*, 37, 338, 1944. With permission.)

AAS) the addition of standards and salt solutions (W + NaCl). The percentage of organochlorinated pesticides found in these artificial aerosols were strongly dependent on the type of pesticide used.

Secondary Agrochemical Aerosols in the Atmosphere

The SAA can contaminate the air in the workplace during the spray application, then the urban air, as well as the air in remote regions, and may be further introduced into the global atmosphere by long-range transport such that they reach even the atmosphere of global regions.

There is clear evidence that the inhalation exposure of spray operators is very high. For example, after an application of pesticides in greenhouses, airborne concentrations of pesticides and fungicides were measured.[27] The measured concentrations ranged between 2 and 30 $\mu g\ m^{-3}$. After application of termicides in private homes, air concentrations in the range of nanograms per cubic meter (1– 300 $ng\ m^{-3}$) were found.[28,29]

The exposures of the general population in urban air are, of course, much lower. DDT concentrations in urban air in some American cities (measured in 1970s) were about 1 to 20 $ng\ m^{-3}$; for parathion values between 20 and 400 $ng\ m^{-3}$ were obtained. Measurements in 1996 of urban air in several California cities indicated that pesticide concentrations were much lower; almost all were less than 1 $ng\ m^{-3}$.[30]

The mean concentrations of lindane in the air of Paris, France measured in 1990 were approximately 1 $ng\ m^{-3}$.[31] Later measurements obtained[32] concentrations in the picogram per cubic meter range. Relatively high concentrations of two pesticides — mecoprop and isoproturon — were found in the air of rural areas near Colmar in 1991/1993 (Figure 27.10).[32] Almost 40 different types of pesticides were measured in the city air of Kitakyushu, Japan, in 1996.[33] Concentrations in the range of 10 $ng\ m^{-3}$ were found. In the remote areas of a Spanish national part in the Pyrenees Mountains, organochlorine pesticides were measured; concentrations between 70 $pg\ m^{-3}$ and 3 $ng\ m^{-3}$ were obtained.[34]

Polar Regions

The SAA contaminate not only the atmospheric environment near application sites, but they are transported across continents and can pollute the global atmosphere, including polar regions.

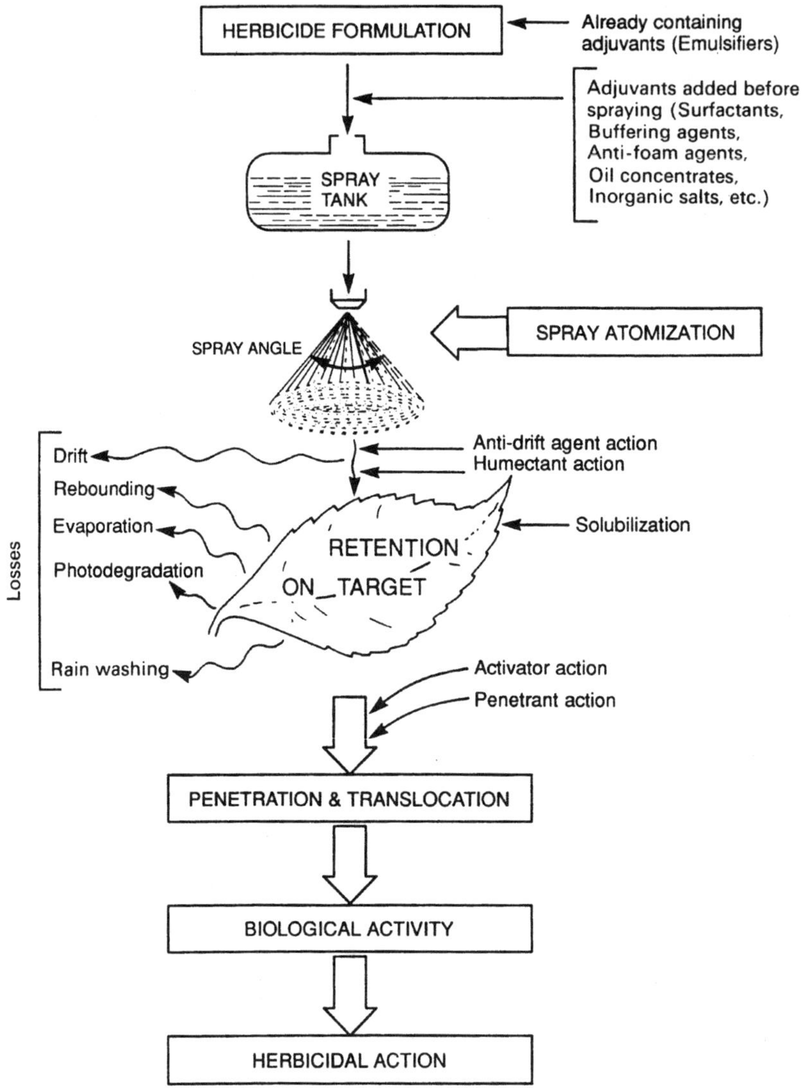

FIGURE 27.6 Schematic illustration of the multiple functions of spray deposition and losses. (From Matthews, G.A. and Hilsop, E.C., Eds., *Application Technology for Crop Protection*, CAB International, Wallingford, U.K., 1993, 1. With permission.)

Different SAA, mainly PCBs and organohalogens, were measured in Canada in 1992.[35] The results showed a large variation of toxics as a function of time and year. The summer values were the highest (Figure 27.11), lying in the range of 20 to 500 pg m^{-3}. Atmospheric transport of SAA to polar regions has been observed since 1960. Chlorinated pesticides in air were measured during ship passage between New Zealand and Antarctica in 1993. The measured concentrations ranged between 0.5 and 60 pg m^{-3}.[36]

Selected polychlorinated compounds and pesticides were determined during winter and summer sampling periods in 1984 at four stations along a south-north profile 2500 km long, starting at Birkenes, Norway, and ending at Ny Ålesund, Spitzbergen. Several long-range transport periods were detected by comparing profiles of selected substances with calculated air transport trajectories to the measuring stations. The data obtained clearly proved that long-range transport of such

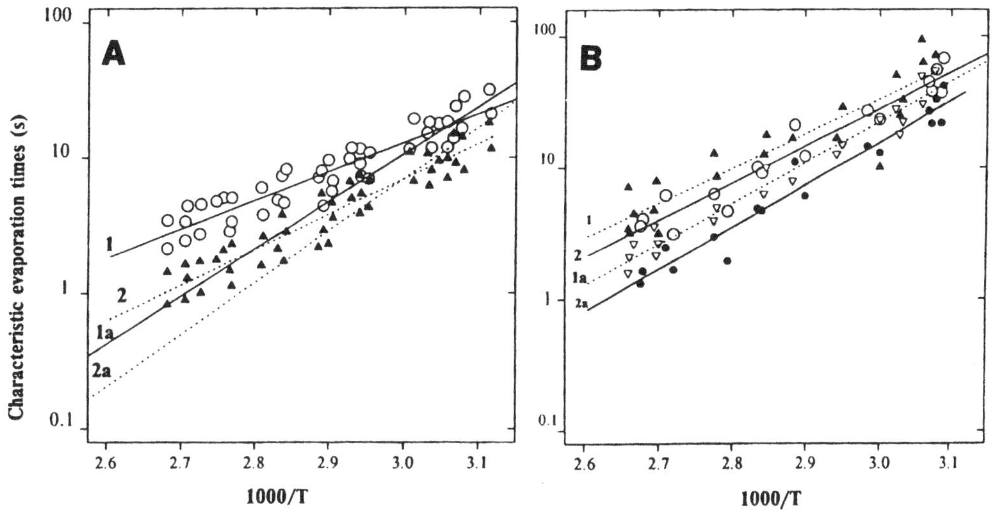

FIGURE 27.7 Experimental data and fitted temperature (T) dependencies of characteristic evaporation times for, (A) the polydisperse system of malathion, and (B) the parathionmethyl aerosols. (From Samsonov, Y.N., Makarov, V.I., and Koutsenogii, P.K., *Pestic. Sci.*, 52, 292, 1998. With permission.)

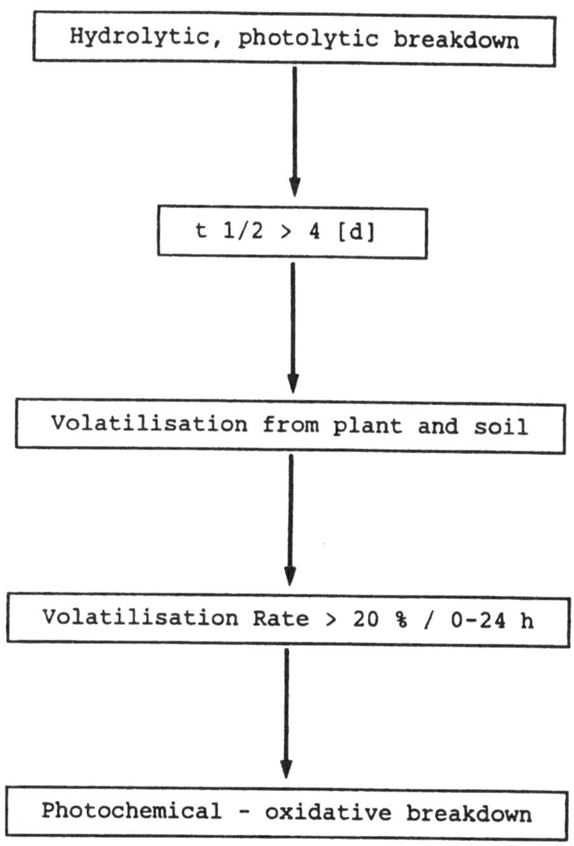

FIGURE 27.8 Schematic table of the fate of pesticides in air. (From Plimmer, J.R. and Johnson, W.E., Pesticide Degradation Products in the Atmosphere, American Chemical Society, Washington, D.C., 1991, 274. With permission.)

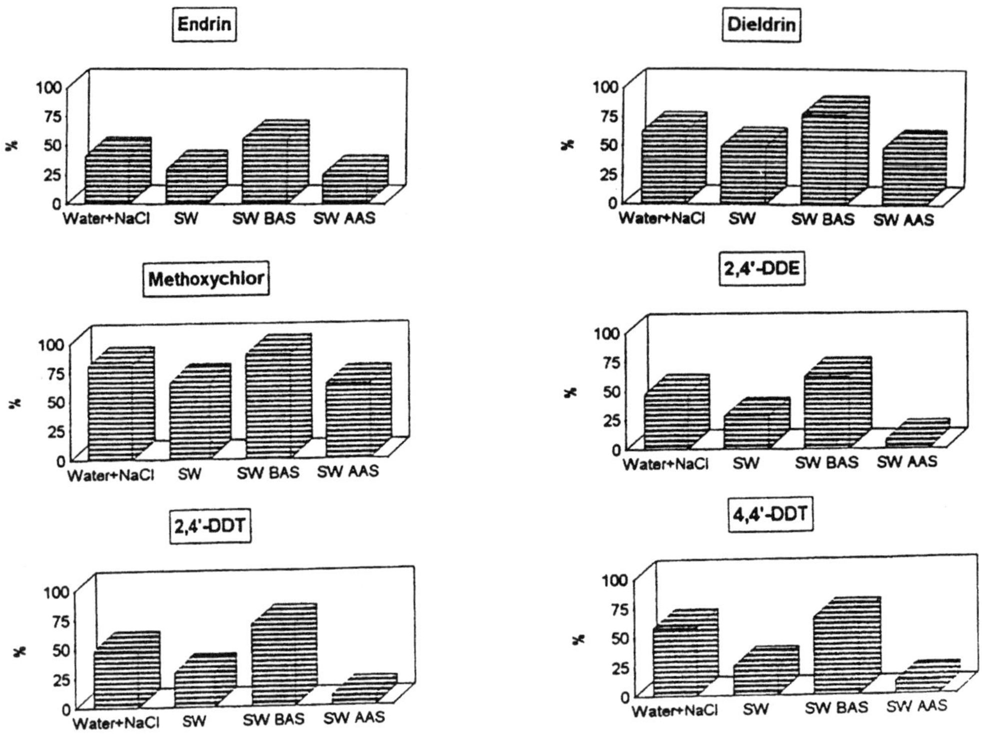

FIGURE 27.9 Percentages of organochlorinated pesticides in the artificial aerosol from different solutions and for different types of pesticides. (From Lepri, L., Desideri, P., Cini, R., Masi, F., and Van Erk, M.S., *Analytica Chim. Acta*, 317, 149, 1995. With permission.)

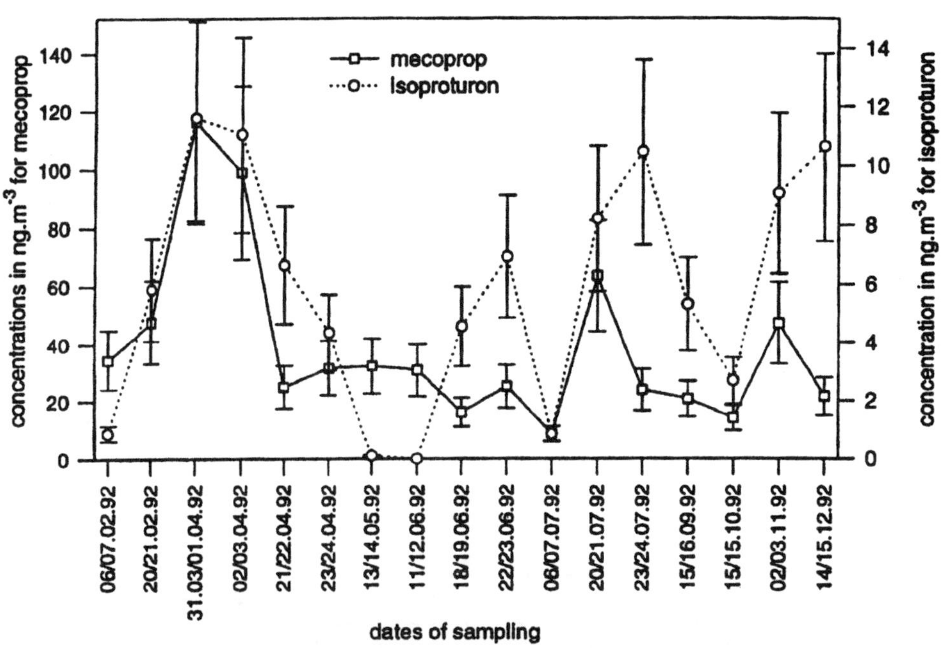

FIGURE 27.10 Annual atmospheric concentrations (gas and particles) of two pesticides — mecoprop and isoproturon — in the air of rural areas near Colmar. (From Millet, M., Wortham, H., Sanusi, A., and Mirabel, P., *Environ. Sci. Pollution Res.*, 4, 172, 1997. With permission.)

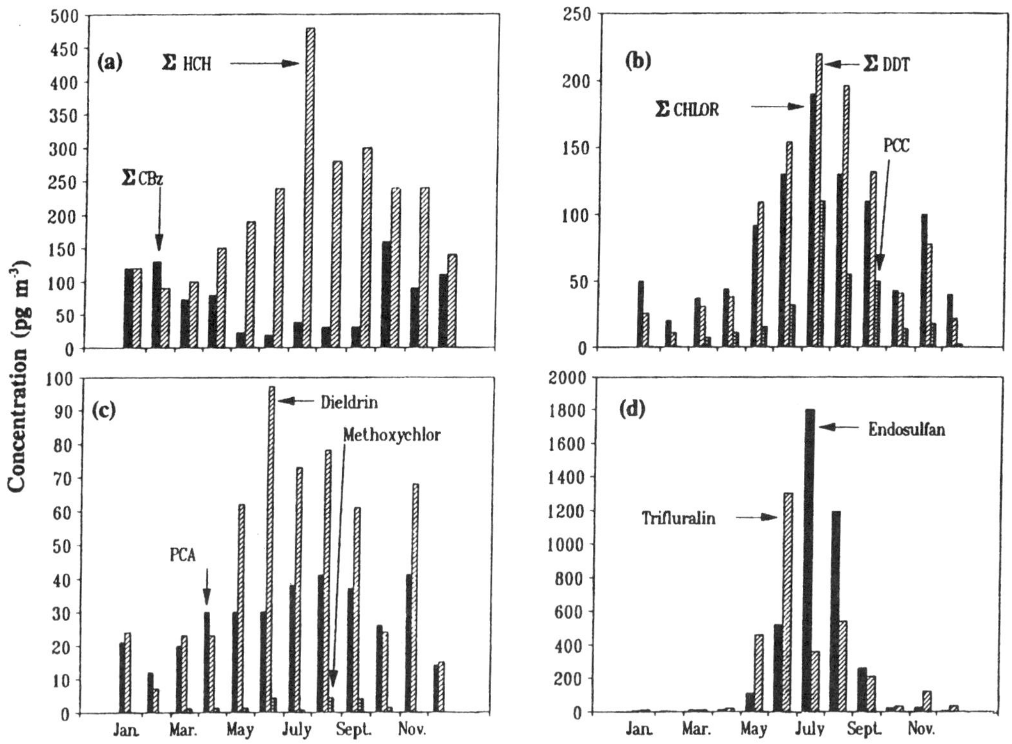

FIGURE 27.11 Organohalogen concentrations vs. time for different types of pesticides in the atmosphere. (From Hoff, R.M., Muir, D.C., and Grift, N.P., *Environ. Sci. Technol.*, 26, 266, 1992. With permission.)

substances is possible from North America and Eurasia to the Arctic.[37] The measured concentrations of selected polychlorinated compounds were in the range between 1 and 500 pg m^{-3}.

Fourteen polychlorinated biphenyl congeners (PCBs) were quantified in air samples of the tropospheric boundary layers across the Atlantic Ocean and on the Cape Verde Islands in the North Atlantic Ocean. Values for the sum of PCBs ranged between 22 and 385 pg m^{-3}.[38]

Organochlorine compounds, including pesticides, herbicides, etc., were monitored in the Arctic in 1992.[39] Atmospheric transport and condensation of these compounds at low-temperature conditions are important factors contributing to the presence of such pollutants in the Arctic. Some 18 compounds representative of different compound classes were measured. During cold periods, the portion of the compounds found in the gas phase decreased, and almost all detected compounds were typically found in the aerosol form. Concentrations of pesticides, herbicides, as well as termicides (such as chordane) were in some cases in the rage of 50 pg m^{-3}, but usually were less than 5 pg m^{-3}. See also Figure 27.12.[39] The atmospheric transport of persistent compounds, including agrochemicals, into the Arctic regions can also be determined by indirect measurement methods, such as chemical analysis of the particulate deposit in the polar snow. In 1990, snow samples were collected at high Arctic weather stations (Figure 27.13) and depositions of PCBs were found to range between 2 and 4 ng m^{-2} day^{-1}.[40]

AIR SAMPLING AND ANALYSIS

SAA occur in the atmosphere in two phases: as aerosols and as vapors.

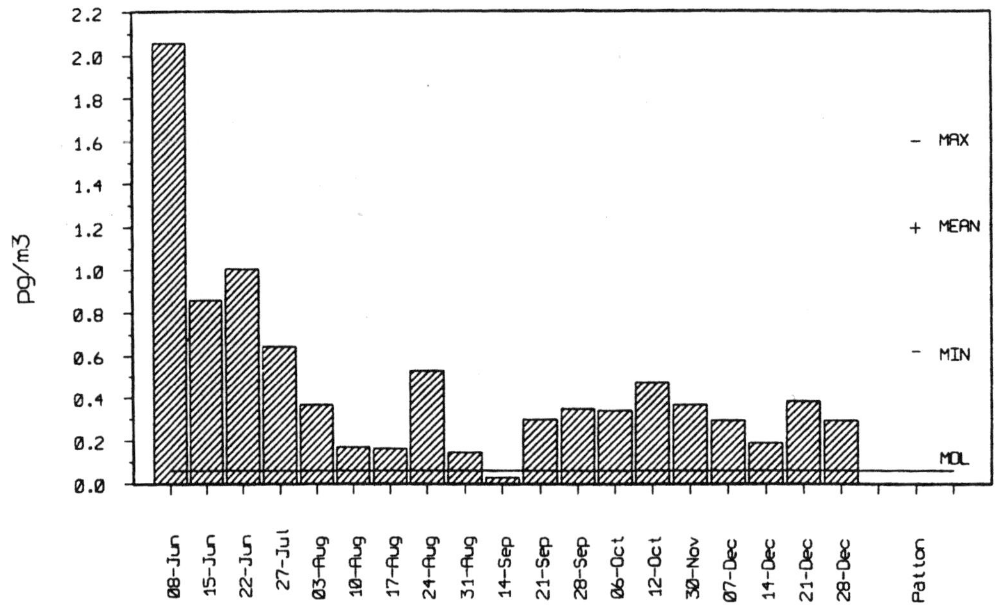

FIGURE 27.12 Time–concentration profiles of measured *trans*-chlordane. (From Fellin, P., Barries, L.A., Dougherty, D. et al., *Environ. Toxicol. Chem.*, 15, 253, 1996. With permission.)

Atmospheric Sampling

Combined sampling devices, consisting of effective aerosol filters and adsorbers, are necessary for analytical sampling. Analytical glass-fiber filters have been successfully used for sampling aerosols. Backing filters prepared by selected adsorbents (polyurethane foam, Chromosorb 102, ORBO 42, ORBO 44, as well as TENAX GC) have proven useful for adsorption of vapors (see also Figure 27.14). In Japan, activated carbon filters are used for combined sampling of pesticide aerosols and vapors in ambient air.[41] For size-selective sampling, special personal samplers for inhalable mixed-phase aerosols are used.[42] For sampling in outdoor measurements, low-volume pesticide samplers with cut-off size at 2.5 μm are recommended.[43]

Analytical Methodology

The analytics of primary as well as secondary agrochemical aerosols and vapors have been well-developed and standardized.[1,44] Sampling by aircraft has become a very suitable technique, making it possible to measure atmospheric concentrations of agroaerosols in large locations and at different elevations, as well as to estimate the flux of agroaerosols. Such an aircraft-based air sampling system using the relaxed eddy-accumulation technique has been developed and successfully tested in Canada.* Atrazine and metolachlor pesticides were detected in the troposphere; concentrations as high as 4.6 ng m^{-3} for atrazine and 9.8 ng m^{-3} for metolachlor were found. Fluxes ranged from 1.1 to 2.5 ng m^{-3} s^{-1} for metolachlor.

Modern analytical methods and equipments are desirable for the evaluation of air samples of SAA. In Figure 27.15, the evolution of analytical methodology for organic agrochemical toxicants in environmental samples is presented.[1] The detection limits lie in the range of low picograms to low nanograms.[45-47]

* Zhu, T., Desjardins, R.L., MacPherson, I.I. et al., Aircraft measurements of the concentration and flux of agrochemicals, *Environ. Sci. Technol.*, 32, 1032, 1998.

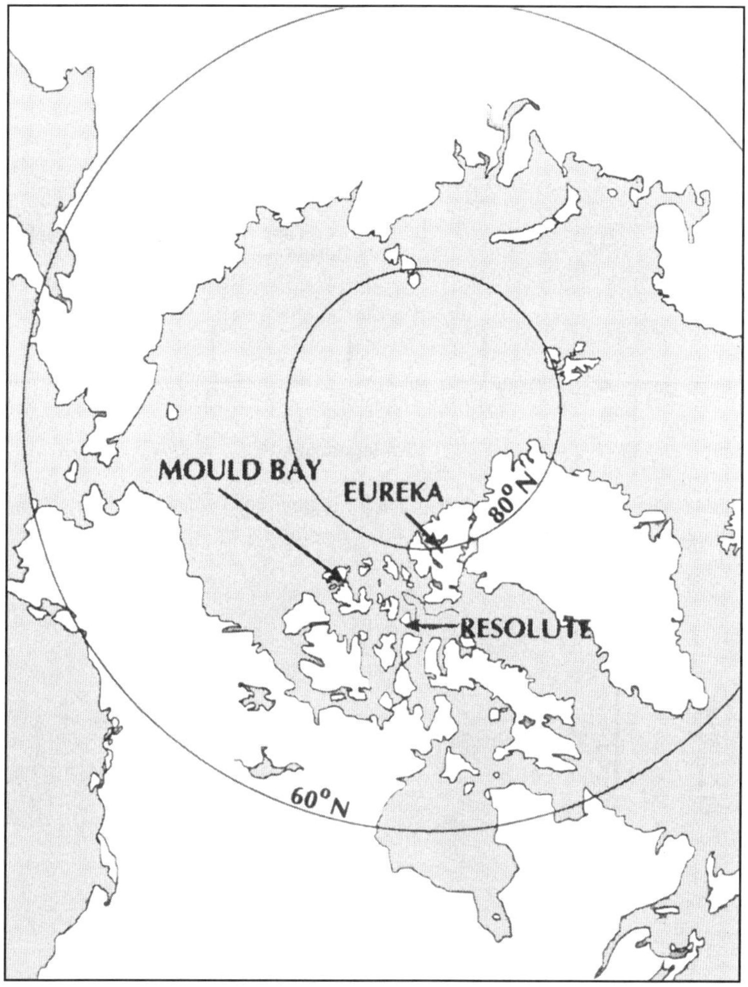

FIGURE 27.13 Circumpolar map showing the sampling locations in the Northwest Territories (Canada). (From Gregor, D., Teixeira, C., and Rowsell, R., *Chemosphere*, 33, 227, 1996. With permission.)

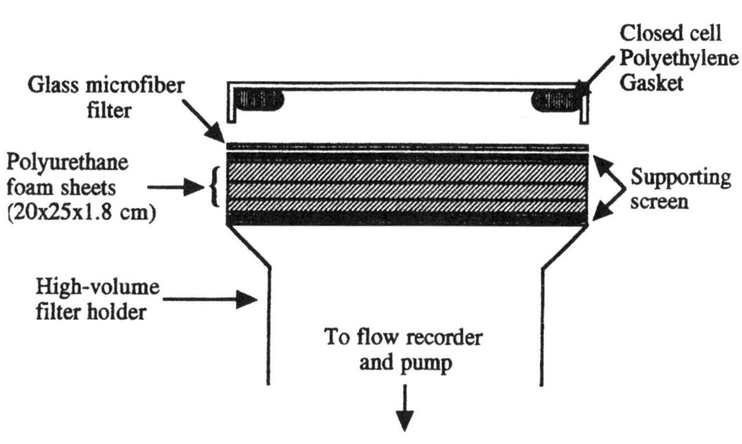

FIGURE 27.14 Scheme of the air filter sampler used for the collection of volatile pesticides. (From Nerin, C., Polo, T., Domeno, C., and Echarri, I., *Int. J. Environ. Anal. Chem.*, 65, 83, 1996. With permission.)

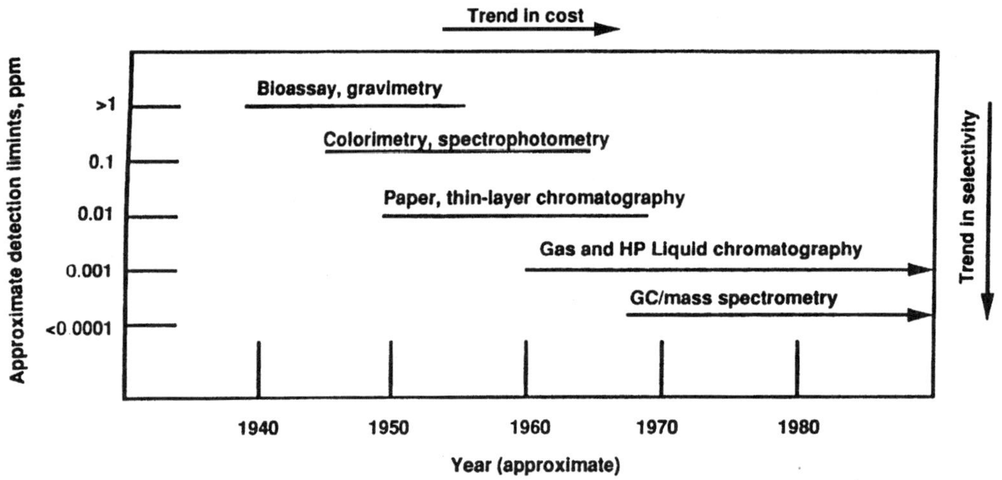

FIGURE 27.15 The evolution of analytical methods for organic toxicants in environmental samples. (From Marco, J.G., Hollingsworth, R.M., and Plimmer, J.R., *Regulation of Agrochemicals*, American Chemical Society, Washington, D.C., 1991, 1. With permission.)

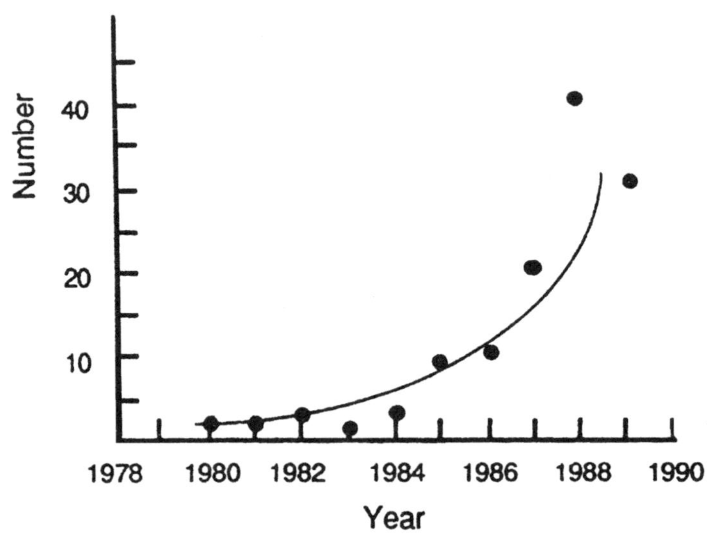

FIGURE 27.16 Publication statistics on enzyme immunoassays for pesticides. (From Hock, B., *Acta Hydrochim. Hydrobiol.*, 21, 71, 1993. With permission.)

Enzyme immunoassays belong to the most sensitive procedures for the detection and analysis of pesticides.[48-50] Enzyme immunoassay profits from the unique properties of antibodies concerning their selectivity and affinity toward ligands (e.g., pesticides). The most commonly applied types belong to the class of competitive immunoassays with phase separtaion, such as ELISA. This procedure can be used for quantitative analysis of several types of herbicides, fungicides, and insecticides (Table 27.2). The growing interest in this techniques is illustrated in Figure 27.16.[50]

TOXICOLOGY AND HEALTH RISK

Before 1960, the development of agrochemicals for plant protection was almost entirely in the hands of entomologists, microbiologists, chemists, and agronomists. Toxicological and epidemiological

TABLE 27.2
Literature Survey on Immunoassays for the Determination of Pesticides

Herbicide	Fungicide	Insecticide
Alachlor	Benomyl and Metabolites	Aldicarb
Atrazine, etc.	Blasticiden S	Aldrin
Triazines	Carbofuran	Benzoylphenyl-urea derivatives
Bentazon	Fenpropimorph	Bioallethrin
Chlorsulfuron	Metalaxyl	Carbamate
Clomazone	Sulfathiazole	Chlordane
Cyanazin	Triadimefon	DDA (2.2-*bis*(*p*-chlorophenyl)acetic acid
2,4-D and 2,4,5-T Diclofop-methyl	Diflubenzuron	DDT
Dinitro-aromates		Dieldrin
Maleic hydrazide		Endosulfan, etc.
Matazachlor		Chlordienes
Norflurazon		Fenitrothion
Paraquat		Heptachlor
Picloram		Iprodion
Thiocarbamate		Malathion
Trifluralin		Paraoxon
		Parathion
		Penfluron
		Permethrin and other pyrethroids

TABLE 27.3
Chronic Adverse Effects of Pesticides

Chronic effects (delayed onset, or protracted, recurrent, or irreversible course)
Peripheral neuropathy
Effects on reproduction
Sensitization
Suspected, but generally unconfirmed effects
- Effects on brain, heart, liver, kidney, lung, blood, reproductive organs
- Accelerated atherosclerosis, hypertension
- Carcinogenesis
- Teratogenesis
- Impaired immunity and immunopathies

studies were developed later.[1,51-55] Both epidemiological studies done on collectives of farmers and animal exposure studies have shown that the majority of the agrochemicals used are toxic, teratogenic, or carcinogenic. Environmental exposure to agrochemicals results in both acute and chronic health effects, including neurotoxicity, lung damage, chemical burns, etc. A variety of cancers — leukemia, Hodgkins disease, sarcoma, brain, stomach, and prostate cancer, lung carcinoma, etc. — have been observed.[51]

By design, agrochemicals are biologically active — in most cases, toxic. Thus, it is clear that they pose potential risk to human beings and other living organisms.

Chronic adverse effects to exposed persons are the most serious ones. They occur as a result of sustained exposures, but are much more difficult to evaluate than acute effects (Table 27.3). Worldwide, according to WHO estimates, there may be as many as 20,000 deaths and 500,000 illnesses per year related to pesticides. In addition, a number of agrochemicals have been shown to affect male reproduction in humans.

REFERENCES

1. Marco, J.G., Hollingworth, R.M., and Plimmer, J.R., *Regulation of Agrochemicals*, American Chemical Society, Washington, D.C., 1991, 1.
2. Mattheus, G.A. and Hislop, E.C., Eds., *Application Technology for Crop Protection*, CAB International, Wallingford, U.K., 1993, 1.
3. Seiber, J.N., Knutson, J.A., Woodrow, J.E., et al., Eds., *Fumigants: Environmental Fate, Exposure and Analysis*, American Chemical Society, Washington, D.C., 1996.
4. Goodhus, L.D., Insecticide aerosols, *J. Econ. Enthomol.*, 37, 338, 1944.
5. Metzener, W.H., Aerosole, *Schädlingsbekämpfung*, 42, 222, 1950.
6. Spurny, K.R., Ed., *Aerosols: Physical Chemistry and Applications*, Publ. House Academia, Prague, 1965, 571.
7. Himmel, C.M., Loats, H., and Bailey, G.W., Pesticide sources to the soil and principles of spray physics, *Pesticides in the Soil Environment*, SSSA, Book Series 2, 1990, 7.
8. Melnikov, N.N., *Chemistry of Pesticides*, Springer-Verlag, New York, 1971.
9. Miyamoto, J., Pearney, P.C., Doyle, P., and Fujita, T., Eds., *Pesticide Chemistry*, Pergamon Press, Oxford, U.K., 1985.
10. Montgomery, J.H., *Agrochemicals desk references*. Lewis Publ., Boca Raton, FL, U.S., 1993.
11. Blancato, J.N., Brown, R.N., Dary, C.C., and Saleh, M.A., Eds., *Biomarkers for Agrochemicals and Toxic Substances*, American Chemical Society, Washington, D.C., 1996.
12. Makropulos, W., Herbicides and pesticides and its potential effects on human health, *Wissenschaft u. Umwelt (Germany)*, 3, 129, 1989.
13. Weisenburger, D.D., Human health effects of agrochemical use, *Persp. Human Pathol.*, 24, 571, 1993.
14. Plimmer, J.R. and Johnson, W.E., *Pesticide Degradation Products in the Atmosphere*, American Chemical Society, Washington, D.C., 1991, 274.
15. Miller, P.C.H., Spray drift and its measurement, G.A. Mattheus and E.C. Hislop, Eds., *Application Technology for Crop Protection*, CAB Int. Wallingford, U.K., 1993, 101.
16. Walklate, P.J., A simulation study of pesticide drift from an air-assisted orchard sprayer, *J. Agri. Eng. Res.*, 51, 263, 1992.
17. Hobson, P.A., Miller, P.C.H., Walklate, P.J., et al., Spray drift from hydraulic spray nozzles, *J. Agri. Eng. Res.*, 54, 293, 1993.
18. Stafford, J.V. and Miller, P.C.H., Spatially selective application of herbicide to cereal crops, *Comput. Electronics in Agricult.*, 9, 217, 1993.
19. Miller, P.C.H. and Hadfield, D.J., A simulation model of the spray drift from hydraulic nozzles, *J. Agri. Eng. Res.*, 42, 135, 1989.
20. Teske, M.E., Bowers, J.F., Rafferty, J.E., and Barry, J.W., FSCBG: An aerial spray dispersion model for predicting the fate of released material behind aircraft, *Environ. Toxicol. and Chem.*, 12, 453, 1993.
21. Teske, M.E., Barry, J.W., and Thistle, W.H., Aerosol spray drift modeling, P. Zannetti, Ed., *Environmental Modeling*, Vol. 2, Computational Mechanics Publ., Southampton, 1994, 11.
22. Samsonov, Y.N., Makarov, V.I., and Koutsenogii, P.K., Physicochemical model and kinetics of pesticide constituent evaporation out of multi-ingredient polydisperse aerosols, *Pestic. Sci.*, 52, 292, 1998.
23. Finizio, A., Diguardo, A., and Cartmale, L., Hazardous air pollutants and their effects on biodiversity, *Environ. Monitoring and Assess.*, 49, 327, 1998.
24. Palm, W.U., Elend, M., Krüger, H.U., and Zetzsch, C., OH radical reactivity of airborne terbuthylazine adsorbed on inert aerosol, *Environ. Sci. Technol.*, 31, 3389, 1997.
25. Lepri, L., Desideri, P., Cini, R., Masi, F., and Ven Erk, M.S. Transport of organochlorine pesticides across the air/sea interface during the aerosol process, *Analytica Chimica Acta*, 317, 149, 1995.
26. Hargrave, B.T., Barrie, L.A., Bidleman, T.F., and Welch, H.E., Seasonality in exhange of organochlorines between Arctic air and seawater, *Environ. Sci. Technol.*, 31, 3258, 1997.
27. Siebers, J. and Mattusch, P., Determination of airborne residues in greenhouse after application of pesticides, *Chemosphere*, 33, 1597, 1996.
28. Dobbs, A.J. and Williams, N., Indoor air pollution from pesticides reused in wood remedial treatments, *Environ. Pollution*, 6, 271, 1983.
29. Wallace, J.C., Brzuzy, L.P., Simonich, L. et al., Case study of organochlorine pesticides in the indoor air of a home, *Environ. Sci. Technol.*, 30, 2715, 1996.

30. Baker, L.W., Fitzell, D.L., Seiber, J.N. et al., Ambient air concentrations of pesticides in California, *Environ. Sci. Technol.*, 30, 1365, 1996.
31. Bintein, S. and Devillers, J., Evaluating the environmental fate of lindane in France, *Chemosphere*, 32, 2427, 1996.
32. Millet, M., Wortham, H., Sanusi, A., and Mirabel, P., Atmospheric contamination by pesticides: Determination in the liquid, gaseous and particulate phases, *Environ. Sci. Pollution Res.*, 4, 172, 1997.
33. Haraguchi, K., Kitamura, E., Yamashita, and Kido, A., Simultaneous determination of trace pesticides in urban air, *Atm. Environ.*, 28, 1319, 1994.
34. Nerin, C., Polo, T., Domeno, C., and Echarri, I., Determination of some organochlorine compounds in the atmosphere, *Int. J. Environ. Anal. Chem.*, 65, 83, 1996.
35. Hoff, R.M., Muir, D.C., and Grift, N.P., Annual cycle of pesticides, *Environ. Sci. Technol.*, 26, 266, 1992.
36. Bidleman, T.F., Walla, M.D., Roura, R., Carr, E., and Schmidt, S., Organochlorine pesticides, *Marine Pollut. Bull.*, 26, 258, 1993.
37. Oehme, M., Further evidence for long-range air transport of polychlorinated aromates and pesticides: North American and Eurasia to the Arctic, *Ambio*, 20, 293, 1991.
38. Schreitmüller, J. and Ballschmitter, K., Levels of polychlorinated biphenyls in the lower troposphere of the North and South Atlantic Ocean, *Fresenius J. Anal. Chem.*, 348, 226, 1994.
39. Fellin, P., Barries, L.A., Dougherty, D. et al., Air monitoring in the Arctic: Results for selected persistent organic pollutants for 1992, *Environ. Toxicol. Chem.*, 15, 253, 1996.
40. Gregor, D., Teixeira, C., and Rowse, I.R., Deposition of atmospherically transported polychlorinated biphenyls in the Canadian Arctic, *Chemosphere*, 33, 227, 1996.
41. Kawata, K., Moriyama, N., Kasahara, M., and Urushiyama, Y., GC determination of pesticides in aerial application using activated carbon fiber paper for sample collection, *Bunseki Kagaku*, 39, 429, 1990; 39, 601, 1990.
42. Brouwer, D.H., Ravensberg, J.C., De Kort, W.L.A.M., and Van Hemmen, A personal sampler for inhalable mixed-phase aerosols. Validation test with three pesticides, *Chemosphere*, 28, 1135, 1994.
43. Lewis, R.G., Evaluation of a low-volume PM-2.5 sampler for pesticides, *4th Int. Aerosol Conf.*, Los Angeles, CA, 1994.
44. Scharf, J., Wiesiolek, R., and Bächman, K., Pesticides in the atmosphere, *Fresenius J. Anal. Chem.*, 324, 813, 1992.
45. Rodante, G., Marrosu, G., and Gatalani, G., Thermal analysis and kinetics study of decomposition processes of some pesticides, *J. Thermal Analysis*, 38, 2669, 1992.
46. Roinestad, K.S., Louis, J.B., and Rosen, J.D., Determination of pesticides in indoor air and dust, *J. AOAC Inter.*, 76, 1121, 1993.
47. Millet, M., Wortham, H., Sanusi, A., and Mirabel, P., A multiresidue method for determination of trace levels of pesticides in air and water, *Arch. Environ. Contam. Toxicol.*, 31, 543, 1996.
48. Schmidt, R.D., Gebbert, A., Kindervater, R., and Krämer, P., Biosensoren zur Analytik von Pflanzenschutzmitteln, *Z. Wasser-Abwasser-Forsch*, 24, 15, 1991.
49. Kaufmann, B.M. and Clower, M., Immunoassay of pesticides, *J. Assoc. Anal. Chem.*, 74, 239, 1991.
50. Hock, B., Enzyme immunoassays for pesticides analysis, *Acta Hydrochim. Hydrobiol.*, 21, 71, 1993.
51. Ragsdale, N.N., Kuhr, R.J., Eds., Pesticides, Minimizing the risks, *ACS Symposium Ser.*, 336, American Chemical Society, Washington, D.C., 1987.
52. O'Brien, R.D. and Yamamoto, I., Eds., *Biochemical Toxicology of Insecticides*, Academic Press, New York, 1970.
53. Henderson, P.T., Brouwer, D.H., Opdah, J.G. et al., Risk assessment for worker exposure to agricultural pesticides, *Ann. Occup. Hyg.*, 37, 499, 1993.
54. Takahashi, H., Yoshida, M., Murao, N., and Maita, K., Different inhalation lethality between micron-sized and submicron-sized aerosols of organophosphorus insecticide, chlorfenvinphos, in rats, *Toxicol. Lett.*, 73, 103, 1994.
55. De Raat, W.K., Stevenson, H., Hakkert, B.C., and Van Hemmen, J.J., Toxicological risk assessment of worker exposure to pesticides, *Regul. Toxicol. Pharmacol.*, 25, 204, 1997.

28 Transport and Chemistry of Pesticides in the Atmosphere

Kai Bester and H. Hühnerfuss

CONTENTS

INTRODUCTION

Pesticides are being used worldwide — with the exception of the polar regions — for crop protection. As many commercial formulations are distributed by spraying them onto the soil, plants, and animals, respectively, farm workers are often faced with considerable health risks. However, the application of pesticides did not only give rise to problems in the vicinity of treated fields, but it is well-documented that long-range atmospheric transport and deposition of organochlorine compounds have resulted in widespread contamination of pristine areas, including Arctic and Antarctic ecosystems. In addition, evidence has been presented that suggests the transformation of

0-87371-829-19/00/$0.00+$.50

xenobiotics in the environment, as well as the formation of potential pollutants from biogenic and/or man-made precursors. As a consequence, many processes, be they transport or transformation processes, may contribute to the worldwide distribution of xenobiotics. This chapter aims to summarize the actual status of these mechanisms, placing emphasis on transport and transformation processes of pesticides in the atmosphere.

MODES OF ATMOSPHERIC TRANSPORT AND SAMPLING PROCEDURES

Basically, the following different modes of atmospheric transport are conceivable: (1) compounds that show vapor pressures of ≥0.0016 Pa, e.g., several chlorinated hydrocarbons may be easily volatilized and thus travel in the gaseous phase in the atmosphere over several thousand kilometers[1]; (2) other xenobiotics, which exhibit low vapor pressures, e.g., polycyclic aromatic hydrocarbons (PAHs) are bound to aerosols and thus also transported in the atmosphere over long distances, for example, to Greenland[2]; (3) while PAHs are known to adsorb on particles, other organic compounds like phenols are more likely solved in droplets (e.g., in clouds or mists)[3,4]; (4) furthermore, a fourth transport mechanism may be encountered which involves transport by ocean currents accompanied by air/sea exchange processes, depending on the meteorological conditions. A comprehensive review on mobilization and long-range transport of pesticides was given by Kurtz.[5]

The question as to whether or not one of these transport mechanisms is encountered can be answered by a combination of different sampling techniques, including gaseous-phase samplers like polyurethane foam (PUF) plugs,[6] XAD,[7] and silica[8] with particulate material samplers like filter materials or impactors and denuders. While filters are not so reliable for droplet sampling, impactors and denuders can be used for this purpose as well.[9,10] As an alternative sampling method for mists or wet aerosols, so-called "cloud harps" have been used.[3,11] Deposition samples are taken by funnels with aperture areas ranging from 0.01 to 1 m^2. In order to gain insight into the historical development of atmospheric transport of pesticides, snow and ice cores from glaciers can be sampled.

MECHANISMS OF MOBILIZATION

Basically, four different mechanisms of mobilization of pesticides are being discussed in the literature: direct transfer (spray) into the atmosphere and three indirect remobilization mechanisms (volatilization, particle-borne remobilization, and marine bubbling processes).

SPRAY

Pesticides are often sprayed using emulsions or solutions of the bioactive compounds. As a consequence, a part of the material does not reach the plants or the soil, but it remains in the air and can thus form a primary source for the atmospheric load of pesticides that may be transported even to non-application sites. In addition to this direct transfer into the air, various remobilization processes may be encountered.

VOLATILIZATION

Direct volatilization of the pesticides that have reached the soil or plants may take place. However, it should be noted that under environmental conditions, an equilibrium state approach on the basis of partition coefficients like Henry constants can rarely if ever be used. The exchange between soil/air, plant/air, and water/air is assumed to obey much more complex laws which, in addition to the usual parameters vapor pressure of the pesticide and temperature, must include at least the

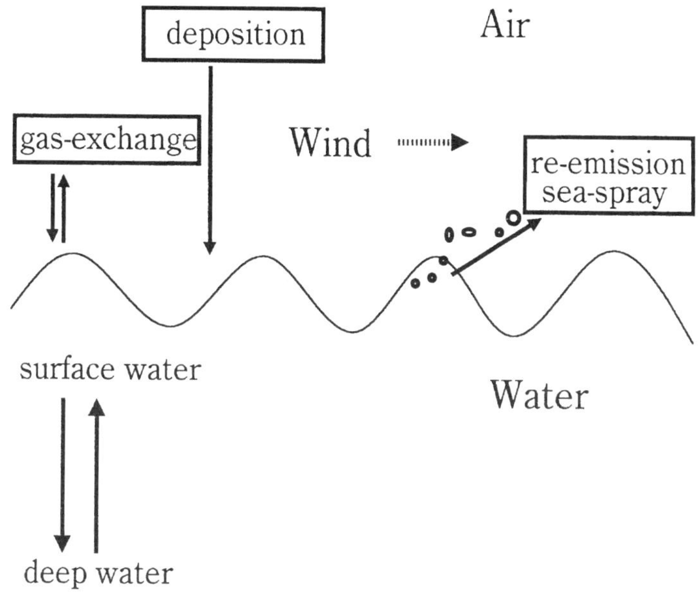

FIGURE 28.1 Exchange processes through the water surface.

meteorological conditions, total organic carbon (TOC), and alkalinity of the soil, as well as the humidity of the air. Furthermore, Glotfelty et al.[12] reported that the humidity of the soil affected the mobilization of pesticides, although their observations did not allow an unequivocal general decision as to whether or not higher soil humidity gives rise to higher or lower mobilization effects. Presumably, lipophilic compounds can be mobilized more easily from wet than from dry soil, while more polar derivatives adsorb more strongly to moist than to dry soil.[5] Similar problems are encountered with regard to the mobilization of pesticides from leaves.

Particle-Borne Remobilization

In addition, under favorable aerodynamic conditions, small particles that are covered with sprayed pesticides may be raised by the wind and thus form aerosols. Some authors[12,13] claim that about 30 to 40% of some pesticides can be transferred into the atmosphere by this mechanism during the first days after application. Boehncke et al.[14] even postulated that most of the pesticides presently applied will be mobilized by one of the mechanisms mentioned above in the course of a few weeks.

As the oceans cover more than 70% of the globe, many investigations have placed emphasis on the exchange of pesticides between the ocean surface and the air, including the determination of Henry's law constants (see Reference 15 and literature cited therein). However, Cotham and Bidleman[16] showed that Henry's law constants thus obtained may vary by about one order of magnitude even at the same environmental temperature. The highest variations were observed for hexachlorobenzene (HCB), chlordane, and toxaphene, while the values for hexachlorocyclohexane (HCH) derivatives, dieldrin, and DDT differed only by about a factor of two. Furthermore, additional parameters have to be taken into account as, for example, the formation of sea spray by means of the so-called "bubbling process" and/or the "jet-drop" effect,[17] and dry and wet deposition (see Figure 28.1). These processes can be modified by the presence of monomolecular surface films ("sea slicks," secreted by plankton and fish) that may accumulate at the ocean surface under favourable wind conditions below about 7 m s^{-1} (for details, see Reference 18). Sea slicks affect air/sea interaction processes by their strong wave damping ability,[19,20] as well as by their potential to extract and concentrate hydrophobic xenobiotics.[21] The latter aspect is schematically depicted

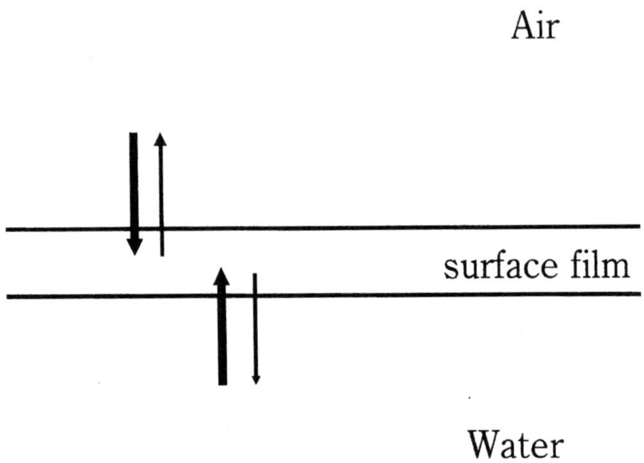

FIGURE 28.2 Extraction of air and water by a (monomolecular) surface film.

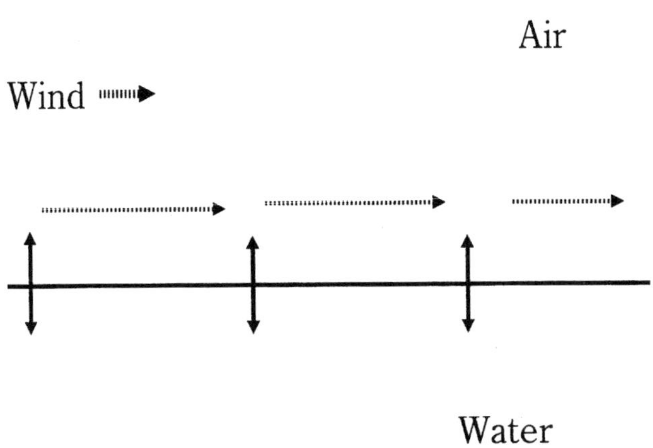

FIGURE 28.3 Long-range transport of organic compounds along the ocean surface.

in Figure 28.2. It is conjectured that some pesticides are subject to several ocean/air emission and deposition cycles when crossing the ocean (see Figure 28.3).

RANGES OF TRANSPORT

SPRAY MISTS

Spray mists are believed to travel only short distances, that is, in the range between about 100 m and a maximum of 100 km from the site of application. They contain high concentrations of the respective pesticides and can thus cause considerable health problems to farmworkers and animals, in particular, if no sufficient precautions against skin contact and inhalation of these mists are being taken. Sprayed pesticides are more mobile if they are applied from airplanes than those distributed from the ground, due to the greater height of application.

MESOSCALE TRANSPORTS

In addition to spray mist formation, in nearly all application cases, very small droplets are either in part leaving the spraying nozzle or in part being produced in the air due to aerodynamic and/or physicochemical effects (formation of wet and dry aerosols). These small droplets or particles are known to remain in the atmosphere for a long period and to be thus transported in the air for some hundred kilometers, until they have agglomerated to a sufficiently large size that will allow them to sink to the ground. In addition, these aerosols may be washed out by rain events. These mesoscale transports can be complemented by secondary soil emissions from agricultural sites.

The pesticide concentrations can in this case, attain relatively high values. Glotfelty et al.,[11] for example, who sampled fog droplets, found parathion concentrations in the range 5800 to 51,000 ng L^{-1}. The respective values for air samples were 0.88 to 7.9 pg m^{-3}. Comparison between the concentrations in droplets and in air samples allowed the calculation of distribution ratios air/water of about 0.2×10^{-6} for parathion. It turned out that the *in situ* coefficients thus determined in some cases were at variance with laboratory values by some orders of magnitude (Henry's law constant: 9.5×10^{-6}). Enrichment factors for ten different pesticides ranged between 0.05 and 3200; for example, for parathion, values between 38 to 63 were found.

Long-Range Transports

In some cases, organic compounds may travel in the atmosphere over several thousands of kilometers. These phenomena were observed in the course of research cruises that started in relatively highly polluted coastal areas and ended in pristine areas like the Arctic and Antarctic regions, where no primary sources exist. Basically, two different reasons may give rise to such long-range transports:

1. The respective pesticide exhibits a high Henry's law constant[22]; that is, the partition between ocean and atmosphere leads to a preferential enrichment in the air, which in turn allows atmospheric transport over long distances.
2. Rain events can also give rise to deposition of those organic xenobiotics showing a high Henry's law constant. However, these kinds of compounds mostly exhibit a strong hydrophobic character and are, as a consequence, preferentially enriched in sea slicks. The mechanisms already discussed above and illustrated in Figures 28.2 and 28.3 lead to a re-emission into the atmosphere and thus to a long-range transport over thousands of kilometers.

This latter mechanism is often assumed to be responsible for long-range transports of polychlorinated biphenyls (PCBs) and many chlorinated hydrocarbons, including pesticides.[1] The Arctic and Antarctic areas are supposed to trap these compounds due to the low temperature usually encountered in this part of the globe. But during the summer period when the temperatures are rising, a moderate remobilization can also be observed in such areas.[1]

In general, long-range transport of particles and thus of xenobiotics adsorbing at particles is assumed to be less probable. By way of contrast, PAHs were recently detected in ice samples from Greenland,[2] which implies that organic compounds adhering to particles of at least certain sizes may be subject to considerable atmospheric transports. However, this aspect requires additional research.

Recent experiments by Knap and Binkley[22] aimed at investigating the vertical distribution of pesticides in the atmosphere. Airborne measurements using an airplane showed that most of the organochlorine pesticides were present in the range between 0 and 3000 m above ground. The concentrations ranged between 10 and 730 pg m^{-3} for HCHs, 50 and 280 pg m^{-3} for HCB, 0 and 390 pg m^{-3} for dieldrin, and 3 and 650 pg m^{-3} for the chlordanes. However, no dependence of the concentrations on the height above ground could be inferred from this data set. The differences observed during this experiment were largely caused by the different origins of the respective air masses.

PRINCIPLES OF TRANSFORMATION IN THE ATMOSPHERE

A comprehensive assessment of air contamination and its environmental risk must include potential transformation pathways. Most compounds can be transformed into the atmosphere by direct or indirect photochemistry.[23] Direct photochemical transformation in general leads to radicals that in turn can react with water or other atmospheric components to form the respective alcohols or phenols (Equations 28.1 and 28.2).

$$R\text{–}H + h\nu \rightarrow R^* + H^* \tag{28.1}$$

$$R^* + H_2O \rightarrow R\text{–}OH + H^* \tag{28.2}$$

Similar transformations can be obtained on the basis of photochemically produced hydroxyl radicals (Equation 28.3).

$$R\text{–}H + OH^* \rightarrow R\text{–}OH + H^* \tag{28.3}$$

From a toxicological point of view, the reaction of organic compounds in the atmosphere with NOx radicals must be considered, because nitro- and nitroso-derivatives of pesticides may be formed (Equation 28.4), which exhibit a highly carcinogenic character.

$$R\text{–}H + No_x^* \rightarrow R\text{–}NO_x + H^* \tag{28.4}$$

In the course of analyses aimed at nitro- and nitroso-derivatives special precautions must be taken; in particular, the application of filters can result in many artifacts because the reactions described by Equation 28.4 can increase on the surfaces of filter materials and thus lead to higher values.[24]

In addition, dehydrohalogenation processes as well as side-chain reactions have to be taken into account. A brief survey of atmospheric photochemistry was given by Woodrow et al.[25]

COMPOUND STUDIES

ORGANOCHLORINES

1,1,1-Trichloro-2,2-bis(4-chlorophenyl)ethane (DDT)

The insecticide 1,1,1-trichloro-2,2-bis(4-chlorophenyl)ethane (DDT), which has been applied since 1942, became well-known because of its overwhelming potential to reduce *anopheles* mosquito larvae in Europe, southeast Asia, as well as other parts of the world. All efforts aimed at developing comparatively potent as well as low-priced successor pesticides have failed thus far. Therefore, DDT is still being used in some parts of Africa, Asia, and possibly eastern Europe,[26] although its severe impact on many environmental food webs has been recognized; it was banned in the U.S (January 1st, 1973), in Germany (1977), in other western European countries, and most countries in South America.

Bidleman and Olney[6,27] were the first to prove that DDT is subject to atmospheric long-range transport and thus distributed worldwide — even in pristine areas like the polar regions, where it still causes problems in Arctic food webs.[28] In 1983 (i.e., several years after the DDT ban), Bidleman et al.[26] still found DDT and the respective transformation products DDE in Swedish air samples at concentrations ranging between 0.8 and 100 pg m^{-3}. No differences between urban and rural sampling sites were detected. As the complete ban in western Europe excluded any primary sources or mesoscale atmospheric transports, the authors interpreted their results on the basis of long-range transports, possibly from eastern European or Asian countries.

In line with these conclusions, Iwata et al.[29] measured DDT in air and surface water samples of the Atlantic, Pacific, and Indian Oceans, in the Arctic and Antarctic regions, as well as in Mediterranean waters in the years 1989 and 1990. In general, the concentrations in the air ranged from 2 to 1000 pg m^{-3}. The highest values were reached in the Arabian Sea, at the west coast of India, where this pesticide is still being applied. The concentrations decreased with distance to that continent, but they were still detectable even within most pristine areas (Figure 28.4). DDT is supposed to be deposited in the Arctic mainly by dry particles; but to some extent, other mechanisms including ocean/air exchange processes as described above might also contribute to this long-range transport.[16]

1,2,3,4,5,6-Hexachlorocyclohexane (HCH)

Because of their ubiquitous nature and their relative persistence, 1,2,3,4,5,6- hexachlorocyclohexane (HCH) isomers still play an important role in environmental analyses of organic pollutants. In western Europe, Canada, and the U.S., the application of the technical mixture (α-HCH: β-HCH: γ-HCH: δ-HCH: ε-HCH = 60: 10: 15: 8: 14) is no longer permitted; but in other parts of Europe (especially in eastern Europe) and in Asia, it may still be entering the environment by way of such formulations. For example, in India, 40,000 tons per year of the technical HCH mixture are applied.[30] In Germany, the insecticidal HCH isomer, γ-HCH, has to exhibit a purity of >99% ("lindane") and may only be applied in this form since 1980. The toxicological risk of farmworkers spraying HCH formulations is considered to be relatively low, α-HCH being more toxic to humans than the γ-isomer.

The HCH isomers are comparatively volatile and water soluble, while the adsorption on particulate matter in the aqueous phase plays a negligible role. As a consequence, a very fast and far-reaching atmospheric transport was reported for these compounds. They were detected in oceanic,[8] Arctic,[30-32] and Antarctic[29] air and water samples. The concentrations in seawater, sufficiently distant from the sources, generally ranged between 1 and 2 ng L^{-1}, while the concentrations in the air depended on seasons and meteorological conditions. Iwata et al.[29] conjectured that high concentrations of the α-isomer in the atmosphere of the Northern Hemisphere are closely related to the application of technical mixtures in the Southern Hemisphere. For example, the concentrations of α-HCH in the gaseous phase in Scandinavian air, published by Oehme,[31] showed a range from 50 to 300 pg m^{-3} at the continental sites; while at marine and Arctic sampling stations, 100 to 800 pg m^{-3} were found. At all sampling stations, concentration maxima were encountered in the middle of May 1984, where at the Arctic stations the maximum peak was observed 5 days later. Calculations of backward trajectories confirmed that the respective air masses moved from eastern Europe to Scandinavia and from America to the stations in the Scandinavian Arctic. Simultaneously, γ-HCH was analyzed at the same stations. The respective concentrations reached from 5 to 60 pg m^{-3}. The concentration maxima mentioned above were also observed, but they were less pronounced than those of the α-HCH concentrations. It is interesting to note that the same maxima were detected for HCB, and in some cases for the chlordanes.

In another study, Oehme[32] compared HCH concentrations encountered in Scandinavia and the Scandinavian Arctic region during the summer period with those of lead and sulfate. It turned out that a significant correlation of the concentration dynamics of these compounds was observed at Scandinavian sampling sites, while no correlations were found at the Arctic stations. Basically, the concentrations found during the summer campaign were higher than those encountered during the spring exercise (α-HCH: 200–1000 pg m^{-3}; γ-HCH: 50–600 pg m^{-3}), and they showed no geographical preference.

During a cruise from the North to the South Atlantic (i.e., 50°N to 50°S) in 1990, Ballschmiter[23] studied the global distribution of some xenobiotics using silica adsorbents for the air samples. For α-HCH, the highest atmospheric concentrations were encountered in the Northern Hemisphere (120 pg m^{-3} at 30°N; Figure 28.5B), while in the Southern Hemisphere, considerably lower con-

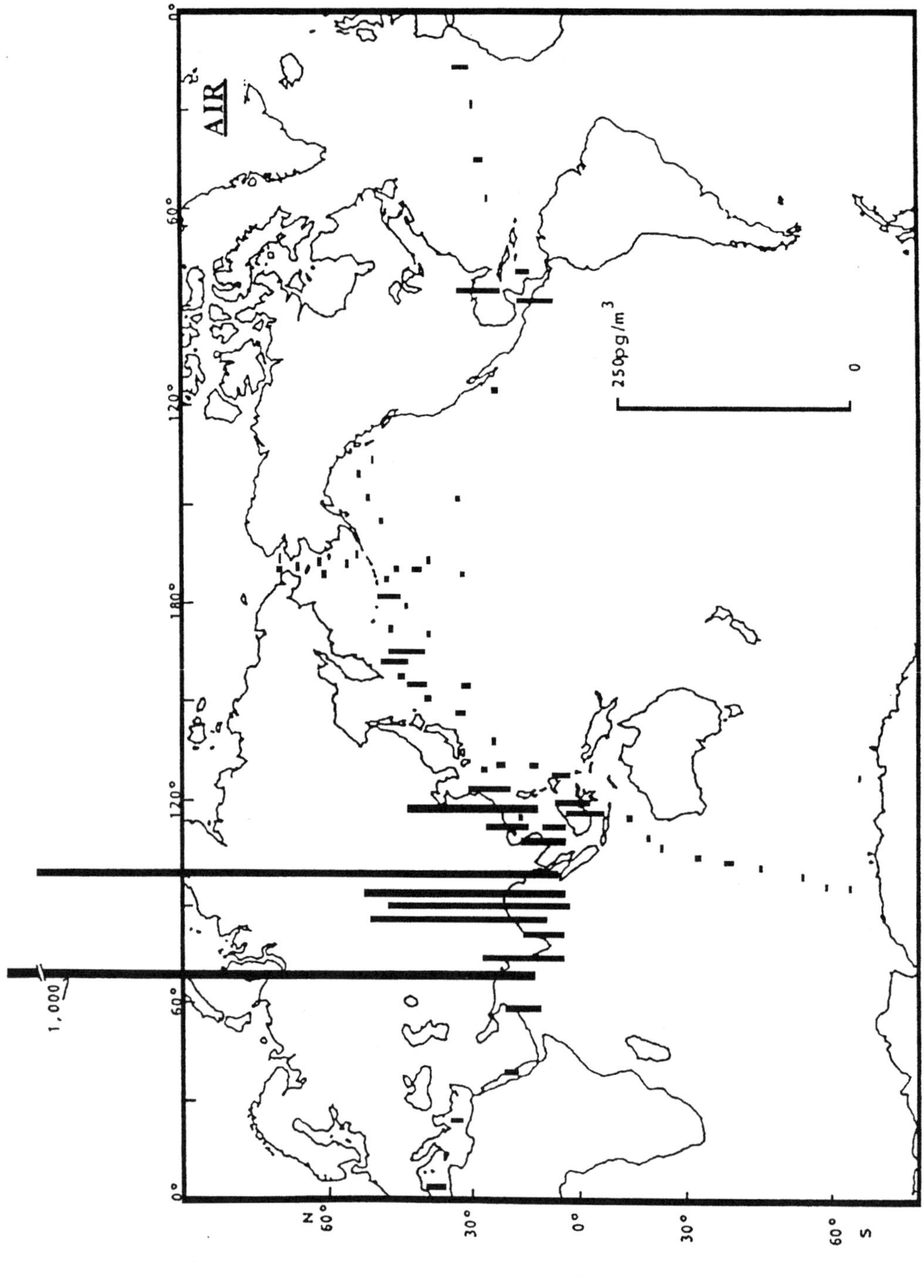

FIGURE 28.4 Distribution of DDT in the atmosphere. (From Reference 29. With permission).

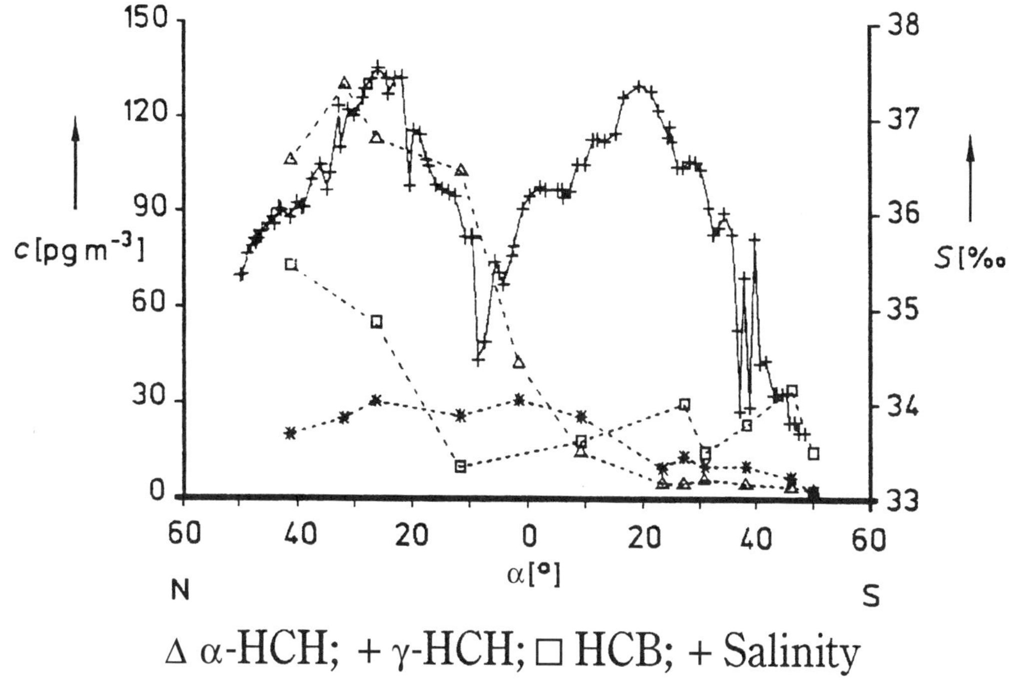

FIGURE 28.5 Distribution of α and γ-HCH as well as HCB in atlantic air in dependence on the salinity (S) of the water and the latitude (α). (From Reference 23. With permission.)

centrations were found (about 20 pg m^{-3}). By way of contrast, γ-HCH largely showed a uniform distribution in both hemispheres (about 30 pg m^{-3}). For comparison, Schreitmüller and Ballschmiter[8] determined the concentrations of the HCH isomers both in the Atlantic water and atmosphere during a cruise east of the previous track in 1991. In this case, the concentrations in the water ranged from 6.7 to 70 pg L^{-1} for α-HCH, and from 21 to 75 pg L^{-1} for γ-HCH, where the maximum values were encountered at the northern stations. The respective concentrations in the air ranged between 2 and 143 pg m^{-3} for α-HCH and between 1.5 and 331 pg m^{-3} for γ-HCH. The authors suggested the higher concentrations in the air to be correlated with higher water temperatures, which would be in accordance with a Henry's law approach, and they supported their conclusions by a comparison of their *in situ* results, both with laboratory and theoretical data. It turned out that good agreement was obtained for areas of higher water temperatures, while the results were less consistent in some areas of the Northern Hemisphere.

Jantunen and Bidleman[30] followed a similar approach for determining ocean/atmosphere fluxes for the HCH isomers. Water and air samples were taken in the course of several cruises in the Bering and Chukchi Seas during the period from 1986 to 1993, using polyurethane foam (PUF) for the air samples. For α-HCH, the concentrations ranged from 1.66 to 2.98 ng L^{-1} in the ocean surface water; while for γ-HCH, concentrations between 0.32 and 0.73 ng L^{-1} were found. The corresponding concentrations determined in the atmosphere were 49 to 128 pg m^{-3} for α-HCH and 13 to 46 pg m^{-3} for γ-HCH. The atmospheric HCH concentrations observed by Jantunen and Bidleman in the Arctic region and by Schreitmüller and Ballschmiter for moderate climates compare quite well; this supports the assumption that HCH isomers may be subject to long-range atmospheric transports. It is worth noting that in the Arctic, there appears to be only a slight dependence of atmospheric HCH concentrations on latitude, although the respective concentrations in the ocean water increase from 50°N to 80°N. Furthermore, it should be noted that between 1979 and 1993, no decline in HCH concentrations in the ocean was observed, but there was a slight decrease in atmospheric concentrations. This is in line with observations by Bidleman et al.,[33] who also found a decrease in α-HCH concentrations in the Arctic air from 1978 to 1993; however, for γ-HCH, these authors reported constant values for this period.

Jantunen and Bidleman[30] also tried to gain insight into deposition rates by using a Henry's law approach. For the year 1988, the authors calculated a flux from the atmosphere to the ocean of 20 ng m^{-2} per day (i.e., 8000 ng m^{-2} per year), while they inferred a volatilization rate of about 30 ng m^{-2} per day (12,000 ng m^{-2} per year) from their results for the year 1993. The respective results for γ-HCH were as follows: during summer 1988, a deposition rate of 35 ng m^{-2} per day was observed, while in the year 1993 no net flux was determined. However, it should be noted that fluxes in the Arctic probably exhibit a strong seasonal variability due to the temperature differences encountered in the course of a year, which complicates the calculation of annual balances. For example, the same group[34] tried to calculate annual fluxes for the Great Lakes area, and obtained a variability of about one order of magnitude. With the exception of August, all fluxes were directed from the atmosphere toward the Ontario lake water, where the depositions ranged from 10 to 60 ng m^{-2} per month and 0 to 1300 ng m^{-2} per month for α-HCH and γ-HCH, respectively. For α-HCH, these results imply an annual deposition of about 6000 ng m^{-2} into the Lake Ontario, while the deposition rates for Lake Michigan were different. However, it should be noted that the fluxes thus obtained were not based on annual measurements, but were calculated from concentrations in the air and the lake water determined during periodic campaigns and by applying theoretical models.

Direct deposition measurements were performed by Lode et al.,[35] who analyzed total deposition samples that were taken in Scandinavia and the European Arctic. Lindane concentrations ranged from <5 to 84 ng L^{-1} in southern Norway, where maximum values were observed in May. Deposition rates inferred from their data ranged from 1200 to 3700 ng m^{-2} (12 to 37 mg ha^{-1}) per year. These values agreed with deposition rates determined in Denmark at four different locations during a period of 3 years. It is worth mentioning that the α-HCH concentrations remained at about 10 ng L^{-1} rainwater throughout the experimental period, while the γ-HCH values showed a seasonal variability between 10 and 150 ng L^{-1} rainwater, with a maximum in April. The authors assumed that the seasonal peak of the γ-HCH concentration was not caused by any sources in Denmark, but they suggest that it reflected applications in Germany and/or other European countries. The deposition rates inferred from these data ranged between 1100 and 3000 ng m^{-2} per year for α-HCH and 9100 to 16,900 ng m^{-2} per year for γ-HCH. On the basis of these results, a total deposition of 0.5 tons of the two HCH isomers per year were estimated for Denmark.

In recent years, α-HCH has received additional attention in process studies because it is the only one of the eight conceivable HCH isomers that is chiral, and can thus be studied by chiral gas chromatography.[36] The cyclodextrin separation phases that have recently been developed have made it possible to distinguish between enzymatic and non-enzymatic processes. This method has been used to study the pathways of microbial and photo-oxidative degradation of α-HCH in marine and terrestrial environments as well as its enzymatic degradation in marine and terrestrial organisms from various trophic levels.[36-39] With regard to air samples, Falconer et al.[40] compared enantiomeric ratios of α-HCH in atmospheric samples from Arctic areas of Canada with those in water samples. While the authors observed a significant enantiomeric excess in water samples, presumably a reflection of enantioselective microbial degradation of α-HCH,[41] the air samples contained the racemic mixture. This result implies that the remobilization of α-HCH from the ocean to the air cannot have been very pronounced at the experimental site. A verification of this conjecture was recently published by Falconer et al.,[42] who found in snow samples of the same area a racemic mixture of α-HCH.

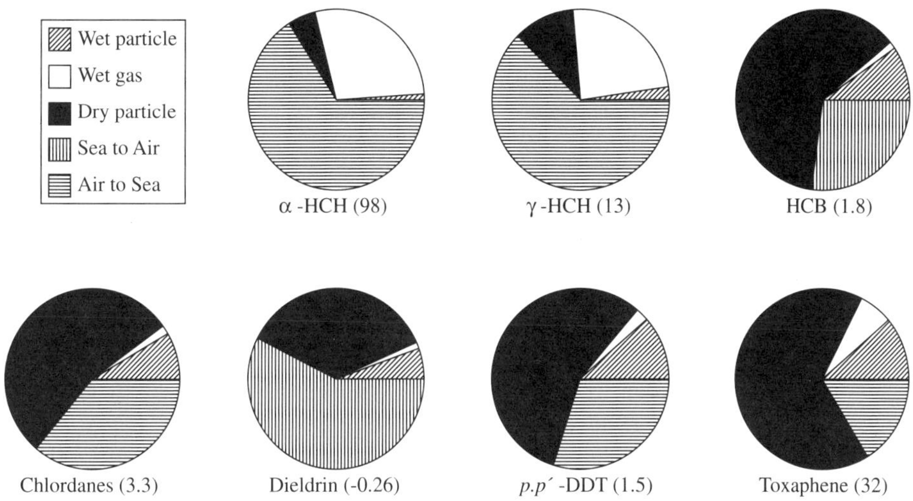

FIGURE 28.6 Deposition of some organochlorine pesticides in the Arctic in tons/year. (Modified from Reference 16. With permission.)

Cyclodienes

Cyclodiene insecticides comprise a group of chlorinated hydrocarbons that are synthesized by a Diels-Alder reaction. The most prominent derivatives are *cis*- and *trans*-chlordane, heptachlor, and nonachlor. About 10,000 tons of chlordane have been applied in the U.S. since 1950.[43]

Bidleman and Olney[6] were the first to analyze chlordane in air samples taken over the Atlantic. They reported a high variability of the chlordane concentrations, which ranged between <5 and 250 pg m^{-3}. By way of contrast, Oehme,[31] who investigated air samples in continental Scandinavian air samples during spring 1984, observed almost constant values for both *cis*- and *trans*-chlordane (both isomers about 2 pg m^{-3}). However, Oehme found elevated concentration levels at the Arctic station Spitsbergen during a short period in March (i.e., an increase from 1 pg m^{-3} to about 5 pg m^{-3}). This temporal development correlated with the α-HCH concentrations at this station for the same period. Another maximum in atmospheric chlordane concentrations (up to 6 pg m^{-3}) was observed at Spitsbergen during summer 1984,[33] where *cis*-chlordane attained slightly higher concentrations than *trans*-chlordane. But also during the summer campaign, the atmospheric chlordane concentrations in samples taken further southward of Spitsbergen were lower. Backward calculations of the respective air mass trajectories revealed that the air samples of Spitsbergen originated in Canada and in the Atlantic, while the Scandinavian air samples originated in eastern Europe. Similar results were obtained in the year 1992.[44] In summary, the results presented by Oehme support the hypothesis by Bidleman et al.[26] that the North American continent is a major source of chlordanes, which may be subject to long-range transport over the Atlantic Ocean. This conjecture is in line with observations that chlordane concentrations in Swedish air samples attained values of 1 to 15 pg m^{-3}, while in American and Canadian air samples, values between 4 and 1260 pg m^{-3} were found. The concentrations in the Canadian Arctic are comparable with the Scandinavian values: chlordanes about 2 pg m^{-3}; nonachlor about 1 pg m^{-3}.[45] The same conclusions can be drawn from the results of Iwata et al.,[29] who determined the atmospheric concentrations of *cis*- and *trans*-chlordane and of *trans*-nonachlor during a cruise on the major oceans. The total concentrations of these three cyclodiene derivatives were as follows: around Asia and America, values from 30 to 100 pg m^{-3} were found; while in the Arctic, 30 pg m^{-3}, and in the Antarctic regions, about 4 pg m^{-3} were determined.

Furthermore, it is worth noting that Cotham and Bidleman[16] have presented experimental evidence that, in the Arctic, chlordane is deposited adsorbed to dry particles. Thus, the deposition mechanism for cyclodiene derivatives appears to be similar to those for DDT and toxaphene (see Figure 28.6). In snow cores from the Canadian Arctic that represented the actual surface, Gregor[46] found concentrations for the chlordanes and heptachloroepoxide ranging from 0.02 to 0.7 ng L^{-1}. It is interesting to note that the values observed in the years 1986/1987 considerably exceeded those from earlier years. The concentration values determined at 12 stations allowed the calculation of an annual flux of about 300 μg m^{-2}. Furthermore, it was concluded that large quantities of pesticides disappear from the snow during the summer season.

Toxaphenes

Insecticide toxaphenes (synonyms: polychlorinated camphenes (PCC); polychlorinated bornanes) are preferentially used in cotton cultures in tropical areas. About 1.9 million tons of the technical mixture of PCCs, which consists of more than 200 congeners, were produced from 1950 to 1993,[47] and thus the potential environmental impact as well as the analytical problems may be compared with those of the PCBs. Because of the high variety of differently halogenated derivatives and isomers, several chromatographic co-elutions of congeners impede unequivocal quantitative assignments, and therefore, in the literature several authors often confined themselves to "total PCC concentrations."

Another problem related to toxaphenes is their extreme stability against atmospheric processes in the course of atmospheric long-range transports. Woodrow et al.[25] claimed toxaphenes to be more stable in the atmosphere than, for example, parathion, DDT, and aldrin. This implies considerable impact on marine and terrestrial ecosystems that are located far away from toxaphene application sites, including so-called pristine areas. For example, as early as 1975, Bidleman and Olney[27] observed toxaphenes in maritime air, reporting at that time concentrations between 40 and 1200 pg m^{-3}. In subsequent investigations of Swedish air samples, Bidleman et al.[26] found "total toxaphene concentrations" in the range from 4 to 225 pg m^{-3}. Comparing these values with atmospheric concentrations determined at American test sites (53–4800 pg m^{-3}), the authors suggest atmospheric long-range transports from the application sites in the southern U.S. toward Scandinavia. Furthermore, additional atmospheric transports of these contaminants from the southern states of the U.S. toward Canada were inferred from a 1-year experiment by Hoff et al.[48] In the latter case, a re-volatilization of toxaphenes from the soil was discussed, at least for the summer period.

Samples taken from the water and the air above an Arctic Russian lake were analyzed by McConnell et al.[47] during the summer of 1991. While the medium concentrations found in the air samples attained values of about 16 pg m^{-3}, in the water samples about 60 pg L^{-1} were determined. On the basis of these results, the authors suggested a flux from the air into the water.

Even in more pristine areas (i.e, in northern parts of the Arctic), Barrie et al.[49] demonstrated the occurrence of toxaphenes in atmospheric samples (1.6–27 pg m^{-3}), where the concentrations during the summer period were significantly higher than those found during the wintertime. It is worth noting that these authors verified the prevalence of C18 toxaphene congeners over those with seven and nine chlorine atoms. The results of Barrie et al.[49] agree well with those of Bidleman et al.[45] who reported for air samples taken in August 1992 concentrations around 7 pg m^{-3}, while in the ocean surface water 0.3 to 0.44 ng L^{-1} were determined. In this case, C17 congeners appeared to dominate those of C18, and only one C19 congener was detectable. According to a Henry's law approach for a temperature of about 272K (–1.4°C), a net flux from the water to the air was proposed. However, this conclusion may be subject to change due to the high temperature variations usually encountered in the Arctic (about 60K). Furthermore, Cotham and Bidleman[16] proposed that toxaphenes are predominantly deposited in the Arctic by dry particles (see Figure 28.6).

Hexachlorobenzene (HCB)

Hexachlorobenzene (HCB) was applied worldwide as a fungicide until it was banned in many countries (e.g., in Germany since 1981). Despite the ban of this relatively volatile pesticide, it is still found in nearly all kinds of environmental samples. Another potential source for HCB is the incomplete combustion of toxic wastes.

In atmospheric samples, a notable distribution of HCB was found by different authors: Bidleman et al.[26] reported concentrations of about 64 pg m^{-3} and 73 to 300 pg m^{-3} for Swedish and for American air samples, respectively. For pristine regions like the Arctic, Oehme et al.[44] published values from 100 to 200 pg m^{-3}. In this case, the samples of the northern station showed the higher concentrations. Oehme et al.[44] also included in their study some stations from southern Norway, where HCB concentrations of 100 to 180 pg m^{-3} were determined. These results led to the conclusion that a largely homogeneous atmospheric distribution of HCB can be assumed.

Schreitmüller and Ballschmiter[8] placed emphasis upon potential transformation products of HCB, in particular, pentachloroanisole (PCA) and tetrachlorodimethoxybenzole (TCDMB). Air samples taken over the Atlantic ocean contained concentrations of these two compounds from 1.8 to 96 pg m^{-3}, where the values decreased from the northern to the southern stations. These values have to be compared with concentrations of the parent compound HCB, which were determined during another cruise in the same area by the same authors (12 to 75 pg m^{-3}). These results allow the tentative conclusion that HCB is subject to atmospheric long-range transport, in part being deposited in the ocean, where it is partly transformed to PCA and TCDMD, which in turn may be re-emitted into the atmosphere.

Organophosphorus Insecticides

In 1990, about 100 derivatives of organophosphorus insecticides were on the market. Together with the carbamates, this class of "pesticides of the second generation" comprised 62% of the world's market for insecticides.[50] These numbers may illustrate the actual importance of these compounds. As pointed out by Zabik et al.,[51] organophosphorus insecticides are, in the meanwhile, being applied in larger quantities than chlorinated hydrocarbons. For example, 24,000 tons per year of malathion are used worldwide.[52]

It is generally assumed that organophosphates are not so persistent as, for example, DDT. On the other hand, most of these compounds are known to be toxic to mammals and humans due to their cholinesterase inhibition effect. Most research activities related to air sampling of such compounds were, therefore, focused on occupational health effects of farmworkers. In this context, it is important to note that the insecticidal thiophosphates can be transformed to the respective "oxons." These oxons are mostly more toxic to humans than the parent compounds.[25,51]

Basically, the equipment used for taking atmospheric samples for the analysis of organophosphorus insecticides resembles that for the investigation of organochlorines.[7] Organophosphorus insecticides were measured in fog and air samples by Glotfelty et al.[11] at four different stations, including agricultural and non-agricultural sites. The atmospheric concentrations were as follows: diazinon 5000 pg m^{-3}, ethyl-parathion 4000 pg m^{-3}, chlorpyrifos 7000 pg m^{-3}. The respective transformation products, the oxon-analogons, were found as well, but their concentrations in the air were relatively low (30–250 pg m^{-3}). By way of contrast, high concentrations of the transformation products were found in the fog samples; the maximum concentration of ethyl-paraoxon encountered was 184,000 ng L^{-1}, while only 51,000 ng L^{-1} of the parent compound ethyl-parathion was determined. However, it should be noted that the air/water ratios observed by Glotfelty et al. were not in line with laboratory results. This disagreement can be explained by oversaturation effects.

A comprehensive study on the atmospheric transport of organophosphorus pesticides was performed in California by Zabik and Seiber,[51] who investigated the transport and transformation of parathion, diazinon, and chlorpyrifos from Californian agricultural sites to the Sierra Nevada

TABLE 28.1
Concentration (in ng L^{-1}) of Some Pesticides in Deposition Samples from the North Sea 54°52′N, 07°00′E in Winter 1992

Compounds	Sampling Interval 11/22/92–11/28/92	11/28/92–12/4/92	12/4/92–12/6/92
Simazine	0.6	94	1.0
Atrazine	0.4	20	0.3
Propazine	<0.3[a]	39	1.0
Terbutylazine	<0.3[a]	40	<0.3[a]
Prometryn	1.0	8	4.0
Parathion	0.2	0.4	0.15

[a] Below the limit of detection.

Source: From Reference 76. With permssion.

National Park (altitude up to 1900 m). In the vicinity of the orchards, high concentrations were found in atmospheric samples (gaseous and particulate phase) during the winter period: parathion 26–13,000 pg m^{-3}, paraoxon (transformation product of parathion) 8–3800 pg m^{-3}, diazinon 13–10,000 pg m^{-3}, diazinon oxon (transformation product) 4–3000 pg m^{-3}, chlorpyrifos 570–3900 pg m^{-3}. The transformation of parathion to paraoxon — a key process from a toxicological point of view — and the increase in the relative concentration of the metabolite in atmospheric samples were verified by also taking samples at an intermediate distance at an altitude of 533 m; at this station, the relative concentration of paraoxon had increased, thus dominating the parent compound by about 400%, the concentration of which in turn had decreased. The final concentrations at an altitude of 1900 m showed an additional (slight) increase in the relative concentration of the transformation product, although the values rarely exceeded 0.7 pg m^{-3} at this location.

Deposition samples obtained during the same experiment allowed the same conclusions: extremely high concentrations (about 6,000 ng L^{-1}) were found in the vicinity of the orchards, and decreasing values in the mountains (7–50 ng L^{-1}). However, additional investigations are required in order to answer the question as to whether or not this decrease in concentrations is an exclusive result of transformation, or if additional effects like dilution and wash-out need to be considered. The range of atmospheric transport of these compounds was at least 200 km. It is worth noting that concentrations in deposition samples in some cases attained values known to be lethal to some fish.

Complementary investigations in Californian orchards were performed by Brown et al.,[53] who monitored malathion, the byproduct (impurity) iso-malathion, trimethylphosphorothionates, and trimethylphosphates, as well as the transformation products malaoxon and malathioncarboxylate in the gaseous phase and in deposition samples. The gaseous phase was sampled by pumping air through an adsorbent; the fall-out load was determined by analyzing filter plates that had been exposed at the beginning of the experiment and then successively analyzed after few hours and some days, respectively. High amounts of malathion and isomalathion were shown to prevail in deposition at several sites close to the orchards and within a distance of few miles directly after an airborne application. The deposition rates were about 90,000 ng m^{-2} (1000 μg per sq. ft.) for malathion and about 450 ng m^{-2} (5 μg per sq. ft.) for isomalathion, where a dependence on wind direction was observed. After the application period, the deposition load of the parent compounds found on the filter material decreased significantly, while the results for the transformation products were mixed: the deposition of malaoxon decreased at one sampling site (7200 ng m^{-2}) after 200 hours, and it increased on two other sampling sites (7200 ng m^{-2} and 290 ng m^{-2}, respectively). In the orchards, the malathion is possibly transformed to malaoxon during the period immediately following application of this pesticide. The malaoxon, in turn, is re-volatilized and deposited again after an atmospheric short-range transport. This conjecture is supported by the observation that the maximum of malaoxon in the gaseous phase appeared at all sampling stations about 24 hours after the malathion peak.[76] Furthermore, the relative content of malaoxon in the air was higher than that in the pesticide formulation applied in the orchards.

With regard to the open problem as to whether or not these compounds are subject to atmospheric medium- or long-range transports, data by Trevisan et al.[54] allow first insight: for parathion, these authors found concentrations in the range 0.36 to 0.84 ng m^{-3} in the Italian air. For comparison, precipitation samples at a remote site in the Italian mountains contained values up to 170 ng L^{-1}, which implies that parathion may be transported in the atmosphere at least over medium distances of some hundred kilometers.

High time resolution sampling of air aimed at the analysis of fenitrothion and diazinon was performed by Kawata and Yasukara.[55] The results show a day/night variation of up to one order of magnitude that was caused by the daily variation of the temperature. The concentrations determined in this study were about 10,000 pg m^{-3} and 100,000 pg m^{-3}, respectively, declining with time after application.

Herbicides

1,3,5-Triazines

The most prominent triazine herbicide, atrazine, has been applied since 1957.[56] During the last three decades, it has become the most popular herbicide for maize cultivation, but it has been used in other cultures (like sorghum) as well. Furthermore, the application of triazine herbicides has been extended to railway tracks, ship antifoulings, and algaecidal treatments as well.[57,58]

In the U.S. about 50,000 tons of atrazine are applied annually,[59] while propazine and cyanazine play a minor role. In Germany, atrazine (April 1st, 1991) and simazine (1992) have been banned, although the employment of all other triazine derivatives (in particular, terbutylazine) is still allowed. In the other EU-countries, except for Italy and some Scandinavian countries, atrazine is still being used; in addition, other important triazine derivatives include terbutylazine, cyanazine, propazine, simetryn, prometryn, terbutryn, and atraton.

Farmworkers' exposure to applications of triazines is considered a problem in some countries, as these compounds can be transformed to genotoxic and carcinogenic N-nitroso-derivatives.[60,61] In particular, in the atmosphere triazines can react with NO_x* radicals, giving rise to the formation of such transformation products.

In addition, other atmospheric transformation pathways are discussed for the triazines. Although the exact mechanism remains unclear, triazines can be dealkylated,[62,63] lead to the formation of the desethyl-, desisopropyl-, or bidesalkyl-derivatives. The importance of this mechanism is stressed by recent results of Berg et al.[63] and Bester et al.,[64,65] who analyzed these compounds in deposition samples. Probably by means of indirect photooxidation, including OH* radicals, hydroxytriazines may be formed. However, as these polar derivatives cannot be directly analyzed by gas chromatographic separation, a verification of this latter transformation mechanism in the envrionment is difficult.

The concentrations of triazine herbicides in air and deposition samples are closely related to maize agriculture; maximum concentrations are found by the end of May/June in the entire Northern Hemisphere.[54,66,67] Another season with elevated triazine levels may be encountered in November, although this period is seldom discussed in the literature.

Atmospheric concentrations as determined in the U.S. reached from <160 to 3400 pg m^{-3}, with air/water distribution coefficients that differed from those in the literature by some orders of magnitude.[11] Extremely high atmospheric concentrations were determined in Japan by Haraguchi et al.,[68] who found 32,000 pg m^{-3} and 58,000 pg m^{-3}, respectively, for atrazine and simazine but were not able to detect triazines in precipitation samples. By way of contrast, Trevisan et al.[54] found no atrazine in Italian air samples, but this may be a result of their detection limit of 1600 pg m^{-3}. With regard to fog samples, Glotfelty et al.[11] used a rotating steel screen collector and were thus able to determine atrazine concentrations between 100 and 1200 ng L^{-1}.

The concentrations found in deposition samples varied by two orders of magnitude and showed a seasonal and latitudinal dependence. Richards et al.[66] sampled rainwater in Ohio (U.S.) during spring and summer 1985 and obtained atrazine concentrations of 50 ng L^{-1} in early April, as well as from July to August; while increased levels up to 1500 ng L^{-1} were determined during the period May/June. The concentrations for cyanazine were significantly lower, ranging from about 20 to 400 ng L^{-1} (maximum values in May). These data imply that the "wash-out effect" for airborne atrazine appears to be slow.

The transformation products of the triazines, desalkyl-triazines, were investigated by Scharf et al.[69] in deposition samples in Germany in 1990 (i.e., shortly before the ban of atrazine). The authors reported for desethylatrazine 40 to 880 ng L^{-1}, and for desisopropylatrazine 28 to 174 ng L^{-1}; while for the parent compounds in general, lower concentrations were determined (atrazine 17 to 135 ng L^{-1}; simazine 10 to 94 ng L^{-1}; propazine 10 to 50 ng L^{-1}; terbutylazine 8 to 34 ng L^{-1}).

In France, atrazine concentrations in rainwater between 10 and 350 ng L^{-1} were found, similar to those determined in the U.S.[67] It should be noted that in this case, the concentrations of simazine (10 to 810 ng L^{-1}) exceeded those of atrazine. The maximum values for atrazine and for simazine were observed in April and May, respectively. Both concentration peaks were assumed to be related to the respective application periods in maize and in vine cultures. The high atrazine concentrations in April may have been a result of exceptionally low amounts of rainfall during that particular month. It should be noted that other triazines, like ametryn, prometryn, terbutryn, and cyanazine, were also found in those deposition samples. Elevated deposition of pesticides was found in winter as well as in spring. The results allowed the calculation of total deposition rates for atrazine (28,000 ng m^{-2}) and simazine (16,000 ng m^{-2}) for the year 1991. Furthermore, the authors concluded that the amount of atmospherically transported triazine herbicides was comparable with the load transported by the rivers. Glotfelty et al.[70] found high concentrations of simazine and atrazine in deposition samples (up to 2,000 ng L^{-1} atrazine) and in the vapor phase (up to 350 and 20,000 pg m^{-3} for simazine and atrazine, respectively). As the concentration maxima of these compounds were observed prior to application near the sampling site, the authors related these maxima to atmospheric transports from the southern states of the U.S. to the experimental site in Maryland.

The question as to whether or not triazines may be subject to medium- and/or long-range atmospheric transport is under considerable debate. In general, it is assumed that high concentrations in air samples are closely related to nearby application sites. However, more recent results by Trevisan et al.,[54] who took rainwater samples running from the trunks of trees in Italian mountain regions at least 200 km away from agricultural sites, showed pronounced maximum atrazine concentrations of 1990 ng L^{-1} during the application period in May, while the normal background levels were generally less than 100 ng L^{-1}. A medium-range transport of atrazine was also reported by Lode et al.,[35] who found concentrations of 80 ng L^{-1} in precipitation samples in southern Norway although these compounds were not applied in this region. The same conclusion was drawn by Bester et al.,[64] who took precipitation samples on the island of Heligoland in the Inner German Bight and on the German marine research platform "Nordsee" (i.e., several hundred kilometers away from agricultural application sites). In the samples taken at Heligoland, atrazine, and terbutylazine as well as the transformation products desethylatrazine and desethylterbutylazine were found (Figure 28.7). It should be noted that atrazine and simazine had been substituted by terbutylazine in Germany at that time, as already mentioned above. The concentrations ranged from <3 to 500 ng L^{-1}, that is, they in part exceeded the levels in seawater,[71] which turned out to affect phytoplankton.[65] The temporal development of the concentrations of these substances as well as their deposition rates as determined on the isle of Heligoland (marine sampling site) were compared with those measured at different distances from the agricultural application sites on the mainland.

During the experiment, southwest wind directions prevailed. Elevated levels of atrazine and simazine (not in Figure 28.7) observed at the westerly stations indicated the importance of medium-range atmospheric transport for these triazine derivatives. Furthermore, gradients for the atmospheric transport and deposition, respectively, were calculated, and assumed to represent in part the dilution effect between the sources (i.e., the application sites in Germany and Heligoland). For terbutylazine, this gradient turned out to be 5:1, while for atrazine the gradient was about 1:1 (Figure 28.7). These different gradients may point to sources other than those in Germany where simazine and atrazine are banned. Calculations of air mass backward trajectories for the experimental periods and for the experimental site Heligoland confirmed these assumptions, suggesting possible inputs from the Netherlands and other western European countries. The total annual deposition to the German Bight was estimated to comprise about 20,000 ng m^{-2} for the triazines investigated by Bester et al.[64] This result is comparable with conclusions drawn by Chevreuil and Gramouma[67] for France. The first observations for "medium-range atmospheric transport of triazine herbicides," as summarized above, were further substantiated by studies of Hühnerfuss et al.,[72] who tried to infer annual balances of the input of various organic pollutants into the German Bight from deposition rates, riverine, as well as lateral in- and output by ocean currents. Additional evidence of atmospheric transports of triazine herbicides is presented in Table 28.1, which summarizes the analytical results of deposition samples taken during the winter of 1992 at the German research platform "Nordsee" located in the North Sea several hundreds kilometers from the application sites. The concentrations in deposition samples ranged from <0.3 to 94 ng L^{-1}, while those in seawater in the same region comprised about 1 ng L^{-1},[71] which implies that a considerable atmospheric input of triazine pesticides into the North Sea is also encountered during the winter period.

Remobilization studies were performed by Gaynor and MacTavish,[13] who presented experimental evidence that simazine can be mobilized as a dry aerosol. They showed a reduction of simazine that had been applied in a peach orchard to 57% within 8 days. These results are consistent with measurements of Glotfelty et al.,[12] who found pesticide losses up to 30% due to wind erosion (aerosol mobilization). They pointed out that the wind velocities necessary for inducing this remobilization mechanism are lower than those usually assumed for the beginning of soil erosion, because the particles involved in pesticide mobilization are known to be smaller. In addition, the authors concluded that the magnitudes of the pesticide input into the environment by atmospheric transports and by runoff water are comparable. The importance of atmospheric transport of triazine herbicides and its potential impact on vegetation was recently demonstrated by de Jong et al.,[73] who estimated that vascular plants on more than one quarter of the area of the Netherlands may be subject to effects induced by airborne atrazine that had been sprayed somewhere else during standard agricultural activities.

Acetanilides

Herbicidal acetanilides such as metolachlor, metazachlor, alachlor, and acetachlor are widely used for corn, rape, and cabbage cultures. Alachlor and metolachlor were found by Glotfelty et al.[11] in fog water (1,500 ng L^{-1} and 2,000 ng L^{-1}, respectively), while the concentrations in gaseous samples turned out to be below the limit of detection (<400 pg m^{-3}).

Similar concentrations (<136 ng L^{-1} to 810 ng L^{-1}) were reported for alachlor in Italian rainwater,[54] while the concentrations in the air were up to 2000 pg m^{-3}. High amounts of acetanilides in rainwater were also recorded by Richards et al.,[66] who observed a seasonal cycle for alachlor and metolachlor (Figure 28.8) during an experimental period of 5 months (alachlor: 100 to 5,000 ng L^{-1}; metolachlor: 100 to 3800 ng L^{-1}). The highest concentrations were determined in May.

In Germany, Scharf et al.[69] investigated rainwater samples (total deposition) and found metolachlor and metazachlor in concentrations ranging from 91 to 330 ng L^{-1} and from 32 and 130 ng L^{-1}, respectively. An indication for the importance of spray applications for the mobilization of pesticides into the air was obtained by Capel et al.[74] In 1994, acetachlor was allowed to be applied in Minnesota

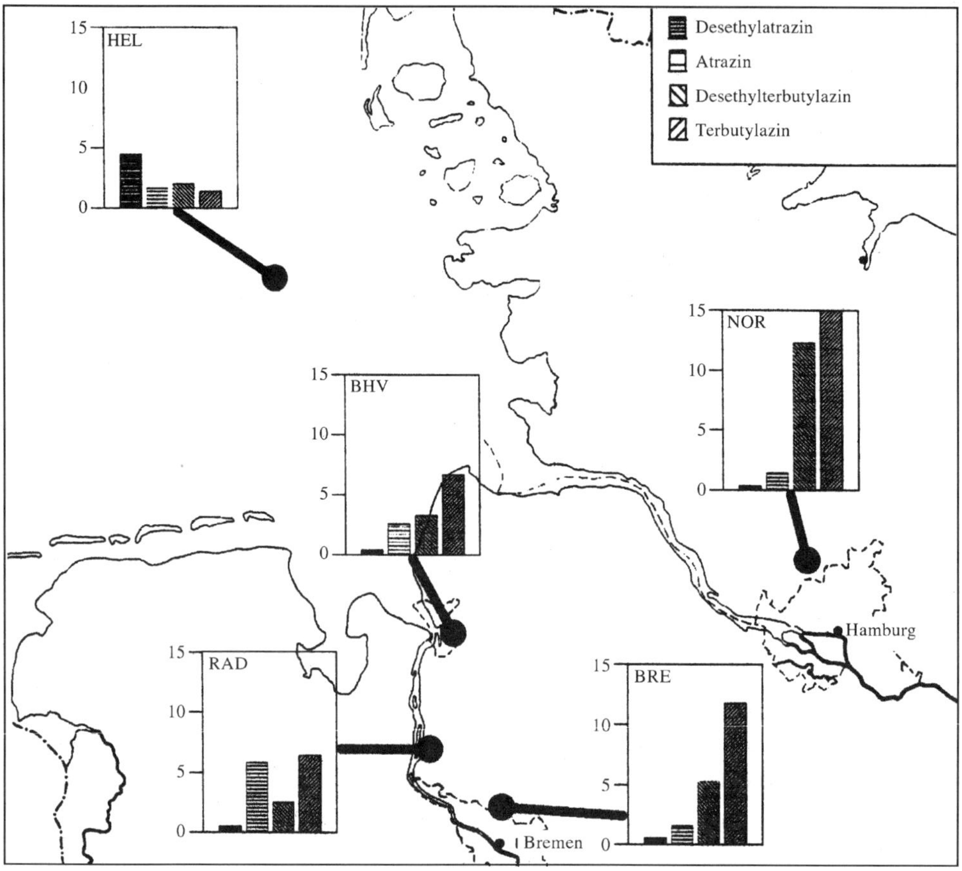

A

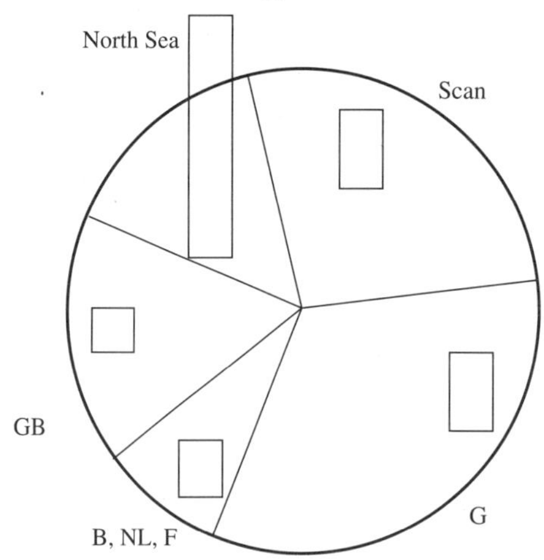

B

FIGURE 28.7 (A) Deposition rates (μg m^{-2})of atrazine, terbutylazine, and the respective transformation products in Northern Germany and the German Bight (24.5–7.7, 1993). (From Reference 64. With permission.); (B) Distribution of trajectories of the respective air masses connected to relevant areas near the North Sea during the sampling period Scan = Scandinavia, G = Germany, GB = Great Britain, NL, B, and F = Benelux and France. (From Reference 76. With permission.)

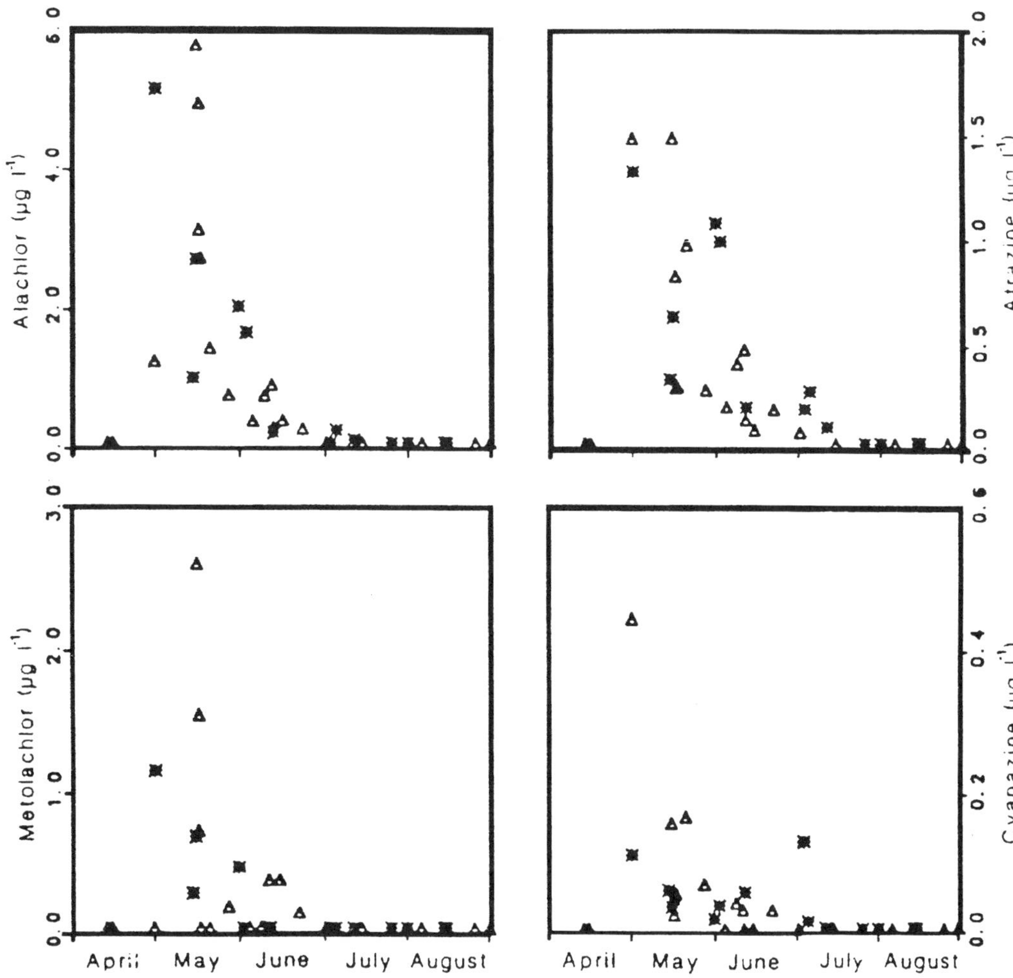

FIGURE 28.8 Temporal development of deposition of alachlor, metolachlor, atrazine and cyanazine in summer 1985; asterix, samples from Indiana; triangles, samples from Ohio. (From Reference 66. With permission.)

(U.S.) for the first time and only with a conditional registration of the EPA (U.S.). The concentrations found for this compound in deposition samples ranged from 50 to 300 ng L^{-1}, and only those of atrazine and metolachlor were higher.

Phenols and Phenoxycarboxylic Acids

Lode et al.[35] analyzed Norwegian total deposition samples with regard to their content of phenoxycarboxylic acid derivatives. In summer 1992, the concentrations of 2-(4-chloro-2-methylphenoxy)propanoic acid [MCPA] and dichlorprop attained values between 19 and 320 ng L^{-1} and between 14 and 250 ng L^{-1}, respectively. During the winter period and in 1993, lower values were encountered. These data allowed the calculation of the following annual deposition rates: MCPA 900–10,000 ng m^{-2} (9–101 mg ha^{-1}) and dichlorprop 50–3500 ng m^{-2} (0.5–35 mg ha^{-1}). It should be noted that MCPA can be transformed in the atmosphere, leading to 4-chloro-2-methyl-phenol.[75]

As some nitrophenols and chlorophenols have also been applied as pesticides, efforts have been devoted to the investigation of their environmental fate. Herterich[4] found for these compounds concentrations of <60 to 300 pg m^{-3} in particulate airborne matter, while their concentrations in cloud-water ranged from <200 to 10,000 ng L^{-1}. These results suggest that phenols are preferentially transported as wet aerosols. But caution should be applied when interpreting these experimental data with regard to sources, because it cannot be excluded that these compounds may in part originate from photosmog processes.

SUMMARY AND CONCLUSIONS

All pesticides discussed herein may be subject to atmospheric transport, although the range and the mechanisms of this transport depends on the chemical and physicochemical characteristics of the respective compounds. Sufficient experimental evidence is available that shows atmospheric long-range transport for several chlorinated hydrocarbon derivatives, thus giving rise to considerable impact even upon pristine areas like the polar regions. By way of contrast, the question as to whether or not more polar pesticides like triazines and acetanilides are being transported in the air some hundred or even thousands of kilometers is still under discussion. The experimental evidence thus far available supports the hypothesis of atmospheric medium-range transports; however, experiments aimed at the investigation of potential long-range transports of these latter polar compounds are lacking. But the medium-range atmospheric transport may also lead to a significant impact on pristine areas (e.g., the Italian mountains, the Alps, and the shelf-seas like the North Sea). With regard to fog and rainwater, considerable concentrations of polar pesticides were found. In several cases, rainwater did not match the quality criteria imposed by the EU for drinking water (100 ng L^{-1}).

In conclusion, atmospheric transport of pesticides cannot be neglected in assessment studies because it is a likely path of man-made organic compounds to fragile ecosystems of remote pristine areas. In addition to occupational health risks, special attention has been drawn to the following effects related to atmospheric transport of pesticides:

1. Muir et al.[28] and Bidleman et al. (References 33 and 45, and literature cited therein) investigated the contamination problem caused by atmospheric transport of organochlorines to the Arctic and its potential impact on the native people who rely on fish and marine mammals for their food supply.
2. Zabik and Seiber[51] showed that the concentrations of organophosphorus compounds in rainwater may exceed the LD_{50} values of some fish.
3. De Jong et al.[73] concluded that atrazine concentrations in soil caused by atmospheric deposition are likely to affect vascular plants on an area as large as a quarter of the Netherlands.

Furthermore, it should be noted that most of the pesticides are subject to transformation processes during atmospheric transport. In the case of organophosphorus pesticides, the respective oxon-analoga are being formed that are more toxic to many organisms, including humans, than the parent compounds. Another class of toxic transformation products, the nitrated derivatives, is assumed to be at least in part a consequence of photo-smog reactions.

ACKNOWLEDGMENTS

This work was supported by the Ministry of Traffic and Transportation of the Federal Republic of Germany through the BSH, Hamburg, Germany (project 532021/93, Organische Stickstoff- und Phosphorverbindungen in der Nordsee).

REFERENCES

1. Wania, F. and Mackay, D., Modelling the global distribution of toxaphene: A discussion of feasibility and desirability, *Chemosphere*, 27, 2079, 1993.
2. Jaffrezo, J.-L., Masclet, P., Clain, M.P., Wortham, H., Beyne, S., and Cahiers, H., Transfer function of polycyclic aromatic hydrocarbons from the atmosphere to the polar ice. I. Determination of atmospheric concentrations at Dye 3, Greenland, *Atmos. Environ.*, 27A, 2781, 1993.
3. Herterich, R., Atrazin–Atmosphärischer Eintrag und Immissions-Konzentrationen, *UWSF-Z. Umweltchem. Ökotox*, 3(4), 196, 1991.
4. Herterich, R., Gas chromatograpic determination of nitrophenols in atmospheric liquid water and airborne particulates, *J. Chromatogr.*, 549, 313, 1991b.
5. Kurtz, D.A., Ed., *Long Range Transport of Pesticides,* Lewis Publishers, Chelsea, MI, 1990.
6. Bidleman, T.F. and Olney, C.E., Chlorinated hydrocarbons in the Sargasso Sea atmosphere and surface water, *Science*, 183, 516, 1973.
7. Lewis, R.G. and Jackson, M.D., Modification and evaluation of a high volume air sampler for pesticides and semivolatile industrial organic chemicals, *Anal. Chem.*, 54, 592, 1982.
8. Schreitmüller, J. and Ballschmiter, K., Air-water equilibrium of hexachlorocyclohexanes and chloromethoxybenzenes in the North and South Atlantic, *Environ. Sci. Technol.*, 29, 207, 1995.
9. Zdrakal, Z. and Vecera, Z., Preconcentration and determination of 2,3,5-trichlorphenol in air using a wet effluent denuder and high performance liquid chromatography, *J. Chromatogr.*, A, 668, 371, 1994.
10. Johnson, N.D., Lane, D.A., Schroeder, W.H., and Strachan, W.M., Measurement of selected organochlorine compounds in air near Ontario lakes: gas-particle relationships, Kurtz, D.A., Ed., *Long Range Transport of Pesticides,* Lewis, Chelsea, MI, 1990, 105.
11. Glotfelty, D.E., Seiber, J.N., and Liljedahl, L.A., Pesticides in fog, *Nature*, 325, 602, 1987.
12. Glotfelty, D.E., Leech, M.M., Jersey, J., and Taylor, A.W., Volatilization and wind erosion of soil surface applied atrazine, simazine, alachlor and toxaphene, *J. Agric. Food Chem.,* 37, 546, 1989.
13. Gaynor, J.D. and MacTavish, D.C., Movement of granular simazine by wind erosion, *Hort Sci.*, 16, 756, 1981.
14. Boehncke, A., Siebers, J., and Nolting, H-G., Verbleib von Pflanzenschutzmitteln in der Umwelt — Exposition, Bioakkumulation, *Abbau — Teil B, Forschungsbericht UBA/BBA,* 89-126 05 008/2, 1989.
15. Kucklick, J.R., Hinckley, D.A., and Bidleman, T.F., Determination of Henry´s Law constants for hexachlorocyclohexanes in distilled water and artificial seawater as a function of temperature, *Mar. Chem.*, 34, 197, 1991.
16. Cotham, W.E., Jr. and Bidleman, T.F., Estimating the atmospheric deposition of organochlorine contaminants to the Arctic, *Chemosphere*, 22(1-2), 165, 1991.
17. Cini, R., Desideri, P., and Lepri, L., Transport of organic compounds across the air/sea interface of artificial and natural marine aerosols, *Anal. Chim. Acta*, 291, 329, 1994.
18. Hühnerfuss, H. and Garrett, W.D., Experimental sea slicks: their practical applications and utilization for basic studies of air–sea interactions, *J. Geophys. Res.*, 86, 439, 1981.
19. Alpers, W. and Hühnerfuss, H., The damping of ocean waves by surface films: A new look at an old problem, *J. Geophys. Res.*, 94, 6251, 1989.
20. Hühnerfuss, H., Gericke, A., Alpers, W., Theis, R., Wismann, V., and Lange, P.A., Classification of sea slicks by multi-frequency radar techniques: new chemical insights and their geophysical implications, *J. Geophys. Res.*, 99, 9835, 1994.
21. Seba, D.B. and Corcoran, E.F., Surface slicks as concentrators of pesticides in the marine environment, *Pest. Monitor. J.*, 3, 190, 1969.
22. Knap, A.H. and Binkley, K.S., Chlorinated organic compounds in the troposphere over the western North Atlantic ocean measured by aircraft, *Atmos. Environ.*, 25A,157, 1991.
23. Ballschmiter, K., Transport und Verbleib organischer Verbindungen im globalen Rahmen, *Angew. Chem.*, 104, 501, 1992, *Angew. Chem. Int. Ed. Engl.,* 31,487, 1992.
24. de Raat, W.K., Bakker, G.L., and de Meijere, F.A., Comparison of filter materials used for sampling of mutagens and PAH in ambient airborne particles, *Atmos. Environ.*, 24 A(11), 2875, 1990.
25. J.E. Woodrow, D.G. Crosby, and J.N. Seiber, Vapor-phase photochemistry of pesticides, *Res. Rev.*, 85, 111, 1983.

26. Bidleman, T.F., Wideqvist, U., Jansson, B., and Söderlund, R., Organochlorine pesticides and polychlorinated biphenyls in the atmosphere of southern Sweden, *Atmos. Environ.*, 21(3), 641, 1987.
27. Bidleman, T.F. and Olney, C.E., Long range transport of toxaphene insecticide in the atmosphere of the western North Atlantic, *Nature,* 257, 475, 1975.
28. Muir, D.C.G., Grift, N.P., Ford, C.A., Reiger, A.W., Hendzel, M.R., and Lockhart, W.L., Evidence for long-range transport of toxaphene to remote arctic and subarctic waters from monitoring of fish tissues, D.A. Kurtz, Ed., *Long Range Transport of Pesticides,* Lewis, Chelsea, MI, 1990, 329.
29. Iwata, H., Tanabe, S., Sakai, N., and Tatsukawa, R., Distribution of persistent organochlorines in the oceanic air and surface seawater and the role of ocean on their global transport and fate, *Environ. Sci. Technol.*, 27, 1080, 1993.
30. Jantunen, L.M. and Bidleman, T.F., Reversal of the air-water gas exchange direction of hexachlorocyclohexanes in the Bering and the Chukchi Seas:1993 versus 1988, *Environ. Sci. Technol.,* 29, 1081, 1995.
31. Oehme, M., Dispersion and transport paths of toxic persistent organochlorines to the Arctic — levels and consequences, *Sci. Total Environ.*, 106, 43, 1991.
32. Oehme, M., Further evidence for long-range air transport of polychlorinated aromates and pesticides: North America and Eurasia to the Arctic, *Ambio*, 20, 293, 1991.
33. Bidleman, T.F., Falconer, R.L., and Walla, M.D., Toxaphene and other organochlorine compounds in air and water at Resolute Bay, N.W.T., Canada, *Sci. Total Environ.*, 160/161, 55, 1995.
34. McConnell, L.L., Cotham, W.E., and Bidleman, T.F., Gas exchange of hexachloro-cyclohexane in the Great Lakes, *Environ. Sci. Technol.*, 27, 1304, 1993.
35. Lode, O., Eklo, O.M., Holen, B., Svenson, A., and Johnsen, A.M., Pesticides in precipitation in Norway, *Sci. Total Environ.*, 160/161, 421, 1995.
36. Hühnerfuss, H., Faller, J., Kallenborn, R., König, W.A., Ludwig, P., Pfaffenberger, B., Oehme, M., and Rimkus, G., Enantioselective and nonenantioselective degradation of organic pollutants in the marine ecosystem, *Chirality,* 5, 393, 1993.
37. Faller, J., Hühnerfuss, H., König, W., and Ludwig, P., Gas chromatographic separation of the enantiomers of marine organic pollutants — distribution of α-HCH enantiomers in the North Sea, *Mar. Pollut. Bull.*, 22/2, 82, 1991.
38. Hühnerfuss, H., Faller, J., König, W.A., and Ludwig, P., Gas chromatographic separation of the enantiomers of marine pollutants. 4. Fate of hexachlorocyclohexane isomers in the Baltic and North Sea, *Environ. Sci. Technol.*, 26, 2127, 1992.
39. Möller, K., Bretzke, C., Hühnerfuss, H., Kallenborn, R., Kinkel, J.N., Kopf, J., and Rimkus, G., Durchlässigkeit der Blut-Hirn-Schranke von Seehunden für das Enantiomer (+) α-1,2,3,4,5,6-hexachlorcyclohexan und dessen absolute Konfiguration, *Angew. Chem.,* 106, 911, 1994.
40. Falconer, R.L., Bidleman, T.F., and Gregor, D.J., Air-water gas exchange and evidence for metabolism of hexachlorocyclohexanes in Resolute Bay, N.W.T., *Sci. Total Environ.*, 160/161, 65, 1995.
41. Ludwig, P., Hühnerfuss, H., König, W.A., and Gunkel, W., Gas chromatographic separation of the enantiomers of marine pollutants. 3. Enantioselective degradation of α-HCH and γ-HCH by marine microorganisms, *Mar. Chem.*, 38, 12, 1992.
42. Falconer, R.L., Bidleman, T.F., Gregor, D.J., Semkin, R., and Teixeira, C., Enantioselective breakdown of α-hexachlorocyclohexane in a small Arctic lake and its watershed, *Environ. Sci. Technol*, 29(5), 1297, 1995b.
43. Puri, R.K., Orazio, C.E., Kapila, S., Clevenger, T.E., Yanders, A.F., McGrath, K.E., Buchanan, A.C., Czarnzki, J., and Bush, J., Studies on the transport and fate of chlordane in the environment, Kurtz, Ed., *Long Range Transport of Pesticides,* Lewis, Chelsea, MI, 1990, 271.
44. Oehme, M., Haugen, J-E., and Schlabach, M., Ambient air levels of persistent organochlorines in spring 1992 at Spitsbergen and the Norwegian mainland: a comparison with 1984 results and quality control measures, *Sci. Total Environ.*, 160/161, 139, 1995.
45. Bidleman, T.F., Jantunen, L.M., Falconer, R.L., Barrie, L.A., and Fellin, P., Decline of hexachlorocyclohexane in the arctic atmosphere and reversal of air-sea gas exchange, *Geophys. Res. Lett.*, 22(3), 219, 1995.
46. Gregor, D.J., Deposition and accumulation of selected agricultural pesticides in Canadian snow, D.A. Kurtz, Ed., *Long Range Transport of Pesticides,* Lewis, Chelsea, MI, 1990, 373.

47. McConnell, L.L., Kucklick, J.R., Bidleman, T.F., and Walla, T.F., Long-range atmospheric transport of toxaphene to lake Baikal, *Chemosphere*, 27, 2027, 1993.
48. Hoff, R.M., Muir, D.C.G., Grift, N.P., and Brice, K.A., Measurements of PCCs in air in southern Ontario, *Chemosphere,* 27(10), 2057, 1993.
49. Barrie, L., Bidleman, T., Dougherty, D., Fellin, P., Grift, N., Muir, D., Rosenberg, D., Stern, G., and Toom, D., Atmospheric toxaphene in the High Arctic, *Chemosphere*, 27(10), 2037, 1993.
50. Naumann, K., Neue Insektizide, *Nachr. Chem. Tech. Lab.*, 42, 255, 1994.
51. Zabik, J.M. and Seiber, J.N., Atmospheric transport of organophosphate pesticides from California's Central valley to the Sierra Nevada mountains, *J. Environ. Qual.*, 22, 80, 1993.
52. Barcelo, D., Porte, C., Cid, J., and Albaiges, J., Determination of organophosphorus compounds in Mediterranean coastal waters and biota samples using gas chromatography with nitrogen — phosphorus and chemical ionization mass spectrometric detection, *Int. J. Environ. Anal. Chem.,* 38, 199, 1990.
53. Brown, M.A., Petreas, M.X., Okamoto, H.S., Mischke, T.M., and Stephens, R.D., Monitoring of malathion and its impurities and environmental transformation products on surfaces and in air following an aerial application, *Environ. Sci. Technol.*, 27, 388, 1993.
54. Trevisan, M., Montepiani, C., Ragozza, L., Bartoletti, C., Ioannilli, E., and Del Re, A.A.M., Pesticides in rainfall and air in Italy, *Environ. Pollut.*, 80, 31, 1993.
55. Kawata, K. and Yasukara, A., Determination of fenitrothion and diazinon in air, *Bull. Environ. Contam. Toxicol.*, 52, 419, 1994.
56. Cai, Z., Ramanujam, V.M.S., Giblin, D.E., Gross, M.L., and Spalding, R.F., Determination of atrazine in water at low- and sub-parts-per-trillion levels using solid-phase extraction and gas chromatography/high resolution mass spectrometry, *Anal. Chem.*, 65, 21, 1993.
57. Büchel, K.H., *Chemistry of Pesticides,* John Wiley & Sons, New York, 1983.
58. Gough, M.A., Fothergill, J., and Hendrie, J.D., A survey of southern England coastal waters for the s-triazine antifouling compound irgarol 1051, *Mar. Pollut. Bull.*, 28, 613, 1994.
59. deNoyelles, F., Kettle, W.D., and Sinn, D.E., The responses of plankton communities in experimental ponds to atrazine, the most heavily used pesticide in the United States, *Ecology,* 63(5), 1285, 1982.
60. Meli, G., Bagnati, R., Fanelli, R., Benfenati, E., and Airoldi, L., Metabolic profile of atrazine and N-nitrosoatrazine in rat urine, *Bull. Environ. Contam. Toxicol.*, 48, 701, 1992.
61. Janzowski, C., Klein, R., and Preussmann, R., Formation of N-nitroso compounds of the pesticides atrazine, simazine, and carbaryl with nitrogen oxides, *IARC Scientific Publications,* 31, 329, 1980.
62. Barcelo, D., Durand, G., deBertrand, N., and Albaiges, J., Determination of aquatic photodegradation products of selected pesticides by gas chromatography-mass spectrometry and liquid chromatography-mass spectrometry, *Sci. Total Environ.*, 132, 283, 1993.
63. Berg, M., Müller, S.R., and Schwarzenbach, R.P., Simultaneous determination of triazines including atrazine and their major metabolites hydroxyatrazine, desethylatrazine and deisopropylatrazine in natural waters, *Anal. Chem.*, 67, 1860, 1995.
64. Bester, K., Hühnerfuss, H., Neudorf, B., and Thiemann, B., Atmospheric deposition of triazine herbicides in Northern Germany and the German Bight (North Sea), *Chemosphere*, 30, 1639, 1995.
65. Bester, K., Hühnerfuss, H., Brockmann, U., and Rick, H.J., Biological effects of triazine herbicide contamination on marine phytoplankton, *Arch. Environ. Contam. Toxicol.*, 29, 277, 1995.
66. Richards, R.P. , Kramer, J.W., Baker, D.B., and Krieger, K.A., Pesticides in rainwater in the northeastern United States, *Nature*, 327, 129, 1987.
67. Chevreuil, M. and Gramouma, M., Occurrence of triazines in the atmospheric fallout on the catchment basin of the river Marne (France), *Chemosphere*, 27, 1605, 1993.
68. Haraguchi, K., Kitamura, E., Yamashita, T., and Kito, A., Simultaneous determination of trace pesticides in urban precipitation, *Atmos. Environ.*, 29, 247, 1995.
69. Scharf, J., Wiesiollek, R., and Bächmann, K., Pesticides in the atmosphere, *Fresenius J. Anal. Chem.*, 342, 813, 1992.
70. Glotfelty, D.E., Williams, D.E., Freeman, H.P., and Leech, H.P., Regional atmospheric transport and deposition of pesticides in Maryland, Kurtz, D.A., Ed., *Long Range Transport of Pesticides,* Lewis, Chelsea, MI, 1990, 199.
71. Bester, K. and Hühnerfuss, K., Triazines in the Baltic and North Sea, *Mar. Pollut Bull.*, 26, 623, 1993.

72. Hühnerfuss, H., Bester, K., Landgraff, O., Pohlmann, T., and Selke, T., Annual balances of hexachlorocyclohexanes, polychlorinated biphenyls, and triazines in the German Bight *Mar. Pollut. Bull.,* 34, 419, 1997.
73. de Jong, F.M.W., van der Voet, E., and Canters, E., Possible side effects of airborne pesticides on fungi and vascular plants in the Netherlands, *Ecotox. Environ. Saf.*, 30, 77, 1995.
74. Capel, P.D., Ma, L., Schroyer, B.R., Larson, S.J., and Gilchrist, T.A., Analysis and detection of the new corn herbicide acetochlor in river water and rain, *Environ. Sci. Technol.*, 29, 1702, 1995.
75. Woodrow, J.E., McChesney, J.E., and Seiber, J.N., Modeling the volatilization of pesticides and their distribution in the atmosphere, D.A. Kurtz, Ed., *Long Range Transport of Pesticides,* Lewis, Chelsea, MI, 1990, 61.
76. Bester, K., Über Eintrag, Verbleib und Auswirkungen von stickstoff- und phosphorhaltigen Schadstoffen in der Nordsee, Shaker Verlag, Aachen, 1996.

Index

A

B

D

E

F

G

H

I

K

L

M

N

O

P

R

S

T

U

V

W

X

Z